DIGITAL TELEPHONY AND NETWORK INTEGRATION
Second Edition

Bernhard E. Keiser
and
Eugene Strange

VAN NOSTRAND REINHOLD
I(T)P A Division of International Thomson Publishing Inc.

New York • Albany • Bonn • Boston • Detroit • London • Madrid • Melbourne
Mexico City • Paris • San Francisco • Singapore • Tokyo • Toronto

I(T)P™ A division of International Thomson Publishing, Inc.
The ITP logo is a trademark under license

Printed in the United States of America

For more information, contact:

Van Nostrand Reinhold
115 Fifth Avenue
New York, NY 10003

Chapman & Hall GmbH
Pappelallee 3
69469 Weinheim
Germany

Chapman & Hall
2-6 Boundary Row
London
SE1 8HN
United Kingdom

International Thomson Publishing Asia
221 Henderson Road #05-10
Henderson Building
Singapore 0315

Thomas Nelson Australia
102 Dodds Street
South Melbourne, 3205
Victoria, Australia

International Thomson Publishing Japan
Hirakawacho Kyowa Building, 3F
2-2-1 Hirakawacho
Chiyoda-ku, 102 Tokyo
Japan

Nelson Canada
1120 Birchmount Road
Scarborough, Ontario
Canada M1K 5G4

International Thomson Editores
Campos Eliseos 385, Piso 7
Col. Polanco
11560 Mexico D.F. Mexico

2 3 4 5 6 7 8 9 10 QEBFF 01 00 99 98 97 96 95

Library of Congress Cataloging-in-Publication Data

Keiser, Bernhard, 1928–
Digital telephony and network integration/Bernhard E. Keiser.
Eugene Strange. 2nd ed.
p. cm.
Includes bibliographical references and index.
ISBN 0-442-00901-1
1. Digital telephone systems. I. Strange, Eugene. II. Title. TK6421.K45 1994
621.385 dc20 94-29054
CIP

Preface

What is "digital telephony"? To the authors, the term digital telephony denotes the technology used to provide a completely digital telecommunication system from end-to-end. This implies the use of digital technology from one end instrument through transmission facilities and switching centers to another end instrument. Digital telephony has become possible only because of the recent and ongoing surge of semiconductor developments, allowing microminiaturization and high reliability along with reduced costs.

This book deals with both the future and the present. Thus, the first chapter is entitled, "A Network in Transition." As baselines, Chapters 2 and 11 provide the reader with the present status of telephone technology in terms of voice digitization as well as switching principles.

The book is an outgrowth of the authors' consulting and teaching experience in the field since the early 1980s. The book has been written to provide both the engineering student and the practicing engineer a working knowledge of the principles of present and future telecommunication systems based upon the use of the public switched network. Problems or discussion questions have been included at the ends of the chapters to facilitate the book's use as a senior-level or first-year graduate-level course text.

Numerous clients and associates of the authors as well as hundreds of others have provided useful information and examples for the text, and the authors wish to thank all those who have so contributed either directly or indirectly.

The first chapter, which is a joint effort of both authors, provides an overview of the field. Chapters 2–4 deal with the subject of speech digitization, while Chapters 5 and 6 are devoted to the use of digital technology in the telephone network and for transmission in general. Chapter 7 deals with the new and rapidly growing field of digital cellular radio. Chapters 8–10 treat three principal facility types for digital transmission: microwave radio, communication satellite systems, and fiber optics.

Chapter 11 deals with generic architectural considerations for digital switching systems, and Chapter 12 describes the architecture of some digital switching systems that are currently operating in the North American network. Network structure, transmission, and signaling of the North American network, both local and long distance, as well as digital network synchronization, are covered in Chapter 13, as the all-digital network continues to evolve. As the network has become more and more digital, initial implementation of the integrated services digital network (ISDN) on a North American basis has become practicable, and is the subject of Chapter 14, along with the beginnings of broadband networking. Chapter 15 is a new chapter concerning the efforts to extend digital telephony to the "last mile," the local loop to subscribers' premises.

Because of the rapidly changing nature of the subjects covered in this volume, the authors invite reader comments, questions, and suggestions for future editions.

The authors acknowledge the useful comments and suggestions provided by the reviewers of the text. A special note of thanks goes to Dr. Stuart Yuill and his students at Johns Hopkins University for their careful review of a prepublication copy of the text, and for their many useful comments and suggestions. Thanks go to Mr. Eric Schimmel of the TIA, Mr. James Mullen of Hughes Network Systems, Inc., Dr. Andrew Viterbi and Dr. Mark Epstein of QUALCOMM, Inc., Dr. Donald Schilling of Interdigital Communications Corporation, Mr. Keith Radousky of Bell South Cellular Corporation, and to Mr. Robert Dixon of Omnipoint Communications, Inc., all of whom were very helpful in providing essential information in the development of Chapter 7. Thanks also go to Jerry Klein of AT&T Bell Laboratories, Pat Maluchnik of AT&T Network Systems, Wanda Baldwin and Kevin Molloy of Northern Telecom, Inc., and Jayne Scott of Siemens Stromberg-Carlson for their kindness in making available product information for Chapter 12. Kathleen Roberts of the National Institute of Standards and Technology (NIST) was most helpful in connection with information concerning the inauguration of nationwide ISDN in the United States, and Dr. Mark Karol of AT&T Bell Laboratories provided information on ATM switching, both covered in Chapter 14.

BERNHARD E. KEISER
EUGENE STRANGE

Vienna, Virginia

Contents

1 A NETWORK IN TRANSITION **1**

1.1 Introduction 1
1.2 Major Historical Milestones 2
1.2.1 Modulation of Direct Current 3
1.2.2 Vacuum-Tube Technology 4
1.2.3 Semiconductor Technology 5
1.2.4 Optical Transmission Technology 6
1.2.5 Wireless Technology 6
1.3 A Futuristic Look 7
1.4 Organization of the Text 8

2 SPEECH DIGITIZATION FUNDAMENTALS **10**

2.1 Introduction 10
2.2 Speech Coding Approaches 11
2.3 Sampling 13
2.4 Quantization 15
2.5 Effect of Digitization on Bandwidth 17
2.6 Speech Digitizer Performance Definitions 18
2.7 Speech-Coding Advantages 20
Problems 21
References 22

3 WAVEFORM CODING **23**

3.1 Introduction 23
3.2 Basic PCM Encoding 25
3.3 Compression and Nonuniform Quantization 31
3.4 PCM Performance 35
3.5 Special PCM Techniques 39

3.5.1 Coarsely Quantized PCM 39
3.5.2 Nearly Instantaneous Companding 42
3.6 Differential PCM 44
3.6.1 Basic DPCM Concepts 46
3.6.2 Adaptive DPCM (ADPCM) 49
3.7 Delta Coding 52
3.7.1 Linear Delta Coding 52
3.7.2 Adaptive Delta Coding 58
3.7.2.1 Continuously Variable-Slope Delta (CVSD) 60
3.7.2.2 Digitally Controlled Delta Coding 62
3.8 Subband Coding (SBC) 62
3.9 Adaptive Transform Coding (ATC) 68
3.10 Harmonic Scaling 70
3.11 Vector Techniques 70
3.12 Performance of Efficient Waveform Coding Techniques 73
Problems 81
References 81

4 PARAMETRIC AND HYBRID CODING 84
4.1 Introduction 84
4.2 Principles of Low-Bit-Rate Speech Encoding 86
4.3 Vocoders 86
4.3.1 Linear Predictive Coding (LPC) 87
4.3.2 Other Vocoders 90
4.3.3 Effect of Model on Speech Quality 93
4.4 Hybrid (Waveform-Parametric) Techniques 93
4.5 Adaptive Predictive Coding (APC) 94
4.6 Multipulse LPC 95
4.7 Residual-Excited Linear Prediction Vocoder (RELP) 95
4.8 Code-Excited Linear Prediction (CELP) 98
4.9 Vector-Sum-Excited Linear Prediction (VSELP) 104
4.10 Regular Pulse Excitation with Long-Term Prediction (RPE-LTP) 106
4.11 Voice Digitizer Comparisons 108
Problems 108
References 109

5 DIGITAL TECHNIQUES IN THE TELEPHONE NETWORK 111
5.1 Introduction 111
5.2 Synchronization 112
5.2.1 Frame Synchronization 112
5.2.2 Timing 115
5.2.3 Timing Recovery 116
5.3 Time-Division Multiplexing 117
5.3.1 T-Carrier Systems 119
5.3.2 The Digital Hierarchy 120

5.3.3 Muldems 122
5.3.4 Transmultiplexers 125
5.4 Scramblers 128
5.5 Channel Coders 130
5.6 Echo Cancelers 135
5.7 Digital Speech Interpolation 137
5.8 Digital Repeaters 139
5.9 Digitization of the Loop Plant 142
5.9.1 Transmission Modes 143
5.9.1.1 Wired Systems 143
5.9.1.2 Pair-Gain Systems 145
5.9.1.3 Discrete Multitone (DMT) Systems 145
5.9.1.4 Radio Systems 146
5.9.2 Line Codes 147
5.9.3 Digital End Instruments 147
5.10 Monitoring and Maintenance 149
5.11 Speech Recognition 150
5.11.1 Speech Production 150
5.11.2 Speech-Recognition Systems 151
5.11.3 Speech-Recognition Techniques 152
5.11.4 Voice-Input Systems 153
5.12 Voice Response 153
Problems 156
References 156

6 DIGITAL TRANSMISSION 159
6.1 Introduction 159
6.1.1 Performance Objectives 159
6.1.2 Signal Impairments in Transmission 159
6.1.3 The Nyquist Theorem 161
6.1.4 The Shannon Limit 166
6.2 Digital Modulation Techniques 167
6.2.1 Phase-Shift Keying 170
6.2.1.1 Binary Phase-Shift Keying (BPSK) 173
6.2.1.2 Quaternary Phase-Shift Keying (QPSK) 174
6.2.1.3 Eight-Phase Shift Keying (8-PSK) 175
6.2.1.4 Sixteen-Phase Shift Keying (16-PSK) 177
6.2.2 Amplitude-Shift Keying and Quadrature Amplitude Modulation 177
6.2.3 Amplitude-Phase Keying 180
6.2.4 Frequency-Shift Keying 181
6.2.5 Correlative Techniques 184
6.2.6 Comparison of Modulation Techniques 188
6.2.6.1 Bandwidth Efficiency 189
6.2.6.2 Modulation Technique Performance 191
6.3 Error Control 191

6.3.1 Error Detection 195
6.3.2 Error Correction 195
6.3.3 Block and Convolutional Codes 196
6.3.4 Bose–Chaudhuri–Hocquenghem (BCH) Code 197
6.4 Multiple Access 197
6.4.1 Frequency-Division Multiple Access (FDMA) 197
6.4.2 Time-Division Multiple Access (TDMA) 198
6.4.3 Code-Division Multiple Access (CDMA) 198
6.5 Packet Transmission 199
6.5.1 Packet Voice 200
6.5.2 Statistical Multiplexing 202
6.5.3 Fast Packet Switching 204
6.5.3.1 Frame Relay 204
6.5.3.2 Cell Relay 204
6.6 Retrofit 205
Problems 208
References 208

7 DIGITAL CELLULAR RADIO 211
7.1 Introduction 211
7.2 Radio-Wave Propagation 213
7.2.1 The Mobile Antenna 213
7.2.2 Signal Fading Characteristics 214
7.2.3 Effect of Fading on Digital Transmission 218
7.2.4 Effect of Propagation on System Design 220
7.2.5 Frequency Reuse 224
7.2.6 Channel Assignments 227
7.3 Cellular System Operation 229
7.3.1 Call Setup 230
7.3.1.1 Fixed-Station Originated Call 232
7.3.1.2 Subscriber-Set Originated Call 232
7.3.1.3 Disconnect 233
7.3.2 Hand-off 233
7.3.3 Roaming 234
7.4 Multiple Access Concepts 235
7.4.1 Time-Division Multiple Access (TDMA) 237
7.4.1.1 Basic TDMA 237
7.4.1.2 Extended TDMA (E-TDMA) 239
7.4.2 Code-Division Multiple Access (CDMA) 245
7.4.2.1 QUALCOMM CDMA 246
7.4.2.2 Broadband CMDA 251
7.4.3 Hybrid Digital System 253
7.4.4 Frequency-Hopping Multiple Access (FHMA) 253
7.5 Data Transmission 253
7.6 Spectrum Efficiency 254

7.7 Summary of Digital Cellular System Characteristics 256
7.7.1 North American Digital Cellular System (TDMA) 256
7.7.2 Extended TDMA (E-TDMA) 257
7.7.3 Japanese Digital Cellular System 257
7.7.4 The Global System Mobile (GSM) 257
7.7.5 Code-Division Multiple Access (CDMA) 258
7.8 The Microcellular Environment 258
7.8.1 The Propagation Environment 259
7.8.2 Personal Communication Networks (PCNs) 260
7.8.3 Indoor Radio Systems 260
7.8.4 System Examples 262
7.9 The Wireless Local Loop 263
7.10 The PCS Interface to the PSTN 266
Problems 267
References 267

8 MICROWAVE TRANSMISSION 270
8.1 Introduction 270
8.2 Characteristics of Microwave Propagation 271
8.2.1 Spreading Loss and Absorption 271
8.2.2 Fading 273
8.2.3 Polarization 275
8.3 Microwave System Engineering 276
8.3.1 Frequency Allocations 276
8.3.2 Link Budget 279
8.3.3 Repeater Siting 280
8.3.4 Diversity 282
8.3.5 Reliability and Availability 282
8.4 Characteristics of Microwave Equipment 284
8.4.1 Amplifiers 284
8.4.1.1 Power Amplifiers 284
8.4.1.2 Small Signal Amplifiers 284
8.4.2 Antennas 285
8.4.3 System Interface Arrangements 289
8.5 Digital Microwave Radio Systems 289
8.5.1 Systems for Intercity and Long-Haul Applications 290
8.5.2 Metropolitan-Area Systems 294
8.5.2.1 Digital Termination Systems 294
8.5.2.2 Private Systems 294
8.5.2.3 Common-Carrier Systems 295
8.5.2.4 SONET-Compatible Radio Systems 295
Problems 296
References 296

9 SATELLITE TRANSMISSION **298**
9.1 Introduction 298
9.2 Characteristics of Satellite Propagation 299
9.2.1 The Satellite Orbit 299
9.2.2 Time Delay on Satellite Paths 300
9.2.3 Atmospheric Attenuation 300
9.2.4 Rain Depolarization 301
9.3 Satellite System Design 302
9.3.1 Frequency Allocations and Usage 302
9.3.2 Link Budgets 305
9.3.3 Earth-Station Siting 307
9.3.4 Multiple Access 309
9.3.4.1 Frequency-Division Multiple Access 309
9.3.4.2 Time-Division Multiple Access 311
9.3.4.3 Code-Division Multiple Access 312
9.3.5 Multiplexing 313
9.3.6 Demand Assignment 313
9.3.7 Echo Cancellation 314
9.4 Characteristics of Satellite System Equipment 314
9.4.1 Space Segment 316
9.4.1.1 Transponders 316
9.4.1.2 Control Subsystem 321
9.4.2 The Earth Station 321
9.5 Major Operational Communication Satellite Systems 324
9.6 Satellite Systems for Direct Service to the User 325
9.7 Future Trends in Communication Satellite Systems 327
Problems 330
References 331

10 FIBER-OPTIC TRANSMISSION **333**
10.1 Introduction 333
10.2 Fiber Transmission Characteristics 334
10.2.1 Attenuation 335
10.2.2 Dispersion 337
10.2.3 Soliton Transmission 337
10.3 Fiber Types 337
10.3.1 Multimode Fibers 340
10.3.2 Single-Mode Fibers 341
10.3.3 Graded-Index Fibers 341
10.4 Optical Sources 342
10.4.1 Types 342
10.4.2 Coherent Sources 343
10.4.3 Modulation 344
10.4.4 Coupling of Sources to Fibers 348
10.5 Photodetectors 348
10.5.1 Noncoherent Detectors 348

10.5.2 Coherent Detectors 349
10.6 Repeaters 350
10.7 Optical Amplifiers 351
10.8 Noise Sources 352
10.9 Transmission Systems and Subscriber Lines 353
10.9.1 Typical Transmission Systems 355
10.9.2 Subscriber Systems 355
10.9.3 Network Compatibility 357
10.10 Wavelength-Division Multiplexing 357
10.11 The Fiber Distributed-Data Interface (FDDI) 359
10.11.1 Basic FDDI 359
10.11.2 FDDI-II 360
10.11.3 FDDI Applications 361
10.12 Synchronous Optical Network (SONET) 362
10.12.1 The SONET Standards 364
10.12.2 Virtual Tributaries 365
10.12.3 Architecture 367
10.12.4 Applications 368
10.13 Atmospheric Optics 368
10.14 Future Optical Telephone Network 369
Problems 369
References 370

11 DIGITAL SWITCHING ARCHITECTURE 373
11.1 Introduction 373
11.2 Terminal Interface Techniques 374
11.2.1 Terminal Interface Functions 374
11.2.2 Implementation Considerations 375
11.2.2.1 Analog Line Interface 375
11.2.2.2 Analog Trunk Interface 377
11.2.2.3 Digital Line and Trunk Interfaces 377
11.2.3 Implementation Trends 377
11.3 Switching Network Considerations 379
11.3.1 Principles of Time-Division Switching 379
11.3.1.1 Time Switching in Memory 380
11.3.1.2 Time Switching in Space 381
11.3.1.3 Sampling and Coding Rates 382
11.3.2 Multistage Digital Switching 382
11.3.2.1 Time-Switching Considerations 382
11.3.2.2 Space-Switching Considerations 383
11.3.2.3 Time-Space-Time (TST) Structure 383
11.3.2.4 Space-Time-Space (STS) Structure 385
11.3.2.5 Combined versus Separated Switching 386
11.3.3 Economic and Traffic Considerations 386
11.3.4 Digital Symmetrical Matrices 387

11.4 Service Circuit Techniques 388
11.4.1 Tone Generation 388
11.4.2 Tone Reception 389
11.4.3 Digital Conferencing 389
11.4.4 Digital Recorded Announcements 391
11.4.5 Digital Echo Suppression 392
11.4.6 Digital Pads 392
11.4.7 Provisioning of Service Circuits 393
11.5 Control Architectures 394
11.5.1 Control Workload Distribution 394
11.5.2 Central Control Systems 395
11.5.3 Shared Control Systems 396
11.5.4 Distributed Control Systems 396
11.5.4.1 Distribution of Control by Function 396
11.5.4.2 Distribution of Control by Block Size 397
11.6 Maintenance Diagnostics and Administration 397
11.6.1 Maintenance Diagnostics 397
11.6.1.1 Maintenance Phases 398
11.6.1.2 Diagnostic Methods 399
11.6.2 Administration 401
11.6.2.1 Database Management 401
11.6.2.2 Generic Program Changes 401
11.6.2.3 Data Collection 402
11.6.3 Traffic Administration 402
11.6.4 Network Management 402
Problems 403
References 404

12 OPERATIONAL SWITCHING SYSTEMS 406
12.1 Introduction 406
12.2 Siemens Stromberg-Carlson DCO 406
12.2.1 DCO System Description 406
12.2.2 DCO System Architecture 407
12.2.2.1 Interfaces 407
12.2.2.2 Local Line Switch 409
12.2.2.3 Common Control 411
12.2.2.4 Switching Network 412
12.2.3 DCO Remote Operation 415
12.2.3.1 Remote Line Group 415
12.2.3.2 Remote Line Switches 416
12.2.3.3 Remote Network Switch 416
12.3 AT&T 5ESS Switch 416
12.3.1 5ESS General Description 416
12.3.2 5ESS Hardware Architecture 417
12.3.2.1 Administrative Module 417
12.3.2.2 Communications Module 417

12.3.2.3 Switching Module 420
12.3.2.4 Remote Switching Module 422
12.3.3 5ESS Software Architecture 424
12.3.3.1 Operating System Software 424
12.3.3.2 Call Processing Software 424
12.3.3.3 Maintenance Software 425
12.3.3.4 Administrative Software 426
12.3.4 5ESS ISDN Architecture 426
12.3.5 5ESS Broadband Architecture 428
12.4 AT&T 4ESS Switch 428
12.4.1 Overview of the 4ESS Switch 429
12.4.2 4ESS Switching Network 430
12.4.2.1 Architectural Concept 430
12.4.2.2 Time-Slot Interchange (TSI) Frame 431
12.4.2.3 Time-Multiplexed Switch (TMS) Frame 433
12.4.2.4 Network Clock 433
12.4.2.5 Call-Switching Process 434
12.4.3 4ESS Terminal Interface Equipment 435
12.4.4 4ESS Control System 437
12.5 Northern Telecom DMS-100 Switch 439
12.5.1 Overview of the DMS-100 Family 439
12.5.2 Evolution of DMS Architecture 439
12.5.3 S/DMS SuperNode Architecture 441
12.5.3.1 DMS-Core 442
12.5.3.2 DMS-Bus 445
12.5.3.3 Link Peripheral Processor 446
12.5.3.4 S/DMS Switching Network 447
12.5.3.5 S/DMS Peripheral Modules 449
12.5.3.6 S/DMS SuperNode ISDN Architecture 450
12.5.4 DMS SuperNode Software Architecture 451
Problems 452
References 453

13 THE EVOLVING SWITCHED DIGITAL NETWORK 455
13.1 Introduction 455
13.2 Intra-LATA Networks 456
13.2.1 Local Access and Transport Areas (LATAs) 456
13.2.2 Intra-LATA Network Switching Plan 456
13.2.3 Intra-LATA Network Transmission Plan 458
13.2.4 Intra-LATA Signaling 458
13.3 The North American Network 459
13.3.1 Network Signaling 459
13.3.1.1 Inband Signaling 460
13.3.1.2 Common-Channel Signaling 460
13.3.2 Digital Network Transmission Considerations 469
13.3.2.1 Loss and Level Considerations 470

13.3.2.2 Digital Network Transmission Plan 471
13.4 Digital Network Synchronization 472
13.4.1 The Need for Synchronization 472
13.4.2 Hierarchical Network Synchronization 474
13.4.3 Private Network Synchronization 476
13.5 Digital Implementation Progress 476
Problems 477
References 477

14 THE INTEGRATED SERVICES DIGITAL NETWORK (ISDN) 479
14.1 The ISDN Concept 479
14.1.1 User Perspective of the ISDN 480
14.1.2 Narrowband ISDN User-Access Arrangements 481
14.1.2.1 The Basic-Rate Access 482
14.1.2.2 The Primary-Rate Access 483
14.1.2.3 The 7-kHz Audio Access 483
14.2 Subscriber-Loop Technology 484
14.3 Narrowband ISDN Network Topology 486
14.4 Narrowband ISDN Signaling Protocols 486
14.5 ISDN Implementation Progress 489
14.5.1 ISDN in Europe 489
14.5.1.1 Deregulation and Competition in Europe 490
14.5.1.2 ISDN in France 490
14.5.1.3 ISDN in Germany 492
14.5.1.4 ISDN in Belgium 492
14.5.1.5 International ISDN Connectivity 492
14.5.2 ISDN in Japan 493
14.5.3 ISDN in North America 494
14.6 Broadband ISDN (BISDN) 496
14.6.1 Overview of BISDN 496
14.6.2 Existing Broadband Services 496
14.6.2.1 Frame Relay Service 497
14.6.2.2 Fiber-Distributed Data Interface (FDDI) 498
14.6.2.3 Switched Multimegabit Data Services (SMDS™) 500
14.6.3 Broadband Access Technology 501
14.6.4 Broadband Transport Technology 504
14.6.5 Broadband Switching Technology 507
14.6.5.1 Switching Fabric Considerations 508
14.6.5.2 A Prototype ATM Switch Architecture 512
14.6.6 Broadband Network Synchronization 514
14.6.7 Congestion Control 518
14.6.8 Assimilation of BISDN 519
14.7 Intelligent Network Functionality 519
14.7.1 AIN Overview 522
14.7.2 Triggers 523
14.7.3 AIN Implementation 524

Problems 524
References 525

15 CLOSING THE LOOP 527
15.1 The Loop Environment in North America 527
15.2 Loop Requirements for Broadband Services 528
15.3 Broadband Services on Copper Loops 529
15.3.1 High-Bit-Rate Digital Subscriber Line (HDSL) 530
15.3.2 Asymmetrical Digital Subscriber Line (ADSL) 532
15.3.2.1 General Description of ADSL 533
15.3.2.2 ADSL Technical Challenges 533
15.3.2.3 Prototype ADSL Transceiver Technology 534
15.3.2.4 Discrete Multitone Technology 535
15.3.3 SONET/ATM Signals on Copper Conductors 536
15.3.3.1 UTP and STP Cables 536
15.3.3.2 Transmission Techniques 537
15.3.3.3 Transmission Performance 538
15.4 Fiber in the Loop (FITL) 540
15.4.1 FITL Configuration Alternatives 541
15.4.2 Powering Alternatives for ONU 544
15.4.3 Spectrum Considerations 545
15.5 Fiber to the Home (FTTH) 545
15.5.1 Biarritz, France Optical Network 545
15.5.2 Heathrow, Florida Optical Network 546
15.6 Wireless-Access Communications System (WACS) 549
15.6.1 WACS System Overview 550
15.6.2 WACS System Architecture 550
15.7 Trends and Issues 552
15.7.1 Multimedia Services via BISDN 552
15.7.2 Multimedia Services via Narrowband ISDN 553
Problems 554
References 554

APPENDIX A. NORTH AMERICAN MIXED-NETWORK TECHNOLOGY 557
A.1 Introduction 557
A.2 The North American Network 558
A.2.1 Network Numbering Plan 558
A.2.2 Network Routing Plan 559
A.3 Transmission Technology 561
A.3.1 Subscriber-Loop Transmission 562
A.3.1.1 Copper-Loop Environment 562
A.3.1.2 Subscriber Carrier Systems 564
A.3.2 Network Transmission 565
A.3.2.1 Frequency-Division Multiplexing 565
A.3.2.2 Transmission Impairments 566

A.3.2.3 Control of Impairments 567
A.3.2.4 Via Net Loss Plan 569
A.3.2.5 Network Transmission Plans 569
A.4 Signaling Technology 570
A.4.1 Supervisory Signaling 570
A.4.1.1 Subscriber Line Supervisory Signaling 571
A.4.1.2 Inband Interoffice Supervisory Signaling 574
A.4.1.3 E & M Lead Control 575
A.4.1.4 Control of Disconnect 577
A.4.2 Address Signaling 577
A.4.2.1 Dial-Pulse Signaling 578
A.4.2.2 Dual-Tone Multifrequency Signaling 579
A.4.2.3 Multifrequency Signaling 581
A.4.2.4 Control of User Address Signaling 583
A.4.2.5 Control of Interoffice Address Signaling 583
A.4.2.6 Glare Detection and Resolution 585
A.4.2.7 Signaling Transients 586
A.4.3 Network Information Signals 587
A.4.4 Inband Signaling Techniques 587
A.4.5 Switched Access for Inter-LATA Carriers 587
A.4.6 Equal-Access Dialing and Signaling Plan 589
A.4.7 Common-Channel Signaling 593
A.4.7.1 Principles of Common-Channel Signaling 593
A.4.7.2 Signaling Link Operation 593
A.4.7.3 Call Setup with CCIS 594
A.4.7.4 Signaling-Message Formats 595
A.4.7.5 Datagram Direct Signaling 597
A.4.7.6 Advantages of Common-Channel Signaling 597
A.5 Switching Technology 597
A.5.1 Basic Switching Functions 597
A.5.1.1 Supervision 597
A.5.1.2 Control 598
A.5.1.3 Signaling 599
A.5.1.4 Switching Network 599
A.5.2 Control Concepts 599
A.5.2.1 Operator Control 599
A.5.2.2 User Control 599
A.5.2.3 Common Control 600
A.5.3 Switching Network Technology 602
A.5.3.1 Space-Division Switching 603
A.5.3.2 Time-Division Switching 611
References 612

APPENDIX B. TRAFFIC CONSIDERATIONS IN TELEPHONY 614
B.1 Introduction 614
B.2 Traffic Assumptions 614

B.3 Traffic Measurements 615
B.3.1 Traffic Data Collection 615
B.3.2 Traffic Analysis Considerations 615
B.3.3 Traffic Loss Probabilities 617
B.3.3.1 Erlang B Formula 617
B.3.3.2 Poisson Formula 617
B.3.3.3 Erlang C Formula 618
B.3.3.4 Comparison of Traffic Formulas 618
B.3.3.5 Nonrandom Traffic Theories 620
B.4 Network Management 621
B.4.1 Principles of Control 622
B.4.2 Principal Controls Available 622
References 625

APPENDIX C. ANALOG CELLULAR SYSTEMS 626
C.1 Introduction 626
C.2 Analog System Operation 626
C.2.1 Call Setup 627
C.2.2 Signal-to-Noise Ratio 627
C.2.3 Signal-to-Interference Ratio 627
C.2.4 FM Deviation 627
C.3 Analog System Control 627
C.3.1 Supervision 628
C.3.1.1 Signaling Tone 628
C.3.1.2 Supervisory Audio Tone 628
C.3.1.3 Locating Function 628
C.3.2 Paging and Access 628
C.3.3 Setup Channels 629
C.3.4 Seizure Collision Avoidance 630
C.3.5 Error Limits 630
C.3.6 Blank and Burst 631
C.3.7 Making Calls 631
C.3.7.1 Call to Mobile 631
C.3.7.2 Call from Mobile 632
C.3.8 Handoff 633
C.3.9 Disconnect 633
C.3.9.1 Mobile-Initiated Disconnect 633
C.3.9.2 System-Initiated Disconnect 634
C.3.10 Summary 634
C.4 Narrowband Advanced Mobile Phone Service (NAMPS) 635

REFERENCES 635

GLOSSARY 636

INDEX 655

1

A Network in Transition

1.1. INTRODUCTION

To say that telephony is in a state of transition is to describe the history of telephony. It always has been in a state of transition and will continue so for the foreseeable future. As the world has been increasingly compressed in size through television and telecommunications, the trend of telephony standards has been toward enabling worldwide application rather than having each national network go its own way. While national networks still are interconnected through gateways, primarily for billing control, international signaling systems are becoming standardized with only minor national differences.

This chapter contains some broad background of major developments in telephony and encapsulates in snapshot form the networks as they prepare to make the transition into the twenty-first century. Since the authors are American and have had more extensive exposure to the networks in the United States, it is only natural to focus on that country. Because the networks in the United States and Canada operate transparently across the international border, we consider telecommunications developments based upon North American standards (i.e., United States and Canada) as applying to both countries.

Solid-state technology has made the use of digital techniques feasible. In particular, very-large-scale-integration (VLSI), with its ability to implement high degrees of circuit complexity reliably, has ushered in the digital era.

What advantages do digital techniques have in telephony, however? First, what do we mean by "digital"? By *digital*, we mean that only a limited number of *signal states*, or *levels* is allowed. A two-level system is called a *binary system*; a three-level system is called a *ternary system*; a four-level system is called a *quaternary system*; and an *M*-level system is called *M-ary*.

With a limited number of levels, the signal is easier to detect, since the detector need not attempt to reproduce an exact replica of the original signal. The detector only needs to decide whether the received signal is on one side or the other of a built-in threshold. This detection simplicity leads to many advantages for digital communications. A key advantage is waveform regeneration. Waveform regeneration allows a noisy distorted signal to be reconstructed at a digital detector if its received level is on the correct side of the detection threshold at the sampling instant.

Another advantage of having a limited number of levels is that the received carrier-to-noise ratio can be significantly lower with digital transmission than with analog. Moreover, *modems* (modulator-demodulators) are not needed for digital transmission because the role of a modem is to convert a digital signal to analog form for transmission on an analog network. Here, we need only make certain that the digital signal is in the proper digital format for transmission.

Digital transmission allows *time-division multiplexing* (TDM) to be implemented readily, since bits or groups of bits need only be properly interleaved to accomplish TDM. TDM offers greater spectral economy than *frequency-division multiplexing* (FDM) since each channel of an FDM system experiences group-delay distortion at its band edges. Even with group delay correction circuits, frequencies near band edges often are not usable in an FDM system, and there are many more band edges in an FDM system than in a TDM system!

From the viewpoint of system implementation costs, single-channel analog systems may be implemented as economically as single-channel digital systems and may even be cheaper because no analog-to-digital conversion may be required. Multichannel signal handling and transmission, however, are a way of life in telephony, and the economies of digital systems over analog systems can be shown readily on a multichannel basis.

Finally, the issue of spectrum economy must be addressed. For many years, digital transmission was felt to be less spectrally efficient than analog transmission. That no longer is the case. With the arrival and widespread implementation of high-level digital modulation techniques, and the use of speech coding at rates only half as great as those conventionally used in the public switched network, digital systems are proving that they can be more spectrally efficient than analog systems. In fact, cellular common carriers are going to digital transmission for the very purpose of obtaining more channels in their limited spectral allocations.

A special advantage of digitized speech is the fact that it can be encrypted readily, thus providing the answer to growing public concerns about privacy in the use of radio telephony.

1.2. MAJOR HISTORICAL MILESTONES

We have identified *five basic technological developments* that have had major impacts on the history of telephony. The *first* was the discovery that direct current on metallic conductors could be modulated by the human voice. The *second* was the invention of the vacuum tube. The *third* was the development of semiconductor technology, beginning with the invention of the transistor, including solid-state crosspoints, and culminating in large-scale integration/very large-scale integration (LSI/VLSI) of solid-state devices. The *fourth* was the development of optical transmission technology, providing almost unlimited bandwidth on glass fibers. The *fifth* was the combination of radio transmission (wireless) technology with signal-processing technology to reduce the severity of multipath problems in the radio local-loop environment. Each of these technological discoveries has enabled telephony to be implemented within the limitations of those developments. We believe that the "optical era," as well as expanded "wireless" usage, will carry telephony well into the twenty-first century.

1.2.1. Modulation of Direct Current

On March 10, 1876, while working on his invention, Alexander Graham Bell spilled battery acid on his clothes and called out, "Mr. Watson, come here, I want you!" Mr. Watson came rushing in, shouting, "Mr. Bell, I heard every word you said, distinctly." The conversion of diaphragm vibrations into electrical energy that could modulate a battery-driven, metallic circuit and be converted back to sound waves at the distant end signaled the beginning of telephony.

The practical application of telephony was limited by the necessity of having each pair of users connected by a single pair of copper wires. That limitation was overcome by terminating the wire pairs in a jackfield into which a switchboard operator could insert plugended cords to connect one user to another. The turning of the hand crank of a magneto on the telephone set activated a metallic drop on a switchboard to alert an operator. Common-battery switchboards soon were developed which enabled operators to be alerted by the simple lifting of a telephone receiver from its hookswitch, thereby lighting a lamp on the switchboard.

In 1889, Almon B. Strowger patented the first workable two-motion stepping switch, controlled by pushbuttons on a telephone. After establishing a connection, the caller rang the called telephone by turning a magneto hand crank. The first commercial version was installed in 1892. Improvements came rapidly, and the rotary dial was invented in 1905, leading to an improved two-motion stepping switch that still is in use in many parts of the world today. Connections were established under the direct control of the user by dialing the digits of the called telephone number. Subsequently, common-control systems were developed, using panel and crossbar switches. Dialed digits were stored in senders, composed of relays, and then used to activate the switches. Common-control systems enabled electromechanical switching systems to grow to about 20,000 lines (see Appendix A).

The transmission medium began with the use of open-wire pairs, followed by the use of paired cable. Transmission on a pair of open wires strung on poles was enhanced by transposing the wires at certain intervals to compensate for capacitance between conductors, thereby reducing crosstalk, and by the use of loading coils at specified intervals to increase transmission range. Speech was transmitted on a single-channel basis as an analog signal. By about 1900, the weak electric currents produced by a telephone transmitter could operate a receiver at a distance of more than 1600 km (1000 mi) over open wire or about 160 km (100 mi) on underground cable. Line repeaters increased these distances somewhat, but the transmission range had approached the practical limits of these media.

1.2.2. Vacuum-Tube Technology

The limiting factor in the practical use of telephony was that of distance. The invention of the multiple-electrode vacuum tube in 1906 was the breakthrough that permitted an eventual quantum jump in transmission efficiency. Development of vacuum-tube line amplifiers in 1913 enabled speech to be transmitted over longer distances, and coast-to-coast transmission was commercially practical by 1915. Transatlantic radiotelephone service was introduced in 1927 using a single-sideband suppressed-carrier technique, and multiplexing systems in the 1930s permitted as many as 12 conversations to be conducted over two pairs of cable wires, using frequency-division multiplexing (FDM). In principle, FDM involves the use of a unique portion of the spectrum for each telephone voice channel. Coaxial cable permitted wideband transmission of telephone signals when the L1 coaxial carrier system, initially providing 480 two-way telephone circuits in a pair of coaxial tubes, was introduced in 1941, increasing both the capacity and quality of transmission systems. Later developments enabled over 32,000 telephone channels to be carried in a single cable with 20 coaxial tubes. The number of channels per system is limited by the amount of spectrum used for each channel and the total amount of spectrum allocated to the system.

The invention of the multiple-cavity vacuum tube in 1940 and experience with radar during World War II resulted in the appearance of the first commercial microwave system (see Chapter 8) used in telephony in 1947. It provided two broadband radio channels in each direction, each capable of handling one television signal or 480 telephone circuits. Since then, the capacity has been increased many times.

The first deep-sea telephone cable system was installed across the Atlantic Ocean in the 1950s; it had a capacity of 36 telephone conversations using two cables—one for each direction of transmission. That capacity was increased by use of Time Assignment Speech Interpolation (TASI) to take advantage of pauses in conversations when one party is listening or stops speaking for a moment. Subsequent cables increased the capacity many times.

1.2.3. Semiconductor Technology

The invention of the transistor in 1948 has revolutionized telephony by making digital telephony practical. In 1962, the *T-carrier* transmission system, using time-division multiplexing, was introduced (see Chapter 5). T-carrier was first installed on short-haul, high-capacity trunk groups for economic and transmission quality reasons. Advances in solid-state technology permitted T-carrier to be multiplexed to higher levels and to be installed on long-distance paths (see Chapter 6). Pulse-code modulation (PCM) (see Chapter 3) enables discrete samples of analog signals to be encoded for transmission with immunity from the noise that normally accompanies analog transmission. A derivative, adaptive PCM (ADPCM), enables encoded analog signals to be transmitted at half the bit rate of PCM.

Other methods of encoding are more efficient than PCM and are used in special applications (see Chapter 4) as well as in mobile telephony (see Chapter 7). Users have sought service while on the move, thus bringing about demands for mobile telephony. Because the radio spectrum is limited in terms of its availability, spectrally efficient speech coding techniques are vital. Parametric speech coding is spectrally very efficient and provides good speech intelligibility, but is poor in terms of speaker recognition. Consequently, hybrid speech coding has been developed, based on both parametric and waveform techniques. Further spectral efficiencies are achieved in mobile telephony by dividing the geography into cells and by having users time-share their channels. This time-sharing is accomplished through the use of time division multiple access (TDMA). Channel sharing is done through code division multiple access (CDMA).

Cellular radio would not have been possible without VLSI technology to enable sufficient miniaturization of components. Though cellular initially was implemented using analog signals, digital cellular is replacing analog cellular as the prevalent technology (see Chapter 7 and Appendix C). Satellite communications, initially deployed in the 1960s, have been enhanced through LSI/VLSI technology as a reliable medium for worldwide digital communications (see Chapter 9).

The transistor was first used in switching systems in the development of electronic common-control systems in analog switches. In the AT&T No. 1 ESS, the electronic common control supported an encapsulated-reed switching network. Later, it replaced the electromechanical common control in the No. 4 Crossbar system. Time-division switching of digital signals began with transistors and has been greatly enhanced through developments in VLSI technology. Digital switching systems (see Chapters 11 and 12) entered the network in the 1970s and, for the first time, enabled completely digital telephone connections from central office to central office. The local loops, however, were still copper conductors but now are being converted to digital technology (see Section 5.9 and Chapter 15). Common-channel signaling has distinct advantages over inband signaling and has enabled many new services to be offered (see Appendix A and Chapter 13).

Standardization within international standards bodies has resulted in the integration of voice and data services in narrowband, circuit-switched, integrated services digital networks (ISDN). Purely data networks have proliferated as user demand has increased. Both private networks and public data networks employ packet switching, predominantly using the CCITT X.25 standard. Local-area networks (LAN) operating at 10 Mb/s are in wide usage, and some are interconnected at 100 Mb/s by wide-area networks (WAN), formerly called metropolitan-area networks (MAN). Video-compression techniques have enabled video conferences to be conducted over increasingly narrower bandwidths (see Chapter 14).

1.2.4. Optical Transmission Technology

This *fourth* major development is in progress currently in the form of optical transmission. The ever increasing demand for more and more circuit capacity was addressed in the 1980s through the implementation of fiber-optic circuit technology, with fibers increasingly spanning more and more of the world. Fiber-optic transmission systems permit extremely high bit rates to be transmitted over long distances (see Chapter 10). Advances in fast packet switching (see Sections 6.5 and 14.6.5) are allowing speech, data, and video signals to be switched by the same equipment, leading to Broadband ISDN (BISDN). The Synchronous Optical Network (SONET) is being implemented along with Asynchronous Transfer Mode (ATM) switching (see Chapters 10 and 14).

1.2.5. Wireless Technology

Radio-wave propagation was known in the earliest days of telephony, and in the early 1900s began to be used in a crude form (spark transmission) for telegraphic communication to and from vessels at sea. The application of the vacuum tube allowed the introduction of radio telephony, and transoceanic circuits were soon established. The 1920s saw the addition of radio equipment for such public service applications as police and fire. The technology of those days, however, allowed only bulky equipment, which had to be vehicle mounted.

World War II saw the fielding of the "walkie-talkie." By this time vacuum tubes had been reduced in size and power consumption to the point that handheld units were feasible. Their actual size and weight, however, left much to be desired. After the war, the 1960s saw the implementation of public mobile telephony by radio common carriers, some of which were associated with the wireline (franchised) carrier for the area, and some of which were independent. These systems took advantage of the development of "static-free" frequency modulation (FM), as well as hardware developments in the VHF and UHF frequency ranges. Many of the early systems provided only half-duplex transmission (one direction at a time), but such limitations were overcome in the 1970s. Service providers attempted to place their antennas as high as possible in major metropolitan centers to maximize their coverage areas. Mobile users used transmitting powers as high as 100 watts. The limited number of available radio channels, however, meant that only people with special emergency needs, such as physicians, police, and fire and rescue personnel, could be provided the service, and waiting times for service approval often were measured in months.

The implementation of the cellular concept in the 1980s, with its frequency reuse within metropolitan areas, finally brought about the widespread availability of radio telephony.

1.3. A FUTURISTIC LOOK

In the first edition of this book, Fig. 1.1 was presented as a possible access arrangement for ISDN. With multiple services beginning to access local switching systems via optical fiber using ATM in SONET format, the local public network configuration is approaching that of Fig. 1.1. User demand is increasing. As BISDN standards development progresses, BISDN will begin to replace existing

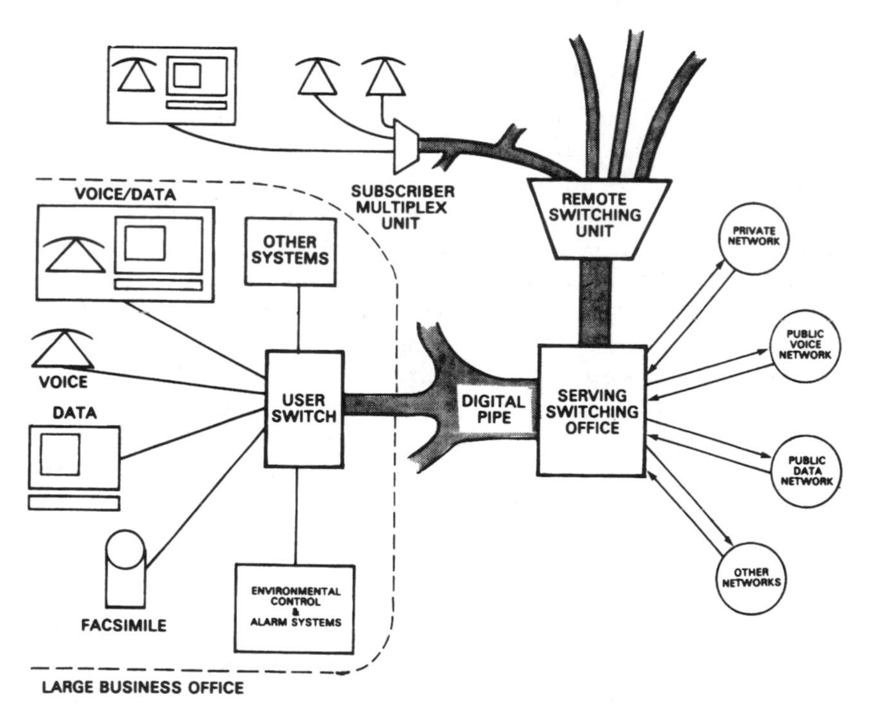

FIGURE 1.1 Potential access arrangement for Broadband ISDN. (Based on Fig. 1, Irwin Dorros, "Telephone nets go digital," *IEEE Spectrum*, April 1983, © 1983 IEEE.)

wideband data services as an overlay network. Then, over the next two decades or so, as older equipment reaches the end of its economic life cycle, most telecommunications may be carried over BISDN. Wireless technology is being integrated rapidly into the network as cellular and personal communications systems (PCS) grow. These systems can be used at home as cordless phones, in the car, and also at work or elsewhere. In addition, they may permit individuals to see one another in real-time motion as they speak, using portable equipment that will function almost anywhere in the world. Already, subscriber-assisted registration is in use on voice units, and plans are being made to further implement the autonomous registration capabilities built into subscriber sets, thus enabling subscribers to make and receive telephone calls from many areas. Users in sparsely populated areas can communicate by satellite, and telephone numbers can be assigned to individual persons in addition to lines at specific locations. Standards bodies are developing standards for a global personal communications network (PCN), and plans to provide global PCS via groups of satellites in low or medium earth orbit are being implemented. The day may arrive soon when the telephone is thought of as a device to be carried everywhere by a person rather than as a device used only in a fixed location.

1.4. ORGANIZATION OF THE TEXT

Since the development and introduction of T-carrier into the North American network in 1962, digital technology has been incorporated gradually into an otherwise analog network. One cannot identify a time at which the network changed from mostly analog to mostly digital. The progression from analog-to-mixed-to-all-digital has not yet been completed. Significant amounts of analog technology are still used in the network, especially in subscriber loops. In the organization of this second edition, the authors have developed three appendices that deal with analog technology. Appendix A is an in-depth coverage of the North American analog network as it has evolved into a mixed analog and digital network. Although most of the coverage pertains to analog technology, there is substantial applicability to the mixed analog-digital network and some applicability to an all-digital network. Appendix B is a brief discussion of traffic considerations and network management applicable to a circuit-switched network. Appendix C covers analog cellular telephony. All three appendices have some degree of applicability to network telephony as it currently exists in North America. This material is not covered in the chapters. Therefore, readers who have little or no background in the network as it has evolved over the last quarter century may wish to familiarize themselves with the appendices before beginning study of the chapters on purely digital technology.

The sequence of the chapters in the text is based, to an extent, on the logical progression of a signal through a telephone network. The words spoken into the telephone transmitter are first digitized (coded). Accordingly, the text begins with speech coding (Chapters 2 through 4). Chapter 2 provides an overview of speech coding; Chapter 3 describes waveform coding, as used in public and nonpublic wireline networks; and Chapter 4 deals with parametric coding, which is significant in mobile telephony.

Since this is a book on telephony, the emphasis is on speech, but since this is a book on *digital* telephony, the ability of the network to handle data and graphics (including video) also is important. Thus Chapter 5 covers many topics dealing with the use of digital techniques in the network.

After the speech has been digitized and related control functions have been described, the text next deals with transporting the speech or other information to another location. Accordingly, Chapters 6 through 10 deal with various aspects of transmission. In Chapter 6, the signal is prepared for transmission by being modulated. Other general transmission topics, such as error control, multiple access, and packet transmission also are included in Chapter 6. This leads naturally to the discussion of the radio link used in the local transmission to and from wireless subscriber sets in Chapter 7. Technology applicable to analog cellular is discussed in Appendix C. Following this are three chapters covering the long haul topics of microwave radio, satellite communication, and optical fiber transmission.

The heart of any telephony system, of course, is the switching function. This vital function is the subject of Chapters 11 and 12. Chapter 11 covers digital switching-system architecture from the viewpoints of line and trunk terminations, switching network, service circuits, control, and overhead for maintenance diagnostics, data base administration, traffic administration, and network management. Chapter 12 contains architectural discussions of four digital switching systems currently in operation in North America.

Chapters 13 and 14 tie all the foregoing chapters together in discussions of networking as it applies to North America. Chapter 13 treats the progressive introduction of digital switching and transmission equipment into the formerly analog network that is described in Appendix A. Chapter 14 discusses the fledgling Integrated Services Digital Network (ISDN) and its evolution into Broadband ISDN. Finally, Chapter 15 deals with various methods being used to bring broadband services into the home by applying digital technology to the local loop.

2

Speech Digitization Fundamentals

2.1. INTRODUCTION

In digital telephony, speech is handled in digital form. Accordingly, the speech from the telephone transmitter must be converted from its analog waveform to a series of ones and zeros. A multitude of speech digitization methods has been devised; these methods can be categorized as waveform, parametric, and hybrid coding. A *waveform coder* takes the actual waveform and produces a series of ones and zeros representative of that waveform according to a set of rules. A *parametric coder*, sometimes called a *source coder*, attempts to detect certain characteristics of speech, such as pitch and amplitude, and produces a series of ones and zeros according to a set of rules that describes the detected speech characteristics rather than the waveform. A *hybrid coder* is one that uses both waveform and parametric principles in its production of a digital version of speech.

A speech waveform is a continuous function of time, but it or its parameters are to be converted into a series of digits which occur at a specific rate. Accordingly, the speech must be sampled periodically. Thus, sampling is an important step in speech digitization.

Each sample conveys a magnitude or a set of parameters based on the type of coding being used. This magnitude must be expressed as a series of digits. The magnitude may be a waveform amplitude or a speech parameter value at some sampling instant or during some sampling period. The process of converting a waveform's magnitude or parameters at a given time into a series of digits is called *quantization*. The processes of speech digitization described in this chapter may result in the need for a bandwidth greater than the original analog speech bandwidth. The reasons for this increase are described in this chapter.

The relative performance capabilities of the waveform, parametric, and hybrid digitization methods are summarized at the close of the chapter, along with the advantages of speech digitization from the viewpoints of overall performance as well as system cost and reliability.

2.2. SPEECH CODING APPROACHES

Waveform and parametric coding constitute the two basic approaches to encoding a source which, in telephony, usually is thought of as producing a voice waveform, but which may produce various signaling and supervision tones such as dial tone, address signaling pulses or tones, or ringing and busy tones.

Waveform coding is performed by transforming the waveform itself into a series of digits. The rules used to achieve this transformation may be applied to the waveform itself, or to a differential version of the waveform which has been formed by subtracting a value based on recent past samples from the present value, as illustrated in Fig. 2.1. As can be seen from the figure, the differential waveform has a smaller amplitude, in general, than does the original waveform. This results from the fact that the original waveform is band-limited and thus its value does not change rapidly over a very short time interval. Recovery of the original waveform from the differential version is accomplished by integration. Chapter 3 describes waveform coding in detail.

The coding parameters may be varied during transmission with respect to their numerical values to accommodate different waveform characteristics. Such variable techniques are called *adaptive coding*. Adaptive coding is very useful in dealing with speech waveforms which may change, for example, from male to female voices, or with waveforms which may change their volume and pitch rapidly.

Parametric coders do not attempt to deal with the speech or signaling waveform at all. Instead, they function by extracting some characteristic of the waveform. Parametric coders are designed to look for speech characteristics, and to reproduce a waveform that sounds like the original one, but which may be different, especially in terms of relative phase values. Since the human ear is relatively insensitive to phase, parametric coders do not attempt to reproduce phase accurately. As a result, any phase-modulated data signals cannot be reproduced suitably. In general, the waveform will consist of speech having a given pitch, amplitude and other spectral characteristics. Such characteristics then are represented by a set of parameters. Other approaches involve taking a Fourier or other transform of the wave and then transmitting the parameters of the transform. Chapter 4 discusses parametric coding in detail.

The concept of *prediction* enters often into source coding. Many speech en-

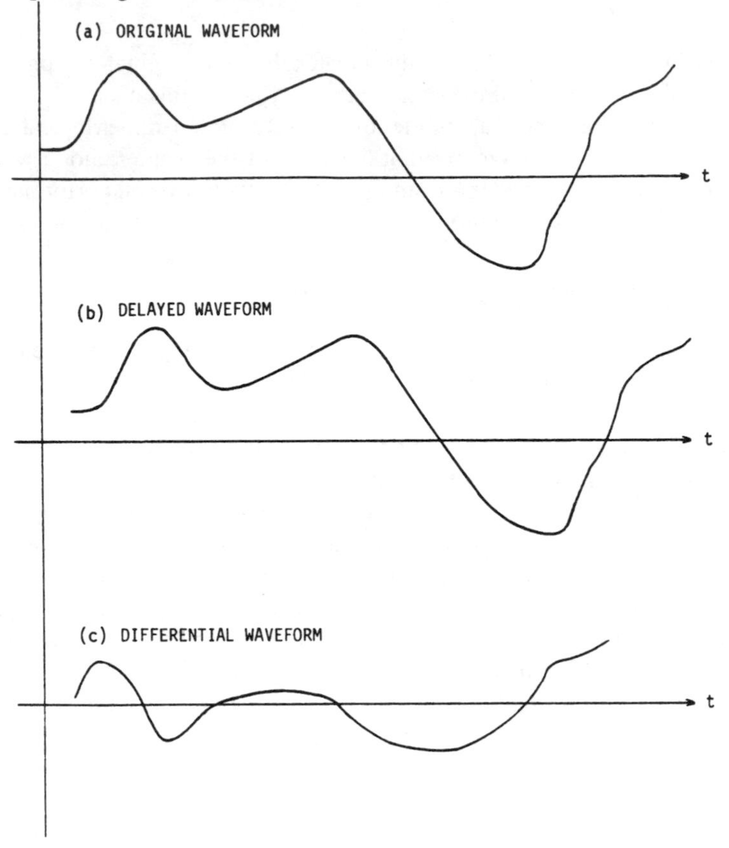

FIGURE 2.1. Formation of a differential version of a waveform: a. original waveform; b. delayed waveform; c. differential waveform.

coders predict the next digital value of the input and combine this predicted value with the actual next value to produce a quantized differential. The prediction is based upon a weighted sum of recent past values. At the receiver, the differentials are summed to obtain the total value of the signal.

The speech-coding technique most commonly used in telephony today is a waveform-coding scheme called pulse-code modulation (PCM); it is detailed in Chapter 3. The coding for PCM currently is done at the central office, in remote units, or in private networks. PCM coders, however, now can be implemented at very low cost on a single chip, thus allowing their placement within the telephone handset. For such techniques to be accepted and used on a widespread basis, however, digital subscriber lines are needed, as outlined in Section 5.7 and in Chapter 14.

2.3. SAMPLING

The process of digitizing an input involves, first of all, making this input discrete in the time domain. A sample, therefore, must be taken each time a change is to be recorded. In the case of waveform coding, each significant change is to be recorded; thus, samples are taken quite frequently. In the case of parametric coding, however, only changes in characteristics of the input are to be recorded. Consequently, parametric samples are taken much less frequently. The discussion that follows describes the sampling that is done in waveform-coding processes. Sampling for parametric-coding purposes is done at a rate consistent with the rate of production of human speech syllables, and is discussed in Chapter 4.

If a waveform contains significant frequencies as high as W Hz, then, according to Nyquist's Theorem, that waveform can be reconstructed perfectly if it is sampled at a rate of $2W$ per second. Actually, if the lowest significant frequency is greater than zero, the sampling rate can be decreased to twice the actual bandwidth. The sampling theorem can be stated as follows:

If a signal that is band-limited is sampled at regular intervals at a rate at least twice that of the highest frequency in the band, then the samples contain all the information of the original signal.

Because frequencies outside the band W may be reproduced inside the band W by the receiving desampler, the sampler must be preceded by a filter to minimize such out-of-band energy.

To illustrate the operation of a sampler, let $f(t)$ be a band-limited signal which has no spectral components above W Hz. Multiply $f(t)$ by a uniform train of time impulse functions $\delta_T(t)$, as illustrated in Fig. 2.2.

The result of this multiplication is the product function $f_s(t)$:

$$f_s(t) = f(t)\delta_T(t) = \sum_{n=-\infty}^{\infty} f(nT)\delta(t - nT) \tag{2.1}$$

The Fourier transform of $f(t)\delta_T(t)$, according to the frequency convolution theorem, is called $F_s(\omega)$ and is given by

$$2\pi f_s(t) \longleftrightarrow F(\omega) * \omega_0 \delta_{\omega 0}(\omega)$$

Letting $\omega_0 = 2\pi/T$,

$$f_s(t) \longleftrightarrow (1/T)F(\omega) * \delta_{\omega 0}(\omega)$$

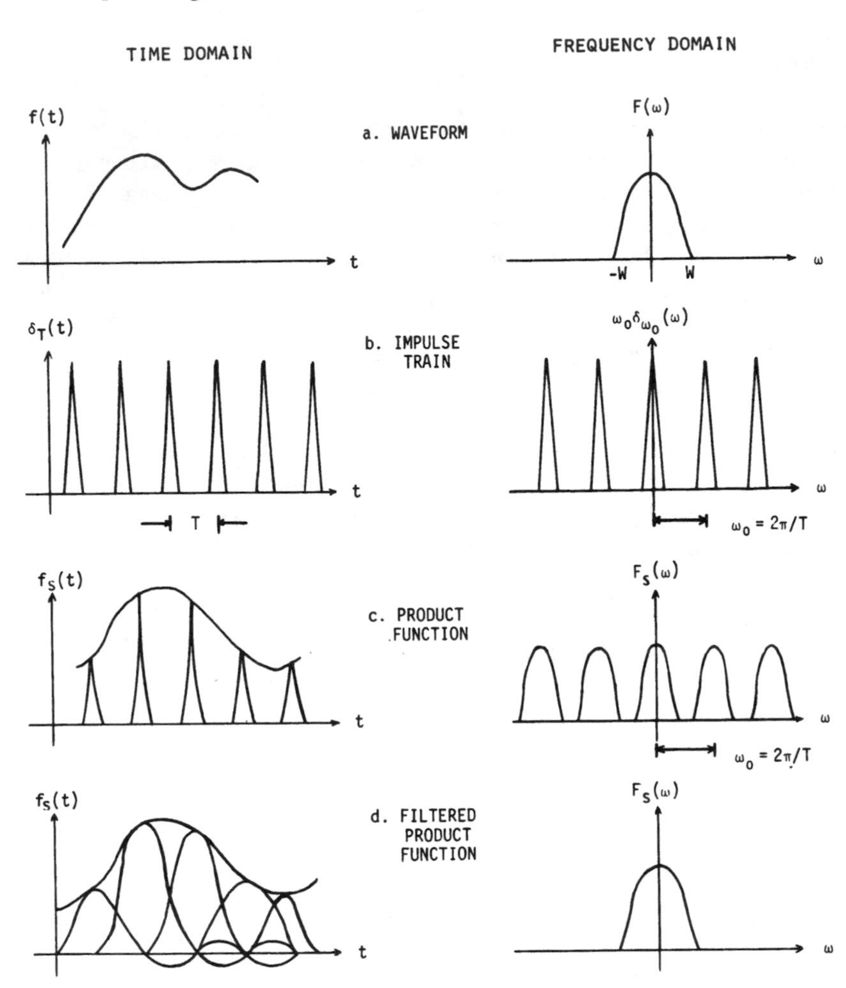

FIGURE 2.2 Illustration of sampling theorem.

In the frequency domain, the corresponding function is a series of harmonics spaced $\omega_0 = 2\pi/T$. The product function $f(t)\delta_T(t)$ is a sequence of impulses located at regular intervals of T seconds each, and having amplitudes equal to the values of $f(t)$ at the corresponding instants. In other words, this is a pulse-amplitude modulated (PAM) sequence. Note that the process of multiplication corresponds, in the frequency domain, to a replication of the original spectrum $F(\omega)$ every ω_0 rad/s.

If $\omega_0 > 2(2\pi W)$, the spectral representations of $F(\omega)$ do not overlap, i.e., $2\pi/T > 2(2\pi W)$, and thus $T \leqslant 1/2W$.

Therefore, if $f(t)$ is sampled at regular intervals less than $1/2W$ seconds apart, the sampled spectral density function $F_s(\omega)$ will be a periodic replica of the true $F(\omega)$. Thus $F_s(\omega)$ will contain all the information of $f(t)$. Removal of all but the spectrum around zero frequency by low-pass filtering then leaves the reconstructed function as the output.

2.4. QUANTIZATION

Each time a signal or one of its characteristics is sampled, its value must be expressed in digital form, that is, as a series of ones and zeros. This is done by quantization, a process in which the continuous range of values of an input signal is divided into nonoverlapping subranges. The presence of the input signal in a particular subrange results in the production of a unique series of bits by the quantizer. In designing a quantizer, the level of detail desired is first established, and then the quantization levels or bins are set up accordingly. Thus, quantization may be said to remove details that the designer regards as irrelevant.

Figure 2.3 illustrates a typical quantizing characteristic. The fact that irrelevant details are removed results in a reconstructed output that can take on only specific values. Thus an error has been generated known as *quantizing distortion*, or *quantizing noise*. A linearly increasing input (ramp function), for example, exhibits the quantizing noise indicated in Fig. 2.4.

For an oscillating input of the type that might be found in a speech waveform, the error exhibits the characteristics of the input near its peaks, as illustrated in Fig. 2.5.

The choice of the level spacing, or step size, may be a difficult one if the input

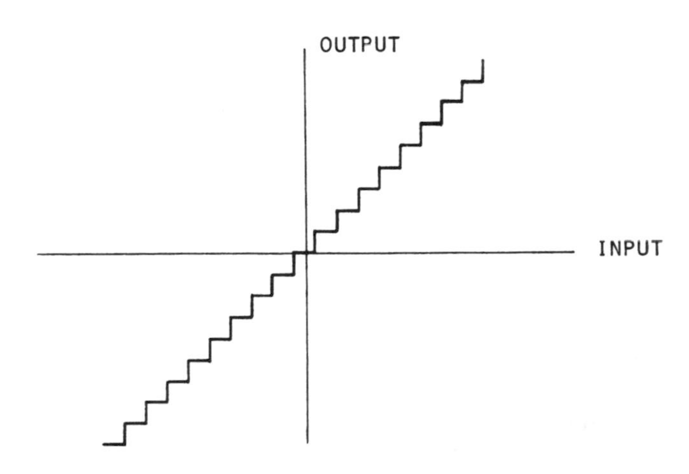

FIGURE 2.3. Typical quantizing characteristic.

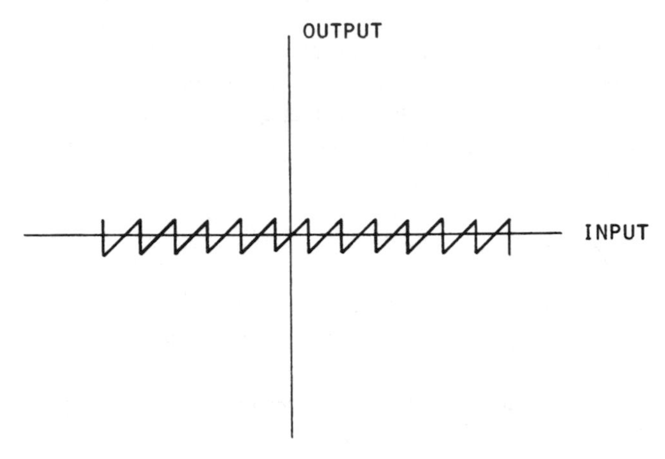

FIGURE 2.4. Errors in quantization of a ramp function.

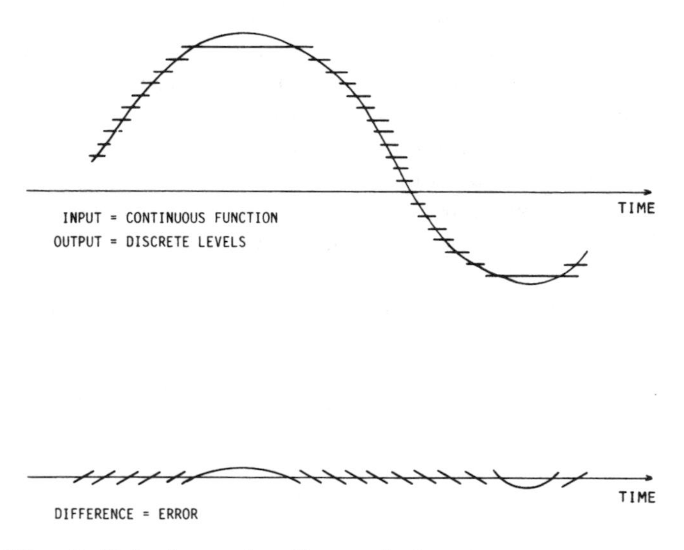

FIGURE 2.5. Oscillating input and resulting quantization error.

is highly variable, as speech generally is. For example, when the speaker is talking loudly, the waveform peaks are large, and large steps are desirable to prevent peak clipping and still not use an unreasonably large number of steps. On the other hand, when the speaker is talking softly, small steps are desirable. One solution to this dilemma is the use of adaptive quantization. The step size is adapted to the signal variance based upon a memory associated with the quantizer. Usually the step size is modified for each new input sample, based upon a knowl-

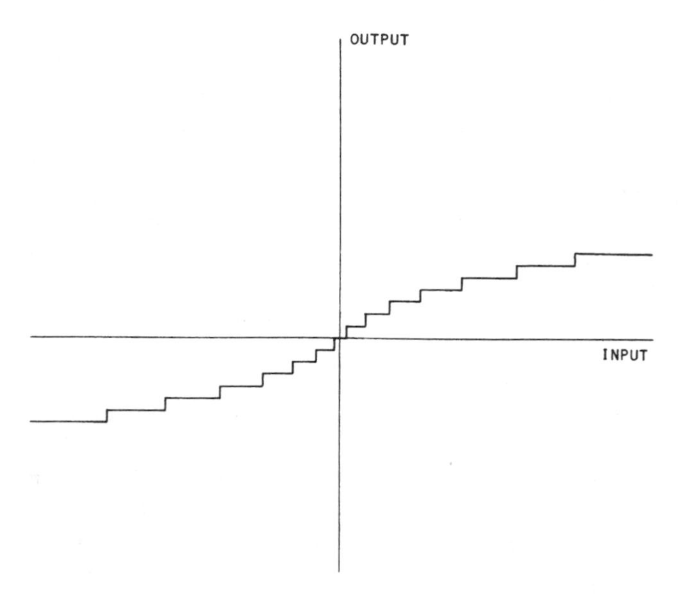

FIGURE 2.6. Nonuniform quantizer characteristic.

edge of which quantizer slots were occupied by the previous samples. Adaptive quantization is discussed further in Chapter 3.

Another approach to the problem of preserving the large dynamic range characteristic of human speech is the use of nonuniform quantization, as illustrated in Fig. 2.6. Here the total number of steps remains the same as in the case of uniform quantization, but low levels are represented by a large number of small steps and high levels are represented by a smaller number of large steps. Note that the use of nonuniform quantization is equivalent to passing the input waveform through an amplitude compressor, quantizing linearly, and then reconstructing the result by dequantizing linearly and passing the result through an amplitude expander. Nonuniform quantization is discussed further in Chapter 3.

Adaptive and nonuniform quantization techniques make use of specific properties of speech. Accordingly, these techniques may be said to be *speech specific.*

2.5. EFFECT OF DIGITIZATION ON BANDWIDTH

Sections 2.2 and 2.3 described two fundamental principles of waveform coding—sampling and quantization. What is the consequence of using these principles on the bandwidth of the transmitted signal?

The digitization of a waveform produces a bit stream whose rate depends upon the sampling rate and the number of bits per sample produced by the quantizer. In the case of waveform coding, Nyquist's theorem states that the sampling rate must be at least twice the maximum significant frequency contained in the input. To keep quantization noise low, quantization with a small step size is needed; this means a large total number of steps and thus many bits per sample.

For current telephone applications, speech most often is sampled 8000 times per second and quantized at 8 bits/sample, resulting in a 64-kb/s stream. If each bit involves a level change (new symbol), then a 64-kBd stream results. The bandwidth of this stream depends on the selectivity of the waveform shaping filters used, but probably will be on the order of 100 kHz. On the other hand, if a highly efficient keying technique providing 4 bits/symbol is used, then a 16-kBd stream is produced, occupying a bandwidth on the order of 25 kHz. In North American practice, however, 24 telephone channels usually are time-division multiplexed together to produce a 1.544-Mb/s stream which then may be transmitted, for example, using a seven-level format, resulting in a 386-kHz bandwidth for the 24 channels, or an equivalent of nearly 16 kHz per channel. European standards time-division multiplex 30 telephone channels plus two housekeeping channels together to produce a 2.048-Mb/s stream, which may be similarly treated. In general, the term b/s per Hz is used as a measure of spectrum efficiency for digital-modulation techniques.

Digital speech now can be transmitted in less bandwidth than is required for analog transmission. Using a standardized 32-kb/s speech-coding technique (described in Chapter 3), and a 256-level digital-modulation technique (described in Chapter 6), the equivalent bandwidth occupancy of a speech signal can be reduced to 4 kHz, and this value is reduced to 2 kHz with the use of a 16-kb/s speech standard.

The foregoing discussion addressed the digitization of a waveform. If hybrid coding is used instead, as discussed in detail in Chapter 4, speech can be represented by a stream rate of 4.8 kb/s or even less. Since some data modems are capable of sending 19.2 kb/s or more over conditioned telephone facilities, one might say that the equivalent bandwidth of such digitized speech channel is only one-fourth that of a 4-kHz telephone voice channel, or about 1000 Hz.

2.6. SPEECH DIGITIZER PERFORMANCE DEFINITIONS

To set the stage for the discussion of waveform coders in Chapter 3 and parametric coders in Chapter 4, speech-coder performance definitions are needed. A wide variety of speech digitizers has been devised, ranging all the way from 64-kb/s *pulse-code modulation* (PCM) systems (see Section 3.2) to very-low-bit-rate *for-*

mant vocoders (see Section 4.3). The speech reproduced by these devices may be described as being in one of the following categories.

1. *Toll quality:* quality based[1] upon a laboratory test in which the signal-to-noise ratio exceeds 30 dB, the frequency response is 200–3200 Hz, and the total harmonic distortion is less than 3%. This quality is acceptable to most members of the general public, but is not necessarily achieved on many connections. Listeners rate toll-quality speech as "good" or better.
2. *Communications quality:* quality which is acceptable to users of cellular telephones, as well as to military and amateur radio operators. Listeners rate communications quality speech as "fair" to "good."
3. *Synthetic quality:* the quality of computer-generated speech, which often lacks human naturalness and the ability to recognize the speaker. Listeners rate synthetic speech as below "fair" to "good."

A fourth quality level exists. It is called "broadcast" or "commentary" quality. This quality level usually is associated with a transmission capability of at least 7 kHz.

Figure 2.7 indicates that waveform coding can provide toll-quality speech at digital rates on the order of 11 to 12 kb/s and higher. At lower rates, down to about 4 kb/s, the quality is judged to be communications quality. At rates below about 4 kb/s, parametric or "source" coders provide better intelligibility than waveform coders, but with synthetic quality. The vertical dashed lines in Figure 2.7 are not intended to be precise lines of demarcation. Some hybrid coders provide very close to toll quality, whereas some waveform coders provide only communication quality.

The bit rate at which parametric coders perform better than waveform coders is not clearly established. Figure 2.8 summarizes the impressions expressed by large groups of listeners to different speech coders.[2] Here the parametric coders

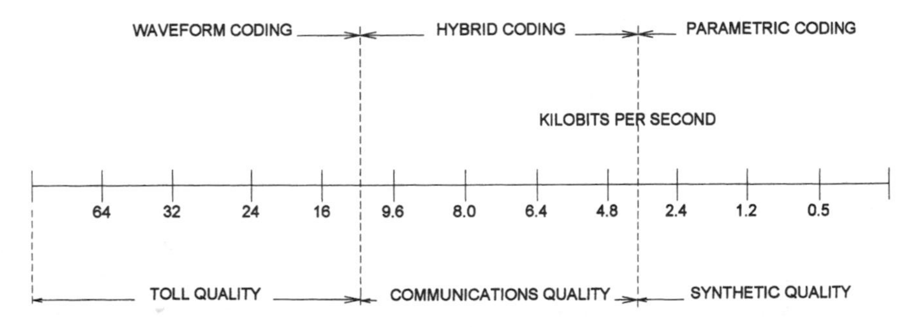

FIGURE 2.7. Quality achievable in speech coding.

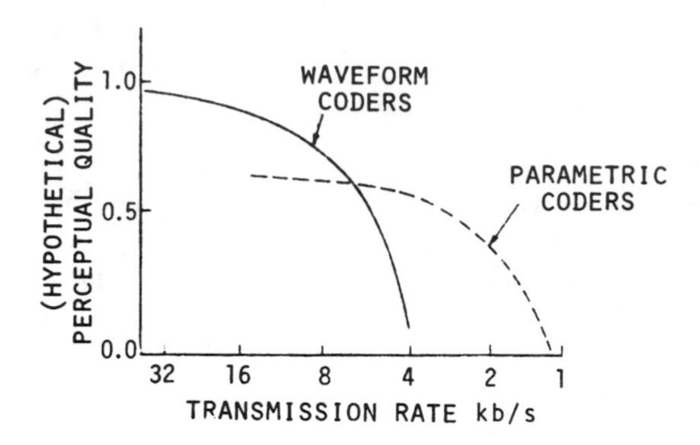

FIGURE 2.8. Perceptual quality comparison of waveform and parametric coders. (Courtesy J. L. Flanagan et al., Ref. 2, © IEEE, 1979.)

are called *vocoders*, a name commonly used for them. Nevertheless, the superiority of waveform coders above about 8 kb/s is clearly evident. Chapter 4 describes the superiority of hybrid (combined waveform and parametric) coders at rates between 4 and 14 kb/s.

Two general criteria are used in judging the performance of a speech coder. One is the ratio of signal to quantization distortion (noise) (SNR), which can be derived analytically. Since quantization noise is generated by the coder, not the channel, the SNR depends primarily on the type of coder used, as well as its bit rate. Alternatively, subjective and perceptual effects (How does it sound?) can be used to judge coder performance. The results can be evaluated only by actual listener tests. They include not only quantization noise effects but also the various types of distortions which a speech coder may produce and encounter.

2.7. SPEECH-CODING ADVANTAGES

Section 2.5, "Effect of Digitization on Bandwidth," has shown that the digitization of speech may increase its bandwidth if standard waveform coding is used. Offsetting this disadvantage, however, are a number of advantages:

1. The coded signal is easier for a detector to recognize than is an uncoded one. As a result, lower carrier-to-noise ratios can be tolerated than for analog signals. Only bits and bit combinations need be detected, rather than the exact values of a continuous waveform.
2. The coded signal can be regenerated efficiently. This allows the elimination of noise and distortion that may have entered between or at the repeaters of a

transmission facility. However, severe noise levels may result in bit errors which are not eliminated or reduced unless special error-correction techniques are used.

3. Coded signals are easily encrypted. This can be achieved using scrambling techniques of the type discussed in Chapter 5. Based on pseudo-noise sequences, these techniques continually change the significance of each bit combination. Thus only a receiver with the proper "key" can recover the transmitted intelligence.
4. Coding makes possible the combination of transmission and switching functions. Since both speech and signaling are in the form of bit streams, such combinations become possible.
5. Coding provides a uniform format for different types of signals. This format is the same whether the input is dial tone, addressing, ringing, speech, or disconnect signal. Everything is in the form of a digital bit stream.

The foregoing technological advantages translate into economic advantages as well. As noted in Chapter 1, digitized speech allows the use of digital techniques throughout the network, with their large-scale economies. In addition, as the point of digitization moves toward the end user, a greater portion of network costs are incurred only when the user is ready to use and pay for them.[3] Moreover, digital techniques require less critical circuit adjustments, thus reducing maintenance costs.

PROBLEMS

2.1 Explain why parametric coding devised for speech will not accurately reproduce switching, signaling and data waveforms from the analog network.

2.2 Can a waveform coder be speech specific? Explain.

2.3 Common carriers provide special lines selected for low noise and low group-delay variation. These lines are called conditioned lines. A corporation leases a conditioned voice-grade line connecting its offices on opposite sides of the country. How many digital voice circuits at 4.8 kb/s can be accommodated on this line, using 9.6-kb/s modems? Using 19.2-kb/s modems?

2.4 Explain why a band-limiting filter is required ahead of the sampling device in a waveform coder. If an attempt were made to improve the speech quality by increasing the passband of this filter to greater than half the standard sampling rate of 8000/s, what would be the result?

2.5 Explain how the use of nonuniform quantization allows a speech sample to be coded using a smaller number of bits than would be required for uniform quantization.

REFERENCES

1. Flanagan, J. L., "Opportunities and Issues in Digitized Voice," *IEEE-EASCON '78 Record*, pp. 709–712.
2. Flanagan, J. L., et al., "Speech Coding," *IEEE Trans. Comm.*, Vol. COM-27, No. 4, p. 729 (April, 1979).
3. Bellamy, J. C., *Digital Telephony*, John Wiley & Sons, Inc., New York, N.Y., 1982.

3

Waveform Coding

3.1 INTRODUCTION

Waveform coding is the type of speech digitization used in the public switched (wireline) telephony network. Waveform coding techniques describe the waveform's instantaneous behavior. This means that the waveform does not have to be speech; in fact it can be analog data or a signaling tone. The simple process of sampling and quantizing a waveform might be called "brute force coding." It is done independently of any special characteristics the input may exhibit. Such coding actually is called linear *pulse-code modulation* (PCM). This chapter begins with a discussion of PCM, but then shows that by making use of certain special characteristics of speech, the required bit rate can be reduced significantly below the 96 kb/s or more of linear PCM, and even below the 64 kb/s of standard telephony PCM. The term PCM refers to the use of a specific set of rules for transforming a waveform into a stream of digits, and vice versa. To clarify the concept, however, the first point to be made is that PCM is coding, but it *is not* the modulation of a carrier. Modulation usually refers to an alteration of a periodic function (often a radio-frequency sine wave) to cause it to convey intelligence. This alteration may be a change in amplitude or frequency or phase of the sine wave. These are the ways in which a carrier can be modulated. The PCM process, in contrast, is a coding technique. The resulting stream of ones and zeros can be used to modulate a carrier in any of the ways mentioned here, or special combinations of them. The coding process known as PCM uses such techniques as sampling and quantization, already discussed in Chapter 2. It also involves the concept of synchronization. Synchronization refers to the timing with which information is transmitted. The receiver must know when to look for a new bit, and which bits constitute bytes or samples, etc. As applied to the transmission of a bit stream, communication may be accomplished either on an asynchronous or a synchronous basis.

Asynchronous communication refers to the transfer of data at nonuniform rates. A stop interval is required to guarantee that each character will begin with a one-to-zero transition, even if the preceding character was entirely zeros.

Synchronous communication is the technique used for PCM transmission, as well as for many other speech coding techniques. Synchronization can be performed on either a digit or a frame basis. *Digit synchronization* uses the timing information contained in the transmitted signal without altering the signal itself. In other words, the receiver simply derives its timing from changes in the bit stream itself. *Frame synchronization*, in contrast, uses a fixed synchronization signal once per frame, where "frame" refers to a block or group of digits transmitted as a unit, over which a coding procedure is applied for synchronization or error control purposes. This is the type of synchronization used in PCM transmission.

A major advantage of synchronous communication over asynchronous is that all, or nearly all, bits are used to transmit data, since stop intervals or bits are not needed for framing.

To determine which groups of bits constitute a character, a "sync" character is used, which is chosen so that its bit arrangement is significantly different from that of any regular character being transmitted. As will be seen in Chapter 5, in a PCM stream of bits a large but specific number of information bits is followed by a single "framing" bit, followed by more information bits, etc. If the framing bits were selected from the overall bit stream, they would constitute a much lower rate stream. The sync character is sent within this lower rate stream. For example, in systems using North American standards, 24 PCM voice channels are transmitted at a rate of 1.544 Mb/s. Within that 1.544-Mb/s stream is a framing bit stream whose rate is 8000 b/s. By contrast, systems using European standards combine 30 PCM voice channels plus two "housekeeping" channels to produce a 2.048-Mb/s stream. These housekeeping channels include framing as well as signaling functions.

The receiver must recognize the sync character in order for it to determine which groups of bits constitute characters. Accordingly, when transmission begins, the receiver goes into its "sync search" mode. Occurrence of a data dropout also activates sync search. In the sync search mode, bits are shifted into the receiver's shift register. The contents of this shift register then are compared continually with the stored sync character which resides in another receiver register. This process continues until a match occurs. When a match is found on two successive sync characters, the receiver raises a "character available" flag every eight bits and normal receiver operation proceeds. In the North American systems, a framing bit appears once every 193 bits, whereas in European systems, two separate "channels" are used to convey such information (see Section 5.2).

3.2. BASIC PCM ENCODING

A continuous waveform representing speech, music or data can be transformed into a bit stream using PCM.[1] The procedure is as follows: the waveform is sampled at a rate of at least $2W$ Hz, where W is the highest significant frequency contained in the waveform. Usually higher frequencies will have been heavily attenuated by a filter so they do not result in spurious output components. The sampling rate can be less than $2W$ if the lowest frequency contained in the continuous waveform is nonzero. Standard PCM as used today on the public switched network operates with a sampling rate of 8000/s. This rate is adequate for telephonic speech, all of whose frequency components are confined to frequencies below 4000 Hz. For music, as stored on digital audio tape or compact disks, the sampling rates are above 40 000/s.

The sampling process produces a pulse train that is amplitude modulated, as illustrated in Fig. 2.2(c).

The continuous waveform now has been made discrete in the time domain by virtue of the sampling process. Next, it must be made discrete in the amplitude domain as well. This is accomplished by the quantization process, in which a large number of possible amplitude levels for the waveform are established. This number is a multiple of 2. For example, a total of 256 possible levels, 128 positive and 128 negative, can be achieved by 8-bit quantization, since $2^8 = 256$. This 8-bit quantization is the standard used in all of the PCM systems being implemented in the public switched network. In general, the PCM process quantizes the amplitude of each signal sample into one of 2^B levels. Thus the amount of information is said to be B bits/sample. Since there are $2W$ samples/s, the information rate is $2WB$ b/s.

The quantization process represents discrete amplitude levels by distinct binary words of length B. For example, if $B = 2$, there are four levels: 00, 01, 10 and 11.

The PCM decoding process maps the binary words back into amplitude levels. The initial result is an amplitude-time pulse sequence which is low-pass filtered, using a filter with a cutoff frequency W.

The discussion in the remainder of this chapter focuses on speech transmission. The statistical characteristics of speech are used in making the speech coding more efficient than it could be if these characteristics were not used. By contrast, music usually is coded without regard to special statistical characteristics because of the wide variety of properties that music may exhibit. Data, on the other hand, tends to be of constant amplitude. Data can be transmitted most efficiently without subjecting it to the type of coding done on speech. Thus a key advantage of digital networks is their direct transmission of data without the need for modem-type conversions.

Figure 3.1 illustrates the use of 8 bits in producing the 256 possible encoded

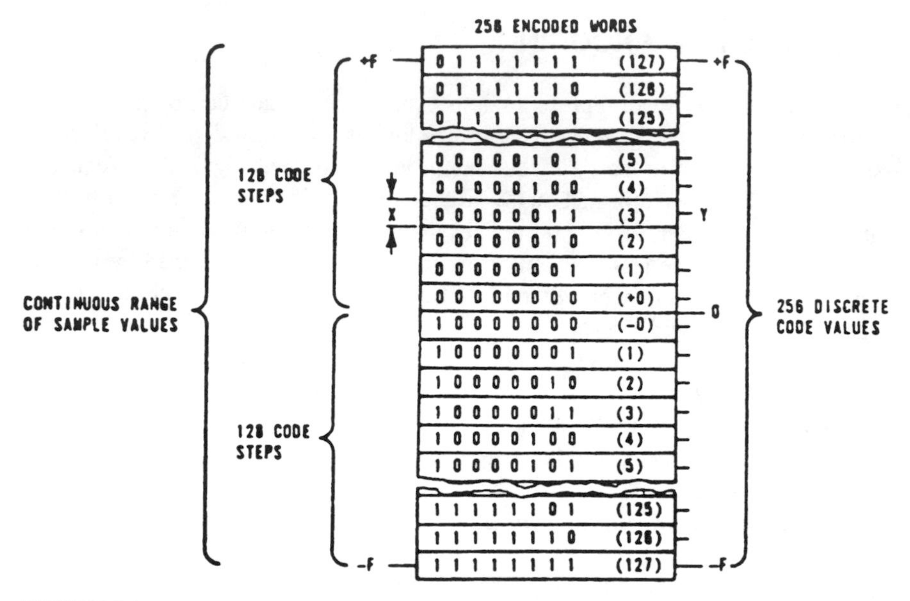

FIGURE 3.1. PCM encoding.

words representing the continuous range of sample values. As shown, however, the use of this code results in the sending of a large number of zeros, since the most probable speech input amplitudes fall in the middle of the code range, where the zeros predominate. As will be discussed in Chapter 5, a digital zero corresponds to a zero level on the digital transmission line. Since timing for systems such as the DS1 (see Section 6.3) and other standard digital arrangements is derived from the data signal, a long sequence of zeros means that there is no data signal from which to derive timing. Accordingly, a mixture of zeros and ones is needed. This mixture can be provided by inverting the code words, thus using what is called *zero code suppression.* Figure 3.2 illustrates the use of this technique. The −27 level corresponds to a sample that produces the all zero code. When this level is reached, a one is placed in the next-to-least-significant bit position, and the word is transmitted as 00000010, corresponding to −125. The use of zero code suppression as illustrated here is important, since on a T1 repeated line no more than 14 consecutive zeros can be transmitted.[2]

The foregoing discussion has described the 8-bit PCM parameters of the voice encoding system currently used on the public switched network. How well does it perform? Is there an incentive to change from $B = 8$ to some other value of B? An increase in B will increase the spectrum occupied, or require higher-level modulation techniques (see Chapter 6). Accordingly, B should be kept as small as possible consistent with meeting performance requirements.

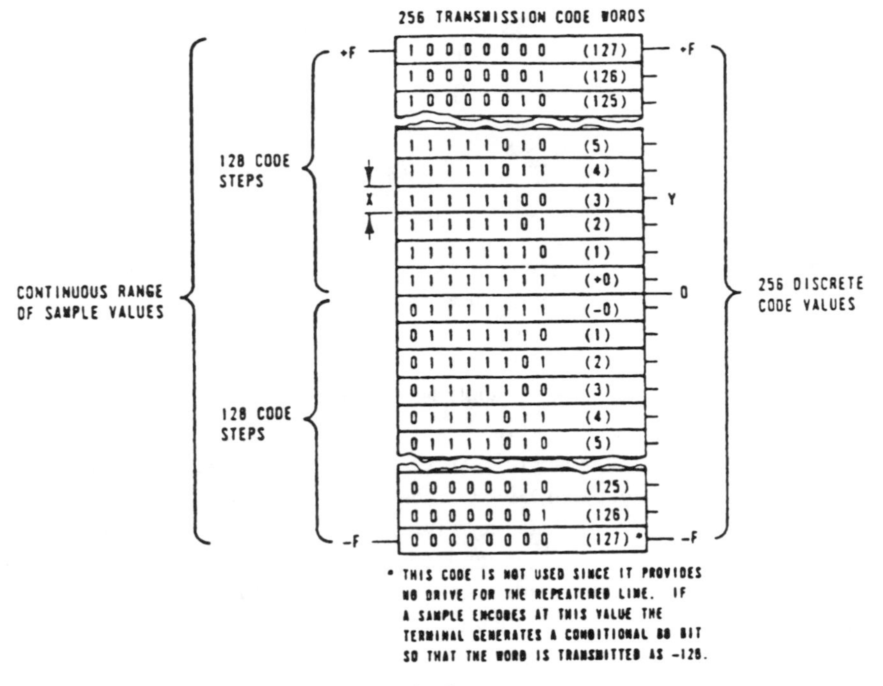

FIGURE 3.2. PCM code words as transmitted.

Small values of B (that is, $B < 8$) mean coarser quantization and a resulting increased quantization distortion (error), also called *quantization noise*. If Δ is the quantizer step size, then the error E will be between -0.5Δ and $+0.5\Delta$ for any given sample. Moreover, the error E is assumed to be uniformly distributed over the interval -0.5Δ to $+0.5\Delta$, so that $\Delta p(E) = 1$, or

$$P(E) = 1/\Delta \tag{3.1}$$

where $p(E)$ is the amplitude probability distribution of the error E. Equation (3.1) is valid provided the signal does not overload the quantizer and provided the signal traverses a large number of steps. To determine the ratio of signal to quantization error (noise) on a power basis, the mean-square error must be calculated. It is

$$\int_{-0.5\Delta}^{0.5\Delta} E^2 p(E)dE = \frac{\Delta^2}{12} \tag{3.2}$$

Then, for an input level having a value X, the signal-to-quantization-noise ratio is

$$\text{SNR} = \frac{X^2}{\left[\frac{\Delta^2}{12}\right]} \tag{3.3}$$

A formula can be developed for performance limited by quantization error, not channel noise. This formula is based upon the input signal being complex enough, or sufficiently finely quantized, that the quantization error has a uniform distribution throughout the interval -0.5Δ to $+0.5\Delta$. Moreover, the quantization is assumed to be fine enough that signal-correlated patterns do not appear in the error waveform, as illustrated in Fig. 2.5. Thus the effect of errors must be measurable in terms of a noise power or error variance. Finally, the quantizer must be aligned with the amplitude range, which means that any peak clipping that occurs will be equal for both positive and negative peaks. In other words, the signal has a zero mean value, that is, no dc component.

Since the number of quantization levels is 2^B, it follows that if the saturation limits of the quantizer are $\pm X_{pk}$, then the step size Δ is

$$\Delta = \frac{2X_{pk}}{2^B} \tag{3.4}$$

Letting A equal the peak amplitude of an input wave, $A < X_{pk}$ if clipping is to be avoided. Let $A/X_{pk} = m$, the fractional portion of the quantizer range occupied by the input waveform. For a sinusoidal input, the rms signal power is

$$\left(\frac{A}{\sqrt{2}}\right)^2 = \frac{A^2}{2}$$

Thus

$$\begin{aligned}\text{SNR} &= 10 \log_{10} [(A^2/2)/\Delta^2/12] = 10 \log_{10} [6A^2/\Delta^2] \\ &= 10 \log_{10} [6m^2 X_{pk}^2/\Delta^2] = 10 \log_{10} [(3m^2/2)\, 2^{2B}]\end{aligned}$$

As a result,

$$\text{SNR} = 1.76 + 20 \log m + 6B \tag{3.5}$$

This is the PCM performance equation for uniform encoding.

For example, suppose a system must be able to provide a 35-dB SNR for inputs whose magnitudes may differ as much as 40 dB. What number of quantization bits B must be used if the encoding is uniform? Using Equation (3-5), one finds that $B = 12.2$. Thus, to meet the stated requirements, 13-bit quantization would be required. Actual PCM telephony systems, however, use only 8 bits or less. This smaller number is feasible through the use of compression and nonuniform quantization, which are discussed in Section 3.3.

The foregoing SNR formulas are valid generally for toll quality speech links except for amplitude alignment. An amplitude alignment problem may arise from the fact that a single encoder-decoder system usually must handle several speakers or circuits, all of which have nonstationary speech signals. The result is a deterioration of quantizer performance, that is, lower SNR values than those predicted by the SNR equations. Solutions to this problem take two forms: nonuniform quantization and adaptive quantization. These two subjects are discussed later in this chapter.

To detect the presence or absence of a PCM (or other) pulse reliably requires a certain signal-to-channel-noise ratio. Two assumptions will be made to determine what this ratio must be. First, the detector is assumed to be ideal, that is, no error is contributed by the detector. (Realistically, this simply means that detector error is negligible compared with the error produced by channel noise.) Second, the channel noise is assumed to have a uniform power spectrum, to have a zero mean value, and to have a gaussian amplitude distribution. (As far as the receiver is concerned, the noise is flat over the receiver passband and has the characteristics of thermal noise within this band.)

The input PCM pulses presented to the detector are assumed to be of short width (less than a time τ, to be defined), and to have been filtered by an ideal low-pass filter of bandwidth W. The pulse centered at time $t = m\tau$, where $\tau = 1/2W$, has the form

$$V = V_0 \frac{\sin (\pi/\tau)(t - m\tau)}{(\pi/\tau)(t - m\tau)} \tag{3.6}$$

and will be zero at $t = k\tau$, where $k \neq m$, as illustrated in Fig. 3.3. By sampling the pulse at $t = m\tau$, however, only the pulse belonging to that time is seen, all others being zero at that instant, assuming the pulses occur at multiples of τ seconds.

As will be seen in Chapter 5, the actual transmission line code usually used is known as *bipolar return to zero* (BRZ) or *alternate mark inversion* (AMI).

For bipolar transmission, detection occurs as follows: if the signal, when sampled, has an output $>V_0/2$, a pulse is said to be present, whereas if the signal when sampled exhibits an output $<V_0/2$, no pulse is said to be present. Thus an

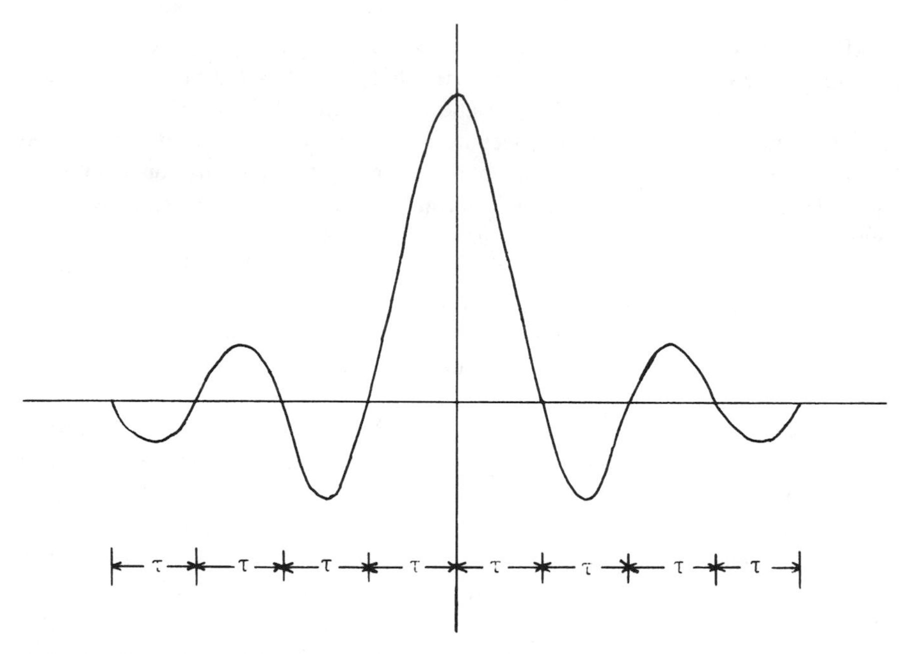

FIGURE 3.3. Pulse of the form. V_0 [sin $(\pi t/\tau)/(\pi t/\tau)$]

error occurs for an instantaneous noise value $>V_0/2$ in the right direction at the sampling time. (The noise is assumed to have a zero mean value.)

Let σ = rms noise amplitude and P_s = signal "power" = V_0^2. The noise "power" in bandwidth W is $N = \sigma^2$. For unipolar transmission (V_0, O), the probability of an error occurring then is

$$p_e = erfc(V_0/2\sigma) = erfc\sqrt{P_s/4N} \tag{3.7}$$

where

$$erfc(x) = \left(\frac{1}{\sqrt{2\pi}}\right)\int_x^{\ell} e^{-(\lambda^2/2)}d\lambda \tag{3.8}$$

Equation (3.8) is plotted in Fig. 3.4. It shows how the channel error ratio varies with signal-to-channel-noise ratio for the case of simple pulse detection.

For bipolar transmission (V_0, O, $-V_0$), the error ratio is somewhat higher (about 40%) because both positive and negative noise pulses can cause unwanted threshold crossings when the zero-level signal is sent. By comparison, unipolar

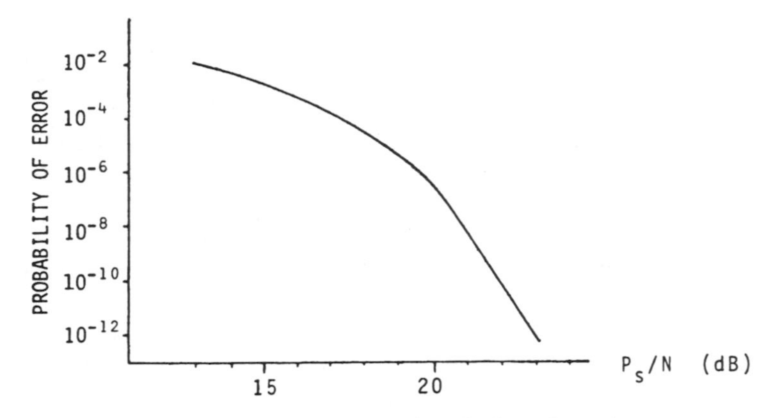

FIGURE 3.4. Probability of detection error for pulse in random noise.

transmission (V_0, O) is influenced only by positive noise pulses for the zero-level signal and only by negative pulses when the V_0 level is sent.

If the AMI detector reproduces the unwanted pulse as a one, then the probability of error is doubled for transmitted zeros. As a result, the overall probability of error is increased by 50% if zeros and ones are equally likely.

3.3. COMPRESSION AND NONUNIFORM QUANTIZATION

As noted in the last section, the nonstationary nature of speech (variations in probability density and talker volume) results in an amplitude alignment problem and a corresponding ineffective use of a uniform quantizer's characteristic. Some speech inputs may use only the lowest levels of the characteristic while others may use primarily the higher levels, and even suffer clipping. Thus it is desirable to provide large end steps for the relatively infrequent large amplitude ranges, while providing a finer quantization for the lowest levels. Alternatively, encoding might be achieved at a lower bit rate for a given quality and dynamic range through the use of nonuniform quantization.

A compressor, as used in analog circuits, is a device that accepts a large dynamic range and reduces the dynamic range according to a predetermined compression law. A companion device, the expander, has the inverse characteristic. It takes the compressed signal and restores it to its original dynamic range. The overall process is called *companding*. It is useful in transmitting speech over channels whose dynamic range (range of minimum to maximum values) is limited. Companding may be syllabic or instantaneous, or somewhere between. Syllabic companding uses the fact that the speech power level varies only slightly from one 125-μs sample to the next. The gain of an amplifier is made to vary automatically to achieve increased dynamic range, whereas in the case of instantaneous companding, an amplifier with a nonlinear input-output characteristic is used. In the case of instantaneous companding, the companding interval is simply the 125-μs interval. "Nearly instantaneous companding," a third technique, is discussed in Section 3.5.2.

For good voice reproduction, the signal-to-quantization distortion ratio should be kept constant over a wide dynamic range. This means that the distortion should be proportional to the signal amplitude for any signal level. A logarithmic compression law achieves this result. The use of a nonuniform quantizer is equivalent to the presentation of a compressed signal to a uniform quantizer, with subsequent expansion of the output. The compression law used in North American telephone networks is the μ-law. It is expressed by the equation

$$|v| = \frac{V\,\ell n\left(1 + \frac{\mu|X|}{V}\right)}{\ell n(1 + \mu)}, \quad \mu > 0 \tag{3.9}$$

where

v = output of speech compressor,
V = overload level of speech compressor designed for nonuniform quantization,
μ = dimensionless quantity in expression for logarithmic compression function,
X = input to speech compressor.

Figure 3.5 illustrates companding characteristics for various values of μ.

The $\mu = 0$ curve corresponds to uniform (linear) quantization. Based upon statistical observations of the dynamic range of speech signals, desirable values of μ are on the order of 100 or more.[3] The gain achieved over uniform quantization is shown in Fig. 3.6 for various values of μ. The gain is seen to be a function of the normalized input level. The areas under the curves measure how well quantizer performance changes as X_{rms} changes. The value $\mu = 255$ has been selected for 8-bit speech quantizers because it can be approximated closely by a set of eight straight-line segments for either polarity (15 segments peak-to-peak, since the two segments just above and below zero are collinear). Where 7-bit quantizers are in use, usually in older designs, $\mu = 100$.

The compression law used in European networks is called the A-law. It is expressed by the equation

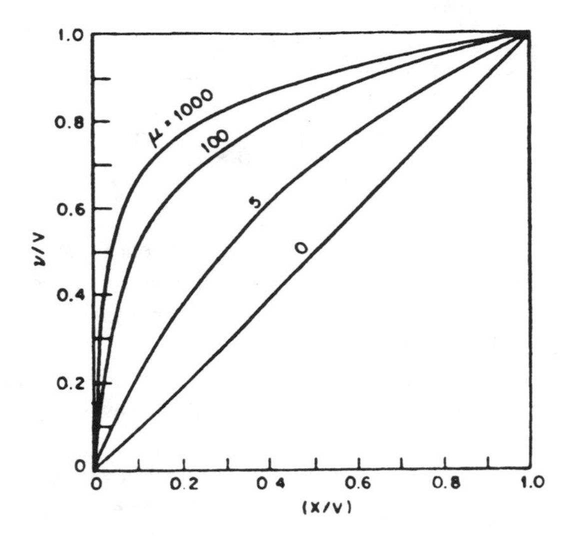

FIGURE 3.5. Companding characteristics using the μ-law. (Courtesy N. S. Jayant, Ref. 3, © IEEE, 1974.)

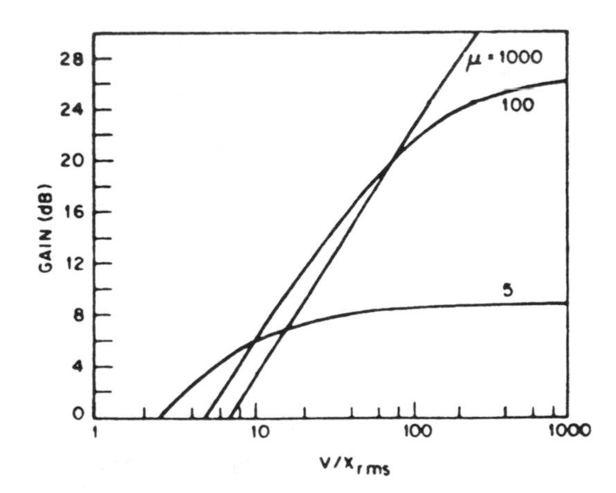

FIGURE 3.6. Logarithmic quantizer performance curves. (Courtesy N. S. Jayant, Ref. 3, © IEEE, 1974.)

$$|v| = V\left[\frac{(1 + \ell n A|X|)}{(1 + \ell n A)}\right] \text{ for } \frac{1}{A} \leq X \leq 1 \tag{3.10a}$$

$$|v| = V\left[\frac{A|X|}{(1 + \ell n A)}\right] \text{ for } 0 \leq X \leq \frac{1}{A} \tag{3.10b}$$

The logarithmic curve is approximated by a linear segment for small signals. The value of X/V after which the curve transitions smoothly to a true logarithmic form is $1/A$. The parameter A is usually set at 87.6. Segmented versions of A-law companding often use 13 segments.

Over the full input dynamic range, A-law companding provides a more nearly constant SNR than does μ-law, as illustrated in Fig. 3.7; however, μ-law provides somewhat more dynamic range.

The minimum step size for A-law is 2/4096, whereas for μ-law the minimum is 2/8159. Moreover, the A-law approximation used does not define a zero-level output for the first quantization interval. In other words, it uses what is known as a mid-riser quantizer. For low-level signals, the companding advantage, that is, the ratio of the smallest step with and without companding, is 24 dB (4096/256) for A-law versus 30 dB (8192/256) for μ-law.

The term "optimum" quantizer refers to one that yields minimum mean-square error when matched to the variance and the amplitude distribution of the signal. The nonstationary nature of speech, however, results in less than satisfactory results with optimum quantizers, especially if the number of quantization levels is small. If a quantization boundary is placed at the zero level, then during idle

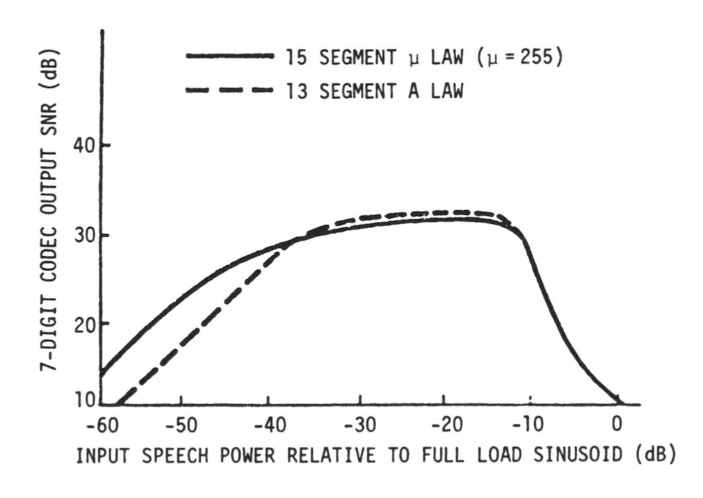

FIGURE 3.7. Signal-to-distortion performance of μ-law and A-law quantizers. (© 1982, Bell Telephone Laboratories. Reprinted by permission.)

channel conditions, the output of the quantizer jumps back and forth between the lowest magnitude quantization levels. If these lowest quantization levels are greater than the amplitude of the background noise, the output noise can exceed the input noise. Thus such a boundary is undesirable in quantizer design.

Quantizers based on the μ-law produce lower idle channel noise than do optimum quantizers. For $\mu = 255$, the ratio of maximum to minimum output level is 4015:1.

Segmented approximations to the logarithmic functions are implemented so that each successive segment changes its slope by a factor of two. Each linear segment then contains an equal number of coding intervals, with bit 1 being determined by the polarity (0 is +; 1 is –); bits 2, 3, and 4 being determined by the segment of the encoder characteristic into which the sample value falls (binary progression); and bits 5, 6, 7, and 8 being encoded using linear quantization within each segment. Figure 3.8 illustrates this coding scheme, which is the encoding-decoding algorithm for $\mu = 255$ PCM.

The advantages of nonuniform quantization increase with the crest factor (peak/rms) of the signal. The values of μ cited relate to the standard 7- and 8-bit quantizations used in the public switched network.

The "optimum" quantizer (whether uniform or nonuniform) is one that maximizes the SNR, with the effect of overload errors included.

3.4. PCM PERFORMANCE

Two methods exist for the evaluation of a speech-coding technique. One is the SNR, the ratio of signal to quantization noise. This is an objective rating, and one that can be calculated. The other is, "How does the result sound?" The response of a listener to the quality of a given coding technique is a subjective but a very important factor in the technique's acceptability. This response is determined by all forms of speech disturbance, not quantizing distortion only.

The SNR has inadequacies as a performance measure because the quantization error sequence has components that depend on the signal, that is, they are signal correlated. These components appear more like distortion than background noise. This distortion does not have the same annoyance value as does independent additive noise of equal variance, provided $B = 7$ or 8, as is the case on the public switched network, or >8, as in high-fidelity digital recording. In Section 3.3.1, however, the reverse will be found to be true for $B < 6$, that is, distortion will be more objectionable than background noise in these lower bit rate systems.

Table 3.1 summarizes the performance characteristics of PCM as used in the currently manufactured D4 and D5 channel banks[4] (8-bit quantization, $\mu = 255$). As can be seen from the table, PCM furnishes toll quality. Shown in Table 3.1 are the end-to-end system requirement (the design objective), the allocation to the coder-decoder (codec), and the allocation to the digital-to-analog converter

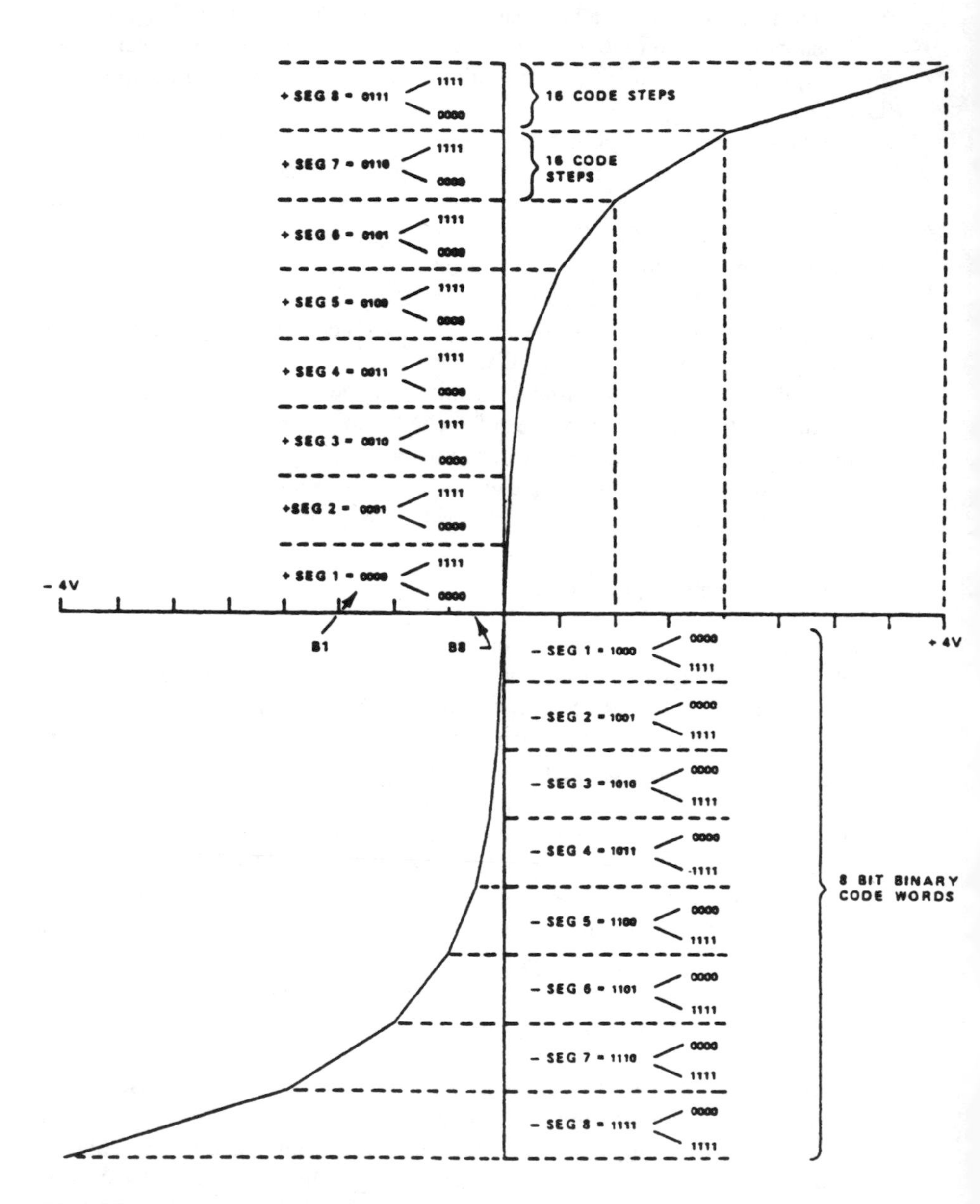

FIGURE 3.8. Segmented approximation to logarithmic function.

TABLE 3.1. PCM Performance Characteristics[a]

Characteristic	1-kHz Level (dBmO)	System Requirement (dB)	D4 Codec Objective (dB)	DAC Objective (dB)
Signal-to-distortion[b]	0	33	33	35
	–10	33	33	35
	–20	33	33	35
	–30	33	33	34
	–40	27	27	30
	–45	22	22	26
Gain tracking[b]	+3	±0.5	±0.25	±0.08
	0 (reference)	—	—	—
	–10	±0.5	±0.25	±0.08
	–20	±0.5	±0.25	±0.08
	–30	±0.5	±0.35	±0.12
	–40	±1.0	±0.50	±0.24
	–45	±1.0	±0.75	±0.35
Harmonic distortion	0 dBmO	–40	–50	–55
Gain stability (0°C–50°C)	0 dBmO	±0.25	±0.20	±0.05 (150 ppm/°C) (0°C–75°C)
Crosstalk coupling loss	0 dBmO			
200 to 3400 Hz		±65	±75	Not applicable
Idle channel noise		23 dBrncO	20 dBrncO	Not applicable

[a] Includes crossover error.

[b] Courtesy American Telephone and Telegraph Company, © 1982. Reprinted with permission.

(DAC). Figure 3.9 shows the transmit and receive filter frequency responses for the D4 channel bank.[5]

While PCM is designed primarily to handle voice, occasions arise in which a subscriber may use a voice line for data transmission via modems. Thus the ability to handle the tones produced by high-speed modems also is important. PCM has been found to pass modem signals satisfactorily at 4800 b/s for up to eight tandem A/D conversions, and at 9600 b/s for up to four tandem A/D conversions. At 14,400 b/s, only one A/D conversion generally can be tolerated. Accordingly, 9600 b/s usually is the maximum rate for switched lines, with higher rates generally using private conditioned lines. Applications information can be found in Bell System Technical Reference PUB41004[6] and in McNamara.[7]

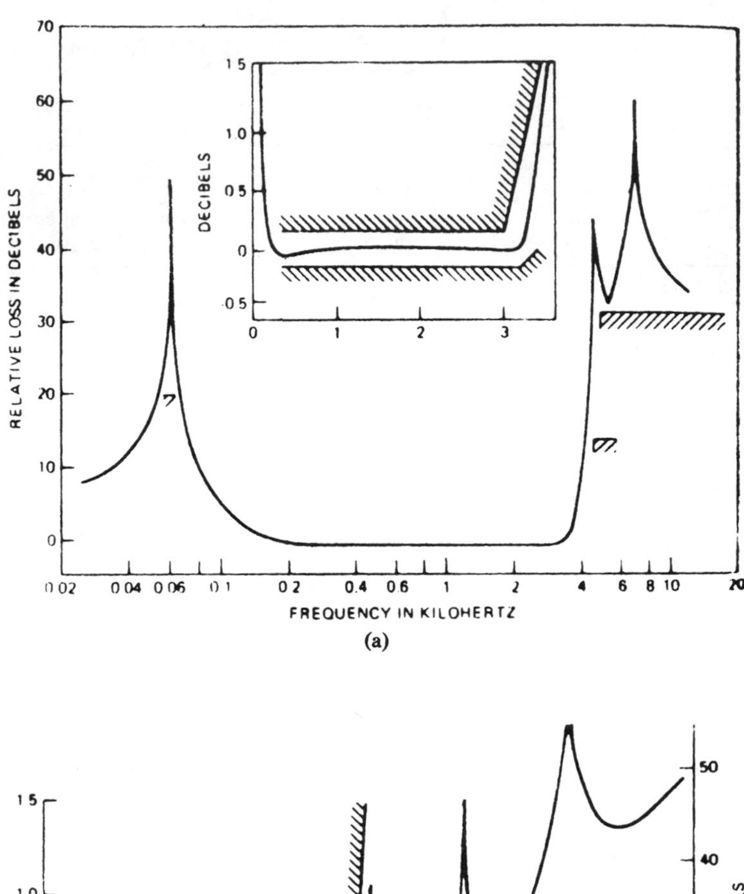

(a)

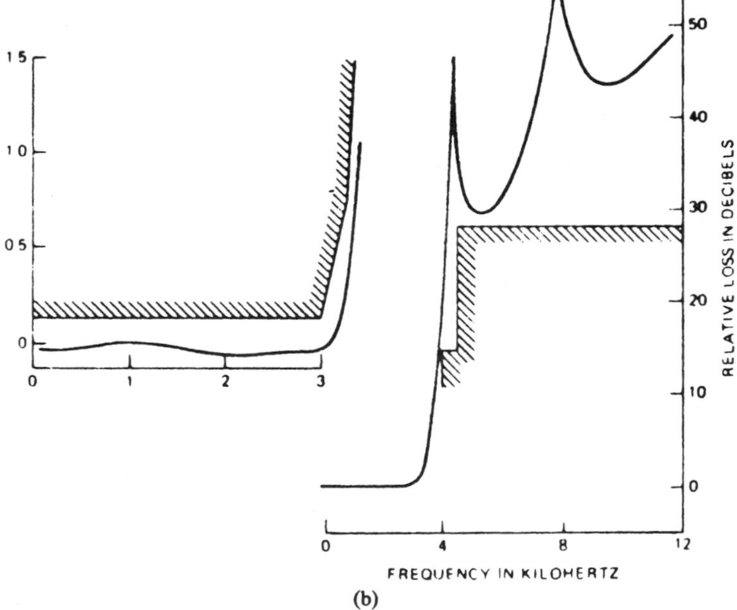

(b)

FIGURE 3.9. a. Transmit and b. receive filter frequency responses of the D4 channel bank (Courtesy AT&T, © 1982. Reprinted with permission.)

Thus far, the effect of channel noise on the bit-error ratio has been shown (see Fig. 3.4), but the tolerance of PCM to a given bit-error ratio has not been discussed. Digital facilities generally are rated in terms of error-free seconds and exhibit a bit error ratio less than 10^{-7} (at worst, 10^{-5} where long-haul analog circuits may be involved in a tandem connection). PCM, however, works quite well at a bit-error ratio of 10^{-4}. This requires only a 17-dB signal-to-channel-noise ratio, according to Fig. 3-4. Consequently, PCM is insensitive to the occasional errors that may exist on a facility.

In conclusion, PCM, as used on the public switched network and on equivalent private-line facilities, provides toll quality voice transmission as well as transparency to other analog signals. Consequently, it can be regarded as providing a "universal" voice channel.

3.5. SPECIAL PCM TECHNIQUES

The desire to conserve spectrum is a very widespread one, prevalent not only with respect to the radiated spectrum, but also significant in terms of the conducted spectrum that exists on any cable transmission facility. Numerous multilevel modulation techniques (see Chapter 6) have been devised to increase the number of bits per second that can be transmitted per hertz of bandwidth. It should be no surprise, therefore, that significant efforts are being made to encode voice at lower information rates (b/s) than required for standard 64-kb/s PCM, as described earlier in this chapter. The fact that 64-kb/s PCM as such generally requires much more bandwidth than does a 3-kHz analog signal is in itself a continuing incentive to seek lower-bit-rate forms of digitized speech.

Several techniques have been applied to PCM to enable the transmission of a voice channel at less than 64 kb/s. These techniques include coarsely quantized PCM, often aided by adaptive quantization, as discussed next in Section 3.5.1, and *nearly instantaneous companding*, discussed in Section 3.5.2. Another technique, *digital speech interpolation* (DSI), is applied to large numbers of channels (often 40 or more). The discussion of DSI is reserved for Section 5.8.

3.5.1. Coarsely Quantized PCM

An examination of the factors that cause standard PCM to require its 64-kb/s rate shows that there are only two ways of reducing the rate. Either the sampling rate or the fineness of quantization or both must be reduced. Reductions in the sampling rate from 8000 to about 6000 can be achieved with only moderate degradation in quality. Below that the result starts to sound very muffled or distorted because its bandwidth is being narrowed appreciably by a low-pass filter or, if not, distortion is occurring due to a problem called aliasing, in which unwanted frequency components are produced in the voice band. Reduction of the fineness of quantization can be done to a greater extent without such a serious lowering of the quality. Moreover, adaptive quantization can be introduced to help maintain the quality.

When quantization levels are present in abundance, as is the case for $B \geqslant 7$, there is no real need to be concerned about adapting the levels to the input. When B is smaller, however, especially for $B \leqslant 4$, adaptive quantization produces a significant quality improvement. By definition, adaptive quantization is step-size adaptation to the signal variance based on quantizer memory. The step size generally is modified for every new input sample, based on a knowledge of which quantizer slots were occupied by the previous samples. The result is called *adaptive PCM* (APCM). Alternatively, a fixed quantizer can be preceded by a time varying gain which tends to keep the speech level constant.

For example, in the case of a one-word (one sample) memory, let the quantizer output be

$$Y_r = H_r \Delta_r / 2 \tag{3.11}$$

where $\pm H_r = 1, 3, 5, \ldots 2_B{-}1$, $\Delta_r > 0$, $B \geqslant 2$. Thus, the various output levels Y_r are governed by the step size, Δ_r. The next step, Δ_{r+1}, is chosen according to the equation

$$\Delta_{r+1} = \Delta_r M(|H_r|), \tag{3.12}$$

that is, a multiplying factor is applied to Δ_r, depending on the magnitude of H_r for the previous step. Thus the adaptation logic matches the step size, at every sample, to an updated estimate of the signal variance.[8]

Figure 3.10 shows a 3-bit quantizer characteristic. Notice that the multiplier M depends upon the step being used by the signal at any given moment, that is, on the magnitude $|H_r|$. The assumption, valid for speech, is made that the input probability density function $P(x)$ is symmetrical about a mean value of zero.

The general advantage of adaptive quantization is an increased dynamic range for a given number of bits per sample.

Table 3.2 lists step size multipliers for the APCM coding of low-pass filtered speech. Note that the negative steps are treated the same as the positive ones. Consequently, there are two M values for $B = 2$, four for $B = 3$ and eight for $B = 4$. The values are not critical. For example, for $B = 3$, adequate choices are $M_1 <, M_2 = M_3 = 1, M_4 > 1$. Step-size increases, however, should be more rapid than step-size decreases, because overload errors tend to harm the SNR more than do granular errors. The general shape of the optimal multiplier function for $B >$

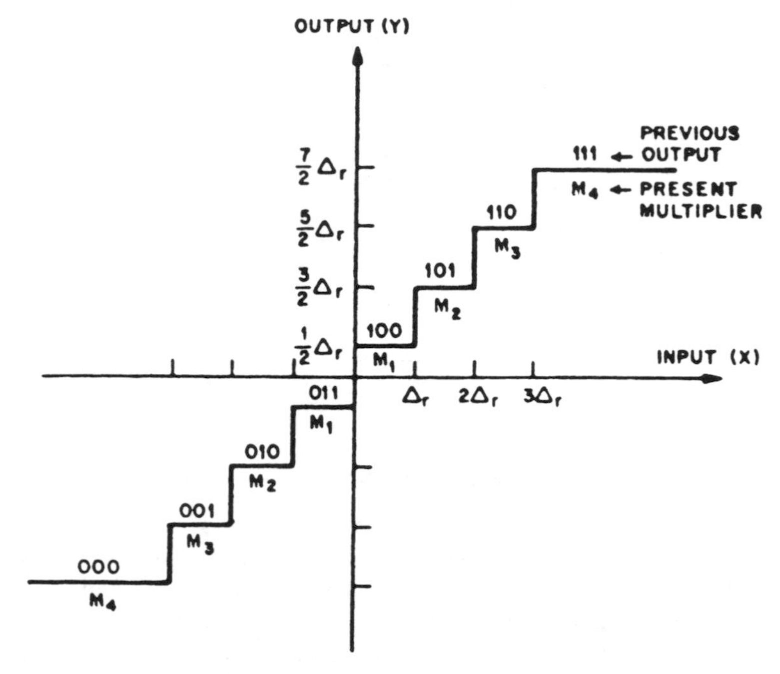

FIGURE 3.10. Three-bit adaptive quantizer characteristic: one-word memory. (Courtesy N. S. Jayant, Ref. 8, © IEEE, 1974.)

2 is shown in Fig. 3.11. The shaded region shows the extent to which M may vary for changes in B as well as in input signal characteristics. The curve of Fig. 3-11 does not include the case $B = 2$ because in this case, the distinction between the expected magnitudes of granular and overload errors is small.

Adapting the quantizer by knowing which quantizer slots were occupied by the previous samples avoids the need to send side information to the receiver. This technique is called *backward adaptive quantization* because estimates are derived from samples of the quantizer output.

Other approaches to obtaining APCM have been devised. The adaptations may be achieved by switching between two invariant quantizers, with the switching being based on the use of quantizer memory. Alternatively, instead of instantaneous adaptation, as discussed thus far, the adaptation may be done at a syllabic rate. In this case, the step size Δ_r is adapted with a time constant of 5–10 ms, rather than every sample. The disadvantage of the use of syllabic techniques is that they are somewhat more complex to implement than is instantaneous adaptation.

TABLE 3.2. APCM Step-Size Multipliers[a]

B	2	3	4
M_1	0.60	0.85	0.80
M_2	2.20	1.00	0.80
M_3		1.00	0.85
M_4		1.50	0.90
M_5			1.20
M_6			1.60
M_7			2.00
M_8			2.40

[a]Courtesy N. S. Jayant, Ref. 1., © IEEE, 1974.

3.5.2. Nearly Instantaneous Companding

Section 3.3 discussed the use of companding in PCM transmission. As described there, the companding is instantaneous in that each speech sample is companded as such. Later, Section 3.7.2 will discuss the use of syllabic adaptation in connection with delta modulation. There, changes in the coding occur at a syllabic rate. In nearly instantaneous companding (NIC), also known as block companding, the companding is done based upon the amplitude of a small block (for example, 10) of speech samples.[9] In general, a block of N speech samples is searched to determine which sample has the largest magnitude. Each sample then is encoded (for example, using six instead of eight bits) with the top of the "chord" of the maximum sample being the overload point. The chord consists of the second, third, and fourth bits of a PCM sample, with the first bit designating the sign of the sample. The encoding of the chord is sent to the receiver together with the sample encoding. Thus the scaling information is sent only once for each block of N samples rather than with each sample. The name *variable quantizing level* (VQL) is used for this technique.

In general, for a block of N samples and n-bit encoding the resulting bit rate is $2\,W(n{-}2 + 3/N)$ b/s. Thus for $W = 4$ kHz, $n = 8$ and $N = 10$, the bit rate is 50.4 kb/s. Some NIC systems (VQL) operate[10] at 32 kb/s.

NIC can be achieved by processing 15-segment μ-law or 13-segment A-law PCM for a reduced bit rate. As such, it is a form of APCM, or digital automatic gain control, with the quantization being nearly uniform and the step size being varied adaptively based upon the short-term signal level. Because speech power varies at a syllabic rate, the transmission of the power level once per block is adequate. The quantization of all samples within a given block is done with similar accuracy, thus avoiding the fine quantization of a small sample when a neighboring sample is much larger and thus more coarsely quantized, as is done in μ-law and A-law PCM.

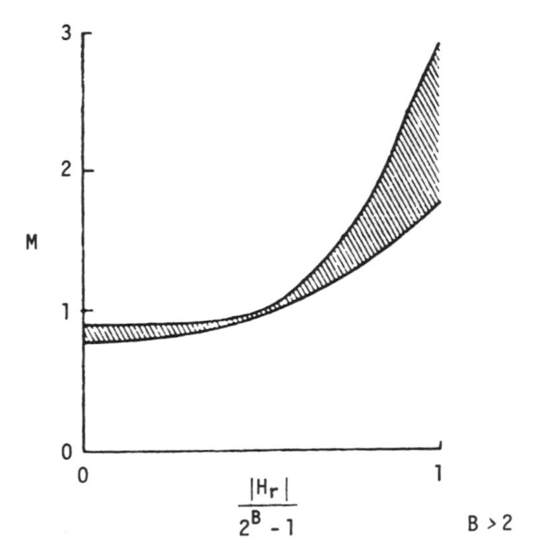

FIGURE 3.11. General shape of optimal multiplier function for adaptive quantization of speech. (Courtesy N. S. Jayant, Ref. 8, © IEEE, 1974.)

The performance of NIC is found to be somewhat better than that of (n–1)-bit companded PCM, and is largely insensitive to the statistics of the input signal, unlike many other low-bit-rate speech-coding techniques. At 6.3 bits/sample, NIC speech exhibits as much as a 6-dB SNR improvement over $\mu = 255$ PCM. An overall SNR $\geqslant$30 dB is achieved. The group delay is less than 50 μs over the 200–3000-Hz band.

Modem transmission at 9600 b/s can be achieved at a 10^{-6} ber on conditioned voice channels when NIC is used.

In the presence of channel noise, the sign bits are the most vulnerable to noise, as is the case with standard PCM. The scaling information is affected less often with NIC, but an error affects an entire block of samples rather than one sample only. As a result, NIC produces fewer "pops" at small amplitudes. Listening tests show that NIC sounds better than standard PCM at a channel bit-error ratio of 10^{-3}.

The achievement of NIC encoding requires an inherent delay of one block at the transmitter. For $N = 10$ and $2W = 8$ kHz, this delay is 1.2 ms. Such a delay may result in echo problems, since echo suppressors generally are installed at lengths of 3000 km, based upon a median incremental round-trip delay of 8 μs per kilometer of trunk length. Thus, a 2.4-ms round-trip processing delay corresponds to a 300-km foreshortening of the nominal 3000-km trunk length.

3.6. DIFFERENTIAL PCM

The formation of a differential waveform was discussed briefly in Chapter 2. It is produced by subtracting from the present value a recent past value or values, or some weighted combination of values. The basis for the quantization of differential waveforms is that once the initial level of a waveform has been established, the information content is expressed by the changes in value of that waveform. Accordingly, only the changes need to be transmitted. As a result, differential PCM (DPCM) can be used at a lower bit rate than PCM with a comparable quality of reproduction. The sampling rate, of course, is presumed to be at the Nyquist rate.

Figure 3.12 shows a typical DPCM system in block diagram form. As can be seen there, instead of direct sampling of the input, as was done in PCM, the difference between the input and a prediction signal (based on past samples) is sampled and coded. For this reason, DPCM is a form of *predictive coding*. The "prediction signal" is derived in the same way as the receiver detects the signal, and is performed in the box labeled "Integrator."

Since the quantizer's input is analog but its output digital, the formation of the

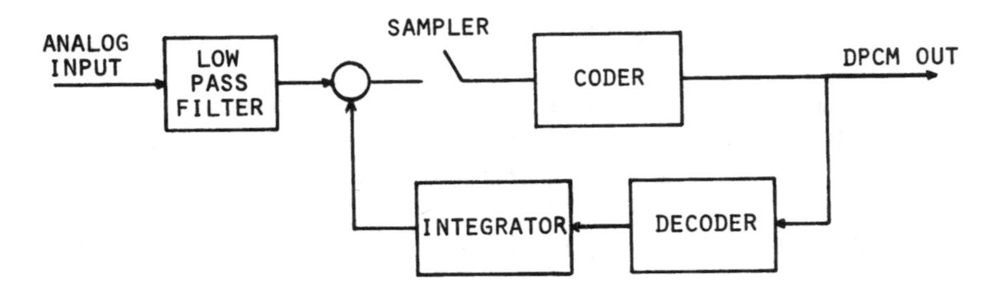

FIGURE 3.12. Block diagram of DPCM system using analog differencing.

differential waveform requires some discussion. Figure 3.12 shows that the quantizer consists of a sampler and a coder. The coder converts sampled amplitude values to specific bit combinations. Thus an analog-to-digital conversion is performed. The decoder followed by the integrator performs the conversion back to analog. The result is the formation of the differential on an analog basis, that is, *analog differencing and integration.* Note that the coder and decoder need handle only the dynamic range of the difference signal.

An alternative approach is illustrated in Fig. 3.13. Here the digital form of the difference signal is summed and stored in a register, thus producing a digital representation of the previous input sample. Following this step, a decoder and integrator produce an analog version of the signal for subtraction from the analog input. The decoders of Fig. 3.13 thus must be able to handle conversions over the entire dynamic range of the input. A third approach is illustrated in Fig. 3.14. In this approach, the coder must handle the full dynamic range of the input. From

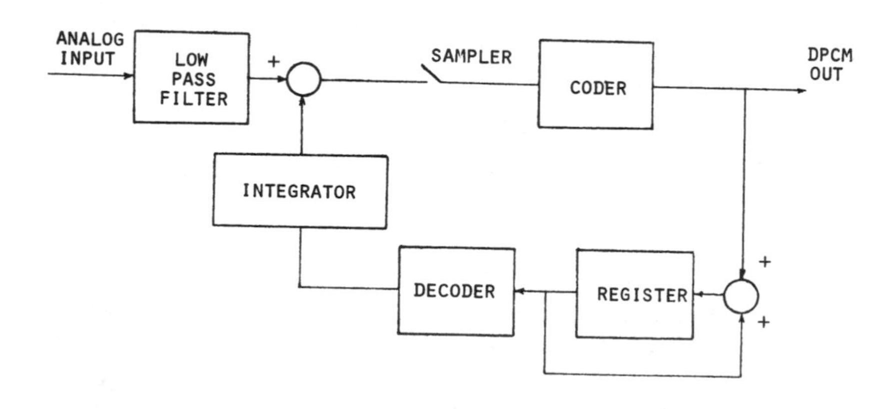

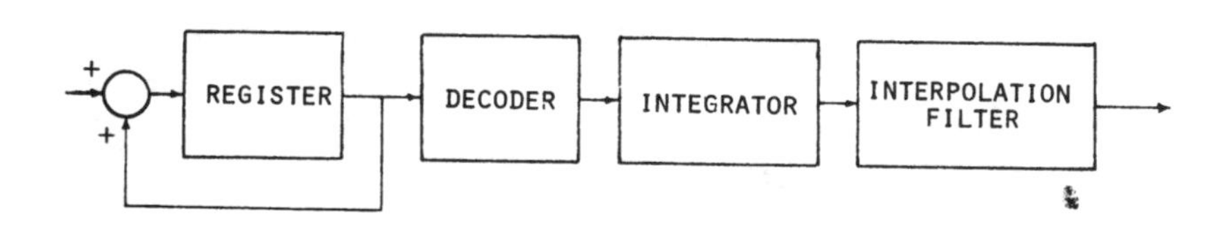

FIGURE 3.13. Block diagram of DPCM system using digital integration.

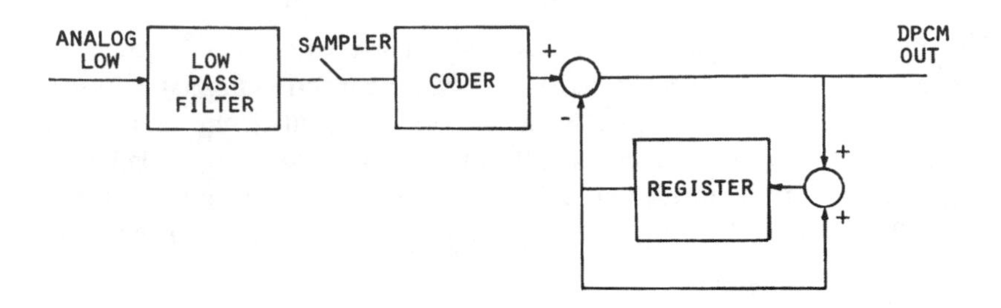

TRANSMITTING SECTION

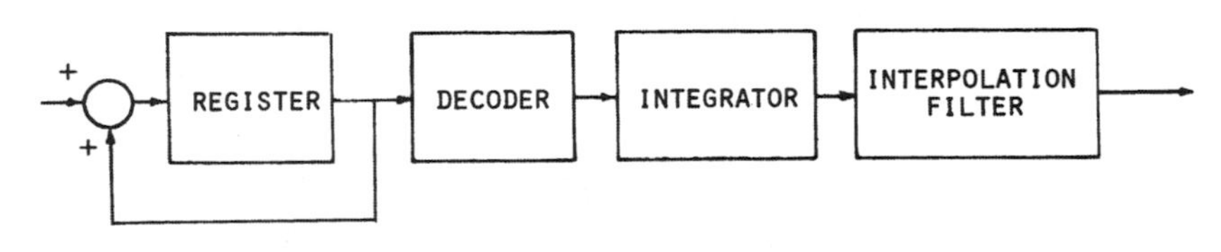

RECEIVING SECTION

FIGURE 3.14. Block diagram of DPCM system using digital differencing.

the coder's output is subtracted a digitally produced approximation of the code representing the previous input amplitude.

The implementations of Figs. 3.13 and 3.14, being digital, are insensitive to minor changes in circuit constants, and are easily reproduced on a large-quantity basis. The advantage of the analog approach of Fig. 3.12 is its simplicity when built using lumped components, as well as its modest cost because of its relatively small dynamic range.

In Figs. 3.12, 3.13, and 3.14, the receiving section consists of components identical to those in the feedback path of the transmitting section. This results from the fact that the feedback path built into the transmitter is intended to predict the receiver's output and will, in fact, in the absence of channel errors.

3.6.1. Basic DPCM Concepts

In treating DPCM, the concept of correlation is useful. The *autocorrelation function* $\phi(\tau)$, usually called the *correlation function*, is a measure of the extent to which two values of a function $f(t)$, separated by a time τ, are related to one another. It is defined as

$$\phi(\tau) = \int_{cl}^{\ell} f(t)f(t-\tau)dt,\ f(t) \rightarrow 0 \ as\ t \rightarrow \infty \tag{3.13}$$

For zero separation in the time domain, $\tau = 0$, $\phi(0)$ expresses the energy of the function. If $f(t)$ does not go to zero for large values of t, then the following alternative expression must be used:

$$\phi(\tau) = \lim \frac{1}{2T} \int_{-T}^{T} f(t)f(t - \tau)dt \tag{3.14}$$

Speech sampled at the Nyquist rate exhibits good correlation between successive samples. Let τ_0 be the time interval between successive samples. For 8000 samples/second, $\tau_0 = 125\ \mu s$. Then the variance of the first difference of the samples, $D_r(1)$, is

$$D_r(1) = X_r - X_{r-1} \tag{3.15}$$

where X_r is the present input and X_{r-1} is the immediately preceding input, i.e., τ_0 seconds ago. Letting $\langle\ \rangle$ denote the mean value,

$$\begin{aligned} D_r^2(1) &= \langle (X_r - X_{r-1})^2 \rangle \\ &= \langle X_r^2 \rangle + \langle X_{r-1}^2 \rangle - 2\langle X_r X_{r-1} \rangle \\ &= \langle X_r^2 \rangle\ [2(1 - C_1)] \end{aligned} \tag{3.16}$$

where C_1 is the value of the correlation function at $\tau = \tau_0$. Thus, if $C_1 > 0.5$, the variance of the first difference $D_r(1)$ clearly is smaller than the variance of the speech signal itself. Actually, a variance reduction can be demonstrated[8] for all values of C_1. The quantization error power is proportional to the variance of the signal present at the quantizer input. Therefore, by quantizing $D(1)$ instead of X and using an integrator to reconstruct X from the quantized values of $D(1)$, the result is a smaller error variance and thus a better SNR. Alternatively, for a given SNR, DPCM permits a smaller B and thus operates at a lower bit rate.

Note that in the three approaches to DPCM (Figs. 3.12, 3.13, and 3.14) the differential is obtained by feeding back the estimate of the receiver's output to the transmitter's input. As a result, quantization errors do not accumulate. If quantization errors should cause the feedback signal to drift away from the input, the next encoding of the difference signal corrects automatically for the drift. Channel errors, of course, will cause receiver output deviations, but these deviations do not persist because of the finite time constants used in the integrator.

How many past samples of a speech waveform should be used in obtaining a predicted value? Certainly some improvements are to be expected through the use of more than one correlation coefficient. The answer depends upon the extent to which significant correlation exists among successive samples. Fig. 3.15 illustrates the autocorrelation functions of speech signals from two male and two female speakers. Sampling was done at 8 kHz and the correlations were done over a 55-second duration. Thus the correlation index $i = 1$ corresponds to 125 μs. The upper curves show the results with a 0 to 3400-Hz filter in use, while the lower curves show the results with a 300 to 3400-Hz filter. As expected, the presence of lower-frequency components causes high correlation up to greater values of the correlation parameter, where $C_1 = \phi(\tau_0)$, $C_2 = \phi(2\,\tau_0)$, . . . $C_i = \phi\,(i\tau_0)$.

The gain G of a DPCM system is the number of dB by which the SNR is improved over a corresponding PCM system. Figure 3.16 shows values of G obtained as a function of the number of predictor coefficients n used in the system. In each case, the maximum, average, and minimum for four speakers are plotted. The low-pass filtered speech (0 to 3400 Hz) provides greater gain than the band-pass filtered speech (300 to 3400 Hz) because of the greater low-frequency energy of the low-pass filtered speech and hence the greater adjacent sample correlations.

The gain advantage achieved by differential encoding is directly related to its redundancy removal. The disadvantage of DPCM is that changing input speech material generally degrades the performance of a DPCM system operating with a predictor that has been designed on the basis of average speech statistics. One rather significant problem is that the voiced segments of speech usually have $C_1 > 0.5$ while noiselike fricatives usually have $C_1 \approx 0$. Voiced segments occur more frequently than unvoiced segments, so the way they are coded tends to control overall speech quality. Fricatives (unvoiced sounds) are relatively tolerant to granular quantization noise.

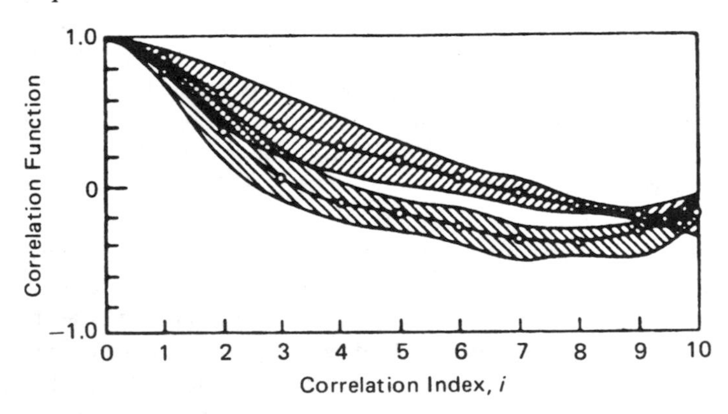

FIGURE 3.15. Autocorrelation function of speech signals. (Courtesy N. S. Jayant, Ref. 8, © IEEE, 1974.)

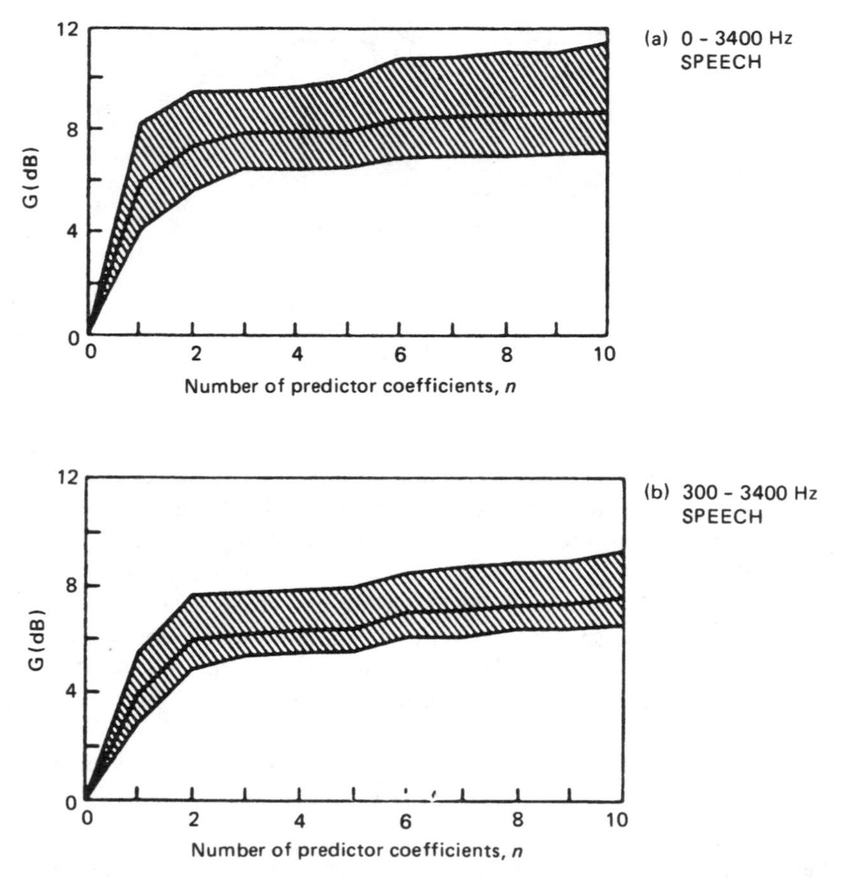

FIGURE 3.16. Gain of DPCM system versus number of predictor coefficients. (Courtesy N. S. Jayant, Ref. 8, © IEEE, 1974.)

The foregoing problems make the use of adaptive quantizers and predictors highly desirable on DPCM systems.

3.6.2. Adaptive DPCM (ADPCM)

The adaptive quantizer follows changes in signal power. In fact, the one-word adaptive quantizer, as described for APCM, is applicable also for ADPCM. The step-size multipliers, however, are slightly different from those for APCM, and are shown in Table 3.3. These ADPCM step-size multipliers are very close to the theoretical optimum for white gaussian signals, for which $C_1 = 0$. The reason is that the input samples of a DPCM quantizer are less correlated than those of a

TABLE 3.3. ADPCM Step-Size Multipliers[a]

B	2	3	4
M_1	0.80	0.90	0.90
M_2	1.60	0.90	0.90
M_3		1.25	0.90
M_4		1.75	0.90
M_5			1.20
M_6			1.60
M_7			2.00
M_8			2.40

[a]Courtesy N. S. Jayant, Ref. 8, © IEEE, 1974.

PCM quantizer because of the differentiating process. In addition to an adaptive quantizer, an adaptive predictor (part of the integrator function) is used in ADPCM. It responds to changes in the short-term spectrum of speech. There are two methods for adapting the predictor coefficients to changes in the short-term spectrum of speech. One, called backward-acting prediction, measures the power level of a section of the speech waveform, often a syllable or less in length. The resulting information is used to determine what gain should be applied to the following speech sections. This method is satisfactory if the power levels change relatively slowly.

The other method for adapting the predictor coefficients is called forward-acting prediction. This method uses the following steps:

1. Store finite sections of the speech waveform (typically a syllable in length).
2. Calculate the correlation function for each section.
3. Determine the optimum predictor coefficients. (The set of coefficients is sometimes called the vector).
4. Update the predictor at intervals based on the length of the stored section.

The predictor coefficients are then sent to the receiver as low bit-rate "housekeeping" information. Because these coefficients can be coarsely quantized and are updated only at a syllabic rate, the amount of channel capacity they require is a very small percentage of a full voice channel.

The advantage of forward-acting prediction is that the encoder and decoder use gain values that are directly related to the speech sections from which they are determined. Such devices as shift registers and charge-coupled devices allow forward-acting prediction to be done in a straightforward manner.

The use of the adaptive predictor means that variable predictor coefficients can be used to achieve a greater SNR gain over standard PCM than would be effective with a fixed predictor. These coefficients are derived from the correlation coefficients. Table 3.4 shows that for only one coefficient ADPCM has nearly the same gain as DPCM, whereas an increased number of coefficients results in an appreciable gain of ADPCM over DPCM.

A gain of about 20 dB can be achieved by means of the Atal-Schroeder adaptive predictor,[11] which exploits the quasiperiodic nature of the speech signal to obtain a more complete signal prediction than is provided by those predictors using ten or fewer coefficients. This requires an increased amount of computation in order to determine the pitch period. In the Atal-Schroeder method, the signal redundancy is removed in two stages. One predictor removes the quasiperiodic nature of the signal while another removes the *formant* information from the spectral envelope (see Section 4.3 for a definition of formant). The first predictor consists of a gain and a delay adjustment. The second provides a linear combination of the past values of the first predictor's output.

Other approaches to exploiting the quasiperiodic nature of speech provide useful speech reproduction for the low-bit-rate applications discussed in Section 4.3. Some involve the replication of an entire pitch-period-long segment of speech. The limitation of this approach is the existence and perceptual significance of very small variations of the pitch period in voiced speech.

Because ADPCM is capable of providing toll-quality speech at 32 kb/s, it is replacing some 64-kb/s PCM for interoffice transmission, not only in the United States, but internationally as well. For this purpose, it must be able to handle both speech and data signals without their being previously identified. In addition, it must be able to handle many tandem encodings. Other requirements include the ability to perform well in the presence of analog impairments, and with minimum coding delay to minimize potential echo problems. Moreover, moderate levels of channel noise must not cause adverse effects. Simplicity is important in keeping costs commensurate with the savings in bandwidth.

To enable the handling of data as well as speech, a device called a *dynamic locking quantizer* (DLQ) has been devised with two "speeds" of adaptation.[12] The rapid power variations of speech cause it to operate in its "unlocked" mode, while the relatively constant power of data causes it to operate in its "locked" mode. In addition, an adaptive transversal predictor with adaptive tap coefficients tracks

TABLE 3.4. Comparison of Nonadaptive and Adaptive Predictors (SNR Gain over PCM, dB)[a]

Order of predictor	1	3	5	10
Nonadaptive DPCM	5.4	8.4	8.6	9.0
ADPCM, 4-ms update	5.6	10.0	11.5	13.0
ADPCM, 32-ms update	5.6	9.8	11.1	12.6

[a]Courtesy N. S. Jayant, Ref. 8, © IEEE, 1974.

TABLE 3.5. Step-Size Multipliers for Network ADPCM

$M_1 = 0.984$
$M_2 = 1.006$
$M_3 = 1.037$
$M_4 = 1.070$
$M_5 = 1.142$
$M_6 = 1.283$
$M_7 = 1.585$
$M_8 = 4.482$

the nonstationary sample correlations of speech waveforms, providing adaptation to speech, data, or tone signals. To avoid the need for sending side information, backward adaptive prediction is used.

Performance tests at 32 kb/s of ADPCM/DLQ show it to have an SNR about 2 dB lower than that of 64-kb/s PCM for tones; however, 32-kb/s ADPCM/DLQ fully meets CCITT Recommendation G.712, Total Distortion, Method I (Narrowband Noise). Listener tests also show good to excellent opinions even after eight encodings for synchronously operated 32-kb/s ADPCM/DLQ. In the presence of channel errors, 32-kb/s ADPCM/DLQ outperforms 64-kb/s PCM for error ratios higher than 1.5×10^{-4}. Table 3.5 shows the step-size multipliers for this version of ADPCM. Figure 3.17 shows the use of backward prediction in both the coder and the decoder.[13]

3.7. DELTA CODING

Delta coding, also known as delta modulation, is a very simple coding technique to implement. It is a one-bit version of DPCM, which means that it uses only a two-level quantizer. The output bits convey only the polarity of the difference signal. A delta coder uses a simple first-order predictor (an integrator). To compensate for the fact that the quantizer has only two levels, the adjacent sample correlation is increased by oversampling, that is, sampling at a rate well beyond $2W$. The *oversampling index* is the ratio of the actual sampling rate to $2W$. This index is designated F.

3.7.1. Linear Delta Coding

The principle of linear delta coding is illustrated in Fig. 3.18. The delta coder approximates the input time function by a series of linear segments of constant slope, hence, the name linear delta coding. If the Y_r denote successive output bits and the X_r denote the corresponding sampled input levels, then

CODER

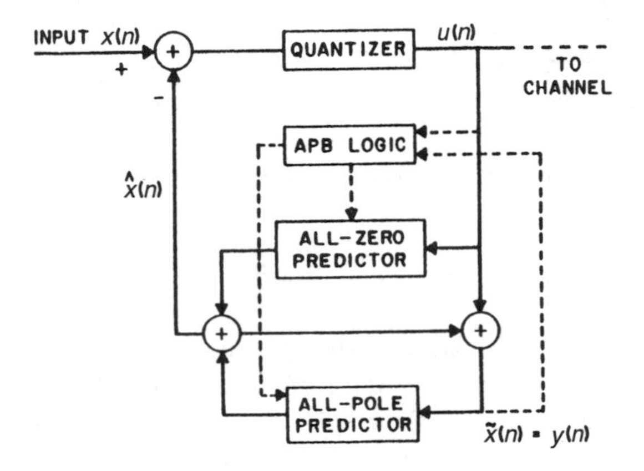

DECODER

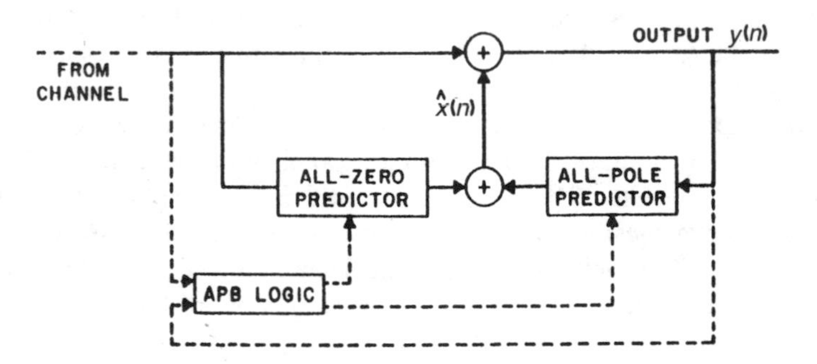

NOTES: Broken lines indicate adaptation processes for predictor coefficients

$\hat{x}$ denotes estimate of x; $\bar{x}$ denotes sampled version of x

FIGURE 3.17. Use of backward prediction in coder and decoder for network ADPCM. (Broken lines indicate adaptation processes for predictor coefficients; $\hat{x}$ denotes estimate of x; $\bar{x}$ denotes sampled version of x.) (Courtesy T. Nishitani et al., Ref. 13, © IEEE, 1982).

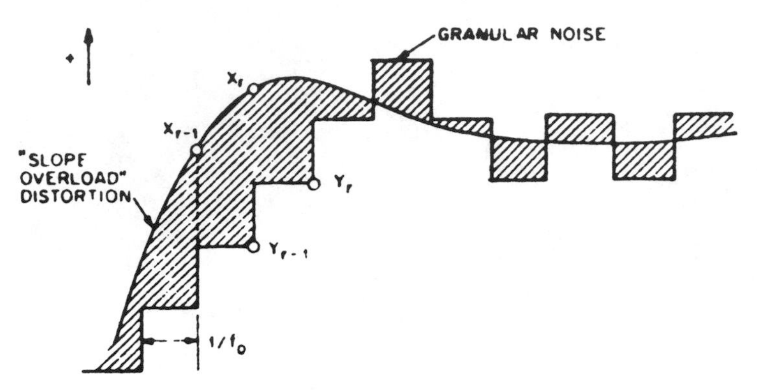

FIGURE 3.18. Principle of linear delta coding. (Courtesy N. S. Jayant, Ref. 8, © IEEE, 1974.)

$$Y_r - Y_{r-1} = \Delta b_r \tag{3.17}$$

where $b_r = \text{sgn}\,(X_r - Y_{r-1})$, that is, b_r, the transmitted channel symbol (either a 0 or a 1), is determined by whether the input is greater than, or less than, the accumulated (integrated) value of the output. Δ is the size of the step.

Two types of problems can arise in the production of delta coding. First, the input can change so rapidly that the output cannot keep pace with it. This is called slope overload distortion. The solution might seem to be that of designing in a larger step size Δ; however, this would aggravate the second problem, granular noise. During relatively flat inputs, the quantizer tends to jump between its two levels. The larger the Δ is, the greater the granular noise.

Figure 3.19 is a block diagram of a linear delta coder. The input signal is sampled at a rate $f_0 >> 2W$. Then a staircase approximation to the input, as in Fig. 3.18, is constructed. This is done in the accumulator, which feeds back to the quantizer input the previously received value of Y_r, that is, Y_{r-1}. Note that the feedback circuit of the transmitter constitutes a replica of the receiver circuits, except for the output filter. This is a characteristic of all of the differential-type digital coding techniques. At the receiver, the value Y_r is incremented by a step Δ in the direction b_r, at each sampling instant. Then Y_r is low-pass filtered to the original signal bandwidth. The low-pass filter rejects out-of-band quantization noise in the staircase function Y_r.

Thus, for each signal sample, the transmitted channel symbol is the single bit b_r. The bit rate equals the sampling rate f_0.

If ΔV is the height of the unit step in volts, the quantized noise power using a single integration[14] is

$$N_0 = \frac{2W(\Delta V)^2}{3f_0} \text{ watts} \tag{3.18}$$

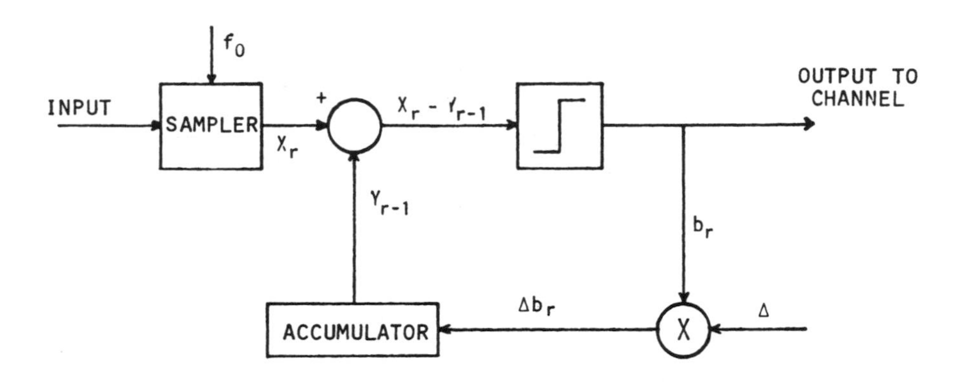

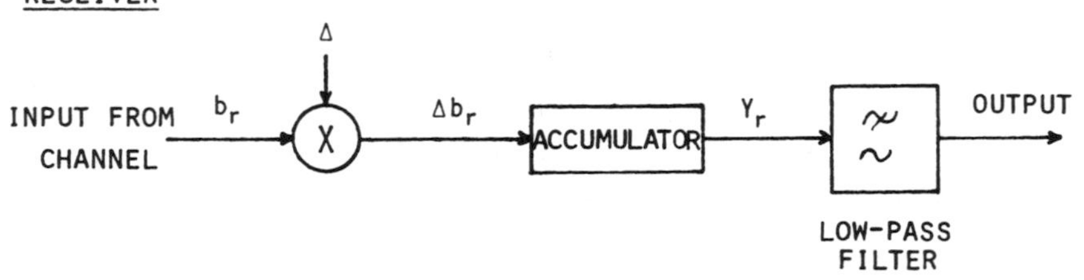

FIGURE 3.19. Linear delta coder block diagram.

The maximum amplitude at frequency f that can be transmitted without overloading in a single-integration coder[15] is

$$A = \frac{f_0 \Delta V}{2\pi f} \text{ volts} \tag{3.19}$$

The average signal power[16] is

$$S_0 = \frac{f_0^2 (\Delta V)^2}{4\pi^2 f^2} \text{ watts} \tag{3.20}$$

If T expresses the sample time, $1/f_0$, then the performance of linear delta coding compared with that of 2-bit DPCM can be displayed as shown in Fig. 3.20.

The SNR of linear delta coding depends significantly on the normalized step size, as indicated in Fig. 3.21. Shown there are the SNR results for an input signal that was gaussian band-limited flat to $1/2T$ Hz with unit power. The values of T

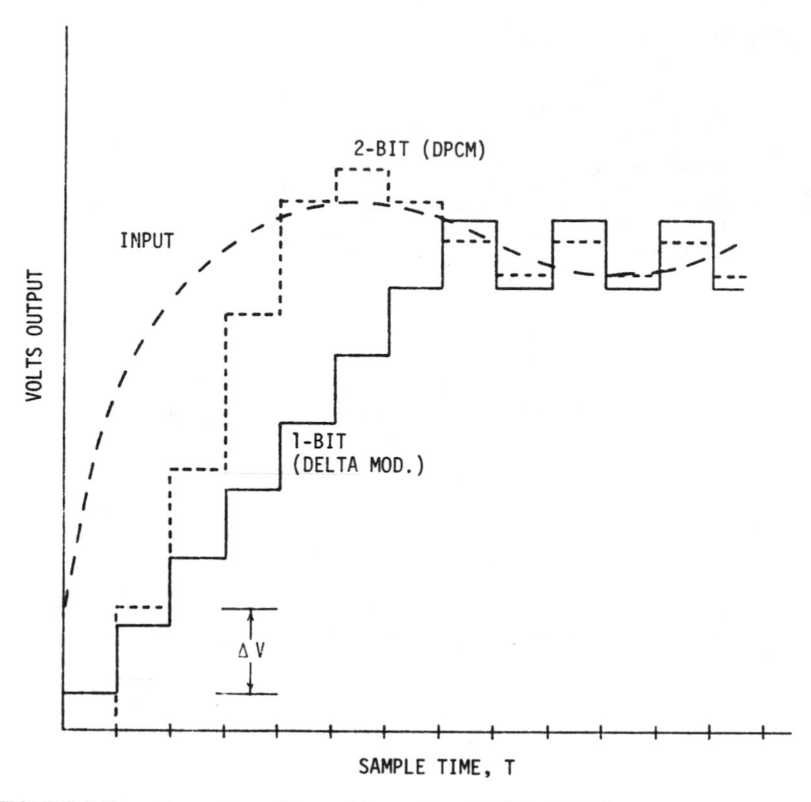

FIGURE 3.20. Comparison of linear delta coding with 2-bit DPCM.

are in time units. This figure shows that by doubling the output digital rate, the SNR is improved by nearly 9 dB (theoretically 9 dB). This compares with an improvement of only 6 dB in the case of PCM.

For a sine wave of frequency f_s Hz, delta coding with a first-order predictor (single integrator) in the feedback loop[17] provides a SNR of

$$\mathrm{SNR} = 10 \log_{10}\left(\frac{f_0^3}{f_s^2 W}\right) - 14 \text{ dB} \tag{3.21}$$

For a sine wave of frequency f_s Hz, delta coding with a second-order predictor (double integrator) in the feedback loop[18] provides an SNR of

$$\mathrm{SNR} = 10 \log_{10}\left(\frac{f_0^5}{f_s^2 f_m^0 W}\right) - 32 \text{ dB} \tag{3.22}$$

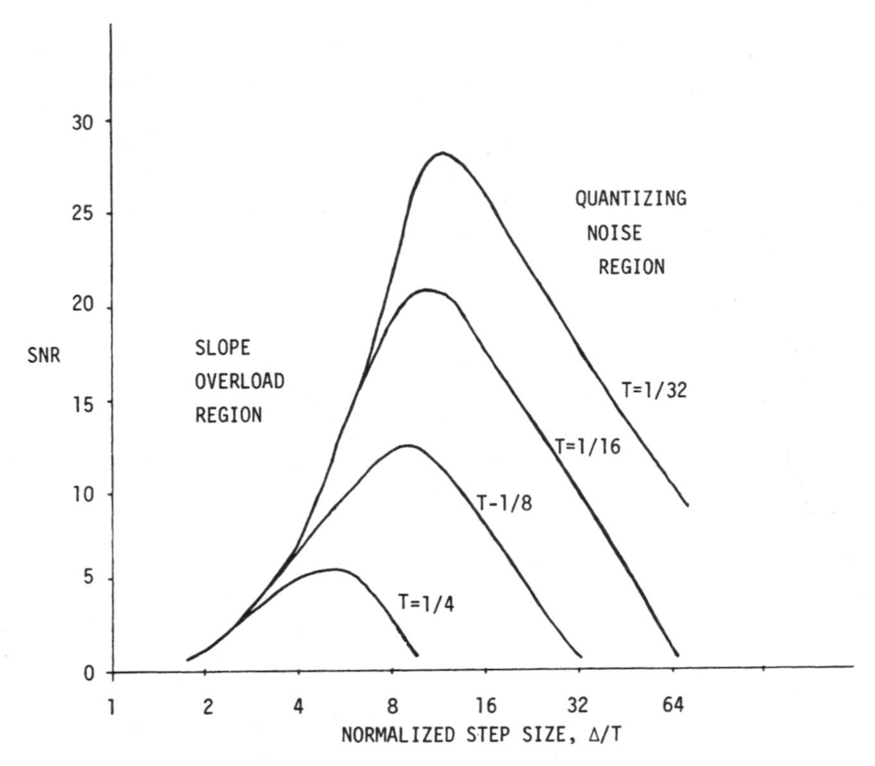

FIGURE 3.21. Performance of linear delta coding. (© 1982, Bell Telephone Laboratories. Reprinted by permission.)

where f_m is the frequency above which second-order prediction takes place in the feedback loop.

The advantage of double integration is that it provides a nearly constant overload probability for all components of the average speech spectrum. In addition, its gain is 15 db/octave. The disadvantage is possible instability when it is used with a very strong instantaneously adaptive logic such as the one-bit memory scheme, as will be discussed in Section 3.7.2. The solution to this problem is to increase the length of the quantizer memory or to use delayed encoding in which the magnitudes of the higher-order predictor coefficients are increased.

The optimum step size Δ_{opt} is given by Abate's Rule:[19]

$$\Delta_{opt} = \langle (X_r - X_{x-1})^2 \rangle^{1/2} \, ln(2F) \tag{3.23}$$

The results of an experimental verification of Abate's Rule are shown in Fig.

3.22. The curves were obtained from a simulation using gaussian signals with a uniform power spectrum. To the left of the curve peaks, the SNR is controlled by slope-overload distortion, whereas to the right of the peaks the SNR is limited by granular noise. The SNR can be seen to increase as F^3, or about 9 dB per octave of bit-rate increase. Companding can be done by varying the step size of a delta modulator. Other variations also can be done to advantage, as will be seen next in the discussion of adaptive delta coding.

3.7.2. Adaptive Delta Coding

A variable step size can be used to improve the dynamic range of a delta coder. The concept is illustrated in Fig. 3.23. The variable step size increases during a steep segment of the input but decreases while slowly varying portions of the input are occurring.

Observing a single sample of quantizer output does not provide an indication of overload or underload because there are only two quantizer levels. Step-size variation is done based on the quantizer output, however, thus no separate housekeeping information needs to be transmitted. This is accomplished as follows: sequences of quantizer outputs of length two or more are inspected to determine how to adjust the step size. If successive bits b_r and b_{r-1} are alike, the assumption is made that slope overload may be occurring. On the other hand, if $b_r \neq b_{r-1}$, the

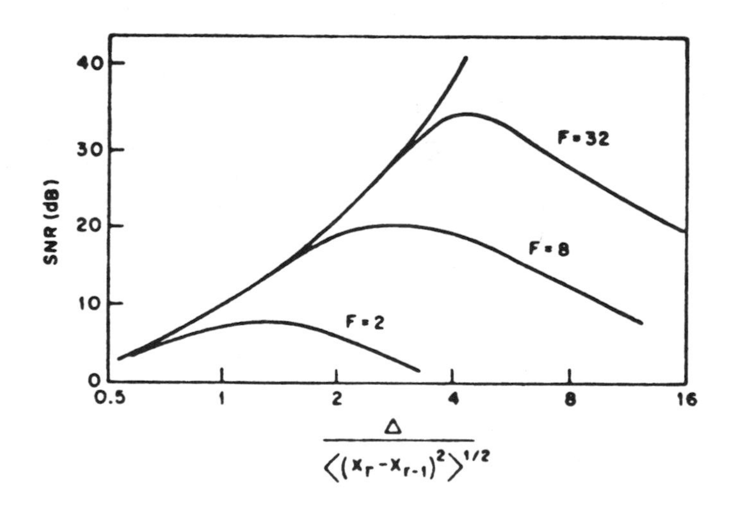

FIGURE 3.22. Experimental verification of Abate's Rule. (Courtesy N. S. Jayant, Ref. 8, © IEEE, 1974.)

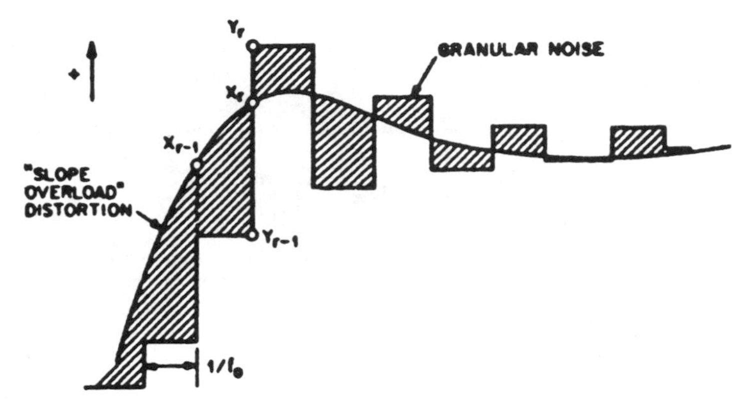

FIGURE 3.23. Principle of adaptive delta coding. (Courtesy N. S. Jayant, Ref. 8, © IEEE, 1974.)

assumption is made that granular noise is being produced. A formula[20] has been devised for adaptation. It is

$$\Delta_r = \Delta_{r-1} P^{b_r b_{r-1}} \; [P \geq 1] \tag{3.24}$$

where P is a factor giving the rate of step-size increase or decrease. For linear (nonadaptive) delta coding, $P = 1$. To minimize the quantization error power for speech encoding, the optimum value is $P_{\text{opt}} \approx 1.5$.

Adaptive delta has about a 10-dB SNR advantage[21] over linear delta for $W = 3.3$ kHz, $f_0 = 60$ kHz, and $P = 1.5$. Typical attainable dynamic ranges for adaptive delta are 30–40 dB.

Figure 3.24 is a block diagram of an adaptive delta coder. Input sampling is done at a rate $f_0 >> 2W$. Then a staircase approximation to the input is constructed as in Fig. 3.23. This is done in the accumulator, which derives the previously received value of Y_r, that is, Y_{r-1}. Unlike linear delta, however, the step-size Δ (here called Δ_r) is not fixed. Instead, it is computed by adaptation logic based upon a formula such as Eq. (3.24). One form of adaptation used for adaptive delta is *syllabic adaptation.* The time constant is on the order of 5–10 ms, providing for smooth changes in step size. The result has been called *continuous delta coding.* A block diagram is shown in Fig. 3.25. The companding is syllabic because of the use of LPF 1, typically a 100-Hz filter. Step-size control is derived from the number of ones in the bit stream. This number reflects the input signal level through the feedforward control provided by the low-frequency envelope signal S_{en}.

Syllabic adaptation decreases the granular noise in the output and thus pro-

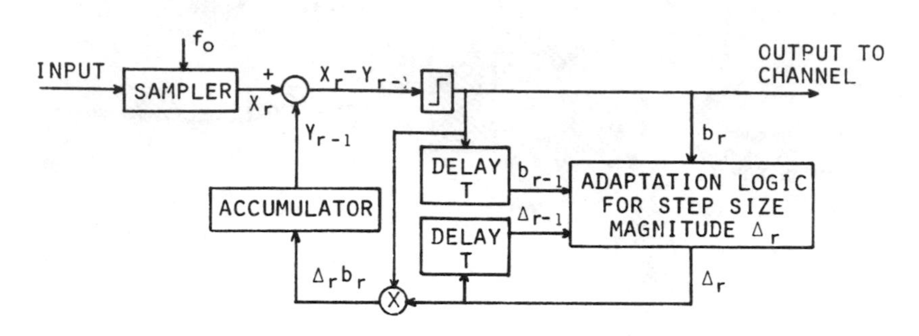

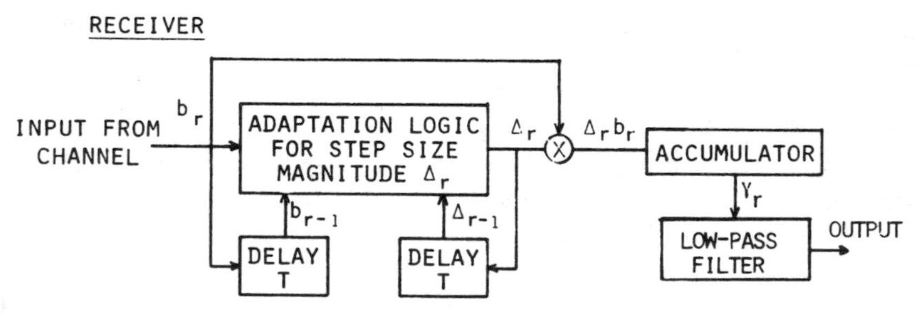

FIGURE 3.24. Block diagram of adaptive delta coder.

duces a clean sound (quiet background), but results in increased slope overload distortion and thus loss of crispness (a tendency to sound muffled). Since changes are produced at a syllabic rate, the system is more tolerant to channel errors than is ordinary PCM. A significant disadvantage of syllabic adaptation is that it is more complex to implement than is instantaneous adaptation, especially for the time-division multiplexing of several speech channels.

3.7.2.1. Continuously variable-slope delta (CVSD)

Variable-slope delta coding is achieved by encoding differences in slope rather than differences in amplitude. The result is a system that is responsive to a wider range of signal voltages than would be the case otherwise. Variable-slope delta coding is implemented by establishing a set of discrete values of slope variations. For example, in the case of 3-bit error encoding, the slopes are determined by the last three transmitted bits. Each of a set of standard slopes (e.g., 1, 2, 4, 6, 8, 16) is available for selection. As usual, the direction (+ or –) is controlled by the current bit, with one bit being transmitted at each sampling interval. Table 3.6 shows the rules used for changing the slope magnitude.

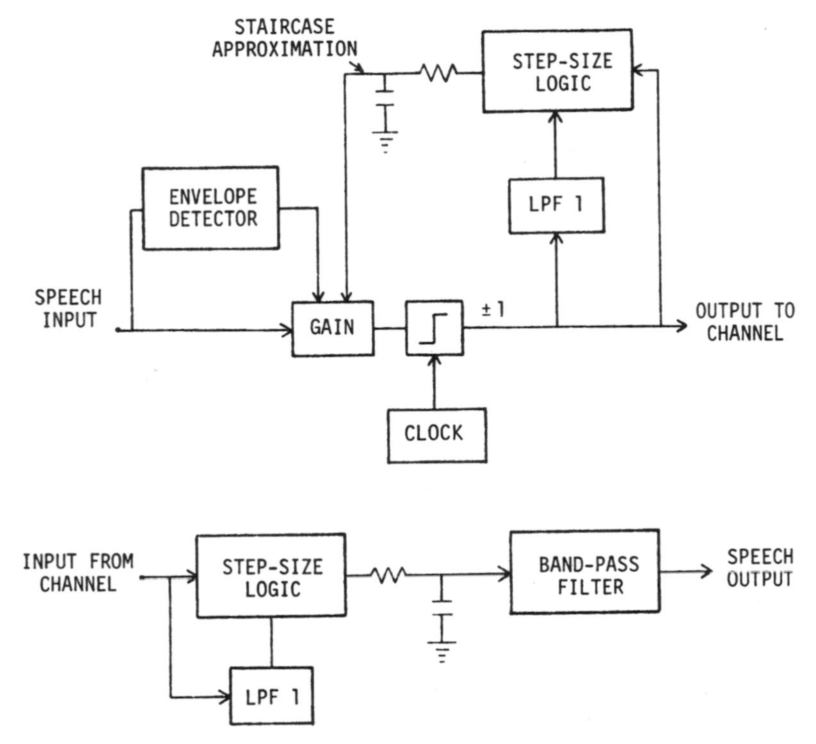

FIGURE 3.25. Continuous delta coder block diagram.

The most commonly used form of variable-slope delta is called continuously variable-slope delta (CVSD). The slope changes usually are done at a syllabic rate, and may result from either 1. a sensing of an overload when a train of ones or zeros appears in the output, or 2. an exponential increase or decrease of the feedback error voltage until the overload pattern ceases to exist, as indicated by a change in bit sense (one to zero or zero to one). A block diagram of a CVSD coder-decoder (codec) is provided in Fig. 3.26.

The CVSD codec senses two or three bit values at a time. The pulse height control function changes the quantization step size of the receiver according to the following rules:

1. If the group (two or three) of successive output bits are of the same sign, the step size is increased.
2. In all other cases the step size is decreased.

Thus the increases and decreases do not occur equally. A syllabic time constant

TABLE 3.6. Variable-Slope Delta-Coding Rules

Current Bit	Previous Bit	Pre-previous Bit	Slope Magnitude Relative to Previous Slope Magnitude
0	0	0	Take next larger magnitude[a]
0	0	1	Slope magnitude unchanged
0	1	0	Take next smaller magnitude[b]
0	1	1	Take next smaller magnitude[b]
1	0	0	Take next smaller magnitude[b]
1	0	1	Take next smaller magnitude[b]
1	1	0	Slope magnitude unchanged
1	1	1	Take next larger magnitude[a]

[a] If the slope magnitude is already at the maximum, this instruction is not obeyed.
[b] If the slope magnitude is at the minimum, this instruction is not obeyed.

(several ms) is used for the step-size decreases. The resulting performance shows lower quantization noise than that of variable-slope delta.

3.7.2.2. Digitally controlled delta coding

Digitally controlled delta coding uses syllabic companding but derives the step-size information directly from the bit sequence,[22] thereby avoiding the need for speech-envelope detection, as in continuous delta coding. For example, a step-size increase may follow the detection of a specified number of consecutive bits of the same polarity. Digitally controlled delta works best above 16 kb/s. At lower rates, the correlation between adjacent signal samples becomes too low to allow observations on the bit stream to be useful for step-size control. At these lower rates, direct monitoring of the speech envelope provides better control results than observations of the bit stream, but toll quality voice is not achieved.

3.8. SUBBAND CODING (SBC)

The various forms of PCM, DPCM, delta coding, etc., described in Sections 3.5–3.7 are time-domain coding techniques in that the speech is coded on a time-domain basis. A frequency-domain coder, on the other hand, does its coding based upon the spectral content of speech. The subband coder is a prime example of frequency-domain coding.[23] The speech signal is divided into a number of separate frequency components, each of which is encoded separately. The frequency-domain approach has two distinct advantages. One is that the number of bits used to encode each frequency component can be selected so that the encoding accuracy is placed where it is needed in the frequency domain. The second advantage is that bands with little or no energy may not be encoded at all. The main differ-

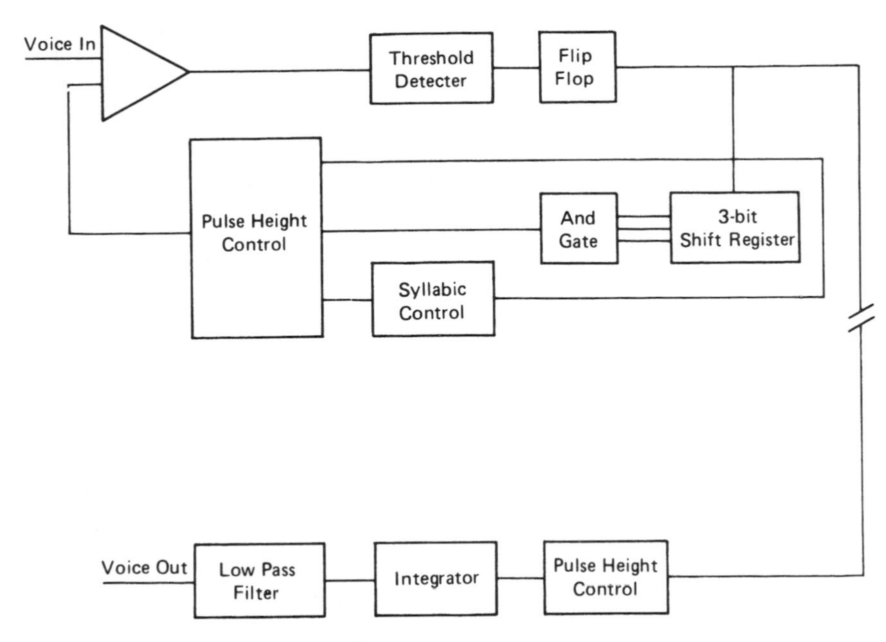

FIGURE 3.26. CVSD codec block diagram.

ences among the frequency-domain techniques usually lie in the degree of prediction used in each.

In subband coding (SBC), as the name implies, the speech spectrum is divided into subbands. This division is based on the fact that the quantizing distortion is not uniformly detectable at all speech frequencies. Thus coding based on subbands offers a possibility for controlling the distribution of the quantization noise. The overall speech band actually is partitioned so that each subband contributes equally to the articulation index (AI), as shown in Table 3.7. The AI indicates

TABLE 3.7. Subbands for Equal Articulation Index[a]

Subband	Frequency Range (Hz)	Contribution to AI
1	200–700	20%
2	700–1310	20%
3	1310–2020	20%
4	2020–3200	20
		80%

Note: 80% AI corresponds to 93% intelligibility. The remaining 20% would come from frequencies below 200 Hz and above 3200 Hz.

[a] Reprinted with permission from *Bell Syst. Tech.* J. © 1979 and 1976, AT&T.

the average contribution of each part of the spectrum to the overall perception of the spoken sound.[24] Figure 3.27 shows the partitioning in the frequency domain. SBC is implemented by translating each of the subbands downward in frequency so that the low end of each is essentially at zero frequency. This facilitates sampling rate reduction. Figure 3.28 shows the sequence of operations in SBC transmission and reception. The term *decimate* refers to a reduction of the sampling rate consistent with the width of the band to be transmitted. At the receiver, the term *interpolate* refers to the smoothing of the decoder output.

The process shown in Fig. 3.28 takes place simultaneously in each of the four bands, as portrayed in Fig. 3.29. APCM is used in each band because it provides low sample-to-sample correlation of the low-pass translated, Nyquist-sampled signals.

The transmission bit rate can be reduced by limiting the subband widths, thus allowing spectral gaps, as shown in Fig. 3.30. If these spectral gaps can be placed in frequency ranges where there is little contribution to intelligibility, their effect will be minimal. Moreover, quantizing distortion is not equally detectable at all frequencies. As the spectral gaps are made larger, however, to reduce the bit rate, the intelligibility begins to suffer. A variable-band SBC coding scheme has been devised[23] which attempts to identify and encode those spectral components (formants) of the signal that are most significant to listeners. The two upper bands are allowed to vary in accordance with the dynamic movement of the vocal tract

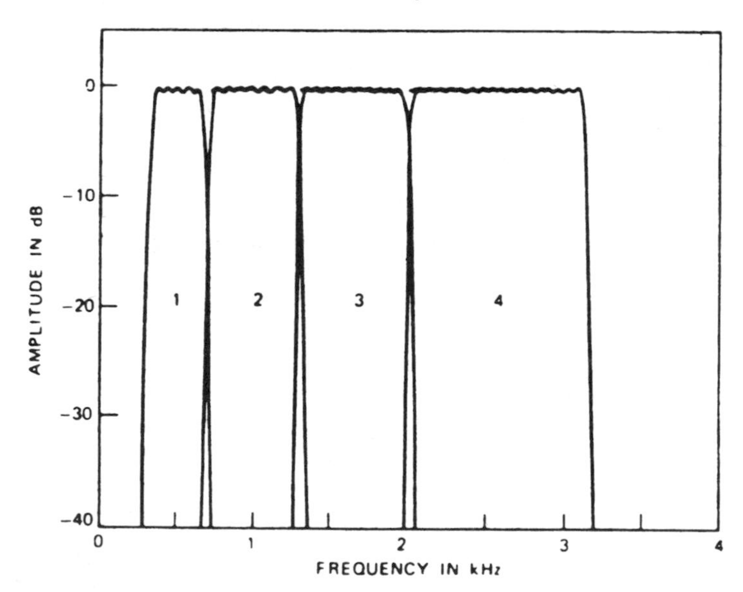

FIGURE 3.27. Subband coding partitioning in the frequency domain. (Reprinted with permission from *Bell Syst. Tech. J.*, © 1983 as a book, AT&T.)

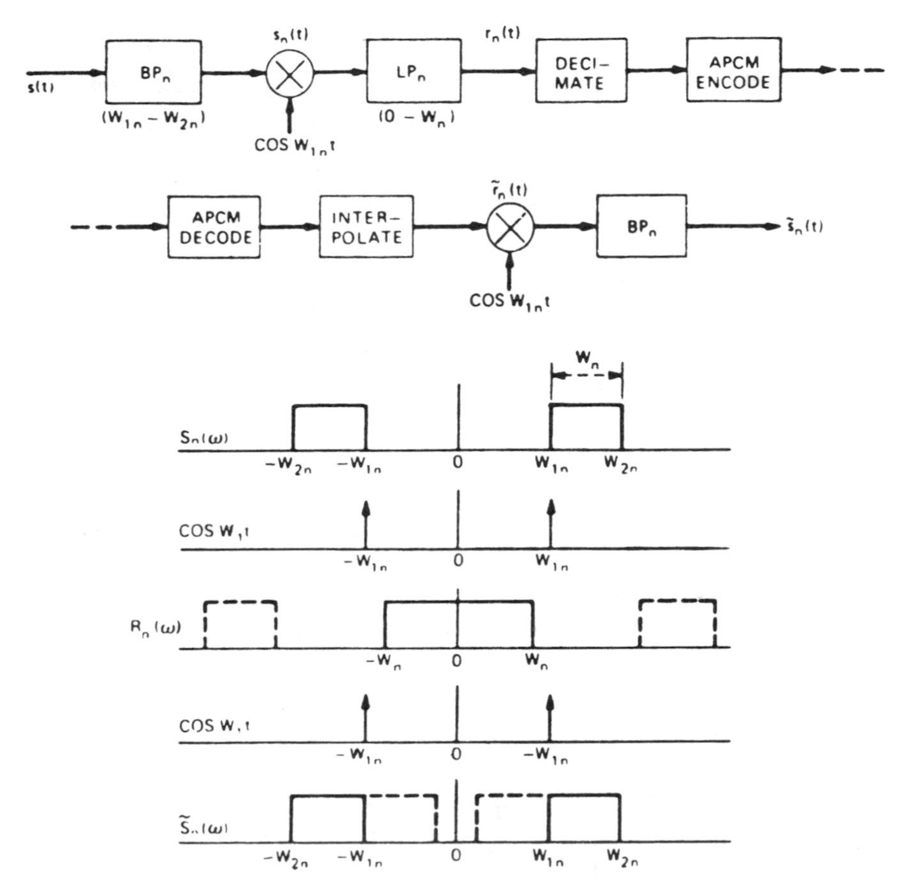

FIGURE 3.28. Subband coding transmission and reception. (Reprinted with permission from *Bell Syst. Tech. J.*, © 1976, AT&T.)

resonances, as measured by a zero crossing technique. The result is a 4800-b/s coder which provides moderate quality intelligible speech.

A technique that eliminates the use of the modulators and achieves a theoretical maximum efficiency in sampling is the integer band sampling technique, illustrated in Fig. 3.31. These advantages are somewhat offset, however, by the fact that the bands no longer can be chosen only on the basis of their contribution to the articulation index.

In integer band sampling, the signal subbands are chosen to have lower cut-off frequencies mf_n and upper cut-off frequencies $(m + 1)f_n$, where m is an integer and f_n is the bandwidth of the nth band. A sampling rate of $2f_n$ is used on each incoming subband. The received signal is recovered by decoding it and then

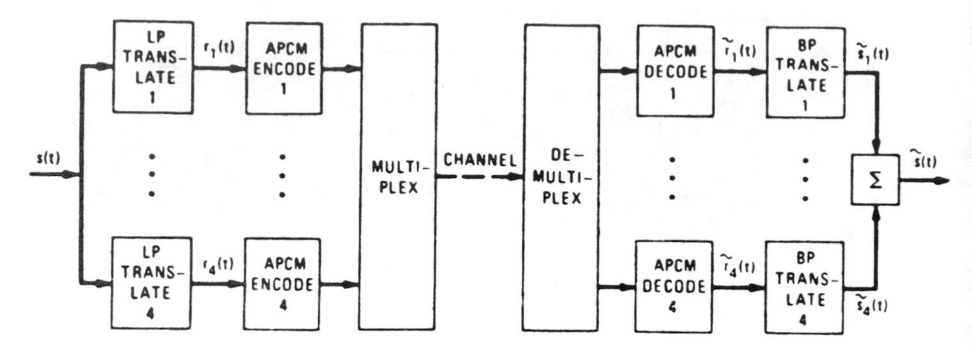

FIGURE 3.29. Four-band SBC system. (Reprinted with permission from *Bell Syst. Tech. J.*, © 1976, AT&T.)

bandpassing it to its original signal band. Table 3.8 shows the frequencies and sampling rates for 16-kb/s SBC, and Table 3.9 shows the same parameters for 9.6-kb/s SBC.

The sampling rate reduction needed to keep the overall bit rate low results in aliasing in each of the subbands. This aliasing is introduced on both sides of each band edge and is called *interband leakage*. The amount of leakage depends upon

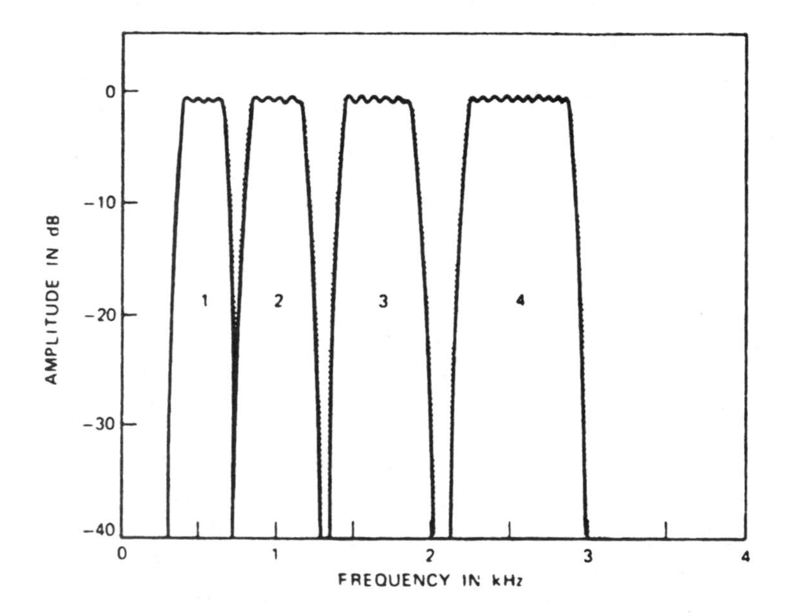

FIGURE 3.30. SBC with limited subband widths. (Reprinted with permission from *Bell Syst. Tech. J.*, © 1979 and 1976, AT&T.)

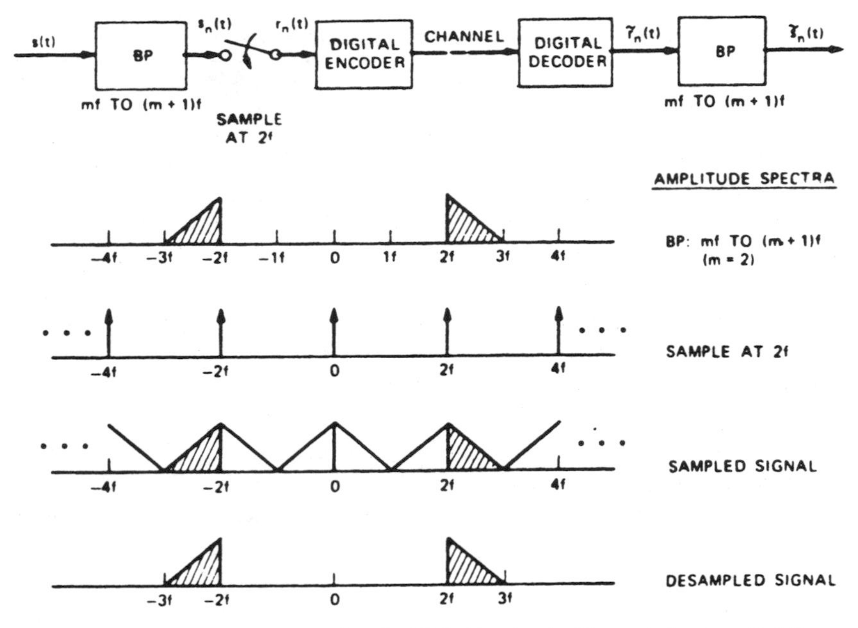

FIGURE 3.31. Integer-band sampling for SBC. (Reprinted with permission from *Bell Syst. Tech. J.*, © 1979 and 1976, AT&T.)

how sharply the subband filters attenuate out-of-band energy. The solution to this problem[26] is the use of the *quadrature mirror filter bank* (QMFB). In the QMFB, the quadrature relationship of the subband signals results in a cancellation of the unwanted images. This cancellation is achieved down to the quantization noise level of the coders.

SBC compares favorably with both ADPCM and adaptive delta coding, often

TABLE 3.8. Frequencies and Sampling Rates for a 16-kb/s SBC[a]

Subband	Center Frequency (Hz)	Sampling Rate (S^{-1})	Decimation[b] from 10 kHz	Quantization (Bits)
1	448	1250	16	3
2	967	1429	14	3
3	1591	1667	12	2
4	2482	2500	8	2

Note: Bandwidth $= \dfrac{1}{2 \times \text{sampling rate}}$

[a]Reprinted with permission from the *Bell Syst. Tech. J.* © 1976, AT&T.

[b]Decimation is the sampling rate reduction by an integer factor *m* (retain only one out of every *m* samples of filter output).

TABLE 3.9. Frequencies and Sampling Rates for a 9.6-kb/s SBC[a]

Subband	Center Frequency (Hz)	Sampling Rate (S^{-1})	Decimation[b] from 10 kHz	Quantization (Bits)
1	448	800	25	3
2	967	952	21	2
3	1591	1111	18	2
4	2482	1538	13	2

Note: Bandwidth $= \dfrac{1}{2 \times \text{sampling rate}}$

[a]Reprinted with permission from Bell System Technical Journal © 1979 and 1976, AT&T.

[b]Decimation is the sampling rate reduction by an integer factor *m* (retain only one out of every *m* samples of filter output).

designated "ADM" for adaptive delta modulation, as shown by the curves of Figure. 3.32. These curves are based on the preferences of 12 listeners, each making 16 comparisons. Two ADPCM systems were used, one (3 bits/sample) at 24 kb/s, and the other (2 bits/sample) at 16 kb/s. An adaptive delta coder was operated at 10.3 kb/s, 12.9 kb/s, and 17.2 kb/s to obtain the lower pair of curves of Fig. 3.32. The curves show SBC to be preferable to ADPCM below about 22 kb/s and to be preferable to adaptive delta (ADM) below 18 to 20 kb/s. Tests of SBC at 7.2 kb/s have shown its quality to be only slightly poorer than at 9.6 kb/s. At 4.8 kb/s, however, SBC's quality was considerably poorer than at 9.6 kb/s because of increased band limiting and the gaps between the bands needed to achieve this lower rate.

SBC is being applied to provide 7-kHz audio at 64 kb/s based upon CCITT Recommendation G.722. The 7-kHz band is divided into a lower subband, 300 to 3500 Hz, which is ADPCM coded using 6 bits. The upper subband, 3500 to 7000 Hz, is ADPCM coded using only 2 bits. Since each subband is less than 4000 Hz wide, the sampling rate can be kept at 8000/s. The overall bit rate then is $(6 + 2) \times 8000 = 64000$ b/s. The result is called subband ADPCM.[27]

SBC may be preceded by *time domain harmonic scaling* (TDHS), which is a method for frequency scaling the short-term spectrum of speech to reduce the bandwidth, and thus to reduce the sampling rate and, therefore, the bit rate.[28] At the receiving end, the spectrum is expanded again to its original form. TDHS also can be used independently of other techniques. One TDHS/SBC coder has been developed using 17-kb/s SBC with 2:1 TDHS. With the addition of side information for pitch and framing, the resulting bit rate is 9600 b/s.

3.9. ADAPTIVE TRANSFORM CODING (ATC)

Adaptive transform coding (ATC) is similar to SBC in that the speech band is divided into a number of subbands. A fast transform algorithm, such as a 128-

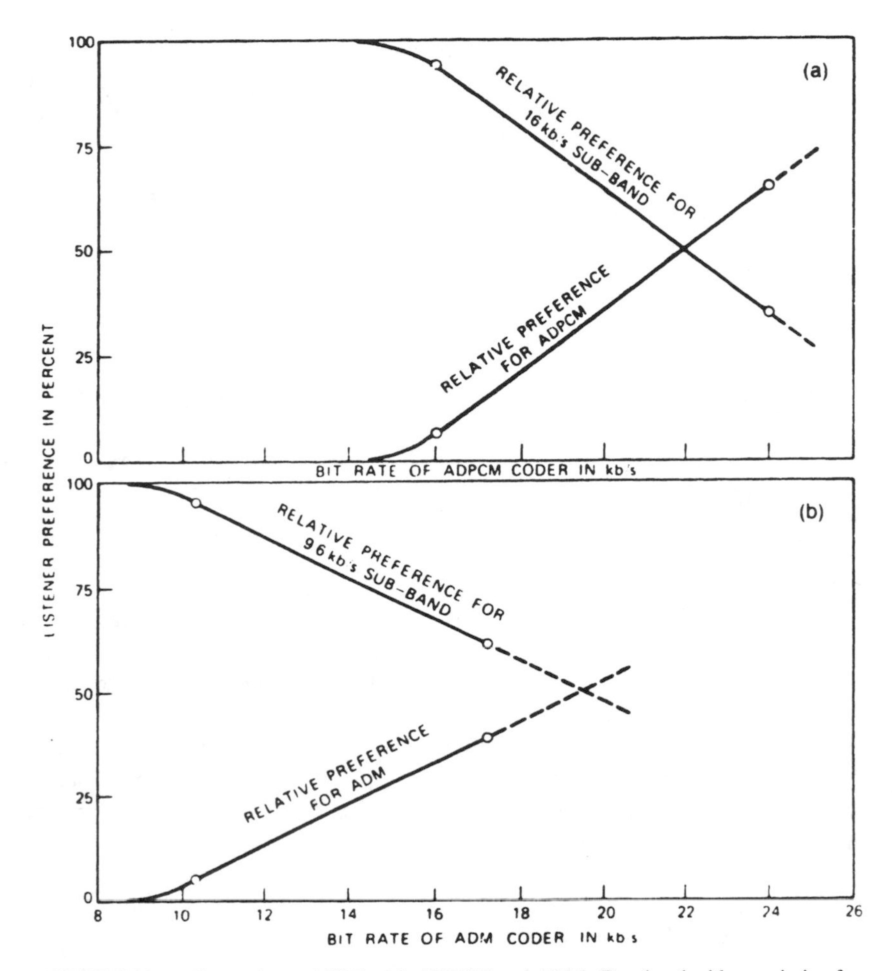

FIGURE 3.32. Comparison of SBC with ADPCM and ADM. (Reprinted with permission from *Bell Syst. Tech. J.*, © 1979 and 1976, AT&T.)

point discrete cosine transform, then translates the input to the frequency domain. The transformed outputs then are smoothed and decimated to 16 values. Finally, the transformed coefficients are encoded using APCM. Figure 3.33 is a block diagram[29] of an ATC system. The complexity of the ATC system is comparable to that of ADPCM with a variable predictor. Based upon talker-listener tests, ATC is claimed to be superior to SBC and ADPCM in the 9.6 to 24-kb/s range.[30]

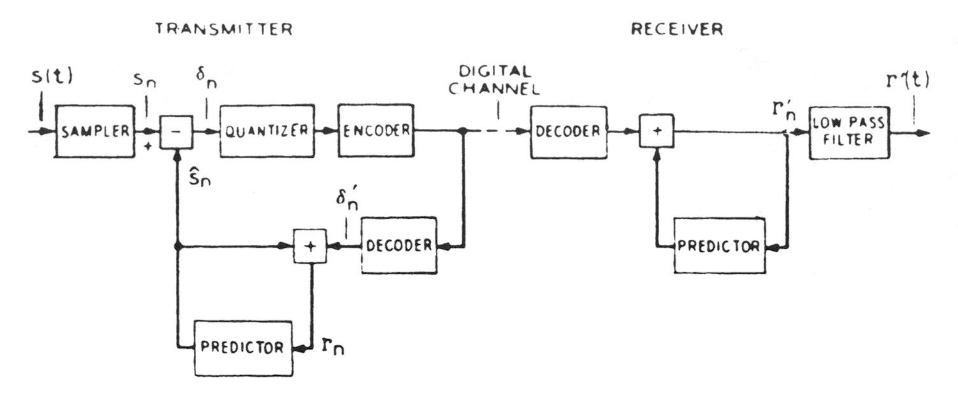

FIGURE 3.33. Adaptive transform coder system. (Reprinted with permission from *Bell Syst. Tech. J.*, © 1979 and 1976, AT&T.)

3.10. HARMONIC SCALING

Harmonic scaling is an *overlay* technique, that is, one that can be applied to other techniques. Harmonic scaling is useful in reducing the bit rate required for transmission; it is a waveform-coding technique that parametrically compresses speech in bandwidth and sampling rate. Methods of harmonic scaling focus on redundancies in speech due to pitch structure and local stationarity. In the frequency domain, the spacing of the pitch harmonics is compressed at the sending end and expanded to the original spacing at the receiving end. In the time domain, harmonic scaling involves selectively eliminating pitch periods in a pitch synchronous interpolative manner at the sending end, and then reinserting them in the same way at the receiving end.

3.11. VECTOR TECHNIQUES

A segment of a waveform can be treated as an entity, that is, its shape can be given a "catalog number," or set of digits that designates it. The shapes, or patterns, are optimized to allow the lowest mean-square error-matching based on a limited number of patterns in a codebook. Each waveform segment then can be encoded by comparing it with a set of stored reference vectors called *codevectors* (patterns). The *vector dimension* is the number of samples per sequence. An example is illustrated in Fig. 3.34, which shows superimposed patterns for 6-dimensional vectors at 1 bit per sample. Each of 64 patterns consists of six samples joined by straight-line segments. The patterns can be superimposed.

To produce *vector PCM*, for example, the number of possible 6-sample sequences, at 8 bits per sample, is $256^6 = 2^{48}$, but not all are equally probable. A small number of the most probable sequences form a code book based on the

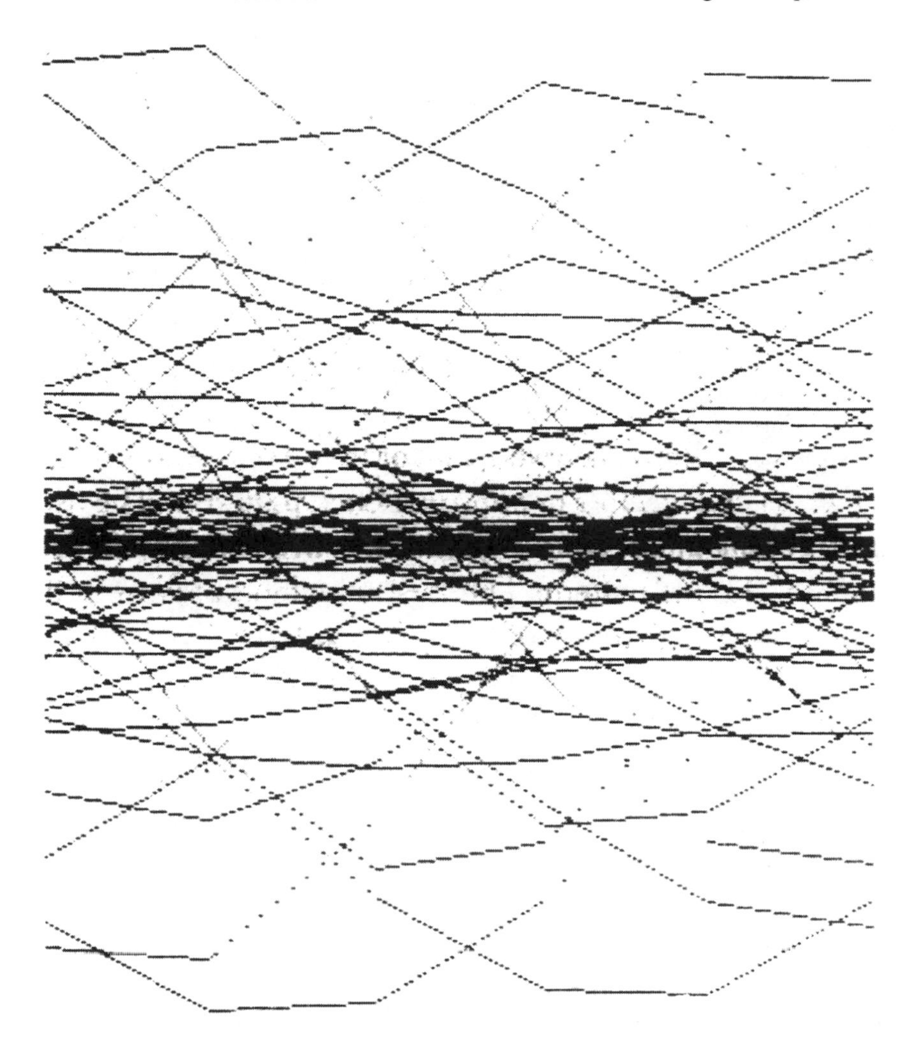

FIGURE 3.34. Superimposed vector patterns (Courtesy Gersho and Cuperman, Ref. 31, © IEEE, 1983.)

average characteristics of many voices. Many of the other possible sequences may never occur. The code book sequences then are stored in ROM, and the best-fit vector is sent instead of the actual sequence. Thus, for a code book of 65 536 vectors (2^{16}), 16 bits can be sent instead of the original 48, for a three-fold data compression.

For vector PCM in general, if N = number of patterns in the codebook, k = dimension of the vectors, and r = bits/sample, then

$$r = \frac{(\log_2 n)}{k} \tag{3.25}$$

Thus, a codebook of 64 6-dimension vectors requires the transmission of only 1 bit/sample.

Short code books mean higher quantization error, but shorter searches, and therefore less delay, in finding the best-fit vector. The procedures for rapid searching include the use of the *tree structure* and *multistage vector quantization.* In the tree-structure approach, instead of searching every code word, the processor goes through a series of binary choices in which the input sequence is compared with the stored vectors. In multistage vector quantization, a small code book is used allowing a fast search, and a second small code book provides for encoding of the residual.

Vector quantization is applicable to the output of any algorithm. Figure 3.35 illustrates the computational complexity as a function of vector dimension. A significant application of vector techniques to public telephony is its use in the speech-coding standard selected for use for North American digital cellular telephony.

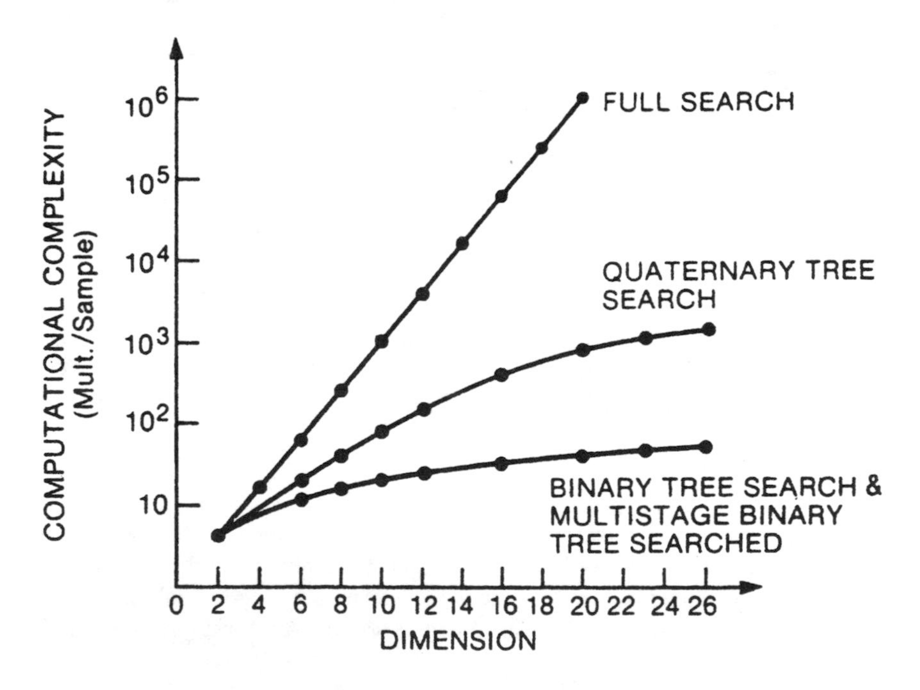

FIGURE 3.35. Vector quantization computational complexity. (Courtesy Gersho and Cuperman, Ref. 31, © IEEE, 1983.)

3.12. PERFORMANCE OF EFFICIENT WAVEFORM CODING TECHNIQUES

Research on efficient coding techniques over the years has produced some reasonably good-sounding results at bit rates well below that of 64-kb/s PCM. Why, then, except for ADPCM, is there no strong move at the present time toward the use of such techniques in the public switched network? Although such reasons as standardization and compatibility with existing plant certainly may be cited, together with questions about the acceptability of slight quality degradations by the general public, probably the strongest impediment to the widespread utilization of efficient coding techniques is their lack of transparency in the case of nonspeech signals. A telephone network, in order to transmit speech, also must handle a variety of in-band signaling and supervision tones. These are handled very well by 64-kb/s PCM, even when many A/D conversions have to be placed in tandem. The various efficient coding techniques, however, do reasonably well if only one or two A/D conversions are involved in a connection, but beyond that their performance may degrade appreciably. In the future, with an expected extensive use of signaling and supervision outside the voice channel, as in CCITT No. 7 signaling, many aspects of the transparency problem will be removed. In particular, data will be transmitted on a direct digital basis without the need for modems. This will be done by bringing the digital stream directly to the subscriber (see Section 5.9) and using digital format converters such as data service units (DSU) or channel service units (CSU) to convert the digital format of the computer to the communications line format.

Numerous evaluations of efficient coding techniques have been performed using both human speakers and listeners and computerized simulations. As explained in Sections 2.6 and 3.4, performance may be compared on either an SNR or a listener preference basis.

Computer simulations have been used to compare ADPCM and ADM with logarithmically companded PCM (log PCM). Figure 3.36 shows the results of these comparisons. SNR was measured in response to the input sentence, "A lathe is a big tool." For log PCM, the sampling frequency was 8 kHz, while for ADPCM it was set at 6.6 kHz. The ADPCM system had an adaptive quantizer and a first-order predictor. The bit rate, Bf_0, was varied by varying B, the number of bits per sample. For ADM, a one-bit memory was used. Since $B = 1$ for ADM, the bit rate was varied by changing the sampling rate.

The curves of Fig. 3.36 show that ADPCM has a constant gain over PCM because of the differential coding advantage. It also is superior to ADM at all bit rates, but has the added complexity of a multibit quantizer. The SNR for ADM increases as the cube of the bit rate, whereas the PCM SNR increases exponentially. ADM is simpler to implement than PCM and provides a better SNR below

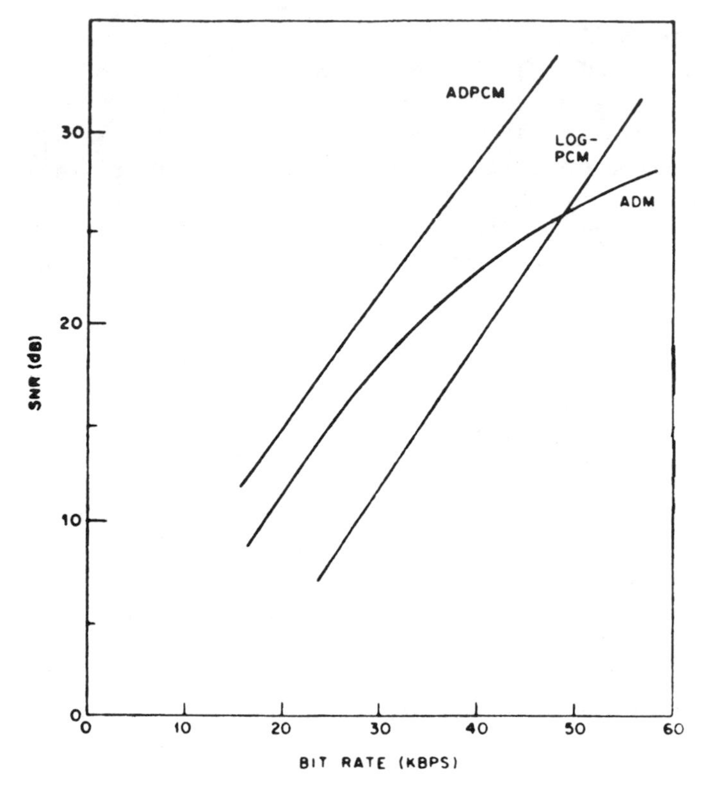

FIGURE 3.36. Comparative SNR performance of ADPCM, ADM and Log PCM. (Courtesy N. S. Jayant, Ref. 3, © IEEE, 1974.)

48 kb/s. APCM is not shown but is better than log PCM for $B \leq 5$. At $B = 3$, the advantage of APCM over PCM is 6 dB.

The adaptive techniques all exhibit distinct advantages over their nonadaptive counterparts. For example, at a sampling rate of 56 kHz, ADM is 10 dB better than LDM. At $B = 3$, ADPCM is 3 dB better than DPCM.

Log companding also produces considerable performance improvements over noncompanded PCM. Measured improvements are 24 dB to 30 dB, the equivalent of four to five quantization bits.

Perceptual and subjective effects are determined from the responses of listeners. Such results are more meaningful than the analytically determined SNR values for several reasons. First, the quantization error sequence has signal-dependent (or signal-correlated) components. In addition, signal-dependent noise (distortion) does not have the same annoyance value as does independent additive noise of equal variance.

The SNR may overstate PCM performance for $B < 6$, where dithering is used to mask the noise. With low B operation, amplitude-correlated noise is more objectionable than additive noise of the same variance. On the other hand, at high B with nonuniform quantization, the quantization error is correlated with the input amplitude. Such amplitude-correlated noise tends to be less annoying than additive noise of equal variance.

A comparison of the objective (SNR) and subjective (preference) ratings of ADPCM and log PCM is shown in Table 3.10. Perceptual effects to which the differences can be attributed[32] are the following:

1. ADPCM errors have more low-frequency (slope overload) distortion, so are more correlated with speech than is PCM noise, which is relatively white.
2. ADPCM noise contains more pitch-related buzziness than PCM.
3. ADPCM suppresses quantizing noise better during silent intervals than does PCM.

Thus, for the same SNR, ADPCM errors have a more signal-correlated distribution in time, making the error variance less objectionable. In the case of delta coding, signal-correlated errors take the form of temporal bursts of slope-overload distortion. The perceptual quality of ADM is determined largely by granular errors. For example, in the case of a 20-kb/s system with a 1-bit memory, referring to Eq. (3.24), a value of $P = 1.5$ maximizes the SNR, but $P = 1.2$ produces some interesting results. They are: (1) a 40% increase in overload noise power, but (2) a 1-dB decrease in SNR, as a result of (3) a 30% decrease in granular noise, which is only 2% of the total noise power. Above 20 kb/s, SNR results are believed to provide a conservative estimate of DM performance.

Performance characteristics of logarithmically quantized DPCM at various numbers of quantization bits and various sampling frequencies[33] are shown in Fig. 3.37. The results consist of preference judgments by a pool of 17 listeners.

TABLE 3.10. Comparison of Objective and Subjective Performance of ADPCM and log PCM[a]

Objective Rating (SNR)	Subjective Rating (Preference)
7-bit PCM	7-bit PCM
6-bit PCM	4-bit ADPCM
4-bit ADPCM	6-bit PCM
5-bit PCM	3-bit ADPCM
3-bit ADPCM	5-bit PCM
4-bit PCM	4-bit PCM

[a]Courtesy N. S. Jayant, Ref. 3, © IEEE, 1974.

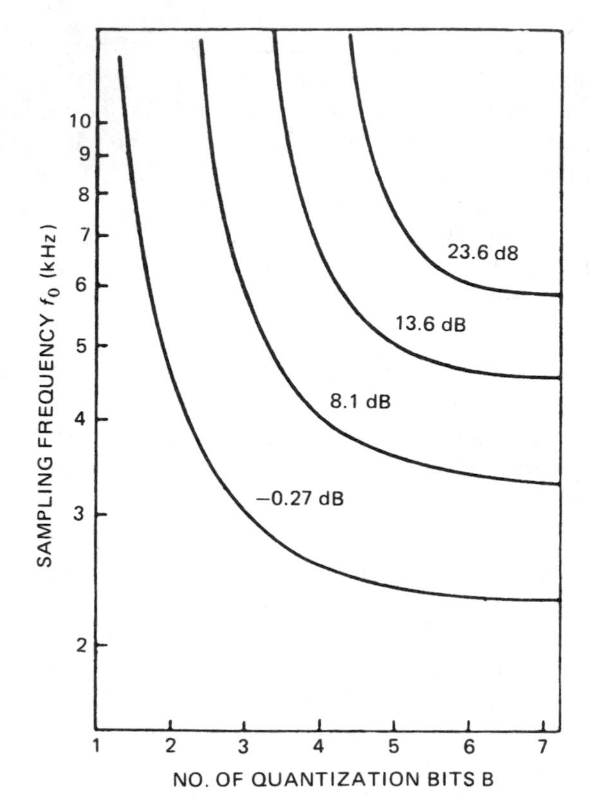

FIGURE 3.37. Performance of logarithmically quantized DPCM: isopreference curves. (Courtesy Chan and Donaldson, Ref. 33, © IEEE, 1971.)

The comparisons were based on listener preferences relative to a white-noise degraded speech signal having the indicated SNR. The isopreference contours connect points of equal subjective quality. These curves clearly show the need for an adequate sampling rate in maintaining performance quality.

The performance of CVSD in noisy channels makes it useful in military environments. The remainder of this section focuses on the performance of CVSD in view of this widespread interest. As an introduction to this matter, Fig. 3.38 shows how the word intelligibility of VSD varies with bit rate as well as with channel error probability, designated "bit-error rate, percent" in the figure.[34] Note that VSD is essentially unaffected by error rates as high as 5%. By comparison, PCM is significantly degraded by only a 0.1% error ratio.

The intelligibility values shown in Fig. 3.38 are word intelligibilities. Sentence intelligibility may be higher. Comparable performance is claimed for "two-chan-

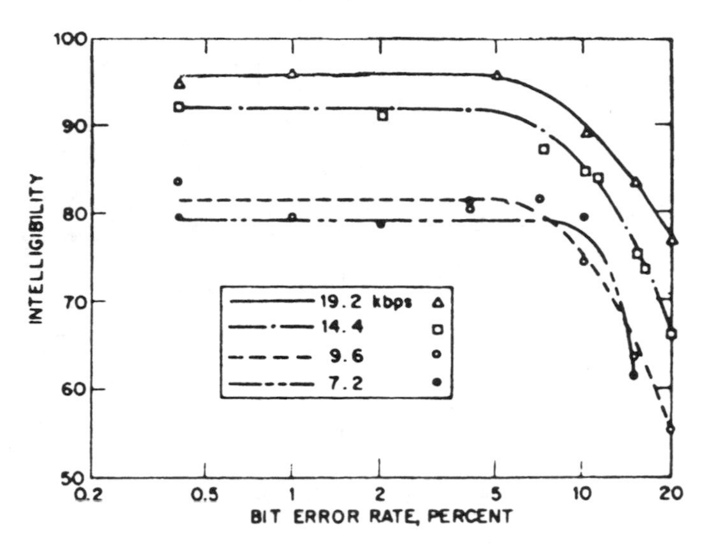

FIGURE 3.38. Intelligibility performance of variable-slope delta coding. (Courtesy M. Melnick, Ref. 34, © IEEE, 1973.)

nel DM," which uses an auxiliary delta coder to transmit a slope-envelope signal as a basis for step-size control.[35]

The performance degradation caused by channel errors is quantizer dependent. DPCM is more tolerant of random bit errors than is PCM because the error spikes are related to waveform differences rather than to the total waveform. Delta coding also can be made more resistant to errors than PCM for equal bit rates and equal channel error ratios. Instantaneously adapting delta coding, however, is more vulnerable to channel noise than is slowly companded delta or linear delta. Syllabically companded delta, designed for use over noisy channels, performs well in low-bit-rate applications (<20 kb/s).

Clustered bit errors may occur in land mobile radio links. Here the slow signal fading causes bit-error patterns in which temporal correlations are usually very obvious. Differential systems such as adaptive delta coding can be designed to provide a high degree of noise suppression. Such designs usually involve operation of the delta coder in a slope-overloading mode.

Figure 3.39 shows the SNR for CVSD at 38.4 kb/s for tones of various frequencies and levels. These curves are useful in evaluating the performance of a CVSD system with respect to signaling and supervision as well as data modem tones. Figure 3.40 compares CVSD at 32 kb/s with 7-bit and 8-bit PCM as used in the D1 and D2 channel banks, respectively. Figure 3.41 displays the overall frequency response of a CVSD encoder/decoder to voice-frequency transmission for a fixed amplitude.

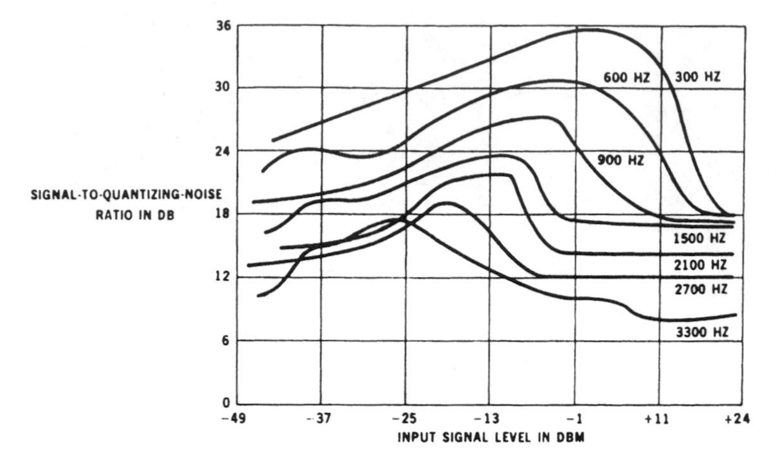

FIGURE 3.39. SNR for CVSD at 38.4 kb/s: tone response.

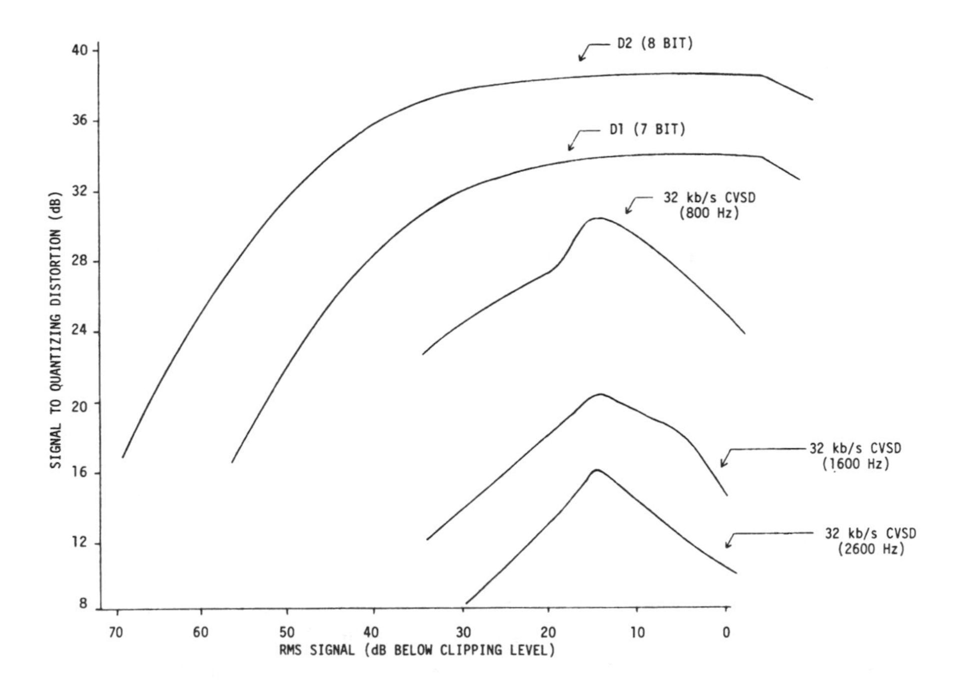

FIGURE 3.40. Comparison of CVSD with PCM for signaling and supervision purposes.

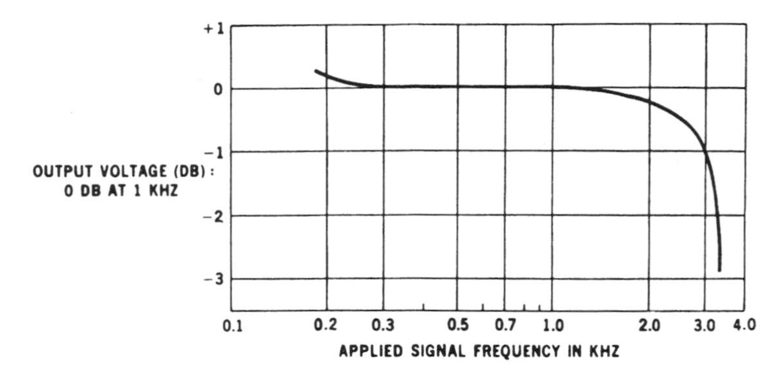

FIGURE 3.41. Frequency response of CVSD encoder/decoder.

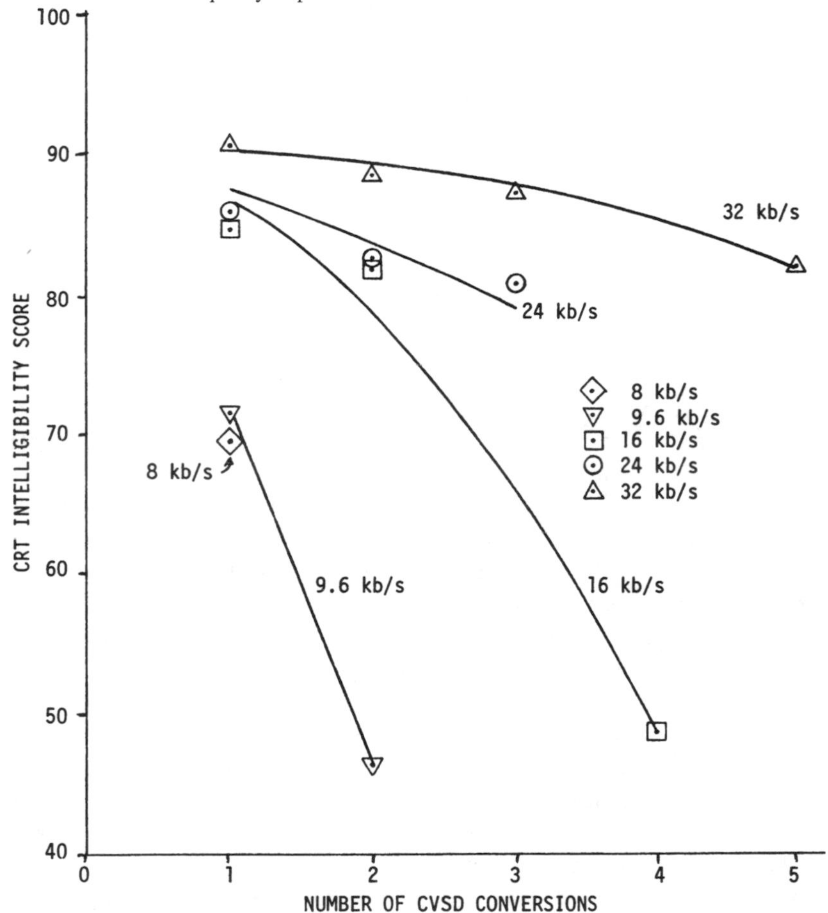

FIGURE 3.42. Intelligibility versus number of CVSD conversions at various bit rates.

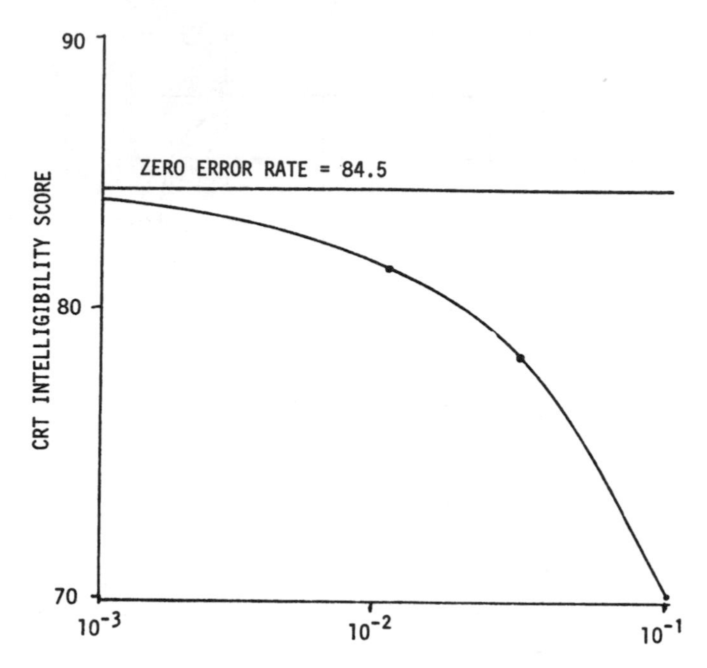

FIGURE 3.43. Intelligibility versus channel error ratio for one CVSD conversion at 16 kb/s.

Repeated encoding of speech may occur by identical or nonidentical quantizers in tandem. Moreover, direct digital conversions among adaptive and nonadaptive formats may be required. While CVSD performs fairly well through a single analog-digital conversion, additional conversions rapidly result in performance degradation that is increasingly severe at the lower bit rates, as shown in Fig. 3.42. These curves are based on the use of the consonant recognition test, which measures how well listeners distinguish one consonant from another. Figure 3.43 displays the way in which CVSD intelligibility varies with channel error ratio. Note that a 10^{-3} bit error ratio, which definitely degrades 7-bit and 8-bit PCM, has an almost negligible effect on CVSD.

In summary, PCM provides toll-quality voice transmission with transparency to other analog signals, that is, a "universal" voice channel. The ADPCM algorithm standardized by the CCITT provides voice quality comparable to that of PCM, provided that no more than 8 tandem asynchronous encodings are required. CVSD provides a voice quality only modestly degraded from that of PCM, but it shows marginal to unsatisfactory performance with nonvoice analog signals, especially under tandem encoding conditions.

A summary of some of the more popular speech digitization and compression methods with respect to their applications is provided at the end of Chapter 4 in Table 4.11.

PROBLEMS

3.1 Calculate the number of quantization bits needed for a linear PCM system to provide a 30-dB SNR with a 48-dB dynamic range.

3.2 Explain why the advantages of nonuniform quantization increase with crest factor.

3.3 Explain the difference between asynchronous and synchronous communication. Why is synchronous communication said to be more efficient?

3.4 Explain the difference between bit and frame synchronization.

3.5 Why can't a truly optimum fixed quantizer be devised for speech?

3.6 If digital transmission is used on interoffice trunks for analog voice service, explain why end users still must use modems to transmit data.

3.7 If the same individuals always use an adaptively quantized system, is the adaptive feature of any use to them, or could they achieve the same results with a properly adjusted fixed quantizer?

3.8 Explain why instantaneously adaptive systems are more sensitive to background (room) noise than are syllabically adaptive ones.

3.9 Explain why differentially coded systems are not as susceptible to channel noise as are nondifferential systems.

3.10 A speech waveform sampled at 8000 times per second is found to have a correlation coefficient C_1 of 0.95. The sampling rate then is doubled to 16 000. Would you expect C_1 to increase or decrease? Why?

3.11 Two codecs are each designed for subband coding at 16 kb/s. Will the two work properly with one another without further specification? In what ways might the two differ?

3.12 List and discuss the reasons why 64-kb/s PCM continues to be installed as the standard digital voice technique by the common carriers. What conditions might change this situation in the future?

3.13 You are required to implement a broadcast-quality circuit for a speech to be given over the radio. You have only a 64-kb/s line available. Which speech coding technique would you use and why?

REFERENCES

1. Oliver, B. M., Pierce, J. R., Shannon, C. E., "The Philosophy of PCM," *Proc. IRE*, Vol. 36, Nov. 1948, pp. 1324–1331.
2. *Transmission Systems for Communications*, Bell Telephone Laboratories, Inc., 1971.
3. Jayant, N. S., "Digital Coding of Speech Waveforms: PCM, DPCM, and DM Quantizers," *Proc. IEEE*, Vol. 62, May 1974, pp. 611–632.
4. Crue, C. R., et al., "The Channel Bank," *Bell Syst. Tech. J.*, Vol. 61, Nov. 1982, pp. 2611–2664.
5. Adams, R. L., et al., "Thin Film Dual Active Filter for Pulse Code Modulation Systems," *Bell Syst. Tech. J.*, Vol. 61, Nov. 1982, pp. 2815–2838.

6. "Data Communications Using Voiceband Private Line Channels," Bell System Tech. Ref. PUB 41004, AT&T Co. 1973.
7. McNamara, J. E., *Technical Aspects of Data Communication*, Digital Equipment Corp., 1977.
8. See note 3.
9. Duttweiler, D. L., and Messerschmitt, D. G., "Nearly Instantaneous Companding for Non-uniformly Quantized PCM," *IEEE Trans. Comm.*, Vol. COM-24, No. 8, August, 1976, pp. 864–873.
10. "The VQL Codec," Aydin Monitor Systems Document 901–0022, Ft. Washington, PA.
11. Atal, B. S. and Schroeder, M. R., "Adaptive Predictive Coding of Speech Signals," *Bell Syst. Tech. J.*, Vol. 49, Oct. 1970, pp. 1973–1986.
12. Petr, D. W., "32 kb/s ADPCM/DLQ Coding for Network Applications," *Globecom '82*, (Miami, FL, Nov. 29–Dec. 2, 1982), pp. A8.3.1–A8.3.5.
13. Nishitani, T. et al, "A 32 kb/s Toll Quality ADPCM Codec Using a Single Chip Signal Processor," *Proc. ICASSP*, April, 1982, pp. 960–963. IEEE.
14. *Reference Data for Radio Engineers*, Howard W. Sams & Co., Inc., Indianapolis, IN, Sixth Edition, 1975.
15. See note 14.
16. See note 14.
17. Schindler, H. R., "Delta Modulation," *IEEE Spectrum*, Vol. 7, Oct. 1970, pp. 69–78.
18. See note 17.
19. Abate, J. E., "Linear and Adaptive Delta Modulation," *Proc. IEEE*, Vol. 55, pp. 298–308 (Mar. 1967).
20. Jayant, N. S., "Adaptive Delta Modulation with a One-Bit Memory," *Bell. Syst. Tech. J.*, March 1970, pp. 321–342.
21. See note 20.
22. Greefkes, J. A., "A Digitally Companded Delta Modulation Modem for Speech Transmission," *ICC '70 Conf. Rec.* June 1970, pp. 7-33 to 7-48.
23. Crochiere, R. E., Webber, S. A., and Flanagan, J. L., "Digital Coding of Speech in Subbands," *Bell Syst. Tech. J.*, Vol. 55, No. 8, Oct. 1976, pp. 1069–1085.
24. Crochiere, R. E., and Sambur, M. R., "A Variable-Band Coding Scheme for Speech Encoding at 4.8 kb/s," *Bell Syst. Tech. J.*, Vol. 56, No. 5, May-June 1977, pp. 771–779.
25. See note 24.
26. Crochiere, R. E., "Sub-Band Coding," *Bell Syst. Tech. J.*, Vol. 60, No. 7, Sept. 1981, pp. 1633–1653.
27. Maitre, X., "7 kHz Audio Coding within 64 kb/s," *IEEE J. on Selected Areas in Comm.*, Vol. 6, No. 2. Feb. 1988, pp. 283–298.
28. Daumer, W. R., "Subjective Evaluation of Several Efficient Speech Coders," *IEEE Trans. Comm.*, Vol. COM-30, No. 4, April, 1982, pp. 655–662.
29. Tribolet, J. M., Noll, P., McDermott, B. J. and Crochiere, R. E., "A Comparison of the Performance of Four Low-Bit-Rate Speech Waveform Coders," *Bell Syst. Tech. J.*, Vol. 58, No. 3, March, 1979, pp. 699–712.
30. See note 29.

31. Gersho, A., and Cuperman, V., "Vector Quantization: A Pattern-Matching Technique for Speech Coding," *IEEE Comm. Mag.*, Vol. 29, No. 9, Dec. 1983, pp. 15–21.
32. See note 3.
33. Chan, D. and Donaldson, R. W., "Subjective Evaluation and Pre- and Postfiltering in PAM, PCM and DPCM Voice Communication Systems," *IEEE Trans. Comm.*, Vol. COM-19, Oct. 1971, pp. 601–612.
34. Melnick, M., "Intelligibility Performance of a Variable Slope Delta Modulator," Proc. Intl. Conf. on Comm., (Seattle, WA, June, 1973), pp. 46-5 to 46-7.
35. Greefkes, J. A., "Code Modulation Systems for Voice Signals Using Bit Rates Below 8 kb/s," Proc. Intl. Conf. on Comm. (Seattle, WA, June, 1973), pp. 46-8 to 46-11.

4

Parametric and Hybrid Coding

4.1. INTRODUCTION

The desire to conserve spectrum is a very widespread one, prevalent not only with respect to the radiated spectrum, but also significant in terms of the conducted spectrum that exists on any cable transmission facility. Numerous multilevel modulation techniques (see Chapter 6) have been devised to increase the number of bits per second that can be transmitted per hertz of bandwidth. It should be no surprise, therefore, that significant efforts are being made to encode voice at lower information rates (b/s) than required for 64-kb/s PCM, as described in Chapter 3. The fact that 64-kb/s PCM, as such, generally requires much more bandwidth than does a 3-kHz analog signal is of itself a continuing incentive to seek lower-bit-rate forms of digitized speech.

Another impetus toward the development of low-bit-rate speech has come from the desire to store speech elements digitally for future synthesis and output in response to information requests. Speech that can be expressed in the smallest number of bits requires the least amount of digital memory, and thus is the least expensive in terms of storage costs.

Perhaps the most significant motivation for the use of low-bit-rate speech coding techniques comes from the spectrum conservation requirements of the portable/mobile telecommunications community. Low-bit-rate coding techniques that are robust with respect to noise and interference in the channel, as well as acoustic noise in the vehicle, are essential; in fact, these techniques must perform well without requiring large amounts of error correction which, itself, uses additional channel capacity.

Digital speech has numerous advantages in portable/mobile system applications. From the user's viewpoint, privacy can be assured through the use of encryption, readily achievable by using bit-scrambling techniques. In addition, multipath fading effects can be mitigated by the use of forward error correction, described in Section 6.3. Moreover, efficient signal regeneration (see Section 5.8) can be achieved, and speech and control signal bits can be time-division multiplexed.

What information rate actually corresponds to the spoken word? One could take a cavalier approach and point out that a person skilled in keyboard entry can convert the spoken word to alphanumeric form, which then can be conveyed by a data circuit at 75 b/s to a receiving terminal where the result can be read. That demonstration might be felt to prove that the information content of the spoken word is only 75 b/s. In fact, based upon the relative frequencies of occurrence in the English language, each set of distinctive sounds (phoneme) conveys 4.9 bits. In conversational speech about ten phonemes per second are produced. On this basis the bit rate for speech is less than 50 b/s.

What is missing in the foregoing arguments is the fact that the human auditory apparatus can scan the multidimensional attributes of the acoustic image and concentrate on minute details which may have emotional content or information about the physical environment of the source. The ability of a listener to recognize a known speaker generally is denied by these very low-bit-rate implementations of speech. The true information rate of the spoken word thus must be higher than 50 to 75 b/s. Nevertheless, enormous economies in bandwidth might be realizable if some deterioration in quality were acceptable and if the resulting reduction in transmission costs justified the expenses of the pre- and post-processing equipment, or if the available bandwidth constituted a restriction in meeting further demands for service.

As noted at the beginning of Chapter 2, speech can be digitized on a wave form, parametric, or hybrid basis. The waveform coder operates in a manner independent of what the waveform is, that is, the waveform may be male or female speech, a signaling tone, dial tone, or busy tone. The parametric coder, however, is designed specifically for one type of input, usually speech. Therefore, for example, it may try to determine at any given moment whether its input is voiced or unvoiced and, if voiced, its pitch. The coder may also determine its amplitude, and the nature of its frequency content. If the input is speech, the parametric coder does what it is intended to do, perhaps well, perhaps poorly. However, the device is only built to respond to speech. It will attempt to treat a dial tone as if it were speech and may produce something that sounds totally different. Accordingly, the parametric coder can be used for speech only. Nonspeech signals on the system must be routed in other ways to achieve proper system operation.

The hybrid coder often is based on a waveform coder whose operation has been tailored or optimized for speech in a specific way. Examples include devices that are built to adapt to the nature of their inputs at a rate that corresponds to the syllabic rate in human speech. Other hybrid coders may contain spectral gaps in their frequency responses, and thus may respond well to tones of some frequencies but not others.

This chapter discusses a variety of coding techniques that have been devised to digitize a speech channel on a parametric basis, that is, without endeavoring to reproduce the original waveform as such. Accordingly, Nyquist sampling need not be used, and the rate can be significantly less than the Nyquist sampling rate.

4.2 PRINCIPLES OF LOW-BIT-RATE SPEECH

Before proceeding further, it is useful to note that there are several principles of low-bit-rate speech coding that form the basis of operation of all speech-coding processes other than linear PCM, which might be regarded simply as "brute force" waveform coding. The use of these principles makes the coding more and more speech specific as the bit rate is diminished.

These principles are as follows:

1. Remove temporal redundancy in the signal. This principle is used in those coders that perform what is called *long term prediction* to remove a significant portion of the redundancy that exists as a result of the periodicity of the pitch component of speech.
2. If a signal is not present in a time interval, do not transmit it. This principle is used in the digital speech interpolation (DSI) systems, as described in Section 5.7. These systems allow other speakers to use the blank intervals.
3. If a signal is not present in a frequency interval, do not transmit it. Adaptive transform coding (ATC), as described in Section 3.9, does this in that the coefficients for missing frequency intervals are zero.
4. Assign long code words to infrequently transmitted messages and short code words to frequently transmitted messages. This is known as the Huffman principle. Examples can be found in the vector quantization techniques used in code-excited linear prediction (CELP), which forms the basis for the speech-coding algorithm that has been adopted for digital cellular radio in North America.
5. Shape the quantizing noise spectrum so it is least objectionable to the user. This principle is found in ATC as well as in tree encoding.
6. Adapt the coding algorithm to the changing statistics of the signal. This principle is used in all of the adaptive techniques, including ADPCM, as well as in many of the parametric techniques to be discussed in this chapter.

4.3 VOCODERS

The actual information rate needed to convey conversational speech has been the subject of much study and discussion. Intelligibility is not the only factor of significance. Also important are speaker recognition and naturalness.

The foregoing considerations have been instrumental in the development of devices called vocoders (voice coders), whose operation depends upon a parametric description of the characteristics of speech rather than the actual coding of its waveform. Vocoders often exhibit a synthetic sound because of difficulties in accurately measuring and coding the pitch or because the spectral energy measurements made do not preserve all of the perceptually significant attributes of the speech spectrum.

Vocoder operation generally depends upon a parametric description of the transfer function of the human vocal tract. The voice provides an acoustic carrier, the pitch, for the speech intelligence, which appears largely in the form of time variations in the envelope of the radiated signal. The three types of sounds with which a vocoder must deal are *voiced sounds, fricatives*, and *stops.*

The voiced sound is a nearly periodic sequence of pulses having a spectrum that falls at 12 dB/octave. Its network equivalent is a current source exciting a linear, passive, slowly time-varying network. A fricative is a sustained voiceless (unvoiced) sound produced from the random sound pressure that results from turbulent air flow at a constricted point in the vocal system. It covers a broad band and exhibits a gentle attenuation at the band edges. Its network equivalent is a series voltage source whose internal impedance essentially is that of the constriction and typically is of large value.

A stop is produced by an abrupt release of air pressure built up behind a complete occlusion. Its spectrum falls as $1/f$. Its electrical equivalent is a step function.

In the present state of the art, the synthetic quality of totally parametric vocoded speech is not appropriate for commercial telephony. Numerous special applications exist already, however, and quality is continuing to improve.

Many types of vocoders have been devised. They include the *channel vocoder*, in which the values of the short-term amplitude spectrum of the speech signal are evaluated at specific frequencies; the *formant vocoder*, based on the frequency values of major spectral resonances; the *autocorrelation vocoder*, which uses the short-time autocorrelation function of the speech signal; the *orthogonal function vocoder*, which uses the coefficients of a set of orthonormal functions that approximate the speech waveform; and the *cepstrum vocoder*, which uses the Fourier transform of the logarithm of the power spectrum. The most popular approach to parametric coding, however, has proven to be the *linear predictive coder* (LPC), which uses linear prediction coefficients describing the spectral envelope. The following paragraphs describe the operation of the LPC vocoder.

4.3.1. Linear-Predictive Coding (LPC)

Since the design of a vocoder depends on speech characteristics, many attempts have been made to model such characteristics. One model that is widely used is the linear speech production model.[1] Its parameters are easily obtained using linear mathematics. The result is a *linear-prediction* model, which forms the basis for the LPC vocoder.

Figure 4.1 shows a digital model of speech production[2] in which, at any moment, the speech is either voiced and simulated by an impulse train having the proper pitch period, or unvoiced and simulated by a random function source. The impulse train or random function source then is multiplied by a gain factor, following which it is filtered by a time-varying device whose parameters are reflection coefficients of an acoustical tube which simulates the human vocal tract.

The foregoing functions are implemented as shown in Fig. 4.2. Here the transmitter function is called *analysis*, whereas the receiver function is called *synthesis*. Figure 4.3 shows[3] the way in which LPC approximates the speech spectrum envelope as a function of the number of predictor coefficients p. Implementations using $p = 8$ and $p = 10$ provide reasonably good quality speech at 2.4 kb/s, a rate at which LPC vocoders often operate. This rate is based on sending 48 bits every 20 ms or 54 bits every 22.5 ms. The bit allocation used for the LPC-10 algorithm established by the U.S. Defense Department as a standard for LPC voice transmission is shown in Table 4.1.

The predictor coefficients define a spectrum, which is the same as that of a vowel. They also define narrow-bandwidth *formants* and continuously follow the movements of formant frequencies. (A formant is a resonance frequency of the vocal tract tube.)

A problem arises in the detection of pitch in telephony. It results from the fact that the fundamental frequency (50–200 Hz for males; 100–400 Hz for females) may be physically suppressed by the low-frequency rolloff of the voice-frequency circuits. Even though the fundamental is suppressed, it is still fully perceived by the listener because of what is called the *residue pitch* effect of hearing. The

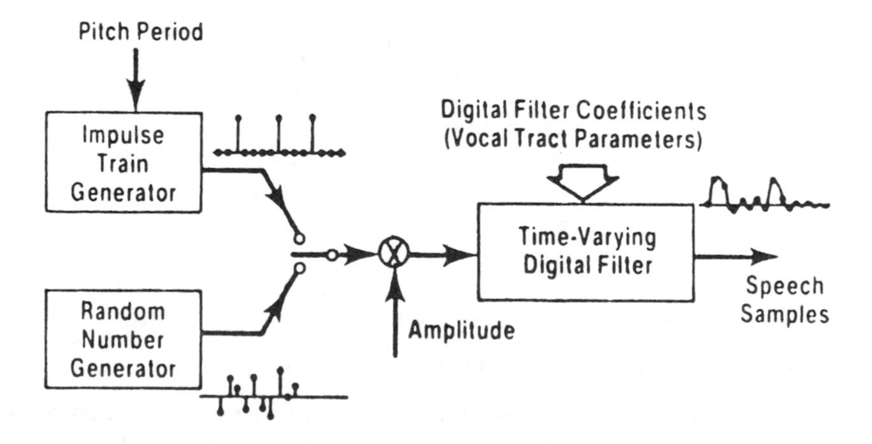

FIGURE 4.1. Digital model of speech production. (Courtesy R. W. Schafer, Ref. 2.)

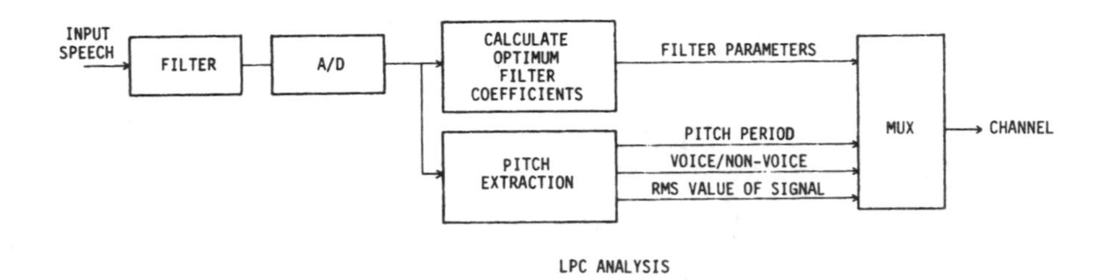

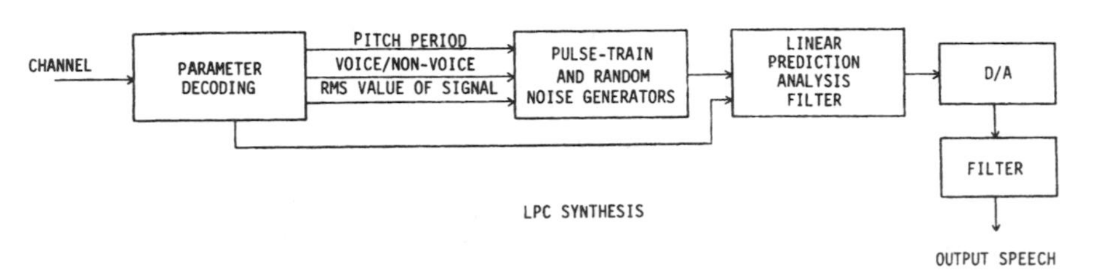

FIGURE 4.2. Implementation of an LPC system.

listener hears the harmonics of the pitch and, as a result, perceives the presence of the fundamental.

Whether a speech signal is from a telephone transmitter or not, it is forced into a simplistic model of voiced and unvoiced sounds. The excitation for voiced sounds, however, is not always well represented by a periodic pulse train, just as the excitation for unvoiced sounds is not always well represented by a noise generator. These differences between real and simulated voice account for the artificial sound often exhibited by LPC and other vocoder techniques.

LPC systems using the LPC-10 algorithm and operating on 2400 b/s channels have been shown to achieve a 92% intelligibility score on the diagnostic rhyme test (DRT)[4] using male speakers. Although the tests were based on an error-free channel, a bit error ratio as high as 10^{-3} has a negligible effect on LPC performance.

LPC offers a speech quality that, while not suitable for telephony in the present state of the art, nevertheless is better than other vocoder techniques requiring a comparable bit rate, usually 2400 b/s. Accordingly, research efforts are aimed at improvements in LPC. Improvements to the basic "simplistic" model that may prove useful include those which provide better duplication of the short-time amplitude spectrum and better duplication of the natural acoustic interactions between the source and the system. An improved model may be one in which the sound source and resonant system are allowed to load one another acoustically, that is, to interact. Such a model would account not only for voiced and unvoiced

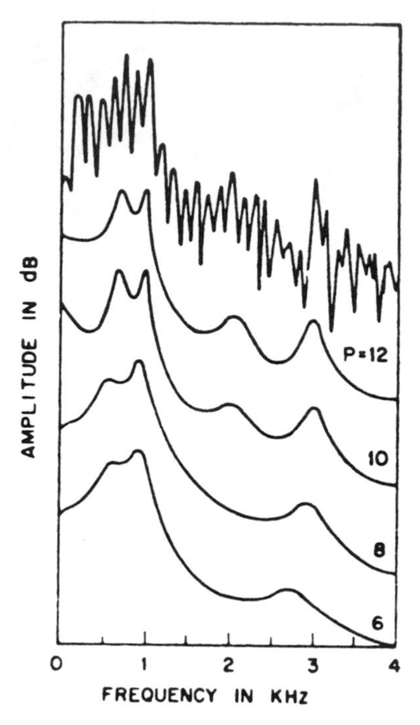

FIGURE 4.3. Actual speech spectrum compared with LPC approximations. (Courtesy J. L. Flanagan et al., Ref. 3, © IEEE, 1979.)

sounds, but also for sound radiation from the yielding side walls of the vocal tract, as well as for detailed loss factors due to viscosity, heat conduction, and radiation resistance. Preliminary results indicate that such an improved model requires no increase in information rate over a simplistic one, but does make more effective use of the information available to it.

4.3.2. Other Vocoders

A vocoder can operate on either a time-domain or a frequency-domain basis. (The LPC vocoder described in Section 4.3.1 is a time-domain vocoder.) The parametric description of the vocal-tract transfer function, on which a vocoder operates, can take a variety of forms, either in the frequency or the time domain. Examples of frequency-domain vocoders include the *channel vocoder* and the *formant vocoder*. These vocoders send spectral coefficients describing the formant frequencies and amplitudes. Time-domain vocoders include the *autocorrelation vocoder*, the *orthogonal function vocoder*, and the *cepstrum vocoder*. These vocoders describe the vocal-tract transfer function in the time domain.

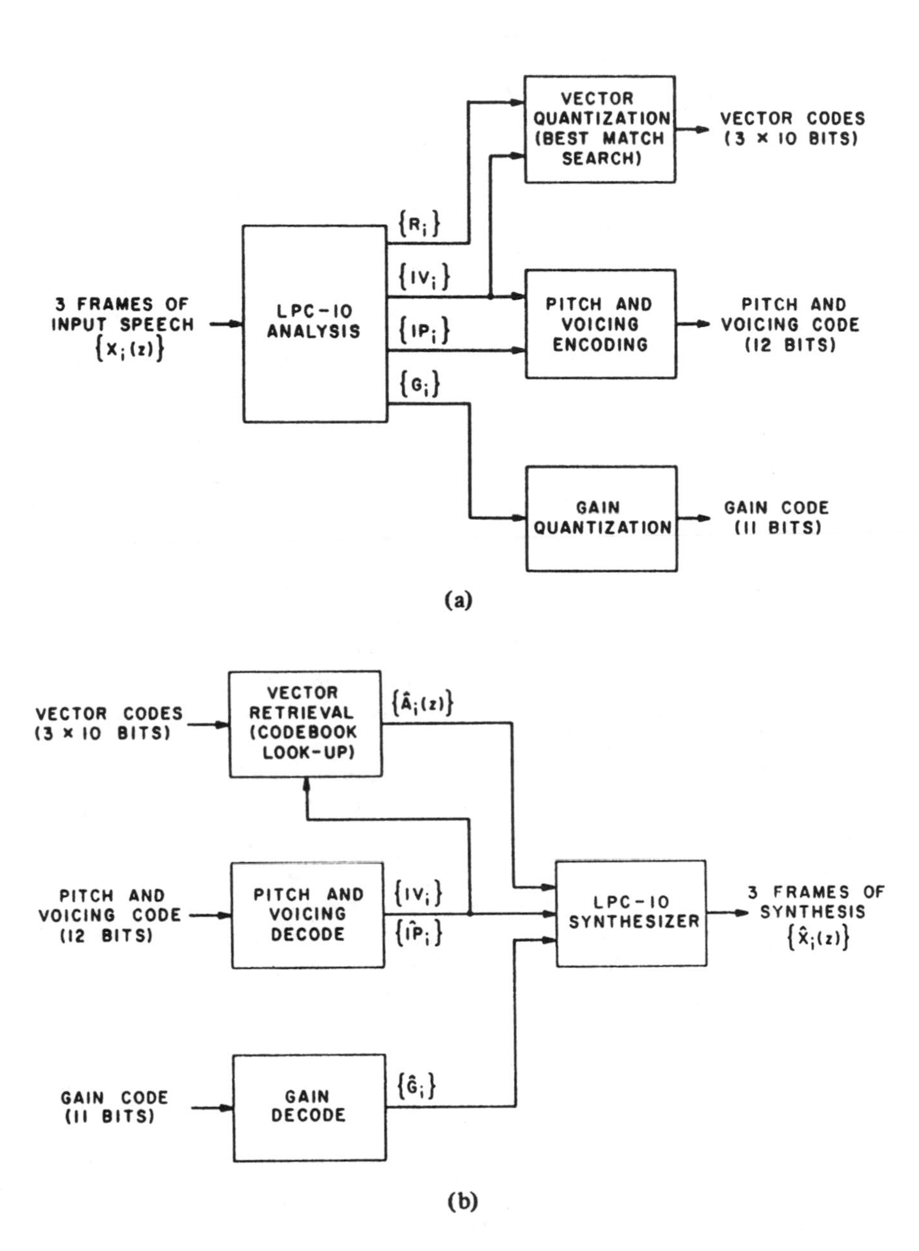

FIGURE 4.4. Vector Quantization LPC block diagram. (Courtesy Wong, Juang and Gray, Ref. 5, © IEEE, 1982).

TABLE 4.1. Bit Allocation for the LPC-10 Algorithm

	Voiced	Unvoiced
Pitch/voicing	7	7
RMS	5	5
Sync	1	1
K(1)	5	5
K(2)	5	5
K(3)	5	5
K(4)	5	5
K(5)	4	0
K(6)	4	0
K(7)	4	0
K(8)	4	0
K(9)	3	0
K(10)	2	0
Total:	54	33
Error correction	0	20

A technique called *vector quantization LPC* (VQ-LPC) allows a reduction from the 2400 b/s rate to 800 b/s. This reduction is achieved by reducing the number of bits per frame from the 54 of LPC-10 to 18, while retaining the 44.4 frame-per-second rate.[5] Table 4.2 illustrates the bit allocation.

Several techniques are used to achieve the bit-rate reduction. Three consecutive frames of LPC parameters are coded. Then a vector quantizer is used on the LPC coefficients. This process is based on a voiced-speech code book, as well as an unvoiced-speech codebook, which includes nonspeech pauses and background noise. Pitch and gain coding are scalar for one of every three frames, and differential for the remaining two frames.

Figure 4.4 is a conceptual block diagram of a VQ-LPC coder and decoder. As can be noted from the figure, VQ-LPC is compatible with LPC-10 because the design differs only in the quantization and encoding algorithms.

The performance of VQ-LPC compared with that of LPC-10, based on DRT scores, is illustrated in Table 4.3.

TABLE 4.2. Bit Allocation for Vector Quantization LPC

	2400 b/s LPC-10	800 b/s VQ-LPC
Bits/frame	54	18
LPC Coefficients	41	10
Voicing, Pitch, Gain, Sync	13	8

TABLE 4.3. Performance Comparison: VQ-LPC and LPC-10

Feature	LPC-10 (2400 b/s)	VQ-LPC (800 b/s)
Voicing	90.6	87.1
Nasality	96.1	86.7
Sustenation	78.9	62.1
Sibilation	98.0	94.5
Graveness	73.8	57.0
Compactness	93.0	92.6
Overall Average)	88.4	80.0

Table 4.4 illustrates the performance of VQ-LPC in the presence of transmission errors.

4.3.3. Effect of Model on Speech Quality

Speech quality limitations in the vocoders discussed up to this point partially result from the simple model used to parameterize speech information. For example, pitch is represented as a periodic pulse train; unvoiced sounds are represented as random noise. Useful model improvements include a better duplication of the short-time amplitude spectrum, as well as better duplication of the natural acoustic interactions between the source and the system. The speech coders discussed in the remainder of this chapter use the LPC principle, but combine it with more sophisticated analysis and synthesis procedures to provide more natural speech at a modest increase in bit rate.

4.4 HYBRID (WAVEFORM-PARAMETRIC) TECHNIQUES

The difficulties encountered in pitch detection have resulted in a number of hybrid techniques, which combine waveform and parametric coding. Some hybrid techniques avoid the problems and complexities of pitch detection by coding the lower

TABLE 4.4. Performance of VQ-LPC in the Presence of Transmission Errors

Bit Error Ratio	DRT Score
0%	80.0
1%	75.3
2%	71.7
5%	55.5

frequencies on a waveform basis but the higher frequencies on a parametric basis. Hybrid arrangements of SBC (see Section 3.8) and LPC have been devised for this purpose. These combinations generally operate at 4.8–9.6 kb/s. Other hybrid techniques of a more sophisticated nature are discussed in the remaining sections of this chapter.

4.5 ADAPTIVE PREDICTIVE CODING (APC)

The term *adaptive predictive coder* (APC) is used to describe a large class of speech coders that use adaptive prediction. Included among them are the ADPCM and ADC coders described in Chapter 3. Predictive coders have been built in which the processed speech has fair voice naturalness and intelligibility even when acoustic background noises are present. In addition, the coding can be done efficiently without the need for large codebook memories.

The design principle on which an APC is built is that the parameters of a linear predictor are optimized to obtain the minimum mean-square error between the predicted value and the true value of the signal. As such, the term APC often is used to describe an adaptive version of an LPC vocoder, as is discussed next.

An implementation algorithm using a fourth-order LPC coefficient analysis has been developed by the U.S. Department of Defense (DOD), and is called APC-4.[6] As can be seen below, use is made of parametric techniques in that sets of bits are used to describe the pitch, the relative pitch amplitude (*pitch gain*), the overall amplitude (*energy scale factor*), and vocal tract resonances (*formants*). The formants are described by the *K* factors, as were presented in Section 4.3.1. The variation of residual energy during a frame is described by the SEGQ parameters.

The system parameters are shown in Table 4.5. Figure 4.5 is a descriptive block diagram of an APC analyzer, and Fig. 4.6 is a corresponding diagram of an APC synthesizer.[7]

As can be seen from the foregoing discussion, APC is primarily a parametric coding technique, but one in which waveform error parameters are sent separately. Actual digitized speech (residual) samples are transmitted, unlike LPC, in which the information is sent on a parametric basis entirely.

Transmission errors $\leq 10^{-3}$ do not degrade the speech perceptibly. The prediction process, however, may produce errors in the predicted pitch resulting in a reverberant sound, although occasional pitch errors do not severely affect voice naturalness. Granular noise results from 1-bit quantization of the error signal. The signal-to-quantizing-noise ratio can be improved through preemphasis at 6-12 dB/octave above 500 Hz before sampling and deemphasis at the output of the synthesizer (receiver). The reason for the choice of 500 Hz is that the spectrum of the human voice tends to fall off beyond 500 Hz, whereas the spectrum of quantization noise tends to be flat.

TABLE 4.5. System Parameters for the APC-4 Algorithm

Rate:	9600 b/s
Frame length:	25 ms
Bits/frame:	240
Sampling rate:	7600/s = 190/frame
Bit allocation:	

Parameter	Bits
Pitch period	6
Pitch gain factor	3
Energy scale factor	5
K(1)	5
K(2)	5
K(3)	5
K(4)	5
SEGQs	20 (10 parameters, 2 bits each)
Sync	1
Error correction	5
Residual (Error signal)	180
	240

4.6 MULTIPULSE LPC

The discussion of LPC in Section 4.3.1 noted the fact that not all sounds are well represented simply by a pulse train or a noise source. Some sounds are semi-voiced. Thus instead of pulse and white noise generators selected by a voiced–unvoiced switch, multipulse LPC provides an excitation which is a sequence of pulses at times t_1, t_2, . . . t_n, . . . with amplitudes α_1, α_2, . . ., α_n, ... Figure 4.7 is a block diagram[8] illustrating the multipulse LPC concept.

The pulse times and amplitudes are obtained by an analysis-synthesis procedure in which the difference between the input speech and synthetic speech (generated at the analyzer) is given a perceptual weighting. This weighting accounts for the way in which human perception treats error, considering the masking and limited frequency resolution of the ear. The error sequence then is used to generate the pulse times and amplitudes which serve as the excitation. Compared with conventional LPC, multipulse LPC provides improved speech quality.

4.7 RESIDUAL-EXCITED LINEAR-PREDICTION VOCODER (RELP)

Instead of sending the parameters of pitch, amplitude, and the voiced–unvoiced condition, residual-excited linear prediction (RELP) sends a prediction error signal called the *residual*. No pitch extraction is done. Encoding of the residual is

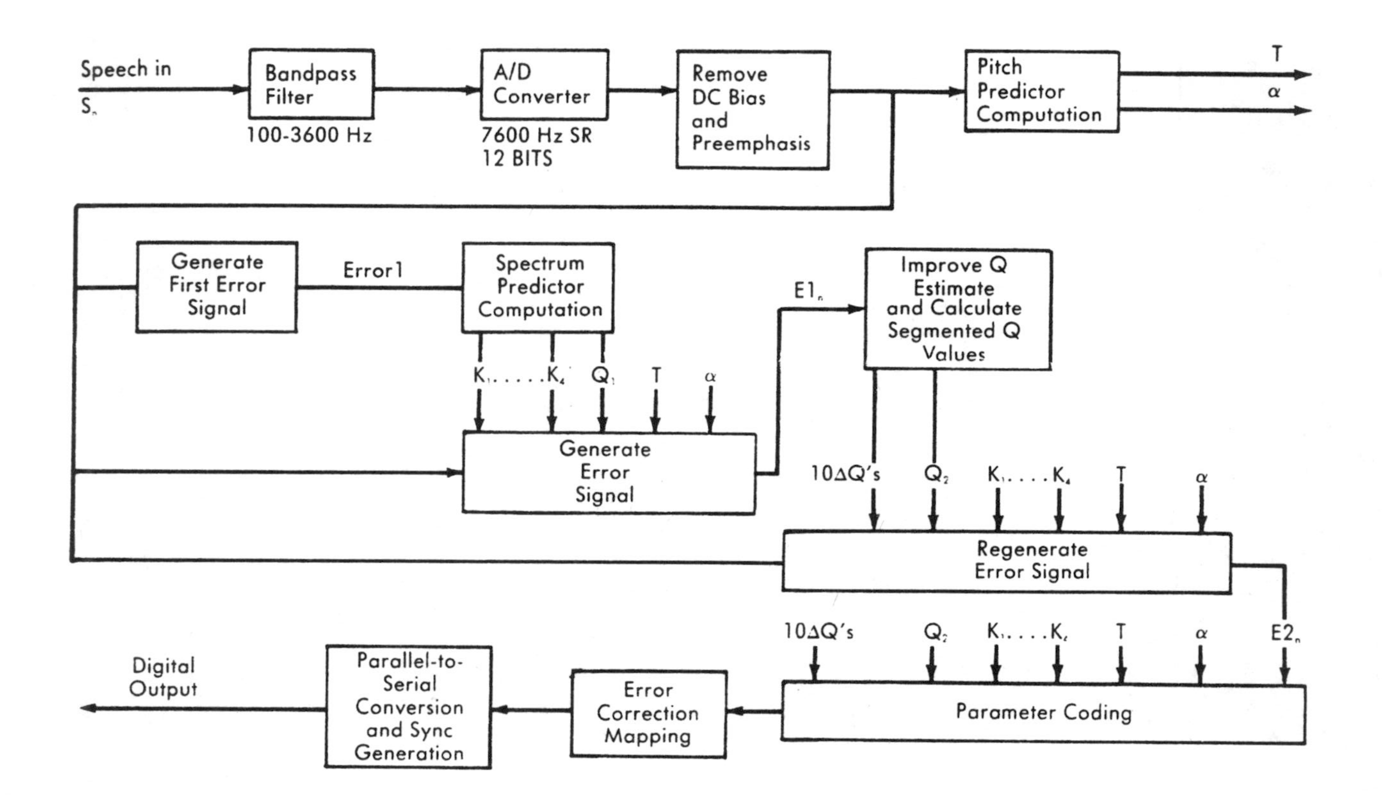

FIGURE 4.5. Adaptive predictive coder analyzer. (Courtesy T. Tremain, Ref. 6, © Media Dimensions, Inc., 1985)

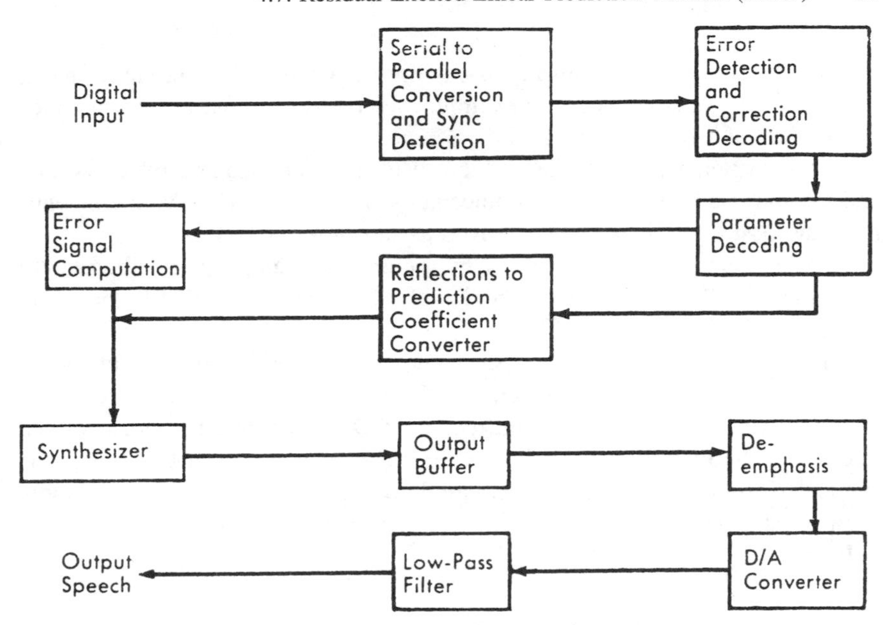

FIGURE 4.6. Adaptive predictive coder synthesizer. (Courtesy T. Tremain, Ref. 6, © Media Dimensions, Inc., 1985).

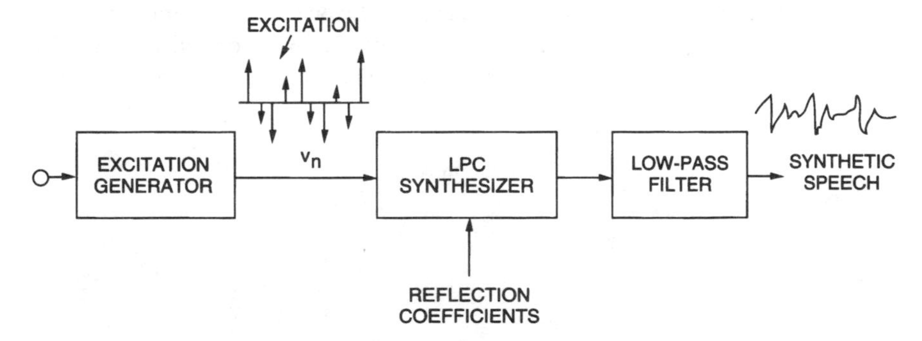

FIGURE 4.7. Multipulse LPC concept. (Courtesy Atal and Remde, Ref. 8, © IEEE, 1982)

based on spectral matching rather than waveform matching. Thus the entire analog band below a given frequency, for example, 800 Hz, is sent on a waveform-coded basis. The residual is sent using ADC.

In one implementation of RELP,[9] the residual is sampled at a 6800 b/s rate, and the analysis frame rate is 50 frames per second. Using 10 LPC coefficients, the transmission rate is 9600 b/s, derived as shown in Table 4.6.

A reduction to 6200 b/s is feasible by reducing the sampling rate allocated to the residual to 3400 b/s. Figure 4.8 is a block diagram[9] of a RELP coder and decoder.

RELP can be used to code the speech from analog end instruments, including signaling and supervision, by extending the waveform coding up to 1600 Hz, thereby covering the dual-tone multifrequency (DTMF) range. The coded speech rate then is 14.3 kb/s. Adding 1.7 kb/s for housekeeping information brings the channel to a 16 kb/s rate, suitable for digital transmission on a wireless local loop, such as is done in the *Basic Exchange Telecommunications Radio Service* (BETRS).

4.8 CODE-EXCITED LINEAR PREDICTION (CELP)

In code-excited linear prediction (CELP), vector quantization (VQ) techniques are used to represent the excitation, that is, the input to the LPC synthesizer. From a stored collection of possible sequences, one sequence is selected that produces the lowest error between the original and the reconstructed signal. If the collection of sequences (stored or generated deterministically) is available at both the encoder and the decoder, only the *index* of the sequence that produces the minimum error has to be transmitted. This index is a set of bits representing the sequence. The receiver is built to recognize this index, and to reproduce the corresponding sequence.

CELP uses tree coding in which the branches of the tree are populated with Gaussian random numbers rather than deterministic ones. A "random" codebook is used because it produces better speech quality than a deterministic one. The term *stochastic* (chance) coding sometimes is used to describe this approach. The

TABLE 4.6. RELP Parameters

Residual	6800 b/s
LPC coefficients	2250 b/s
Gain	200 b/s
Normalized energy	200 b/s
Frame synchronization	150 b/s
	9600 b/s

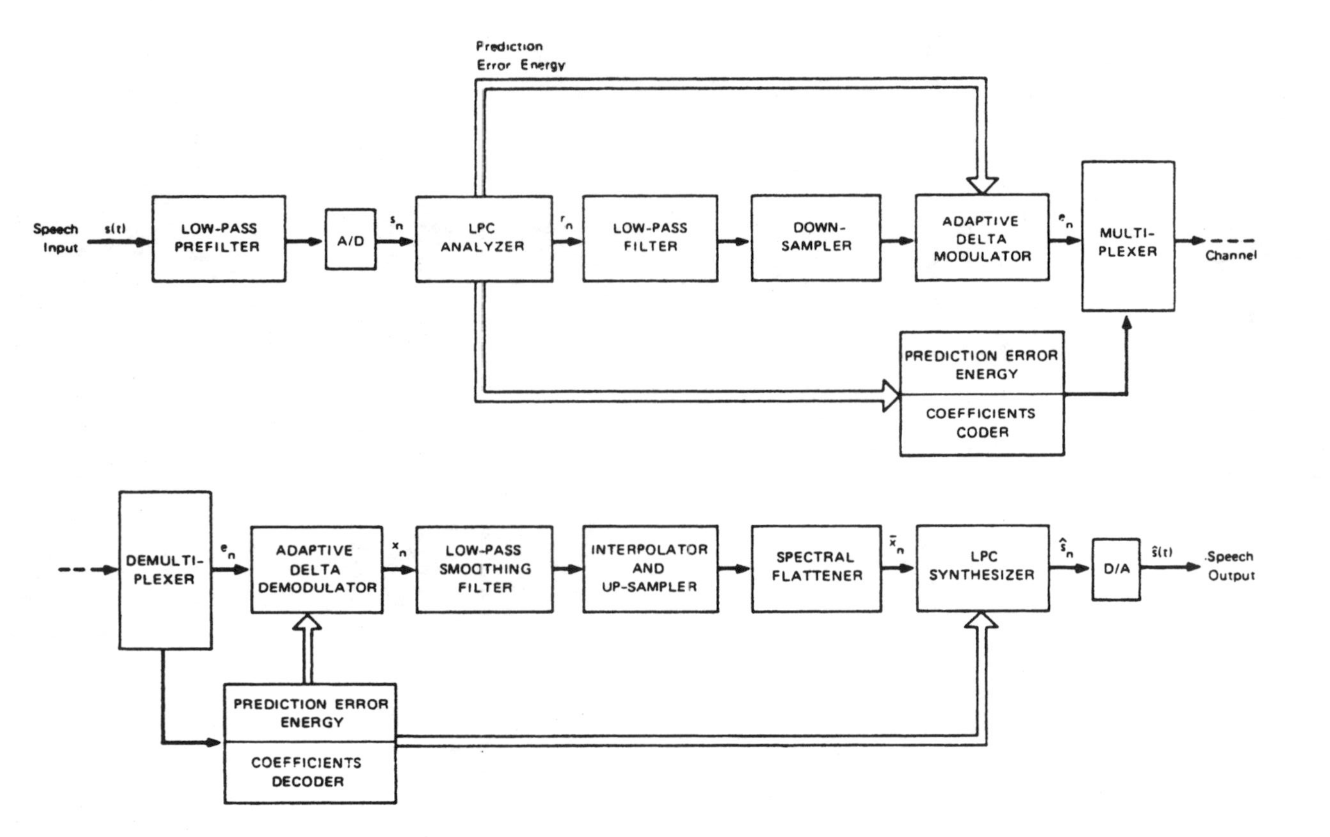

FIGURE 4.8. Block diagram of RELP coder and decoder. (Courtesy Un and Magill, Ref. 9, © IEEE, 1975).

optimum excitation (or *innovation*) sequence is selected from a codebook of stored sequences to optimize a given fidelity criterion.

Figure 4.9 illustrates two CELP analyzer implementations, conventional and low-delay. Conventional CELP is forward adaptive. In the version shown,[10] the 20-ms segment of speech to the right of the "current speech vector to be coded" is used to establish the adaptation to be done. Thus conventional CELP is forward adaptive.

For low-delay CELP (LD-CELP), the 20-ms segment of speech to the left of the "current speech vector to be coded" is used to establish the adaptation to be done. Thus LD-CELP is backward adaptive. In LD-CELP the only source of coding delay is the forward adaptation of the shape of the excitation signal. LD-CELP can provide toll quality (MOS $= 4.0$) at 16 kb/s with a delay of ≤ 5 ms. The algorithm uses a 50-tap all-pole synthesis filter whose coefficients are updated every 5 ms. A full duplex (FDX) implementation can be done using two *digital signal processing* (DSP) devices.

Figure 4.10 is a conceptual block diagram of a CELP synthesizer. As can be seen from Fig. 4.10, each sample is filtered sequentially through two time-varying linear filters, one with a long delay predictor (related to the pitch period) in the feedback loop, and one with a short-delay predictor (related to the formants, or spectral envelope) in the feedback loop.

The long-term filter includes a long-term predictor with one to three taps. A single-tap long-term predictor in the filter gives it the transfer function $B(z)$, where

$$B(z) = \frac{1}{1 - \beta z^{-L}} \tag{4.1}$$

In Eq. (4.1), L is the *lag*, meaning the pitch period or a multiple thereof, while β is the long-term predictor coefficient.

The short-term filter has 8 to 12 taps (usually 10), and is the traditional LPC synthesis filter. The short-term parameter update rate is the frame rate, 20 ms. There is a subframe rate of 5 ms, which is the code update rate. The gain and long-term parameters may be updated at the frame or subframe rate, or in between, depending on the CELP coder design.

To reduce the search times, use is made of the fact that different sequences have several common elements. Thus, tree and trellis coding can be used to advantage. Each sequence forms a path through the tree or trellis, and the transmitted code provides information about how to trace through the tree or trellis.

At a sampling rate of 8000/second, speech is coded using 5-ms blocks, thus resulting in 40 samples per block. Each block is described in terms of one of 1024 (2^{10}) possible *innovation sequences*. (The innovation sequence is the input to the prediction filter.) Accordingly, a block is described by 10 bits, so the result is just 0.25 bit per original waveform sample.

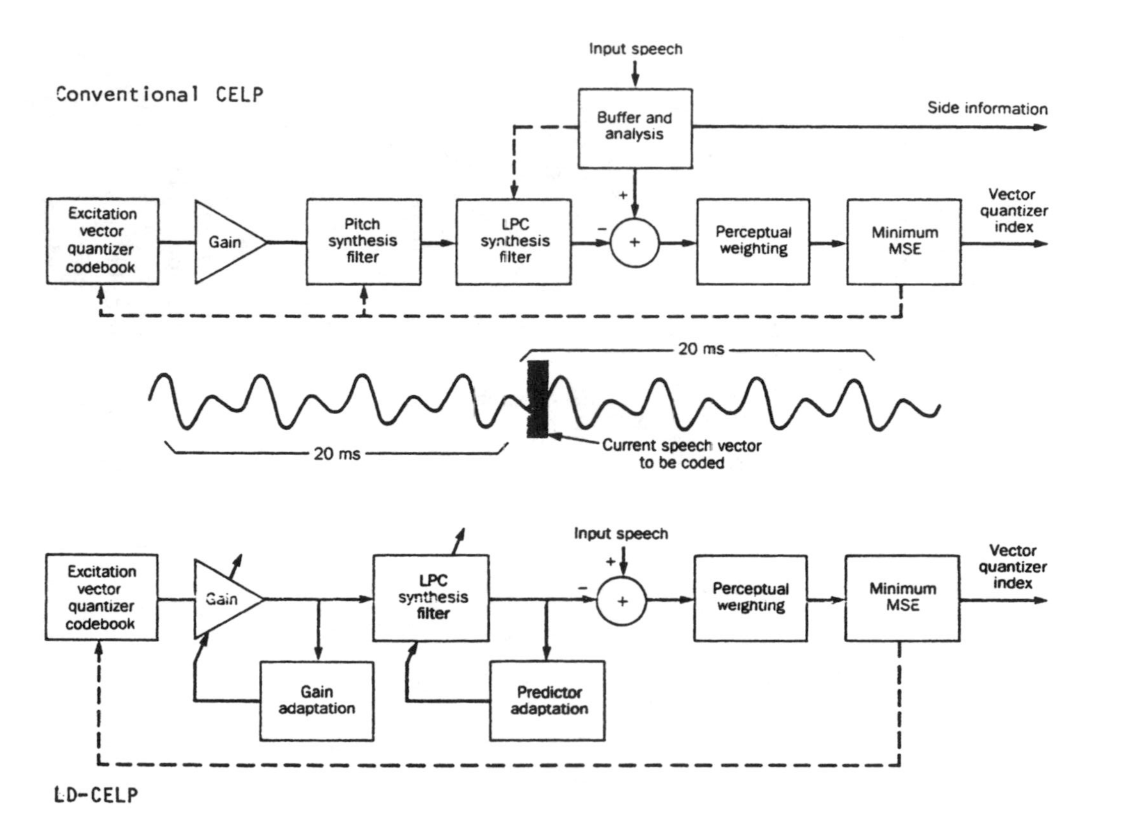

FIGURE 4.9. CELP analyzers (Courtesy Jayant, Lawrence, and Prezas, Ref. 10, © IEEE, 1990).

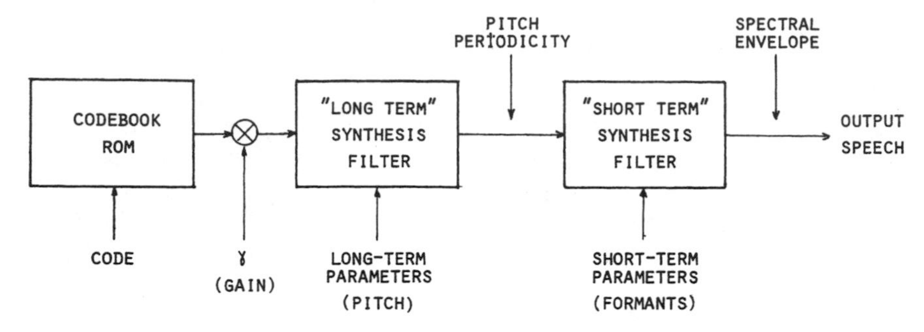

FIGURE 4.10. CELP synthesizer block diagram.

An alternative set of CELP parameters has been developed for U.S. Government applications.[11] It is known as FED-STD-1016. A 4.8-kb/s transmission rate and a 30-ms frame period (7.5-ms subframe period) allow a 144-bit frame to be constructed containing the bit allocations shown in Table 4.7. An adaptive codebook is searched to find the pattern that best approximates the voice's pitch characteristics.

As has been noted previously, the total processing delay required by a speech-coding technique becomes significant when the total system delay, including transmission delays, is long enough to become noticeable to users while in conversation. Accordingly, the delay budget for FED-STD-1016 CELP is held to following the limits of Table 4.8.

While a 162-ms delay may seem reasonably small, note that in satellite service (see Chapter 9), an additional 270-ms propagation delay will be incurred in each direction of transmission. An additional processing delay for buffering initial speech will occur if a speaker starts to talk during the end of a previous burst. Such a delay will consist of the following components:

Delay Source	Duration (ms)
Burst header	63.4
Burst tail end overhead	20
Voice-operated control circuits	50 (estimate)
Total	133.4 ms

The total outbound delay on a satellite circuit thus would be 565 ms.

At this point it is useful to compare CELP with conventional LPC, as well as with multipulse LPC. Figure 4.11 provides this comparison.[12] Note that in all three cases, the spectral envelope information is provided by a short-delay correlation filter. The pitch information, however, is provided by a long-delay correlation filter in the case of multipulse LPC and CELP, whereas conventional

TABLE 4.7. FED-STD-1016 Bit Allocations for CELP

Information	Bits
Code index	36
Code gain	20
Adaptive index	28
Adaptive gain	20
Line spectral parameter	34
Algorithm version bit	1
Frame synchronization	1
Parity (15,11 Hamming)	4
	144

TABLE 4.8. Delay Budget for FED-STD-1016 CELP

Delay Factor	Duration (ms)
Sample one frame of speech	30
Sample next two subframes	15
Process frame	27 (typical)
Receive frame	30
Decode processing	10 (estimate)
Circuit delays	50 (estimate)
	162 ms

LPC uses simply a voiced–unvoiced switch. All three differ with respect to their excitation, as has been described earlier in this chapter.

4.9. VECTOR-SUM-EXCITED LINEAR PREDICTION (VSELP)

Vector-sum-excited linear prediction (VSELP) is a variation of CELP using a codebook which has a predefined structure such that the computations required for the codebook search process can be significantly reduced. The decoder uses two codebooks, each with its own gain. A conceptual block diagram of a VSELP synthesizer[13] is shown in Fig. 4.12. Echo control is needed because the delays may exceed acceptable telephony limits.

The basic VSELP parameters are shown in Table 4.9. Table 4.10 lists the VSELP parameter codes per frame. The *Lag* parameters describe the pitch characteristics. The *Code* parameters are used to describe the characteristics of the excitation. The GSPO parameters are gain parameters.

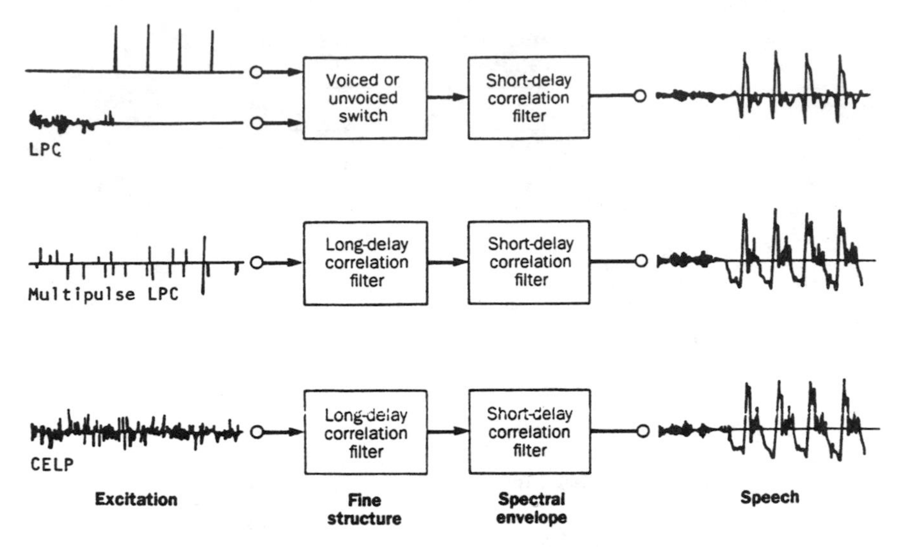

FIGURE 4.11. Comparison of coders. (Courtesy Jayant, Lawrence, and Prezas, Ref. 10, © IEEE, 1990).

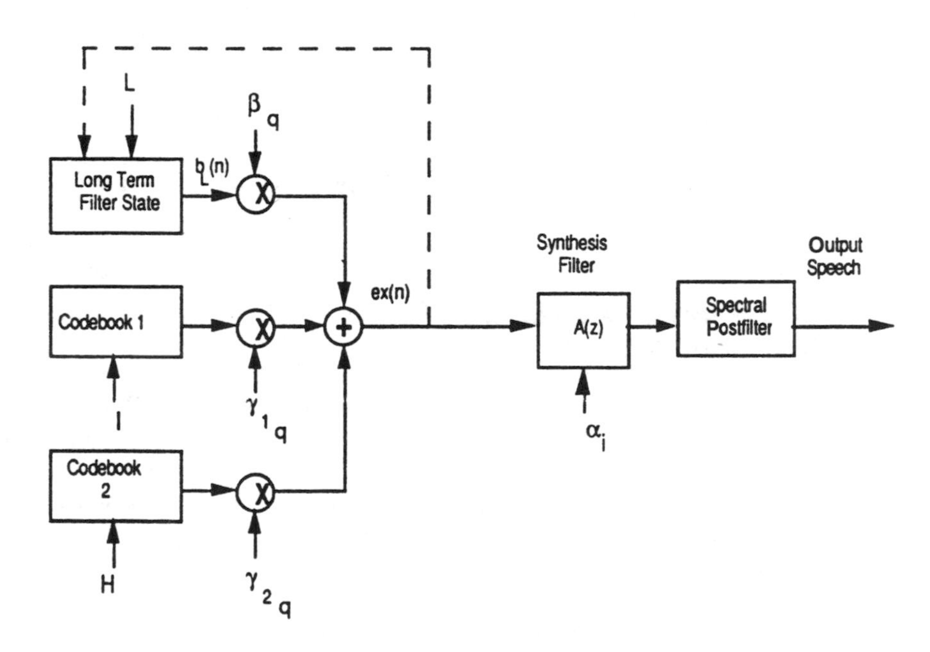

FIGURE 4.12. Vector-sum-excited linear predictive coder synthesizer. (Courtesy Telecommunications Industries Association).

TABLE 4.9. Basic VSELP Parameters

Basic data rate	7950 b/s
Frame length	20 ms
Frame contents	
Short term filter coefficients, α_i	38 bits
Frame energy, $R(0)$	5 bits
Lag, L	28 bits
Codeword I	28 bits
Codeword H	28 bits
Gains β, γ_1, γ_2	32 bits
Total	159 bits
Subframe length	5 ms
Short-term predictor order	10
Long-term predictor taps	1

TABLE 4.10. VSELP Parameter Codes per Frame[a]

Code	Name	Bits
R0	Frame energy	5
LPC1	1st reflection coefficient	6
LPC2	2nd reflection coefficient	5
LPC3	3rd reflection coefficient	5
LPC4	4th reflection coefficient	4
LPC5	5th reflection coefficient	4
LPC6	6th reflection coefficient	3
LPC7	7th reflection coefficient	3
LPC8	8th reflection coefficient	3
LPC9	9th reflection coefficient	3
LPC10	10th reflection coefficient	2
LAG-1	Lag for 1st subframe	7
LAG-2	Lag for 2nd subframe	7
LAG-3	Lag for 3rd subframe	7
LAG-4	Lag for 4th subframe	7
CODE 1-1	1st codebook code, I, for 1st subframe	7
CODE 1-2	1st codebook code, I, for 2nd subframe	7
CODE 1-3	1st codebook code, I, for 3rd subframe	7
CODE 1-4	1st codebook code, I, for 4th subframe	7
CODE 2-1	2nd codebook code, H, for 1st subframe	7
CODE 2-2	2nd codebook code, H, for 2nd subframe	7
CODE 2-3	2nd codebook code, H, for 3rd subframe	7
CODE 2-4	2nd codebook code, H, for 4th subframe	7
GSP0-1	(GS, P0, P1) code for 1st subframe	8
GSP0-2	(GS, P0, P1) code for 2nd subframe	8
GSP0-3	(GS, P0, P1) code for 3rd subframe	8
GSP0-4	(GS, P0, P1) code for 4th subframe	8

[a]From EIA/TIA IS-54A, courtesy Telecommunications Industries Association, Washington, DC

VSELP is the standard for North American TDMA digital cellular telephone systems. Further details on its use in this application are provided in Chapter 7. A single-codebook version of VSELP known as QSELP (QUALCOMM Sum-Excited Linear Prediction) is used as the speech-coding technique in the CDMA system being developed for digital cellular systems by QUALCOMM, Inc. It is a variable-rate technique, with a peak rate of 8 kb/s during maximum speech activity and a standby rate of 1 kb/s during speech pauses. The average rate during normal speech activity is about 3.4 kb/s.

4.10 REGULAR-PULSE EXCITATION WITH LONG-TERM PREDICTION (RPE-LTP)

The speech-coding technique that has been devised for use in the pan-European digital cellular telephone system known as the *Global System Mobile* (GSM) is *regular-pulse excitation with long-term prediction* (RPE-LTP). This technique is a special case of multipulse excited LPC.[14]

In RPE-LTP, linear-predictive techniques remove the short-time correlation (structure) from the vowel sounds. The residual information then is modeled by a low-bit-rate reduced excitation sequence that produces a signal close to that of the original speech. The excitation signal is found by solving strongly coupled sets of linear equations. During an encoding delay, the best version of the input speech is found. The prediction residual is modeled by a signal that resembles an upsampled sequence and thus has a "regular" (in time) structure. The samples are found by a least-squares analysis-by-synthesis fitting procedure described in terms of matrix arithmetic.

The residual modeling process performed in the encoder uses the time varying filter, $A(z)$, to produce the residual, $r(n)$, from the signal sequence, $s(n)$. $A(z)$ is a pth-order LPC filter,

$$A(z) = 1 + \sum_{k=1}^{p} a_k z^{-k} \qquad (4.2)$$

Fig. 4.13 illustrates the relationship of these functions. The difference between $r(n)$ and the modeled residual $v(n)$ is fed through the shaping filter

$$\frac{1}{\left[A\left(\frac{z}{\gamma}\right)\right]}$$

$$\frac{1}{A}(z/\gamma) = \frac{1}{1 + \sum_{k=1}^{p} a_k z^{-k}}, \; 0 \le \gamma \le 1 \qquad (4.3)$$

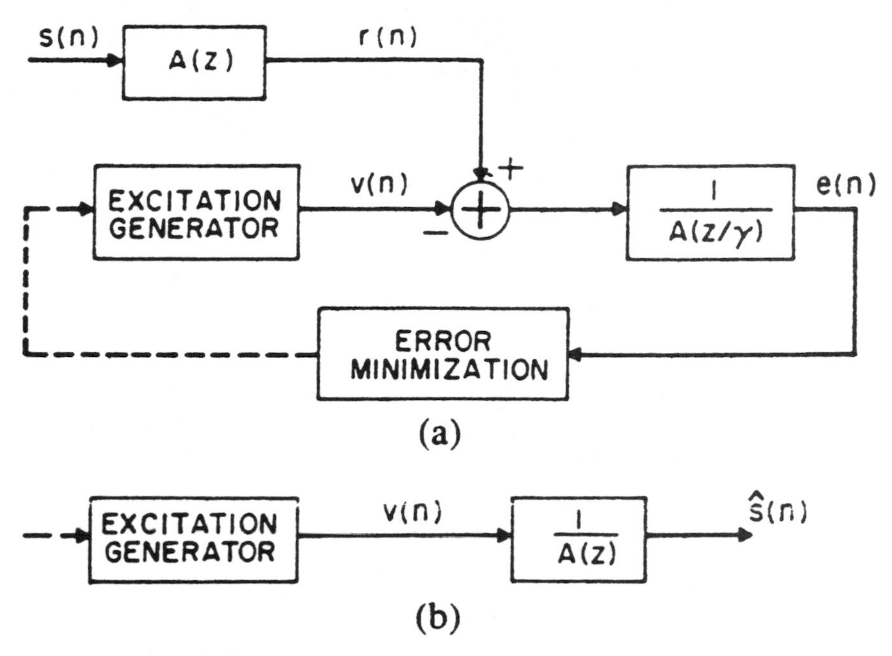

FIGURE 4.13. Block diagram of the regular-pulse excitation coder: (a) encoder, (b) decoder. (Courtesy Kroon, Deprettere, and Sluyter, Ref. 14, ©. IEEE, 1986).

This filter provides an error-weighting function, and serves to provide the analysis-by-synthesis function as was described under multipulse excitation. The resulting weighted difference, $e(n)$, is squared and accumulated, and is used as a measure for determining the effectiveness of the model $v(n)$ of the residual $r(n)$.

The excitation sequence $v(n)$ is determined for adjacent frames consisting of L samples each. The duration of a frame is 5 ms.

Speech is processed in 20-ms blocks, each of which consists of 260 bits. Thus, the basic codec rate is 13.0 kb/s. Figure 4.14 shows that the 182 most significant of these bits, in terms of their effect on overall speech quality, are provided forward error correction by adding a redundant bit for each information bit, as well as by adding some special check bits and tail bits. (See Chapter 6 for a discussion of error control.)

The resulting 456 bits then are interleaved. This involves distributing them over eight TDMA time slots, with consecutive bits being mapped into different slots, but every eighth bit being mapped into the same slot. Consecutive frames are sent in an interlaced manner. The overall channel rate is 22.8 kb/s per channel. A half-rate standard has been developed.

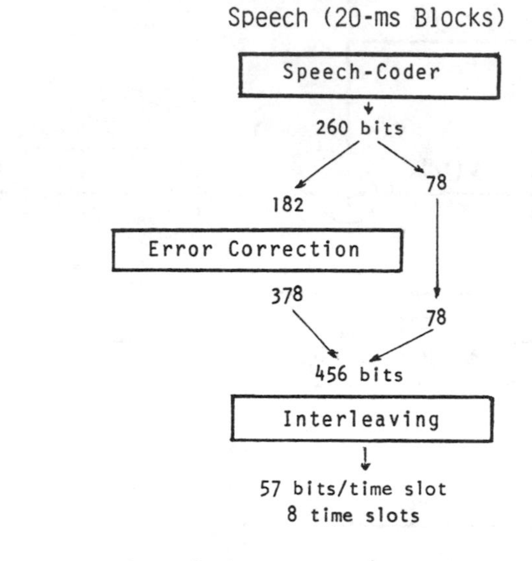

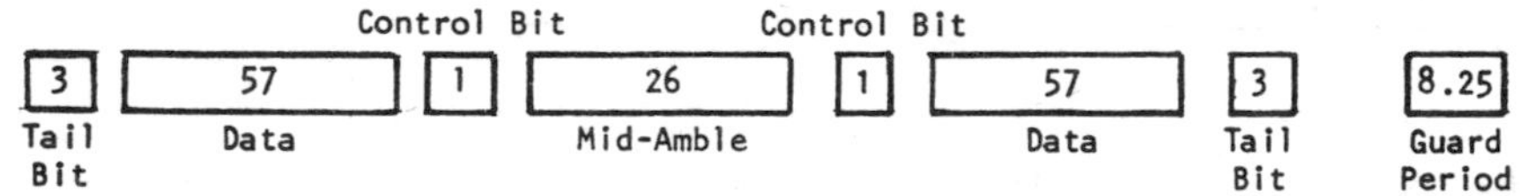

FIGURE 4.14. Data structure for transmission of RPE-LTP speech in GSM.

4.11. Voice Digitizer Comparisons

Numerous speech-coding techniques have been discussed—waveform techniques in Chapter 3 and parametric and hybrid techniques in Chapter 4. Table 4.11 lists the present and anticipated future uses of these techniques in digital telephony.

PROBLEMS

4.1 Explain why a true information rate for speech cannot be calculated readily.

4.2 Parametric coders are said to be useful for speech, but not for other sounds. Explain, then, how RELP-based end instruments can be used in the PSTN on analog subscriber lines (BETRS) on which signaling and supervision tones must be transmitted.

4.3 What is the lower limit on the sampling rate required in parametrically coded speech?

4.4 In view of the frequency-response characteristics of the telephone channel, explain why a higher-pitched voice may sound more natural than a lower-pitched voice over the telephone.

TABLE 4.11. Voice Digitizer Comparisons

Bit Rate (kb/s)	Speech Coder Type	Standard for	Present Use	Future Use
64	PCM	PSTN	CO to CO	ISDN to Subs.
32	ADPCM	PSTN	CO to CO	CO to CO
32	ADC		Satellite	Satellite
32/16	CVSD	Mil./Govt.	Mil./Govt.	Mil./Govt.
16	LD-CELP	PSTN		CO to CO
14.3	RELP	BETRS	BETRS	BETRS
13	RPE-LTP	GSM	Euro. Cell.	Euro. Cell.
9.6	APC	Govt.	Govt.	Govt.
7.95	VSELP	TIA	N.A. Cell.	N.A. Cell
6.5	?	GSM		Euro. Cell.
4.8	CELP	Govt.		AMSS
4+	?	TIA		N.A. Cell.
2.4	LPC	Comcl./Govt.	Priv. Voice/Govt.	Priv. Voice/Govt.

PSTN, Public Switched Telecommunications Network; CO, Central Office; Mil., Military; Govt., Government; Euro. Cell., European Cellular; N.A. Cell., North American Cellular; Comcl., Commercial; Priv. Voice, Private Voice

4.5 Explain why sentence intelligibility generally is higher than word intelligibility. Why is word intelligibility higher than phoneme intelligibility?

REFERENCES

1. Markel, J. D. and Gray, A. H., Jr., *Linear Prediction of Speech*, Springer-Verlag, Berlin, 1976.
2. Schaefer, R. W., "A Survey of Digital Speech Processing Techniques", *IEEE Trans. on Audio and Electroacoustics*, Vol. AU-20, No. 1, Mar. 1977, pp. 28–35.
3. Flanagan, J. L., et al., "Speech Coding," *IEEE Trans. Comm.*, Vol. COM-27, No. 4, April 1979, pp. 710–737.
4. Voiers, W. D., "Diagnostic Evaluation of Speech Intelligibility," in *Speech Intelligibility and Speaker Recognition*, M. Hawley (ed.), Dowden Hutchinson Ross, Stroudsburg, PA, 1977.
5. Wong, D. Y., Juang, B-H, and Gray, A. H., "An 800 bit/s Vector Quantization LPC Vocoder," *IEEE Trans. ASSP*, Vol. ASSP-30, No. 5, Oct. 1982, pp. 770–780.
6. Tremain, T., "The Government Standard Adaptive Predictive Coding Algorithm APC-04," *Speech Technology*, February/March, 1985, pp. 52–62.
7. See note 6.
8. Atal, B. and Remde, J. R., "A New Model of LPC Excitation for Producing Natural Sounding Speech at Low Bit Rate," *ICASSP '82*, pp. 614–617.
9. Un, C. K., and Magill, P. T., "The Residual-Excited Linear Prediction Vocoder with Transmission Rate Below 9.6 kb/s," *IEEE Trans. Comm.*, Vol. COM-23, No. 12, Dec. 1975, pp. 1466–1473.

10. Jayant, N., Lawrence, V., and Prezas, D., "Coding of Speech and Wideband Audio," *AT&T Tech.*, Vol. 69, No. 5, pp. 25–41 (Oct. 1990).
11. "Telecommunications Analog to Digital Conversion of Radio Voice by 4800 Bit/Second Code Excited Linear Prediction (CELP)," FED-STD-1016, General Services Administration, Washington, DC, February 14, 1991.
12. See note 9.
13. "Cellular System: Dual-Mode Mobile Station–Base Station Compatibility Standard," EIA/TIA Standard IS-54 (Revision A), Telecommunications Industries Association, Washington, DC (January, 1991).
14. Kroon, P., Deprettere, E. F., and Sluyter, R. J., "Regular-Pulse Excitation—A Novel Approach to Effective and Efficient Multipulse Coding of Speech," *IEEE Trans. Acoustics, Speech and Signal Processing*, Vol. ASSP-34, No. 5, Oct. 1986, pp. 1054–1063.

5

Digital Techniques in the Telephone Network

5.1. INTRODUCTION

This chapter deals with many of the functions that are needed to make a digital telephone system operate. It begins with the subjects of synchronization and time-division multiplexing and then covers the basic functions of multiplexing, scrambling and channel coding. Figure 5.1 relates some of these functions to one another as well as to source coding as they might be found in a generalized digital transmission system.

Echo cancellers are covered next, followed by digital speech interpolation, digital repeaters, and digitization of the loop plant. Then monitoring and maintenance is covered. The chapter concludes with a discussion of speech recognition and voice response.

Error control is discussed in Section 6.3. Signaling and supervision are covered in Appendix A.

The parts of the telephone system that remain analog in most areas are the subscriber set and the local loop. Progress is being made, however, in the digitization of the local loop, and a section of the chapter is devoted to this topic.

An *encryption* function is needed by certain government agencies in their transmission of classified material, by industry in the transmission of proprietary information, and by members of the general public who wish privacy for their conversations. To achieve encryption, the bit stream is altered in such a way that only recipients with the correct descrambling algorithm can decipher the message.

With reference to Fig. 5.1, the *source coder* performs the conversion of analog inputs, such as voice, to a digital form, usually PCM, but possibly ADPCM or another form in the case of wired telephony, or VSELP in the case of porta-

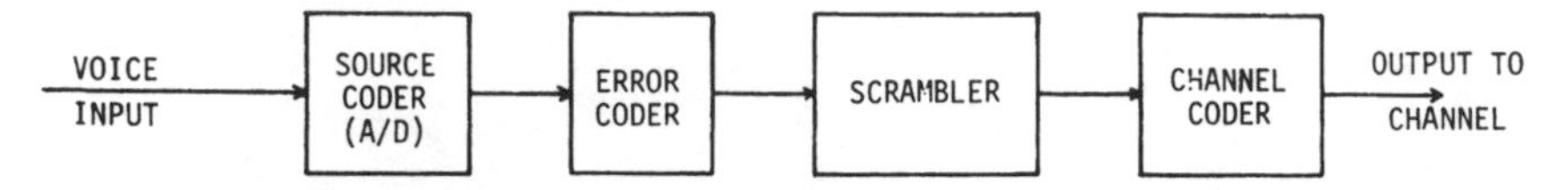

FIGURE 5.1. Digital transmission system functions.
Note: The source coder usually is 8-bit PCM in standard telephony systems.

ble/mobile telephony. Chapters 3 and 4 discussed such conversions in detail. The *error coder* computes and adds certain redundant bits that will be used at the receiver in the correction of those errors that may have occurred in the transmission process. Error control is discussed in Section 6.3.

The *scrambler* alters the bit stream in such a way as to prevent long sequences of zeros that might make timing recovery difficult at the receiver, or unwanted periodicities that might result in interference.

The *channel coder* processes the binary input into a multi-level or modified binary signal.

Other important functions include *signaling and supervision*, that is, addressing, ringing, and on- or off-hook supervision. Additional functions are performed by information signals and test signals. Signaling on loops often still is done on an analog basis, but this is changing with the digitization of the local-loop plant, as described in Section 5.9, and in Section 14.2. Signaling on trunks is done digitally, where the trunks have been implemented to carry a digital stream.

5.2. SYNCHRONIZATION

Synchronization is the process whereby the receiver is made to sample the incoming bit stream so that each bit is properly identified, and so that the bits constituting a particular channel are combined to provide the proper output. The process is called *frame synchronization* or *alignment*. The timing signal is made available to the receiver from the received bit stream itself through the process of timing recovery.

5.2.1. Frame Synchronization

Framing is the determination of which groups of bits constitute quantized levels (characters) and which quantized levels belong to which channels. Framing allows the separation of one signal from another by counting pulse positions relative to the beginning of a frame. Thus code words representing the elements of each signal can be extracted from the combined bit stream and reassembled into a single stream of pulses associated with a particular channel's signal.

The framing of each character is accomplished by defining a synchronization character, commonly called *sync*. The bit arrangement of the sync character is significantly different from that of any of the regular bit combinations being transmitted. In the formation of a DS1 digital stream (1.544 Mb/s), a framing bit is added after each sequence of 24 8-bit words to supply synchronizing information for the receiver system. This is known as the *added digit* framing strategy, that is, a dedicated digit position in the frame is used. It represents a negligible reduction in the information rate.

A frame thus contains the bits from one sample of each channel being transmitted, plus signaling and frame alignment bits. Figure 5.2 illustrates the frame structure for the CCITT 24-channel PCM system providing two signaling channels, as commonly used in the United States. The corresponding signal is designated DS1. Note that 12 frames constitute a multiframe, lasting 1.5 ms. Since each frame contains one sample from each channel, the 12 frames at 1.5 ms provide an 8000/s sampling rate on each of the 24 channels. Frame alignment assures that Channel 1 from the transmitting terminal is interpreted correctly as Channel 1 at the receiving terminal, etc. The frame-alignment signal, as shown in Fig. 5.2, is 101010. It is conveyed by Bit 1 in the odd-numbered frames. This bit is separate from those used to convey speech, and thus does not detract from

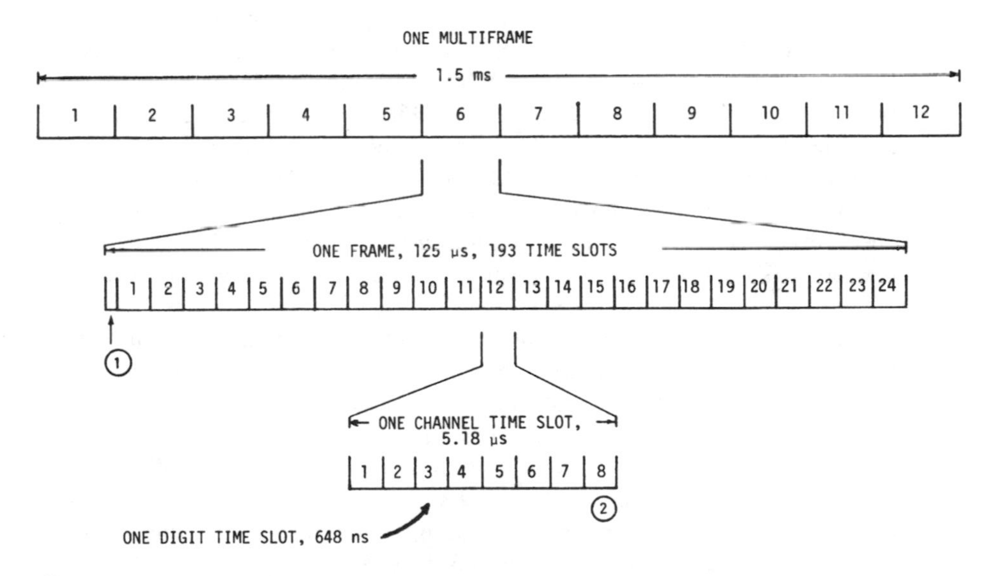

FIURE 5.2. Frame and multiframe structure for 24-channel PCM system (DS1).

speech quality. It is generated in the transmitting terminal and used as a reference for all the following digits up to the next frame alignment signal.

Address and supervisory bits are assigned as indicated in Fig. 5.2 for Signaling Channel A and Signaling Channel B. Thus, the least significant of the eight quantization bits is robbed every sixth frame. This means that each of the 24 channels is quantized on an 8-bit basis 5/6 of the time and on a 7-bit basis 1/6 of the time. The receiving portion of the system performs the inverse operations. The incoming signal from the digital line is converted to binary form. A framing circuit then searches for, and synchronizes to, the framing bit pattern. This provides assurance that the locally generated timing pulses are in synchronism with the incoming pulse train. The signaling digits then are sorted out and directed to the individual channels.

The multiframe alignment signal, 001110, identifies the sixth and twelfth frames, which carry signaling states in the least significant bit position of each time slot. Thus, the composite framing sequence is 100011011100.

At the receiving terminal, the frame alignment sequence is recognized by the receiving terminal logic. This sequence becomes the fixed reference from which the logic determines the position of the character signal for Channel 1 and all the following channels. An extended framing format[1] is being implemented in new designs of DS1 level equipment which frame on a pattern contained in the framing bit position of the DS1 signal. It is used where both the transmitter and the receiver have the hardware and software capabilities to handle it. The extended framing format extends the multiframe structure of Fig. 5.2 from 12 to 24 frames and redefines the 8 kb/s-framing bit pattern into 2 kb/s for mainframe and robbed bit signaling synchronization (instead of 8 kb/s as before). As a result, it provides 2 kb/s for a *cyclic redundancy check* code, called CRC-6 because it contains six check bits, and 4 kb/s for a terminal-to-terminal data link. The advantage of using this format is that it allows in-service monitoring of the channel without disturbing the traffic.

The CRC-6 code can detect most errors in the DS1 signal, and can be used for such functions as false framing protection, protection switching, end-to-end performance monitoring, automatic restoral after alarms, line verification, and the determination of error-free seconds. Thus, if false framing should occur, parity checks at the receiver would reveal it. Protection switching refers to the replacement of a malfunctioning chain of repeaters with a properly operating spare chain. Line verification refers to the determination that the line is operating properly. These functions are discussed further in Section 5.9.

The 4-kb/s data link may be used for protection switching, alarms, loop-back, received-line performance monitoring, supervisory signaling, network configuration information and general maintenance information.

Table 5.1 shows the extended multiframe (superframe) structure. The F-bit is the first bit of the 193-bit frame, as shown in Fig. 5.2.

TABLE 5.1. Extended Superframe F-Bit Assignments.1[a]

		F Bit Assignments			
Frame Number	Bit Number	FPS[b]	FDL[c]	CRC[d]	Robbed Bit Signaling
1	0		m		
2	193			CB_1	
3	386		m		
4	579	0			
5	772		m		
6	965			CB_2	A
7	1158		m		
8	1351	0			
9	1544		m		
10	1737			CB_3	
11	1930		m		
12	2123	1			B
13	2316		m		
14	2509			CB_4	
15	2702		m		
16	2895	0			
17	3088		m		
18	3281			CB_5	C
19	3474		m		
20	3667	1			
21	3860		m		
22	4053			CB_6	
23	4246		m		
24	4439	1			D

[a]Courtesy American Telephone and Telegraph Company
[b]FPS = Framing Pattern Sequence (2 kb/s)
[c]FDL = Facility Data Link Message Bits, *m* (4 kb/s)
[d]CRC = Cyclic Redundancy Check (Check Bits $CB_1, \ldots CB_6$) (2kb/s)

Figure 5.3 shows the frame structure for another 24-channel PCM system approved by CCITT and providing common-channel signaling. All bits are usable for speech or information, that is, full 8-bit quantization occurs at all times. It is based on CCITT Recommendation G.733.

5.2.2. Timing

In addition to the maintenance of frame alignment, the timing signals within both the transmitting and the receiving terminals must occur at the same average rate. The transmitting clock governs the rate at which the overall digital signal is produced. If the receiving terminal were to use its own internal clock, any slight

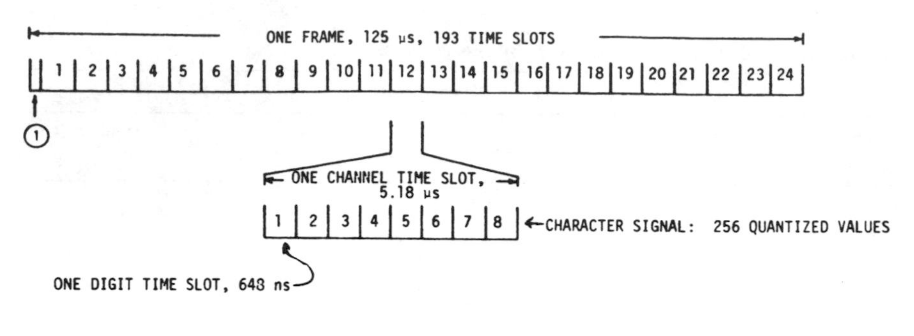

FIGURE 5.3. Frame structure for 24-channel PCM system (DS1) with common-channel signaling.

difference in rate would cause a loss of frame alignment. Thus the receiving terminal is slaved or synchronized to the transmitting terminal.

For conversational voice transmission, two directions of transmission are involved, so each terminal must receive as well as transmit. While each direction of transmission could have independent timing, the need to time-division multiplex numerous channels together makes it imperative that there be only one timing source in an all-digital network. Section 13.4 discusses digital synchronization in the public switched network.

5.2.3. Timing Recovery

Timing information can be recovered from a digital signal provided a sufficient number of transitions exists in the received signal. The timing extractor is a circuit tuned to the timing frequency. This circuit has a sufficiently high Q to provide an adequate output during a sequence of zeros in the received signal. The desired output is amplified and limited to produce a square wave at the signaling rate. This square wave then controls a clock-pulse generator that generates narrow pulses which are alternately positive and negative at the zero crossings of the square wave. Then a delay circuit in the timing path adjusts the timing pulses so they occur at the middle of each signal-pulse interval. This technique, called *forward-acting timing*, is used in self-timed digital repeaters.[2]

Note that ordinarily a string of zeros is produced as the binary number corresponding to the smallest quantized value. Thus this code would be produced by many of the channels when traffic is light. In order that this condition not have an adverse effect on timing recovery, special measures must be taken. One approach is to have the coder generate a binary output in which alternate digits are inverted (ADI). In other words, a string of all zeros becomes 01010101. Another approach, called *zero code suppression*, is to place a one in the next-to-least-significant bit position. Thus 00000000 becomes 00000010.

Two conditions may cause the receiver to go into its "sync search" mode: (1) when transmission begins, and (2) when a bit-stream dropout occurs. The "sync search" function operates as follows: with reference to Fig. 5.2, beginning with an arbitrary bit position, every 193rd bit is sent into a shift register. When twelve such bits have filled the register, its contents are compared with the frame-alignment character stored in another register. Actually, the frame-alignment signal appears only in the odd frames (see Figs. 5.2 and 5.3) with the multiframe alignment signal appearing in the even numbered frames when common-channel signaling is not being used. The foregoing process is repeated by examining each of the 193 bit positions within the frame until the match is found. When a match is detected on two successive characters, the receiver raises a "character available" flag every eight bits, thus delineating the eight-bit characters of the system. Each "character" corresponds to a sample of the speech waveform of one of the 24 channels being transmitted. Further details of this system, known as the T1 system, are provided in Section 5.3.1.

The occurrence of a bit-stream dropout, as mentioned in the preceding paragraph, indicates either system failure, or a very high error ratio. Usually three consecutive frame-alignment signals must be received incorrectly before the system is assumed to have lost alignment.

5.3. TIME-DIVISION MULTIPLEXING

The concept of time-sharing a channel is a very old one in communications technology. In the early days of telephony, the party-line concept was commonly used. Now, scarce radio-telephone channels are shared among many users. Another sharing concept, digital speech interpolation (DSI), uses the fact that one party to a conversation usually is listening at least half the time, and that hesitations, pauses and other silent intervals also occur. Accordingly, for example, 100 two-way channels can be used to carry over 200 conversations by assigning channels in either direction only to those users who are talking at the moment. Time-division multiplexing differs from the foregoing time-sharing concepts in that it involves time sharing among continuous users. Continuous band-limited signals are sampled at discrete times and the full channel is assigned to each input for a short interval, Fig. 5.4 illustrates the concept. Here, each of four analog signals are sampled in accordance with Nyquist's theorem. The sampling is timed so that the samples can be combined on an interleaved basis at the output of the sampler. This type of arrangement, with 24 channels instead of four, forms the basis for the frame structure illustrated in Fig. 5.2.

The channel bank time-division multiplexes a number of voice channels (for

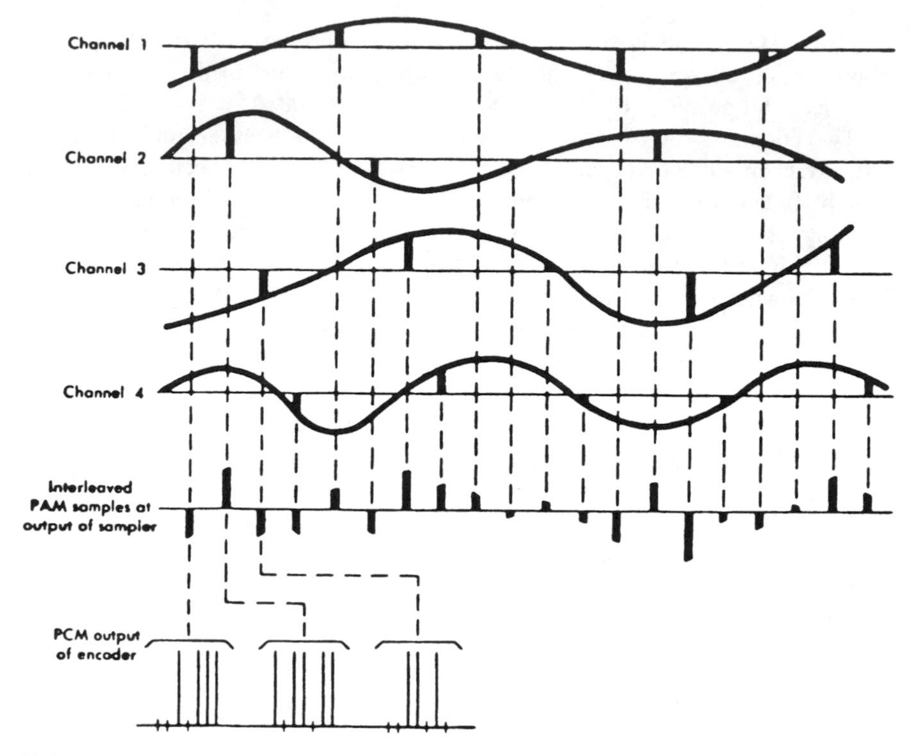

FIGURE 5.4. Time-division multiplexing. (© 1982, Bell Telephone Laboratories. Reprinted by permission.)

example, 24) together into a digital stream (for example, at 1.544 Mb/s) based on sampling each 3.4-kHz voice channel at 8000 times per second. The samples are time-interleaved, allowing them to be combined on a TDM basis. Companding, using $\mu = 255$, is done to achieve a 40-dB dynamic range. Thus, the functions of these banks include filtering, sampling, compressing, coding, multiplexing, synchronizing, and framing at the transmitter, with the inverse of these processes at the receiver. The various channel banks differ to some extent in detail, but all perform the same basic functions. Each of one or more 24-channel digroups (North American standards) is processed into one or more DS1 pulse streams. The resulting signal is binary (unipolar) within the channel bank itself, but is converted to a bipolar format for digital transmission, as will be defined in Section 5.5. Within the channel banks, plug-in channel units provide the interface between analog signal transmission circuits and digital transmission circuits. Separate interfaces are used for voice and for signaling. Various types of channel units provide for the various types of signaling arrangements such as common channel and channel-associated, as well as for the many types of trunks and special-service circuits, and for 2-wire or 4-wire subscriber-loop operation.

5.3.1. T-Carrier Systems

The T1 Digital Transmission System is used for the baseband wireline transmission of PCM voice between the central offices of major metropolitan areas, and between cities for distances up to about 80 km. Multipair exchange cables of a variety of standard gauges (24- and 26-gauge copper pair, as well as 22-gauge in some cases) are used. Regenerative repeaters (capable of waveform reshaping) are placed at nominal spacings of 1.6 km, with closer spacings (for example, 0.8 km) near offices. Digroups consisting of 24 voice channels, each at 64 kb/s, are arranged in the frame structure of Fig. 5.5. As shown there, a framing bit is added to each digroup. The T1 system thus transmits a digital stream which is made up as follows:

```
(8 × 24 + 1) 8000 = 1.544 Mb/s
|    |    | |
|    |    | samples/second
|    |    framing
|    |
|    channels
8 bits/channel
```

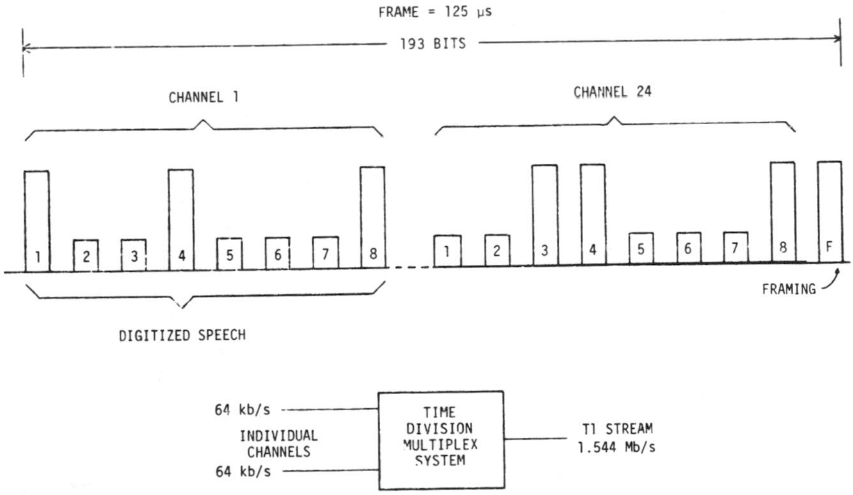

FRAME = 125 μs
193 BITS
CHANNEL 1
CHANNEL 24
1 2 3 4 5 6 7 8 1 2 3 4 5 6 7 8 F
FRAMING
DIGITIZED SPEECH
64 kb/s
INDIVIDUAL CHANNELS
64 kb/s
TIME DIVISION MULTIPLEX SYSTEM
T1 STREAM 1.544 Mb/s

FIGURE 5.5. T1 carrier frame structure. (From D. R. Doll, *Data Communications*, © John Wiley & Sons, Inc., 1978. Reprinted by permission of John Wiley & Sons, Inc.)

In heavily populated areas, use is made of metropolitan area trunk (MAT) cable,[3] designed to minimize crosstalk where large numbers of circuits are required between central offices. For the much longer systems needed between major cities, the T1 Outstate (T1/OS) System has been developed.[4] It can function with as many as 200 repeaters in tandem. Repeater spacings are determined based upon the need to meet well-established performance objectives. A *failure time* is said to occur whenever the b.e.r. $\geqslant 10^{-6}$. Failure time is not to exceed 0.01% per year (52 minutes) over a two-way 400-km path based upon propagation outages. In Canada, only 0.02% per year (104 minutes) is allowed for all causes of unavailability (propagation plus hardware causes).

Of critical importance to users is the number of error-free seconds (EFS) or the error-free interval (EFI) which the transmission facility achieves. This is the time duration, or the number of bits, between error bursts. An error burst, in turn, is a series of bits which starts and ends with an erroneous bit and which is separated from other error bursts by a number of error-free bits. An error burst longer than 300 ms is called an *outage*.

The derivation of T-carrier transmission channels through the retrofit of existing analog transmission systems is discussed in Section 6.6.

T-carrier systems for rates in excess of 1.544 Mb/s are discussed in Section 5.3.2. North American digital stream rates are designated by the letters DS. In addition, a hierarchy for the Synchronous Optical Network (SONET), has been established, as described in Section 10.12.

5.3.2. The Digital Hierarchy

The T1 Digital Transmission System for the transmission of 24 PCM voice channels at an overall 1.544-Mb/s rate is the most commonly used digital transmission system in the United States, Canada and Japan, and is the basic building block in the digital hierarchy. Both this system and the 30-channel European system have been included in the CCITT Recommendations. The 30-channel system (actually 32 channels if the two channel time slots for frame alignment and signaling are counted) operates at a 2.048-Mb/s rate and is used extensively in Europe as well as in the Latin American countries. Higher level digital multiplex rates are listed in Table 5.2.

The types of multiplexing equipment used to translate the signals of the North American system between levels are shown in Fig. 5.6. The multiplexers must be capable of providing the proper interface parameters in terms of transmitted signal format as well as bit stream organization. These parameters include transmission rate, signal format (polar, bipolar or multilevel), pulse amplitude, parity-bit locations and allowable number of consecutive zeros. In addition, signal compatibility with the terminal equipment used at the ends of the facility must be maintained. Thus the message, framing and signaling formats must be maintained.

TABLE 5.2. Digital Multiplex Hierarchies

Level	North American	Japanese	European (CEPT)[a]
First	1.544 Mb/s (DS1) (24 ch)	1.544 Mb/s (24 ch)	2.048 Mb/s (E1) (30 ch)
Second	6.312 Mb/s (DS2) (96 ch)	6.312 Mb/s[b] (96 ch)	8.448 Mb/s (E2) (120 ch)
Third	44.736 Mb/s (DS3) (672 ch)	32.064 Mb/s (480 ch)	34.368 Mb/s (E3) (480 ch)
Fourth	274.176 Mb/s (DS4) (4032 ch)	97.728 Mb/s (1440 ch)	139.268 Mb/s (E4) (1920 ch)
Fifth		400.352 Mb/s (5760 ch)	565.148 Mb/s (E5) (7680 ch)

Note: Many systems are being implemented using multiples of the these levels. They include DS1C (3.152 Mb/s), twice the DS3 rate (≈90 Mb/s), and three times the DS3 rate (≈135 Mb/s). Some military systems operate at six times DS1, or 9.696 Mb/s, and at eight times DS1, or 12.928 Mb/s. Another example is the 36 DS3 rate (1.7 Gb/s) for North American light guide systems. In addition, fiber-optic transmission systems are being designed to operate at the rates of the Synchronous Digital Hierarchy, as described in Section 10.12.

[a]Conference of European Postal and Telecommunications Administrations

[b]Alternative: 7.876 Mb/s (120 ch).

At the DS1 rate, 1.544 Mb/s, the most common application is the channel bank handling 24 PCM voice channels. Other applications include data terminals, often with a composite of lower-bit-rate data streams, and some limited-motion video terminals. Some high-speed facsimile machines also operate at the DS1 rate. T1 lines generally have a maximum length of about 80 km with repeater spacings not exceeding 1.6 km. An exception is the T1/OS system[5] described in Section 5.3.1. Increases in system length well beyond 80 km have been achieved through improved equipment reliability and maintenance procedures, as well as increased cable shielding to reduce *near-end crosstalk* (NEXT), which will be discussed in Section 5.8. Improved fault location procedures as well as protection switching also are used to maintain service standards.[6]

The DS1C rate provides for the transmission of two DS1 signals which generally are not synchronized with one another. A 64-kb/s stream is added for synchronization and framing by a process called pulse stuffing to produce a 3.152-Mb/s rate. Pulse stuffing involves the addition of enough time slots to each signal so that the DS1C signal rate is that of the clock circuit in the transmitter. The stuffed pulses carry no information, but the signal is coded so that they can be recognized and removed at the receiving terminal. Details of the multiplexing process are discussed in Ref. 3. Generally transmissions at the DS1C rate use the same transmission media and length limits as those described for the T1 system. The DS2 transmission rate, 6.312 Mb/s, can handle 96 PCM voice channels. DS2 lines may extend to distances as great as 800 km. Stuff pulses are used in combining four DS1 lines together to provide the input to a DS2 line, since the DS1

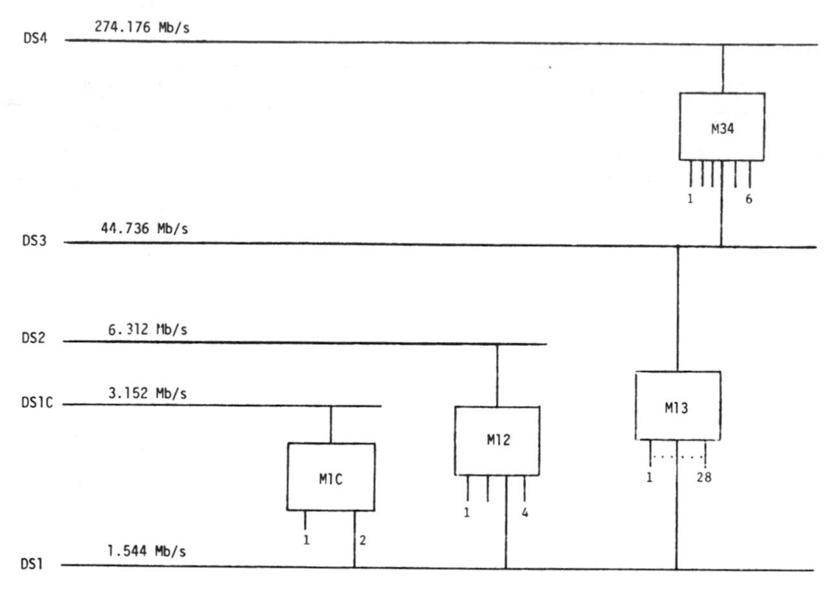

FIGURE 5.6. North American digital hierarchy and multiplexing plan. (© American Telephone and Telegraph Company, 1977.)

lines may have been timed by independent (unsynchronized) clocks. In addition, the necessary control and framing bits must be added.

As shown in Fig. 5.6, an M13 multiplexer combines 28 DS1 lines into a DS3 line. Since the DS3 line can handle 672 PCM voice channels, it can readily transmit a 600-channel frequency-division multiplexed mastergroup. A system capable of twice the DS3 rate can handle full 90-Mb/s composite color video.

The highest line rate in the North American digital hierarchy, the DS4, is 274.176 Mb/s, which corresponds to six DS3 lines; these lines are applied to it through an M34 multiplexer. Systems designed for higher rates use multiples of the lower rates.

5.3.3. Muldems

Multiplexing is accomplished by two classes of devices, *muldems* and *transmultiplexers*. The muldem, a contraction for multiplexer–demultiplexer, is used to combine message channels at one level of the digital hierarchy to produce a higher level and, at the receiving end, to reproduce the channels at the original level. A transmultiplexer is used to convert TDM signals to FDM, and vice versa, and thus serves as an interface device between digital and analog networks.

Figure 5.6 showed what muldems are used in the digital hierarchy. In the M12 multiplexer, two DS1 signals in bipolar format are converted to unipolar signals. The second input signal is inverted (all zeros are changed to ones and all ones are changed to zeros). Pulse stuffing then is done, and the two bit streams next are interleaved bit-by-bit. The multiplexed bit stream then is scrambled such that each output bit is the modulo-2 sum of the corresponding input bit and the preceding output bit.[7] A control bit sequence then is multiplexed into the stream to permit the correct demultiplexing of the two DS1 streams, as well as the deletion of the stuff bits at the receiving end.

The M12 multiplexer combines four DS1 streams[8] together with the needed control, framing and stuff bits, since the original streams may have come from sources with separate clocks. Prior to multiplexing, the second and fourth input streams are inverted. At the M12 output, the combined stream is unipolar and is converted to bipolar with a 50% duty cycle. The format, known as B6ZS, is described in Section 5.5. The M13 multiplexer (see Ref. 3) contains two multiplexing steps. First, up to four DS1 streams are combined to produce a DS2 stream. Then, up to seven DS2 streams are combined to produce the DS3 stream. Internally, all streams are in the polar format, so there are no bipolar violations. The output DS3 stream uses the B3ZS format (see Section 5.5) with a 50% duty cycle. Within the DS3 stream, a frame length of 4760 bits is established. Each frame consists of seven 680 bit subframes, corresponding to the fact that a DS3 stream contains the information from seven DS2 streams. Each subframe is divided into eight blocks of 85 bits, with the first bit used for control purposes and the remaining 84 bits containing message information.

The M34 multiplexer (see Ref. 3) combines six DS3 streams into a single DS4 stream. The resulting DS4 stream contains polar binary signals in which the ones are 100% duty cycle positive voltage pulses, measured from the center conductor to the outer conductor of the coaxial cable and the zeros are correspondingly negative voltage pulses. A DS4 superframe is 4704 time slots. Each superframe is divided into 24 frames having 196 time slots each. Each frame, in turn, contains two subframes of 98 time slots. The first two of these 98 slots are used for control bits, while the remaining are for information bits.

At the various levels of the digital hierarchy, the bit stream rates must conform to the tolerances shown in Table 5.3. The term *subrate* means any rate below 64 kb/s.

The timing of one bit stream relative to another in the multiplexing process must be controlled carefully. This timing is achieved by pulse stuffing. Upon demultiplexing, the stuffed pulses are removed. To prevent gaps, and consequent jitter, upon removal of the stuffed pulses, elastic stores are used. These are specially designed buffers whose output repetition rate can be controlled precisely using phase-locked loop techniques. When pulse stuffing is performed at the correct rate, the elastic stores never become completely filled or completely empty. Figure 5.7 illustrates the basis upon which jitter is removed in a regen-

TABLE 5.3. Summary of Repetition Rates and Codes in the Digital Hierarchy[a]

Signal	Repetition Rate (Mb/s)	Tolerance (ppm)	Format	Duty Cycle
Subrate	0.0024	—	bipolar	50
Subrate	0.0048	—	bipolar	50
Subrate	0.0096	—	bipolar	50
Subrate	0.056	—	bipolar	50
DSO	0.064	[b]	bipolar	100
DS1	1.544	±130	bipolar	50
DS1C	3.152	± 30	bipolar	50
DS2	6.312	± 30	B6ZS	50
DS3	44.736	± 20	B3ZS	50
DS4	274.176	± 10	polar	100

[b]Expressed in terms of slip rate.

erative repeater using an elastic store. The clock rate is recovered from the received stream and also is used to write bits from the receiver into the elastic store. The bit level in the elastic store is monitored continuously. Short duration changes in the level are smoothed via the low-pass filter. An increase in level beyond the medium range causes the frequency of the voltage-controlled oscillator to increase. This, in turn, means that bits will be read out of the elastic store at an increased rate, thus reducing the level. The voltage-controlled oscillator also serves as the transmit clock. A decrease in the elastic-store level correspondingly causes a decrease in the frequency of the voltage-controlled oscillator. This decrease allows the level in the elastic store to increase, since bits then are read out of it more slowly.

The combination of a muldem and a codec is called a *channel bank*. Channel banks are described in Section 5.5.

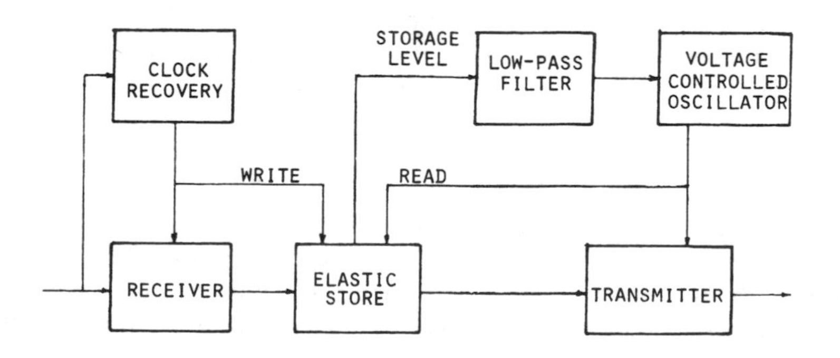

FIGURE 5.7. Use of elastic store in removing jitter at regenerative repeater.

5.3.4. Transmultiplexers

Both digital and analog transmission facilities now exist together in telephone networks. A given connection may involve a tandem combination of both facility types, such as may occur at a junction or at a transmission node where digital facilities connect to analog facilities. Accordingly, the need exists for the conversion of TDM signals to and from FDM signals. The device which performs this function is called the transmultiplexer. It is equivalent to a back-to-back connection between a digital demultiplexer and an analog multiplexer, or between an analog demultiplexer and a digital multiplexer. An example is the interfacing of five DS1 streams with two analog supergroups (120 voice channels), or the interfacing of two European 30-channel systems with a single analog supergroup (See Fig. 5.8). The latter is a type of application that is significant in providing international connections between systems using North American and European standards.

In Fig. 5.8(a), two European primary (30 channel) bit streams, each at 2.048

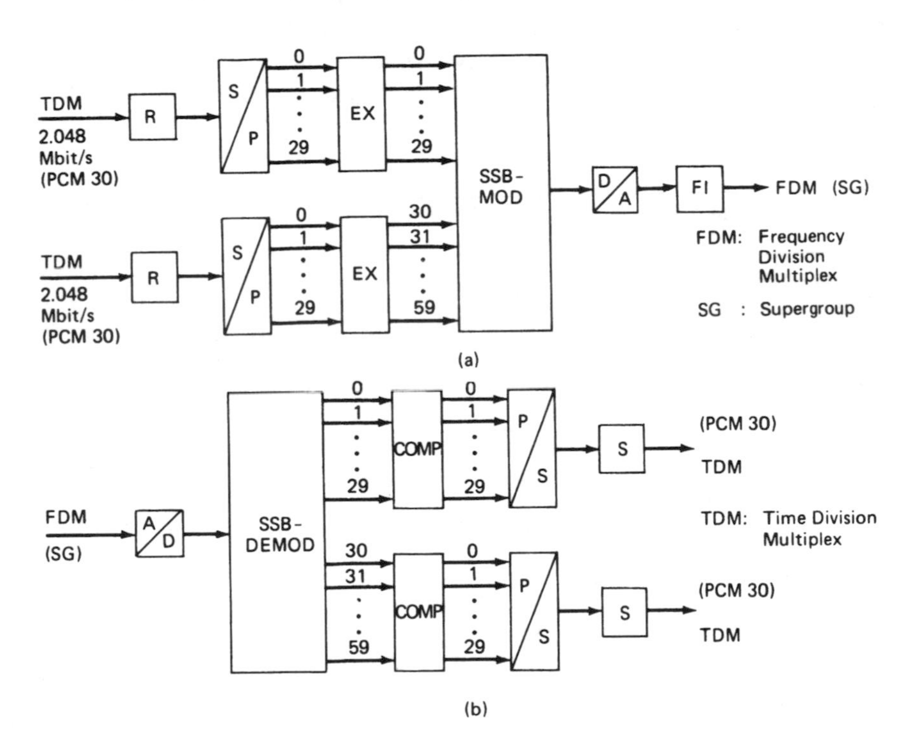

FIGURE 5.8. Block diagram of a 60-channel transmultiplexer. (Courtesy Scheuerman and Göckler, Ref. 9, © IEEE).

Mb/s and designated PCM 30, are applied to receivers R, which provide for their synchronization and signal level adjustment. Serial–parallel (S/P) converters then convert the TDM signals from serial to parallel form. The expander (EX) then converts the signals from their compressed (A-law) form to linearly encoded samples. Following this step, a bank of single-sideband modulators (SSB-MOD) produces the FDM signal using one of several digital signal-processing methods. Finally, a digital-to-analog (D/A) converter is used, followed by an analog filter (FI) to smooth the output, consisting of an FDM supergroup (SG).

Figure 5.8(b) illustrates the reverse process of conversion of an FDM SG to two European primary streams. Here the box labelled COMP is a compressor. Requirements for a 60-channel transmultiplexer based upon CCITT Recommendations G.792 and G.793 are shown in Table 5.4. This table shows the allowable variation of the amplitude response A and group delay distortion τ within the passband, as well as specification limits on crosstalk, and channel noise and distortion. Also shown are the signaling and pilot frequencies.

Digital signal-processing techniques can be used to implement transmultiplexers. Four major approaches[9] of this type are called (1) the bandpass filter bank, (2) the low-pass filter bank, (3) the Weaver structure method, and (4) the multistage modulation method. Two additional techniques[10] are the polyphase method

TABLE 5.4. Requirements for 60-Channel Transmultiplexers

Amplitude response, A	
600 Hz to 2400 Hz:	-0.5 dB $\leqslant A \leqslant 0.6$ dB
Maximum allowance at band edges	
300 Hz:	-0.6 dB $\leqslant A \leqslant 1.7$ dB
3400 Hz:	-0.6 dB $\leqslant A \leqslant 2.4$ dB
Group delay (absolute value)	$\tau \leqslant 3$ ms
Distortion	
1000 Hz to 2600 Hz:	$\Delta\tau \leqslant 0.5$ ms
600 Hz to 1000 Hz:	$\Delta\tau \leqslant 1.5$ ms
500 Hz to 600 Hz, 2600 Hz to 2800 Hz:	$\Delta\tau \leqslant 2.0$ ms
Minimum crosstalk attenuation between any two channels	
Intelligible crosstalk:	65 dB
Unintelligible crosstalk:	58 dB
Maximum idle channel noise with all channels loaded except the one measured, relative to peak signal level:	-80 dB
Maximum in-band rms nonlinear distortion, relative to peak signal level:	-40 dB
Out-of-band signaling frequency:	3825 Hz
Pilot frequency:	3920 Hz

Note: Levels and delays are those of the multiplexed analog signal with the digital parts looped.
Basis: CCITT Recommendations G.792 and G.793.
Source: Scheuermann and Göckler [Ref. 9]

and the Classen–Mecklenbräuker method. Numerous papers have been published on transmultiplexers[11] in Europe, Japan, and North America.

Because the FDM hierarchy uses single sideband for each voice channel, a major factor in transmultiplexer implementation is single sideband modulation and demodulation. These operations can be viewed as forms of digital sample rate *interpolation* and *decimation.* Digital interpolation consists of taking one sequence $a(n)$ and producing another sequence $b(m)$ whose samples occur r times as fast. Interpolation can be used to extract any portion of the spectrum of Fig. 5.9(a), including single sidebands.[12] Decimation is the process of sample rate reduction, that is, starting with the sequence $b(m)$ and producing a sequence $a(n)$ from it. Thus let $a(n)$ be a sequence whose sample period is T seconds. The spectrum of this signal is shown in Fig. 5.9(a). It is periodic with a period of $1/T$ Hz. Let $H(f)$ be a low-pass digital filter with a cutoff frequency of $1/2T$ Hz and operating at a sampling rate of r/T Hz [see Fig. 5.9(b)]. If the low-rate signal is applied to this filter (considering intervening input samples to be zero), the filter will produce an r/T rate output whose samples are the same as those that would be obtained in the interpolation process. For group band operations, a suitable r

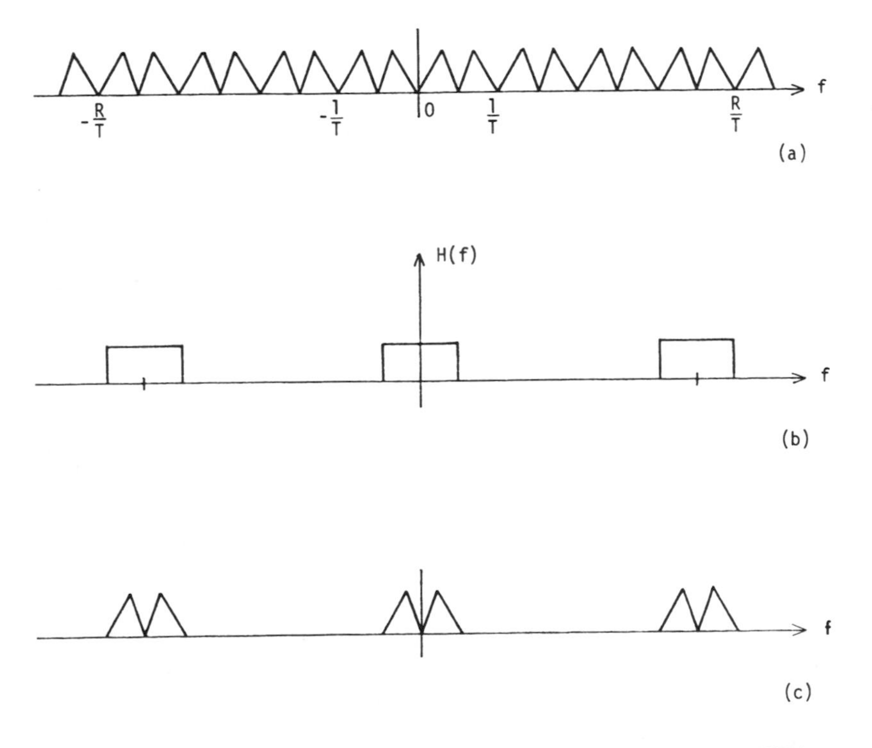

FIGURE 5.9. Interpolation spectra. (Courtesy S. L. Freeny, Ref. 10, © IEEE, 1980.)

value is 14. The reason for this choice is that since $1/T = 8$ kHz, $r/T = 112$ kHz, which is suitably above the top group band frequency of 108 kHz. (The group band of 60 to 108 kHz contains 12 voice channels, each occupying 4 kHz.) As indicated in Fig. 5.9(c), the filter removes the unwanted signal images. The result is a spectrum that is the same as would have resulted from sampling an analog signal at the high rate initially, to the extent that $H(f)$ approximates an ideal low-pass filter.

The result of using interpolation is known as the bandpass method of digital single-sideband generation. Rather than multiplying by a sine wave, as in the analog case, the "carrier" is built in because of the periodic nature of the digital spectrum. For the bandpass demodulation process, the composite FDM stream is applied simultaneously to twelve bandpass filters (one for each channel in the group spectrum) whose outputs then are decimated by a factor of $r = 14$. The per-channel computation rate for demodulation thus is the same as for modulation. To prevent the resulting FDM signal in the modulation process from having a mixture of upper and lower sidebands, the sign of the input sequences of every other channel must be alternated. This shifts the spectra of each of the inputs by half the sampling rate, or 4 kHz. Thus in any given frequency slot, an upper sideband is exchanged for a lower one or vice versa. For the bandpass demodulator, a similar sign alternation must be done on half of the output channels after decimation. The Weaver method is described in Refs. 9 and 10, the multistage modulation method in Ref. 9, and the polyphase and the Classen-Mecklenbräuker methods in Ref. 10.

5.4. SCRAMBLERS

In the waveform encoding of speech, the most probable speech amplitudes are the lowest ones, because of the large number of pauses in conversational speech. In addition, the use of waveform encoding means that the frequent zero crossings of any waveform will produce many zero amplitudes to be encoded. The result then may consist of appreciable trains of zeros in the transmitted stream. Since the receiver must derive its clock from the received data stream, it must receive enough ones on a regular basis to allow its timing circuits to function as intended, as discussed in Section 5.2.3. Figure 3.2 illustrated one method of minimizing the long train of zeros caused by speech pauses. This method is called *zero-code suppression.* Scramblers also may be used to prevent the occurrence of long runs of zeros.

Another reason for the use of scramblers is that a data stream, such as one produced by signaling and supervision functions, may have a repetitive pattern with high discrete frequency components. If such a pattern were transmitted, it could constitute a serious source of interference to other DS1 streams using the same facilities because of the line spectrum it produces. Scrambling can be accomplished by adding a pseudorandom digital sequence to the transmitted stream, and then subtracting the same sequence from the received stream. This can be accomplished by the use of pseudorandom generators that are built alike, started in synchronism, and clocked at the same rate. The clocking rate, of course, is derived directly from the received digital stream itself. Figure 5.10 illustrates the scrambling and subsequent descrambling of a digital stream.

With reference to Fig. 5.10, a pseudorandom sequence is added to the stream to be scrambled. This is done using modulo-2 addition, in which the rules are: $0 + 0 = 0$, $0 + 1 = 1$, $1 + 0 = 1$, and $1 + 1 = 0$. The sum then constitutes the scrambled stream which is transmitted. The received stream is assumed to be identical to the transmitted stream. For purposes of illustration, the channel has been assumed to be error free. At the receiver a locally generated pseudorandom stream is added, again modulo-2, to the received stream. The resultant descrambled stream is found to be identical to the original stream to be scrambled. Note that the received stream, on which the receiver's timing circuits function, bears no resemblance to the descrambled stream. Figure 5.11 illustrates the way a pseudonoise generator is built. The period of such a generator is

$$T = \frac{(2^n - 1)}{F_o}$$

where n is the number of shift register stages and F_o is the clock frequency.

A simpler alternative that is commonly used derives the added digital sequence by inverting the original message sequence (as in zero-code suppression) and then using modulo-2 addition of the bits of the original and inverted sequences.

By eliminating periodicities, the scrambler reduces the amplitude of any discrete spectral lines (that otherwise would exist) to negligible levels. Because the

Stream to Be Scrambled:	0 0 0 0 0 0 0 1 0 0 0 0
Pseudo-Random Stream:	1 1 0 0 0 1 0 1 1 0 1 0
Transmitted Stream:	1 1 0 0 0 1 0 0 1 0 1 0
Received Stream:	1 1 0 0 0 1 0 0 1 0 1 0
Pseudo-Random Stream:	1 1 0 0 0 1 0 1 1 0 1 0
Descrambled Stream	0 0 0 0 0 0 0 1 0 0 0 0

FIGURE 5.10. Scrambling and descrambling of a digital stream.

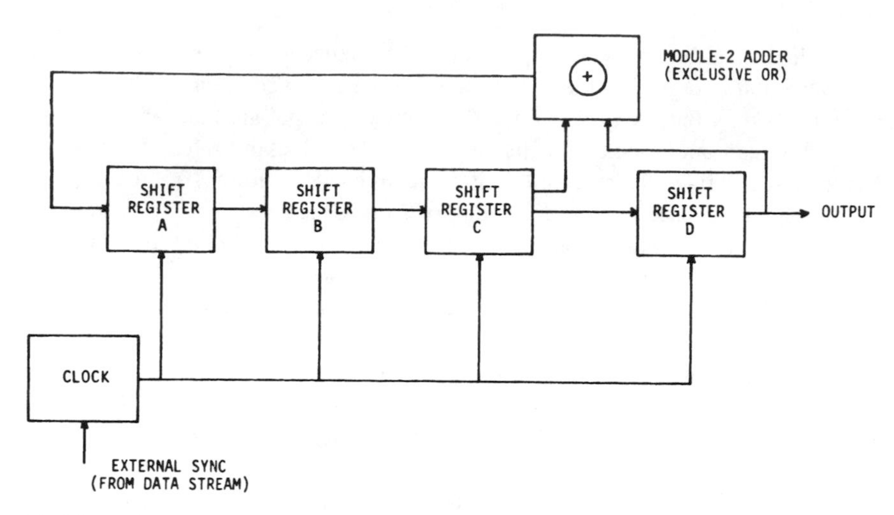

FIGURE 5.11. Pseudo-noise generator.

scrambling process removes spectral lines, interchannel interference between digital systems can be kept to a low level.

5.5. CHANNEL CODERS

The channel coder is the device that processes the binary input into a multilevel or modified binary signal. This processing may alter the spectral shape of the signal since, in general, converting a two-level signal to three or more levels tends to decrease the amount of spectrum it occupies.

The formats which have been selected for use on the public switched network provide reliability together with efficiency of operation in terms of spectrum utilization and power consumption.

Local-loop digital transmission at subrate speeds (2.4, 4.8 and 9.6 kb/s) is accomplished using a bipolar return-to-zero (BRZ) signal with a 50% duty cycle, as illustrated in Fig. 5.12. Conversion from the customer's format to BRZ occurs at the customer's interface with the local loop. Note that zero is sent as 0 volts whereas the first one is sent as a positive voltage (for example, 3 volts) during 50% of the bit period. The second one is sent as a negative voltage of equal magnitude to the positive voltage. Subsequent ones continue to alternate in polarity. Accordingly, the transmitted signal has no zero frequency energy.

A departure from the alternating positive and negative pulse rule for ones is called a *bipolar violation* (BPV). Noise on the channel thus can cause a BPV to

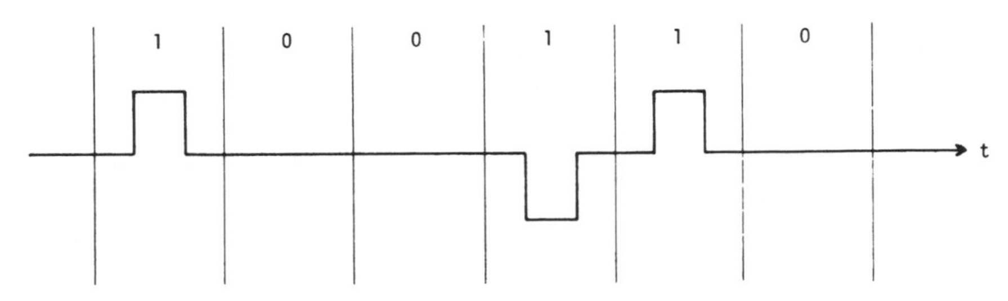

FIGURE 5.12. Bipolar return-to-zero signal with 50% duty cycle.

be produced. However, BPVs are produced intentionally for several purposes. They are used when a trouble condition exists on the facility, when an idle code or more than six zeros (seven for 56 kb/s) is being sent, or when equipment tests are being done.

If a digital sequence from a customer contains too many sequential zeros, the transmitting channel coder inserts BPVs, thus breaking the sequence. These BPVs are detected by the receiving equipment, which replaces them by the proper number of zeros before forwarding the sequence to the customer's receiving terminal equipment. As a result, no restriction needs to be placed on customer digital sequences.[13]

At the central office, various subrate signals may be combined together to produce a 64 kb/s stream known as the DS0 signal. It is a nonreturn-to-zero (NRZ) signal, that is, it has a 100% duty cycle, and is used for intraoffice transmission. To the extent that the subrate signals add to less than 64 kb/s, *pulse stuffing* is used in combining them. Pulse stuffing is a process in which enough time slots are added to an incoming signal to make this signal operate at the specific rate controlled by the clock circuit of the transmitter.[14] Pulses inserted into these new time slots carry no information. The signal must be coded so that these noninformation bits can be recognized and removed at the receiving terminal. The European term for this process is *justification.*

The use of pulse stuffing requires that the output channel rate be higher than the aggregate input rate to allow for the addition of the *stuff* (also called *null*) bits. The stuff bits are added in a prescribed manner so that they are identifiable at the receiver, which must remove them.

For interoffice transmission, 24 DS0 signals are combined to produce a DS1 signal. Because of possible minor differences in the DS0 signals, pulse stuffing is used in this combining process.

The various factors that affect the design of these signals include (1) interconnection and transmission requirements and (2) compatibility with terminal equipment. Under interconnection and transmission requirements are included transmission rate, signal format (polar, bipolar, or multilevel), pulse amplitude, number of consecutive zeros allowable, and the location of parity bits in the bit stream. Compatibility requirements include the multiplexing method, framing, and signaling formats.

As noted in Section 5.3.2, two DS1 signals can be combined using an M1C multiplexer to form a DS1C signal at 3.152 Mb/s. The DS1C uses a BRZ 50% duty-cycle format, with an additional 64 kb/s for synchronization and framing. In this process, time slots are added to the received signal (pulse stuffing) so that the signal produced operates at a rate controlled by the transmitter's clock.

As described in Section 5.3.2, the DS2 signal (6.312 Mb/s) is produced by multiplexing four DS1 signals together in an M12 multiplexer. Synchronization of the four incoming signals is obtained through the addition of stuff bits since the DS1 signals may originate from different sources. Control and framing bits also are added. The DS2 format is called *bipolar with six-zero substitution* (B6ZS). For sequences of five or fewer zeros the stream is bipolar. If six or more zeros occur in a row, the output depends on the polarity of the one pulse that preceded them. If the polarity was positive, the output produced is $0+-0-+$, where $+$ and $-$ denote the polarity of the digit one. Thus two bipolar violations have been caused. However, if the polarity was negative, the output produced is $0-+0+-$. Again, two bipolar violations have been produced. These violations occur in the second and fifth bit positions of the sequence. The receiver recognizes them and substitutes zeros instead.

The DS3 signal (44.736 Mb/s) is produced by an M13 multiplexer, as noted in Section 5.3.2. Within this multiplexer four DS1 streams are combined to produce the DS2, and seven DS2s then are combined to produce the DS3. The polar[15] format is maintained internally, which means there are no bipolar violations. The output stream is bipolar with a 50% duty cycle, and is modified using three-zero substitution (B3ZS). Each group of three consecutive zeros is replaced by *BOV* or by *00V* where *B* is a one pulse that adheres to the bipolar rule, whereas *V* is a one pulse that violates the bipolar rule. The selection between *BOV* and *00V* is made so that the number of *B* pulses between consecutive *V* pulses is odd.

The DS3 signal is partitioned into frames, each of which is 4760 bits long. Each frame, in turn, is divided into seven subframes, each of which is 680 bits long. The number of subframes corresponds to the number of DS2 signals formed within the multiplex unit. Each subframe, in turn, is divided into eight blocks of 85 bits. The first bit of each block is used as a control bit; the remaining 84 bits are available for information.

The initial bits in successive subframes are designated as *X, X, P, P, M0, M1,* and *M0.* The first time slot in each of the first two subframes, designated as an *X* bit, may be used for alarm or other operation and maintenance purposes; however, the two *X* bits in a frame must be the same, either *00* or *11*.

The first time slots in the third and fourth subframes are designated as *P* bits. These bits are used to convey parity information relating to the 4704 information time slots following the first *X* bit in the previous frame. If the modulo-2 sum of all information bits is 1, $PP = 11$, and if the sum is 0, $PP = 00$.

The first time slots in subframes 5, 6, and 7 are designated as M bits. These three time slots always carry the time code *010*, which is used as a multiframe alignment signal. In each subframe, blocks 2, 4, 6, and 8 carry *F* bits. These *F* bits are transmitted in the first time slot of each of these blocks as a *1001* code that is used as a frame alignment signal to identify all control bit time slots.

The first time slots in subframes 3, 5, and 7 carry bits to indicate the presence or absence of a stuff pulse in the subframe. The bits that are designated *Ci1, Ci2*, and *Ci3* are the stuffing indicator bits for the *i*th subframe, where *i* is any number from 1 to 7. In the C-bit position, a *111* code indicates that a stuff pulse has been added, while a *000* code indicates that no stuff pulse has been added.

One stuff pulse per subframe may be added in the eighth block. The stuffing time slot is the first information time slot in that block for the DS2 signal that corresponds numerically to the subframe, that is, the *i*th time slot in the eighth block of the *i*th subframe. The nominal and maximum stuffing rates per 6.312 Mb/s input are 3671 b/s and 9398 b/s respectively. The 6.312 Mb/s signals appear internally in the multiplex unit. Each is a DS2 signal, made up of four multiplexed DS1 signals in a manner similar to that used in the M12.

The DS4 signal (274.176 Mb/s) is the result of multiplexing six DS3 signals together in an M34 multiplexer, as noted in Section 5.3.2. Pulse stuffing is used, as at the lower levels. The DS4 is polar binary, which means that one is positive (100% duty cycle) and zero is negative (100% duty cycle), as measured from the center conductor to the outer conductor of the coaxial cable used for transmission.

European systems time-division multiplex 32 channels together, but use only 30 of them for actual speech or user data transmission. The other two serve to provide frame alignment (Channel 0) and signaling (Channel 16). Standard digital levels in the European hierarchy are shown in Table 5.5.

The first three European levels use the *High Density Bipolar-3* (HDB-3) code which is basically BRZ, but in which each group of four consecutive zeros is replaced by B00V or 000V. The selection between B00V and 000V is made so that the number of *B* pulses between consecutive *V* pulses is odd. The fourth

TABLE 5.5. European Digital Hierarchy

Level	Rate	Capacity
First	2.048 Mb/s	30 PCM channels
Second	8.448 Mb/s	120 PCM channels
Third	34.368 Mb/s	480 PCM channels
Fourth	139.264 Mb/s	1920 PCM channels
Fifth	565.148 Mb/s	7680 PCM channels

European level uses *coded mark inversion* (CMI), also known as biphase-space. This format always has a transition at the beginning of an interval. For a 1, there is no transition in the middle of the interval, whereas for a 0, there is a transition in the middle of the interval.

Table 5.3 summarized the description of the United States standards and showed the timing tolerances in parts per million. Figure 5.13 illustrates how the various levels of the digital hierarchy are related. In that figure, the CSU is a *channel service unit*; the DSU is a *data service unit*. These are devices that convert the customer's digital data signals to the format used by the transmission facilities. The DSU converts the user's two-level data signals into the standard bipolar format. It operates at the desired data rate, usually 2400, 4800, 9600, 56 000, or 64 000 b/s, and provides a loopback point for test purposes. The CSU usually operates at the DS1 rate. For the CSU, the user's signal must be in the bipolar format and comply with the digital encoding rules. The CSU provides loop-around testing capability, regeneration of received-line signals, and protection of the carrier network. Local loop arrangements for digital voice are discussed in Section 5.9.

Figure 5.14 compares the unipolar, bipolar, and polar binary waveforms. The foregoing discussion indicated that signals may require frequent format conversions in digital transmission. Noteworthy in this regard are the signal format differences indicated in Fig. 5.14, which shows (a) the format associated with transistor-transistor logic (TTL), (b) the BRZ format, and (c) the polar binary format used for the DS4 signal. For the unipolar signal, the zero is nominally 0 volts and the one is nominally 5 volts; this format must be converted to the BRZ

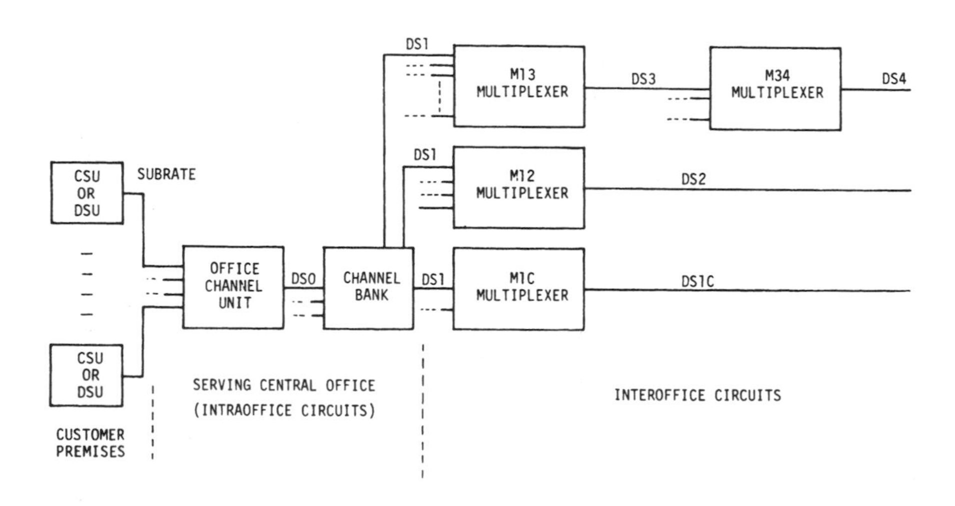

FIGURE 5.13. Relationships among levels of the digital hierarchy.

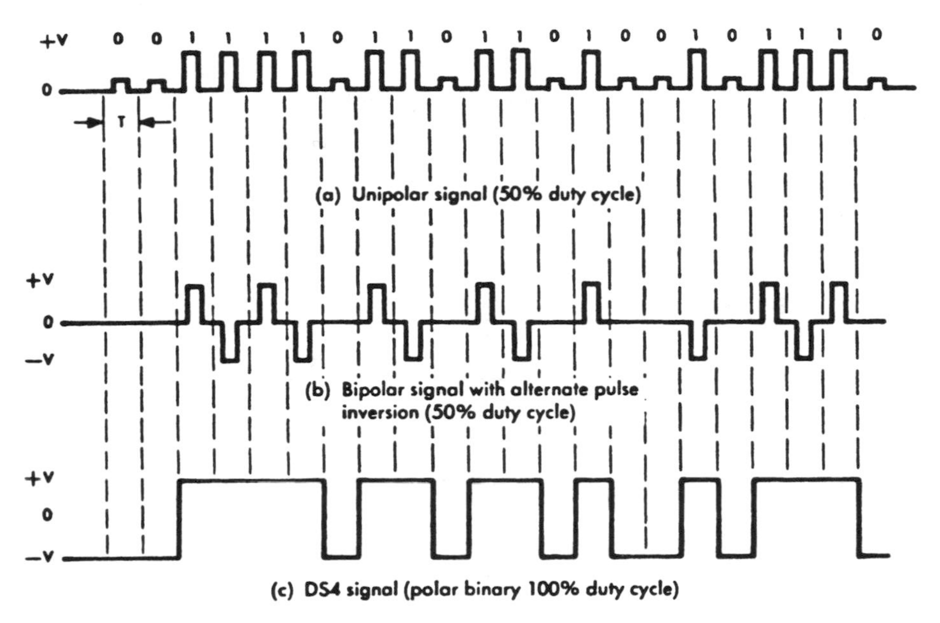

FIGURE 5.14. Signal waveform comparison. (© American Telephone and Telegraph Company, 1977.)

line format for transmission. In the case of TTL levels, tolerances of ±2 volts may be placed around the nominal values.

A conversion from a 2-level to a 3-level format is required in one method of implementing *quadrature partial response signaling* (QPRS), (see Section 6.2.5) to achieve bandwidth compression, as shown in Fig. 5.15. In this case, the input is a binary signal while the output is obtained by combining the present and the immediately preceding input symbols.

5.6. ECHO CANCELERS

Echoes arise from impedance mismatches in terminating equipment where the 2-wire–4-wire conversion takes place. Often echoes cause as much subjective annoyance on long-distance telephone calls as do noise and low speech level. The subjective annoyance of echoes increases with both delay and level. Echo problems have been controlled through the judicious use of circuit loss as well as by the use of the echo suppressor, which opens the transmission path from the listener to the talker.[16]

For circuit lengths that require echo control, the use of loss, however, results in unacceptably low received levels.[17] Echo suppressors, on the other hand, cannot

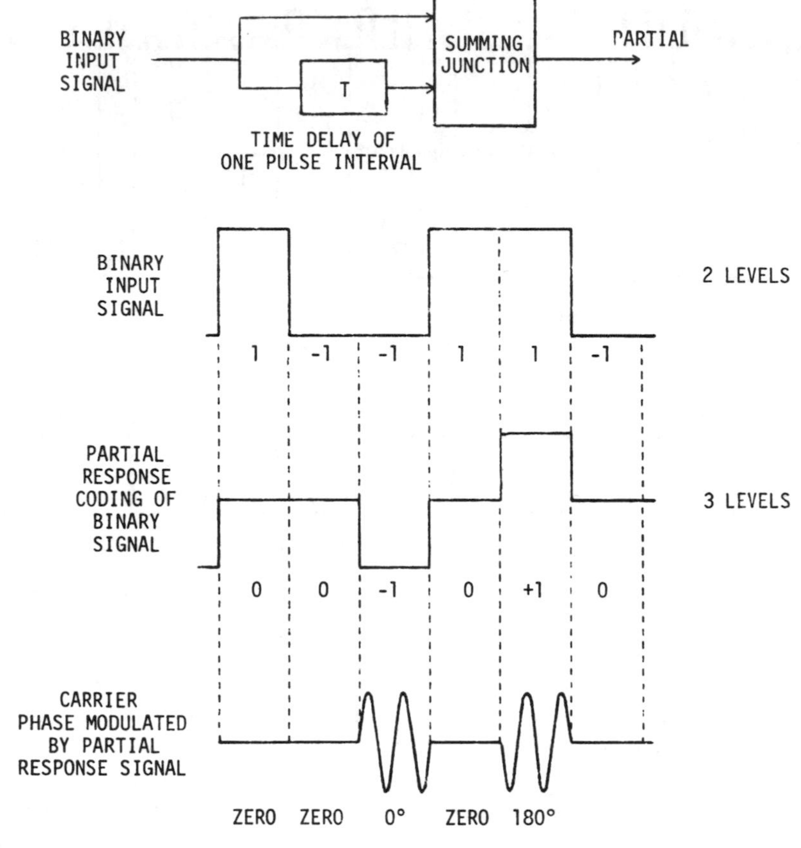

FIGURE 5.15. Conversion from 2-level to 3-level format.

provide satisfactory service under conditions in which both parties attempt to speak simultaneously (double-talk), and also may chop or clip low speech levels. In addition, while echo suppressors often are used on one-way single-hop satellite circuits, their performance has been found to be unacceptable[18] on two-way (full-hop) satellite circuits, where the echo delay is a full 540 ms or more. The foregoing problems of excess loss, chopping, and clipping are best solved through the use of the echo canceller. Echo cancellation was proposed in the early 1960s, but could not be implemented on a widespread basis at that time because of its high cost. The use of very-large-scale integration (VLSI), however, has allowed an entire echo canceller to be placed on a single chip,[19,20] thus rendering the concept economically practical.

Figure 5.16 illustrates the basis on which echo cancelers are implemented. An

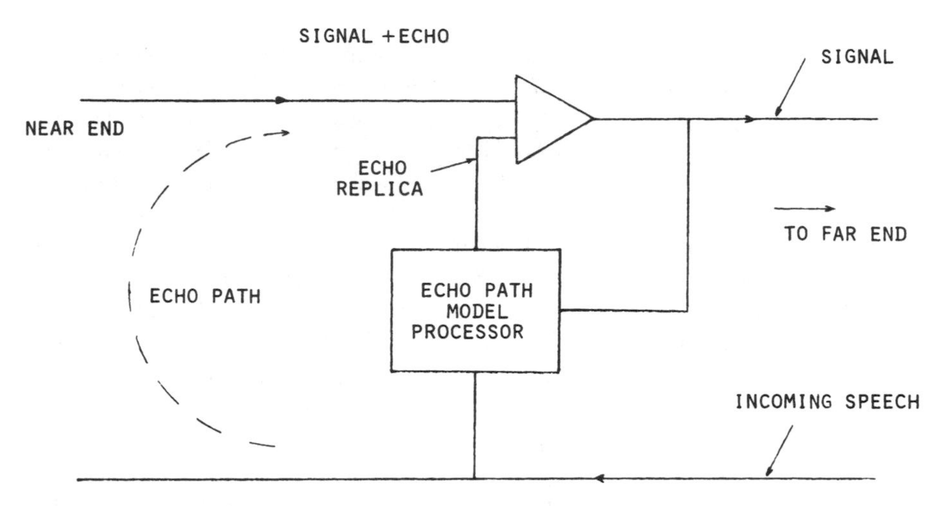

FIGURE 5.16. Principle of echo cancellation. (Courtesy Suyderhoud et al., Ref. 16, © IEEE, 1976.)

adaptive model of the echo signal is built using tapped delay-line principles. The modelled echo then is subtracted from the actual echo, thereby leaving the speech without echo. Accordingly, the failings of the echo suppressor technique do not arise. Adaptive feedback processing is used to obtain a model of the echo-path response. This model then is stored in memory. Following this, the incoming signal is processed by convolution with the stored impulse response, thus providing a close replica of the actual echo signal. This replica then is subtracted from the actual echo signal on the sending side. As a result, the echo is removed, leaving the speech unaffected.

5.7. DIGITAL SPEECH INTERPOLATION

During ordinary telephone conversations, the transmission path in each direction is used only about 40% of the time.[21] The usage is this low because only one person usually talks at a time and because there are natural pauses in speech. This fact has been used to advantage in a technique called *time assignment speech interpolation* (TASI), which has been implemented for many years on long-haul analog channels such as those of transoceanic cable systems. TASI allows an approximate doubling of channel capacity because each talker uses the circuit less than half the time. A digitized version of TASI is one type of *digital speech interpolation* (DSI). In the case of PCM or ADPCM voice, a special TASI technique known as *channel augmentation by bit reduction* can be used to absorb overloads that would otherwise result in clipping.

In operation, a TASI-type system serves a number of trunks N via a smaller number of channels n by connecting, at any moment, only those trunks on which speech activity (a speech "spurt") is present. Special TASI common-signaling channels, whose propagation delay is the minimum possible for a given transmission medium, are used to notify the far end which trunk is connected to which channel. A voice-path delay of 25–50 ms is used to allow time for the connection information to be acted upon.[22] A delay of 50 ms eliminates processing clip caused by the switching not being complete in time for the speech spurt.

Another form of clipping, called *competitive clip*, occurs when all channels are active and an additional trunk has a speech spurt to be connected. Such a new spurt must await the availability of the next channel and thus becomes clipped, that is, its start is cut off. Generally a clip of 50 ms or less is not detectable by the user.[23]

Because data tones normally cannot tolerate clipping, data transmission is provided service called a *digital noninterpolated interface* (DNI) on which TASI is not being used.

Under conditions such that the number of trunks being used actively for voice exceeds the available number of channels, the least significant bits of the PCM channels can be robbed to form additional voice channels. In such a system used by INTELSAT[24] (channel augmentation by bit reduction), the probability of a clip longer than 50 ms is limited to less than 2% by serving 240 trunks with 127 normal channels and 16 overload channels derived by robbing least significant bits. To assure the continuous achievement of <2% clips of >50-ms duration, a 1000-Hz square wave is sent one second out of 10 on one of the trunks. At the receiving end, this square wave's on-off durations are measured to detect the extent of any clipping. If clipping beyond preestablished limits occurs, an alarm message is generated and sent back to the originating encoder.

DSI gain is defined as (Number of Trunks)/(Number of Channels, including assignment channel). Figure 5.17 shows how DSI gain varies with the number of incoming trunks.[25] The parameter P is the "freeze-out" percentage, which is defined as

$$100x \frac{\text{(Number of trunks)}}{\text{(Average of talk spurt time)}}$$

for a given channel.

To date, DSI has been used largely on satellite circuits; however, its use on a significant number of terrestrial circuits now is beginning. The use of DSI offers the possibility of appreciable increases in the number of voice channels that can be carried by digital radio systems, which are discussed in Chapter 8. For example, long-haul systems at 6 GHz can provide 2016 digital voice channels in a 30-MHz radio channel bandwidth through the use of 64-QAM. The use of DSI

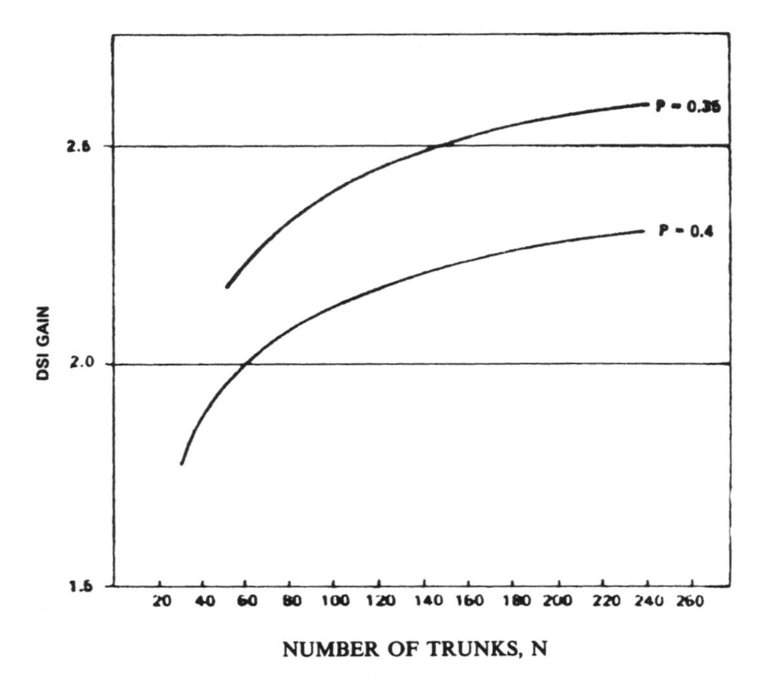

FIGURE 5.17. DSI gain versus number of trunks, *N* (Courtesy Verma, Ramasastry and Monsees, from NTC '78 Conf. Record © IEEE, 1978).

could double this number of voice channels. In addition, DSI forms the basis of the Extended-TDMA (E-TDMA) digital cellular system described in Chapter 7.

5.8. DIGITAL REPEATERS

A major advantage of digital transmission is its ability to perform well at much higher channel noise levels than are feasible for analog transmission. This advantage results directly from the use of regenerative repeaters in digital transmission. The regenerative repeater senses the state of the incoming noisy and distorted signal and reconstructs a clean output signal.

The functions performed within a regenerative repeater are illustrated in Fig. 5.18. The amplifier compensates for signal-level losses prior to the repeater and provides a uniform output level through the use of an automatic line buildout circuit,[26] which corrects for the effects of temperature changes as well as for variations in repeater spacing. The filter performs an equalizing function, optimizing the attenuation versus frequency characteristic for pulse transmission. This keeps pulse distortion low while suppressing out-of-band noise.

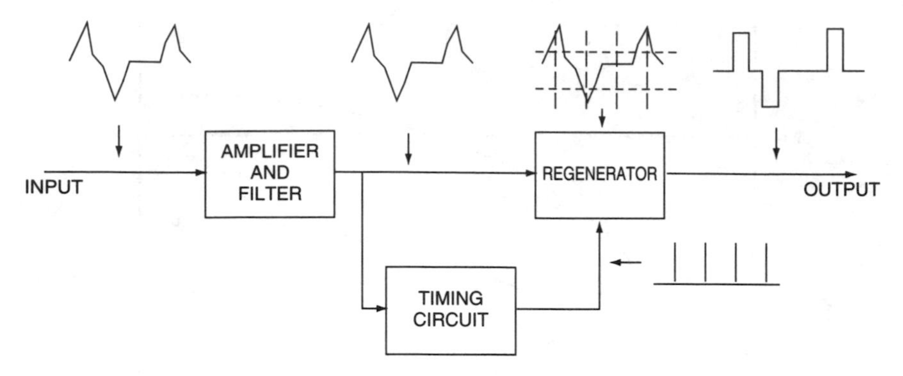

FIGURE 5.18. Regenerative repeater functions.

Timing information is obtained from the signal after the amplification and equalization have been done. A fixed delay is used in the timing path to cause the timing pulses to occur at the middle of each signal pulse interval. The regenerator thus looks at the incoming signal at the middle of each pulse interval. Positive clock pulses gate the incoming pulse stream into the regenerator. Negative clock pulses turn the regenerator off. In this manner, the width of the output pulses is controlled.

Threshold circuits are used to determine whether a positive, zero, or negative signal state exists at the sampling times. When either the positive or negative threshold is exceeded, a pulse of the proper polarity, duration and amplitude is produced as the output. Otherwise, the output remains zero. The tolerable channel error ratio places a limit on repeater performance. Errors are produced both by near-end crosstalk and by receiver front-end noise, which has a gaussian characteristic.

The channel error ratio produced by receiver front-end noise places a limit on the maximum distance between repeaters. Thus, if transmission is to be done at an error ratio not exceeding a specified level, the signal-to-noise ratio must be at least a minimum value based on the equation[27]

$$\text{b.e.r.} = 0.5\, erfc\left(\frac{V_p}{\sqrt{2}\,\sigma_n}\right) \tag{5.1a}$$

where

V_p = peak signal level
σ_n = rms level of gaussian noise.

Equation (5.1a) is based on simple two-level transmission. In the event of multilevel transmission, for example, $m > 2$, a more general equation is used:

$$\text{b.e.r.} = \left[\frac{(m - 1)}{m}\right] erfc \left(\frac{V_p}{(m - 1)\sqrt{2}\,\sigma_n}\right) \tag{5.1b}$$

Figure 5.19 is a plot of Eq. (5.1b). Even for bit-error ratio values on the order of 10^{-10}, the signal-to-noise ratio requirements are very modest compared with those of analog transmission. Note that as the number of levels increases, the signal-to-noise ratio must be increased to maintain a given bit-error ratio.

At each digital regenerative repeater, the signal is received and detected, whereupon a new output waveform is produced. This process is called *waveform regeneration*. It is a primary advantage of digital transmission. To achieve an end-to-end bit-error ratio of 10^{-7} in transmission through a large number of repeaters, an adequate signal-to-noise ratio must be present at the input to each repeater. In the case of three-level (polar ternary) transmission, Fig. 5.20 shows the input signal-to-noise ratio requirement both with and without waveform regeneration. The advantages of regeneration are quite clear from this figure. (Note

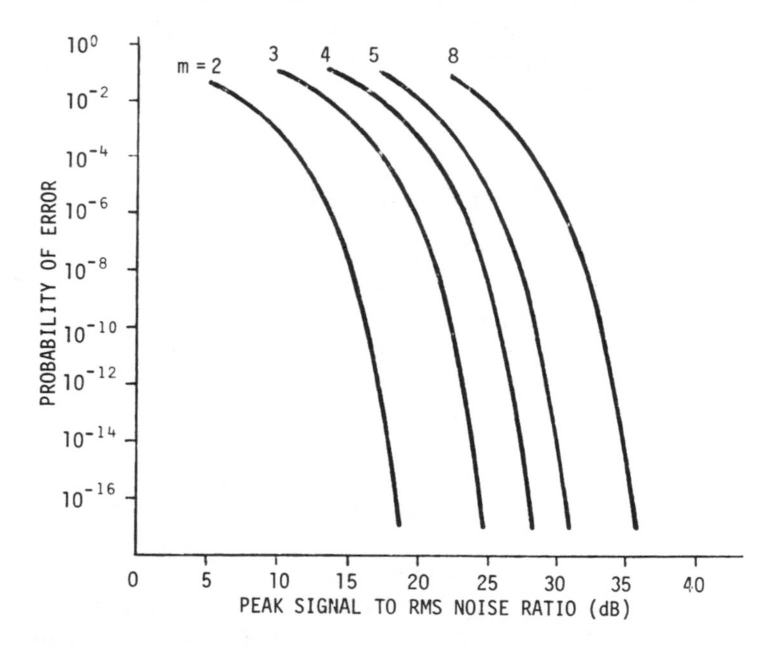

FIGURE 5.19. Bit-error ratio for multiple-decision thresholds. (© 1982, Bell Telephone Laboratories. Reprinted by permission.)

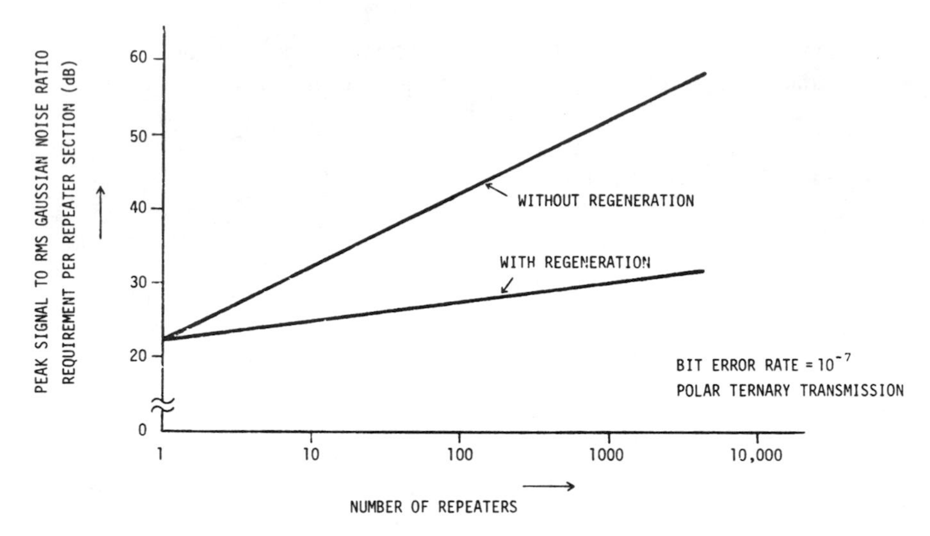

FIGURE 5.20. Required signal-to-noise ratio versus number of repeaters. (© 1982, Bell Telephone Laboratories. Reprinted by permission.)

that for 3-level transmission, Fig. 5.19 calls for about 22 dB for b.e.r. = 10^{-7}. This is consistent, to within drafting accuracy, with the value shown for one repeater in Fig. 5.20.)

5.9. DIGITIZATION OF THE LOOP PLANT

The transition to digital technology within the telephone industry started in the exchange plant, and was extended to long-haul transmission. The final part of the transition, within the loop plant, now is materializing. The description of the digital loop plant (subscriber lines) contained in this section is based on work being done by the common carriers, as well as by manufacturers, to develop the systems to be used.[28,29] Further discussion can be found in Section 15.4, which describes the use of fiber in the local loop.

Because of the considerable investment in subscriber lines by wireline common carriers everywhere, the utilization of that investment to a maximum extent is vital. Accordingly, the use of two-wire transmission between the subscriber's instrument and the serving central office is a significant factor. Important design considerations are compatibility with related systems, crosstalk, and impulse noise from the central office. System parameters include transmission rate, transmitted power, and the choice of line codes. Design uniformity is important with respect to component interchangeability, but the design must be flexible enough to accommodate variations in the loop plant and local implementation details, as well as manufacturing variations in components.

5.9.1. Transmission Modes

Although many end instruments operate on a totally wired basis, the use of radio transmission for part or all of the local loop is gaining in importance. In addition, the use of fiber optics (see Chapter 10) is beginning to be seen primarily in connection with the integrated services digital network (ISDN) (see Chapter 14). This section includes discussions of wired systems (Section 5.9.1.1), pair-gain systems (Section 5.9.1.2), discrete multitone (Section 5.9.1.3), and radio systems (Section 5.9.1.4).

5.9.1.1. Wired Systems

The three basic transmission modes that can be used for two-wire operation are *hybrid, frequency-division multiplexing* (FDM), and *time compression multiplexing* (TCM). The hybrid system achieves isolation through the use of echo cancellers, with transmission in both directions occurring at the same time. The principle of echo cancellation is described in Section 5.6. The hybrid approach is being used in the public implementation of ISDN. Dependable hybrid systems require high-quality echo cancellers.

The FDM system uses separate frequency bands for the two directions of transmission, thus accomplishing them simultaneously. In the analog environment, FDM finds use in FDX modem operation, but it has not been applied to the digital environment.

TCM separates the two directions in the time domain, and is being implemented in private branch exchange (PBX) applications of ISDN. In TCM, the line rate must be more than twice the bit rate. Thus for standard 64 kb/s PCM, a line rate greater than 128 kb/s must be used. Design parameters for TCM are the block length transmitted, the line length (and thus pulse transit time), and the amount of idle time needed for line transients to decay. The TCM system, however, is readily implemented with digital components, and is not sensitive to gauge discontinuities and bridged taps. Because it may not be familiar to the reader, it is described next.

Figure 5.21 illustrates the components of a TCM system. The "bit stream in" is a steady stream, for example, 64 kb/s. The "buffer" converts this stream to bursts at a higher rate, for example, >128 kb/s. These bursts are sent through the "T/R switch" to the subscriber line during the times when no bursts are traveling in the opposite direction. The received bursts are regenerated and then smoothed into a steady 64 kb/s stream at the output. Bursts traveling in the opposite direction are timed so that they do not collide with the bursts just described.

The maximum line length is limited by both propagation time and crosstalk.

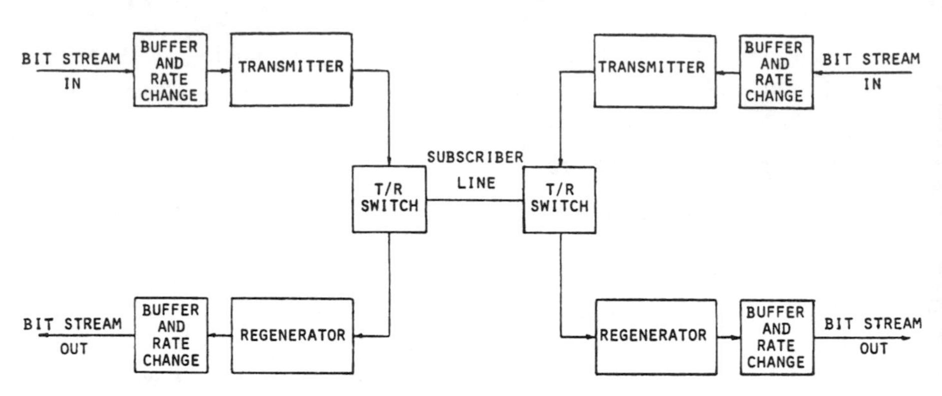

FIGURE 5.21. Basic components of time-compression multiplex (TCM) local-loop transmission system. (Courtesy Ahamed et al., Ref. 28, © IEEE, 1981.)

The propagation time limitation has been shown[30] to be related to the system parameters as follows: Assume that the line is to be used for 64-kb/s PCM plus 8-kb/s data in each direction, and that one additional bit for every nine is to be used for housekeeping purposes. This results in 80 kb/s in each direction. Let each burst in each direction have a length of $k \times 10$ bits. A burst then will have a duration of $k \times 125$ μs.

$$\text{Let } \tau_g = \text{group delay per unit length of line in } \mu\text{s/km}$$
$$\tau_1 = \text{length of a } k \times 10 \text{ bit burst in } \mu\text{s.}$$

Allowing a total time of $k \times 125$ μs for two-way transmissions, the maximum distance that can be served by the line is

$$d_{max} = \frac{[125k - 2\tau_1]}{2\tau_g} \text{ km} \tag{5.2a}$$

$$\text{Let } \zeta = \text{line bit rate/input bit rate}$$
$$= \text{compression factor.}$$

Then $\tau_1 = 125k/\zeta$ μs. Since $\tau_g \approx 5$ μs/km for typical local line conductors,

$$d_{max} = \frac{12.5k(-2)}{\zeta} \text{ km} \tag{5.2b}$$

For example, at a line rate of 256 kb/s, $\zeta = 3.2$. For 10 bit bursts, $k = 1$, so $d_{max} = 4.7$ km, generally more than adequate for most users of such systems. This limit only takes pulse travel time, usually the most significant limit, into account. Crosstalk and impulse noise also may limit some of the longer systems.[31]

5.9.1.2. Pair-Gain Systems

Pair-gain systems, sometimes called *subscriber-loop systems*, provide savings in wire pairs by combining a number of voice channels using time-division multiplexing. Their use has been fostered not only because of wire-pair savings, but also because they enable service additions in areas of rapid growth as well as in areas where emergency conditions have caused system outages. Pair-gain systems also are useful in serving small communities remote from a central office.

The first digital pair-gain system[32] was the *subscriber-loop multiplex* (SLM). These systems are still in operation and serve up to 80 subscribers via a 24-channel T1 line with concentrators. The channel code rate is 57.2 kb/s (delta coding). The system uses two pairs of 22-gauge wire, and can function with remote powering to achieve distances up to 80 km. As many as six remote terminals can be placed along the line to interconnect with individual station lines along the route, with up to 40 lines being served by a remote terminal. Signaling and supervision are handled within the digital bit stream. The SLM system interfaces with both the end-office switch and the subscriber on an analog basis, thus providing system transparency.

The *Subscriber Loop Carrier-40* (SLC-40) system serves 40 subscribers via a DS1 line using 38 kb/s adaptive delta coding without concentration. As does the SLM, the SLC-40 uses two pairs of 22-gauge wire and works up to 80 km using remote powering; however, all subscriber line interfaces must be at a common point. Analog interfaces are provided to the end-office switch as well as to the subscriber.

The SLC-96 was placed into service in 1979.[33] It uses PCM, the same as do the T-carrier systems. Accordingly, the SLC-96 is fully compatible with other digital switching and transmission equipment used in the public switched network.

5.9.1.3. Discrete Multitone (DMT) Systems

Requirements for asymmetric digital subscriber-line (ADSL) systems (see Section 15.3.2) have led to the development of *Discrete Multitone* (DMT) modulation.[34] In DMT, the full transmission capability of the subscriber line is used by probing it for its frequency response and noise versus frequency, and then sending as many as 256 carriers or "tones", each occupying a 4-kHz portion of the line's spectrum. The duration of each tone (symbol) is 250 μs, and each tone carries from 0 to 11 bits. (A zero-bit carrier simply is not present.) Limited bandwidth lines use more bits/symbol, up to 11.

Those portions of the spectrum that are noisy, or whose attenuation is high, simply are not transmitted. DMT thus optimizes the performance of each subchannel by allocating the data based on the subchannel signal-to-noise ratios. Forward error correction (see Section 6.3) is used, and echo cancellation (see Section 5.6) probably will be added. In addition, use may be made of *trellis coding*, in which added error-correction bits are used to select the sequence of signal-state points in a digital modulation system (see Section 6.2).

A digital stream rate of nearly 7 Mb/s has been demonstrated over line lengths of 3658 meters, with rates in excess of 1.544 Mb/s to 5486 meters. Application details are presented in Section 15.3.2.4.

5.9.1.4. Radio Systems

Numerous factors are contributing to the rapid increase in the use of radio to replace the wired local loop in given situations, either partially or entirely. Many present systems operate on an analog basis, but digital implementations are gaining in usage. Partial replacement of the wired local loop is seen in the popularity of the cordless analog end instrument. Cordless systems to date have been wired to the user's location, usually on a 2-wire local-loop basis. The end instrument base unit contains a low-power FM transmitter and receiver, as does the handset. Full duplex operation is achieved through the use of frequency pairs in the 46- and 49-MHz bands, or the 49- and 1.7-MHz bands in older systems, with separate frequency pairs being available to minimize interference between nearby cordless systems. The power levels used allow operation to distances of about 250 m between the handset and the cradle. When placed in the cradle, the handset's self-contained battery is recharged. Extension telephones are implemented as part of the wired portion of the system. Radio common-carrier systems use wired local loops from the common carrier's switching exchange to a mobile base station which covers an entire community, often on a competitive basis, and serves numerous mobile units, most of which are installed in automobiles, but some of which are handheld. Existing systems use analog FM modulation, but work is being done toward the use of digital modulation in such systems[35] as described in detail in Chapter 7.

Cellular radio systems, like the radio common-carrier systems, have been using analog FM modulation in providing service to both mobile and handheld units, but now are changing to digital voice transmission. Because cellular systems make extensive use of space-diversity reception, optimum received-signal-combining techniques using PSK signals have been studied, looking toward the implementation of digital cellular radio systems.[36] Optimum combining is important not only in handling fading problems, but also those of cochannel interference as it occurs in cellular radio systems.

Other uses of radio transmission to the end instrument are found in the single-channel-per-carrier (SCPC) systems used in satellite transmission, as discussed in Section 9.3.4.1, and the digital termination (bypass) systems discussed in Section 8.5.2.1.

The remaining discussion of this section deals with two-wire transmission to digital end instruments.

5.9.2. Line Codes

The line code selected for the user interface for North American systems is called the 2B1Q code. Each two bits constitute one quaternary (1Q) level. Figure 5.22[37] illustrates this line code (a), along with the interface designations in the ISDN architecture (b) for the basic rate interface (BRI). A detailed discussion of ISDN appears in Chapter 14.

5.9.3. Digital End Instruments

A fully digital telephone system is one that is completely digital all the way to the end instrument. This means that speech is encoded in PCM, ADPCM, or some other form before it leaves the calling subscriber's set and, likewise, arrives at the called subscriber's set in digital form, to be decoded there. The technical feasibility of such a set is not new, but its economic feasibility has come into focus only with the appreciable cost reductions that have occurred with advances in very-large-scale integrated circuits (VLSI). The ease with which other subscriber digital services can be integrated into such an all-digital telephone helps to make it even more attractive.

Digital end instruments producing 2.4 kb/s LPC are in use for secure communication purposes. Four such instruments can be time-division-multiplexed to produce a 9.6 kb/s stream, which can be handled by a modem on a single conditioned voice-bandwidth line. Alternatively, one or more such end instruments can operate in locations where 2.4 kb/s digital lines are available. The widespread appearance of digital end instruments is occurring in business systems, in which the private branch exchange (PBX) and the end instruments all are provided by a single supplier. Elsewhere, the change to digital end instruments is proceeding more slowly because of economic reasons, especially the availability of digital subscriber lines. Simplification and uniformity in the design of the subscriber set are paramount factors in saving overall system cost. To this end, the central switch is given the processing functions, as well as the automatic testing of the end instruments attached to it.

An important feature of any end instrument is *sidetone*, which gives the speaker a replica of his own voice in his receiver. Sidetone can be achieved by receiving and decoding the instrument's own transmitted words at a reduced level.

Power consumption and packaging considerations also play an important role in the design of digital end instruments. Transducers made of *polyvinylidene fluoride* (PVDF) can be used for both transmitters and receivers. PVDF is a plastic piezoelectric material. For telephony use it is aluminum-coated on both sides. The

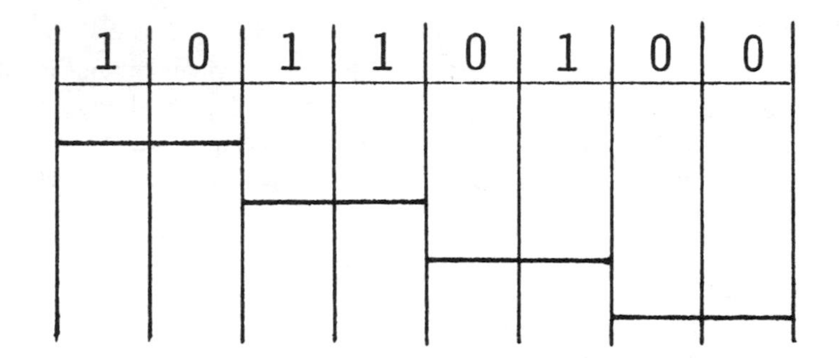

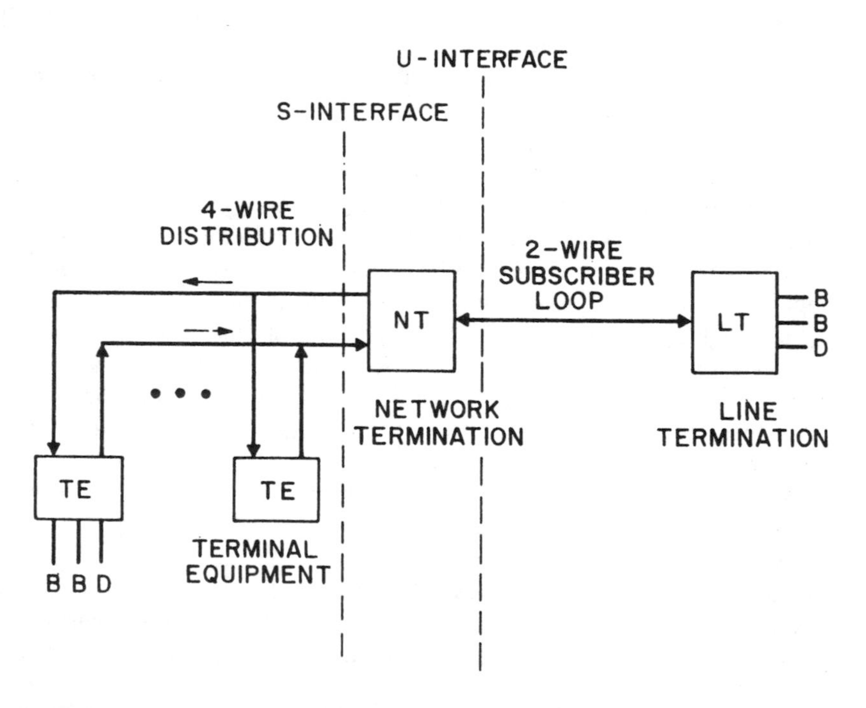

FIGURE 5.22. 2B1Q Line Code and ISDN Interface Designations (Courtesy Lin and Tzeng, Ref. 37, © IEEE, 1988.)

transducer functions as either a transmitter or a receiver, depending on how its associated amplifier is connected. Excellent frequency response is available, well beyond telephony demands. Its impedance is on the order of 100 ohms at voice frequencies. CMOS voltage amplifiers can be used for an overall low-power-drain instrument. Monolithic codecs, filters and LSI logic all help to minimize the parts count of the instrument, and thus to provide high reliability.

5.10. MONITORING AND MAINTENANCE

To maintain service quality, a terminal is removed from service if a failure occurs, or if the quantizing noise exceeds the maintenance limit. The maintenance limit and the time allowed for restoration of service depend on the required grade of service as well as the number of message channels affected. Terminals for 672 channels (DS3) or more must be restored rapidly, thus requiring automatic switching to spares, whereas smaller terminals may be restored by manual connection to spares or by repair.

Received framing information shows how the clock circuits are performing at each end of the circuit, as well as the performance of the digital facility. Frequent misframing initiates alarms at the terminal so that proper action can be taken. Thus, some maintenance features and functions are incorporated within the transmission systems, whereas others are provided by external arrangements that may include data recording and display, as well as operational features.

A *red alarm* (red light and audible alarm, usually appearing at a video display terminal) is initiated when a loss of signal or framing occurs in one direction. This alarm is triggered by a circuit in the receiving channel bank that is affected. Along with the initiation of this alarm, the associated transmitting channel bank is alerted. The result is the initiation of the *yellow alarm* at the transmitting bank. The existence of such alarms may start certain trunk processing functions. For example, those busy network trunks that are involved in the failure are removed from service until repairs can be completed. A resupply (*blue*) or *alarm indication signal* (AIS) is substituted for a failed signal. It satisfies the line format at the bit rate at which it is inserted, but does not carry message or framing information for lower levels in the digital hierarchy. Its purpose is to prevent or minimize protection switching or the initiation of alarms, especially on downstream multiplex equipment and channel banks. The monitoring of repeatered lines is done from offices at the ends or along the routes of the lines. The signal is examined for code format violations, such as bipolar violations. In addition, violations of successive zeros restrictions and loss of signal also are monitored. Defective repeaters are identified by a built-in fault location system.

A line monitor[38] is used on DS1 lines to perform pulse-quality measurements, and to sense bipolar violations, as well as the presence of more than 15 consecutive zeros. The redundancies inherent in partial response coding also can be used in performance monitoring, as described in Section 6.2.5. Excessive violations cause an office alarm to sound, and may initiate protection-line switching. A fault-location system identifies which repeater section is causing trouble. To accomplish fault location, the line under test must be removed from service. A special test signal then is sent from a fault-location test set.[39] Other DS1 line tests include the measurement of transmission pair losses and tests of repeater performance.

Lines operating at the DS1C and the DS2 rates use fault-location systems similar to those used for DS1 lines. Portable battery-operated monitors provide measurements at span terminating frames for DS2 lines, as well as at various access points along the lines, including the output of each regenerator.

5.11. SPEECH RECOGNITION

The remainder of this chapter (Sections 5.11 and 5.12) is devoted to several special techniques which are facilitated by having speech in digital form. Their implementation thus can be handled readily within an overall digital telephone system. Significant application areas for these techniques are expected to develop in the integrated voice-data networks of the ISDN (see Chapter 14). The special techniques are known, respectively, as speech recognition, voice input, and voice response.

5.11.1. Speech Production

Speech recognition by a piece of digital equipment requires that speech be treated in accordance with the nature of speech production. Several significant definitions follow:

diphthong A gliding monosyllabic speech item that starts at or near the articulatory position for one vowel and moves to or toward the position for another. (Examples: bay, boat, buy, how, boy, you.)

formant A frequency at which the vocal tract tube resonates.

phoneme One of a set of distinctive sounds (vowels, diphthongs, semivowels, consonants).

plosive sound A sound resulting from making a complete closure (usually toward the mouth end), building up pressure behind the closure, and abruptly releasing it (Example: tsh, j).

semivowel A sound characterized by a gliding transition in vocal tract area function between adjacent phonemes (Examples: w, l, r, y).

A human sound source provides an acoustic carrier, known as the pitch, for the speech intelligence, which appears largely as time variations in the envelope of the signal produced. Sounds can be characterized as *voiced sounds, fricatives,* and *stops.*

A voiced sound is produced by forcing air through the glottis with the tension of the vocal chords adjusted so that they vibrate in a relaxation oscillation, thereby producing quasiperiodic pulses of air which excite the vocal tract (Examples: u, d, w, i, e). A voiced sound is made up of a nearly periodic sequence of pulses. Its network equivalent is a current source exciting a linear, passive, slowly time-varying network. Its spectrum falls off with frequency at a rate of 12 dB/octave.

A fricative is an unvoiced sound, generated by producing a constriction at some point in the vocal tract (usually toward the mouth end), and forcing air through the constriction at a high enough velocity to produce turbulence (Example: sh). The result is a sustained random sound pressure. Its network equivalent is a series voltage source whose internal impedance essentially is that of the constriction and typically is of large value. Its spectrum is a broad band with gentle attenuation at the band edges.

A stop is a transient sound which may be either voiced (Examples: b, d, g) or unvoiced (Examples: p, t, k). A stop results from an abrupt release of air pressure built up behind a complete occlusion. Its network equivalent is a step function. Its spectrum falls off with frequency f as $1/f$.

5.11.2. Speech-Recognition Systems

Speech recognition systems are of two levels of sophistication. The simpler ones are capable of *isolated word recognition* (IWR), while the more complex ones can achieve *continuous word recognition* (CWR). Applications for isolated word recognition include data entry (via speakerphone) under conditions such that the use of a keyboard is impractical since the user's hands are otherwise occupied. Other (speakerphone) applications include the programming of numerically controlled machine tools, with the user's hands and eyes busy with blueprints, or voice commands to industrial robots where shop floor conditions make the use of buttons difficult or impossible. Continuous word recognition refers to the transcription of any spoken sentence from any speaker. It has the potential for eliminating many clerical tasks as well as improving man's speed of communication through the use of programming as well as data entry. Laboratory work has resulted in the recognition of artificial, limited vocabulary languages with predictable structures for ease of recognition.

One characteristic of a language significant to CWR is the *perplexity* of the language. The perplexity is the average number of words which theoretically could follow a given word. For natural language, the perplexity is greater than 50 words.

5.11.3. Speech-Recognition Techniques

Isolated word recognition (IWR) systems have been devised by splitting each word into a number of equal time segments, for example, 16. Each segment then is analyzed for a number of spectral components, for example, 32. These components are processed to extract a set of key features. The resulting matrix (16 $\times$ 32 using the example numbers) then is matched against a number of "word templates" stored in memory. Figure 5.23 illustrates this sequence of operations in block diagram form.

Continuous word recognition (CWR) involves the use of a technique called *nonlinear time warping*. This technique compensates for the variations in time used by various speakers to pronounce different words. Figure 5.24 illustrates time warping and shows that it amounts to a variable delay implemented by the computer as it expands or contracts the lengths of various parts of words to improve their match with a given word template. The computer tries each of a

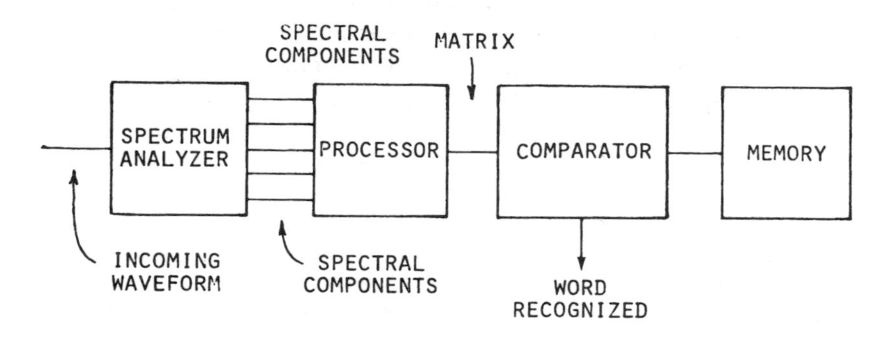

FIGURE 5.23. Isolated word recognition system.

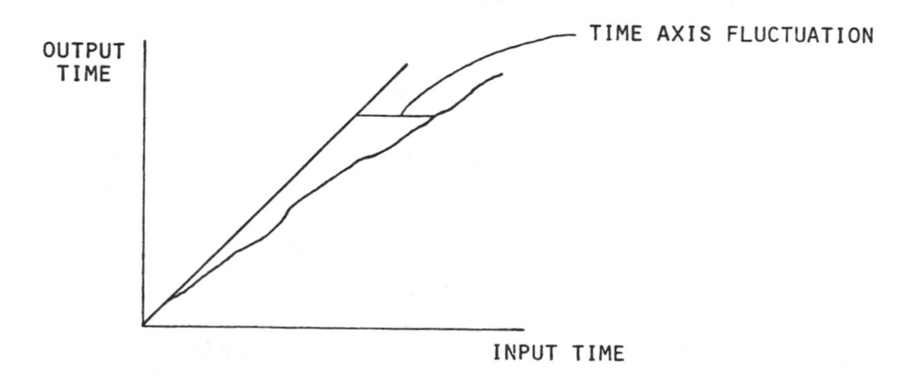

FIGURE 5.24. Time warping for continuous word recognition.

number of path slopes and calculates the cumulative difference between the template and the signal to that point.

5.11.4. Voice-Input Systems

Word recognition systems are used in providing voice input to computers. In many cases this voice input can best be transmitted to the computer via a digital subscriber line, as described in Section 5.9. Numerous applications exist in manufacturing, and include quality control and inspection, automated material handling, and parts programming for numerically controlled machine tools. In many cases, the reason for using voice input is that the inspector's hands and eyes are occupied in the inspection task. Applications include television faceplate inspection, automobile assembly-line inspection, and various types of receiving inspections in manufacturing plants and service facilities. In the case of parts programming for numerically controlled machine tools, the programmer speaks each machine command into a speakerphone or headset. The system eliminates the need for a separate conversion of the information into computer-compatible format. The programmer's hands remain free to handle prints or perform calculations.

5.12. VOICE RESPONSE

The general principle underlying voice response is illustrated in Fig. 5.25. Words or phrases are stored in a read-only memory called the vocabulary. Messages then

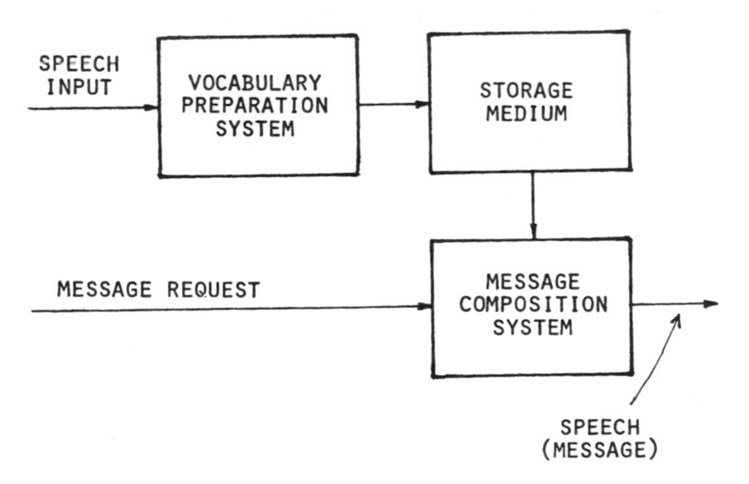

FIGURE 5.25. Voice response system.

are created by retrieving the required words and phrases and reproducing them in the proper sequence. The choice of the digital coding method has a major impact on the amount and type of digital memory required, and therefore on its cost. Thus LPC at 2.4 kb/s requires less than 4% of the memory required by 64-kb/s PCM. While LPC may not be satisfactory from a quality viewpoint, CELP at 9.6 kb/s or CVSD at 16 kb/s may be quite cost-effective while allowing satisfactory voice quality as well.

Voice response requires the preparation of a suitable vocabulary by a trained speaker, its digitization and insertion into memory, and the establishment of a directory of word addresses. The absence of an input is used to establish the start

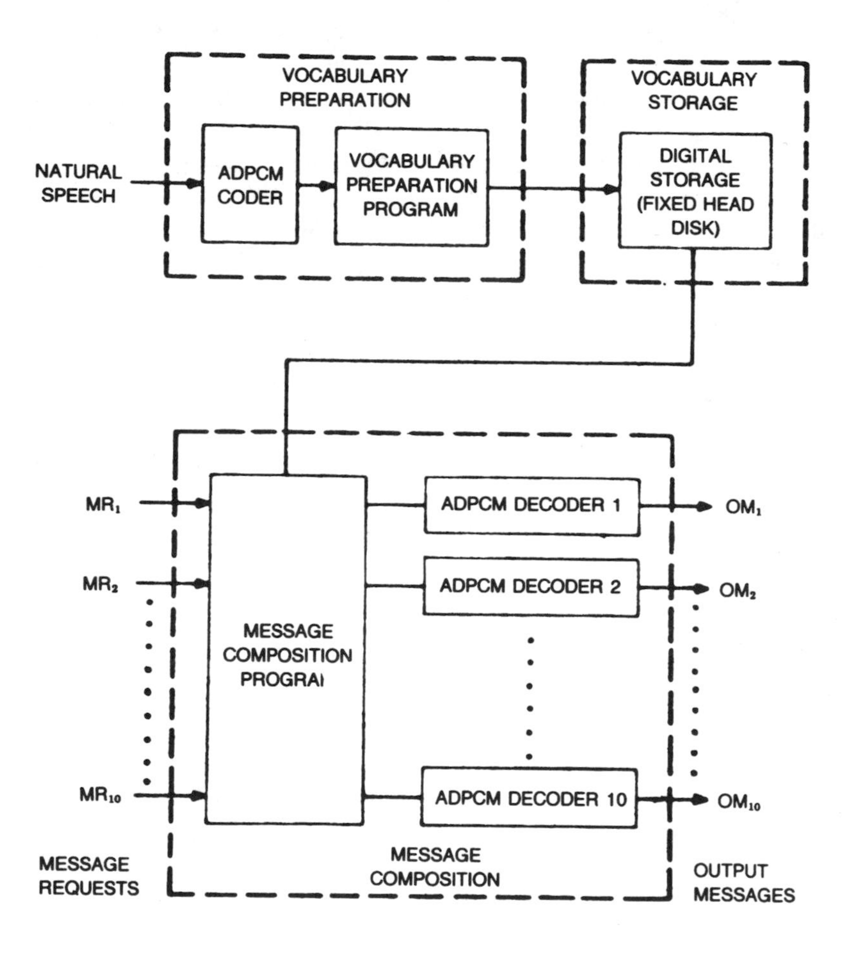

FIGURE 5.26. Ten-line digital voice response system using ADPCM. (Courtesy L. H. Rosenthal et al., Ref. 40, © IEEE, 1974.)

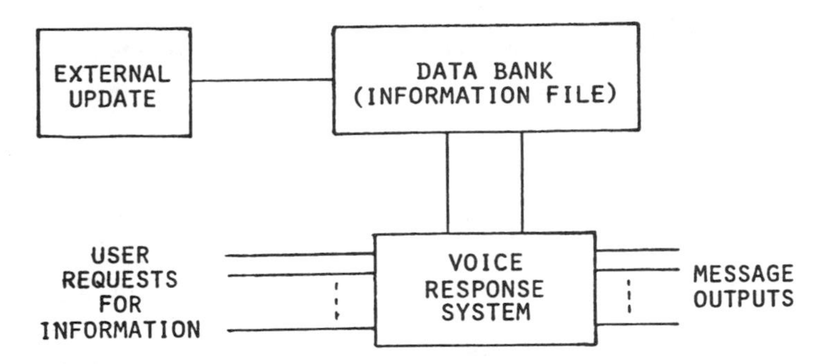

FIGURE 5.27. Voice response system for information retrieval. (Courtesy L. R. Rabiner and R. W. Schafer, Ref. 41, © IEEE, 1976.)

and end of each word. Message composition then involves logical operations followed by the proper data transfers. A single processor, with different programs, can be used for both vocabulary preparation and message composition.

Figure 5.26 shows a voice response system capable of handling ten lines simultaneously.[40] The vocabulary preparation portion is used only initially, or when changes must be made in the vocabulary. A variety of applications exists for systems such as this one. They include computer-aided voice wiring and information retrieval systems. In the case of computer-aided wiring, a wire list is recorded and the wireman actuates the system by a foot switch. He then hears the wiring instructions, one at a time, through a speakerphone. Usually a relatively small vocabulary and frequent modifications make computer design of the list desirable.

Information-retrieval systems using voice response generally operate on the principle illustrated[41] in Fig. 5.27. A data bank is updated from an external source. This data bank then is searched by the voice-response system to provide outputs as requested. Information-retrieval systems include directory assistance systems, in which the user can key in requests on a telephone equipped for dual-tone multifrequency (DTMF) calling. (DTMF is discussed in Section A.4.2.2.) Other applications include a stock-price quotation system and a flight-information system. Enroute weather information for flyers also can be provided. A data-set testing system is programmed so that, in response to an installer's request, a computer applies a custom test signal to determine if the data set, and all its options, are performing correctly. In addition, the computer advises the installer about the expected causes of any trouble.

In a speaker verification system, the user enters his claimed identity, speaks a verification phrase, and requests a transaction. A processor then performs a feature analysis on the voice, based on the verification phrase, compares the result with recorded reference data on the speaker's voice, and determines whether to accept or reject the speaker's request.

PROBLEMS

5.1. A circuit experiences a 10-μs noise burst which causes loss of synchronization. A subsequent portion of the bit stream is lost. If this loss is 1 ms long, what will be the effect on voice communication? on data communication?

5.2. The following stream is to be scrambled: 0 1 1 0 0 0 1 0. Compose three arbitrary pseudorandom streams. In each case, what is the transmitted stream? Assuming no channel errors, derive the recovered stream at the receiver's output.

5.3. Why is two-wire service important in digital subscriber loops?

5.4. Why can digital end instruments not be paralleled directly for extension service as is the case with analog end instruments?

5.5. Under what conditions does channel error coding become very important?

5.6. Why do DSI systems generally require a large number of trunks (e.g., 50 or more) to be effective?

REFERENCES

1. "The Extended Framing Format Interface Specifications," Technical Advisory No. 70, Issue 2, American Telephone and Telegraph Company, Basking Ridge, NJ, Sept. 29, 1981.
2. Aaron, M. R., "PCM Transmission in the Exchange Plant," *Bell Syst. Tech. J.*, Vol. 41, Jan. 1962, pp. 99–141.
3. *Telecommunications Transmission Engineering*, Vol. 2, *Facilities*, Bell System Center for Technical Education, 1977.
4. Haury, P. T. and Romeiser, M. B., "T1 Goes Rural," *Bell Laboratories Record*, Vol. 54, July/Aug., 1976, pp. 178–183.
5. See note 4.
6. Anderson, B. C., "Testing Long Haul Carrier Systems Automatically," *Bell Laboratories Record*, Vol. 54, (July/Aug 1990), pp. 212–216.
7. Booth, T. L., *Digital Networks and Computer Systems*, John Wiley & Sons, Inc., New York, 1971, pp. 41–52.
8. Moore, J. D., "M12 Multiplex," *1973 Conference Record, IEEE Int. Conf. on Comm.*, vol. 1, pp. 22–20 to 22–25.
9. Scheuermann, H. and Göckler, H., "A Comprehensive Survey of Digital Transmultiplexing Methods," *Proc. IEEE*, Vol. 69, No. 11, Nov. 1981, pp. 1419–1450.
10. Freeny, S. L., "TDM/FDM Translation as an Application of Digital Signal Processing," *IEEE Communications Magazine*, Vol. 18, No. 1, Jan. 1980, pp. 5–15.
11. Special Issue on Transmultiplexers, *IEEE Trans. Comm.*, Vol. COM-30, No. 7, Part 1, July, 1982.

12. See note 10.
13. *Telecommunications Transmission Engineering*, Vol. 2, *Facilities*, Bell System Center for Technical Education, Western Electric Company, Winston-Salem, NC, 1977, p. 168.
14. Bellamy, J. C., *Digital Telephony*, John Wiley & Sons, New York, NY, 1982, pp. 337–343.
15. *Telecommunications Transmission Engineering*, Vol. 2, *Facilities*, Bell System Center for Technical Education, Western Electric Company, Winston-Salem, NC, 1977, p. 559.
16. Suyderhoud, H. G., Onufry, M., and Campanella, S. J., "Echo Control in Telephone Communications," *1976 National Telecommunications Conference Record*, pp. 8.1-1 to 8.1-5.
17. Hatch, R. W. and Ruppel, A. E., "New Rules for Echo Suppressors in the DDD Network," *Bell Laboratories Record*, Vol. 52, pp. 351–357, (December, 1974).
18. Helder, G. K. and Lopiparo, P. C., "Improving Transmission in Domestic Satellite Circuits," *Bell Laboratories Record*, Vol. 55, September, 1977, pp. 202–207.
19. Duttweiler, D. L. and Chen, Y. S., "A Single-Chip VLSI Echo Canceler," *Bell Syst. Tech. J.*, Vol. 59, No. 2, Feb. 1980, pp. 149–160.
20. Messerschmitt, D. G., "Echo Cancellation in Speech and Data Transmission," *IEEE Journal on Selected Areas in Communications*, Vol. SAC-2, No. 2, March, 1984, pp. 283–297.
21. Brady, P. T., "A Statistical Analysis of On-Off Patterns in 16 Conversations," *Bell Syst. Tech. J.*, Vol. 47, No. 1, Jan. 1968, pp. 73–91.
22. Easton, R. L., et al., "TASI-E Communications System," *ICC '81 Conference Record*, IEEE Document No. 81 CH1648-5, pp. 49.3.1–49.3.5.
23. See note 22.
24. Reiser, J. H., Suyderhoud, H. G., and Yatsuzuka, Y., "Design Considerations for Digital Speech Interpolation," *ICC '81 Conference Record*, IEEE Document No. 81 CH1648-5, pp. 49.4.1–49.4.7.
25. Verma, S. N., Ramasastry, J., and Monsees, W. R., "Digital Speech Interpolation Applications for Domestic Satellite Communications," *NTC '78 Conference Record*, IEEE Document No. 78 CH1354-0 CSCB, pp. 14.4.1–14.4.5.
26. *Telecommunications Transmission Engineering*, Vol. 2, *Facilities*, Bell System Center for Technical Education, Western Electric Company, Inc., Winston-Salem, NC, 1977, pp. 542–546 and 605.
27. *Transmission Systems for Communications*, Bell Telephone Laboratories, Inc., prepared for publication by Western Electric Company, Inc., Technical Publications, Winston-Salem, NC, 1971, pp. 627–631.
28. Ahamed, S. V., Bohn, P. P., and Gottfried, N. L., "A Tutorial on Two-Wire Digital Transmission in the Loop Plant," *IEEE Trans. Comm.*, Vol. COM-29, No. 11, Nov. 1981, pp. 1554–1564.
29. Abraham, L. G. and Fellows, D. M., "A Digital Telephone with Extensions," *IEEE Trans. Comm.*, Vol. COM-29, No. 11, Nov. 1981, pp. 1602–1608.
30. Brosio, A., et al., "A Comparison of Digital Subscriber Line Transmission Systems Employing Different Line Codes," *IEEE Trans. Comm.*, Vol. COM-29, No. 11, Nov. 1981, pp. 1581–1588.

31. See note 28.
32. *Telecommunications Transmission Engineering*, Vol. 2, Facilities, Bell System Center for Technical Education, Western Electric Company, Inc., Winston-Salem, NC, 1977, pp. 85–88.
33. Brolin, S., Cho, Y. S., Michaud, W. P., and Williamson, D. E., "Inside the New Digital Subscriber Loop System," *Bell Laboratories Record*, Vol. 58, April 1980, pp. 110–116.
34. Cioffi, J. M. and Bingham, J. A. C., "DMT Specification Overview for ADSL," In *ANSI T1E1.4 Committee Contribution No. T1E1.4/93/083*, Chicago, April, 1993.
35. Feher, K., *Digital Communications: Microwave Applications*, Prentice-Hall, Englewood Cliffs, NJ, 1981.
36. Winters, J., "Optimum Combining in Digital Mobile Radio with Co-Channel Interference," *ICC '83 Conference Record*, IEEE Document No. 83 CH1874-7, pp. B8.4.1–B8.4.5.
37. Lin, N.-S. and Tzeng, C.-P.J., "Full-Duplex Data Over Local Loops," *IEEE Communications Magazine*, Vol. 26, No. 2, Feb. 1988, p. 32.
38. Blair, R. W. and Burnell, R. S., "Monitors Take the Pulse of T1 Transmission Lines," Bell Laboratories Record, Vol. 51, Feb. 1973, pp. 55–60.
39. Lender, A., "Correlative Digital Communication Techniques," *IEEE Trans. on Comm. Technol.*, Dec. 1964, pp. 128–135.
40. Rosenthal, L. H., et al. "A Multiline Computer Voice Response System Utilizing ADPCM Coded Speech," *IEEE Trans. Acoustics, Speech, and Signal Proc.*, Vol. ASSP-22, No. 5, Oct. 1974, pp. 339–352.
41. Rabiner, L. R. and Schafer, R. W., "Digital Techniques for Computer Voice Response: Implementations and Applications," *Proc. IEEE*, Vol. 64, No. 4, pp. 416–433, (April, 1976).

6

Digital Transmission

6.1. INTRODUCTION

The objective of digital transmission, as applied to telephony, is to convey digital voice and nonvoice signals at sufficiently low bit-error ratios (b.e.r.) to assure satisfactory performance from the users' viewpoints. The bit error ratio is defined as the ratio of erroneous bits in a bit sequence to the total number of bits in that sequence.

6.1.1. Performance Objectives

For voice, the bit error ratio generally should not exceed 10^{-4}, since 64-kb/s PCM performance is quite good if the b.e.r. is kept to this value or lower. Other voice coding techniques, such as ADPCM, can tolerate higher b.e.r. values, as was described in Chapters 3 and 4. The transmission of signaling and supervision data, however, generally requires lower b.e.r. values, achievable with error correction techniques, to prevent wrong numbers, false alarms, and other undesirable occurrences.

6.1.2. Signal Impairments in Transmission

Bit errors result from signal impairments during transmission. Such impairments can be categorized in a variety of ways. One of these ways is shown in Fig. 6.1. Note that the vertical dimension on the figure denotes *level.* The horizontal dimension denotes bandwidth. As used here, the term level of a signal denotes its modulating effect on a sine-wave carrier, and thus, may be an amplitude, a frequency difference, or a phase shift. Too large a signal level results in *intermodulation* (IM) as repeater characteristics are pressed to their limits. Intermodulation

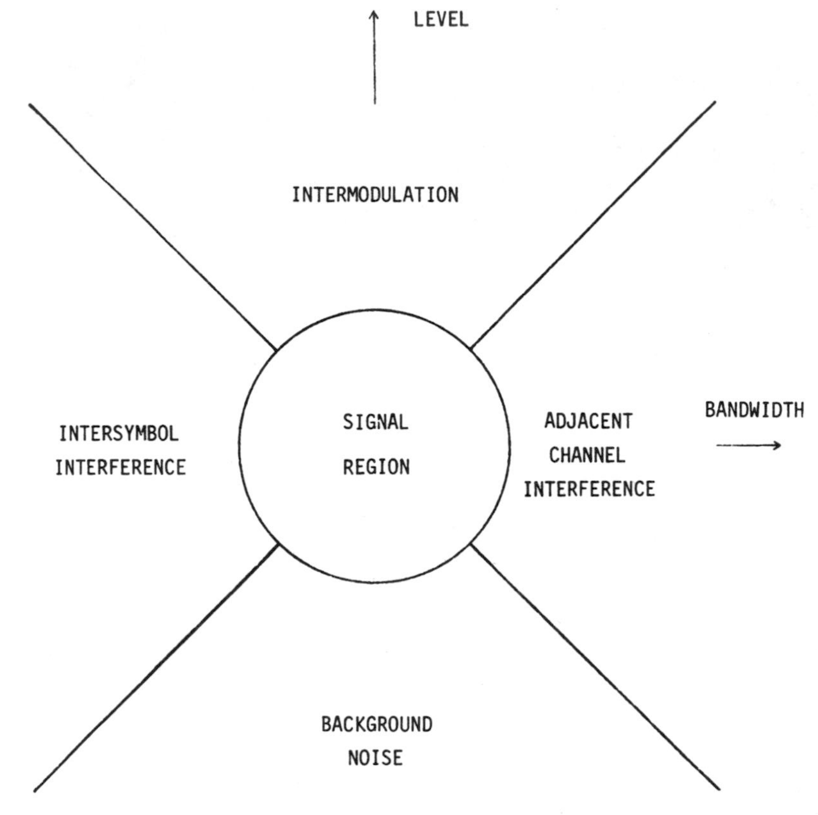

FIGURE 6.1. Categories of transmission impairments.

consists of the production of unwanted frequency components as a result of a nonlinearity within the system. Accordingly, it can produce interference between digital and analog transmissions that use different portions of the spectrum on a given facility. At low signal levels, background noise causes impairments because it alters the amplitudes as well as the zero crossing times of the signals. Background noise includes thermal and impulse noise as well as crosstalk (*co-channel interference*, abbreviated CCI). Looking next at the bandwidth axis of Fig. 6.1, *adjacent-channel interference* (ACI) and *intersymbol interference* (ISI) occur on facilities whose resources are shared on a time- or frequency-division basis. Attempts to place transmissions too close to one another in the frequency domain result in adjacent-channel interference if channel passbands are large enough to encompass some of the adjacent channel. Conversely, if relatively narrow filtering is done, intersymbol interference is the result. Intersymbol interference is a response in a given time slot produced by a symbol in another slot, usually the preceding one. Intersymbol interference also can result from echoes (see Section 5.6).

All forms of impairment result in bit errors. Bit-error ratios as low as 10^{-7} can disturb the accuracy with which signaling and supervision are accomplished. A bit-error ratio in excess of 10^{-4} can degrade PCM voice, while bit-error ratios in excess of 10^{-3} may degrade other forms of digital voice.

6.1.3. The Nyquist Theorem

Digital transmission results in rapid changes from one level to another. Such changes constitute pulses in the time domain. The response of circuits and systems to such pulses, however, usually is described in terms of the frequency domain, which is the complex amplitude (magnitude and phase) versus frequency characteristic of an electrical signal. The frequency spectrum is defined by the Fourier transform of the pulse. For a pulse of duration T_s centered at $t = 0$, and of magnitude A volts, the Fourier spectrum is

$$A(\omega) = \left(\frac{AT_s}{2\pi}\right)\left[\frac{\sin\left(\frac{\omega T_s}{2}\right)}{\left(\frac{\omega T_s}{2}\right)}\right] \tag{6.1}$$

This function exhibits spectral energy to infinite frequency, so for transmission, the spectrum must be limited. Usually, transmission of the components up to $\omega_s = 2\pi/T_s$ provides a sufficiently good representation of the pulse that it can be detected correctly.

The effect of band limiting has been described by the Nyquist theorems,[1] which specify the minimum channel bandwidth required to send a given *symbol* rate without intersymbol interference. (A symbol is one or more bits; each symbol is conveyed as a specific level in a digital transmission system.) Nyquist states that synchronous impulses can be sent at a rate of $2W$ per second through a channel whose bandwidth is W hertz, and that these impulses can be received without intersymbol interference. Note that the use of *impulses* is specified. The time domain response of a network to an impulse is called its *impulse response*. An impulse has an infinite amplitude but zero duration; its energy is defined to be finite. From a practical viewpoint, however, the impulse is viewed as having a duration that is small compared with a cycle at the highest frequency under consideration. Moreover, Nyquist's "channel whose bandwidth is W hertz" has a flat response and linear phase shift within the passband, but its response drops to zero at the edge of the passband. The term *brick-wall channel* often is used to describe such a response. Figure 6.2[2] illustrates the characteristics of the brick wall channel. Its impulse response, $h(t)$, is the inverse Fourier transform of its transfer function, $H(f)$, where

$$H(f) = \begin{cases} T_s, \ |f| \leq \dfrac{1}{2T_s} \\ 0, \ |f| > \dfrac{1}{2T_s} \end{cases} \tag{6.2}$$

$$h(t) = \int_{-\infty}^{\infty} H(f) \exp(j2\pi f T_s) df \tag{6.3a}$$

$$= \frac{\left[\sin\left(\dfrac{\pi t}{T_s}\right)\right]}{\left(\dfrac{\pi t}{T_s}\right)} \tag{6.3b}$$

The phase shift through $H(f)$ is zero for all frequencies.

Note that Eq. 6.3b describes the response to an impulse, Fig. 6.2(a), of a brick-wall channel, Fig. 6.2(b). This response, plotted in Fig. 6.2c, calls for an output before the input impulse occurs at $t = 0$! For this reason, plus the infinitely steep drop at $f = 1/2T_s$, as shown in Fig. 6.2(d), the channel cannot be built. It can be

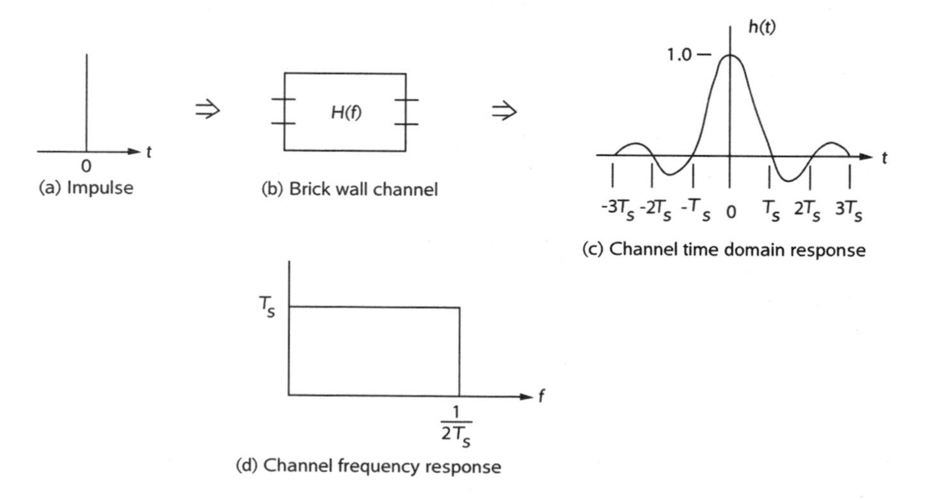

FIGURE 6.2. Characteristics of brick wall channel (Ref. 2).

closely approximated, however, by allowing a smooth roll-off at the band edge in place of the brick wall characteristic. Those filters that are advertised as "brick wall" actually have a very small value of α; not zero, however. The smooth roll-off also eliminates the high degree of sensitivity of ISI to filter bandwidth and symbol rate. The frequency domain response to the impulse, Fig. 6.2d, now becomes that of Fig. 6.3a,[3] while the response to the rectangular pulse becomes that of Fig. 6.3b.

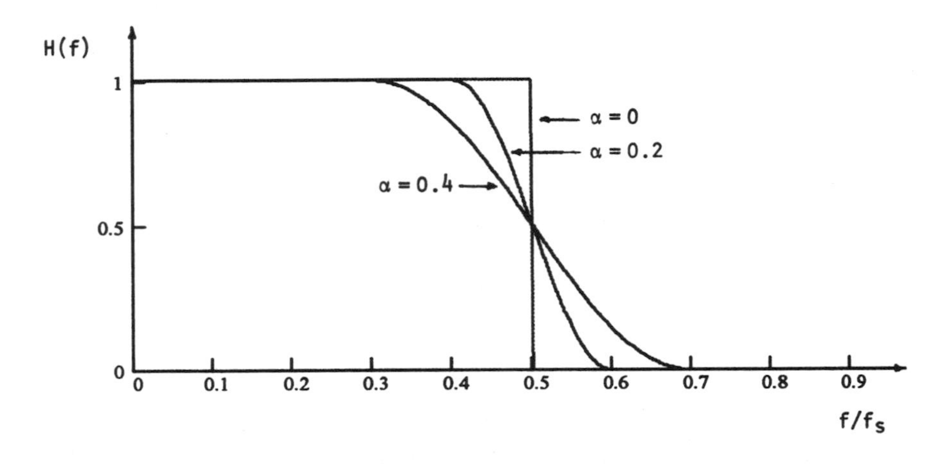

a

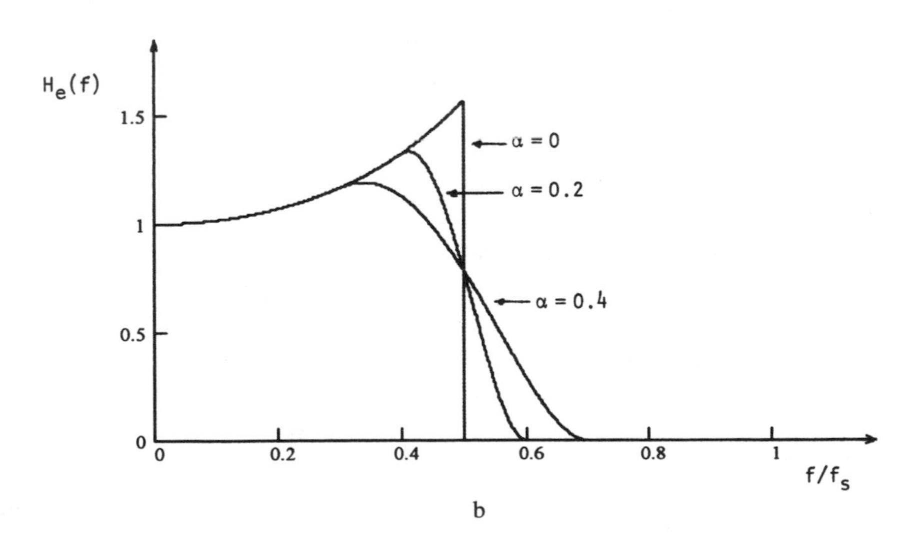

b

FIGURE 6.3. Amplitude versus frequency response of the Nyquist channel.

The equations expressing the channel characteristics for impulse transmission are:

$$H(f) = \begin{cases} 1 \text{ for } 0 \leqslant f \leqslant \left(\dfrac{f_s}{2}\right)(1 - \alpha) \\ \cos^2\left\{\left(\dfrac{1}{4\alpha f_s}\right)[2\pi f - \pi f_s(1 - \alpha)]\right\} \\ \text{for } \left(\dfrac{f_s}{2}\right)(1 - \alpha) \leqslant f \leqslant \left(\dfrac{f_s}{2}\right)(1 + \alpha) \\ 0 \text{ for } f > \left(\dfrac{f_s}{2}\right)(1 + \alpha) \end{cases} \tag{6.4}$$

where

α = channel roll-off factor, a measure of the excess bandwidth built into the channel.
f_s = symbol rate.

This characteristic maintains the zero axis crossings of the impulse response, and thus provides transmission free of ISI. Its shape is that of the raised cosine function, $(1 + \cos x)/2$; accordingly, it is called the raised cosine spectrum. The phase response is linear with respect to frequency.[4]

For finite-width pulse transmission, the equations are:

$$H_e(f) = \begin{cases} \dfrac{\left(\pi \dfrac{f}{f_s}\right)}{\sin \pi \dfrac{f}{f_s}} \text{ for } 0 \leq f \leq \left(\dfrac{f_s}{2}(1 - \alpha)\right) \\ \left[\dfrac{\left(\pi \dfrac{f}{f_s}\right)}{\sin\left(\pi \dfrac{f}{f_s}\right)}\right] \cos^2\left\{\dfrac{1}{4\alpha f_s}[2\pi f - \pi f_s(1 - \alpha)]\right\} \\ \text{for } \left(\dfrac{f_s}{2}\right)(1 - \alpha) \leq f \leq \left(\dfrac{f_s}{2}\right)(1 + \alpha) \\ 0 \text{ for } f > \left(\dfrac{f_s}{2}\right)(1 + \alpha) \end{cases} \tag{6.5}$$

The maximum transmission rate is $2/(1 + \alpha)$ baud/hertz, where one baud is one symbol/second. Note that baud is the symbol rate, whereas bit/second is the information rate. Note also that if the data stream modulates a carrier such that two sidebands are produced (doubling the occupied spectrum), the maximum double-sideband transmission rate is $1/(1 + \alpha)$ baud/hertz.

Double sideband transmission is used in modems to permit the implementation of quadrature modulation techniques, with their improved noise immunity compared with the amplitude only techniques available with single-sideband transmission.

Fig. 6.4 shows a raised cosine spectrum using a logarithmic scale on the right side of the figure. Each horizontal increment of the spectrum corresponds to the symbol rate. The pulse is illustrated in the upper left of the figure. Here each horizontal increment is a symbol period. The lower left of the figure is an *eye pattern*. This is a pattern in which the horizontal axis is time, and spans a total of two bits, that is, is expanded by four times compared with the individual bit. The vertical axis is amplitude. An oscilloscope display of the eye pattern is useful in evaluating the quality of a received signal.

Although the pulse producing the raised cosine spectrum theoretically has infinite duration, it has negligible amplitude beyond $t = \pm 2$ symbols. Thus discarding the tails has little effect other than to create some very minor spectral

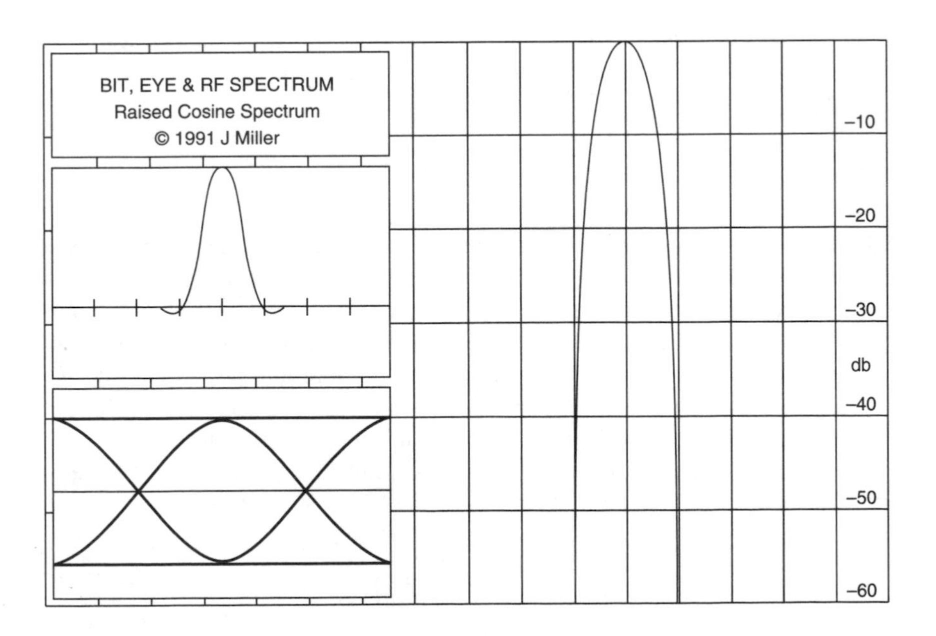

FIGURE 6.4. Raised cosine spectrum. (Courtesy J. Miller, Ref. 5, © J. Miller, 1991.)

sidelobes. In addition, the pulse of Fig. 6.4[5] has the necessary property for zero ISI, namely, a zero value at the instants reserved for the peaks of adjacent bits. This is a direct consequence of the symmetries of the cosine shape. Accordingly, the raised cosine shape is widely used in data transmission system design.

6.1.4. The Shannon Limit

An important aspect of digital transmission is the Shannon-Hartley Law, which states the limits on channel capacity C in terms of the signal-to-noise power ratio ($S/N = P/N_0W$) in the channel.[6] This theorem states that the upper limit on error-free transmission is:

$$C = W \log_2\left(1 + \left[\frac{P}{N_0 W}\right]\right) \text{ b/s} \tag{6.6}$$

where

P = received average signal power
N_0 = single-sided noise power spectral density
Thus N_0W = received average noise power

Note that the single-sided noise power spectral density is the total average noise power N divided by the receiver noise bandwidth which, for the Nyquist channel, is the Nyquist bandwidth, $1/2T_s$, as illustrated in Fig. 6.2(d). Thus

$$N_0 = 2NT_s \tag{6.7}$$

According to Shannon, up to C bits/second can be sent error free through such a channel. However, equipment becomes increasingly complex as the Shannon limit is approached. Shannon's theorem also quantifies some well known trends which are found in practice. The data rate that can be sent over a channel increases in proportion to the bandwidth of the channel. In addition, an increase in S/N allows an increase in data rate for a given bandwidth, although an increase in equipment complexity may be required to realize the higher data-rate.

The Nyquist theorem limit on symbol rate is $2/(1 + \alpha)$ baud/hertz in a memoryless (zero ISI) channel. The Shannon–Hartley Law states that

$$W \log_2\left(1 + \left[\frac{P}{N_0 W}\right]\right)$$

is the maximum number of bits per second that can be sent error-free in a band-

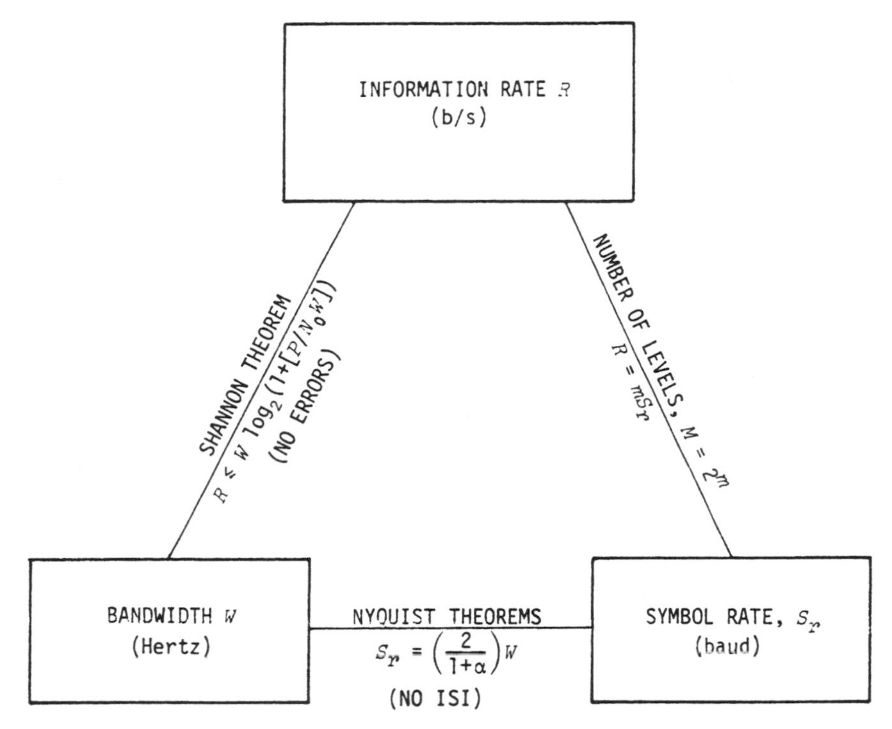

FIGURE 6.5. Relationships between information rate (b/s), symbol rate (baud) and bandwidth (Hz) (Ref. 2).

width W. The symbol rate (baud) and information rate (b/s) are related by the number of levels M used in the system, where $M = 2^m$; thus M is a power of 2. The parameter m thus becomes a multiplier on the spectral efficiency of the transmission. Figure 6.5 shows the interrelations among the foregoing factors.[7]

6.2. DIGITAL MODULATION TECHNIQUES

Modulation is defined as the alteration of a carrier in order to cause it to convey information. The characteristics of a carrier can be expressed in the form $A \sin(\omega t + \phi)$, where A = amplitude, ω = radian frequency, and ϕ = phase. Thus a carrier can be modulated in amplitude, frequency, or phase. Digital modulation refers to the use of a limited set of discrete values of A, ω, and ϕ. The type of digital modulation used affects the relationship between b.e.r. and S/N. In addition, the type of modulation affects the amount of spectrum occupied by a given transmission. Moreover, the equipment complexity and cost depend upon the modulation technique selected.

Modulation produces *sidebands*, that is, spectral components above and below the carrier frequency. In the case of amplitude modulation, the sidebands differ from the carrier by an amount equal to the modulating frequency. These two sidebands convey the same information, and are in phase with one another. As a result, one of them can be eliminated without altering the information content of the signal.

Modulation of the amplitude A and the phase ϕ of a carrier can be shown on a signal-state space diagram, as illustrated in Fig. 6.6. This is also called a constellation. The detector must determine which of the four allowable A, ϕ combinations have been sent at any given time. Since four states are allowed, each state can convey a unique combination of bits. For example, the following assignment might be made:

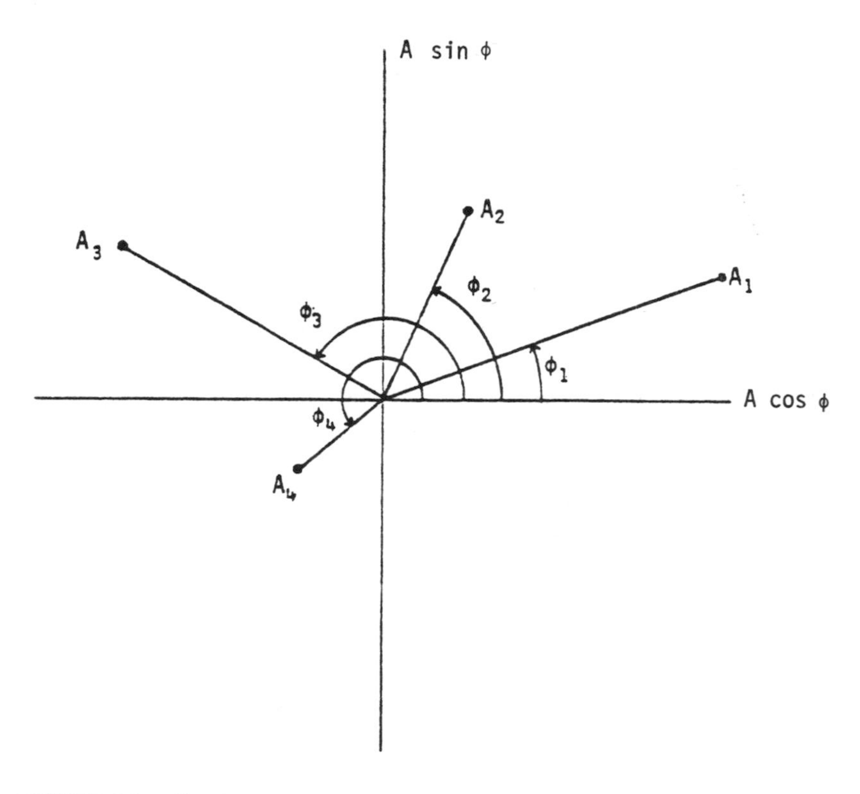

FIGURE 6.6. Signal-state space diagram for amplitude-phase modulation.

Signal State	Bits Conveyed
$A_1\phi_1$	00
$A_2\phi_2$	01
$A_3\phi_3$	10
$A_4\phi_4$	11

For pure phase modulation, $A_1 = A_2 = A_3 = A_4$, while for pure amplitude modulation, $\phi_1 = \phi_2 = \phi_3 = \phi_4$, but the A's may take discrete negative as well as discrete positive values. In the case of phase modulation, the power of the signal remains constant with modulation, unlike amplitude modulation, in which the power increases with modulation. As a result, as phase modulation sideband energy increases, the carrier magnitude decreases. In addition, the phases of the odd order sidebands (first, third, fifth, etc.) are opposite to one another. Both sidebands of a phase modulated signal must, therefore, be transmitted.

Modulation of the frequency of a carrier produces frequency-shift keying, which is discussed in Section 6.1.4. Both sidebands of a frequency-modulated signal must be transmitted for the preservation of its characteristics.

The amount of spectrum occupied by a digitally-modulated signal can be shown to depend on the type of modulation and coding used to produce that signal. Thus, a signal that is only amplitude modulated can be transmitted on a single sideband, or a *vestigial sideband* basis, while an angle- (frequency- or phase-) modulated signal must have both of its sidebands transmitted. Vestigial sideband transmission involves the full transmission of one sideband, usually the upper, while only a vestige of the other sideband is transmitted. An example of vestigial sideband transmission is found in the transmission of television to homes in most parts of the world.

The effect of coding on bandwidth is shown next. For example, Fig. 6.7 shows a digital sequence which is coded on a two-level basis and which thus undergoes changes every bit period. Alternatively, with coding on a four-level basis, a change occurs only every other bit period. The four-level coding is based on the following rules: $00 = -3; 01 = -1; 10 = +1; 11 = +3$. Four-level coding thus requires less frequent level changes and, correspondingly, occupies less spectrum.

Spectrum crowding has become a serious problem in the United States, and in other countries with large volumes of telecommunications traffic. This crowding has made the use of multilevel modulation techniques important.

In the modulation technique descriptions which follow, use is made of the ratio of the energy per bit, E_b, to the noise power spectral density ratio, N_0, which is the noise power in a 1-Hz bandwidth within the receiver's passband. This ratio can be related to the carrier-to-noise ratio, C/N, at the receiver's input in the following way.

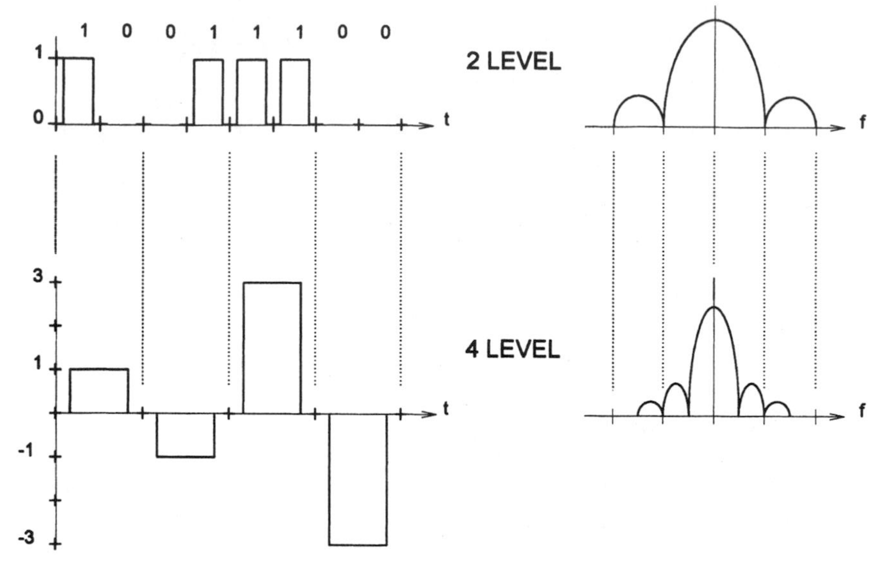

FIGURE 6.7. Use of coding in reducing spectrum occupancy.

Let C = average carrier power (watts) and let N = noise power (watts) in the receiver's bandwidth. If T_b = duration of a bit (seconds), then $E_b = CT_b$ (watt-seconds or joules). The bit rate (b/s) then is $f_b = 1/T_b$. Letting BW = receiver noise bandwidth (Hz), and noting that $N_0 = N/BW$,

$$\frac{E_b}{N_0} = \left[\frac{C}{N}\right]\left[\frac{BW}{f_b}\right] \tag{6.8}$$

Equation (6.8) expresses a numerical ratio. Most commonly, E_b/N_0 is expressed in decibel terms by taking

$$10 \log_{10}\left[\frac{C}{N}\right]\left[\frac{BW}{f_b}\right]$$

6.2.1. Phase-Shift Keying

Phase-shift keying involves the shifting of the carrier's phase among several discrete values. If only two values of phase, e.g., 0° and 180° are used, the result may be called *binary phase-shift keying* (BPSK). Each phase represents one bit and the coding is done at 1 b/s per baud. If four values of phase are used, for example, 45°, 135°, −135°, and −45°, the result is called *quaternary phase-shift keying*

(QPSK). Each phase can represent two bits, so the coding is done at 2 b/s per baud. Systems using 8-PSK (3 b/s per baud) and larger numbers of phases also are in use.

To detect phase changes, the receiver must have a phase reference. There are two ways of obtaining this reference, known, respectively, as the *differential* and the *coherent* methods. The detection efficiency of the differential forms of phase-shift keying is known to be about 1 dB below that of the coherent forms for two-level modulation, and approaches 3 dB for multilevel modulation. Accordingly, coherent detection is preferred in those applications in which small losses in signal-to-noise ratio are significant, as is the case in satellite transmission, especially the downlink (see Chapter 8). Coherent detection, however, requires the production and extraction of a local carrier phase reference at the receiver. Usually a coherent phase estimate is obtained through the use of *phase-locked loop* (PLL) techniques.[8] Because both the transmitter and receiver have inherent frequency instabilities and phase jitter, the bandwidth of the carrier recovery loop cannot be made arbitrarily small. As a result, a somewhat noisy phase estimate is obtained, and only partially coherent reception actually can be claimed.

Differential detection avoids the carrier-phase-recovery problem and thus is immune from slow carrier-phase fluctuations. Under noisy phase estimation conditions and significant intersymbol interference, the detection efficiency of differential detection may approach that of coherent detection. Excellent comparative treatments of differential and coherent detection have been provided by Prabhu and Salz,[9] and by Spilker.[10]

In the case of differentially detected phase-shift keying, the previous phase is used as the reference. Thus, for differential QPSK, the various bit pairs might be conveyed in accordance with the list and diagram of Fig. 6.8. While receivers using differential-phase detection are relatively simple to build, better performance can be achieved through the use of coherent receivers with built-in carrier-recovery circuits, as illustrated in Fig. 6.9. The improvement is equivalent to up to 2.5 dB in E_b/N_0, or to as much as two orders of magnitude in bit-error ratio. In Fig. 6.9, showing a coherent QPSK receiver, the bandpass filter minimizes interference and noise not actually in the band being received. The multipliers (balanced mixers) provide the function of coherent linear demodulation. The carrier-recovery circuit may be any of several types of circuits[11] designed with internal feedback loops to track carrier frequency or phase. A phase splitter then provides outputs at the carrier frequency that are separated by 90°. Harmonics of the carrier wave produced by the multipliers are attenuated by the low-pass filters.

Some modems, such as the General Telephone and Electronics (GTE) Multi-rate Microprocessor Modems, can be programmed to provide either differential or coherent detection.

The symbol-timing recovery circuits[12] perform nonlinear signal processing to derive the symbol-rate clock, and thus to determine the symbol sampling times

TO SEND	CHANGE PHASE BY
00	+45°
01	+135°
10	-135°
11	-45°

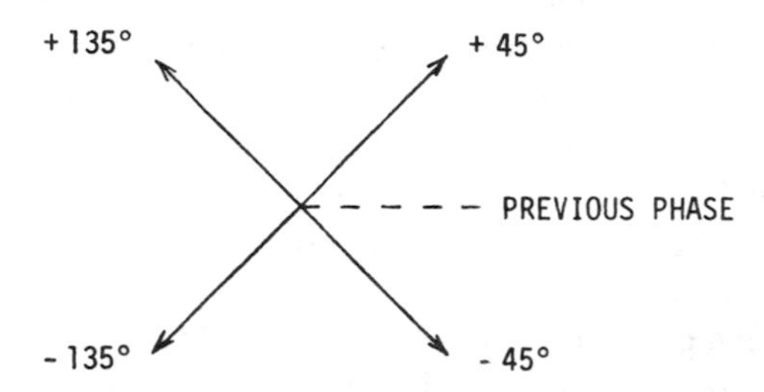

FIGURE 6.8. Differential QPSK. (From D. R. Doll, *Data Communications*, © John Wiley & Sons, Inc., 1978. Reprinted by permission of John Wiley & Sons, Inc.)

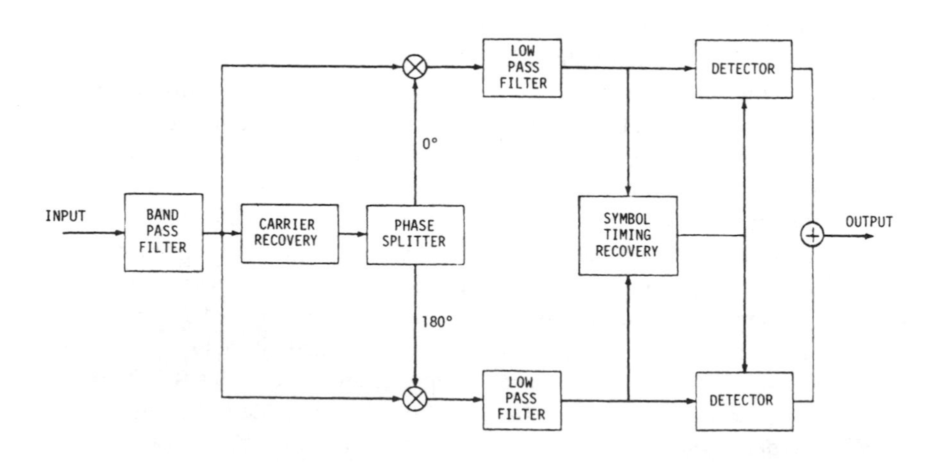

FIGURE 6.9. Coherent receiver for QPSK.

for the detectors. The detectors are threshold comparators whose outputs are logic one or zero states. The output summation circuit performs a parallel-to-serial conversion, thus providing the demodulated output bit stream.

Individual descriptions of M-PSK with $M = 2, 4, 8$ and 16 follow next. The b.e.r. performance of an M-PSK system for any value of M for additive white gaussian noise is given[13] by the expression:

$$\text{b.e.r.} = \left(\frac{1}{n}\right) \mathit{erfc}\left[\sqrt{\left(n\frac{E_b}{N_0}\right)} \sin\left(\frac{\pi}{M}\right)\right] \qquad n > 1 \tag{6.9}$$

where

$$\mathit{erfc}\ x = \left(\frac{2}{\sqrt{\pi}}\right) \int_x^\infty \exp(-y^2)dy$$

E_b = energy per bit
N_0 = noise power spectral density
n = number of bits per keying interval
M = number of phase states. (Thus $2^n = M$.)

Equations (6.13) and (6.14), which follow later, are derived from Eq. (6.9) by substituting $n = 3$ and 4, respectively, into them.

All of these equations for bit-error ratio are based upon thermal noise, that is, noise having a gaussian amplitude distribution. This is the most common type of noise encountered in telecommunication systems.

6.2.1.1. Binary Phase-Shift Keying (BPSK)

In binary phase-shift keying, only two phase states, 180° apart, are used. One state thus represents the digit zero, and the other represents the digit one. Since the transmitter must be keyed once for each digit, the system provides 1 b/s per baud. The Nyquist theorem states that one baud can be transmitted in a bandwidth of 0.5 Hz, but BPSK signals are sent using two sidebands, so the result is, at most, 1 b/s per Hz. This is a theoretical limit. Its practical implications are discussed in Section 6.2.6.1.

The bit-error ratio (b.e.r.) performance of BPSK in thermal noise using coherent detection is

$$\text{b.e.r.} = 0.5\ \mathit{erfc} \sqrt{\frac{E_b}{N_0}} \tag{6.10}$$

Note that Eq. (6.10) has a value only 0.5 as great as might be expected from Eq.

(6.9). This is because half of the noise power that contributes to errors is in quadrature (thus out-of-phase) with the phase reference. Therefore, it is not sensed by the single-phase detector required for BPSK.

6.2.1.2. Quaternary Phase-Shift Keying (QPSK)

In quaternary phase-shift keying, four phase states are used, the adjacent ones being separated by 90°. Each phase state is made to represent a pair of bits or a *symbol*, that is, 00, 01, 10, 11. Thus, a pair of bits is sent each time the transmitter is keyed; accordingly, its theoretical limit is 2.0 b/s per Hz (See Section 6.2.6.1).

A form of QPSK called $\pi/4$-DQPSK has been selected for use in the North American TDMA digital cellular system, as will be described in Section 7.4. Its signal states are the same as those shown in Fig. 6.8. Each new symbol produces at least a 45° phase shift relative to the previous one. The baseband filters ideally have linear phase and square root raised cosine frequency response of the form

$$|H(f)| = \begin{cases} 1 & 0 \leqslant f \leqslant \dfrac{(1-\alpha)}{2T} \\ \sqrt{\dfrac{1}{2}\left\{1 - \sin\left[\dfrac{\pi(2fT-1)}{2\alpha}\right]\right\}} & \dfrac{(1-\alpha)}{2T} \leqslant f \leqslant \dfrac{(1+\alpha)}{2T} \\ 0 & f > \dfrac{(1+\alpha)}{2T} \end{cases} \tag{6.11}$$

where T is the symbol period. The roll-off factor, α, determines the width of the transition band, and is 0.35.

The b.e.r. performance of QPSK using coherent detection is

$$\text{b.e.r.} = 0.5\, erfc \sqrt{\frac{E_b}{N_0}} \tag{6.12}$$

In this case, the energy per symbol $E_s = E_b \log_2 4$, where M is the number of levels, or bits/symbol. Thus

$$E_s = E_b \log_2 4 = 2E_b$$

Its performance in the presence of thermal noise thus is the same as that of BPSK with respect to E_b/N_0. In the presence of such impairments as continuous wave (cw) interference and linear delay distortion, however, Eq. (6.12) no longer is valid, and QPSK is found to degrade more rapidly than does BPSK.

6.2.1.3. Eight-Phase Shift Keying (8-PSK)

In eight-phase shift keying, also called *octonary phase-shift keying* (OPSK), eight phase states are used, the adjacent ones being separated by 45°. Each phase state is made to represent a symbol consisting of a sequence of three bits, that is, 000, 001, 010, etc. Thus three bits are sent each time the transmitter is keyed; accordingly, the technique provides a theoretical limit of 3-b/s per Hz (see Section 6.2.6.1). The b.e.r. performance of 8-PSK using coherent detection is

$$\text{b.e.r.} = \frac{1}{3}\, erfc\left[\sqrt{\left(\frac{3E_b}{N_0}\right)}\sin\left(\frac{\pi}{8}\right)\right] \tag{6.13}$$

The 8-PSK technique is used in a number of digital microwave radio systems operating in the 6- and 11-GHz bands. Although newly implemented terrestrial microwave repeater systems (see Chapter 8) often use 64- or 256-state modulation systems (see Section 6.2.3), satellite systems make frequent use of 8-PSK. A total of 1344 PCM voice channels (90 Mb/s) can be transmitted within a standard 30-MHz channel allocation in the 6-GHz band. Figure 6.10 is a block diagram of the transmitter (providing a 70-MHz IF output) and Fig. 6.11 shows the corresponding receiver.

For transmission, the 90-Mb/s stream is split into three 30-Mb/s streams. The transmit logic circuits produce two 4-level (amplitude) streams, which are used to modulate two quadrature carriers, using double sideband suppressed-carrier amplitude modulation, as shown in Fig. 6.12. The power combiner, accordingly, produces an 8-PSK output that is centered at 70 MHz, and occupies the 55- to 85-MHz spectrum. The inverse functions are accomplished in the receiver (Fig. 6.11), which shows the carrier (70-MHz VCXO) and symbol-timing recovery

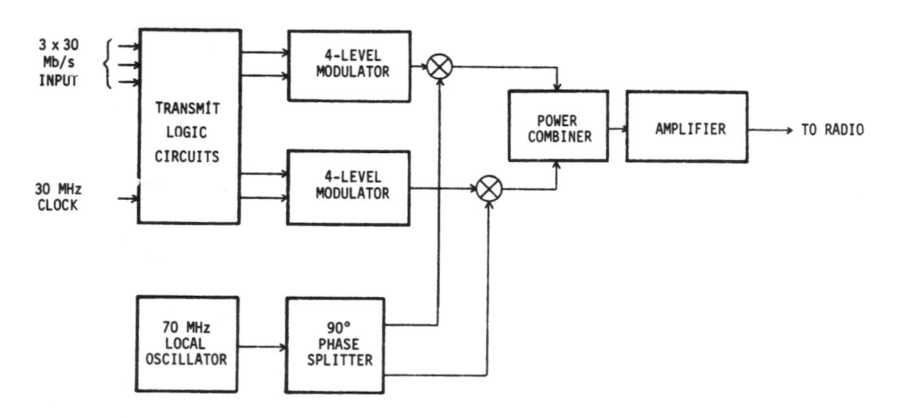

FIGURE 6.10. Block diagram of 8-PSK transmitter.

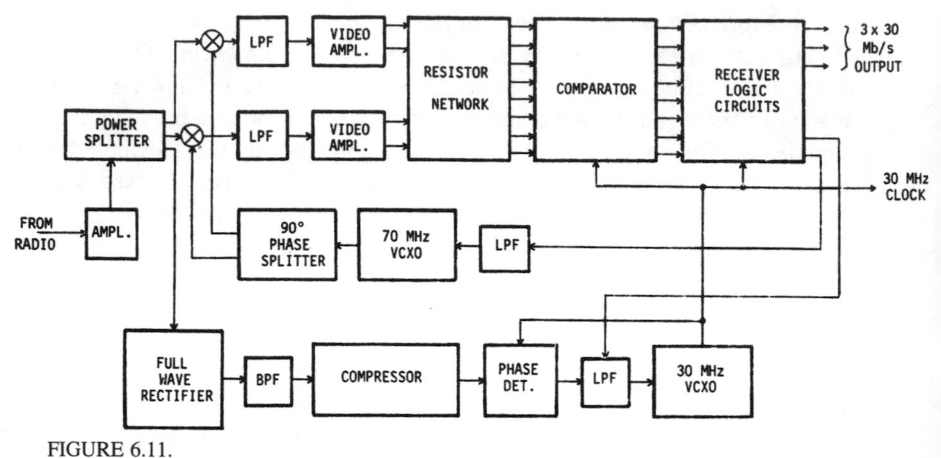

FIGURE 6.11.
Block diagram of 8-PSK receiver.

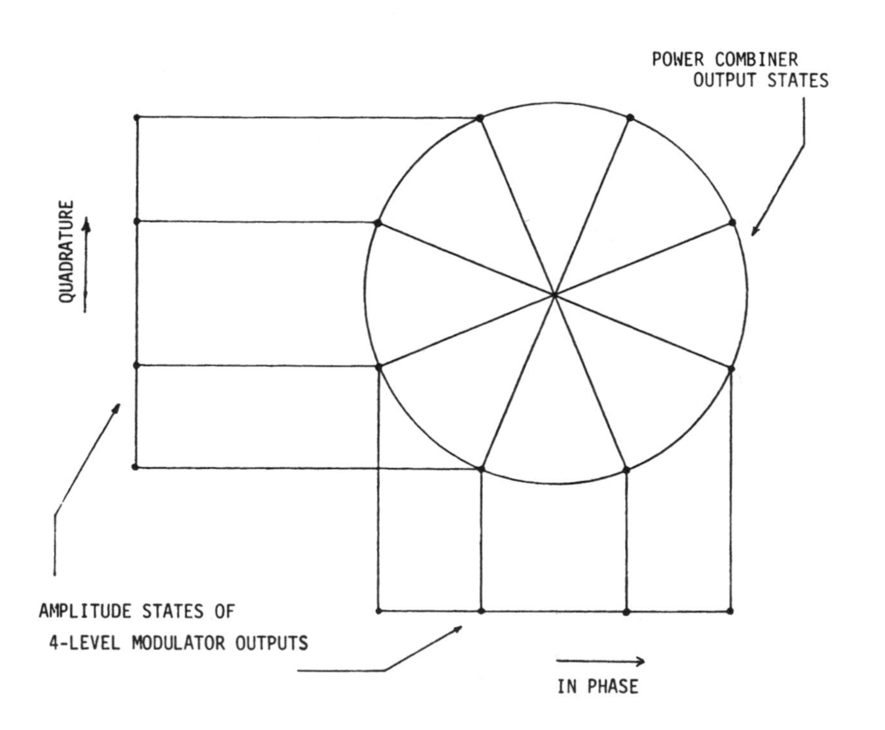

FIGURE 6.12. Amplitude levels used to produce 8-PSK.

(30-MHz VCXO) chains, with the VCXOs being voltage-controlled crystal oscillators.

6.2.1.4. Sixteen-Phase Shift Keying (16-PSK)

In sixteen-phase shift keying, sixteen phase states are used, the adjacent ones being separated by 22.5°. Each phase state is made to represent a symbol consisting of a sequence of four bits, that is, 0000, 0001, 0010, 0011, etc. Thus four bits are sent each time the transmitter is keyed; accordingly, 16-PSK provides a theoretical limit of 4-b/s per Hz (see Section 6.2.6.1). The b.e.r. performance of a 16-PSK system is given by the expression

$$\text{b.e.r.} = \frac{1}{4}\, erfc\left[\sqrt{\left(\frac{4E_b}{N_0}\right)}\sin\left(\frac{\pi}{16}\right)\right] \tag{6.14}$$

No 16-PSK systems are in commercial use at present because of the superior performance of 16-APK, as described in Section 6.2.3.

6.2.2. Amplitude-Shift Keying and Quadrature Amplitude Modulation

The phase-shift keyed systems provide good performance (b.e.r. versus E_b/N_0) provided phase jitter can be kept adequately low. They are insensitive to amplitude jitter; however, channel noise actually consists of both amplitude and phase fluctuations. Accordingly, phase jitter begins to cause significant performance limits when attempts are made to implement more than eight phases, as may be desired for efficient use of the spectrum. While many practical 8-PSK systems are being implemented, the achievement of 4 bits per symbol (16 signaling states) is done most effectively using amplitude- as well as phase-shift keying. In fact, comparisons of 16-PSK with 16-state techniques involving amplitude as well as phase shift, show that 16-PSK may require on the order of 5 to 6 dB more carrier-to-noise ratio; moreover, the intersymbol interference produced by filtering causes greater degradation to 16-PSK and higher order PSK systems than to systems that also involve amplitude shift, as will be seen later from Figs. 6.22, 6.23, and 6.24.

The concept of amplitude-shift keying was introduced already in the discussion of 8-PSK, where it may be used as part of the process of generating the 8-PSK signal. In fact, all of the M-PSK outputs can be generated by quadrature combinations of amplitude-shift keyed signals. The term quadrature amplitude modulation (QAM) is used to describe the combining of two amplitude-shifted streams in quadrature. One of these two can, for example, be a digitally modulated sine wave, while the other is a digitally modulated cosine wave of the same frequency. Thus, in its simplest form, QAM can be produced by the quadrature combination of two two-level signals (4-QAM), as illustrated in Fig. 6.13. The resulting am-

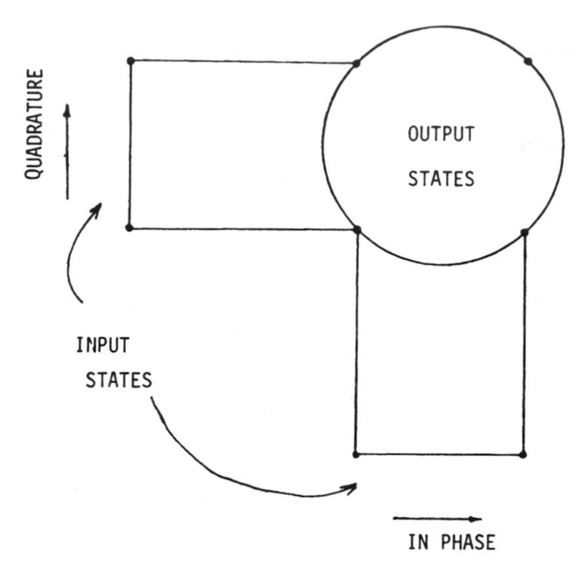

FIGURE 6.13. Quaternary amplitude-modulation states.

plitude-phase diagram will be recognized as being the same as that of QPSK. The difference[14] is that the QAM system uses premodulation and post-detection low-pass filters while the QPSK system has post-modulation and predetection band-pass filters. Accordingly, 4-QAM and QPSK systems have identical transmitted spectra and identical b.e.r. performance, as expressed by Eq. (6.12).

The diagrams of Figs. 6.12 and 6.13 illustrate the positions taken by the transmitted amplitude and phase as the transmitter is keyed by various symbols (bit combinations). Each symbol corresponds to a specific point on such a diagram. Sometimes the diagram is called the *signal-state space diagram*, or the *constellation.* On a normalized basis, the distance between the points may be referred to as the Euclidean distance. The greater this distance, the greater will be the extent to which a given modulation technique is noise resistant. This is why 16-PSK is not commonly used. A constellation with greater Euclidean distances can be devised, for example, using 16-QAM. For comparison, Fig. 6.14 shows 16-PSK (a) and 16-QAM (b), both of which are of equal peak amplitude. Note the much closer spacing of the signal states in 16-PSK, compared with those of 16-QAM. For the 16-QAM, the spacing is $0.47A$, whereas, for the 16-PSK, the spacing is only $0.39A$, where A is the unmodulated carrier amplitude. As shown in Fig. 6.14(b) for QAM, two quadrature 4-amplitude signals are combined, but their timing and magnitudes produce sixteen well-spaced constellation points, using various amplitudes as well as various phases of the carrier.

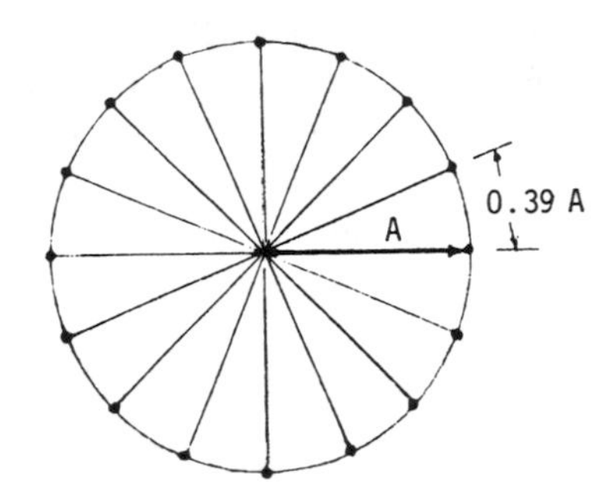

(a) Constellation for 16-PSK

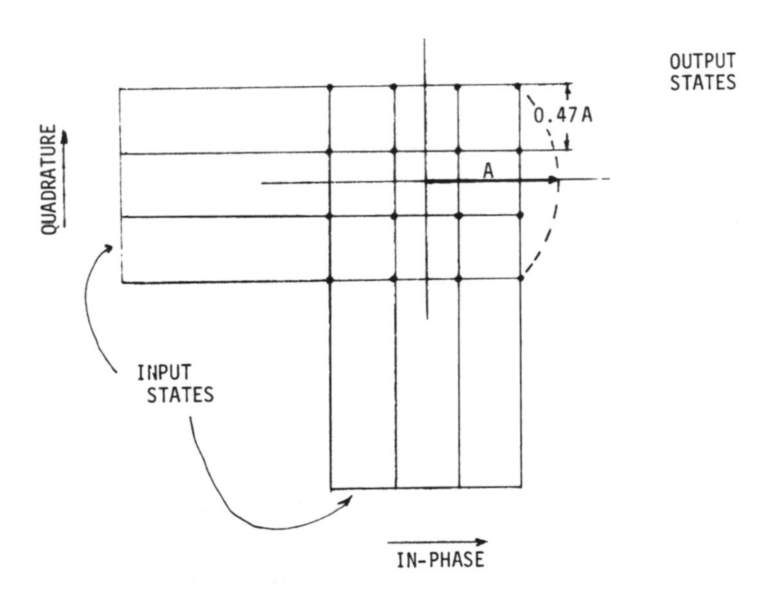

(b) Constellation for 16-QAM and its derivation

FIGURE 6.14. Constellations for 16-PSK and 16-QAM.

6.2.3. Amplitude-Phase Keying

Amplitude-phase keying (APK) is a general term used for any digital modulation system in which both the amplitude and the phase of the carrier are altered to produce the various symbol states. Thus, QAM is one form of APK. Figure 6.15 is a generalized block diagram of an APK modulator using suppressed-carrier QAM, while Fig. 6.16 shows the corresponding demodulator. In these figures, f_b refers to the bit rate of the two-level digital stream being transmitted. In the modulator, M amplitude levels are produced and combined in phase as in Figs. 6.13 and 6.14. The digital stream rate thus decreases by a factor $\log_2 M$ at this point. The receiver derives its carrier using a carrier-recovery circuit similar to the type used for the coherent demodulation of phase-shift keying. Conversion from M to two levels then is achieved by individual threshold comparators, each

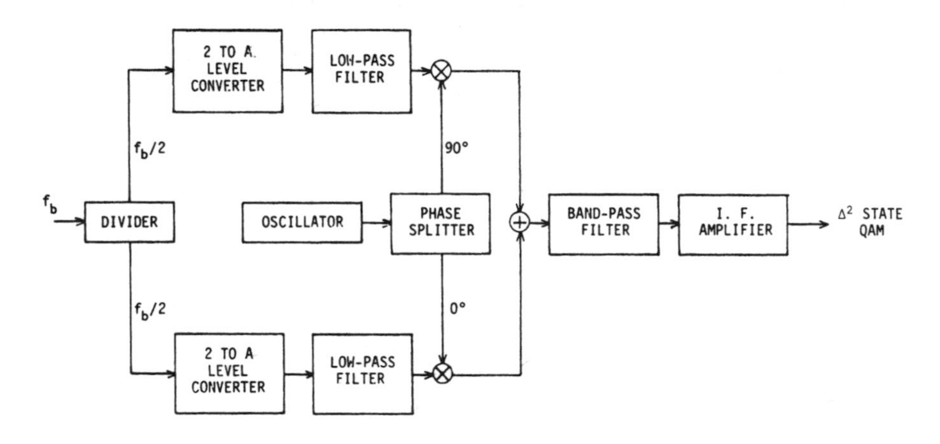

FIGURE 6.15. Generalized QAM modulator.

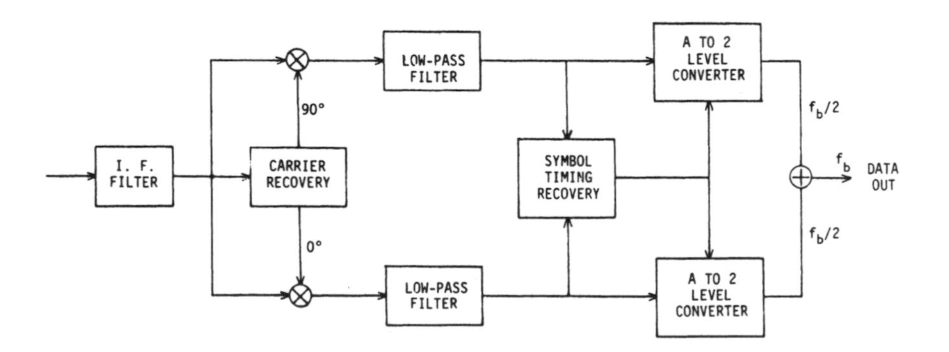

FIGURE 6.16. Generalized QAM coherent demodulator.

with a separate preset threshold level. The outputs of these threshold comparators then are combined in digital logic to produce a stream at a rate of $f_b/2$. Finally, the $f_b/2$ streams from the in-phase and quadrature parts of the demodulator are combined to reproduce the output bit stream f_b.

The b.e.r. performance of a 16-QAM system is given[15] by the expression

$$\text{b.e.r.} \approx \frac{3}{8}\, erfc \sqrt{\frac{2E_b}{3N_0}} \tag{6.15}$$

while that of a 64-QAM system is given[16] by the expression

$$\text{b.e.r.} \approx \frac{7}{24}\, erfc \sqrt{\frac{E_b}{7N_0}} \tag{6.16}$$

and for 256-QAM, the expression is[17]

$$\text{b.e.r.} = \frac{19}{64}\, erfc \sqrt{\frac{E_b}{170N_0}} \tag{6.17}$$

The transmission of digital bit streams through relatively linear analog voiceband facilities can be done using large numbers of signal states. Voiceband modems using high-level APK constitute a major application area for such techniques. Fig. 6.17 for example, is a photograph of the constellation used by the Codex SP14.4 Data Modem. A data rate of 14 400 b/s keyed at 2400 baud requires 6 bits per symbol and thus $2^6 = 64$ signal states.[a] Accordingly, the carrier-to-noise ratio required for a given bit-error ratio is significantly higher (see Eqs. (6.15) and (6.16)) than that of a system using 16 or fewer states. On the conditioned telephone facility,[b] this modem operates with a b.e.r. $\leqslant 10^{-6}$.

A large number of APK constellations have been devised and numerous studies have been done relative to their performance.[18]

Table 6.1 summarizes modem development, showing the symbol rates and constellations used by the first modem to be marketed at each respectively higher data rate. Modems incorporating data compression are not included in this listing.

6.2.4. Frequency-Shift Keying

Frequency-shift keying (FSK) generally uses two frequencies to represent two binary states, that is, zero and one. Noncoherent detection is accomplished with

---

[a] Modems operating at rates as high as 24 kb/s are available.

[b] Four-wire 3002 D1 conditioned or CCITT M 1020 lines.

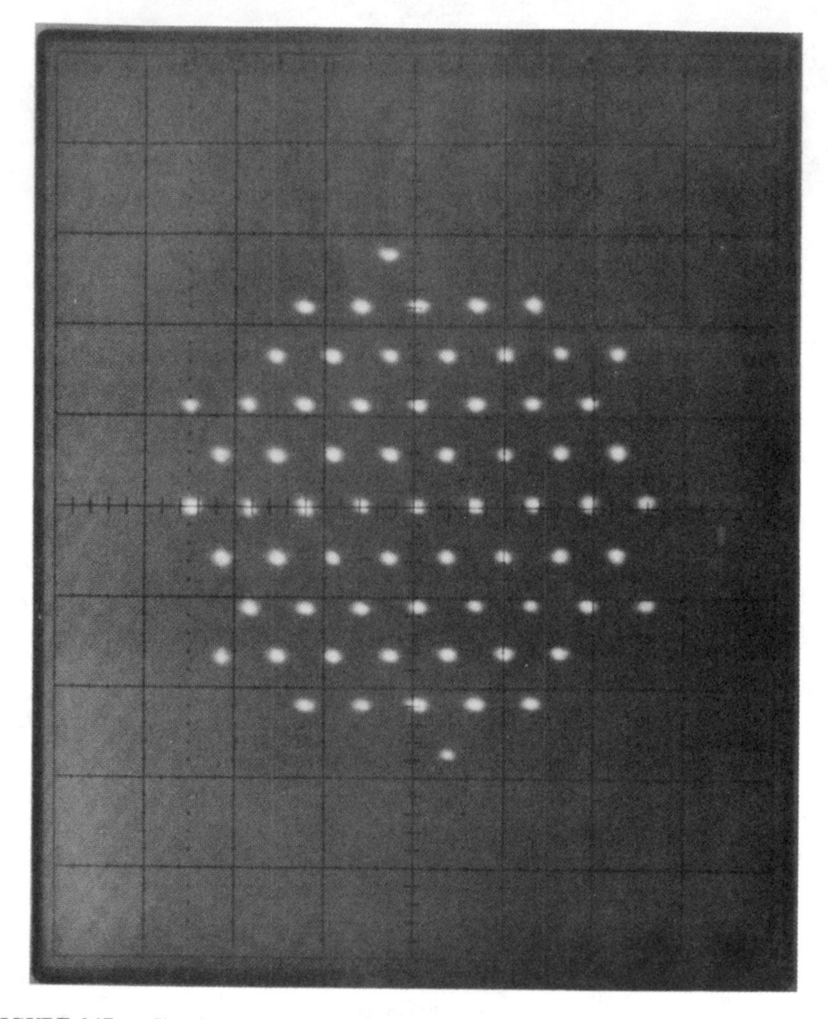

FIGURE 6.17. Signal state space array for 64-state modulation used in 14.4-kb/s modem. (Courtesy Codex Corp.)

circuits tuned to each of these frequencies. The implementation of FSK thus can be simple, and limiters can be used to reduce the effects of selective fading, which is common on the high-frequency (HF) transmission circuits on which FSK often is used. FSK is a commonly used technique for transmission at and below 1200 b/s. The deviation ratio d of an FSK signal is defined as $d = \Delta f/b$ where Δf = peak-to-peak frequency deviation and b = bit rate. With noncoherent detection, d must be at least 1.0 to prevent significant overlap of

Table 6.1. Modem Milestones

Year	Model	Data Rate	Symbol Rate	n	Constellation	Type	CCITT Standard
1962	Bell 201	2400 b/s	1200 Bd	2	QPSK	Fixed eq.	V.26 (1968)
1967	Milgo 4400/48	4800 b/s	1600 Bd	3	8-PSK	Manual eq.	V.27 (1972)
1971	Codex 9600C	9600 b/s	2400 Bd	4	16-QAM	Adaptive eq.	V.29 (1976)
1980	Paradyne MD 14400	14.4 kb/s	2400 Bd	6	64-QAM	Rectangular constellation	V.33
1984	Codex 2680	19.2 kb/s	2400 Bd	8	256+ -QAM	Trellis	V.24, V.28
1990	Codex	24.0 kb/s	Variable	8	256+ -QAM	Trellis + line probing	

the filter passbands. This constraint on d can be eliminated by using a discriminator to convert the frequency variations to amplitude variations, whereupon envelope detection is used.

Continuous-phase FSK (CP-FSK) avoids abrupt phase changes at the bit transition instants by using observation intervals greater than one bit period in length.[19] The result is rapid spectral roll-off and thus narrower filter bandwidths than would be possible otherwise. With coherent detection, $d = 0.715$ is the optimum deviation ratio with respect to transmitted power and occupied bandwidth in the presence of thermal noise if decisions are limited to one bit interval. A useful special case of CP-FSK is called *minimum shift keying* (MSK). For MSK, the peak frequency deviation is equal to $\pm 0.25b$ and coherent detection is used. Thus $d = 0.5$ for MSK. MSK achieves performance identical to that of coherent PSK with the efficient spectral characteristics of CP-FSK, by extending the decision interval to two bit periods for binary detection.[20] An additional advantage of MSK over coherent CP-FSK is the possibility of a relatively simple self-synchronizing implementation.[21]

In the absence of bandwidth or amplitude limiting between the modulator and the demodulator, the bit-error ratio performance of MSK is identical to that of BPSK,[22] as expressed by Eq. (6.10), that is, for a given ratio of signal-to-thermal noise, the two techniques yield the same bit-error ratio. In the presence of the amplitude and bandwidth limitations that often occur in a transmission system, however, MSK suffers less degradation than BPSK because MSK exhibits less out-of-band power.

A special form of MSK is used in the Global System Mobile (GSM), the European digital cellular system. In GSM, the MSK signal, to reduce its spectral occupancy, is filtered using a gaussian-shaped filter. The result is called gaussian MSK (GMSK).

6.2.5. Correlative Techniques

Correlative techniques use a finite memory to change the baseband digital stream to a new form.[23,24] This form provides improvements in coding efficiency from a spectrum occupancy viewpoint. Correlative techniques produce intersymbol interference on a planned and controlled basis, and are sometimes called *partial response* techniques[25] because the resulting filtered pulses no longer exhibit zero outputs at those sampling instants reserved for adjacent pulses. Alternatively, one might say that the output responds only partially to the input. Figure 6.18 illustrates the response of a bandwidth-limited system to a rectangular pulse. The result ideally is zero intersymbol interference; however, if the pulse tails have finite amplitude at the times T, $2T$, . . ., nT reserved for adjacent pulses, the result is intersymbol interference.

Correlative techniques are called duobinary, modified duobinary, and polybinary. The name *duobinary* means doubling the speed of binary. This is done by

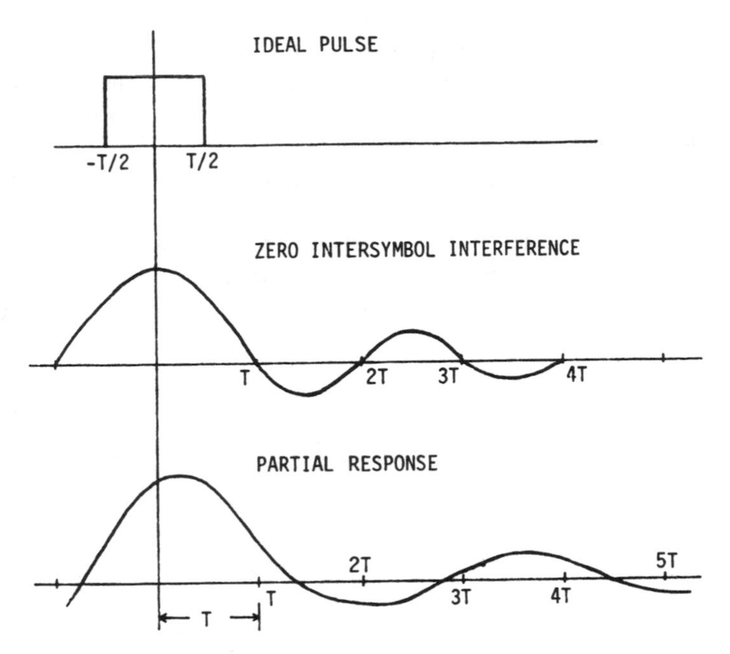

FIGURE 6.18. Illustration of intersymbol interference.

converting an ordinary two-level binary sequence into one that uses three levels. This conversion involves intersymbol interference extending over one bit interval, as will be seen soon. If the intersymbol interference extends over two bit intervals, the technique is called *modified duobinary*. Polybinary is the term used to denote the fact that the correlation span includes more than three bits. The greater the correlation span, the more extensively the energy of the sequence is concentrated at the lowest frequencies. As a result, the number of bits/second per hertz of bandwidth can be increased significantly over that of systems not using such techniques. In addition, certain unique patterns are produced among the correlative pulse trains. These patterns can be monitored at the receiver, thus allowing error detection without the addition of redundant bits at the transmitter.

Figure 5.15 showed the introduction of controlled interference between input bits. The "partial response" is produced by adding the binary input signal to itself delayed by one bit interval using the following rules: $1 - 1 = 0$; $-1 - 1 = -1$; $1 + 1 = 1$. The resulting three-level signal is found to occupy significantly less bandwidth than the original two-level signal.

Correlative techniques have been developed in which the processes of coding and modulation are combined in a single step. An example is duobinary AM-PSK modulation, as illustrated in Fig. 5.15. Here the -1 level corresponds to a

0° carrier shift, the 0 level corresponds to zero output, and the +1 level corresponds to a 180° carrier shift. Because the carrier reverses phase according to duobinary rules, the bandwidth is less by a factor of two than it would be for simple on-off keying.

A decision feedback receiver is used to remove the intersymbol interference. Shown in Fig. 6.19 is the concept for such a receiver for a polybinary signal in which the intersymbol interference extends over four bit intervals.

Quadrature partial response signaling (QPRS) is achieved with two 3-level duobinary signals phase-modulated in quadrature, as illustrated in Fig. 6.20. This

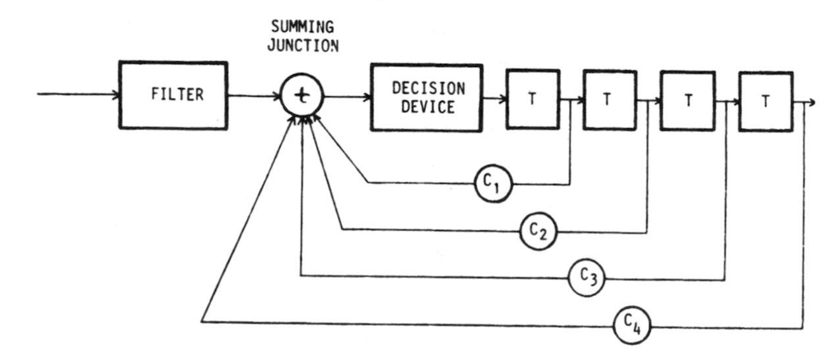

FIGURE 6.19. Decision feedback receiver for polybinary signal with correlation span extending over four bit intervals.

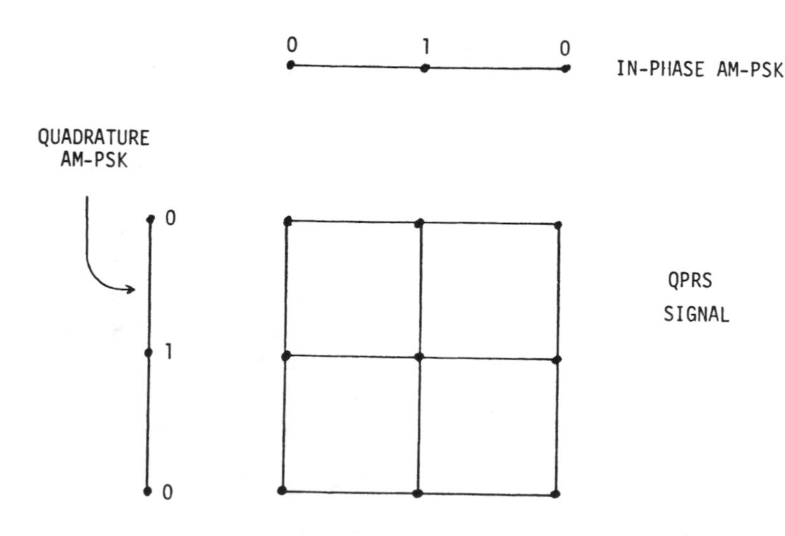

FIGURE 6.20. QPRS modulation.

technique, also called *restricted bandwidth,* is being used in several microwave radio systems that operate in the 2-GHz band. The advantage of using correlative techniques in such systems is that they are relatively simple to implement, being comparable to a four-phase modulation system, while being less complex than the eight-phase systems. With this simplicity, they can still achieve the spectral efficiency (96 voice channels, or 6.312 Mb/s, in a 3.5-MHz bandwidth) required for operation at 2 GHz. This is a 2-b/s per hertz efficiency.

A seven-level modified duobinary signal processor also is used in some 2-GHz radio systems. Three positive, three negative and a zero amplitude level are obtained digitally. The binary input at 6.312 Mb/s (96 PCM voice channels) is converted to 3.156 Mbaud (2 b/s per baud), which occupies a baseband of 1.578 MHz. As applied to a standard FM radio terminal, the resulting double-sideband signal occupies 3.156 MHz, for an overall 2 b/s per hertz spectral efficiency. It thus meets the regulatory requirements for DS2 transmission at 2 GHz. A block diagram of this radio system is shown in Fig. 6.21. There is no zero-frequency energy in the baseband spectrum, and a negligible amount of energy at low frequencies. Accordingly, a voice-frequency order-wire channel is placed into the first 8 kHz of the baseband spectrum. Note the presence of the error detector at the output. This detector monitors the bit patterns from the binary threshold detector and thus achieves error detection without the need for redundant bits in the input bit stream.

Error detection is accomplished using the following rules, based upon observation of the behavior of the "extreme," or top and bottom, levels of the data stream. Thus for a $(+, 0, -)$ system, the $(+)$ and $(-)$ levels are the extreme levels. Violation of the rules below indicates the presence of an error.

Rule 1. For a duobinary $(+, 0, -)$ system, the polarities of two successive bits

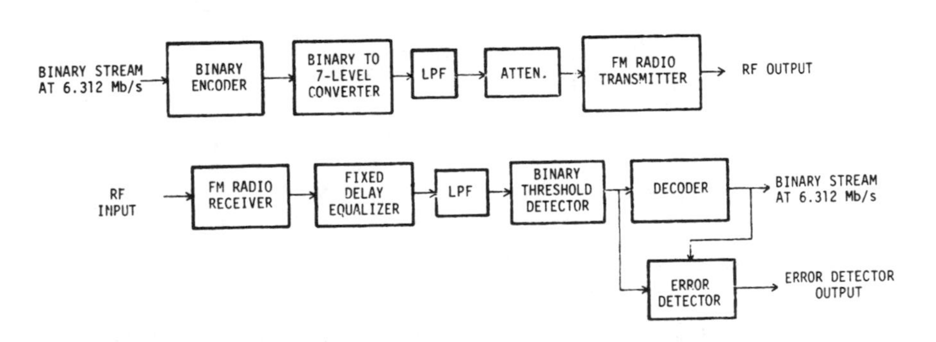

FIGURE 6.21. Seven-level modified duobinary digital transmission system. (Copr. GTE Lenkurt, Inc., San Carlos, CA.)

at the extreme levels are opposite if the number of intervening bits at the zero level is odd; otherwise, the extreme bits have the same polarity.

Rule 2. For a modified duobinary (+, 0, −) system, divide the pulse train into even and odd trains. Both trains follow the same pattern. For either train, two successive bits at the extreme levels must have the opposite polarity.

Note that only the extreme levels are usable for error detection. This also is true for systems with larger numbers of levels, that is, (−3, −1, +1, +3), etc. Intermediate levels may be formed in more than one way, and thus are not suitable for error detection.

Example: Given the following modified duobinary stream:

$$0 - - + 0 0 0 - 0 + 0 - 0 + + 0$$

The "even" and "odd" streams are

$$- + 0 - + - + 0 \text{ (even)}$$
$$0 - 0 0 0 0 0 + \text{ (odd)}$$

Since any two successive bits at the extreme levels have opposite polarity, no bit errors are apparent.

A discussion of the various partial response classes, as well as their performance, can be found in Ref. 4.

6.2.6. Comparison of Modulation Techniques

Three basic questions arise in the choice of a digital modulation technique:

1. What is its bandwidth efficiency (b/s per hertz)?
2. How well does it perform in the presence of:

- channel noise?
- filter limitations?
- other impairments?

3. How complex is it to implement?

This section provides quantitative answers to the first two categories of questions. The third, complexity, can be answered only in a qualitative fashion. General comments are provided below, where appropriate.

6.2.6.1. Bandwidth Efficiency

Bandwidth efficiency is expressed in bits per second (b/s) per hertz of bandwidth. To determine it for a given modulation technique, the number of bits/symbol, or b/s per Bd, m, must be established first. As noted in Section 6.1, the number of logic levels is $M = 2^m$. For a zero-memory (noncorrelative) coding technique, the maximum symbol rate per unit bandwidth is 2 Bd/Hz. This is valid in baseband, or at RF for single-sideband modulation techniques. Most modulation, however, involves either frequency or phase shift, and thus double sidebands. The result is a maximum symbol rate of 1 Bd/Hz. This implies the use of filters with infinitely steep cut-off characteristics, sometimes designated *brick wall* ($\alpha = 0$) filters, where α denotes the filter's roll-off factor. In general, a filter with a roll-off factor α will have an amplitude versus frequency characteristic for impulse transmission[26] that begins to roll off at a frequency of $(1 - \alpha)f_c$, where f_c, is the cut-off or -3-dB frequency. The response is very low (for example, -80 dB or lower) at and beyond a frequency of $(1 + \alpha)f_c$. Accordingly, the maximum double-sideband transmission rate is $1/1 + \alpha$ Bd/Hz. Thus, if $\alpha = 0.25$, the maximum symbol rate is 0.8 Bd/Hz.

Table 6.2 compares various digital modulation techniques on the basis of their bandwidth efficiency, as established by the foregoing principles. The filter is assumed to have an ideal roll-off factor, $\alpha = 0$, since some systems in the field approximate $\alpha = 0$ very closely. For example, some of the operational digital radio systems in the 6-GHz microwave band (see Chapter 8) manufactured by Raytheon, Collins, and Nippon Electric use 8-PSK and achieve a 90-Mb/s rate in a 30-MHz radio-channel bandwidth by using very sharp cut-off filters. This de-

TABLE 6.2. Bandwidth Efficiency of Digital Modulation Techniques

			Maximum Achievable Efficiency	
Type of Modulation	Number of Logic Levels	b/s per Baud	Baseband (b/s per Hertz)	Modulated Carrier (b/s per Hertz)
AM	2	1	2	1
FSK	2	1	2	1
7-Level mod. duobin.[a]	7	2	4	2
BPSK	2	1	2	1
QPSK	4	2	4	2
8-PSK	8	3	6	3
16-PSK	16	4	8	4
16-QAM	16	4	8	4
64-QAM	64	6	12	6

[a]Three of the four bit combinations can each be represented by either of two logic levels. The fourth combination is represented by only one level.

sign, however, enables them to meet the regulatory requirements for spectral efficiency in that band.

Figure 6.22[27] shows that increased bandwidth efficiency carries with it a requirement for increased signal-to-noise ratio (*S/N*), but that the techniques that use only phase shift require a better *S/N* than do those that involve both amplitude and phase changes. Fig. 6.22 is based upon a b.e.r. of 10^{-6}. The numerals refer to the number of signaling states required in each case. Equipment complexity is not displayed. The Shannon limit is based upon a zero b.e.r. Equipment complexity increases substantially, however, as one attempts to get closer to this limit.

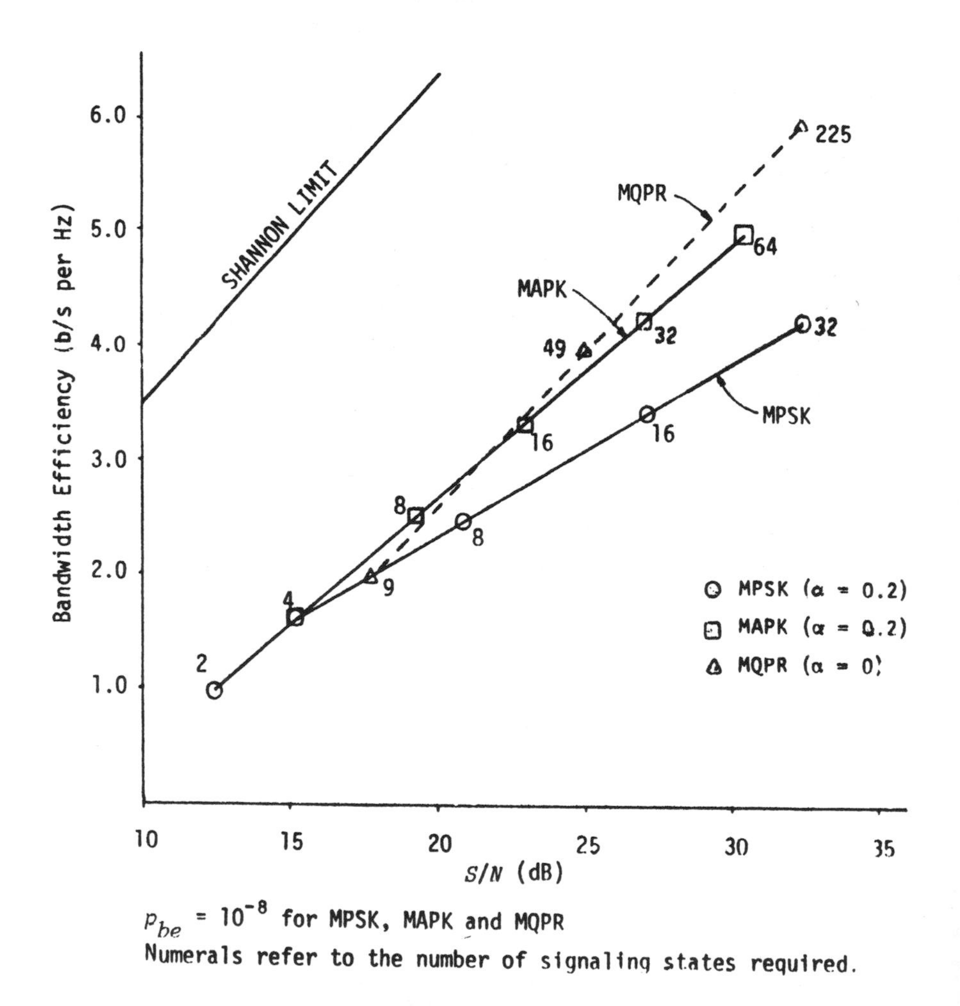

FIGURE 6.22. Bandwidth efficiency comparison of modulation methods.

6.2.6.2. Modulation Technique Performance

Modulation techniques can be evaluated based upon their performance in the presence of various impairments. A detailed discussion of this subject is beyond the scope of this text, but two major types of impairments will be considered: channel noise and filter bandwidth limitations.

Figures 6.23 and 6.24 show how several of the systems of this chapter perform in the presence of channel noise, with performance[28] being expressed in terms of b.e.r., designated as probability of error. The rms *C/N* is specified in the double-sided Nyquist bandwidth in each case.

Figure 6.25 shows the effect of filter bandwidth on several of the systems in terms of *C/N* degradation.[29] In this figure, B is the 3-dB double-sided bandwidth of a gaussian filter, and T_s is the duration of a symbol.

6.3. ERROR CONTROL

Figure 5.1 showed that an error coder follows the speech coder in an overall digital speech transmission system. The error coder is used in those systems in which the detection and/or correction of channel errors is important. In telephony, one major requirement for error coding is in mobile radio applications, especially land mobile, where the problem of rapid fades justifies the complexity of error coding, as well as the resultant increase in transmitted-bit rate. As described in Chapter 4, U.S. cellular systems will use VSELP at 7.95 kb/s, while European cellular systems will use RPE-LTP at 13 kb/s. Such speech-coding techniques can tolerate bit-error ratios on the order of 10^{-2}. Higher error ratios, however, may result from the rapid fading that accompanies unfavorable propagation conditions. Accordingly, forward error correction is used to maintain voice quality. Forward error correction involves the addition of redundant digits to a bit stream according to preestablished rules. The receivcr then examines all received bits, using the redundant ones to determine what errors may have occurred and to correct as many of these errors as possible, or to eliminate the blocks in which they occur. In performing such coding, the speech signal may be handled in blocks or groups of blocks whose length does not exceed the duration of a syllable. *Block coding* is the term used to describe an arrangement in which the redundant bits in a given block relate only to actual speech or information bits of the same block. If the redundant bits also check speech or information bits in preceding blocks, however, the code is called *convolutional.* While forward error correction can provide appreciable improvements in minimizing speech distortion during rapid fades or dropouts less than a syllable in length, it cannot aid in the case of longer-term signal dropouts, such as are experienced in fringe areas relative to a mobile base station.

In addition to its use in improving speech quality under difficult transmission conditions, such as those encountered in mobile radiotelephony, error coding has

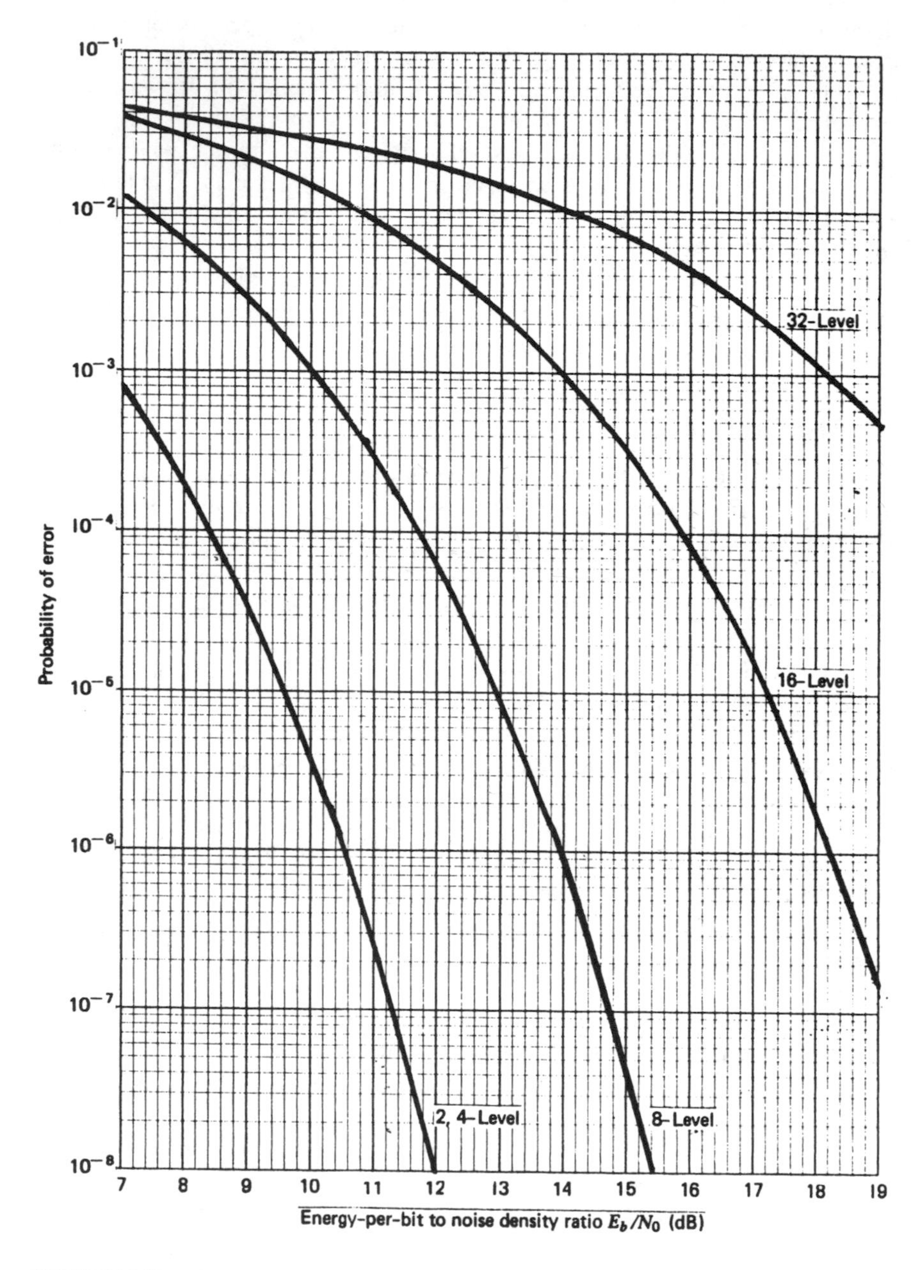

FIGURE 6.23. Performance of *M*-ary PSK modulation systems. (Courtesy J. C. Bellamy, Ref. 28, © Wiley, 1982.)

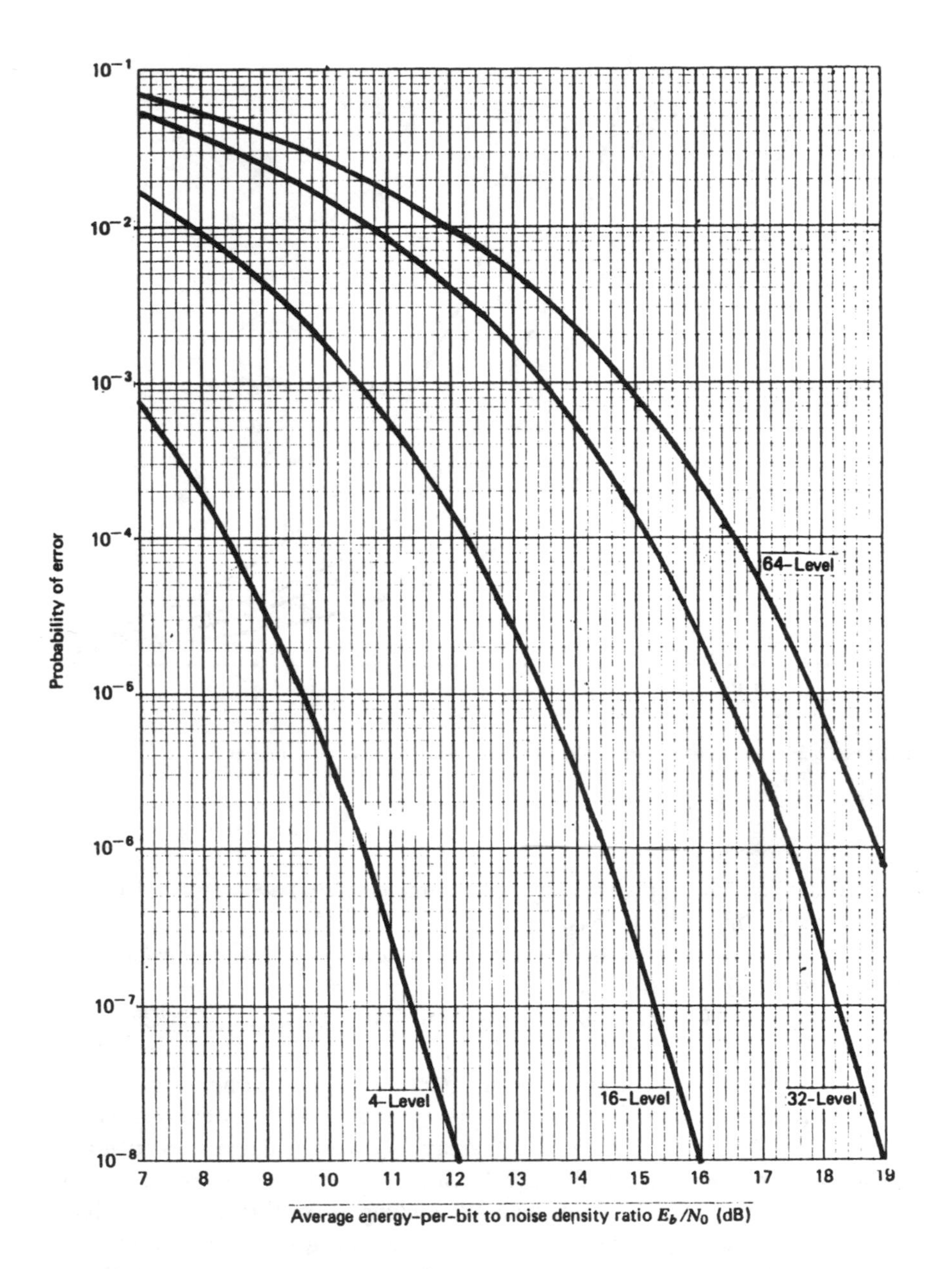

FIGURE 6.24. Performance of M-ary QAM modulation systems. (Courtesy J. C. Bellamy, Ref. 28, © Wiley, 1982.)

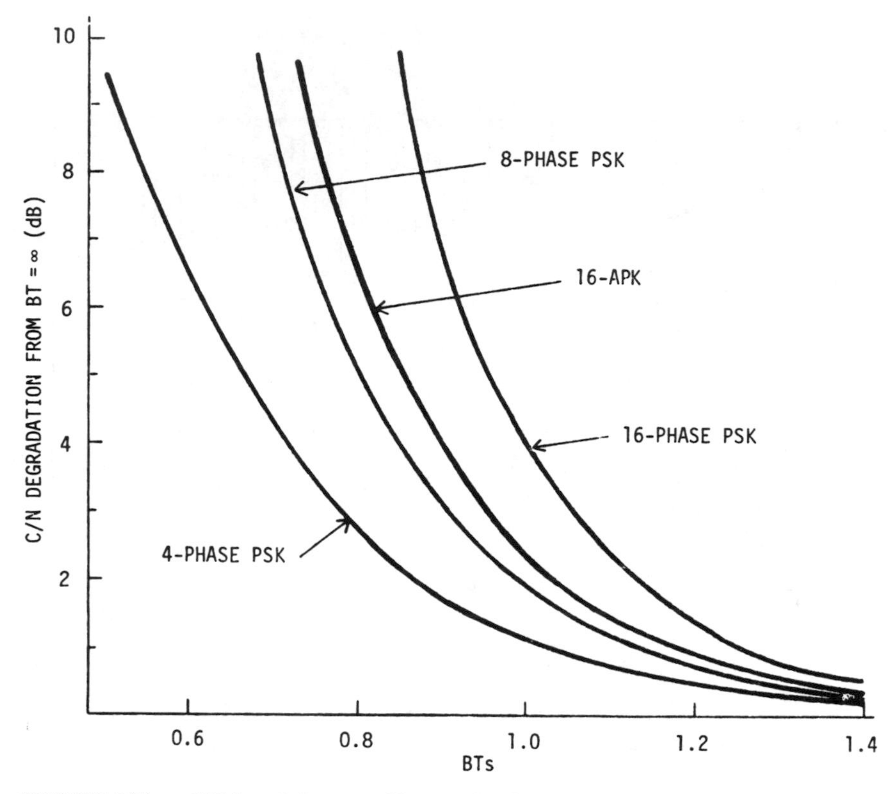

FIGURE 6.25. *C/N* degradation caused by gaussian filter band limitation. (Courtesy Ishio, et al, Ref. 23, © IEEE, 1976.)

obvious applications in maintaining the extremely low-bit-error ratios (10^{-7}) essential in signaling and supervision.

Chapter 3 noted that PCM encoded speech does not demand unusually low error ratios for good performance. For PCM, a bit-error ratio of 10^{-4} or less is quite satisfactory. The other voice-coding techniques, as discussed in Chapter 4, perform quite well at even higher bit-error ratios, such as 10^{-3} or even 10^{-2}. Why, then, should concerns arise about detecting and correcting errors in such environments? First, error detection is important as a means of monitoring facility performance. An increased bit-error ratio may indicate line degradation that may get worse with time. Second, although PCM and ADPCM voice do not demand ultra-low-bit-error ratios, such functions as signaling, supervision, and voice-band data need to be accomplished accurately. For example, if wrong numbers caused by channel and equipment errors are to be minimized, a means of error correction may be desirable to assure a bit-error ratio of 10^{-7} or less.

6.3.1. Error Detection

Many error detection schemes are based on a parity-check code. A simple parity-check code for detecting a single error involves the addition of a computed bit to a sequence of information bits. For example, suppose the information bits are 1 0 1 1 0 1 1. To these seven bits, an eighth bit is added so that the modulo-2 sum of the bits equals zero (or one, if the designer so chooses). Thus, the sequence would become 1 0 1 1 0 1 1 1. This is also called a row check. At the end of a block of bits, a row of parity check bits, each checking its own column, also could be added.

The information transmission rate, when using the simple row check on seven information bits, will be only 7/8 of the value it would have if all eight bits were information bits. Thus, this is called a *rate 7/8 code*, or the code is said to have an *efficiency of 7/8*. Alternatively, it may be designated as an (8,7) code. The lower the efficiency, the more effective the error detection can be in terms of the number of errors that can be detected. Note that a single parity bit detects an odd number of errors, but an even number of errors is undetected by this simple approach.

6.3.2. Error Correction

In land mobile radiotelephone as well as cordless or portable telephone applications, the radio channel may exhibit severe phase distortions for various positions of the end instrument. In addition, momentary signal drops below the threshold may occur. Such circumstances call for error correction, illustrated by the following example[30] of what is known as a Hamming code for single error correction.

Consider information to be grouped into blocks of four binary digits. Let three parity digits be added to each block, for a total block length of seven digits. Let the information digits be labeled I_1, I_2, I_3, and I_4, and let the parity digits be labeled P_1, P_2, and P_3. Each parity digit will be selected for even parity (an even number of ones) among itself and a selected subset of the information digits. The rules for determining each parity digit are given in the encoding-decoding table of Fig. 6.26. The top row in the first column of the table is labeled P_1. The P_1 value is selected so that there will be an even number of ones among I_1, I_2, I_3, and P_1. In the next row, the P_2 value is selected so there will be an even number of ones among I_1, I_3, I_4, and P_2. In the bottom row, the P_3 value is selected so there will be an even number of ones among I_2, I_3, I_4, and P_3. For example, assume the information digits are $I_1 = 1$, $I_2 = 0$, $I_3 = 0$, and $I_4 = 1$. Select P_1 so there will be an even number of ones among I_2, I_3, I_4, and P_1, that is, among 1, 0, 1, and P_1. Therefore, P_1 must be 0. Similarly, the rules indicate that $P_2 = 0$ and $P_3 = 1$. Thus the code word, complete with its parity digits, is 1 0 0 1 0 0 1. Correspondingly the rate is 4/7.

To illustrate the use of this code in error correction, assume that the above code word, 1 0 0 1 0 0 1, is sent over a communication system. At the receiving

	I_1	I_2	I_3	I_4	P_1	P_2	P_3
P_1	X	X		X	X		
P_2	X		X	X		X	
P_3		X	X	X			X

FIGURE 6.26. Encoding-decoding table for a simple error correcting code. (From L. Lewin, ed., *Telecommunications, An Interdisciplinary Survey*, Artech House, Inc., Dedham, MA, 1979.)

end of the system, the parity checks indicated by the encoding-decoding table are made. More specifically, the number of ones among I_1, I_2, I_4, and P_1 is determined. This is *parity check* number 1, and if the number of ones is even, the parity check is said to be correct; otherwise, it is said to have failed. Similarly, parity checks numbers 2 and 3 are made according to the definitions given by the second and third rows of the encoding-decoding table. For example, if an error occurs, the block may be received as 1 0 1 1 0 0 1. At the receiver the first parity check is correct, but the second and third both fail. The assumption then is made that only one error has occurred, and reference to the encoding-decoding table shows the location of the error.

If the error were in I_1, the table shows that I_1 enters into parity checks 1 and 2 but not 3. Thus the first two parity checks would fail, while the third one would be correct. This is not the result of the parity checks obtained in the example, so the error cannot be in I_1. Each of the other possible positions is considered; the only position in which an error could cause the first parity check to be correct and the other two to fail is I_3. Accordingly, the error must be in digit I_3. I_3 then is changed, producing 1 0 0 1 0 0 1, which was the block transmitted. This procedure is called error correction.

For situations in which the probability of double errors is significant, this code is unsuitable, but other codes are available.

6.3.3. Block and Convolutional Codes

A block (or frame) is a grouping of bits established by a sending station. The example in the previous section illustrated a block code, in which the redundant bits relate only to the information bits of the same block. If, on the other hand, the redundant bits also check the information bits in previous blocks, the code is called *convolutional.* With reference to the propagation problems of mobile and portable radiotelephone systems, note that nothing can handle a complete signal dropout; however, a very short-duration signal impairment (e.g., less than a syllable in length) may be bridged by a suitable convolutional code.

6.3.4. Bose–Chaudhuri–Hocquenghem (BCH) Code

The Bose–Chaudhuri–Hocquenghem (BCH) code is used widely in telecommunications systems. Its applications include signaling in cellular radio systems, as well as data transmission in satellite telecommunications systems. The BCH code is a *cyclic* or *polynomial* code, which means that it is a block code in which a data message of b bits is represented in terms of a $b - 1$ degree polynomial in the variable x. Thus, if the data message is given by $a_{b-1}a_{b-2} \ldots a_1a_0$, the corresponding polynomial is

$$M(x) = a_{b-1}x^{b-1} + a_{b-2}x^{b-2} + \ldots + a_1x^1 + a_0$$

A key characteristic of the cyclic code is that any cyclic permutation or end-around shift of a code word results in another code word. The advantage of such codes is that simple feedback shift registers and modulo-2 adders can be used to perform the encoding and decoding operations. Thus, encoding consists of operations on the message by feedback shift registers and modulo-2 adders.

The BCH codes have words of length $n = 2^m - 1$, $m = 3, 4, 5, \ldots$ and can correct any pattern of e or fewer errors using no more that me parity check digits. BCH codes are especially valuable in detecting and correcting randomly occurring multiple errors, that is, errors which affect successive symbols independently.

6.4. MULTIPLE ACCESS

The subject of multiple access is discussed in terms of its general principles in this chapter. Specific discussions applicable to cellular telephony are presented in Chapter 7, and to satellite transmission in Chapter 9.

6.4.1. Frequency-Division Multiple Access (FDMA)

A frequency-division multiple-access (FDMA) system is simply a channelized system. Since international regulations demand that all telecommunications systems using radio be limited in terms of their bandwidth, all radio systems may be said to be channelized. The designation FDMA, however, usually is reserved for those systems in which one channel is distinguished from another solely by the frequency band it occupies. FDMA systems thus can utilize simple terminals requiring no digital addressing, since a signal in the receiver's designated channel is presumed to convey a message for that terminal.

Those FDMA systems handling multiple channels using common electronic amplifiers are subject to intermodulation (IM) problems, and must be designed to keep their operating levels within suitable limits.

6.4.2. Time-Division Multiple Access (TDMA)

Time-division multiple-access (TDMA) systems share a portion of spectrum, usually of substantial bandwidth, with one another. Terminals must take information from a constant rate source and convert it to bursts at a much higher rate, but at a relatively low duty cycle. The bursts must be timed so they do not interfere with one another as a result of propagation delays within the user environment. Thus burst timing depends on terminal distance from the central repeater (cell site or satellite) within the system.

Since signals from two sources generally do not pass through a common electronic amplifier simultaneously, the IM problem is reduced appreciably. Only IM among the sidebands of a given signal now remains of concern.

The demultiplexing of TDMA signals is done by time gating. The receiving terminal generally must be able to recognize its own digital address.

TDMA transmission is said to be more efficient than pure FDMA since wider bandwidths often can be used; consequently, spectrum is not wasted on such factors as receiver phase distortion near band edges, which limits overall spectrum utilization.

6.4.3. Code-Division Multiple Access (CDMA)

A code-division multiple-access (CDMA) system is a major form of spread-spectrum multiple access (SSMA), which is defined as transmission in which the occupied bandwidth is significantly greater than twice the baseband width. Generally excluded are such common systems as wideband FM. In SSMA, each signal occupies all of the bandwidth all of the time. Separation of the signals at the receiver depends upon each one being carried by an underlying waveform that is very nearly *orthogonal* to all of the other signal waveforms.[31] Two waveforms are orthogonal if their product, integrated over some fixed period, is zero. A CDMA system is one in which digital codes are used to form the underlying quasi-orthogonal waveforms.

Each user of a CDMA system has a unique long period digital sequence called a *pseudorandom* sequence, so called because it appears random to the casual observer who does not examine it for a sufficiently long period. This sequence usually is used in either of two ways: (1) it may be combined directly with the information stream to be sent, as in *direct sequence* (DS), or (2) it may be used to select pre-planned spectrum channels among which the transmission is hopped, as in *frequency hopping* (FH). FH may be either slow (many symbols per hop) or fast (more than one hop per symbol). The hopping rate also is called the *chip* rate. The bandwidth is established by the chip rate (FH), or by the rate of the pseudorandom sequence (DS).

Spread spectrum provides a degree of protection against frequency-selective fading since the overall spectrum often is much wider than the faded portion of the spectrum. Correspondingly, the effects of multipath interference can be reduced significantly, since signals that arrive late at the receiver will not match the portion of the code currently being used to decode the signal. Thus, such multipath components are rejected as interference.

The basic detection process in a CDMA system is *correlation*, which measures how alike two signals are to one another. A correlator may consist of a mixer where the two signals to be compared are multiplied together. A match produces a high output value, whereas a lack of match results in zero, or nearly zero. A low-pass filter following the mixer performs an averaging function. One of these signals, of course, is the input, while the other is the receiver's own replica of the code. Interfering signals that are not synchronous with the receiver's local code are spread to a bandwidth equal to their own bandwidth plus that of the local code. The filter, which may be at intermediate frequency (IF), rejects all of the (undesired) signal power lying outside its bandwidth.

CDMA is excellent in discriminating against noise, but all users in the spectrum other than the one of interest look like noise to the one of interest. Accordingly, performance gradually degrades as more and more users attempt to communicate via the given spectrum. The ability of a CDMA system to discriminate against noise (including other users) is called its *processing gain*. Processing gain equals occupied spectrum width divided by information bandwidth.

In DS systems, collapsing the spread-spectrum signal by removing the effects of the spreading sequence is called *despreading*. Since the signal may be viewed as having two types of modulation on it, one for spreading, and the other containing the information, what remains after despreading is the modulation sequence.

CDMA is used in both the FH and the DS forms in digital cellular telephony (see Chapter 7) and also in satellite telecommunications (see Chapter 9).

6.5. PACKET TRANSMISSION

With the development of data transmission has come the realization that the requirements for data differ appreciably from those of voice. Not only must (1) group delay be held within a tight tolerance over the transmission band, and (2) bit-error ratio be held to a value on the order of 10^{-6} or less, but (3) a full-period circuit is wasteful of resources, since data usually is generated and sent in bursts rather than nearly continuously, as is voice. Voice is served best by a *circuit-switched* environment, in which a circuit is available between two parties for the duration of a call. Correspondingly, telegrams can best be handled in a *message-switched* environment, in which messages are sent as soon as transmission facilities are available, but in which delays of minutes or hours may be incurred.

Packet switching, on the other hand, parcels data to be sent into *packets*, each consisting of a header containing address information, a field of data bits, and a set of closing bits called the *trailer*. Compared with circuit switching, packet switching shares many of the advantages of message switching, including the facts that blocking is not noticeable to the user, and that terminal compatibility (code, protocol, and speed) are not required. Perhaps most important to the economy-minded user is the high efficiency of line utilization that packet shares with message switching.

Packet switching has further advantages in a data transmission environment of network flexibility, and the ability to provide for *logical multiplexing* (sharing a high-speed access line with numerous lower-speed devices). A significant disadvantage of packet switching is its requirement for many small switches and processors, along with its need for complex routing and control procedures.

Three types of routing control are used in packet networks. In *dynamic routing*, the network nodes examine the address of each packet and select the outgoing line that will provide minimum delay. Thus the system can respond quickly to changes in network topology or traffic conditions. The packets, however, may arrive out of sequence, so they must be numbered. The second type of routing control is called *virtual circuit routing*. Here all packets of the message follow the same route through the network, but other packets may be interspersed among them for efficient use of the channel. Virtual circuit routing entails the possibility of greater transmission delay than that of dynamic routing. In *fixed-path routing*, successive connections between any two end points always use the same path. This approach is vulnerable to node or link failures.

6.5.1. Packet Voice

A network that has been established primarily for data transmission purposes using packet technology may be required to transmit voice at times. Thus the requirement may exist to send the bits representing voice on a packetized basis. Moreover, conversational voice, with its requirement for delay well under one second, may be needed. The most significant design issue in the packet switching of voice thus is delay, with 250 ms being regarded as the maximum allowable one-way delay for conversational voice.

The design parameters affecting delay are described by the equation:

$$t_{Dmax} = t_{qmax} + t_p + spt_t \tag{6.18}$$

where

t_{Dmax} = maximum delay
t_{qmax} = maximum time a packet may wait in all output queues
t_p = time to build a packet
t_t = transmission time

Then

$$t_p = \frac{\ell_i}{r_i} \tag{6.19}$$

and

$$t_{\mathrm{qmax}} = \frac{n(\ell_i + \ell_n)m}{s}$$

where

ℓ_i = length of packet information field
ℓ_n = length of header
r_i = source symbol rate
n = number of nodes (queues) through which packet travels
m = depth of each queue
s = transmission speed of facility

Delay can be minimized by using a virtual-circuit-based architecture, that is, a fixed route during a given session.

A second design issue in packet voice is echo control. This is a function of the echo return loss at each central office 2-wire-to-4-wire conversion, as well as overall path delay. Delays in excess of 12 ms require echo suppressors, while significantly greater delays require the use of echo cancelers.

The variable waiting times encountered by packets produce random arrival times at the destination. Thus voice synchronization must be achieved. The techniques available for voice synchronization are:

Blind delay. The receiver makes a worst-case assumption.
Round-trip measurement. A special initial packet is used to make the measurement, and the results are sent to the receiver.

Absolute timing. A common master clock is used to establish timing. The receiver then computes a target playback time, based on knowing when each packet was produced.

Added variable delay. Each network element adds its delay to a "delay stamp" as the packet passes through it.

Adaptive delay. A delay is calculated at the receiver based upon the amount of speech in the buffer, or based on repeated round-trip delay measurements.

With reference to the open system interconnect (OSI) model discussed in Section 13.3.1.2, packet voice synchronization is at the *transport layer* or above, while the distribution of the time reference is at the *network layer.*

Since speech consists of talk spurt and silence states, the output queues of the packet switch will be highly variable in length, but nevertheless the required traffic must be supported. Computer modeling may be required to determine if a given system architecture can support the expected traffic. Congestion control involves a tradeoff between queue length (delay vs. speech clipping) and degradation. The most effective solution to the congestion-control problem appears to be restricting the number of channels that can access a given packet link.

The bandwidth efficiency is proportional to the ratio $\ell_i/(\ell_i + \ell_h)$. This indicates the desirability of maintaining a long information field relative to the header; however, the time to build a packet, $t_p = \ell_i/r_i$. Time t_p must be kept short to keep the delay within limits.

Errors in the header should not cause errors in other channels. Thus, error detection must be done on the headers so that erroneous information is not routed to the wrong channel.

6.5.2. Statistical Multiplexing

The discussion thus far has dealt with an environment in which a time slot is assigned for the duration of a call, or at least of a packet of fixed length! This arrangement also is called *synchronous time-division multiplexing* (STDM). On the other hand, if the frame length is adjusted to the activity of the source, the result is called *asynchronous time-division multiplexing* (ATDM). Note that synchronous bit stream transmission and detection are implied, and usually used, with both STDM and ATDM. The framing formats also usually are the same. The difference lies in the channelization within the frame, as well as in the fact that ATDM uses a variable frame length.

The differences between message switching, packet switching, and statistical multiplexing (also called *stat-mux*) are illustrated in Fig. 6.27.[32] As can be seen there, the message switch sends each message on a first-come first-served basis. Packet switching breaks up the messages to allow packets from other sources to be interleaved. Each stat-mux time slot (containing a set of bits called a *cell* or

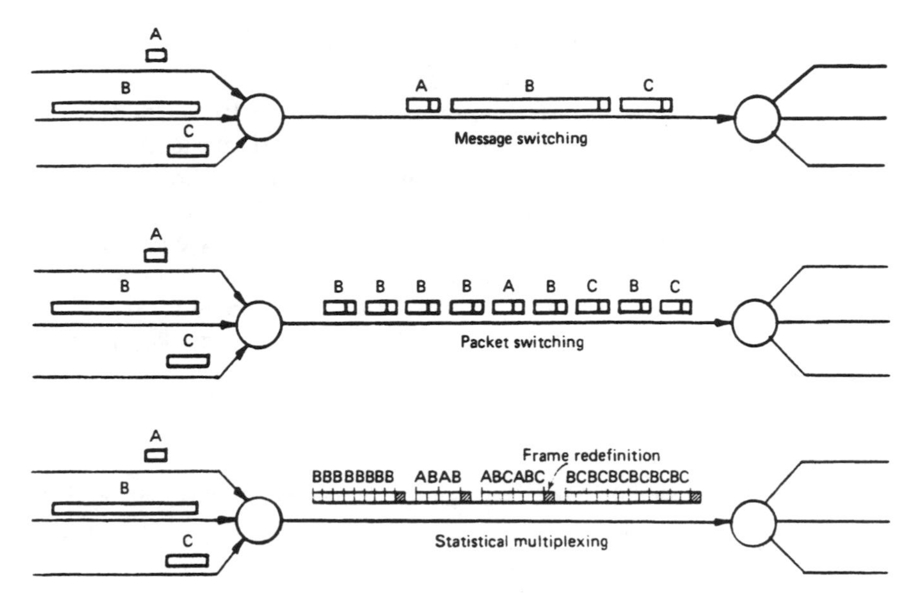

FIGURE 6.27. Comparison of message and packet switching with statistical multiplexing. (Courtesy, J. C. Bellamy, Ref. 28, © Wiley, 1982.)

word) is shorter than the packet slot, and contains only source data. Thus the packet switch transmits larger blocks of data with a header in each block. Stat-mux breaks the messages into even finer blocks and adds periodic frame definition messages to allow the receiver to identify the individual time slots and to switch the incoming data correctly. Note in this regard that Fig. 6.27 shows a new frame redefinition each time the cell addressing information needs to be changed.

Applications for stat-mux include line sharing for multiple interactive terminals communicating with a host computer. A high degree of flexibility is possible, with a relatively high channel data rate for each of a few active sources, or a relatively low rate for a large number of active sources.

The term *fast stat-mux* refers to the use of a data compressor built into the stat-mux. Possible alternatives include the use of fixed-table data compression, based on Huffman principles, or adaptive data compression, which builds a compression table based on the frequency with which various characters appear in the data stream. An example is the Micom fast stat-mux, which uses per channel adaptive data compression. A table is built for each channel through the mux at each end of the path. The chief disadvantage of data compression is the processing delay which it entails.

6.5.3. Fast Packet Switching

Facilities such as fiber-optic systems with a very low (e.g., 10^{-7}) bit-error ratio do not require that error correction be done at every node. Fast packet switching achieves a transmission time reduction via overhead reduction. Error correction may be done only at the ends of the circuit, as part of the application layer. The header can be shortened correspondingly, and there is no delay due to retransmission and recovery. A major advantage of fast packet is its dynamic "bandwidth" (bit rate) allocation on such backbone facilities as T3. Fast packet switching uses the asynchronous transfer mode. The two types of fast packet are frame relay and cell relay.

6.5.3.1. Frame Relay

Frame relay is based on the relaying of packets at the data-link layer (Layer 2), rather than at the network layer (Layer 3) as is done in the packet switching standard, X.25. Frame relay, however, can be added to X.25 equipment. The frame length may be up to 1000 bytes or more.

Public frame-relay services are provided by the local exchange carriers, as well as by the interexchange carriers. Through its use, the number of leased circuits in a local-area network/wide-area-network (LAN/WAN) connection can be decreased since the topology becomes a star rather than point-to-point. Frame relay service entails less installation cost for new service or additions than conventional packet transmission, and the waiting time for new or added service generally can be decreased as well. Frame relay is compatible with a wide variety of data communications equipment, operates with low delay, provides high throughput, and retains the packet-switching advantages of dynamic "bandwidth" allocation, inherent survivability, and dynamic signaling.

The frame relay standard established by the CCITT is I.122, and is an extension of the ISDN standards. In this standard, the address field is called the *data-link connection identifier* (DLCI). Fig. 6.28 compares the LAN packet format with that of frame relay for LAN interworking.

Fast packet switching uses a fixed frame size, in contrast to fast frame switching, which uses a variable frame size. The frame relay standard defines how to switch packet frames within the 64 kb/s D-channel of a 1.544 Mb/s primary rate interface (PRI) line.

6.5.3.2. Cell Relay

Cell relay is the basic transport mode for broadband ISDN (B-ISDN). Cell relay supports very high speeds ($\geqslant$150 Mb/s) to serve multimedia applications. Each cell (also called a *slot*) is 53 bytes long. Fig. 6.29 illustrates the cell relay format.

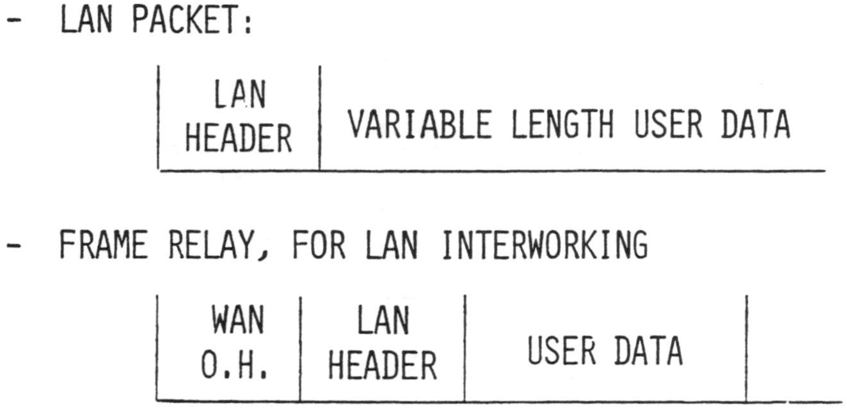

- LAN PACKET:

LAN HEADER	VARIABLE LENGTH USER DATA

- FRAME RELAY, FOR LAN INTERWORKING

WAN O.H.	LAN HEADER	USER DATA

FIGURE 6.28. Comparison of LAN packet format with frame relay format.

WAN O.H.	LAN HEADER	USER DATA	WAN O.H.	USER DATA	WAN O.H.	USER DATA
5	48		5	48	5	48

FIGURE 6.29. Cell relay format. The communications path is established during call setup. An example of a cell relay system is the Stratacom IPX, which is equivalent to the Codex 6290.

6.6. RETROFIT

Digital transmission systems often have been added to existing analog systems on a retrofit basis. The combination thus is a hybrid system, providing the analog channels as before, but adding digital channels in a portion of the baseband spectrum that previously was unused. Often this is more economical than building completely new digital systems. The most commonly used arrangement is known as data under voice (DUV), and is shown in Fig. 6.30. Here the baseband spectrum below 564 kHz on a microwave radio system is utilized. None of the FDM mastergroups occupy this portion of the baseband spectrum. To achieve the needed spectral efficiency of over 3 b/s per hertz, a seven-level partial-response coder is used, as was described in Section 6.2.5. This confines T1 spectrum to the band 0-470 kHz. The resulting baseband signal is used in the 1A-RDS system. Figure 6.31 is a block diagram of the transmitting system. The elastic store performs the function of removing timing jitter. Scrambling is performed next to reduce the level of any discrete spectral components that otherwise might interfere with the

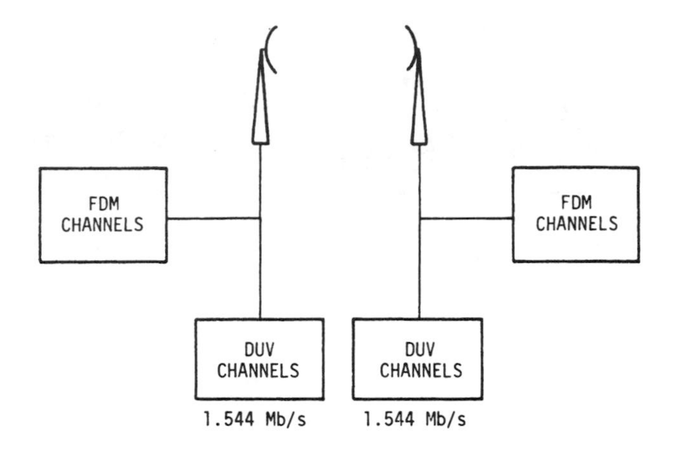

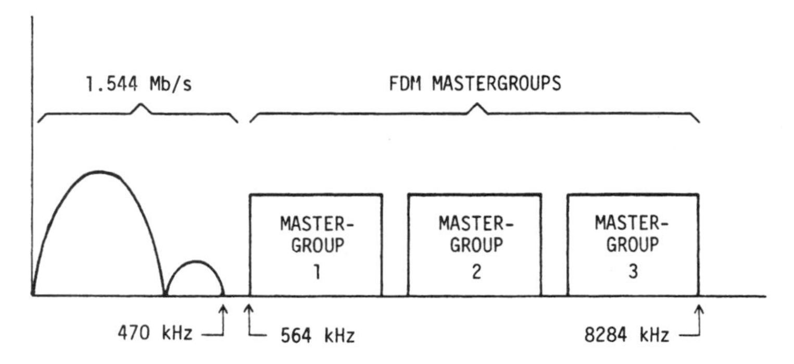

FIGURE 6.30. Data under voice (DUV) system concept.

FDM transmissions. The low-pass filter performs the necessary spectrum shaping to suppress unwanted energy above 386 kHz. The monitoring and status-indicator circuits perform in-service monitoring, as well as failure indications. If the input signal should be lost, a "DS1 substitution signal," consisting of all ones, is sent instead. The 386-kHz pilot is transmitted for synchronization of the receiver circuits. Preemphasis is added because the resulting multilevel output becomes part of the overall baseband (see Fig. 6.30), which is frequency modulated onto the microwave carrier. In any FM system, preemphasis is important in allowing the receiver to attenuate (deemphasize) the background noise, which increases in level with baseband frequency.

Figure 6.32 shows the DUV receiving system, providing an output signal that is a replica of the DS1 input at the transmitting end. The various processes are the inverse of those of the transmitter. The signal monitoring and status indicators show how the receiver is performing.

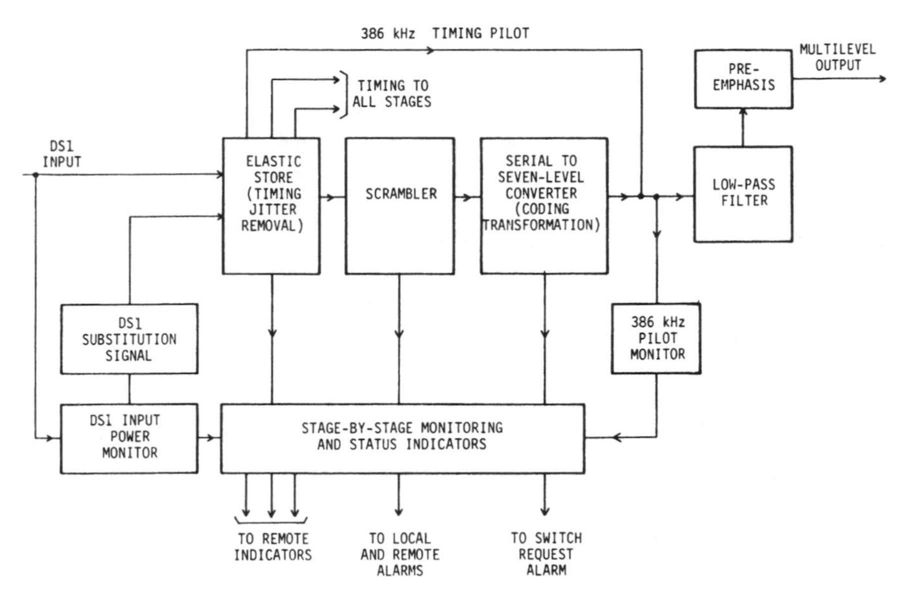

FIGURE 6.31. Block diagram of data under voice (DUV) transmitting system.

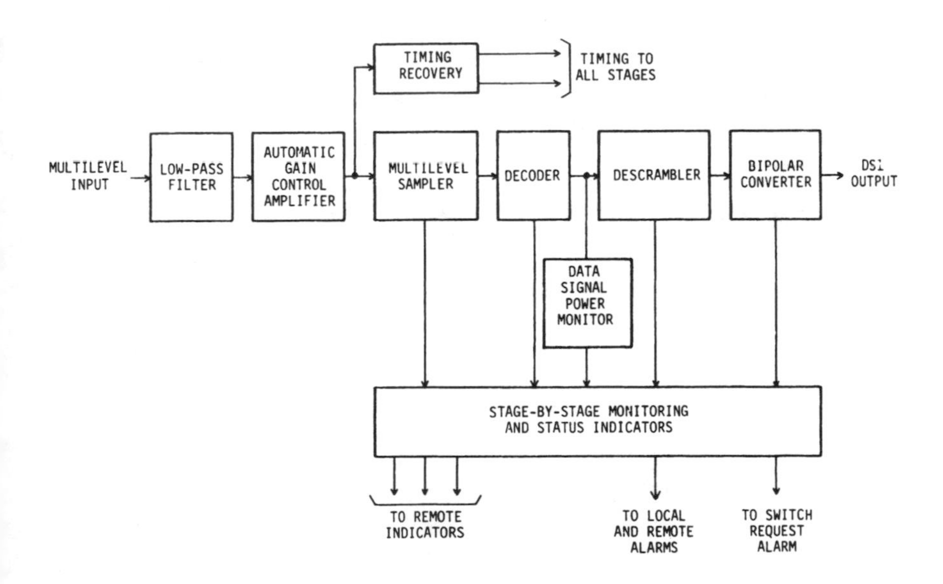

FIGURE 6.32. Block diagram of data under voice (DUV) receiving system.

Several alternative techniques for adding digital transmission to an analog facility have been developed.[33] These are known as *data above voice* (DAV) and *data above video* (DAVID). They allow up to 3.152 Mb/s to be sent above the top of the baseband of an FDM message or video signal on an existing analog facility. The bands used for digital transmission normally cannot be used for analog transmission. Both DAV and DAVID are sent by converting the bipolar data to a binary stream, scrambling, and then translating to the desired frequency, at which the modulation form is QPSK. A bandpass filter keeps the QPSK separate from the lower frequency analog transmissions.

PROBLEMS

6.1. What is the maximum (theoretical) data rate that can be sent error-free over a voice bandwidth channel with a 30-dB ratio of signal to channel noise?

6.2. Explain why BPSK cannot be used in the 3.7–4.2-GHz band. Would the result be any different if dual polarization were used to double the number of voice channels?

6.3. Why are the higher stream rates of the digital hierarchy not integral multiples of the lower stream rates?

6.4. Explain why the tolerable channel-noise level decreases as the number of modulation levels increases, assuming the bit-error ratio is to be held constant.

6.5. A digital transmission engineer has a 20-kHz bandwidth channel in which to transmit a 56-kb/s digital stream. Can QPSK be used to meet this requirement? If not, would 8-PSK be feasible? Would 16-QAM be preferable?

6.6. Why can a 9.6-kBd transmission rate not be sent on a voice-bandwidth line? What is the maximum rate according to Nyquist's theorem?

REFERENCES

1. Nyquist, H., Certain Topics in Telegraph Transmission Theory, *Trans. AIEE*, 47, April 1928, pp. 617–644.
2. Keiser, B. E., *Broadband Communications Systems*, Course 537 Class Notes, © 1988, B. E. Keiser.
3. Feher, K., *Digital Communications: Microwave Applications*, Prentice-Hall, Englewood Cliffs, NJ, 1981
4. Keiser, B. E., *Broadband Coding, Modulation, and Transmission Engineering*, Prentice-Hall, Englewood Cliffs, NJ, 1989.
5. Miller, J., "The Shape of Bits to Come," *The AMSAT Journal*, Vol. 14, No. 4, July 1991, pp. 12–17, 30.
6. See note 4.
7. See note 2.

8. D'Andrea, A. N., and Russo, F., "Noise Analysis of a PSK Carrier Recovery DPLL," *IEEE Trans. Comm.*, Vol. COM-31, No. 2, Feb. 1983, pp. 190, 199.
9. Prabhu, V. K., and Salz, J., "On the Performance of Phase-Shift-Keying Systems," *Bell Syst. Tech. J.*, Vol. 60, No. 10, Dec. 1981, pp. 2307–2343.
10. Spilker, J. J., *Digital Communications by Satellite*, Prentice-Hall, Englewood Cliffs, NJ, 1977, pp. 295–335.
11. Spilker, J. J., *Digital Communications by Satellite*, Prentice-Hall, Englewood Cliffs, NJ, 1977.
12. Feher, K., *Digital Communications: Microwave Applications*, Prentice-Hall, Englewood Cliffs, NJ, 1981.
13. Bellamy, J. C., *Digital Telephony*, John Wiley & Sons, New York, 1982, pp. 496–497.
14. Feher, K., "Digital Modulation Techniques in an Interference Environment," *Multivolume EMC Encyclopedia, Vol. 9*, Don White Consultants, Inc., Gainesville, VA.
15. Bic, J. C., Duponteil, D., and Imbeaux, J. C., "64-QASK Sensitivity to Modem Imperfections and to Interferences," *IEEE Globecom '82 Conference Record*, Miami, FL, Nov. 29–Dec. 2, 1982, IEEE Document CH1819-2/82-0000-0322, pp. B3.3.1–B3.3.6.
16. See note 15.
17. Saito, Y., Komaki, S., and Murotani, M., "Feasibility Considerations of High-Level QAM Multi-carrier System," Proceedings, *IEEE International Conference on Communications, 1984*, pp. 665–671, © IEEE, 1984.
18. Thomas, C. M., Weidner, M. Y. and Durrani, S. H., "Digital Amplitude-Phase Keying with M-ary Alphabets," *IEEE Trans. Comm.*, Vol. COM. 22, No. 2, pp. 168–180 (Feb. 1974).
19. Osborne, W. P., and Luntz, M. P., "Coherent and Noncoherent Detection of CPFSK," *IEEE Trans. Comm.*, Vol. COM-22, pp. 1023–1036 (Aug. 1974).
20. deBuda, R., "Coherent Demodulation of Frequency-Shift Keying with Low Deviation Ratio," *IEEE Transactions on Communications*, Vol. COM-20, No. 3, pp. 429–435 (June, 1972).
21. Weinberg, A., "Effects of a Hard Limiting Repeater on the Performance of a DPSK Data Transmission System," *IEEE Trans. Comm.*, Vol. COM-25, pp. 1128–1133, (Oct. 1977).
22. Mathwich, H. R., Balcewicz, J. F., and Hecht, M., "The Effect of Tandem Band and Amplitude Limiting on the E_b/N_0 Performance of Minimum (Frequency) Shift Keying (MSK)," *IEEE Trans. Comm.*, Vol. COM-22, No. 10, pp. 1525–1540 (Oct. 1974).
23. Lender, A., "The Duobinary Technique for High Speed Data Transmission," *IEEE Transactions on Communication and Electronics*, Vol. 82, pp. 214–218 (May, 1963).
24. Lender, A., "Correlative Digital Communication Techniques," *IEEE Transactions on Communication Technology*, Dec. 1964, pp. 128–135.
25. Kretzmer, E. R., "Binary Data Communication by Partial Response Transmission," *1965 IEEE Annual Communication Conference, Conference Proceedings*, pp. 451–455.
26. Feher, K., *Digital Communications: Microwave Applications*, Prentice-Hall, Englewood Cliffs, NJ, 1981, pp. 47–50.

27. See note 2.
28. Bellamy, J. C., *Digital Telephony*, Wiley-Interscience, New York, NY, 1982, pp. 295, 299.
29. Ishio, H., et al., "A New Multilevel Modulation and Demodulation System for Carrier Digital Transmission," *Proceedings of the IEEE International Conference on Communications, ICC-76*, Philadelphia, PA, June, 1976, pp. 29.7 to 29.12.
30. Maley, S. W., "Telecommunications Systems," Chapter 10 of *Telecommunications: An Interdisciplinary Survey*, ed. by L. Lewin, Artech House, Inc., 1979.
31. Scales, W. C., "Potential Use of Spread Spectrum Techniques in Non-Government Applications," Report MTR-80W335, The MITRE Corporation, Metrek Division, McLean, VA 22102, December, 1990.
32. See note 28.
33. See note 12.

7

Digital Cellular Radio[a]

7.1. INTRODUCTION

Mobile and portable telephone systems present some unique engineering challenges. One of these is their use of radio wave propagation, with its highly variable path characteristics. Another is the fact that the switching system no longer can count on a given subscriber set simply being at the end of a corresponding wire pair. The subscriber set may be anywhere within a telephone company's serving area, or perhaps elsewhere, as what is called a *roamer*.

Numerous wireless personal communication markets have been defined. These include the mobile/portable telephone, personal communication networks, cordless end-instrument type services (such as Telepoint), the wireless local loop, satellite services, and the wireless PBX. Mobile/portable telephony is covered in Sections 7.2 through 7.7. Section 7.8, dealing with the microcellular environment, includes the concepts of personal communication networks and cordless end-instrument type services. The wireless local loop is the subject of Section 7.9, and the PCS interface to the PSTN is discussed in Section 7.10. Satellite services are outlined in Chapter 9.

Wireless telephony systems function through the use of repeaters. A system may use simply a single repeater that serves a city, a portion of a city, or merely a building. If a multiplicity of contiguous repeaters, with automatic hand-off from one to the other, is used, the system is said to be *cellular*. Either type of system may be operated by a common carrier, or by a private carrier. Many of the private carrier systems are programmed for automatic channel assignment. These systems are called *trunked* systems. Subscriber sets may be mobile, portable, or cordless.

---

[a] Much of the material of this chapter is taken from B. E. Keiser, Digital Cellular Radio, Course Notes, © 1992.

The mobile set usually is mounted in a vehicle, and generally has the greatest maximum transmit power of the subscriber sets. Mobiles in North American cellular systems are limited to a transmitter power of 3 watts; some noncellular mobiles, however, have been built with transmitter powers as high as 100 watts. Portables in North American cellular systems are limited to a transmitter power of 0.6 watts, with some noncellular portables below 450 MHz operating at power levels as high as 7 watts. Both battery life between charges and user safety considerations serve to limit portable power.

Cordless end instruments have output powers in the milliwatt range and may be used either as actual cordless telephones, or as end instruments that are part of a cordless PBX system. Other units can be used as part of *personal communication networks* (PCNs) that are connectable to the public switched network, but not actually part of it.

The type of modulation used by mobile radio systems historically has been FM. In a few cases, certain private, industrial and law enforcement systems have been using forms of digital modulation. Digital mobile systems in North America are being built using relatively low-level forms of phase-shift keying (specifically $\pi/4$-DQPSK for public-switched cellular service). A similar modulation type, minimum-shift keying, is being used for European public-switched cellular service. The speech coding for digital mobile service is predictive, syllabically companded coding to provide adequately robust performance in the mobile propagation environment.

Major technical issues in digital cellular radio are spectrum conservation, the highly variable propagation path, and the access technique used by the system. All of these issues are explored extensively in this chapter.

Numerous standards have been devised for mobile telephony. This author views analog as first-generation mobile telephony. It has a long and well known history of service. In addition to the standard FM analog system, a narrowband analog system, using 10 kHz, rather than 30 kHz, channels has been devised by Motorola.

The second generation is time-division multiple access (TDMA). Separate approaches to TDMA are being implemented. Within the US a TDMA standard has been developed under the auspices of the Telecommunication Industries Association (TIA), and the Cellular Telecommunications Industry Association (CTIA) has endorsed TDMA as a standard for North American systems; revisions to that standard will be used to improve the capabilities of digital service. Within Europe, a different TDMA system has been developed by the Group Spéciale Mobile (GSM) of the European Telecommunications Standards Institute (ETSI). An extended TDMA (E-TDMA) system based upon the TIA standard, but providing better spectrum utilization through the use of DSI, is being built by Hughes Network Systems, Inc.

The third approach to cellular telephony is code-division multiple access

(CDMA). A cellular CDMA system is being built based upon research and design efforts led by QUALCOMM, Inc. In addition, a broadband ($\approx$15 MHz) CDMA system is being developed by Inter Digital Communications Corp. for use in the 1800-MHz band. A hybrid TDMA/CDMA system for use in several bands has been developed by Omnipoint Communications, Inc.

7.2. RADIO WAVE PROPAGATION

As will be described in Chapter 8, spreading loss is the major factor causing the signal strength from a source to diminish with distance. As applied to propagation in a terrestrial mobile radio environment, however, typically at lower frequencies than those of microwave relay systems, shadowing, diffraction and multipath reflection play significant roles as well.

7.2.1. The Mobile Antenna

Convenience and economy dictate that the mobile antenna be vertically oriented. This results in a vertically polarized signal, since any monopole or dipole antenna produces, and is sensitive to, signals whose electric vector is parallel to the antenna. Since the direction of the electric vector is defined as the direction of polarization, the result is vertical polarization. Fig. 7.1 shows the basic relationships between the electric field, $\vec{E}$, the magnetic field, $\vec{H}$, and the power flux density, $\vec{P}$. This is called linear polarization.

A basic relationship in discussions of radio-wave propagation is the fact that the product of frequency and wavelength is the velocity of propagation. Thus, in free space, and approximately in air, if f = frequency in Hz and λ = wavelength in meters, then $c = f\lambda$, where c is the speed of light. $c \approx 3 \times 10^8$. A more practical relationship at cellular radio frequencies is $f_{MHr}\lambda_{cm} = 30\,000$.

The antenna pattern will be the same for transmitting as for receiving, and is determined by the current configuration in the antenna element. For omnidirectional coverage to and from a terrestrially located repeater, the pattern in the vertical plane should be maximum at the lowest elevation angles. For this reason, the cellular mobile antenna has a 5/8 λ length above the phasing coil, as illustrated in Fig. 7.2.

Other vertical antenna types also are used in portable/mobile service. Fig. 7.3 illustrates the patterns obtained with various vertical combinations. The loaded quarter wave primarily finds mobile use at frequencies in the 30–47-MHz band. The quarter wave is used at higher frequencies (for example, in the 148–174-MHz band) for mobile service, and for personal portable service in the 450–512-MHz band. The half wave is used for personal portable service in the 800 MHz and higher frequency bands where a highly directional elevation pattern is not desirable. The 5/8 λ over 1/2 λ and the 5/8 λ over 1/4 λ are commonly used

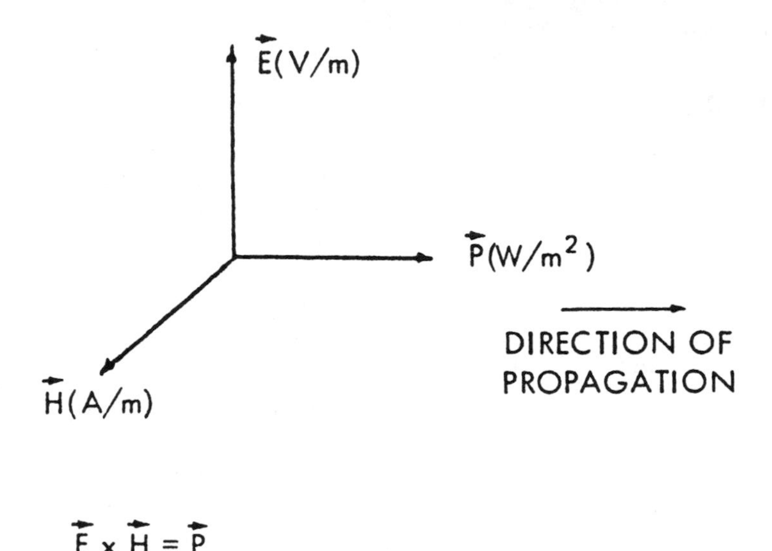

FIGURE 7.1. Basic vector relationships in a propagating wave.

cellular antennas for vehicular service. The 5/8 λ over 5/8 λ antenna is used for applications over 1200 MHz.

Antenna gain relative to a dipole is defined as the extent to which an antenna produces a field strength that is greater than that produced by a simple dipole. The 5/8 λ antenna commonly used in mobile cellular service has a gain of 3 *decibels relative to a dipole* (dBd). The *effective radiated power* (ERP) is defined as the power at the antenna terminals multiplied by the antenna gain (times 2 for a gain of 3 dB).

7.2.2. Signal Fading Characteristics

The fading characteristics of the land mobile environment are different from those of the personal portable environment. This section and the following five sections deal with the land mobile environment (primarily, but not exclusively, in the 800-MHz band), while the personal portable environment is treated in Section 7.8.1.

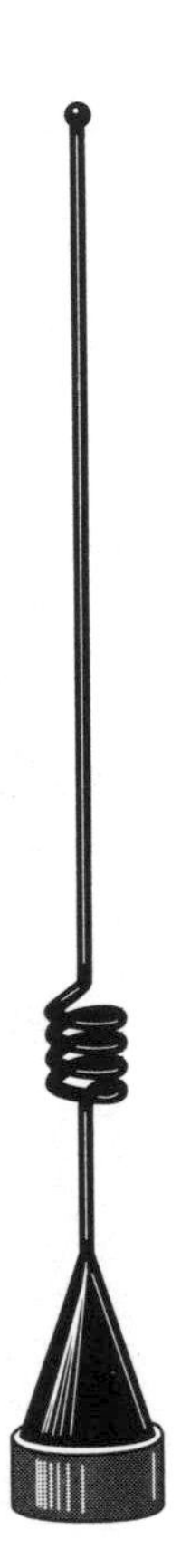

FIGURE 7.2. Cellular mobile antenna.

In the land mobile environment, the motion of the vehicle creates a continuously changing path. The corresponding fluctuations in the diffraction, shadowing, and multipath conditions along the path create rapid amplitude and delay variations, accompanied by Doppler shift due to the motion of the vehicle. Fig. 7.4, taken from measurements in downtown Philadelphia, PA at 836 MHz, illustrates the signal variations present in the absence of any vehicular motion.

Shadowing from buildings and hills is a significant factor in determining received field strength. The received field may arrive via some or all of the following: (1) direct transmission, either with no blockage, or with partial blockage, (2) diffraction, (3) reflection. Each component of the received field has its own time-fluctuating amplitude and phase, the total field being the vectorial sum of its components in time as well as in space.

Shadowing from buildings and hills readily produces fluctuations on the order of 20 dB. The average signal strength at a given distance from the transmitter

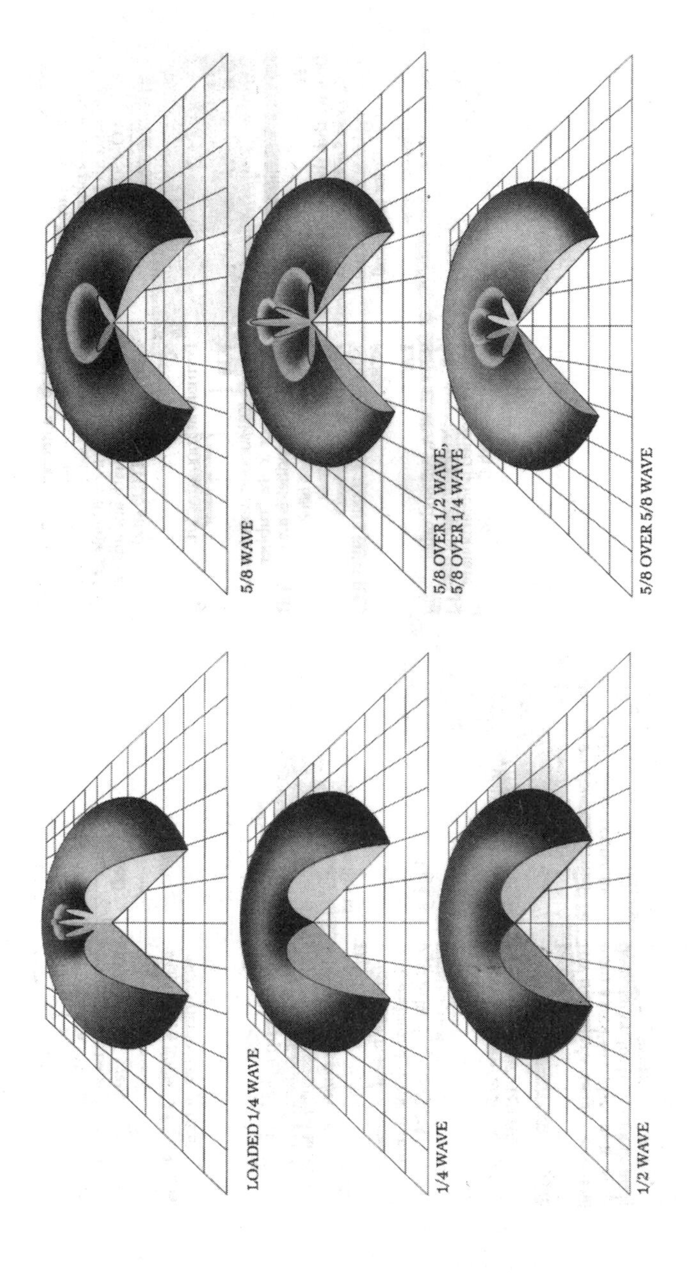

FIGURE 7.3. Patterns of vertical antennas. (Courtesy Larsen Electronics, Inc.)

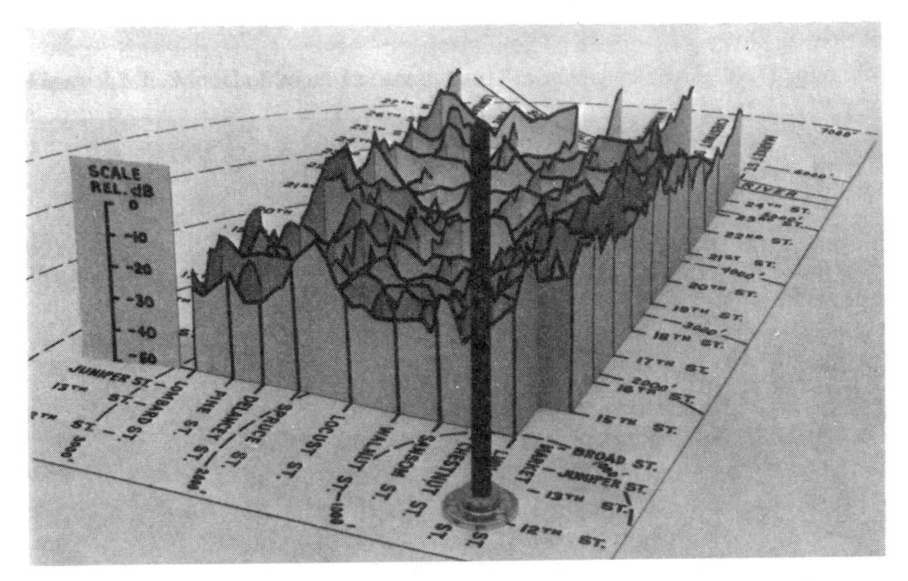

FIGURE 7.4. Signal variations in a segment of downtown Philadelphia (836 MHz). (Courtesy Jakes, *Microwave Mobile Communications*, © AT&T, 1974. Reprinted by permission.)

typically is 10 dB higher in suburban areas than in urban areas because the shadowing is somewhat less in the suburban areas. Correspondingly, open rural areas often exhibit a 20 dB higher signal level than that found in suburban areas.

Foliage attenuation in the 800-MHz band in suburban areas is about 10 dB greater in the late summer than in the winter. In fact, many cellular operators report that they receive their greatest number of service complaints each August.

Reflections can occur in a great many ways. Reflectors may be not only hills and buildings, but even trucks, airplanes and atmospheric discontinuities. Propagation into tunnels is experienced at 800 MHz, a typical tunnel serving as a waveguide. The effectiveness with which a signal propagates into a tunnel depends on the signal's angle of arrival at the entrance to the tunnel. Correspondingly, a signal exiting a tunnel does so with a degree of directionality, and thus may or may not arrive with an adequate strength at the cell site.

The fading environment includes a multipath problem of special significance in digital transmission. This is delay spread. Delay spread basically is the result of echoing, and produces intersymbol interference, depending on the symbol rate. Such interference is quite severe if the symbol time approximates the delay spread. As an example, a 5-μs delay spread causes a 20% symbol overlap at 40 kBd. This places an upper limit on the symbol rate per carrier.

Statistically, the mobile fading environment is described by the Rayleigh distribution. The depth and physical spacing of the fades depends on frequency. The higher the frequency, the shorter the wavelength, and the more closely spaced the fades are. Moreover, the location of the fades differs for each transmit and receive path. As will be seen later, in a cellular system, the transmit and receive frequencies differ from one another to allow full-duplex operation, so the fade patterns for the two directions of transmission also differ. The fade rate increases with vehicular speed. For a vehicular speed of 100 km/h, for example, the fading is rapid enough at 800 MHz to sound like a clicking, rather than the smoother sounding noise bursts that one hears at lower frequencies or lower vehicular speeds.

The equation expressing the Rayleigh cumulative distribution is:

$$p[r \leq R] = 1 - \exp(-R^2/2b) \tag{7.1}$$

where

$p[r \leqslant R]$ = probability that the amplitude r is less than R,
and b = mean power

Figure 7.5 illustrates one author's concept[1] of the mobile fading environment, as described by the Rayleigh distribution.

Doppler shift causes the received frequency to be different from the transmitted frequency. If v = vehicle speed in m/s, f = carrier frequency in Hz, $\omega = 2\pi f$ radians/second, $\beta = 2\pi/\lambda$, and α is the angle between the vehicle direction and the direction of the incoming wave, then

$$\omega = \beta v \cos \alpha \tag{7.2}$$

As an example, at 35 km/h and 900 MHz, the median Doppler offset is approximately 30 Hz, including multipath effects.

7.2.3. Effect of Fading on Digital Transmission

Studies of radio-wave propagation in the 800–950-MHz band have revealed several characteristics especially significant to digital transmission. One of these is the correlation bandwidth, which is defined as the maximum frequency difference for which signals fade together (to within 3 dB of one another). This correlation bandwidth ranges from 150 kHz to over 2 MHz, depending on terrain features, both natural and man made.

The range of signal levels during fading may be as much as 40 to 50 dB. Delay spread typically is 0.5 to 20 μs in urban environments.

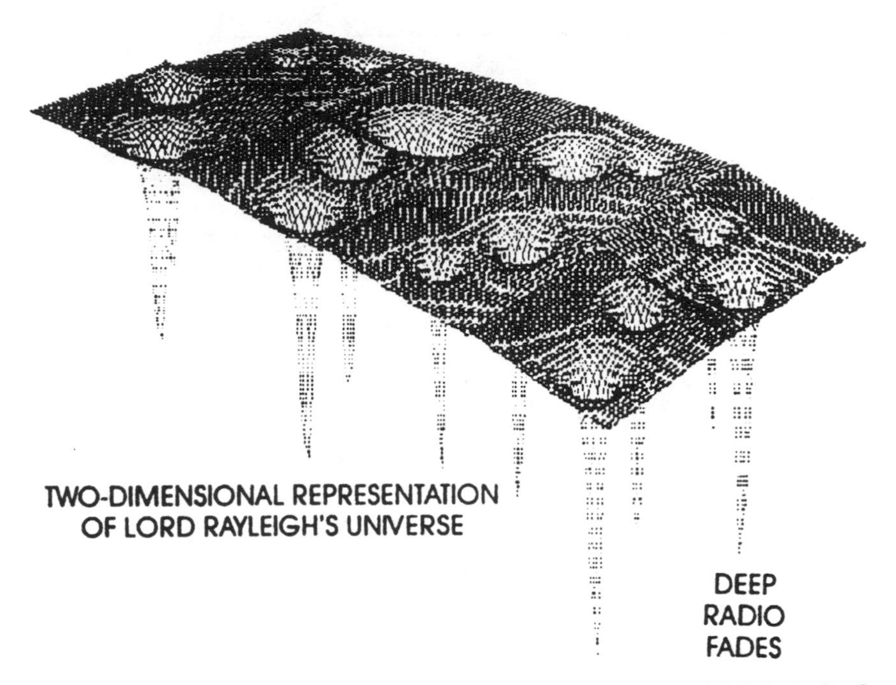

FIGURE 7.5. The mobile fading environment. (Courtesy G. Calhoun, *Digital Cellular Radio*, © Artech House, 1988.)

Delay spread expressed as a percentage of symbol time is shown in Fig. 7.6 for a 200-kb/s rate, and in Fig. 7.7 for a 64-kb/s rate. A comparison of these figures makes the effect of delay spread quite clear, especially as bit rate increases.

The probability density of the Rayleigh distribution describes the short-term amplitude of mobile radio signals. It is

$$p(r) = (r/\sigma^2)\exp(-r^2/2\sigma^2),\ r \geqslant 0 \tag{7.3}$$

where σ = standard deviation. Under the assumption that the time delay spread is less than the reciprocal of the bandwidth, a generally valid assumption using North American TDMA standards, the digital performance characteristics of $\pi/4$-DQPSK are displayed in Fig. 7.8 for a vehicle moving at 100 km/h, and in Fig. 7.9 for a vehicle moving at 8 km/h.[2] Fig. 7.10 shows the performance of Gaussian minimum-shift keying (GMSK), the modulation technique used by the European Global System Mobile (GSM) for a vehicle moving at 40 km/h.

In Figs. 7.8, 7.9, and 7.10, Class 1 bits are those that have been given error correction coding, whereas Class 2 bits are those that must be detected without error correction decoding.

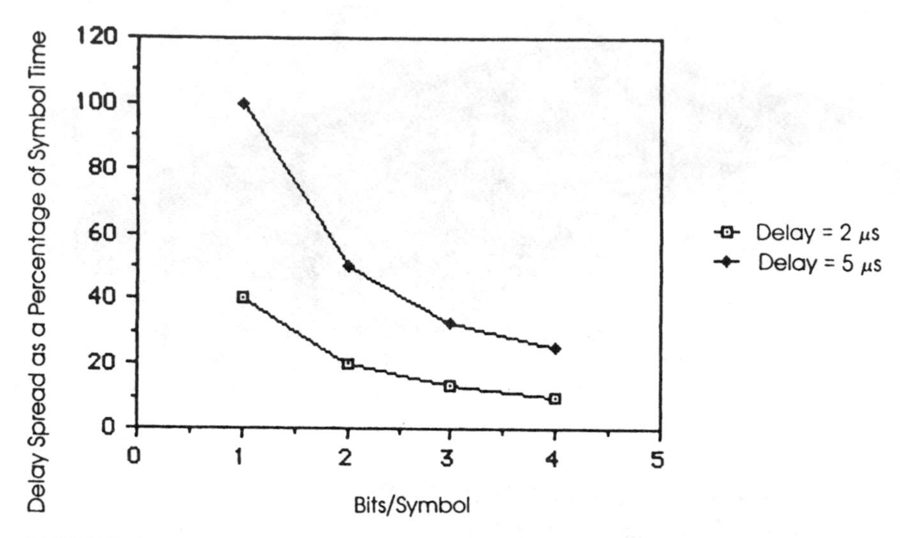

FIGURE 7.6. Delay spread as a percentage of symbol time, 200 kb/s. (Courtesy G. Calhoun, *Digital Cellular Radio*, © Artech House, Inc., 1988.)

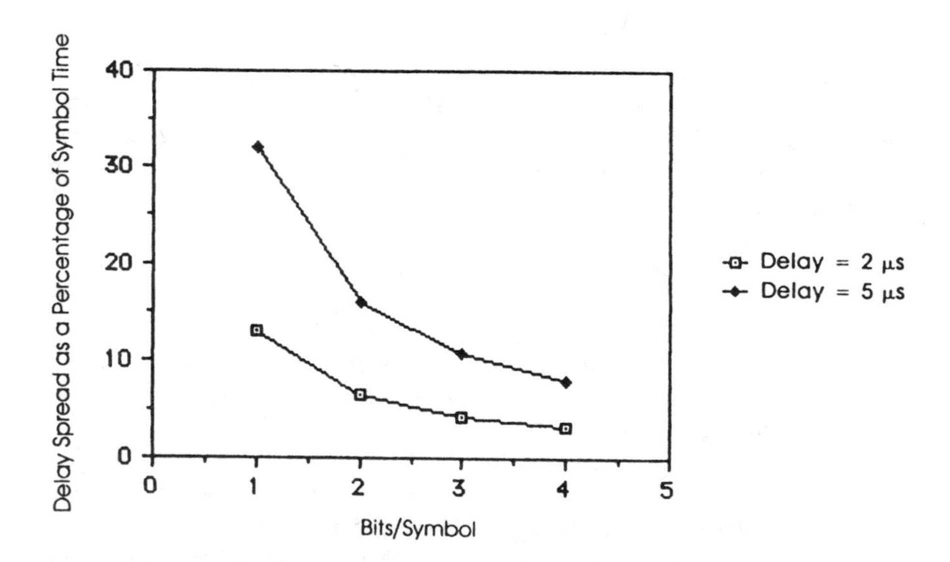

FIGURE 7.7. Delay spread as a function of symbol time, 64 kb/s. (Courtesy G. Calhoun, *Digital Cellular Radio*, © Artech House, Inc., 1988.)

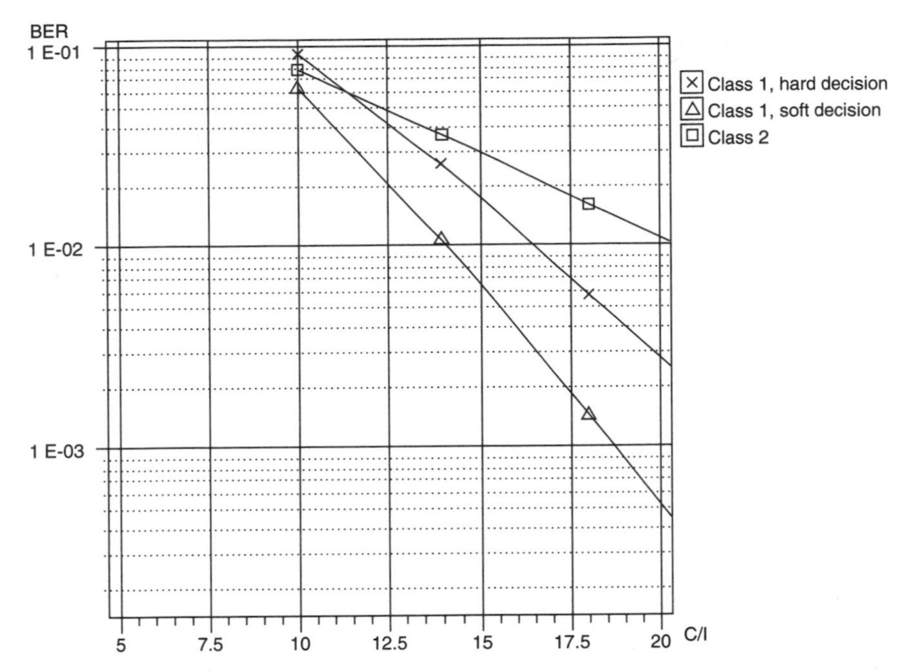

FIGURE 7.8. Performance of π/4-DQPSK in a Rayleigh fading channel at 100 km/h. (Courtesy K. Raith and J. Uddenfeldt, Ref. 2, © IEEE, 1991.)

Figures 7.8, 7.9, and 7.10 clearly show the adverse effect of fading on the mobile signal and the need for speech coding techniques that are robust in a bit error ratio environment on the order of 10^{-2}. The noticeably superior performance of the GMSK signal is attributable, in part, to the fact that the Class 1 bits are given a cyclic redundancy check plus forward error correction at rate 1/2, and $K = 5$ convolutional decoding with interleaving over 8 frames. The comparative performance of the North American and European systems also depends on other factors, such as speech coding, as discussed in Ref. 2, so overall conclusions cannot be drawn from the information displayed here alone.

7.2.4. Effect of Propagation on System Design

Radio-wave propagation has a significant effect on the design of any cellular system. First, the speech coding technique used must be one that can provide satisfactory quality in the presence of the bit errors caused by multipath. Second, the modulation technique must be selected for use in this same bit error environment.

The phase-shift techniques generally are preferred because they are less se-

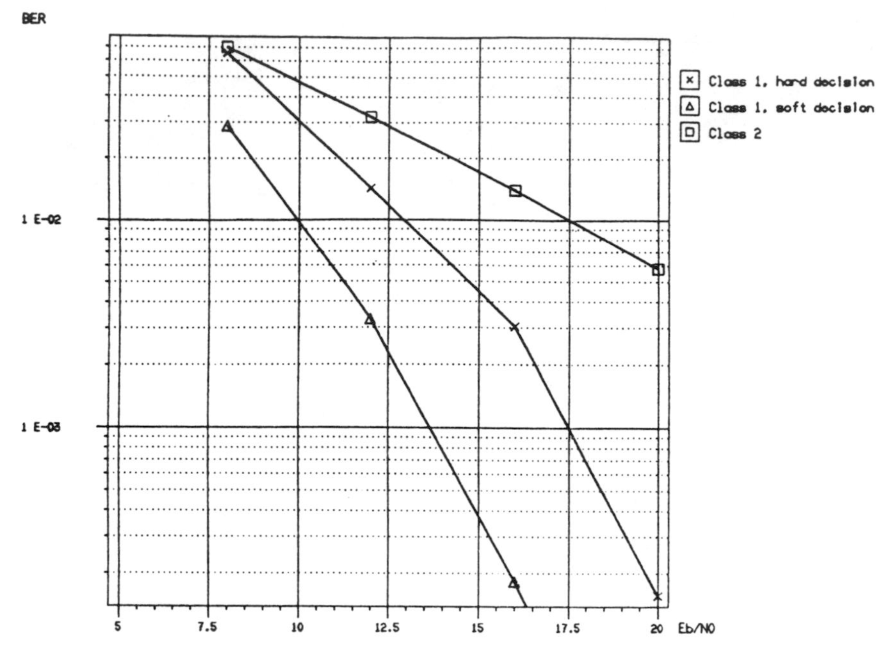

FIGURE 7.9. Performance of π/4-DQPSK in a Rayleigh fading channel at 8 km/h. (Courtesy K. Raith and J. Uddenfeldt, Ref. 2, © IEEE, 1991.)

verely affected by fading. In addition, they occupy less spectrum. With respect to the fading problem, the adjacent channel selectivity that would be required in the absence of fading is only 16 dB. Fading, however, may cause the desired signal to be as much as 22 dB low, and the undesired adjacent-channel signal to be as much as 22 dB high. Accordingly, the requirement on adjacent-channel selectivity becomes 60 dB.

Noncoherent phase demodulation can circumvent the need for fast carrier recovery, which may be difficult to achieve in a fast fading environment. While spectrum economy desires might lead to interest in relatively high level modulation techniques, the use of more than 16 levels has not been found feasible in the mobile fading environment.

Spectrum efficiency requirements dictate that the minimum bit rate for satisfactory quality be used. This makes the more robust coding and modulation techniques especially important in that they require less forward error correction, and thus use the channel more efficiently.

Table 7.1 summarizes the major mobile propagation problems and the system approaches used to mitigate them.

Several types of diversity can be used for the mitigation of fading: time, fre-

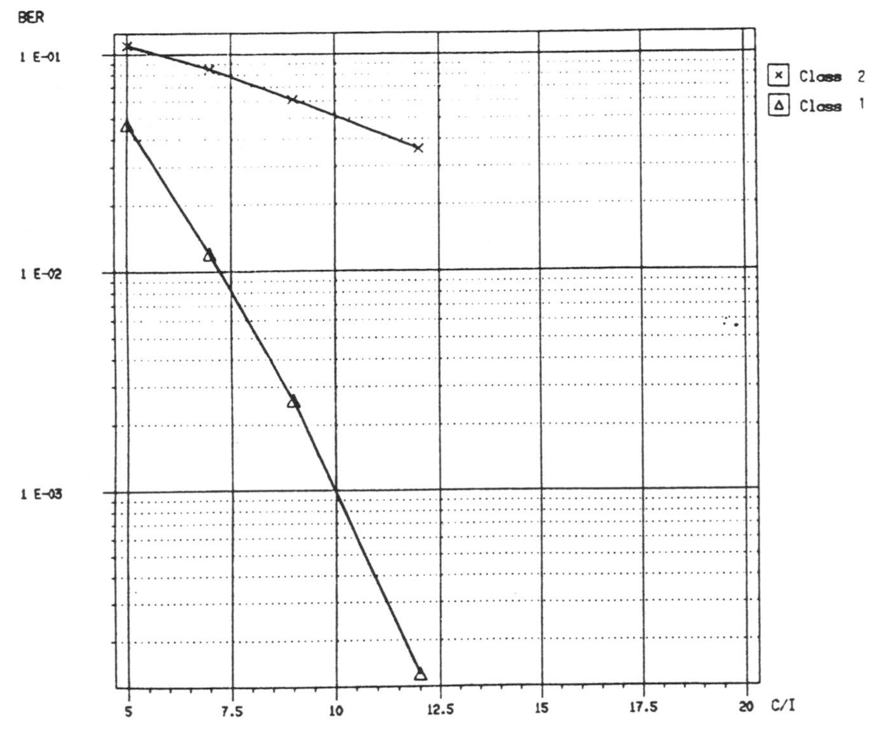

FIGURE 7.10. Performance of Gaussian minimum-shift keying in a Rayleigh fading channel at 40 km/h. (Courtesy K. Raith and J. Uddenfeldt, Ref. 2, © IEEE, 1991.)

quency, and space. Time diversity can be used in the form of interleaving, as is done in the EIA/TIA standard for North American digital cellular systems,[3] currently designated IS-54C. An alternative type of time diversity is error-correction coding, as was described in Chapter 6. Frequency diversity is implemented in the use of spread spectrum techniques; the basic conceptual description of these techniques was provided in Chapter 6, while their utilization in the CDMA cellular system is described in Section 7.4.2.

One form of space diversity involves the reception of multiple signal paths.

TABLE 7.1. Mobile Propagation Problems and Their Solutions

Problem	Cause	Solution	Effect
Delay spread	Multipath	Adaptive equalization	$\leqslant 20\ \mu s$
Flat fading	Weak signal	Diversity	5 to 10 dB
Flat fading	Weak signal	Error-correction coding	3 to 5 dB

This can be done through the use of diversity receiving antennas at a cell site. A totally different type of multiple signal path use is found in the rake receiving circuits of the CDMA system. In this case, the different signal paths are given different delays so they combine additively at the receiver. The hand-off process which occurs as a mobile moves from one cell to another is yet a different type of space diversity.

7.2.5. Frequency Reuse

The ability to reuse frequencies within a metropolitan area distinguishes a cellular system from its predecessors. An analog or TDMA digital cellular system will use one of, for example, 4, 7, or 12 different frequency sets at each cell site. Cell repeating patterns have been devised, as illustrated in Fig. 7.11, so that no two adjacent cells will use the same frequencies.

As was noted previously, however, not all cells have the same size. This size difference is illustrated in Fig. 7.12, in which small cells are found in the urban high-density areas, and large cells are found in the rural low density areas.

Cellular system architects begin their work by drawing their cells as neat looking hexagons. Nothing, however, constrains the radio waves to behave in such a way. Quite to the contrary, the boundaries of a cell are rather irregular, as illustrated in Fig. 7.13. The actual coverage depends on terrain, with multipath producing dropout spots, and the motion of other vehicles causing signal flutter.

The maximum distance reached by a given cell-site transmitter depends upon its height relative to the surrounding terrain, while higher levels of transmitter power help to fill in the dropout spots.

From the foregoing description, one can see that a mobile is not simply handed off to the next cell when it reaches the hexagonal boundary. Instead, near cell boundaries, short-term fluctuations of each signal may greatly exceed the difference in average signal levels, as illustrated in Fig. 7.14. In some locations, both cells may provide good signals, whereas in others, neither one may provide a strong signal. Thus the pattern becomes one in which the coverages are intermixed. The interference zone is wide because of the statistical and constantly changing nature of the mobile environment.

The separation between cells using the same frequency set is governed by the *carrier-to-interference ratio, C/I,* required to achieve the desired transmission quality, as well as the fade margin needed to handle the expected signal level fluctuations. The use of diversity reduces the required fade margin.

The amount of reuse achievable depends on the separation D relative to the cell radius R, where D is the cell-center to remote-cell-center distance, assuming the two cells are using the same frequency set. For hexagonal cells using a 7-cell repeat pattern (six interferers) and for an assumed fourth power decrease of signal

4-cell repeating pattern

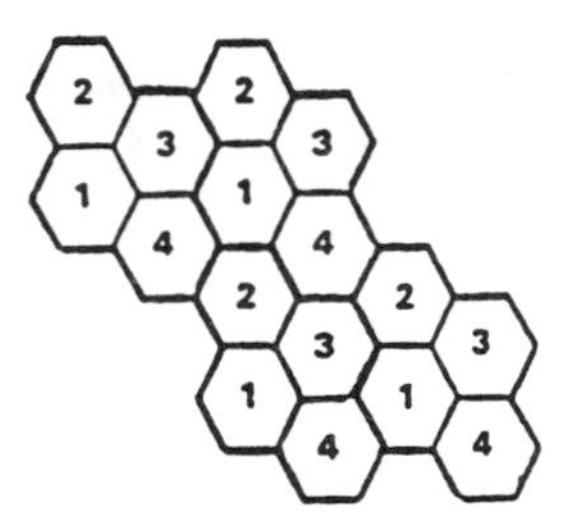

7-cell repeating pattern

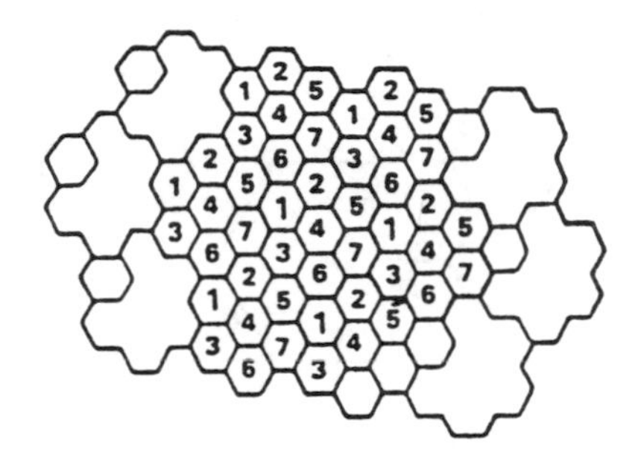

12-cell repeating pattern

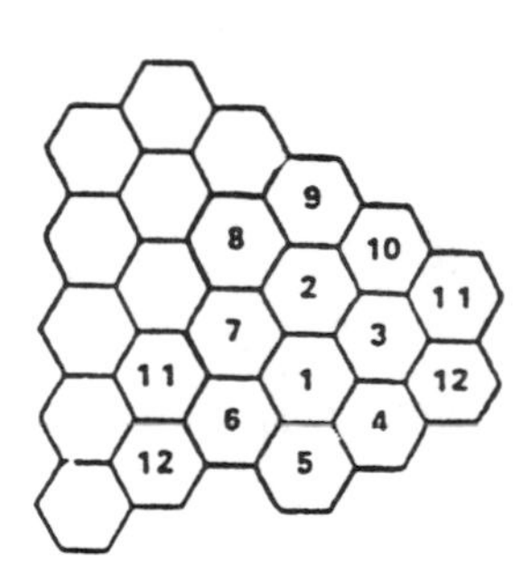

FIGURE 7.11. Cell repeating patterns. (Courtesy Frost & Sullivan, Inc. Reprinted by permission.)

strength with distance (a function of frequency and terrain), the D/R ratio is given[4] by the equation:

$$\frac{D}{R} = \sqrt[4]{6(C/I)},\ R \geqslant 4 \text{ km} \tag{7.4}$$

where C/I is the carrier-to-noise ratio that is exceeded 99% of the time. As an example, if the required C/I is 20 dB and R is 10 km, find D. The solution: $D =$

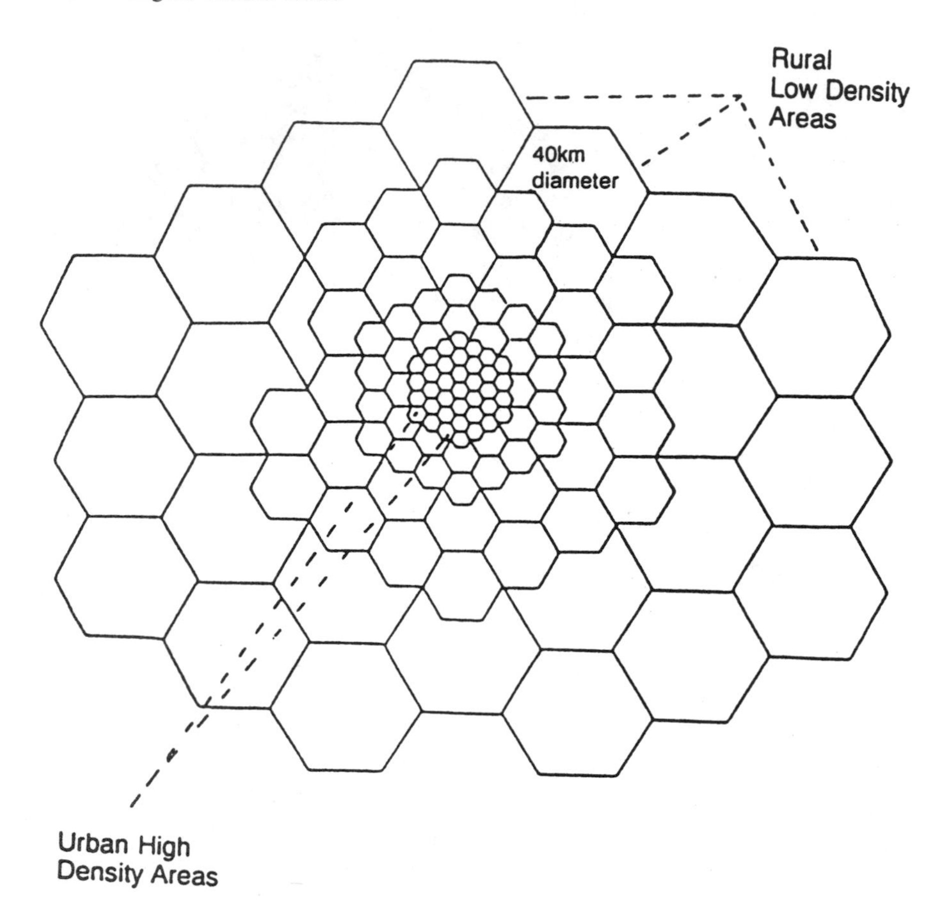

FIGURE 7.12. Cell sizes. (Courtesy Frost & Sullivan, Inc. Reprinted by permission.)

$10 \text{ km} \times \sqrt[4]{6 \times 100} = 49.5$ km. This means that cells closer than D must use other frequency sets. Bear in mind, however, that hexagonal cells have been assumed. In reality, such factors as terrain features and differing cell sizes will alter this value of D.

As can be seen from Eq. (7.4), the lowest C/I yields the lowest D/R, and thus allows frequency reuse at the closest possible distance. This results in the greatest number of circuits per MHz per km^2. Since digital transmission generally can be accomplished at a lower C/I than analog transmission, a 4-cell reuse pattern might be substituted for a 7-cell reuse pattern by using TDMA rather than FM (analog). For a CDMA system, full reuse is achievable, as will be explained in Section 7.4.2.

The 12-cell site repeat pattern uses an omnidirectional site antenna at each cell

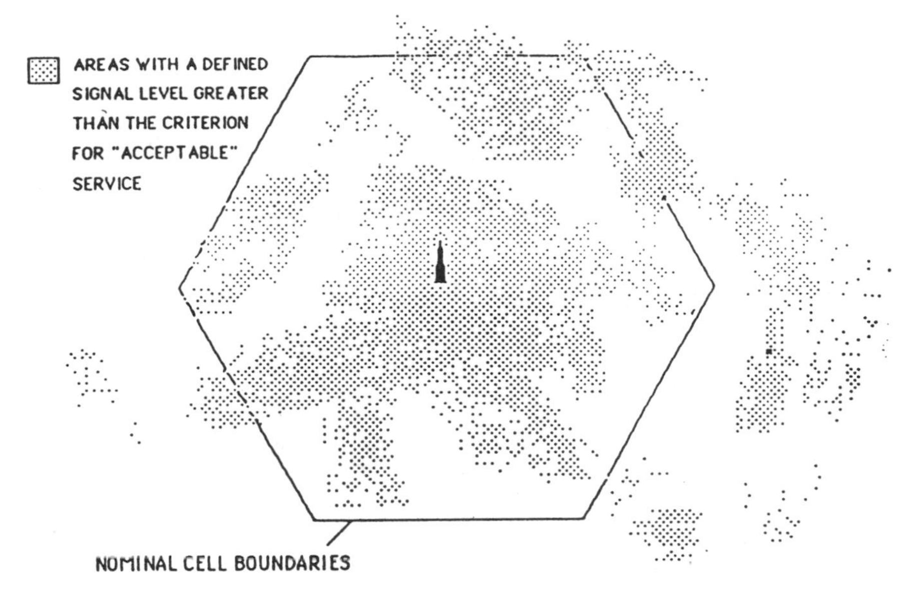

FIGURE 7.13. Cell coverage characteristics. (Courtesy G. Calhoun, *Digital Cellular Radio*, © Artech House, Inc., 1988.)

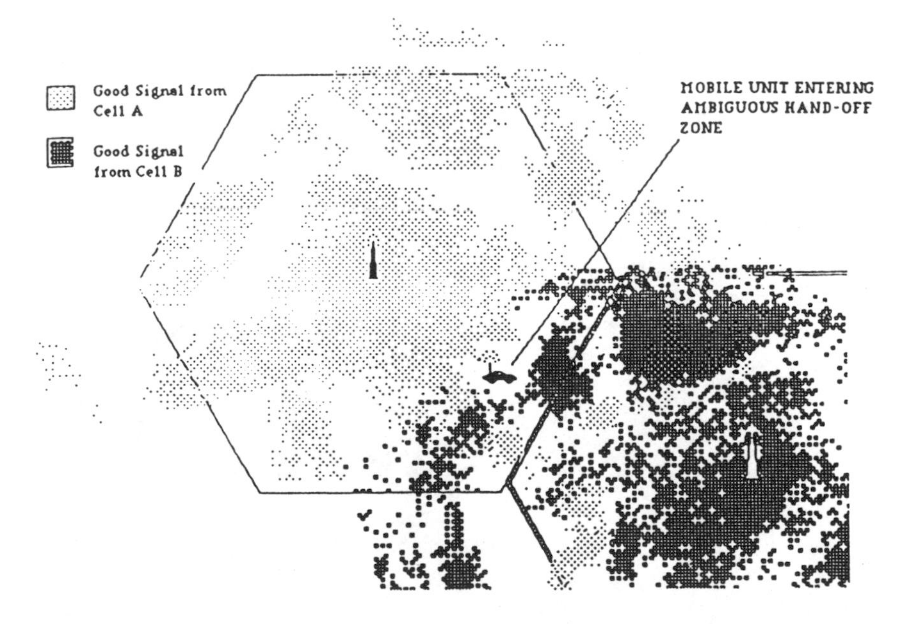

FIGURE 7.14. Coverage of two adjacent cells. (Courtesy G. Calhoun, *Digital Cellular Radio*, © Artech House, Inc., 1988.)

site. The cell's coverage area for analog service is defined as having a $C/I \geqslant 17$ dB over at least 90% of the area. A 7-cell site repeat pattern may use 120° beamwidth antennas. A 4-cell site repeat pattern may use 60° beamwidth antennas. In this arrangement, each primary cell is divided into six *sector cells*, a sector cell being defined as an area with a unique frequency set. A vehicle leaving a sector requires a hand-off to a new channel assignment.

7.2.6. Channel Assignments

North American cellular systems based on analog and TDMA digital transmission use the channel assignment arrangement shown in Fig. 7.15. As can be observed from this figure, there is a 45-MHz spacing between the transmit and the receive frequencies. The mobile-to-cell-site path (sometimes called the *uplink*) always is the lower of the two frequencies, while the cell-site-to-mobile path (sometimes called the *downlink*) always is the upper of the two frequencies. The channel spacing is 30 kHz for systems using TDMA. CDMA (See section 7.4.2.) use channels that are 1.25 MHz wide. The wireline carriers are allocated the B channels, while the nonwireline carriers are allocated the A channels.

All carriers are retaining a pool of analog channels (30 kHz each) to serve those users not capable of operating with the carrier's selected digital standard.

7.3. CELLULAR SYSTEM OPERATION

Communications to the subscriber set in a cellular system is provided by routing all traffic through a *mobile switching center* (MSC), which is a cellular central office. Each subscriber set accordingly has a three-digit exchange number which corresponds to that of its MSC. In North American systems, the MSC often is called a *mobile telecommunications switching office* (MTSO), or an *electronic*

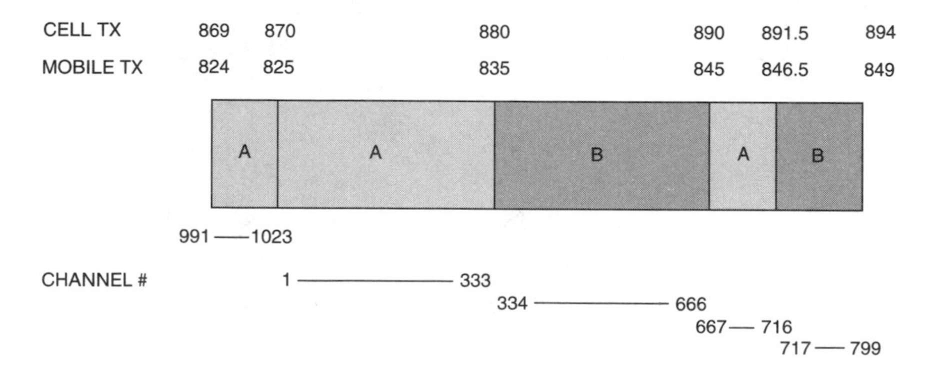

FIGURE 7.15. North American cellular channel assignments.

mobile exchange (EMX). Fig. 7.16 shows the signal path from the MTSO or EMX via a cell site (also called a *base station*) to the mobile. The path from the MTSO or EMX to the cell site is a fixed terrestrial link, such as a microwave or fiber-optic link, while the path from the cell site to the mobile is the radio link in the 800-MHz band. (See Section 7.2.6).

Large metropolitan-area operating systems may use multiple MSCs interconnected by microwave or wireline facilities to achieve the needed traffic handling capability, as well as for system redundancy.

MSCs not only must perform the switching functions, but also must contain the subscriber data bases. The industry now is deploying *home location registers* (HLRs) to handle subscriber data bases, thus removing that burden from the MSC. Typical HLRs can store one million subscriber numbers, whereas MSCs often can store only a fraction of that (16 000 to 100 000).

Typical cellular systems may have cell radii as small as 1 km, and as large as 50 km. Cells smaller than 1 km generally are called *microcells*; they are found in busy urban areas that experience high levels of traffic density. Microcells are discussed in Section 7.8. The large-diameter cells serve the sparsely populated, lightly travelled areas, and sometimes are called *boomer* cells.

The remainder of this section describes cellular system operation using the TIA TDMA standard. Many similarities exist with respect to the CDMA standard, but the reader is referred to Section 7.4.2.1 for further information regarding the CDMA standard.

Mobile end-instrument transmitter power in North American systems is a max-

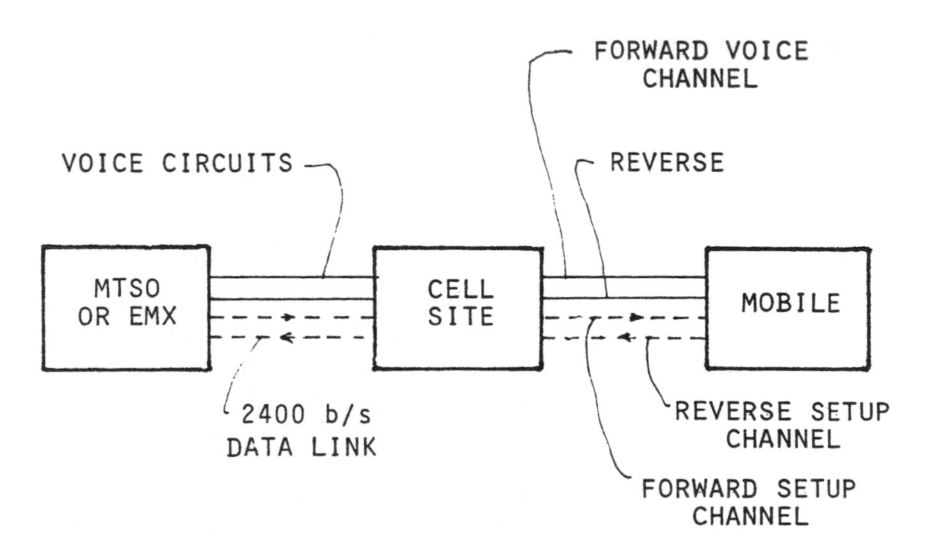

FIGURE 7.16. Signal path between switching center and mobile.

imum of 3 watts to a 3 dBd antenna (5/8 λ). Portable end-instrument power is a maximum of 0.6 watt to a λ/2 antenna. The typical environment in which the portable set is used, however, is often felt to result in a disadvantage averaging about 3 dB, for an effective value of −3 dBd.

Cell-site transmitter power may range from a maximum of 7 to 45 watts per channel. Considering combiner and feed-line losses of about 6 dB, but an antenna gain of 5 to 17 dBd, the ERP typically is on the order of 50 to 100 watts. The FCC limit is 500 watts ERP for an antenna not exceeding 500 feet height above average terrain (HAAT). The function of the combiner is to allow from 5 to as many as 70 channels to be transmitted from a single antenna. Receiving diversity often is used at each cell site.

Signaling is done by using a 10-kb/s stream; the actual data rate is 1.2 kb/s, since 3:1 voting and BCH coding are used.

7.3.1. Call Setup

One or more channels at each cell site serve as setup channels, depending upon the cell configuration (omnidirectional or sectored—discussed later in this section), as well as the infrastructure vendor. Setup channels transmit and receive the continuous digital messages needed to set up calls. When a subscriber set (assumed to be portable or mobile in the discussions of this section) is turned on, the set scans for the best setup channel (in terms of received signal strength), and selects that channel to monitor. Monitoring consists of listening for its *mobile identification number* (MIN), receipt of which usually indicates an incoming call. The MIN is a 34-bit number derived from the subscriber set's 10-digit telephone number. It is not to be confused with the subscriber set's *electronic serial number* (ESN), which is a 32-bit number uniquely identifying that subscriber set to any cellular system.

In some cases, the system may direct the subscriber set to tune to a secondary set of setup channels. The secondary set typically is used for digital overlay systems that have been supplied by a different infrastructure vendor.

Paging refers to broadcasting a message from all cell sites within a system to locate a given subscriber set, and thus to determine that the set is available to receive an incoming call. Directional antennas at the cell site may aid in obtaining a general indication of the set's location to aid in hand-off to the correct *neighbor*, or adjacent cell, when the time for hand-off arrives. Prior to the subscriber set responding to a page attempt or initiating a call, it rescans the setup channels to make certain that the call is set up on the best serving cell site.

Both the North American and the European (GSM) system provide for a subscriber set to identify itself periodically to permit the automatic transfer of its service from its home system to a remote serving system. This arrangement may be viewed as automatic *follow-me roaming*, or *autonomous registration*. This automatic procedure also can reduce the load on the paging channel by providing an indication as to the subscriber set's general location at all times.

Whether autonomous registration is used or not, the subscriber set may continue to monitor its selected setup channel until the data stream directs it to sample another channel, or until the signal strength of the setup channel drops below a specified threshold. Subscriber sets using the North American system have an internal timer that triggers a setup channel rescan every five minutes. In some cases, the system may send out global messages over the forward setup channels directing all subscriber sets to rescan.

Most cellular systems began service with omnidirectional cell sites. As they matured and required more capacity, they converted to sectorized cell sites which increase capacity by improving the amount of frequency reuse being done. When a cell site is sectored, the voice channels usually are split into 3 or 6 sectors, thus allowing the frequencies to be reused more often. The setup channel may remain omnidirectional, or may be sectorized as well. The advantages of directional setup channels have resulted in that feature being implemented on a widespread basis.

Setup channels use two bits to distinguish call attempts and page responses occurring in cofrequency cell sites. These two bits result in four *digital color codes* (DCCs). The system inserts the bits on the forward setup channel. When the subscriber set uses the corresponding reverse setup channels it transponds the DCC. If the cell site does not receive the correct DCC, it does not process the call. This minimizes call setups on the wrong cell site, as well as system reorders.

If all voice channels in a cell are occupied during call setup, the system operator may define directed retry calls. Directed retry allows up to 6 specific setup channels (of other cells) to be sent to the subscriber set after it has made an access attempt. The set will scan those 6 channels, select the strongest one, and make one more attempt to obtain service. If successful, the call is set up on what is the second best server, and its quality may become degraded. The call thus must be handed off to the best server as quickly as possible. Some operators have disabled directed retry.

A digital control channel with increased capacity for setup and other advanced features often associated with digital cellular is in the standardization process, and is expected to be issued in 1994 as TIA standard IS-54D.

7.3.1.1. Fixed-Station Originated Call

A dialed call from a fixed telephone is routed to the called subscriber set's home switching office (MSC). The MSC converts the digits to the subscriber set's MIN, and then instructs the cell sites to page the subscriber set. This paging must be broadcast over the entire service area unless specific subscriber set location information is available from a just completed call, or from autonomous registration by the subscriber set. Upon receipt of the page, the subscriber set attempts seizure of the reverse setup channel if it is idle. The subscriber set then sends a seizure precursor telling the system which cell site it is calling. The subscriber set then looks for the response from that cell site, and repeats the try if it is not received. If the subscriber set is unsuccessful, the caller receives a recorded message stating that the subscriber set is not available for service. If successful, the subscriber set sends its identification to the cell site. The cell site then notifies the MSC of the subscriber set's response via the reverse data link, whereupon the MSC assigns an idle voice channel for the subscriber set's use.

The MSC then directs the serving cell site to send a data message over the forward setup channel to ring the subscriber set and to direct it to tune to the assigned voice channel. In responding, the subscriber set sends a *coded digital verification color code* (CDVCC) back to the cell site. The CDVCC is used to distinguish the communications from CCI that might be present on the radio path due to signals from other cells. When the subscriber answers, the cell site recognizes the answer, and changes the terrestrial trunk to the off-hook state. The MSC then removes audible ringing and establishes the talking connection.

Voice-mail service now is offered by most operating companies, allowing the caller to leave a message for an available but unanswered subscriber set.

7.3.1.2. Subscriber-Set Originated Call

The subscriber dials the desired number while on-hook, and then pushes the *SEND* button to place the call. This is called preorigination dialing. It allows a mobile subscriber to place a call while driving without the distraction of having to dial ten or more digits just before placing the call. Pressing the *SEND* button sends a digital message over the reverse setup channel to the cell site. This message consists of the subscriber set's MIN, the called number, and a *pro-forma* request for a voice channel. The cell site relays this information via the data link to the MSC. In addition, the cell site tells the subscriber set which voice channel to use.

The MSC routes the call through the wireline network, determining the routing and charging information based on the dialed digits, and the subscriber set's chosen long distance carrier, if applicable. When the called party answers, the conversation begins. Upon completion of the call, the MSC informs the cell site, and the voice channel is closed. For a subscriber set-to-subscriber set call, the switching is done within the MSC if both subscriber sets are served by the same MSC.

7.3.1.3. Disconnect

If the subscriber set goes on-hook, it transmits the CDVCC and turns off the transmitter. After the cell site receives this message, it places an on-hook signal on the appropriate wireline trunk. In response to the on-hook signal, the MSC idles all switching office units associated with the call and sends disconnect signals through the wireline network. The MSC then commands the serving cell site to shut down the cell transmitter associated with the call.

If the land party goes on-hook, the MSC receives the disconnect signal from the wireline network, idles all switching office units associated with the call, and sends a release order message over the data link serving the cell site. The cell site then transmits the release command over the voice channel to the subscriber set. The subscriber set confirms receipt of the message by using the same sequence as described above for subscriber-set disconnect. The MSC recognizes release by the subscriber set and uses the data link to command the serving cell to shut down the transmitter used for the call.

7.3.2. Hand-off

In an analog cellular system, control programs at the cell sites monitor the received signals of all calls in progress to obtain general location information. This location information is sent continuously to the MSC. Scanning receivers at each cell site monitor the signals of subscriber sets that are using neighbor cells, thus learning when such subscriber sets might be served better by the monitoring site than the currently serving cell. The MSC thus has signal strength information on a given subscriber set from not only its serving cell, but usually from one or more neighbor cells as well, especially if the subscriber set is anywhere near a place where hand-off might be appropriate.

Whereas in analog systems, hand-off decisions are based on uplink signal strength, the North American TDMA systems use downlink signal strength instead. The subscriber set measures the signal strength and quality from the available cell sites and forwards that information to its serving cell site for evaluation and possible hand-off, if warranted. In addition, the signal strength and quality information is sent to the MSC, and may be used by it in the hand-off process. This is done using a function called *mobile assisted handoff* (MAHO), which consists of the mobile *received signal-strength indication* (RSSI) and the BER for either the serving cell's signal or that of any other channel, as may be requested by the cell site.

If the MSC decides that hand-off is needed, the procedures required to transfer the call to an idle voice channel in another cell are initiated. Since the mobile can send and receive only on a single channel pair at a time, however, the cell site may have to send any quick-action commands in place of the bits being used for voice. This procedure involves use of the *fast associated control channel* (FACCH). The bits used to transmit the voice are interrupted very briefly to allow data to be sent instructing the subscriber set to change to a different channel.

In the EIA/TIA standard for TDMA (see Section 7.4.1), FACCH bits are sent in place of a block of speech bits, allowing "fast" action compared with what can be accomplished using the *slow associated control channel* (SACCH), which operates continuously, but at a rate of only 6 information bits every 20 ms, or 300 b/s. SACCH normally is used for the transmission of control and supervision messages between the cell site and the subscriber set.

When the mobile has switched frequencies, the MSC verifies that the hand-off has been accomplished successfully. Since the hand-off is completed in less than 0.2 second, the conversation is not interrupted; ordinarily the user is unaware that hand-off has occurred, although performance that was beginning to sound somewhat poor may be observed to improve suddenly!

If no channel is available for hand-off, the subscriber set must await an available channel in the new cell. Meanwhile, conversation continues via the current cell during the waiting period. Emergency calls (for example, to 911) may be placed in a high-priority queue for hand-off to the new cell.

7.3.3. Roaming

Roaming refers to the operation of a subscriber set through other than its home carrier. Roaming within a home carrier's service area may occur if none of the home carrier's channels are available; consequently, the subscriber set can obtain service only from the competing carrier or from an outside carrier that may be providing a good signal path to the immediate vicinity of the subscriber set.

Follow-me roaming (FMR) was introduced by General Telephone & Electronics (GT&E) in 1988. In using FMR, the subscriber enters a two or three digit code (for example, *19 or *32) upon entering a new carrier's service area. The code sets up a temporary registration with the new carrier and includes the forwarding of calls made to the subscriber via the home system and the selected long distance carrier. FMR thus allows full service in the new area—both incoming and outgoing calls.

Intersystem roaming allows hand-off among differing companies with differing switch types. EIA/TIA standard IS-41[5] for intersystem signaling provides for call hand-off between foreign systems, and allows for proprietary base stations, hand-off algorithms, control protocols and operational features.

Automatic roaming allows users to roam while automatically receiving and placing calls. It eliminates the need for manual registration and roamer access ports. The calling party generally does not know if the subscriber is in the home area or across the country. The use of automatic roaming is increasing rapidly.

Automatic roaming utilizes the information from subscriber set registration. The two types of registration are *autonomous* and *forced*. Autonomous registration requires the subscriber set to transmit a message periodically, usually every 10 to 60 minutes. The registration identification (REG ID) counter on the forward setup channel is stepped by the MSC. The subscriber set registers every time the REG ID counter has increased by 450 units.

The resulting location and status information can be used to determine if a set has been turned on. If the subscriber set appears to be off, the system operator may decide not to page the subscriber, but instead to route the incoming call directly to a "not available" recording or to voice mail. Autonomous registration results in the incoming call being handled promptly, and reduces the load on the cell-site data links and the forward setup channel.

Forced registration occurs after the subscriber set rescans and monitors a setup channel from another system. This is accomplished easily since the new MSC has a different system identification (SID) from the previous one. The new MSC receives the registration and informs the home MSC to route incoming calls to it. This occurs again when the subscriber either returns to the home area or another roaming area.

7.4. MULTIPLE ACCESS CONCEPTS

The resources provided by a given amount of the radio spectrum can be measured by the parameters power, bandwidth, and time. Accordingly, the spectrum's use as a resource can be shared in the power, frequency, or time domains, as shown in Fig. 7.17.[6] In *frequency-division multiple access* (FDMA), each user has some of the bandwidth and some of the power all of the time. In TDMA, each user has all of the power and all of the bandwidth part of the time. A third way of dividing the spectrum resource is called CDMA. Here, each user has all of the bandwidth all of the time, but only part of the power. CDMA is a digital form of *spread-spectrum multiple access* (SSMA), and may take the form of *frequency hopping* (FH), *time hopping* (TH) or *direct sequence* (DS), as delineated in Chapter 6.

The transmission being done using the TIA standard thus is VSELP–(π/4) DQPSK–TDMA–FDMA. This means that the speech is digitized using VSELP, then modulated using π/4-DQPSK, then burst onto a channel in a preassigned time slot (TDMA). The time slot is one of many channels; hence the designation FDMA. For brevity, however, it is called simply TDMA.

The choice of (π/4)-DQPSK was based on the fact that it produces a phase change for each new symbol, and carries its phase reference with it, thus requiring no special form of timing recovery. With this feature, it maintains satisfactorily low adjacent-channel energy.

The sections to follow discuss TDMA (Section 7.4.1), CDMA (Section 7.4.2) and hybrid TDMA/CDMA systems (Section 7.4.3).

7.4.1. Time-Division Multiple Access (TDMA)

In a system using TDMA principles, each participating subscriber set sends one or more traffic bursts to the cell site. The bursts from the various subscriber sets must be synchronized so that they occupy nonoverlapping time slots in a TDMA frame. The cell site replies with traffic bursts back to the subscriber set. Each

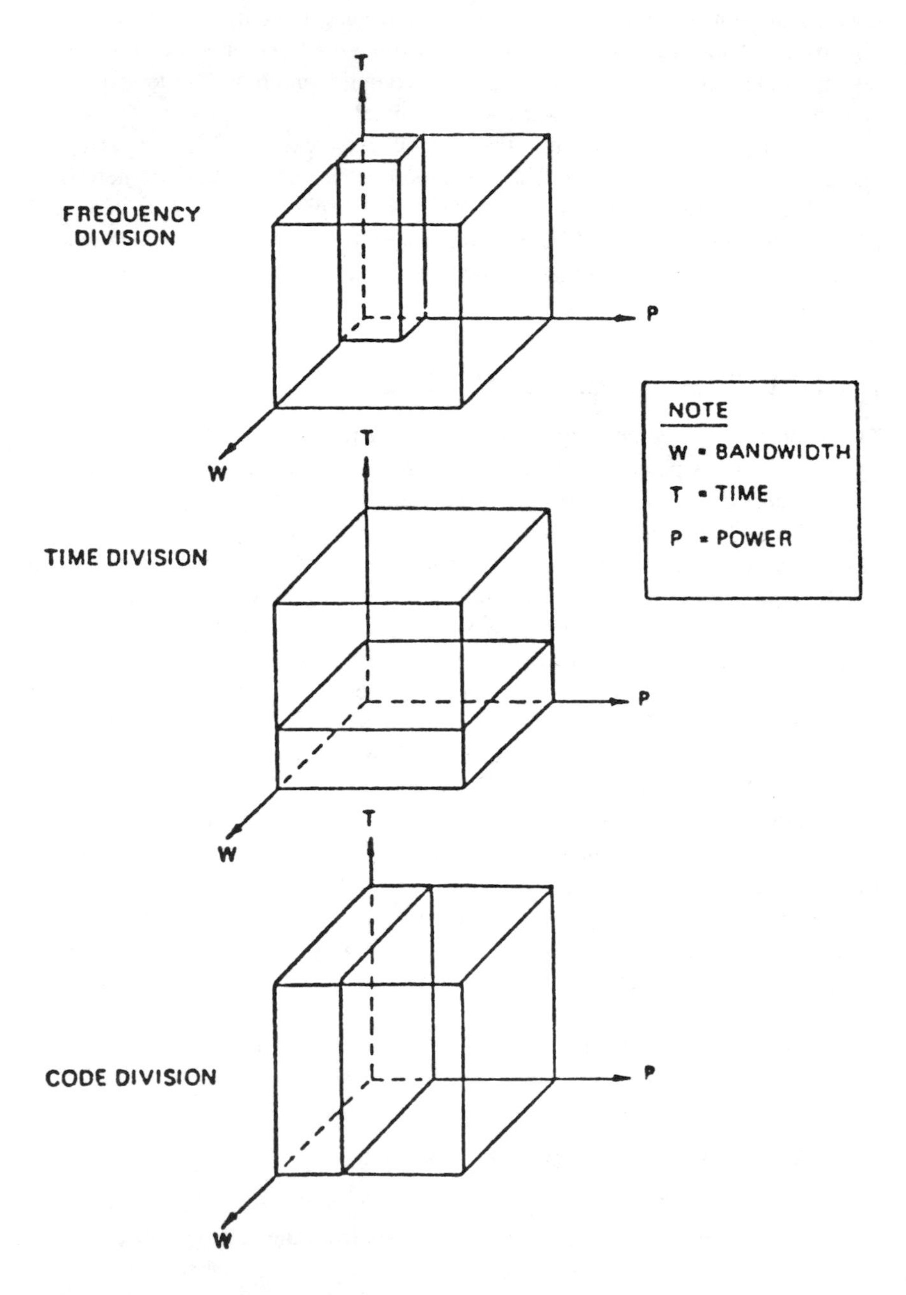

FIGURE 7.17. Multiple access concepts.

burst must be timed such that at its time of arrival at the cell site, it is the only signal present in its channel. In this way, no collision occurs with the traffic bursts of any other source. Thus, multiple access means that a multiplicity of participating subscriber sets can use the network simultaneously by occupying time slots in a time frame. To assure collision-free operation, at call setup the cell site measures the round-trip time to the subscriber set and back, and sends a timing advance message to the subscriber set. Timing advance may be adjusted during the call if necessary.

TDMA exhibits a number of advantages over pure FDMA. First, selectivity is accomplished in the time domain. Digital signals can be stored and processed readily, resulting in numerous performance improvements over what can be obtained using analog systems. Demand assignment can be implemented by adjusting the traffic bursts to accommodate the demand, as is done in extended TDMA (E-TDMA). A decision-feedback equalizer can reduce the frequency-selective effects of multipath fading, while a fractional interval equalizer can reduce adjacent-channel interference effects.

7.4.1.1. Basic TDMA

The principles of TDMA as standardized by the TIA for North American digital cellular systems[7] begin with the establishment of a digital traffic-channel structure consisting of 6 time slots. These 6 time slots occupy 40 ms, and are called a frame. Thus 25 frames per second are sent. A frame contains 1944 bits. Accordingly, the bit rate is 48 600 b/s. The slots are numbered 1 through 6, as shown in Fig. 7.18.

In Fig. 7.18, each full-rate voice channel uses two equally spaced time slots, that is, 1 and 4, 2 and 5, or 3 and 6. Each half-rate channel uses only a single time slot. With no timing advance, the offset at the subscriber's set between the reverse and forward frame timing is 207 symbol periods, where a symbol is 2 bits. Thus 207 symbol periods = 414 bits = 8.5185 ms. Stated another way, time slot 1 of frame N in the forward direction occurs 207 symbol periods after time slot 1 of frame N in the reverse direction.

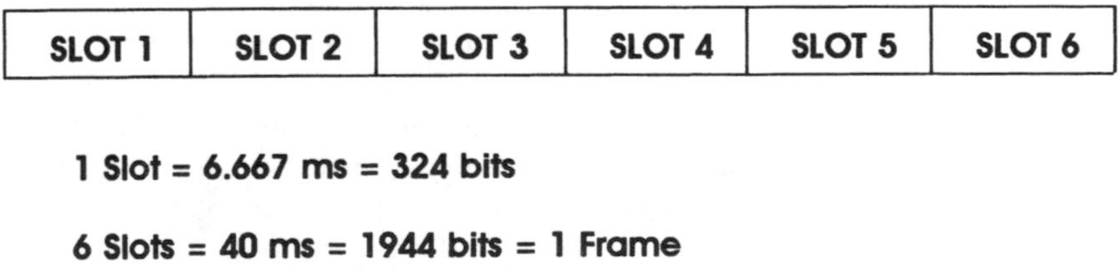

SLOT 1	SLOT 2	SLOT 3	SLOT 4	SLOT 5	SLOT 6

1 Slot = 6.667 ms = 324 bits

6 Slots = 40 ms = 1944 bits = 1 Frame

FIGURE 7.18. Digital traffic channel structure.

The slot format for the subscriber set to the cell site is as follows:

Guard Time	6 bits
Ramp Time	6 bits
Data (User Information or FACCH)	16 bits
Synchronization and Training	28 bits
Data (User Information or FACCH)	122 bits
SACCH	12 bits
CDVCC	12 bits
Data (User Information or FACCH)	122 bits

The *ramp time* is the time needed for the transmitter to reach full output from a quiescent condition. *User information* refers to speech bits or user data bits.

The slot format for the cell site to the subscriber set is as follows:

Synchronization and Training	28 bits
SACCH	12 bits
Data (User Information or FACCH)	130 bits
CDVCC	12 bits
Data (User Information or FACCH)	130 bits
Reserved	12 bits

The CDVCC consists of an 8-bit *digital verification color code* (DVCC) plus four error-correction bits.

FACCH is a signaling channel for the transmission of control and supervision messages between the cell site and the subscriber set. It is sent in place of user information whenever needed, and is used where rapid action is required, for example, for hand-off.

SACCH is a signaling channel that is sent continuously for the transmission of control and supervision messages. It is used for functions that can be performed over a longer period of time than the brief FACCH message, for example, channel quality information for MAHO during continuous transmission.

FACCH and SACCH constitute the *forward digital traffic channel* (FDTC) in the case of transmission from the cell site to the subscriber set, and the *reverse digital traffic channel* (RDTC) in the case of transmission from the subscriber set to the cell site.

The subscriber set derives the timing for its transmit symbol, TDMA frame and slot clocks from a common source which tracks the cell-site symbol rate as observed at the subscriber set. Frequency tracking must be maintained over all specified operating conditions.

Time alignment consists of controlling the time of occurrence of the time-slot burst from the subscriber set by advancing or retarding it. If an overlap with the burst of another subscriber set should be detected, the cell site sends a time-alignment message to the subscriber set using the appropriate forward signaling channel.

If an overlap with the burst from another subscriber set is detected at the head or tail of a time slot, the cell site must send an appropriate time-alignment message to the subscriber set.

Emission limits for subscriber sets are as follows:

Adjacent channel	−26 dB
Alternate channel	−45 dB
Remote channel	−60 dB or −43 dBW, whichever is greater

The alternate channel is defined as one centered ±60 kHz from the center of the reference channel. A remote channel is any channel that is ±90 kHz or more from the center of the reference channel. Total emission power is defined using a passband that matches that of the π/4-DQPSK modulated carrier.

All subscriber sets must be capable of reducing or increasing their power levels upon command from the cell site, to minimize co-channel and adjacent-channel interference, while maintaining good signal quality for the subscriber.

Subscriber sets are made on a *dual mode* (TDMA and analog, or CDMA and analog) basis to allow subscribers who travel into an area without their type of digital service to receive analog service instead.

Because of space limitations, further details of the TIA TDMA standard are not included here, but the interested reader can find such details in the referenced documents.[8,9,10]

7.4.1.2. Extended TDMA (E-TDMA)

Extended TDMA (E-TDMA), also known as *enhanced TDMA*™ (E-TDMA), is a development of Hughes Network Systems, Inc.[11] It utilizes half-rate speech coding and the TIA (IS-54) TDMA format. It achieves a spectrum efficiency improvement over other techniques through the use of digital speech interpolation (DSI). In addition, it provides silence-interval detection on the speech channel. The objective is 10 voice conversations per RF channel and, typically, a 15-fold improvement in capacity. It is designed for backward compatibility with IS-54.

Initially, the operator can deploy half-rate channels without DSI (*continuous mode*) on the same channel as IS-54. Alternatively, a complete channel can be set aside for half-rate operation. Fig. 7.19 shows an example of continuous-mode operation.

E-TDMA works by treating each slot of an IS-54 frame as being available for either speech, forward, or reverse control, packet data, or special applications. A DSI pool is formed using a set of channels, for example, three, as illustrated in

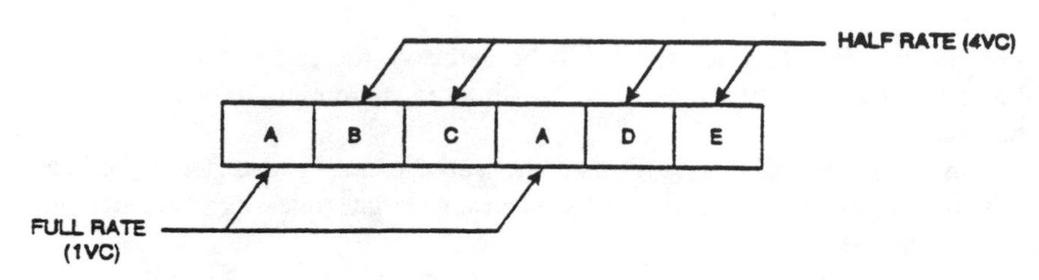

FIGURE 7.19. Example of continuous mode E-TDMA operation. (Courtesy Hughes Network Systems, Inc. Reprinted by permission.)

Fig. 7.20. The channels are assumed to be operating on a half-rate basis. They need not be contiguous; in fact, they may be separated widely.

Because of the DSI, this pool supports more traffic than the 18 channels supported in the continuous mode. The improvement results from the fact that a typical speech pattern involves only a 0.40 (or lower) voice activity factor, as illustrated in Fig. 7.21. Since speech activity is a statistical process similar to trunk loading, statistical distributions are useful in predicting the performance of trunking systems reliably. Speech-spurt lengths are exponentially distributed. Likewise, speech activity statistics can predict the behavior of DSI systems. The 0.40 value has been found to be applicable to a mobile-to-land path. For the land-to-mobile path, a 0.30 to 0.35 factor is more likely to be found. Statistics applicable to portable sets may be more like those of fixed-subscriber paths.

The DSI concept is illustrated in Fig. 7.22. In the left half of the figure, 6 typical speech streams are shown on 6 typical channels. Since the voice activity factor is assumed to be 40%, the 6 speech streams can be accommodated on 3 channels. The resulting channel utilization then is 80% rather than only 40%.

As was shown in Fig. 7.18, the IS-54 format provides 6 time slots. In E-TDMA, the frequency channel/time slot assignments are made by a *base-station controller* (BSC) based on speech activity. The BSC function is performed at the

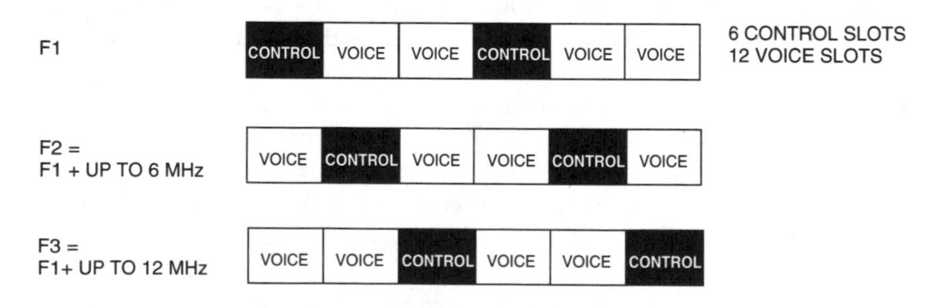

FIGURE 7.20. DSI pool example. (Courtesy Hughes Network Systems, Inc. Reprinted by permission.)

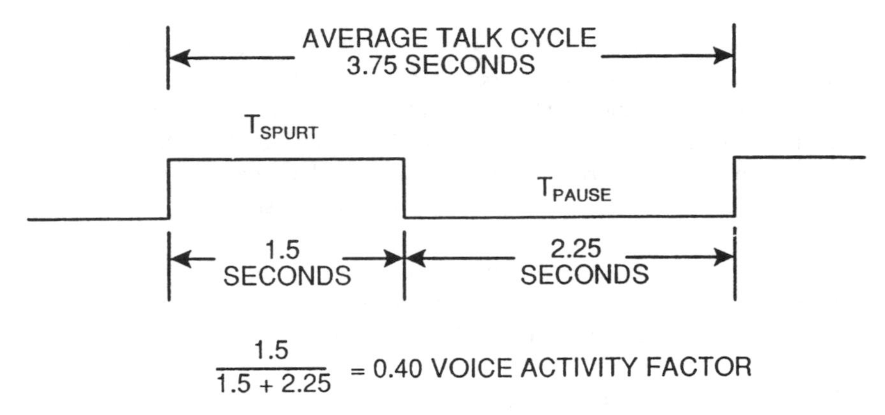

FIGURE 7.21. Typical speech pattern. (Courtesy Hughes Network Systems, Inc. Reprinted by permission.)

MSC in a North American system. Whereas, in the continuous (basic TDMA) mode, a channel is assigned for the duration of a call, in the E-TDMA mode, a channel and time slot are assigned at the beginning of each speech spurt. During speech pauses, there is no assignment.

In the forward direction, *voice-activity detection* (VAD) at the BSC requests a channel/slot assignment. The assignment is made in a DSI control slot. The

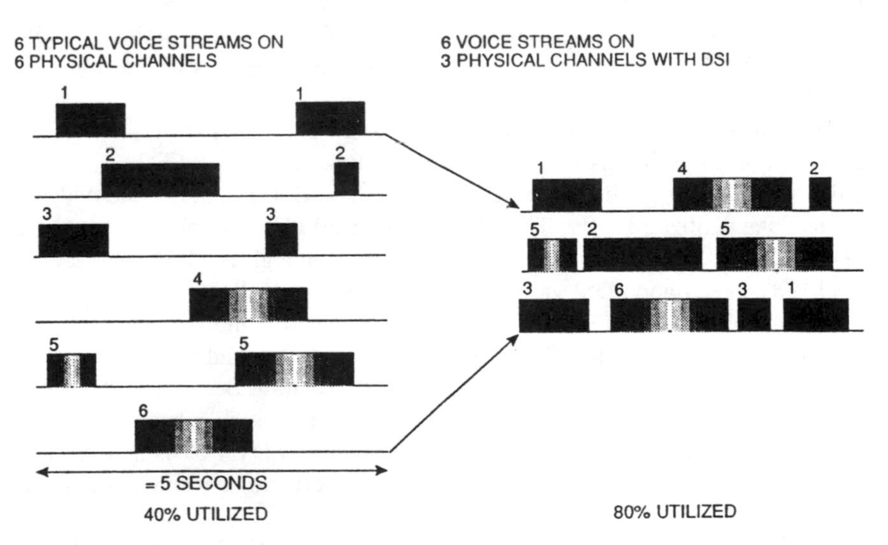

FIGURE 7.22. Digital speech interpolation (DSI) concept. (Courtesy Hughes Network Systems, Inc. Reprinted by permission.)

subscriber set then tunes to the assigned channel/slot. During the idle period (between speech spurts), the subscriber set retunes to the assignment slot.

In the reverse direction, VAD at the subscriber set makes the request. The request is sent to the cell site in one of the DSI control slots. The BSC then sends the assignment to the subscriber set in a DSI control slot. The subscriber set tunes to the assigned channel/slot. During the idle period, the subscriber set again retunes to the assignment slot.

Fig. 7.23 provides an example of a DSI channel pool consisting of 12 RF channels (72 slots). Nine of these slots are control slots, with the remaining 63 being traffic (usually speech) slots. Such an arrangement can support more than 120 calls.

The way in which the slots of a channel may be used is illustrated in Fig. 7.24. The 6 slots are used by 10 different subscriber sets in this particular example. These subscriber sets are designated as MS (mobile station) numbers 1, 2, 4, 8, 11, 15, 19, 21, 22, and 32.

The timing of the channel assignments is done as illustrated in Fig. 7.25. As shown in the upper portion of the figure, the initiation of a speech frame causes VAD to occur, and processing to be initiated. The assignment (AS) is made, whereupon tuning is accomplished. The actual time required is seen to be less than half the speech-frame time. Speech is buffered at the base station until the mobile is ready to receive it, so no speech is lost. In the example, the assignment was to slot 3, so the letters "SP" appear in the following slots 3.

The lower portion of the figure illustrates the reverse channel assignment. Here VAD at the subscriber set causes a request (RQ) to be sent to the BSC. Following processing at the BSC, an AS is sent to the subscriber set (MS). In this example, the assignment was to slot 2, so the designation "SP" appears in the following slots 2.

The transition from basic TDMA to E-TDMA is facilitated by the fact that E-TDMA uses the TDMA frames, modulation type, data rates, FEC, synchronization words and signaling methods of IS-54. The E-TDMA half-rate codec is the one standardized by the TIA. The existing 30-kHz channels are used. E-TDMA can be intermixed with TDMA, provided enough E-TDMA capacity is furnished to get a good DSI gain, as defined in Section 5.10.

Additional benefits of E-TDMA include the fact that the inherent slow frequency hopping reduces the effects of frequency-selective fading if the assigned channels are spaced by at least the correlation bandwidth. Because of the use of DSI, inherent *discontinuous transmission* (DTX) exists at the subscriber set, and the transmissions from the cell site inherently are on a DTX basis as well, thus improving channel reuse, as well as minimizing portable subscriber-set battery drain.

Any DSI system will experience some degree of speech clipping. Except during periods of peak usage, this clipping usually is not noticeable. Even at 2%

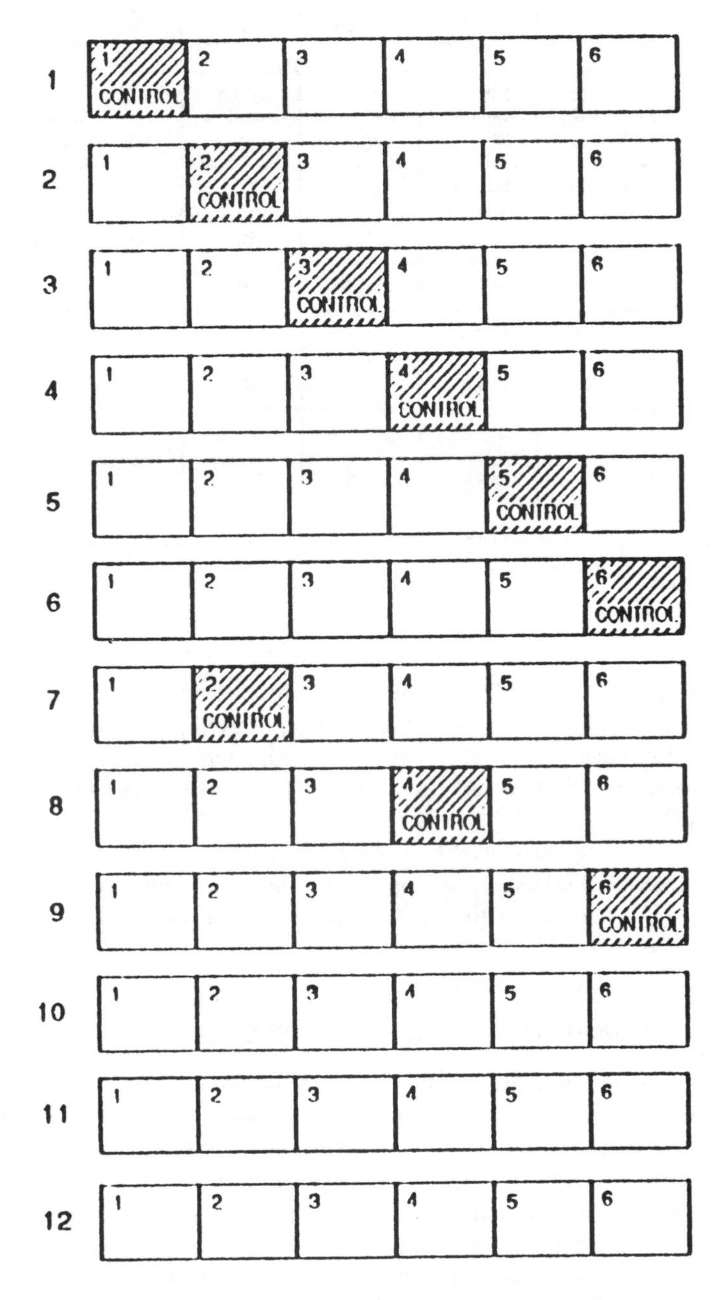

FIGURE 7.23. DSI channel pool example. (Courtesy Hughes Network Systems, Inc. Reprinted by permission.)

FRAME (TIME ↓)	SLOT 1	SLOT 2	SLOT 3	SLOT 4	SLOT 5	SLOT 6
1	MS 15	MS 11	MS 21	MS 4	MS 1	MS 22
2	MS 15	MS 11	MS 21	MS 4		MS 22
3	MS 15	MS 11	MS 21	MS 4	MS 19	MS 22
4	MS 15		MS 21	MS 4	MS 19	MS 22
5	MS 15	MS 8	MS 21	MS 4	MS 19	MS 22
6	MS 15	MS 8		MS 4	MS 19	
7	MS 2	MS 8		MS 4	MS 19	
8	MS 2	MS 8	MS 32	MS 4	MS 19	
9	MS 2	MS 8	MS 32	MS 4	MS 19	MS 15

FIGURE 7.24. Example of E-TDMA™ slot assignment. (Courtesy Hughes Network Systems, Inc. Reprinted by permission.)

clipping, a user will experience one 40-ms overload event once per conversation (about 3 minutes), but this is well under the typical clipping loss on an analog (FM) telephone using a squelch circuit. A reduction in speech loss can be achieved by scheduling FACCH between or at the end of speech spurts as often as possible.

As is the case with basic TDMA subscriber sets, the E-TDMA sets are made on a dual mode (E-TDMA and analog) basis, allowing users who travel into an area without TDMA service to obtain analog service instead. An E-TDMA set will function on basic TDMA as well.

7.4.2. Code-Division Multiple Access (CDMA)

Code-division multiple access (CDMA) is a form of SSMA, as discussed in Chapter 6, which also covers the subjects of FH, both fast and slow, and DS modulation. In CDMA, each circuit is assigned a unique randomized code sequence. In FH, the code generates the frequency hops. In DS, the code generates a noiselike high bit rate (or chip rate, see Section 6.4.3) signal that spreads the spectrum of the information signal. There is no hard limit on the number of circuits per carrier. Each user simply raises the background noise level for all the others. Consequently, quality for all users degrades when a CDMA cell becomes overloaded.

A CDMA standard for use in the 800-MHz cellular-frequency allocation has

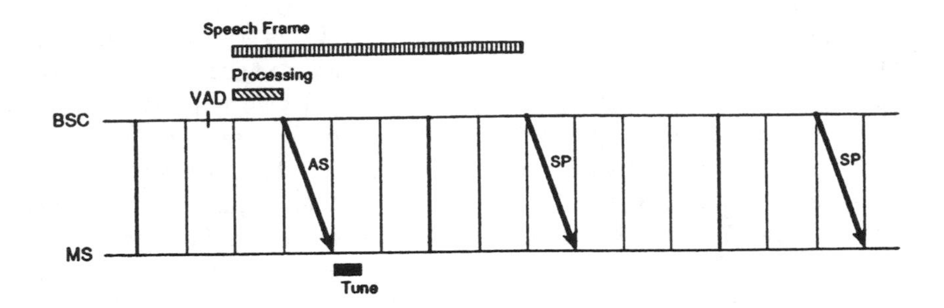

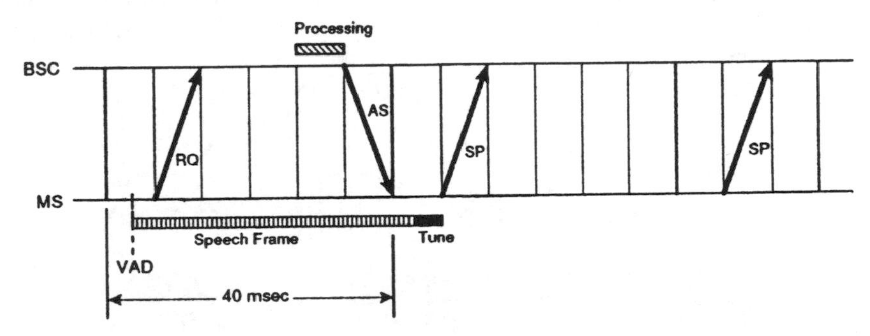

FIGURE 7.25. Example of slot assignment timing. (Courtesy Hughes Network Systems, Inc. Reprinted by permission.)

been developed by QUALCOMM, Inc., and is discussed in Section 7.4.2.1. The corresponding TIA standard is designated IS-95. An alternative CDMA system, developed by the SCS Telecom division of InterDigital Communications Corporation, for use in the 1860–1990 MHz band, is discussed in Section 7.4.2.2.

7.4.2.1. QUALCOMM CDMA

The QUALCOMM CDMA standard[12] uses a 1.25-MHz channel width, this amount being 10% of the spectrum allocated to each carrier in any given serving area. Each carrier thus could have 9 or 10 CDMA channels; accordingly, the FDMA principle applies to CDMA in the sense that it is a channelized system, the same as is TDMA. The 45-MHz separation between the uplink and downlink directions is the same as for other cellular systems. DS is used, with a pilot carrier consisting of a 32 768-bit beacon code being sent from each cell site. Time shifts in increments of about 200

μs distinguish the cells from one another. The pilot carriers provide the synchronization signal for the PN streams. The modulation is DQPSK.

While any speech coder can be used, the QUALCOMM variable-rate coder (QSELP), as noted in Chapter 4, is specified for use in the CDMA system. The rate varies from an idling (no speech activity) rate of 1 kb/s to a maximum of 8 kb/s, with the average being about 3.4 kb/s. The idling stream provides information on the acoustical background noise level. Using the variable-rate system, the receiver must be told what the rate is on a continuing basis. Forward error correction (FEC) is incorporated into the transmission in each direction.

The system capacity is determined by (1) the signal-to-interference ratio and fade margin, (2) the speech duty cycle, and (3) the number of sectors at the cell site. Studies in a variety of terrain conditions have shown that the interference contribution from neighboring cells is about one-half that of sources in one's own cell.[13] This concept is illustrated in Fig. 7.26, which portrays the average subscriber set in an adjoining cell, $k1$, contributing only 6% of the interference pro-

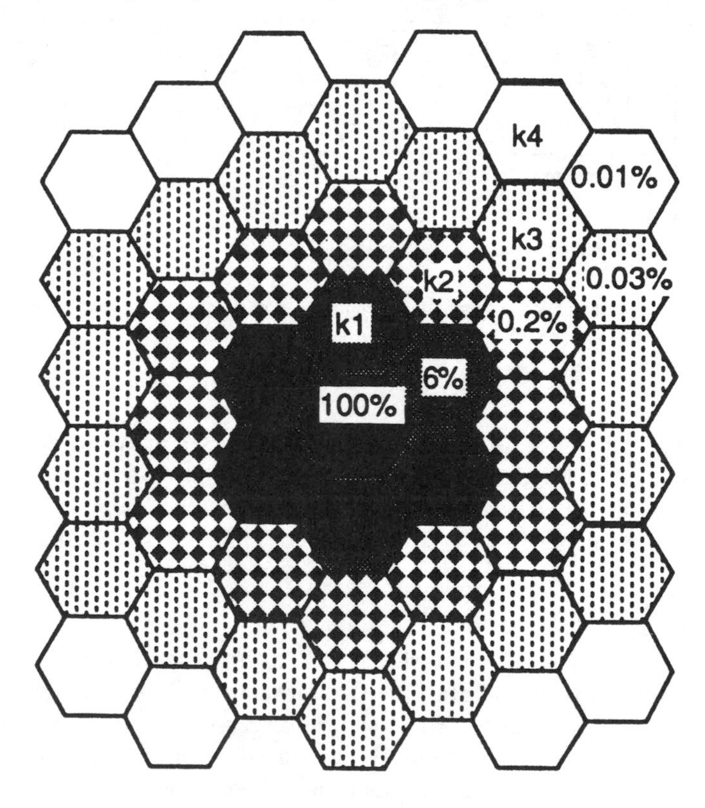

FIGURE 7.26. Percentage of interference contributions from neighboring cells. (Courtesy, QUALCOMM, Inc., Ref. 13.)

duced by a set in the cell under consideration. A subscriber set in the next ring of cells outward, $k2$, contributes only 0.2%, etc. On this basis, the reuse efficiency is $1/1.5 = 0.67$. Admittedly, idealized hexagonal cells were used for the analysis. The results of field tests in over-the-air situations, however, have shown actual results to be close to the results achieved with the idealized model. Thus the D/R considerations applicable to TDMA are not applicable to CDMA, since in CDMA all sources transmit simultaneously in the same spectrum.

Discrimination against multipath can be achieved through the use of a *rake receiver*, in which the multipath echoes are summed so they contribute to the desired signal. Signals whose delay spread exceeds 1 μs are found to be uncorrelated with the code, thus eliminating the need for an equalizer.

System capacity is a function of the following parameters:

N = number of users (communication paths) on the channel.
W = channel bandwidth = 1.25 MHz
R = data rate (Thus W/R = processing gain)
E_b = energy per bit
N_0 = thermal noise + noise generated by other CDMA users
V = speech activity factor
F = frequency reuse factor (Note: Full frequency reuse means that no frequency planning is needed if all the radios are properly synchronized)
G = gain from sectionalizing the cells. (Note: For two or more sectors, this gain is about 70% of the ideal value because of cell-site antenna sidelobe characteristics)

The system capacity then is:

$$N = \left(\frac{W/R}{E_b/N_0}\right) \times \frac{FG}{V} \tag{7.5}$$

Note that, in CDMA, manmade and thermal noise are treated alike. In addition, the occupied bandwidth is appreciably larger than the information bandwidth. Consequently, the designation E_b/N_0 is used instead of C/I, as is done for TDMA. As an example of the application of Eq. 7.5, using R = 9.6 kb/s, assuming a 7-dB value for E_b/N_0, $V = 0.35$, and a 3-sector cell with $G = 2$, the number of users is $N \approx 100$. This is the number that can be accommodated in a 1.25-MHz channel. Of course, operation at a lower E_b/N_0 allows a significant increase in the number of users. Since such a channel is equivalent to about 42 of the 30-kHz channels, the utilization corresponds to 2.4 simultaneous users in a 30-kHz channel with full reuse. Since an analog system usually must be designed on the basis of a 7-cell reuse pattern, a 17-fold increase in utilization over analog can be claimed.

Cell site hardware includes a single redundant output amplifier per 1.25-MHz channel. No power combiner is needed. An ERP of 45 W is sufficient for pilot, setup, and 36 users, compared with the 100 W required for analog per user. Other hardware is an RF rack and a modem rack.

The cell site sends power-control commands to each subscriber set on both an open-loop and a closed-loop basis. Open-loop control is based on the subscriber set's reception of the signal from the cell site. Closed-loop control is based on the cell site's reception of the signal from the subscriber set. Closed-loop power control is achieved over an 80-dB dynamic range. Through use of this 80-dB loop, the cell site attempts to maintain all received signal energies (from the individual subscribers) so that they arrive at the cell site with approximately equal energies. An 800-b/s stream is used to achieve this control with a 2.5-ms response time. Such quick response is vital in the rapidly fluctuating propagation environment.

Hand-off is initiated by the subscriber set. The new cell switches to the code the subscriber is using, while the subscriber still is receiving signals from the old cell. A subscriber set near a cell boundary is served by the two cells simultaneously, thus improving performance in an area which might be served inadequately otherwise, through what amounts to a form of diversity combining. In fact, a subscriber set equidistant from three cell sites experiences a threefold E_b/N_o increase, by being served simultaneously by all three. The type of hand-off achieved is called a *soft hand-off*. Soft hand-off reduces lost calls appreciably. The term *softer hand-off* is used to describe a hand-off between the sectors of a single cell.

Fig. 7.27[14] shows the architecture of the cell site and MSC equipment needed to support the handoff process.

Elevation angle discrimination can provide near/far cells, all on the same frequency.

A CDMA system can readily accommodate unequal cell loading. This results from the fact that lightly loaded cells contribute less interference than their neighbors, allowing more use of the heavily loaded cells.[14] Fig. 7.28 illustrates a situation in which the cells along a major road experience a higher than average demand, while more remote cells have only half as much demand. Because of the low level of interference from the lightly loaded cells (operating at only 30% of capacity), the cells along the road can operate at 120% of their normal capacity. This load flexibility occurs naturally, provided that 1. the heavily loaded cells are equipped with an adequate number of modems and 2. the computer equipment that manages the system is programmed to coordinate the variable-capacity assignment.

Registration can be accomplished on an automatic basis. Upon turn-on or entry into a new cell, a subscriber set signals the cell site. This procedure prevents excessive paging load on the system. The cell site sends the subscriber set such basic information as the identification of the neighboring cells.

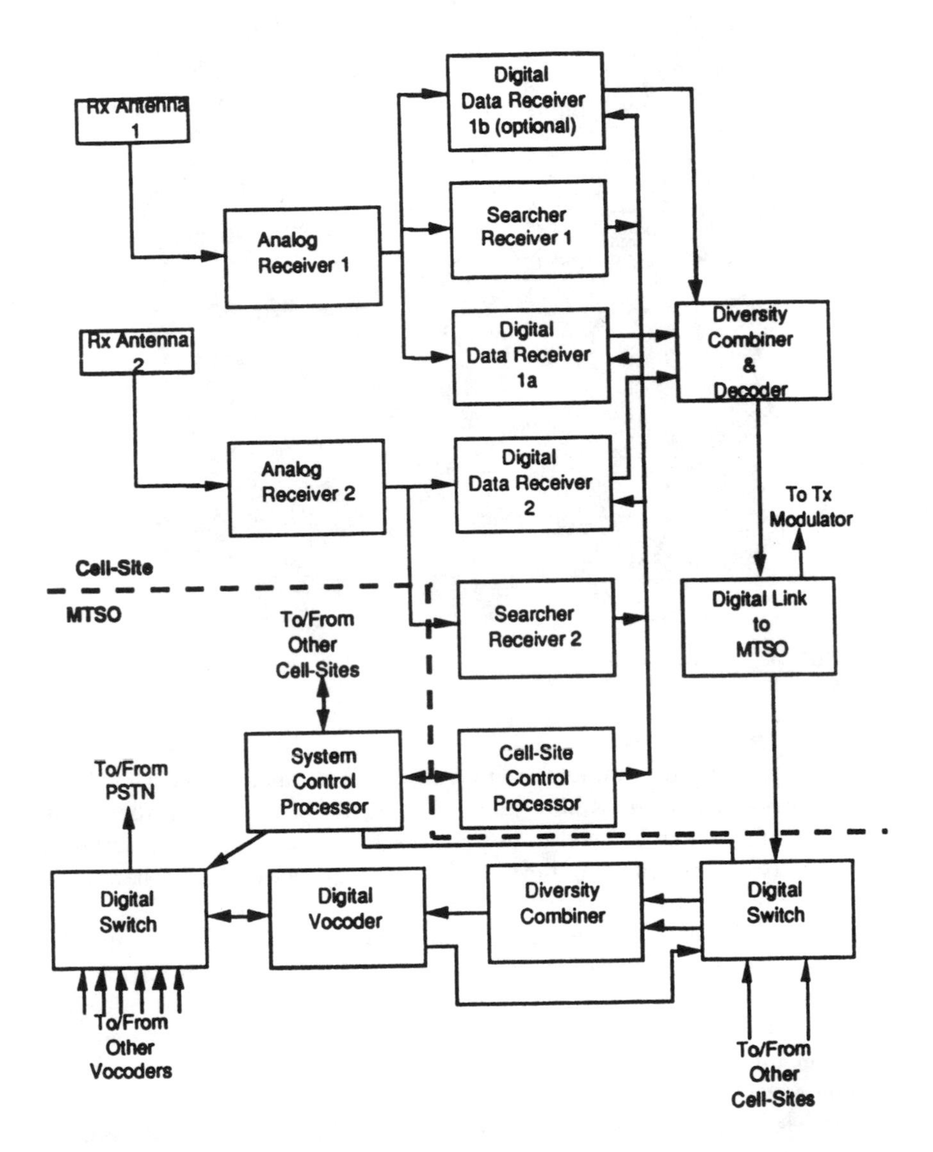

FIGURE 7.27. Cell site and MSC architecture for hand-off. (Courtesy QUALCOMM, Inc., Ref. 13.)

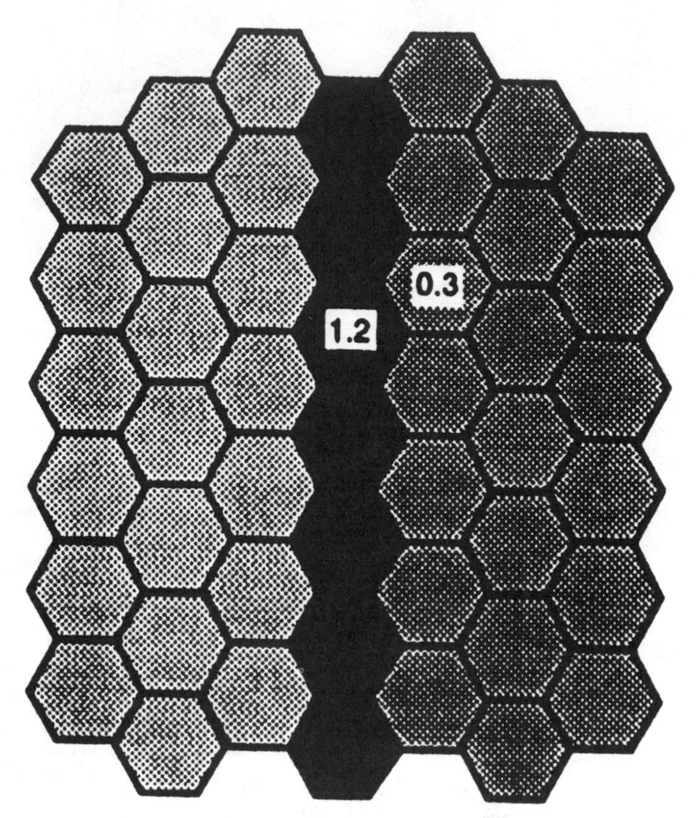

FIGURE 7.28. Unequal cell loading. (Courtesy QUALCOMM, Inc., Ref. 13.)

The transmit power required from the subscriber set can be appreciably lower than in analog service, thus allowing fewer, larger cells in low-traffic-density areas. The transmitter is found to average a 10-mW level under typical urban operating conditions, with a 300-mW level seldom being exceeded. The cell site typically allocates about 1.25 W to a single mobile.

Subscriber sets are made on a dual mode (CDMA and analog) basis, allowing subscribers who travel into an area without CDMA service to obtain analog service instead. Fig. 7.29[16] is a simplified block diagram of a CDMA subscriber set, and Fig. 7.30[17] illustrates, on a more detailed basis, the functions that are performed in a dual-mode subscriber set.

With respect to a given user, additional users merely raise the noise level slightly. Users whose radios do not power down properly may produce many times their share of interference. Although this is an unlikely occurrence, tests involving failed units show that traffic simply migrates to neighboring cells. This

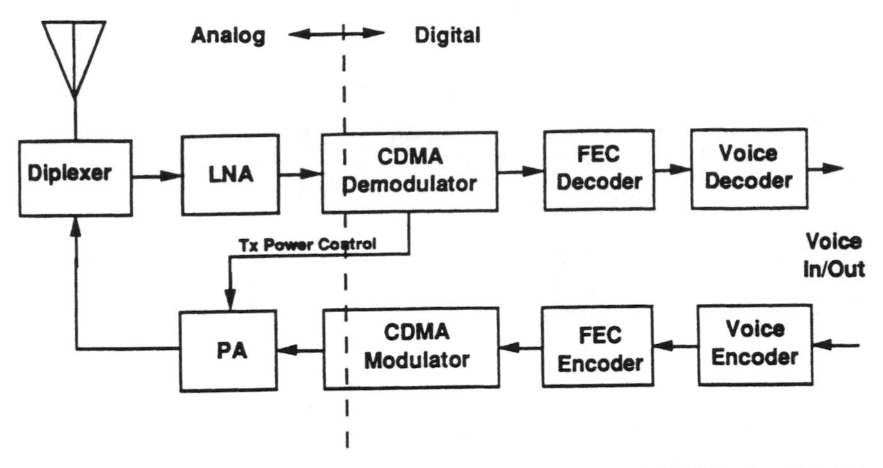

FIGURE 7.29. CDMA subscriber set block diagram. (Courtesy, QUALCOMM, Inc., Ref. 13.)

is especially true of subscriber sets near the cell boundary, which can migrate most readily.

7.4.2.2. Broadband CDMA

A CDMA system for use in the 1850 to 1990 MHz band has been developed by the SCS Telecom division of InterDigital Communications Corporation.[18,19] This system uses a 10-MHz bandwidth (in each direction) to minimize the effects of fading, based upon the fact that the correlation bandwidth often is 8 MHz or less in the hand portable environment in which many of the system's users will operate. Multipath delays in excess of the chip duration are significantly attenuated. Thus at 9 megachips/s, the chip duration is 110 ns $\approx$ 33 m. Accordingly, multipath greater than 33 m is attenuated. On the other hand, for a system using a 1.25-MHz bandwidth, the chip duration is 0.8 μs $\approx$ 240 m. Thus multipath must be greater than 240 m to be attenuated significantly in such a system.

Power control is "open loop," meaning that the power received by the handset indicates the power to be radiated. The techniques used to minimize interference to and from terrestrial microwave systems include (1) notch filtering (less interference in a CDMA system means less power radiated), (2) voice-activity detection (VAD), resulting in lower average radiated power, and (3) dynamic capacity allocation (DCA), which monitors the signal-to-noise (S/N) ratio in a microwave receiver, and directly controls the capacity of the cell in which the PCS users can affect the (S/N) of the microwave receiver.

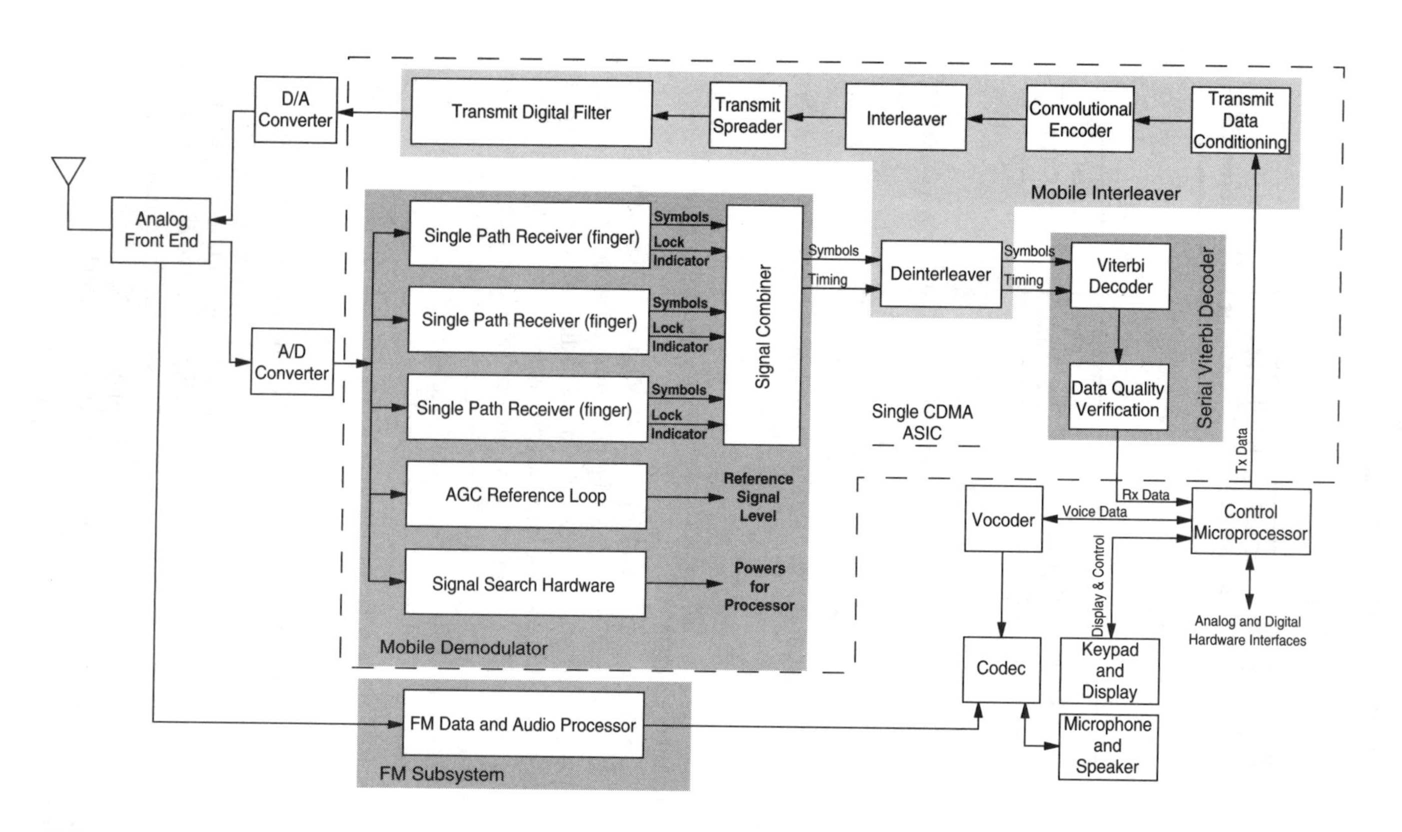

FIGURE 7.30. Dual-mode subscriber set block diagram. (Courtesy, QUALCOMM, Inc., Ref. 15.)

7.4.3. Hybrid Digital System

A system using both TDMA and CDMA concepts has been developed by Omnipoint Communications, Inc.[20] This system uses 9-MHz bandwidth channels with frequency agility. It is optimized for microcell and in-building use. Within each CDMA channel, a 400-kb/s stream carries the traffic of 8 users, each having a 32-kb/s ADPCM stream. Each user operates on a time-division duplex (TDD) or a frequency-division duplex (FDD) basis for full FDX capability within the channel being used.

The subscriber sets are operable in the 902–928 MHz unlicensed (FCC Part 16) band. In addition, subscriber sets that are switchable between two 2 GHz bands have been made. These two bands are 1. 1850–2200 MHz, in which operation is licensed, and will be coordinated with fixed microwave systems using exclusion zones, and 2. 2400–2483.5 MHz on an unlicensed basis (FCC Part 16).

7.4.4. Frequency-Hopping Multiple Access (FHMA)

A frequency-hopping multiple access (FHMA) system has been developed by Geotek Industries, Inc. for use in a proprietary digital network for managing fleet and mobile communications.[21] Short bursts are transmitted over numerous channels, using special hopping algorithms. Frequencies in the 900-MHz band have been assigned. Complex algorithms coordinate and control the assignment of hopping sequences and the level of interference, allowing the reuse of the same set of frequencies in each sector of a macrocell.

Services provided by Geotek through its wireless network, GeoNet™, include telephone service; facsimile and computer data transmission; mobile data for dispatch, two-way messaging, electronic mail, and file transfer; and advanced network operations for coordination of delivery and inventory control functions. Vehicle location also is provided.

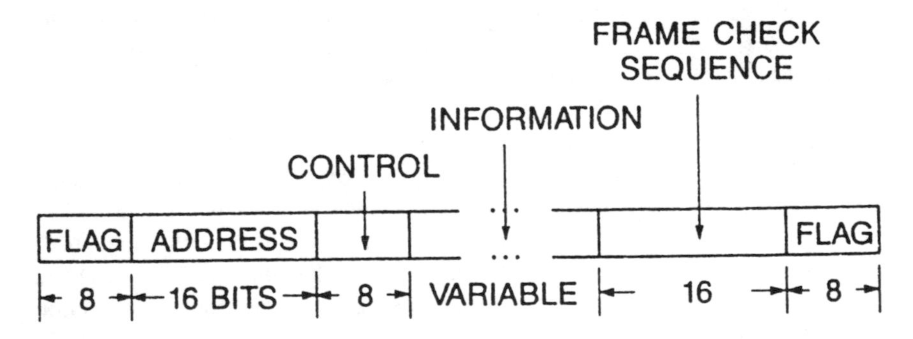

FIGURE 7.31. A possible data link packet structure. (Courtesy Siew and Goodman, Ref. 17, *IEEE Transactions on Vehicular, Technology*, © IEEE, 1989.)

7.5. DATA TRANSMISSION

User desires for the operation of facsimile machines and computers on a portable/mobile basis lead to the need to provide nonvoice service for subscriber sets. Data transmission at rates of 9.6 or 16 kb/s on a packet basis in the Rayleigh fading environment lead to the question of the optimum packet length for the maximum rate of information transfer. Short packets are less likely to encounter fades, but still must carry overhead.

Factors in the establishment of a data-link packet structure are the packet length, the speed of movement of the subscriber set, the length of the packet header, and the required fade margin. Studies[22] have shown that at 16 kb/s, the optimum packet size is about 17 bytes, corresponding to a duration of about 8.5 ms. The exact optima depend on the speed of movement of the subscriber set, the header size, and the required fade margin. Fig. 7.31 shows a possible data link structure based on *high-level data-line control, link access protocol C* (HDLC/LAPC).[23]

7.6. SPECTRUM EFFICIENCY

The advent of cellular telecommunications was initially believed to be a way to solve the continuing spectrum scarcity problems perceived by the land mobile radio community. The concept was that of continuing to make cells smaller and smaller as demands grew. While the day of the microcell certainly has arrived, the end to spectrum scarcity problems has not appeared. Two significant factors in this regard are the cost of adding new cells, and the propensity of radio waves to appear beyond the bounds intended for them. Maximum use of the available spectrum thus is vital to the ability of a carrier to continue to serve the public satisfactorily, as well as the ability of private and government operators to meet their needs.

Several spectrum measures have been defined by Calhoun.[24] The first is *information density*. Information density is the number of b/s per Hz provided by a system, based on the modulation technique it uses. This term sometimes is called *bandwidth efficiency*.

Circuit spectrum efficiency is the number of circuits per MHz. It depends upon the speech coding and the modulation techniques used, as well as adjacent channel restrictions, including guardbands. If C is the number of circuits per carrier, B is the channel bandwidth, including guardbands, in kHz, and A_r is the number of restricted adjacent channels per usable channel, then the number of circuits/MHz is $(C/B[A_r + 1])(1000/2)$, since a separate channel is needed for each direction of transmission.

As an example, for analog cellular, $C = 1$, $B = 30$, and $A_r = 2$. The circuit spectrum efficiency then is 5.5 circuits/MHz. For basic TDMA cellular using half-rate channels, $C = 6$, $B = 30$, and $A_r = 2$. The circuit spectrum efficiency then is 33 circuits/MHz. For CDMA, assuming the results of an example in Section 7.4.2, $C = 100$. Since $B = 1250$, and $A_r = 0$, the circuit spectrum efficiency is 40 circuits/MHz.

Geographical spectrum efficiency is measured in circuits per MHz per km^2, and includes frequency reuse. Its definition is

$$\frac{\text{(no. of cells)(circuits/cell)}}{\text{total area covered}}$$

The number of circuits/cell is $S_a/2R(B/C)A$, where S_a = total spectrum allocation (MHz), and R = reuse pattern. Then

$$\text{geographical spectrum efficiency} = \frac{NS_a}{2R(B/C)A} \tag{7.6}$$

where N is the number of cells, and A is the total area covered by them.

For example, for analog cellular, $S_a = 25$ MHz (12.5 MHz in each direction), $R = 7$, and $A = 7850$ km^2 for an assumed 100 km diameter service area. $N = A/A_c$, where A_c = area per cell. For an assumed 10 km cell radius, $A_c = 314$, so $N = 25$. Each cell may be assumed to have $1/R$ of the channels for the purposes of this example, or 57 channels. Allowing for setup channels, each frequency is used about 6 times. Then the geographical spectrum efficiency is 0.16 circuits/MHz/km^2.

TDMA with its 3 or 6 circuits per carrier, and E-TDMA with even more circuits per carrier, will have correspondingly greater values of geographical spectrum efficiency.

The reuse pattern is a major factor in geographical spectrum efficiency. Thus CDMA, with its 67% reuse efficiency, has an $R = 0.67$. Assuming the other parameters to be the same as in the examples above, the geographical spectrum efficiency of CDMA is 4.75 circuits/MHz/km^2.

Economic spectrum efficiency is defined as the cost per circuit/MHz/km^2. This measure addresses the limited availability of the spectrum. The cost per circuit for cell splitting is high compared with the cost per circuit for expanding into additional spectrum. In the absence of additional available spectrum, however, cell splitting must be implemented.

Communication efficiency is expressed in erlangs/MHz/km^2. Possible improvements may lie in providing for the interchangeability of data and voice, as well as in increasing the use of queuing and sophisticated circuit allocation techniques, as is done in E-TDMA, for example. An acceptable alternative in some cases may be that of shifting traffic away from the busiest hours, for example, through the use of voice messaging and voice mailbox techniques. A further approach is the use of flexible bandwidth allocation for data transmission or for user-selected speech quality.

7.7. SUMMARY OF DIGITAL CELLULAR SYSTEM CHARACTERISTICS

As noted earlier in this chapter, cellular systems using analog voice are designated as first-generation cellular systems. This designation refers primarily to the FM system known formally as the *advanced mobile phone system* (AMPS). A narrowband system using advanced digital control techniques, but analog voice, is the Motorola *narrowband AMPS* (AMPS). Its standards are similar to those of AMPS, but the system operates in 10-kHz channels rather than the 30-kHz channels of the AMPS system.

TDMA systems are designated here as second-generation systems. Included in this category are the EIA/TIA North American digital cellular system (TDMA), the E-TDMA system developed by Hughes, the Japanese digital cellular system, and the European *Global System Mobile* (GSM). The GSM abbreviation also stands for the organization that developed the system, the Groupe Spéciale Mobile of the European Telecommunications Standards Institute (ETSI).

7.7.1. North American Digital Cellular System (TDMA)

The system developed by the TIA can be summarized as follows:

Carrier separation	30 kHz
Users per carrier, full-rate voice	3
half-rate voice	6
Transmission rate in channel	48.6 kb/s
Transmission rate per user, full rate	16.2 kb/s
half rate	8.1 kb/s
Modulation:	π/4-DQPSK
Full-rate speech codec	VSELP
Note: Full rate = 7.95 kb/s + FEC = 13.0 kb/s	
Half rate = Speech rate + FEC = 6.5 kb/s	
Channel equalization: Adaptive, to 40-μs delay	
Channel coding: Convolutional	

The base station may be built to operate on either a single mode or dual mode (analog and digital) basis.

7.7.2. Extended TDMA (E-TDMA)

The E-TDMA system is intended to be compatible with the North American digital cellular system with half-rate voice, but to provide for more than 6 users per carrier through the use of DSI. The number of users per carrier depends on the level of speech activity at any time. A DSI gain of 2.5 times thus implies an average of 15 users per carrier. Other basic parameters are the same as those listed in Section 7.7.1.

7.7.3. Japanese Digital Cellular System

The Japanese Ministry of Post and Telegraph (MPT) has established a digital cellular standard based on the United States TIA IE-54 system (see Section 7.7.1), but adapted to the 25-kHz channels in use in Japan. Its characteristics can be summarized as follows:

Carrier separation:	25 kHz
Users per carrier:	3
Transmission rate in channel:	42.0 kb/s
Transmission rate per user:	14.0 kb/s
Modulation:	π/4-DQPSK
Speech codec: (6.7 kb/s + FEC = 11.2 kb/s)	VSELP

Mobiles are required to use diversity to allow operation at $C/I = 12$ dB.

7.7.4. The Global System Mobile (GSM)

The GSM system was devised with interoperability across national borders as a paramount requirement. Accordingly, the GSM system is designed to allow automatic intersystem hand-off. The architecture uses a *home location register* (HLR) and a *visited location register* (VLR) to facilitate such intersystem operation. Each subscriber set is assigned to an HLR. Through automatic registration, a subscriber set also is listed in the appropriate VLR. GSM characteristics can be summarized as follows:

Carrier separation:	200 kHz
Users per carrier, full rate voice:	8
half rate voice	16

Transmission rate in channel:	271 kb/s
Transmission rate per user, full rate:	33.88 kb/s
half rate:	16.94 kb/s
Modulation:	GMSK
Speech codec:	RPE-LTP
Full rate: 13.0 kb/s + FEC = 16.0 kb/s	
Half rate: 6.5 kb/s + FEC = 8.0 kb/s	
Channel equalization: Adaptive, to 10 μs delay	
Channel coding: Convolutional	
Frequency hopping: Mandatory capability for subscriber sets	
Optional capability for cell sites	

Frequency hopping, where implemented, is at a rate of 217 hops/second; this classifies it as slow frequency hopping.

7.7.5. Code-Division Multiple Access (CDMA)

Standards for CDMA systems have been developed by an industry group headed by QUALCOMM, Inc.; the group includes Pacific Telesis, Ameritech, and NYNEX. The corresponding TIA standard is designated IS-95. The basic characteristics of this system can be summarized[25] as follows:

Channel width:	1.25 MHz
Users per channel:	≈100 (see Sect. 7.4.2)
Transmission rate per user:	9.6 kb/s
Modulation:	QPSK
Speech coding	QUALCOMM CELP (QCELP)
Variable Rate = 1.0 to 8.0 kb/s + FEC = 9.6 kb/s max.	

7.8. THE MICROCELLULAR ENVIRONMENT

A microcell is defined as a cell of limited coverage area for use by a variety of subscriber sets such as personal portables, mobiles, and suitably designed cordless telephones. The limited coverage area is achieved by using antenna heights on the order of 5 to 20 m for coverage distances of 100 m to 1 km. Transmitters may be located, for example, at street intersections at street lighting level. An overlay macrocell (normal size) may be used to act as a safety net for traffic hot spots, propagation dropout spots, and to aid in the hand-off process as necessary. Alternatively, a microcell may serve a single building.

7.8.1. The Propagation Environment

Multipath tends to be less severe in the microcellular environment than for larger cells. This is especially true of the in-building environment. The fading is characterized as *Rician*, which is produced by a combination of a Rayleigh faded (random or scatter) component and a direct (nonfaded or specular) component. The Rician distribution describes the short-time amplitude of the signals. Rician fading typically exhibits a larger correlation bandwidth than Rayleigh fading. Thus, it is less frequency-selective than Rayleigh fading, and the fades tend to be slower.

The probability density for the Rician distribution is:

$$p(r) = (r/\sigma^2) \exp\{-(r^2 + A^2)/2\sigma^2\}\, I_0(rA/\sigma^2) \quad r \geqslant 0, A \geqslant 0$$
$$K = A^2/2\sigma^2$$

where

I_0 = modified Bessel function
A = amplitude of specular component
K = specifies the Rician distribution:
As $K \rightarrow 0$, the distribution becomes Rayleigh
As $K \rightarrow \infty$, no fading occurs.

The indoor radio channel may be characterized by severe multipath and large propagation losses. Losses of up to 60 dB have been experienced on the floor of a building whose interior walls are wood-studded plasterboard. Even higher attenuations have been experienced on floors with metallic partitions. At 2 GHz, fades are found to be frequency selective and slowly varying, lasting for seconds to minutes.

Delay spreads are much shorter in the microcellular environment than in the macrocellular environment. Delay spreads range from 25 ns in medium-sized office buildings to 125 ns in large office buildings.

Solutions to indoor microcellular radio propagation problems may take the form of distributed antenna systems or leaky feeders in each hallway.

7.8.2. Personal Communication Networks (PCNs)

The PCNs have an architecture that allows operation independent of, but connectable to, the *public switched telecommunications network* (PSTN). PCN base stations typically serve a radius on the order of 100 m. CCITT signaling system No. 7 may be used. Most PCNs use CDMA technology, with attempts being made to share the spectrum with microwave point-to-point networks or other users.

Experimental licenses have been issued to operators in many metropolitan areas for operation in the bands 1850–1990, 2400–2483.5, and 5725–5850 MHz. The lowest frequency range is used by the majority of the experimenters.

The cordless booster allows cordless service to mobile users roaming up to several hundred meters from their vehicles. The cordless portion operates outside the cellular frequency bands. Thus a call is routed to the vehicle-mounted mobile system, which includes a cordless base feature, thereby extending service to the cordless portion of the system.

The acronym PCN has a second meaning: *personal calling number*. While different from the personal communications network concept, the two bear a close relation. The objective of numerous efforts now underway in the world of telecommunications development is to assign a number to a person rather than to a place. Combinations of smart switching and transmission systems then route calls to and from the person anywhere in the world. This implies a highly portable end instrument that can access both macrocellular and microcellular as well as satellite services. Automatic intersystem roaming is one step in this direction. Remaining to be solved are such problems as the frequency band(s) to be used, as well as numerous other design parameters, such as access technique, speech coding, signal format, possible antenna directivity, etc.

7.8.3. Indoor Radio Systems

Indoor radio systems provide for subscriber set use not only for telephony, but also for data, facsimile, interactive graphics, and video conferencing. Indoor systems may be found in office buildings, hospitals, factories, warehouses, convention centers, and apartment buildings. Their advantages include the reduction of wiring needs in new buildings, as well as ease of changes without requiring rewiring.

Indoor radio systems may operate on an FDMA, TDMA, or CDMA basis. FDMA can allow frequency diversity at the user equipment, to overcome fades by sending the same information simultaneously over several channels.

TDMA system users share a broadband channel (for example, 1 MHz) using a time-division protocol. ISI is not a problem at rates up to 1 to 2 Mb/s in small to medium-sized buildings. For higher rates and in large buildings with central antennas, ISI requires the use of adaptive equalizers and, thus, coherent signaling. A distributed antenna often eliminates the need for an adaptive equalizer.[26]

CDMA is effective in combatting multipath and CCI. It can share an otherwise lightly loaded spectrum as an overlay on other communication services. CDMA requires tight adaptive power control, as was noted in Section 7.4.2. Such power control is essential in compensating for the near-far problem, an especially difficult matter for portables. A suitable channel access protocol, such as Aloha,[27] can be helpful in such situations.

CDMA chip rates on the order of 300 megachips/second are not uncommon for Ethernet bandwidths (10 MHz); overlay on other services may be required because of the transmitted bandwidth requirements. With 10 MHz of information bandwidth, 100 000 Erlangs per system ($\approx$10 000 E/floor) can be provided with 10 times the E/MHz/km^2 of a cellular system for a given *C/I*, because of the small cell sizes. Frequency reuse every 3 floors has been found feasible.[28]

RF powers on the order of milliwatts are adequate for in-building systems, compared with the 100-mW levels often needed for macrocellular systems. Illumination from inside the building is preferable to illumination from the outside, which tends to produce spotty coverage. The use of frequencies near 800–900 MHz allows maximum use to be made of readily available components.

Indoor systems can use high-level digital modulation to advantage. Modulation at 6 b/s per Hz allows ISDN basic rate service (2B + D) in the equivalent of a 30-kHz channel.

An indoor radio system may use a microcell to serve PSTN subscriber sets while they are in the building. Alternatively, or in addition, a cordless (or wireless) PBX may be used to serve specific end instruments that operate only via the PBX. An interesting concept usable with CDMA is the distributed antenna with intentional delay,[29] as illustrated in Fig. 7.32. Neighboring antennas have time delays inserted into their feeds so that the signals received from any two antennas can be distinguished from one another by the receiver's time domain processing circuits. In some situations, the needed delay is provided by the natural delay of the cable. Additional delay is obtained most easily by use of a coiled section of cable between antennas. While overlapping coverage is not required, it is desirable since it provides diversity to all subscriber sets in the overlap area.

In CDMA cellular service, three adjoining sectors of a cell can be connected to a channel unit. This channel unit provides diversity combining of signals from all three sectors at the symbol level, thus providing a very high level of diversity

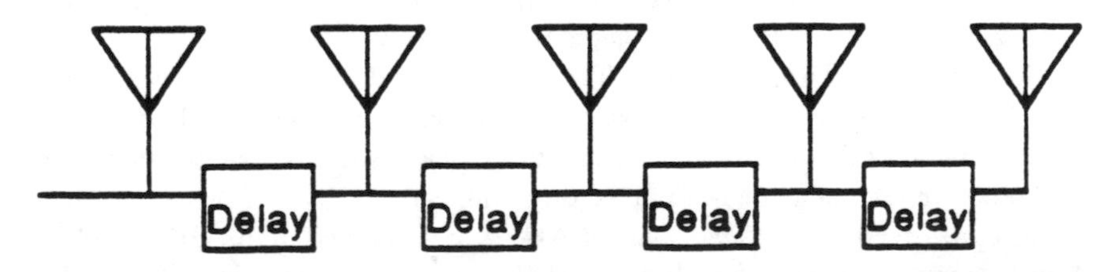

FIGURE 7.32. Distributed antenna with intentional delay. (Courtesy QUALCOMM, Inc., Ref. 13.)

combining. In the wireless PBX application, three antenna strings serving adjoining service areas could be connected to these three buses, resulting in three "pseudo-sectors." This would allow a soft hand-off without switch intervention to be accomplished between any of the antennas in the three strings. This mode of operation has the advantage of hiding the hand-off process from the switch, thus allowing the switch to be a generic PBX. With ten 1.25 MHz channels in use in three "pseudo-sectors," as many as 1200 simultaneous calls could be processed. Accordingly, on the order of 15,000 lines could be served, corresponding to the capacity of a large central office.

7.8.4. System Examples

Work has been done by Bellcore on *universal digital portable communications* (UDPC).[30] In this system, a radio network interface to the system (port) communicates with portables using TDM on the downlink. The portables reply using TDMA, with channel assignments made by the port. The architecture assumes CCI because of the limited number of channels. The other major impairment is AWGN. The portable radio initiates the access protocol, attempting channel selection based on the relative received power levels on the various available channels. A multiple block error check is used to improve performance in CCI limited situations. Hand-off is used.

The *Digital European Cordless Telephone* (DECT) was established by the Swedish Televerket (PTT) and adopted by ETSI for development as an international standard. It uses TDMA for a "seamless" hand-off between cell sites, and features dynamic channel allocation. DECT is designed for use in the microcellular environment (30- to 50-m cells). The handset optimizes the use of the voice channels independently of the central switch to achieve good spectrum efficiency. There is no need for prior frequency planning or for changes as the network expands. Two-way call initiation is achieved via a built-in paging feature. Speech encryption is provided for privacy.

One alternative to DECT is the French *Pointel*, a two-way *telepoint* system. A telepoint is a special microcellular base station. Usually, a telepoint allows only outgoing calls from specific areas in public places. Other alternatives are the CT-2 based cordless PBX by Helsinki Telephone, the Norwegian Telecom CT-2, and the Hutchinson *Rabbit* in the UK. CT-2 refers to cordless telephone generation two. CT-2 is an advanced digital cordless telephone that can be used either as a home cordless, or that can be taken to a public place, such as a shopping center, and used with telepoints there to make outgoing calls.

The *cordless local area network* (CLAN) was developed in the UK on behalf of the Department of Trade and Industry (DTI), the British equivalent of the FCC. The CLAN is intended to provide PC interconnection, as well as links for controls to manage laboratory test equipment. It uses traditional LAN protocols, such as token ring and Ethernet. It is purely data oriented. The *Advanced Radio Data Information Service* (ARDIS) is a radio packet network for portable laptop and handheld computers. It runs at 4800 baud, with a 240-character packet length. ARDIS started as a joint venture between Motorola and IBM. It covers 97% of United States metropolitan locations, using 1000 base stations at 855.8375 MHz with 40-W output, providing a 16-km range. The subscriber terminals transmit at 810.8375 MHz with 4 W out.

7.9. THE WIRELESS LOCAL LOOP

Subscribers in locations remote from switching centers, or in places that cannot be wired readily to a switching center, may be faced with excessive service costs for their subscriber lines. The problem is that of the man-hour installation costs over the distances required. In the case of remote areas, often only light use is made of mobile radio frequencies because of the sparse population. A wireless local loop using such frequencies thus presents an economically attractive alternative to the use of either fiber or copper over a very lightly loaded path.

Wireless local loops operate in the 150-, 450-, and 800-MHz bands. One type provides *Basic Exchange Telecommunications Radio Service* (BETRS). These local loops serve small numbers of widely scattered subscribers. With point-to-point wireless technology, all subscribers share the costs for a large portion of the system; the only cost allocated entirely to the individual subscriber is that of the remote terminal at or near the subscriber premises.[31] Fig. 7.33 is a block diagram of a BETRS system. The *central office terminal* (COT) interfaces with

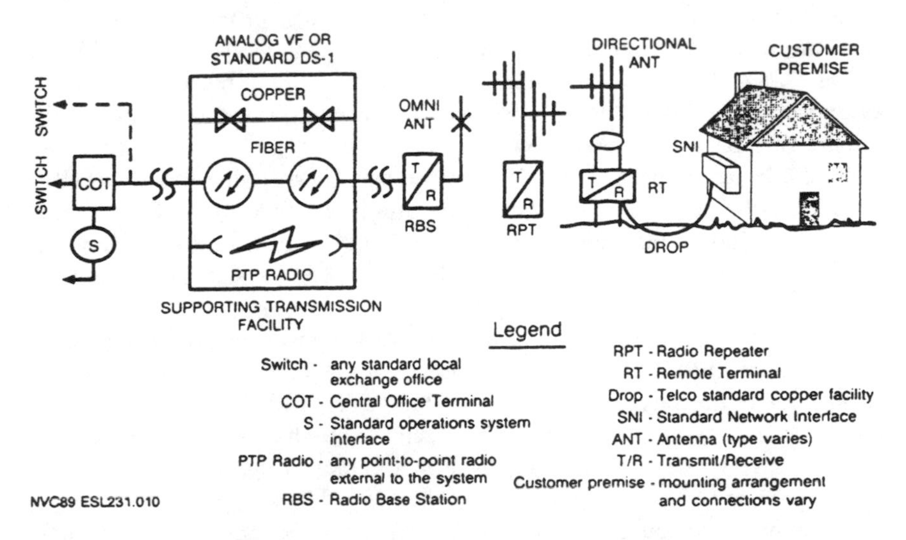

FIGURE 7.33. Basic exchange radio system. (Courtesy Lin and Wolff, Ref. 26, © IEEE, 1990.)

the local exchange switch. The *radio base station* (RBS) serves as the main radio distribution point. The *remote terminal* (RT) provides the connection to the subscriber through a *standard network interface* (SNI). As needed, a repeater (RPT) may be used between the RBS and the RTs. The COT usually is located near the local exchange switch. The RBS may be collocated with the COT, or may be at a remote high-elevation site to reach many scattered subscribers.

Speech coding techniques such as RELP or LD-CELP can be used, along with high level digital modulation techniques, for spectrum conservation. *Demand assigned multiple access* (DAMA) allows each RT to access any of the 26 RF channels in the 450-MHz band, and any of the individual voice circuits multiplexed within an RF channel. The number of available channel pairs in the 150-MHz band is 18; in the 800-MHz band, 50 channel pairs are available.

The typical coverage radius over relatively flat terrain is about 55 km from the RBS.

Not all wireless local loops need serve remote subscribers. There may be a need for improved telephone service to a built-up area with small cost and ease of installation of the necessary infrastructure.[32] In this case, the wireless local-loop equipment could be located with the central office switch serving the area. A distributed antenna system as illustrated in Fig. 7.32 could be used. Fixed end instruments would function in the same way as cordless units, but would be powered from the 117 V AC supply, rather than from an internal battery. To the extent that hand-off is not required, the system is simpler than a cellular type system.

Another concept is being developed by Bellcore under the name *Wireless Access Communications Systems* (WACS).[33,34] This concept visualizes access to the *local exchange carrier* (LEC) network for *fixed-wire applications* (FWA), as well as PCS. FWA includes service in new buildings, loop rehabilitation, temporary service for conventions, fairs, etc., and backup service for critical applications. The connection may be analog or digital (ISDN).

The Bellcore approach to WACS is to provide TDM from a *radio port* (RP) to the subscriber set, with TDMA from the subscriber set back to the RP. Channel assignment is done automatically, with the RP choosing the best frequency, the subscriber unit choosing the RP, and the RP assigning a time slot to the subscriber set.

Preliminary plans call for the use of 4-QAM with differential encoding, a symbol rate of 250-kBd, and a channel spacing of 400 kHz. One 200-μs time slot contains 64 bearer bits, 9 system control bits, and 12 FEC bits, resulting in a (85,73) code. Ten time slots constitute a frame. Speech coding is at 32 kb/s or less, with a 32-kb/s user being designated as "full rate." Link security techniques include encryption, authentication, and subscriber-set lock.

Subscriber-set output power is 10 to 20 mW per active time slot, controllable in 0.75-dB steps (by the RP) over a 30-dB dynamic range. The receiver sensitivity is better than -101 dBm. The antenna is encapsulated in the subscriber set.

Interfaces include a standard air interface (SAI) between the subscriber sets and the RP, a network-to-network interface, and an RP-to-network interface. The RP-to-network interface consists of several RPs connected to an RPMUX, and several RPMUXs connected to a *radio port control unit* (RPCU) at the central office. The RPCU functions may be at an RP or at the central office or at a separate location. Several RPCUs may connect to the switching system. Since the channels are 500 kb/s each, three of them are combined into a DS1 for transport.

The RPCU performs numerous functions. These include:

- Multiplexing and demultiplexing between DS1s and 500-kb/s channels
- Transcoding speech to/from 64 kb/s
- Encryption/decryption
- Echo cancellation as required
- Error-handling functions
- Relay of registration messages to allow location of subscriber sets
- Exchange of signaling information with subscriber sets for customer-initiated calls
- Coordination of multiple RPs
- Determination of power control for subscriber sets
- RP channel allocation
- Dynamic call transfer and rerouting
- Performance of *time-slot transfers* (TST) via local *time-slot interchange* (TSI) hardware
- Transfer of signaling information network components and subscriber sets via RPs.

Another type of wireless local loop is found in the *wireless information network*, or *wireless in-building network* (WIN), providing portable radio handsets and fixed radio ports, thus freeing end-users from the tether of the wireline-based telephone.[35] Systems using infrared transmission are capable of data rates limited to about 1 Mb/s, but are easy to deploy. The use of systems operating at 18 GHz allows not only ease of deployment, but also data rates in excess of 10 Mb/s. Moreover, the propagation characteristics at 18 GHz are such that microcells can be established that are confined to single floors within buildings. A cluster of six or eight sector antennas centrally located on a floor are used to provide the desired coverage.

7.10. THE PCS INTERFACE TO THE PSTN

The PCS services to be offered vary widely in terms of the capabilities of the potential service providers to perform many of the functions of a telephone system. In particular, some of the PCS systems under discussion will be independent of, but connectable to, the PSTN. Those access services that the telephone company (telco) is capable of providing include signaling, transport, switching, and intelligent network capabilities.

Bellcore and the BOCs have defined several ways in which a telco might interface with a PCS. The network (N) interface is one in which telco does the least. The telco provides signaling, transport, switching (routing), and advanced intelligent network (AIN) capabilities. The PCS system provides call control processing, radio management, and subscriber information storage.

With the controller (C) interface, the telco also does call control and all switching, and adds communication between RPCUs. The PCS operator provides radio channel control, local data storage, and RPCU interfaces with the telco for hand-off purposes.

With the port (P) interface, the PCS operator simply provides the air interface to the subscriber sets, that is, the RPs. The telco provides communication between the RPs, radio management, hand-off, and all other functions of the N and C interfaces.

The data (D) interface consists of a set of interface services. The PCS provider may subscribe to any or all of them. They allow the PCS provider to utilize the telco's AIN capabilities, that is, they allow the PCS to access centralized data in the telco network, such as customer location updates, customer authentication, and customer service profiles.

Discussions of PCS systems lead to the realization that many different types of handsets may be required of a user who wishes a truly portable calling number (PCN). Ideally, one would eventually have what this author views as a *universal telephone handset*. No such handset has appeared "on the drawing board" yet, although several manufacturers are taking a step in that direction through the development of multimode handsets.

What must the truly universal handset be capable of doing? Aside from being acceptably small, light weight and acceptably priced from the consumer's viewpoint, it must be capable of the following:

With respect to transmission, it must be capable of operating via satellites, terrestrial cells, microcells, cordless PBXs, cordless bases or twisted pair or fiber. With respect to frequency ranges, it must be capable of functioning in the bands 800–960 MHz, 1535–1665 MHz, 1710–2200 MHz, 2400–2485 MHz, and possibly other older frequency ranges, such as 148–174 MHz and 450–512 MHz. Additional frequency ranges, perhaps higher than those listed here, also may be required. With respect to access, the handset must be capable of operation using FDMA (SCPC), TDMA or CDMA, or combinations of these techniques. Modulation techniques required include π/4-DQPSK (for use in North America, Japan, etc.), GMSK (for use in the GSM countries), and analog FM for use where digital services are not available.

PROBLEMS

7.1. A TDMA cellular system is to be designed so that $C/I = 14$ dB is exceeded 99% of the time. The terrain is such that the fourth-power law describes signal strength variations with distance. For cells whose radius is 12 km, and for a 4-cell repeat pattern, what is the minimum center-to-center distance between cells that use the same frequency set?

7.2. "All present and proposed cellular systems use FDMA." Is this statement true or false? Explain.

7.3. A CDMA cellular system uses omnidirectional cells, each of which is equipped for two channels. Assuming $C/I = 7$ dB, and assuming average speech statistics as outlined in this chapter, how many users can be served?

7.4. Determine the geographical spectrum efficiency of a TDMA system that has 20 cells, each of which handles 240 circuits. The system covers an area of 10,000 sq. km.

7.5. Multipath propagation has been identified as a cause of frequency selective fading, a problem that is especially troublesome in metropolitan areas with slow-moving or stopped vehicles. Frequency hopping is a technique for avoiding the propagation dropout spots. Must the frequency hopping be done on a pseudorandom basis, or will regular short-period hopping suffice? Discuss.

7.6. A given metropolitan area covers 30,000 sq. km. The terrain is flat. Assume $R = 7$ for analog, 4 for TDMA, and 1 for CDMA. Make suitable assumptions for set-up channels in each case, as well as adjacent channel limitations, and state your assumptions. Compute the geographical spectrum efficiency for analog, TDMA, E-TDMA and CDMA.

REFERENCES

1. Calhoun, G. *Digital Cellular Radio*, ©Artech House, 1988.
2. Raith, K. and J. Uddenfeldt, Capacity of Digital Cellular TDMA Systems, *IEEE Transactions on Vehicular Technology*, Vol. 40, No. 2, May 1991, pp. 323–332.
3. "Cellular System Dual-Mode Mobile Station—Base Station Compatibility Standard," EIA/TIA Project Number 2398, Standard IS-54 (Revision C), Engineering Department, Electronic Industries Association, Washington, DC 20006.
4. Lee, W. C. Y., *Mobile Communications Design Fundamentals*, Howard W. Sams, 1986.
5. "Cellular Radiotelecommunications Intersystem Operations," Standard IS-41, Engineering Department, Electronic Industries Association, Washington, DC 20006.
6. Keiser, B. E., *Broadband Coding, Modulation and Transmission Engineering*, ©Prentice Hall, Inc., 1989.

7. See note 3.
8. See note 3.
9. "Recommended Minimum Performance Standards for 800 MHz Dual-Mode Mobile Stations," EIA/TIA Project Number 2216, Standard IS-55, Engineering Department, Electronic Industries Association, Washington, DC 20006.
10. "Recommended Minimum Performance Standards for 800-MHz Base Stations Supporting Dual-Mode Mobile Stations," EIA/TIA Project Number 2217, Standard IS-56, Engineering Department, Electronic Industries Association, Washington, DC 20006.
11. The contents of this section are based largely on presentation material provided by Hughes Network Systems, Inc., Germantown, MD.
12. The contents of this section are based largely on material presented by QUALCOMM, Inc., San Diego, CA.
13. QUALCOMM, Inc. Submission to the Cellular Telecommunications Industry Association (CTIA) "CDMA Technology Investigation Subcommittee," Jan. 7, 1991.
14. See note 13.
15. See note 13.
16. See note 13.
17. Epstein, M., "CDMA Digital Cellular Communications," presented to the IEEE Communications Society, Washington Section, December 19, 1991.
18. Schilling, D. L., et al, "Broadband CDMA for Personal Communication Systems," *IEEE Communications Magazine*, Vol. 29, No. 11, November, 1991, pp. 86–93.
19. Schilling, D. L. Private communication to author, September 21, 1992.
20. Request of Omnipoint Communications, Inc. for a Pioneer's Preference in the Licensing Process for Personal Communications Services (Gen. Docket 90-314), before the Federal Communications Commission, May 4, 1992.
21. Robinson, T. S., "From Analog to Digital: TDMA, CDMA, FHMA, FDMA," *Business Radio*, Vol. 29, No. 8, Oct. 1993, pp. 15–18.
22. Siew, C. K., and Goodman, D. J., "Packet Data Transmission Over Mobile Radio Channels," *IEEE Transactions on Vehicular Technology*, Vol. 18, No. 2, May, 1989, pp. 95–101.
23. Keiser, B. E., *Broadband Coding, Modulation, and Transmission Engineering*, Prentice Hall, Inc., Englewood Cliffs, NJ, 1989, pp. 117, 440.
24. Calhoun, G., See note 1.
25. See note 13.
26. Saleh, A. A. M., and Cimini, L. J., "Indoor Radio Communications Using Time-Division Multiple Access with Cyclical Slow Frequency Hopping and Coding," *IEEE Journal on Selected Areas in Communications*, Vol. 7, No. 1, Jan. 1989, pp. 59–70.
27. Keiser, B. E., op. cit., p. 436.
28. See note 23.
29. See note 13.
30. Bernhardt, R. C., "User Access in Portable Radio Systems in a Co-Channel Interference Environment," *IEEE Journal on Selected Areas in Communications*, Vol. 7, No. 1, Jan. 1989, pp. 49–58.
31. Lin, S. H. and Wolff, S. H., "Basic Exchange Radio—From Concept To Reality," *IEEE Intl. Conf. Comm., 1990*, Paper 206.2, © IEEE, 1990.

32. See note 13.
33. "Generic Framework Criteria for Version 1.0 Wireless Access Communications Systems," Framework Technical Advisory FA-NWT-001318, Issue 1, June 1992, Bellcore, Morristown, NJ 07962-1910.
34. "Generic Criteria for Version 0.1 Wireless Access Communications Systems," Technical Advisory TA-NWT-001313, Issue 1, July 1992, Bellcore, Morristown, NJ 07962-1910.
35. Freeburg, T. A., "Enabling Technologies for Wireless In-Building Network Communications—Four Technical Challenges, Four Solutions," *IEEE Communications Magazine*, Vol. 29, No. 4, April, 1991, pp. 58–64.

8

Microwave Transmission*

8.1. INTRODUCTION

Microwave radio systems still provide a portion of the long-haul message circuit distance within the United States, and are responsible for an appreciable portion of the world's circuits as well. While some of this message-circuit distance still is on an analog basis, digital service constitutes an increasing portion of microwave transmission. Microwave systems not only provide feeder service to long-haul fiber backbone routes but also provide cross-country service through the use of multiple repeater sites.

Thus, individual circuit lengths range from less than 30 km to over 6400 km, with route cross-sectional capacities from less than 60 to more than 22,000 circuits. Applications include not only telephone-message channels but also wideband data and television transmission.

In microwave system terminology, a short-haul system is one covering a distance up to about 400 km between end points. Such applications include intrastate and feeder service. Long-haul systems exceed 400 km and are used for interstate and backbone routes.

While fiber-optic systems are replacing microwave radio routes on a long-haul basis, microwave systems are finding new applications in the metropolitan-area market, where their ability to provide service quickly without the need to drill through concrete is making them popular. Most new microwave systems use digital modulation.[1]

This chapter presents an overview of microwave radio systems with an emphasis on their use in digital telephony applications.

---

* Much of the information in this chapter is summarized from Bernhard E. Keiser, "Broadband Communications Systems," Course Notes, © 1992.

8.2. CHARACTERISTICS OF MICROWAVE PROPAGATION

Several factors contribute to the degradation of a microwave radio signal as it travels from one repeater to another. These are spreading loss and absorption, fading (caused by refraction and multipath effects), and polarization shifts. They are described next, in that order.

8.2.1. Spreading Loss and Absorption

All forms of electromagnetic radiation involve the emission of power from a source, and the spreading of that power in many directions. Thus if p_t is the power radiated from an isotropic source, and if d is the radius of an imaginary sphere centered on the source and representing the locus of all points a distance d from the source, then the power density at a distance d from the source is $P_t/4\pi d^2$ W/m^2. The spreading of power from the source can be exemplified by the spreading of light flux from a flashlight bulb. The greater the distance, the less is the power received.

The actual power P_r captured by a receiving antenna with an effective area A_r on the surface of the sphere is

$$p_r = \frac{p_t A_r}{4\pi d^2} \text{ watts} \tag{8.1}$$

The area A_r may be thought of as the area of a reflector associated with the receiving antenna, but diminished by the overall efficiency of the antenna. Figure 8.1 shows the geometrical relationships involved in this concept. Figure 8.1(a) illustrates the isotropic source just discussed. This source can be made directional, as illustrated in Fig. 8.1(b), by placing a reflector having an effective area A_t behind it. In this case, the source exhibits an on-axis gain of

$$g_t = \frac{4\pi A_r}{\lambda^2} \tag{8.2}$$

where λ is the wavelength of radiation and A_t is the physical area of the transmitting antenna multiplied by its efficiency.

The value of λ in meters can be determined from the operating frequency f in hertz by the relationship $\lambda = c/f$ where c = speed of light = 3×10^8 m/s. Such an antenna concentrates the radiation into a solid angle Ω steradians where

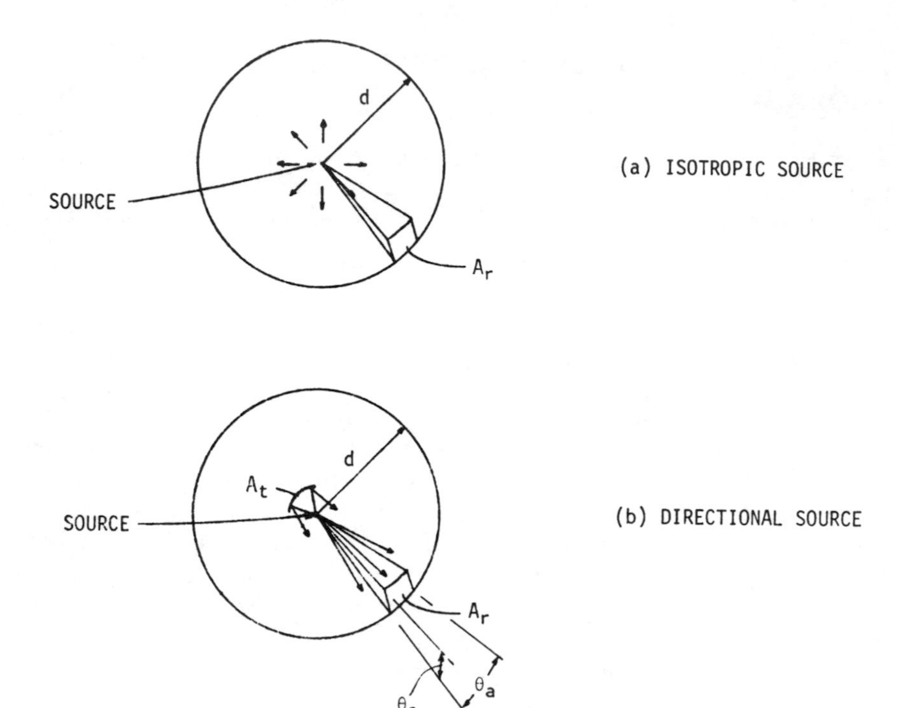

FIGURE 8.1. Geometrical relationships resulting in free-space path loss.

$$\Omega = \frac{\lambda^2}{A_t} \tag{8.3}$$

The expressions for gain and solid angle can be combined to yield the result

$$g_t = \frac{4\pi}{\Omega} \tag{8.4a}$$

This equation, of course, is idealized in that it assumes a 100% antenna efficiency. In reality, an efficiency factor, η, must be included, giving

$$g_t = \frac{4\pi\eta}{\Omega} \tag{8.4b}$$

Thus, the smaller the beam angle of an antenna, the greater its gain. The solid angle Ω is proportional to the product of an azimuthal beam width θ_a and an elevation beam width θ_e.

For a directional antenna, Eq. (8.1) becomes

$$p_r = p_t g_t \left(\frac{A_r}{4\pi d^2}\right) \tag{8.5}$$

Use of Eq. (8.2) yields

$$p_r = p_t \left(\frac{4\pi A_t}{\lambda^2}\right)\left(\frac{A_r}{4\pi d^2}\right) \tag{8.6}$$

Equation (8.6) can be rearranged to express the received power in terms of the transmitting and receiving antenna gains as follows:

$$p_r = p_t \left(\frac{4\pi A_t}{\lambda^2}\right)\left(\frac{4\pi A_r}{\lambda^2}\right)\left(\frac{\lambda}{4\pi d}\right)^2 \tag{8.7}$$

The reciprocal of the last term in Eq. (8.7), $(4\pi d/\lambda)^2$, is the *spreading loss* or *free-space path loss*. Generally it is expressed in decibel form as $20 \log_{10}(4\pi d/\lambda)$, or in the more convenient form

$$\text{Free-space path loss} = 32.5 + 20 \log_{10} d_{km} + 20 \log_{10} f_{MHz} \tag{8.8}$$

where

d_{km} = path length in km
f_{MHz} = frequency in MHz.

The free-space path loss is the most significant loss between the microwave transmitter and receiver of a link; however, it is a fixed loss that is known in advance and can be accounted for in design.

An additional loss between the transmitter and receiver is *absorption*, produced primarily by water vapor in the air. Figure 8.2 shows that absorption increases with frequency and also with rain intensity. Typical microwave path lengths of 30 to 45 km are essentially unaffected by rain, but at frequencies of 11 GHz and higher, path lengths may have to be kept shorter, especially in areas that experience significant amounts of precipitation.

8.2.2. Fading

Fading on a microwave path is caused by atmospheric refraction changes and by multipath propagation. The atmosphere tends to bend microwaves slightly downward toward the earth. In other words, the earth appears to be slightly flatter than

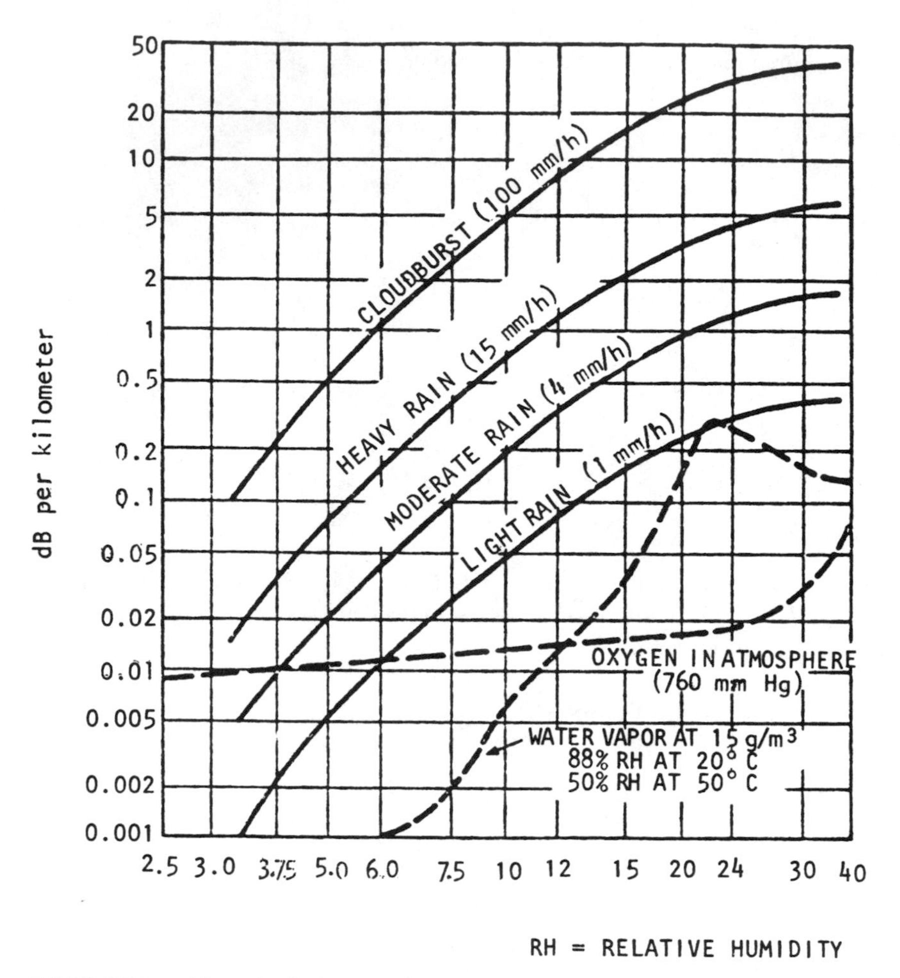

FIGURE 8.2. Absorption in the atmosphere. (© 1982, Bell Telephone Laboratories. Reprinted by permission.)

it really is. The earth nominally appears to have a radius that is about 4/3 of its true radius. Thus the distance to the horizon becomes somewhat greater than the simple geometric distance. Light rays experience a similar bending, but for them the effective earth radius is about 6/5 of the true earth radius. The ratio of the effective to the true earth radius usually is designated by the letter K.

Refraction is not constant. When heavy ground fog is present, or when there is very cold air over a warm earth, atmospheric refraction may become substandard, that is, $K < 1$, in which case the rays tend to be bent upward away from the earth. Other conditions may exist in which *K* becomes infinite, that is, the rays bend along with the curvature of the earth for distances of 100 km or more, resulting in a situation called *overreach*, in which the rays appear at repeaters more distant than intended.

Ray interference is produced when a ray reaches its destination by more than one path. Often one path is the direct one and the other path is one in which a ray has been reflected from the ground. Ray interference also can result from an irregular variation in dielectric constant with height. Unlike atmospheric refraction, ray interference tends to be fast and frequency selective. Figure 8.3 illustrates the statistics of ray interference at 4 GHz. Fortunately, the deep fades are of only brief duration.

In the design of microwave facilities (see Section 8.4), the probability of fades below a given level must be ascertained and a fade margin must be incorporated into the power budget corresponding to the percent availability for which the system is being designed. Space diversity is helpful in combatting fading.

8.2.3. Polarization

In order to minimize adjacent-channel interference along a given microwave path, alternate polarizations, vertical and horizontal, are used. In addition, the alternation of polarizations along a path from one repeater to another can aid in minimizing the overreach interference problem. Although precipitation tends to convert linear polarization into elliptical, especially at frequencies above 10 GHz, the undesired, or cross-polarized, signal generally is 25 to 30 dB below the desired signal, making polarization a very useful technique in microwave radio systems. In fact, separate digital channels can be sent on the two polarizations.

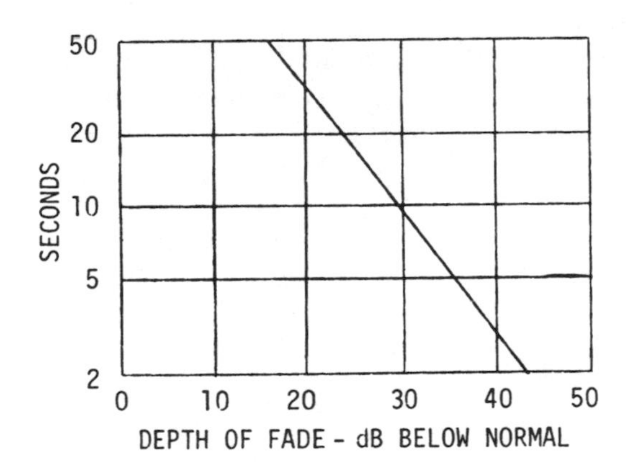

FIGURE 8.3. Median duration of fast fading on 50-km paths. (© 1982, Bell Telephone Laboratories. Reprinted by permission.)

8.3. MICROWAVE SYSTEM ENGINEERING

Numerous factors affect the performance of a digital radio system. Some of the key factors are discussed next.

8.3.1. Frequency Allocations

The most heavily used bands available for long-haul microwave radio relay by the common carriers within the U.S., and their usage, is as follows:

Frequency Bands (MHz)	*Usage*
3700–4200	Long haul
5925–6425	Long haul; short haul
10700–11200	Short haul

Additional common-carrier bands are available in some countries. For example, in Canada, additional available bands are 3550–4200 MHz, 5925–8275 MHz and 10 700–11 700 MHz. Higher frequencies also have been allocated in the U.S. and Canada, but are not in common use for transmission beyond metropolitan areas because of propagation problems during precipitation.

The Radiocommunications Sector (RS), an arm of the International Telecommunication Union (ITU), headquartered in Geneva, Switzerland, has divided the common-carrier bands into radio channels, and has specified the minimum number of voice channels per radio channel, as is shown in Table 8.1.

In addition, to minimize problems of adjacent-channel interference, the mean output power in any 4-kHz band must be attenuated below the mean total output power of the transmitter. This attenuation must be such that the transmitter's output spectrum fits within the regulatory mask. For operation below 15 GHz, this mask is shown in Fig. 8.4 and is defined by the equations

TABLE 8.1. Minimum Number of Voice Channels in Microwave Bands Below 12 GHz

Frequency Range (GHz)	Minimum No. of Voice Channels	Radio Channel Bandwidth (MHz)
2.11–2.13	96	3.5
2.16–2.18	96	3.5
3.70–4.20	1152	20
5.925–6.425	1152	30
10.70–11.70	1152	40

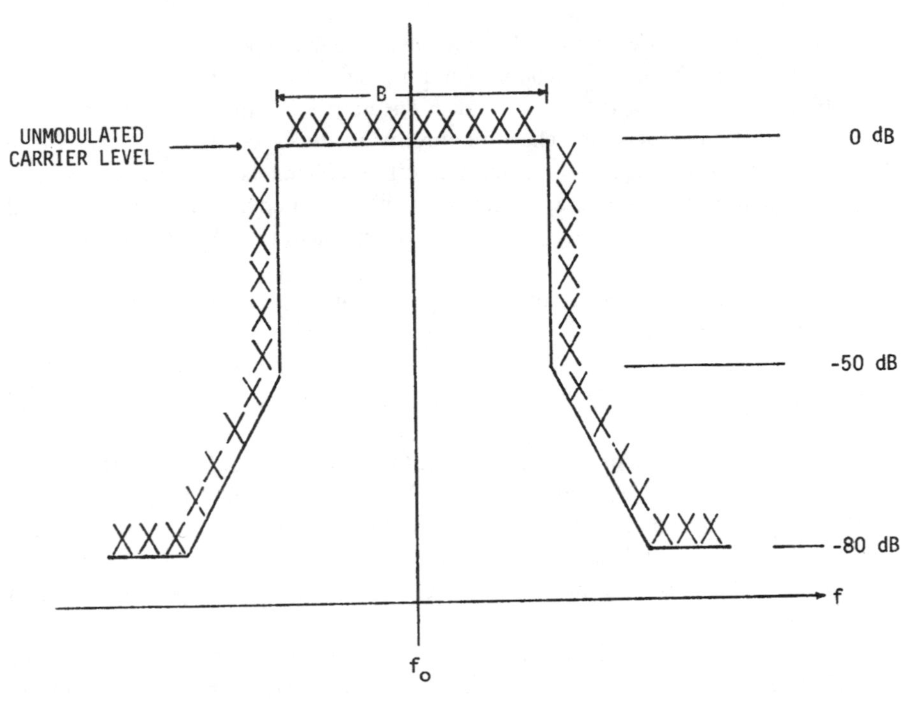

FIGURE 8.4. ITU/RS mask.

$$A(f) = \begin{cases} 0 \text{ dB}, & \text{for } 0 < b < 50 \quad (8.9a) \\ 35 + 0.8\,(b-50) + 10 \log_{10} B \text{ dB}, & \text{for } b > 50 \quad (8.9b) \end{cases}$$

with the provision that the maximum required value of $A(f)$ is 80 dB. In Eq. (8.9),

$A(f)$ = attenuation below the mean wideband output power level (dB), measured in a 4-kHz band
B = radio channel bandwidth (MHz)
f_0 = carrier frequency (MHz)
f = frequency at which attenuation is being specified (MHz)
$b = 100 \dfrac{|f - f_0|}{B}$

Above 15 GHz,

$$A_M(f) = \begin{cases} 0 \text{ dB}, & \text{for } 0 < b < 50 \quad (8.10a) \\ 11 + 0.4(b-50) + 10 \log_{10} B \text{ dB}, & \text{for } b > 50 \quad (8.10b) \end{cases}$$

where $A_M(f)$ is the attenuation below the mean wideband power output level (dB), measured in a 1-MHz band. The maximum required value of $A_M(f)$ is 56 dB.

Filtering is a major technique used to ensure the meeting of the mask requirements in a given system. These requirements are especially important, since otherwise the use of analog transmission in radio channels adjacent to those carrying digital transmission may be seriously affected. Other combinations, digital to adjacent digital and analog to adjacent digital, also result in degradation, but of a less serious nature.

Several system-design factors are closely related to the subject of frequency allocation.[2] These include the proper choice of radio-frequency (RF) carrier, intermediate frequency (IF) and beating oscillator frequencies, as well as the likely severity of selective fading, which may alter the stability of the carrier to sideband phase relationships.

A major factor affecting the selection of the frequency band is expected system growth and thus the required number of channels. The higher-frequency bands, especially 11 GHz and above, tend to be preferable from this viewpoint because more channels generally are available along any route at these higher frequencies. Protection channels also must be provided for use during deep fades or when channel equipment outages occur.

Cost factors also are significant in frequency choice. In general, equipment for the higher frequency bands (for example, 11 GHz) tends to be more expensive, but this may be partially offset by the fact that the number of voice channels per megahertz must be larger at the lower frequencies (for example, 4 GHz), and thus expensive filters and signal-shaping techniques may be needed at these frequencies.

System costs in terms of numbers of channels consist of per site costs and variable costs. The fixed (per site) costs are independent of the number of channels carried. They include roads, buildings, towers, and antennas. Thus, the fixed costs per voice channel decrease with the number of channels carried. The variable costs per channel tend to be constant for low to moderate numbers of channels, but then rise as attempts to fully load the route are made, because of increasingly severe filter requirements, tight tolerances on RF and IF frequencies, and the resulting overall system complexity and cost. The optimum mix of fixed and variable costs depends upon both the present and the projected demand for route capacity.

8.3.2. Link Budget

A link budget is an essential element of any microwave radio design. Its purpose is to assure the reception of adequate signal strength through the use of adequate transmitter power and antenna sizes, while not using excessive power and thereby creating interference problems. The use of decibel expressions is convenient in structuring link budgets because the decibel values can be simply added or subtracted from one another. An illustration of a link budget problem follows.

Problem. Determine the receiver input power for a 4-GHz radio hop given the following conditions:

Transmitter output power = 5 watts:	37.0 dBm
Gain of each antenna (3 m diameter):	39.6 dB
Waveguide loss, each end:	2.1 dB
Network losses, each end:	1.9 dB
Path length:	46 km

Solution. The free-space path loss is computed using Eq. (8.8). The result is 137.5 dB. The received power then is found by adding the antenna gains and subtracting the losses. Thus,

$$\begin{aligned}\text{Receiver input power} &= 37.0 - 1.9 - 2.1 + 39.6 - 137.5 + 39.6 - 2.1 - 1.9 \\ &= -29.3 \text{ dBm}\end{aligned}$$

Assuming a 20-MHz radio-channel bandwidth, as is standard for the 4-GHz band, the receiver noise level must be determined next. The noise figure F of a receiving system is defined as

$$F = 10 \log_{10}\left[\left(\frac{[T_a + T_e]}{290}\right) + 1\right] \tag{8.11}$$

where T_a = noise temperature associated with antenna in Kelvin, K. T_e = noise temperature associated with receiver, K.

Noise figure is discussed further in Section 8.4.1.2. The system noise temperature $T = T_a + T_e$. The receiver noise level then is

$$N = kTB \text{ dBW} \tag{8.12}$$

where k = Boltzmann's constant = –228.6 dBW/HzK.

Thus, for $T_a = T_a = 290\ K$, $T = 580\ K = 27.6$ dBK, corresponding to F = 6.0 dB; in a 20-MHz bandwidth, N = –128.0 dBW = –98.0 dBm.

For n hops, assuming identical repeaters and identical received power levels at each receiver, the noise adds as $10 \log_{10} n$ dB. Thus, for nine hops, N = –98.0 + 9.5, or –88.5 dBm. This obviously is well below the –29.3-dBm signal level;

however, numerous factors must be accounted for between these two levels. Microwave systems are subjected to propagation fades caused by multipath transmission and by rain attenuation. Thus, if a common carrier wants to keep propagation outages as low as equipment outages, a two-way annual fading allocation of 0.01% over a 400-km route (53 minutes per year) may be specified.[3] This is equivalent to seven minutes per year over a 50-km hop. The achievement of such a reliability may require a fade margin on the order of 40 dB, thus leaving 19.2 dB carrier-to-noise-plus-interference ratio in order to ensure a bit error ratio of 10^{-6} with the modulation technique used, including a margin for modem losses and other forms of circuit degradation. On this basis, such modulation techniques as QPR-AM and M-PSK ($M \geq 8$) can be utilized. Several commercial digital radio system types using these techniques are in operation in the 6- and 11-GHz bands.[4,5]

For operation in the 11-GHz band, where heavy rain (15 mm/h) produces an attenuation of 1.0 dB/km, a higher fade margin may be required than would be used in the lower-frequency bands. Alternatively, space diversity may have to be used. Space diversity involves the use of separate receiving sites with sufficient spacing that the probability of a deep fade to both sites simultaneously is extremely low. The signals received at the two sites then are compared, and the stronger of the two is used. Some diversity systems combine the two received signals instead. Section 8.3.4 provides a further discussion of diversity.

8.3.3. Repeater Siting

Microwave repeaters must be placed so that they are within radio line of sight of one another. The term radio line-of-sight implies the fact that the waves tend somewhat to bend with the curvature of the earth during normal refraction conditions. This bending is expressed by viewing the earth's radius as being multiplied by a factor K, which often is taken to be 1.33 for radio waves. Under unusual conditions, K may drop to as low as 0.66 or increase to infinity. The distance d in km to the radio horizon over smooth earth is approximately $d = 3.55\sqrt{Kh}$, where h is the height in meters above the earth.

Repeater antenna heights must be sufficient that the microwave beam will clear all obstacles in the path. The clearance of the beam over an obstacle is expressed in terms of Fresnel zones. These zones are defined such that the boundary of the nth Fresnel zone consists of all points from which the reflected wave is delayed $n/2$ wavelengths. The distance in meters, m, from the line-of-sight path to the boundary of the nth Fresnel zone, H_n, is

$$H_n = \sqrt{\frac{N\lambda d_1(d-d_1)}{d}} \tag{8.13}$$

where

d_1 = path length from reflecting point to one antenna, m
d = distance between the antennas, m.

Figure 8.5 illustrates the Fresnel zone concept by showing the first Fresnel zones[5] at 100 MHz ($\lambda = 3$ m) and at 10 GHz ($\lambda = 3$ cm). Notice that point C is beyond the first Fresnel zone for $\lambda = 3$ cm, but well within the first Fresnel zone for $\lambda = 3$ m. The indirect path d_1 will result in a received signal that may interfere with the signal received via the direct path d. The type of interference will depend on the overall loss and delay in a path d_1 compared with the loss and delay via the direct path d.

For free-space transmission, a clearance of at least 0.6 of the first Fresnel zone is required, but substantially greater clearances usually are provided because of the variability of the refraction effects.

Repeater spacing thus depends upon line of sight path clearance, as well as upon fading, including rain attenuation, as will be appreciated from Section 8.3.2. Other factors affecting repeater spacing are tower costs, interference to and from other systems, and system requirements in terms of the points to be served along the route. Tower costs can be traded versus repeater spacing, with taller (costlier) towers allowing fewer repeaters overall. System interference may occur when any two routes are in proximity to one another. Thus, parallel paths should be avoided and, where routes must cross one another, the crossing should be done at right angles, if possible.

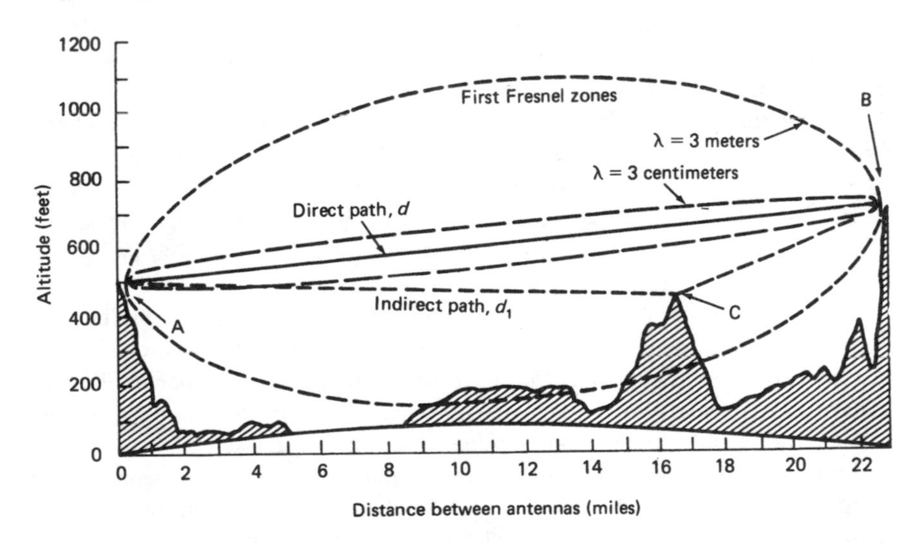

FIGURE 8.5. Fresnel zones.

8.3.4. Diversity

Diversity can be implemented on either a frequency or a space basis. Frequency diversity requires two separate frequencies and thus does not make efficient use of the radio spectrum. Accordingly, tight regulations exist regarding frequency diversity, and industrial users are not allowed it in many countries.

Vertical space diversity is especially effective against ground-or water-reflective fading, as well as against atmospheric multipath fading. Although space diversity is more expensive than frequency diversity because of the additional antennas and waveguides it requires, it generally can provide better protection than frequency diversity, especially where the latter is limited to small frequency spacing intervals. Suitable spacings[7] are 20 m at 2 GHz, 15 m at 4 GHz, 10 m at 6 GHz, and 5–8 m at 11 GHz.

Horizontal space diversity has been found useful in digital radio systems through the use of antennas having differing beamwidths or beam cross-sections. Spacings are comparable to those used for vertical space diversity. The success of horizontal space diversity in digital radio systems is attributed to the fact that digital radio is affected more strongly by multipath dispersion than by total power-fade depth, as is the case in analog radio.

8.3.5. Reliability and Availability

The terms *reliability* and *availability* sometimes are used interchangeably. Reliability, however, often denotes the percentage of time during which the equipment performs its intended function, whereas the actual system availability is less because of propagation outages. Thus, if equipment failures disrupt service 0.01% of the time and if propagation outages also disrupt service 0.01% of the time, then the overall system availability is 99.98%.

Availability objectives for U.S. common-carrier short-haul systems (<400 km) specify that two-way service failure be less than 0.01% per year. This is equivalent to 53 minutes per year. Industrial microwave systems, especially those belonging to electric, oil, and gas utilities, allow no more than 0.01% outage during the worst month, because of the potentially serious effects of outages on their operations.

The major factors related to equipment downtime are hardware reliability, redundancy, spares availability, and power-source reliability. Hardware reliability has increased appreciably with the move to all-solid-state components. Redundancy can be applied to both the radio equipment, in the form of duplication, and to the power source, in the form of a diesel backup generator.

Propagation outages can be minimized through the use of an adequate fade margin as well as space diversity. A useful equation[8,9,10] for determining the required fade margin M_f has been developed from the work of W. T. Barnett and A. Vigants at Bell Telephone Laboratories:

$$M_{\mathrm{f}} = 30 \log_{10} d + 10 \log_{10} (6ABf) - 10 \log_{10} (1 - R) - 70 \text{ dB} \quad (8.14a)$$

where

d = path length, km
R = reliability objective (one way) for a 400 km route
A = roughness factor
B = conversion factor, worst-month to annual probability
f = carrier frequency, GHz.

Values of A are:

4 for very smooth terrain, including over water
1 for average terrain with some roughness
0.25 for mountainous, very rough terrain

Values of B are:

0.5 for lake areas, especially if hot or humid
0.25 for average inland areas
0.125 for mountainous or very dry areas

To obtain the fade margin to be used for the worst month, set B = 1. For example, for $R = 0.9999$ during the worst month,

$$M_{\mathrm{r}} = 30 \log_{10} d + 10 \log_{10} (6Af) - 30 \quad (8.14b)$$

8.4. CHARACTERISTICS OF MICROWAVE EQUIPMENT

Microwave equipment has certain characteristics that distinguish it from equipment for other frequencies. This section discusses these characteristics as well as the limitations and advantages of microwave system operation.

8.4.1. Amplifiers

Microwave amplifiers are built to provide adequate transmitter power-output levels, as well as to amplify weak incoming signals. Such characteristics as linearity and noise level constitute significant system design factors.

8.4.1.1. Power Amplifiers

Microwave power amplifiers exhibit nonlinearities which manifest themselves in the generation of intermodulation products, as well as errors in the reproduction of the input amplitude. In addition, a phase shift is produced that depends on the input power level. These two factors tend to alter the signal-state space diagram (see Chapter 6) in terms of both amplitude and phase. Accordingly, compensation for these nonlinearities must be achieved for the transmission of such multiamplitude digital modulation types as 16-QAM, 64-QAM, and 256-QAM.

Operation in the most linear portion of a power amplifier's characteristic also is important in minimizing adjacent-channel interference. Unwanted spectral sidelobes, which are inherent in the digital modulation process, may be restored by amplifier nonlinearities.

8.4.1.2. Small Signal Amplifiers

The amplifier used at the input of a microwave repeater generally is operated in the linear portion of its characteristic. Its noise figure or noise temperature, as well as gain and bandwidth, are its most important performance characteristics.

The sensitivity of a microwave receiver can be expressed in terms of the effective noise temperature T of the receiver front end, measured in Kelvin, the bandwidth, expressed in hertz, and Boltzmann's constant, $k = 1.38 \times 10^{-23}$ W/HzK. This sensitivity, kTB watts, is the noise level at the receiver input.

The noise factor F is given by

$$F = 1 + \left[\frac{(T_r + T_e)}{290}\right] \tag{8.15}$$

A more commonly used term is the noise figure, which is given by 10 $\log_{10}F$, and which is defined by the expression

$$NF = 10 \log_{10} \left[\frac{P_i/N_i}{P_0/N_0}\right] \tag{8.16}$$

where

P_i = available input signal power
P_0 = available output signal power
N_i = available input noise power
N_0 = available output noise power.

The foregoing terms must all be in the same units, that is, all in watts, milliwatts, microwatts, etc.

8.4.2. Antennas

Antennas used for microwave radio-relay purposes usually are either the parabolic or the horn-reflector type. Parabolic apertures normally range from 0.6 to 4.6 meters in diameter, and can be operated on a multiband basis by using a separate feed for each band of operation. The gain G of such an antenna at a wavelength λ meters ($\lambda = c/f$ where $c = 3 \times 10^8$ m/s and f = frequency in Hz) is

$$G = \frac{4\pi A\eta}{\lambda^2} \tag{8.17}$$

where A is the reflector area and η the efficiency of the antenna, typically 55 to 70%. The beamwidth θ of the parabolic antenna to the –3-dB points is approximately

$$\theta = \frac{70\lambda}{D} \text{ degrees} \tag{8.18}$$

where D is the diameter of the reflector. Such antennas have patterns with front-to-back response ratios of 40 to 70 dB, although sidelobes may be only 13 to 20 dB down from the main lobe. The polarization is that of the feed, and is usually either horizontal or vertical. Alternate radio channels often are transmitted using alternate polarizations to help reduce adjacent channel interference. A standard parabolic antenna is pictured in Fig. 8.6.

Even though parabolic reflector antennas can be built with quite good characteristics, situations arise in which even lower sidelobe levels must be achieved than the parabola can provide. Such situations call for the use of a horn-reflector antenna, which uses a vertically mounted horn under a section of a parabolic surface. Figure 8.7 shows several of these antennas on a repeater tower. Such antennas not only have front-to-back ratios of 80 to 90 dB, but also can be built to have a beam pattern that continues to drop in amplitude off boresight rather than exhibiting the prominent sidelobes characteristic of the parabola. Figure 8.8. shows the radiation pattern envelope of a horn-parabola antenna, Andrew Type Number SHX10B, which has a 3-meter aperture and provides a 42.7 dBi gain in the 6-GHz band.[11] The envelope shown is for a horizontally polarized antenna. The actual antenna pattern, except at 0°, is equal to, or lower than, the levels described by this envelope. Minor variations may occur in the pattern itself as a result of reflections from nearby objects. The manufacturer includes a full set of digital data for each pattern in order to save time and maintain accuracy when

FIGURE 8.6. Parabolic antenna, PXL-series standard microwave antenna. (Photo courtesy of Andrew Corporation, © 1982.)

interpreting from the analog pattern envelopes. One reason for the excellent characteristics of the horn reflector is the fact that the feed does not block the main path of radiation from the reflector, but rather is offset from this direct path. The gain and –3-dB beamwidths of the horn reflector depend on its aperture area, as do those of the parabola, so Eqs. (8.17) and (8.18) are applicable to it. Horn reflectors are suited to locations where frequency usage is heavy. Usually they allow the use of identical radio channels in the repeating and branching directions of a repeater site.

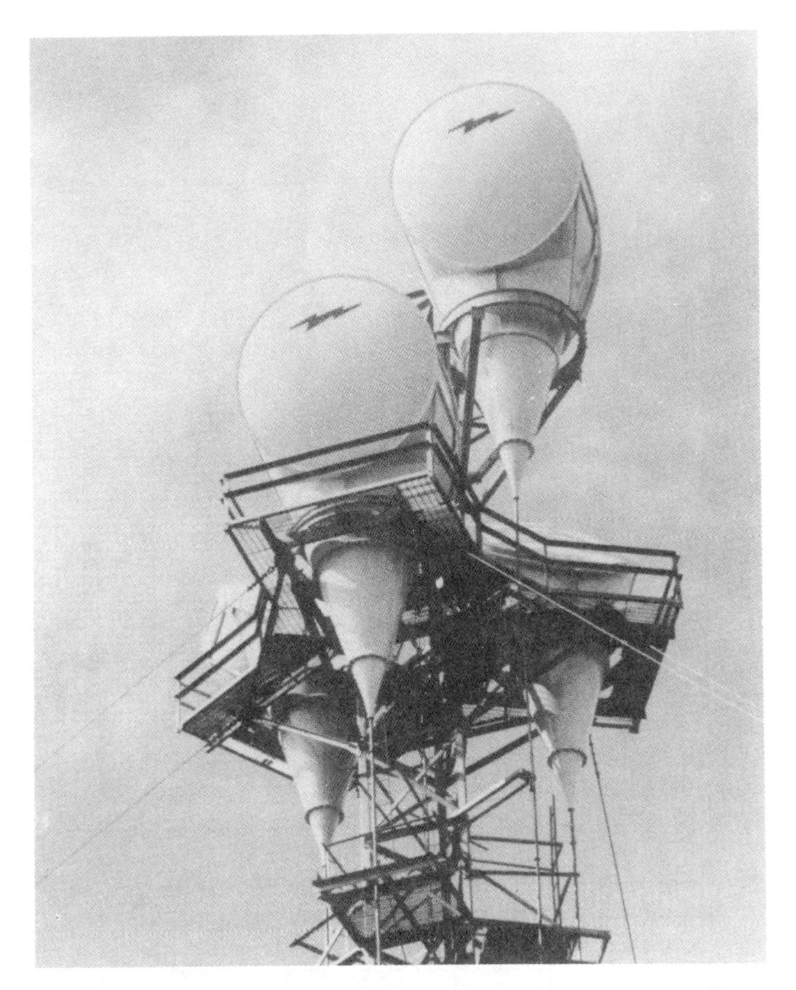

FIGURE 8.7. Horn-reflector antennas, SHZ™ super high performance antennas. (Photo courtesy of Andrew Corporation, © 1982.)

8.4.3. System Interface Arrangements

Digital radios may be built for a DS1, DS2, or DS3 interface with terminal equipment or other repeaters. In addition, some 90-Mb/s digital radios provide an interface consisting of three 30-Mb/s streams. These arrangements allow paths that use the 11-GHz band in the vicinity of large cities and that convert to 6 GHz for the long-distance rural routes, along which the 6 GHz band is more readily available than in the large metropolitan areas.

Digital transmission via the *data-under-voice* (DUV) mode and similar tech-

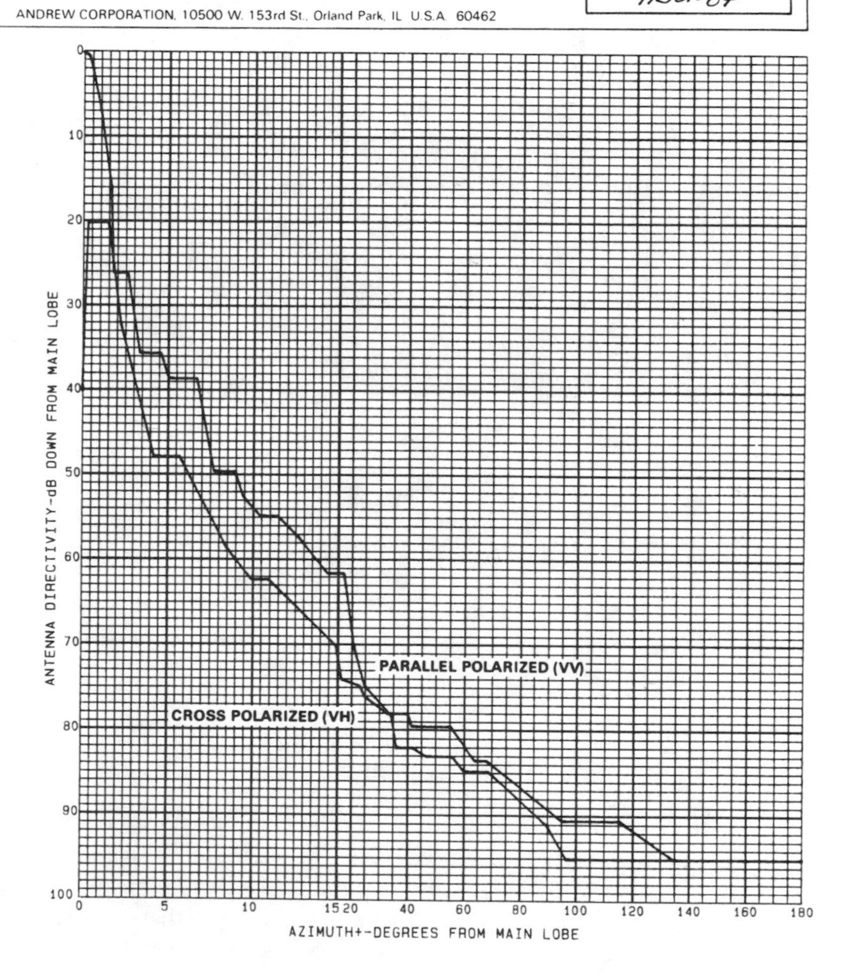

FIGURE 8.8. Radiation pattern envelope of a 3 m horn-parabola antenna Type SHX10B at 6.175 GHz. (Courtesy Andrew Corporation © 1984).

niques is handled by FDM *baseband* repeaters, which provide a conversion to baseband at each repeater at which trunks are dropped. Baseband repeaters are common in the 6- and 11-GHz bands.

Intermediate-frequency (heterodyne) FDM repeaters avoid the FM-to-baseband and baseband-to-FM steps of the baseband repeater, and thus avoid the noise associated with demodulation and remodulation. Moreover, they reduce misalignment problems because the signal deviation does not change through the repeater. Where trunks must be dropped at a heterodyne repeater, an FM terminal is used to perform the required demodulation and remodulation functions.

8.5. DIGITAL MICROWAVE RADIO SYSTEMS

A digital microwave radio is one whose instantaneous RF carrier can assume one of a discrete set of amplitude, frequency, or phase levels as a result of the modulating signal. Since digital modulation techniques produce very broadband signals, filters must be used to limit the bandwidth of the modulated signal to the assigned radio channel while still allowing transmission with a minimum amount of intersymbol interference.

A "digital radio" may include a radio that transmits a signal whose informational content is at least partly digital; that is, the baseband may include both analog and digital signals. In the case of an FM-digital hybrid system, digital modulation techniques are considered to be used if digital modulation contributes 50% or more to the total power of the transmitted RF carrier. Thus the term "digital radio" may refer to any microwave radio that transmits PCM or other digital carrier signals, regardless of how or at what point the signals are inserted into the radio equipment.

Table 6.1 showed the minimum number of voice channels required per radio channel by the FCC; however, this number of channels does not correlate with the digital multiplex hierarchies, as portrayed in Table 6.3. As a result, more voice channels per radio channel than the FCC requires may be designed into digital microwave radios in some bands.

A digital channel bank interfaces the voice channels to the digital radio. This channel bank is followed by the channel encoders, whose primary function is to encode one or more bits of the digital input signal into a symbol signal whose bit content determines the modulating effect it is to have on the amplitude, frequency, or phase of the carrier. Each combination of bit values that may be encoded into a symbol signal corresponds to a logic level within the encoder. The logic level the encoder assigns to a particular bit combination defines the discrete amplitude, frequency, or phase produced in the carrier at the moment of modulation.

Figures 8.9 and 8.10 show, respectively, the transmit and receive block diagrams[12] of a 135-Mb/s system for operation at 6 GHz using 64-QAM and achieving 4.5 b/s/Hz. This system is the Rockwell DST-2300. The input consists of

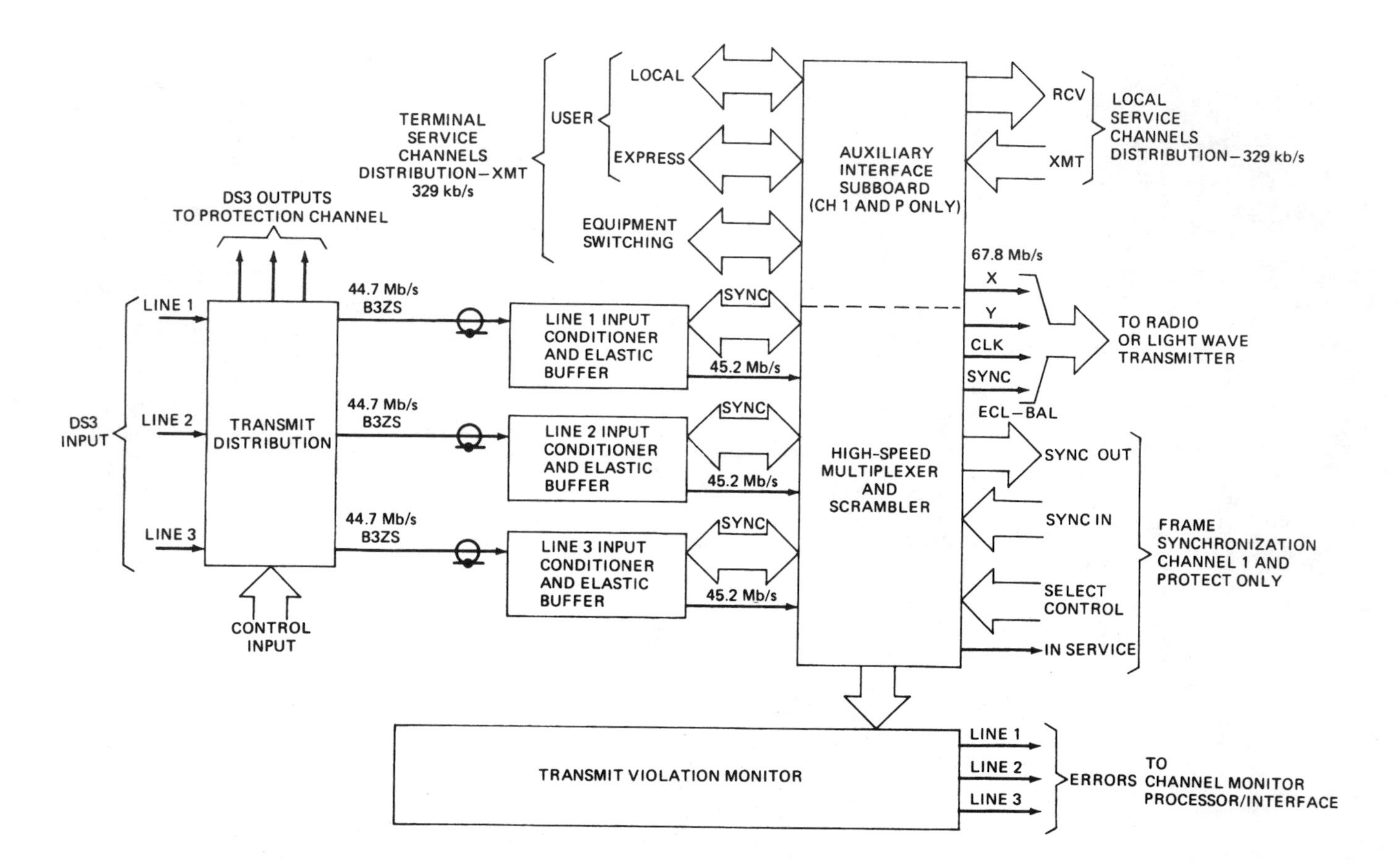

FIGURE 8.9. Transmit block diagram for Rockwell LST-2300. (Courtesy Hartmann and Crossett (Ref. 12). © IEEE 1983.)

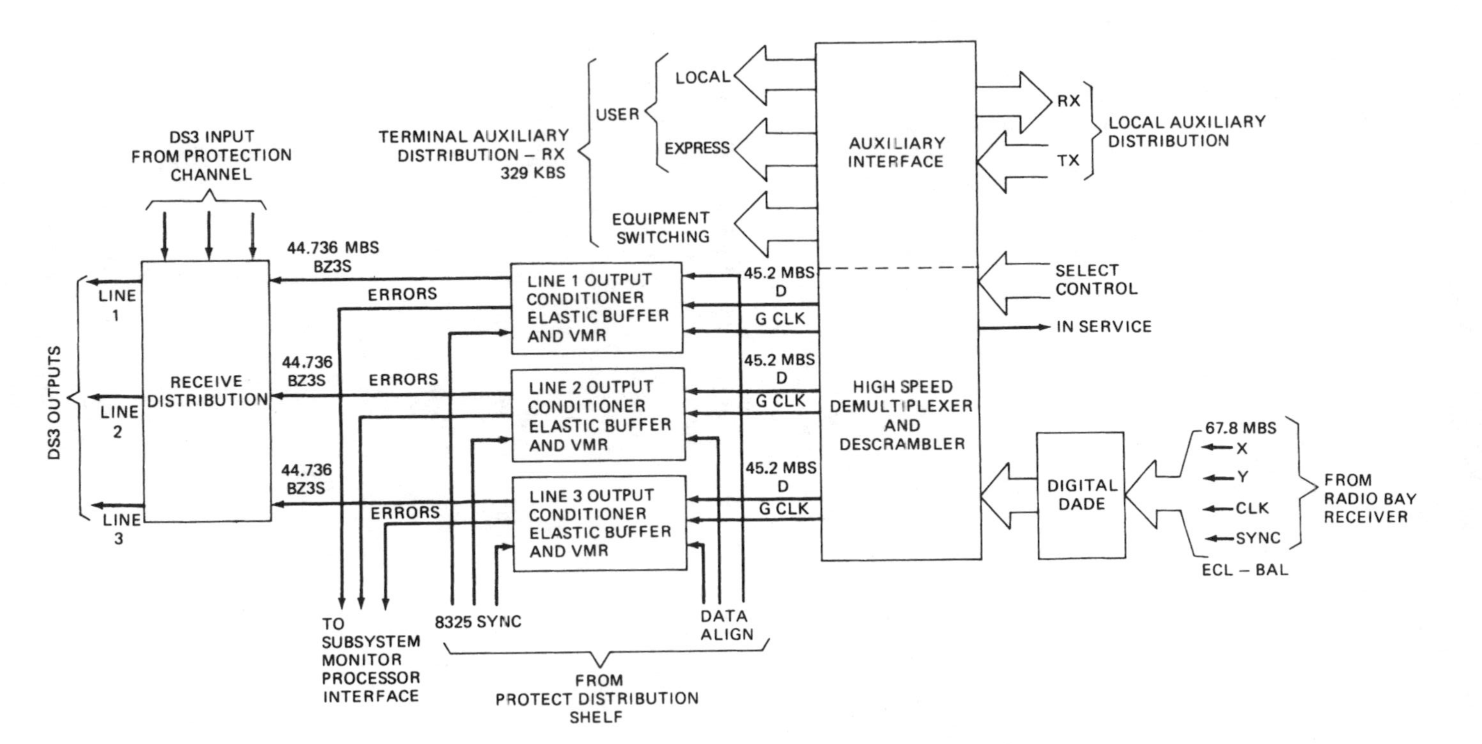

FIGURE 8.10. Receive block diagram for Rockwell DST-2300 (Courtesy Hartmann and Crossett (Ref. 12). © IEEE 1983.)

three DS3 lines, each at 44.7 Mb/s. Input conditioners and elastic buffers are used to obtain uniform pulse quality and to allow stream synchronization. A total of 329 kb/s is added for service-channel functions. Scrambling is performed to allow the data stream transitions needed for receiver clock recovery. An interface with protection channel lines is provided.

In Fig. 8.10, the system monitoring and violation monitor and removal (VMR) function is performed in the output conditioner.[13] In this unit, DS3 framing is recovered and DS3 parity is compared with parity calculated from the DS3 data. The detection of errors above a preset threshold causes switching to be initiated. The receiving equipment also contains the system diversity antenna delay equalization (DADE) module, which equalizes the absolute delay between the various RF channels on a DS3 sectional basis. This helps in minimizing error bursts during switching. A data alignment capability is included to align the protection channel data to the regular channel data prior to the completion of switching. The result is found to be 10 or fewer DS3 errors per switch operation provided framing is maintained on the regular channel.

8.5.1. Systems for Intercity and Long-Haul Applications

One digital microwave radio for intercity and long-haul applications is the Northern Telecom DRS-8. Its basic characteristics[14] are as follows:

Data rate:	91.04 Mb/s
Bit error ratio:	10^{-4} at C/N = 22 dB (usually achieves $<10^{-8}$ at somewhat higher C/N)
Operating frequency:	7.725-8.275 GHz (or other band providing 40 MHz radio channels)
Channel bandwidth:	40.74 MHz
Spectrum utilization:	2.25 b/s per Hz (QPRS)

The DRS-8 has provisions for space diversity and can be overbuilt on existing structures or built as a standalone system. A 40-dB fade margin can be provided over 40- to 50-km paths. With proper repeater spacings, the availability over a 6500-km path is 99.98%.

The Rockwell MDR-6 has the following characteristics:[15]

Data rate:	90.258 Mb/s
Bit error ratio:	10^{-6} at C/N = 29.2 dB
Operating frequency:	5.925 to 6.425 GHz
Channel bandwidth:	29.65 MHz
Spectrum utilization:	3.0 b/s per Hz (8-PSK)

The MDR-6 provides two asynchronous DS3 streams using the B3ZS line code. A similar radio, the MDR-11, operates in the 10.7–11.7-GHz band and also uses 8-PSK, but its radio channel bandwidth is 40 MHz wide, corresponding to the allocated channel widths at 11 GHz. As a result, its channel filtering is less stringent. Its spectrum utilization thus is 2.26 b/s per Hz.

While systems such as the Rockwell MDR-6 actually achieve a spectrum utilization of 3.0 b/s per Hz, they do so at a significant carrier-to-noise ratio penalty. For example, for b.e.r. $= 10^{-6}$, 8-PSK theoretically requires $E_b/N_0 = 14.0$ dB. This corresponds to $C/N = 14.0 + 10 \log_{10}(3) = 18.8$ dB. The filtering required to meet the demands of the mask, as outlined in Section 8.3.1, however, is largely responsible for the additional 10.4 dB of C/N needed by this system.

Accordingly, digital radio designers have turned to 16-QAM and 64-QAM, which are capable of 4 b/s per baud and more. For example, 16-QAM enables the achievement of a 90.258 Mb/s rate at 22.565 MBd. With a filter $BT_s = 1.0$ in a 30-MHz wide radio channel, the roll-off factor α of the channel filters can be 0.33, which is achievable readily.

For operation in the 4-GHz band, where the radio channels are only 20 MHz wide, Fujitsu has designed a 64-QAM system which achieves the 90-Mb/s rate at 15 MBd. In this case, the filter roll-off factor is 0.50. This system suffers only a 2.0-dB degradation in C/N at b.e.r. $= 10^{-6}$, thus requiring 28.7 dB.

The 30-MHz radio channel bandwidth available in the 6 GHz band allows a yet higher transmission rate if 64-QAM is used, as is done in systems designed by Western Electric, Rockwell, and Fujitsu. Accordingly, 3 DS-3 streams, or 135 Mb/s, corresponding to 2016 voice channels at 64 kb/s each, can be transmitted by such systems, one of which was shown in block-diagram form in Figs. 8.9 and 8.10. Systems using 256-QAM also have been implemented, providing 140 Mb/s in the 20 MHz radio channels of the 4-GHz band.

8.5.2. Metropolitan-Area Systems

Several categories of digital radio systems have been developed for the transmission of various stream rates over relatively short-distance, high-traffic-density routes within metropolitan areas. They include *digital termination systems* (DTS), private systems, and common-carrier systems.

8.5.2.1. Digital Termination Systems

Digital termination systems are provided by competing common carriers using radio local loops. Each user has a small rooftop antenna, typically a 0.6-m dish, directed toward the service provider's location. Data rates are from 2400 b/s to 1.544 Mb/s or 2.048 Mb/s at distances as great as 10 km through moderate rain.

The frequency band from 10.55 to 10.68 GHz is allocated for this service both to common carriers and to private users. Additional common carrier bands are: 18.36 to 18.46 GHz and 18.94 to 19.04 GHz. A 10^{-9} to 10^{-11} bit-error ratio is achieved using 500-mW transmitters and 4-level FSK for distances less than 10 km.

The two available classes of service are called "extended" and "limited." Extended-service carriers serve 30 or more cities and have seven 5-MHz channel pairs in each city. Limited-service carriers serve fewer than 30 cities and have six 2.5-MHz channel pairs in each city. The applications include a *digital electronic message service* (DEMS), high-speed facsimile (to 1.544 Mb/s or 2.048 Mb/s) and teleconferencing using motion-compensated video at 1.544 Mb/s or 2.048 Mb/s. Satellites often link the various cities.[16]

In operation, a central control station transmits digital traffic to the remote user sites, with the data for the various users being time-division multiplexed. Each remote site monitors the signal and processes only the data addressed to it, and then responds to the central station by sending bursts of pre-assembled packets of data for a predetermined or controlled time interval on another allocated frequency.[17]

8.5.2.2. Private Systems

Digital radio systems available for private (industrial or commercial) use within metropolitan areas include those made by Telesciences, Digital Microwave Radio, and M/A-Com. Some of these systems provide DS1 or DS2 service[18] using FSK in the 18.36- to 19.04-GHz band. The transmitter output typically is 100 mW. Bit-error ratios of 10^{-6} are achievable over distances from 5 to 16 km, depending on the local climate. Antenna diameters are 61 cm or 122 cm. Other systems provide DS1 or DS1C service using 2-level AM (p-i-n diode modulator) in the 21.2- to 23.6-GHz band. The transmitter output typically is 20 mW. A 10^{-7} bit error ratio is achievable 99.95% of the time over a 4-km distance in an average U.S. mid-latitude climate. The antennas are on the order of 40 cm $\times$ 40 cm $\times$ 25 cm deep.

The M/A-Com MA-23DR is capable of not only DS1 or DS1C service, but also DS2 (6.312 Mb/s), CEPT 1 (2.048 Mb/s), and CEPT 2 (8.448 Mb/s). The system handles the standard AMI and zero-substitution line codes and provides waveform regeneration at the receiver.[19] The modulation is FSK with a ± 4 MHz deviation. The transmitter uses a Gunn diode and has a 66-mW output. The antenna diameter is 0.6 m or 1.2 m. The receiver noise figure is 12 dB. At a 1.544 Mb/s-rate, a 10^{-9} b.e.r. is available 99.99% of the time at distances from 4 to 13 km within the U.S. depending upon climate.

8.5.2.3. Common-Carrier Systems

Digital radio facilities have been designed for DS3 transmission in metropolitan areas.[20] The Western Electric 3A-RDS operates at the DS3 stream rate, using QPSK in the 40-MHz radio channels between 10.7 and 11.7 GHz. Eleven two-way streams are obtained within eleven radio channels using dual linear polarization. Regenerative repeaters are used and the spacing is 16 to 20 km in average climates and 26 to 32 km in dry climates. The 3.2-W transmitter output power provides better than a 17.2-dB C/N ratio at the receivers for a 10^{-8} bit-error ratio. Many 3A-RDS systems thus transmit to distances well beyond metropolitan areas.

8.5.2.4. SONET-Compatible Radio Systems

Section 10.12 describes the synchronous optical network (SONET). As noted there, SONET is to be used for new telecommunication services worldwide. SONET-compatible digital radio systems have been developed as broadband short-haul radio systems.[21] Such systems may find use in subscriber line applications in which 1. cable installation is difficult or costly, or 2. a temporary connection must be established sooner than cable can be installed. Another use is as spurs off backbone fiber routes.

Figure 8.11 illustrates the concept. Here the central office is connected via a 2.3-Gb/s fiber-optic cable (see Chapter 10) to a remote electronics (RE) subsystem, which provides links to users in the 37- to 39.5-GHz band, and performs such necessary functions as multiplexing and demultiplexing, as well as switching and routing control.

Distances up to 4 km should be possible using 100-mW transmitters into 40-dB gain antennas (30-cm diameter dishes) with receiver noise figures on the order of 11.5 dB or better.

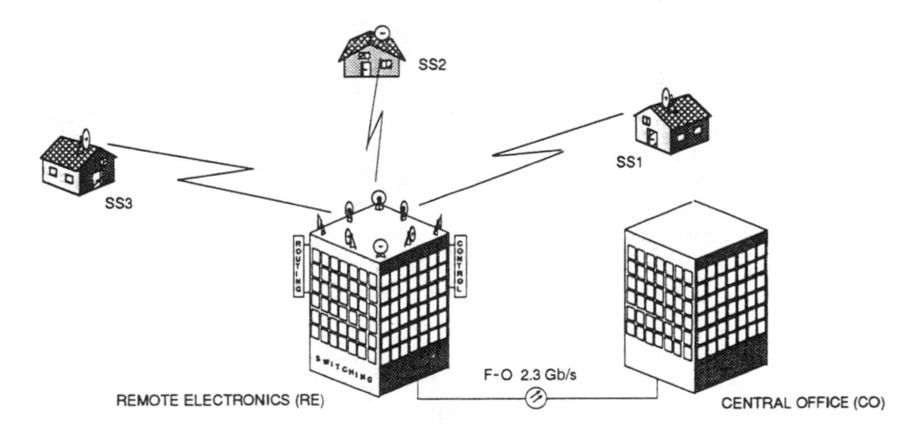

FIGURE 8.11. Broadband local radio system architecture. (Courtesy D1 Zenobio and Russo (Ref. 21) © IEEE, 1991.)

PROBLEMS

8.1. Explain why the propagation spreading loss equation shows loss increasing with frequency.

8.2. A digital radio operates in the 5.925 to 6.425 GHz band using 1.0-W transmitter powers. The antenna diameters are 3 m and the path length is 50 km. The modulation technique requires C/N = 20 dB to maintain an adequately low bit-error ratio. The receiver noise figure is 5.0 dB. Allow 3.0 dB at each end for waveguide and network losses. Assume a 55% antenna efficiency at each end. What is the fade margin?

8.3. How does the fade margin computed in Problem 8.2 compare with the level allowable during the worst month, assuming average inland terrain and a 0.999 reliability requirement?

8.4. During cloudburst conditions over a 5 km section of the path length of Problem 8.2, what additional path attenuation is encountered?

8.5. Microwave propagation in coastal regions often extends over greater distances than inland. Why does this not allow repeaters to be spaced farther apart in such areas?

8.6. The horn-reflector antenna is more commonly found at terminals in large metropolitan areas, whereas the simple parabolic is more frequently found in rural locations. Why?

REFERENCES

1. Keiser, B. E., "Wideband Communications Systems," Course Notes, © 1992.
2. White, R. F., *Engineering Consideration for Microwave Communications Systems*, GTE Lenkurt, Inc., San Carlos, CA, 1970.
3. Feher, K., *Digital Communications: Microwave Applications*, Prentice-Hall, Inc., Englewood Cliffs, NJ, © 1981.
4. Bell-Northern Research, *Telesis*, Vol. 5, No. 6, Special issue: DRS-8 Digital Radio System, Dec. 1977.
5. Rockwell International, Collins Technical Data Sheet, MDR-6, 6-GHz Microwave Digital Radio, Commercial Telecommunications Group, Dallas, TX.
6. *Telecommunications Transmission Engineering*, Volume 2, *Facilities*, Bell System Center for Technical Education, Western Electric Company, Inc., Winston-Salem, NC, pp. 48–51, © 1977.
7. See note 2.
8. Barnett, W. T., "Occurrence of Selective Fading as a Function of Path Length, Frequency, Geography," *ICC '70 Conference Record*, © IEEE, 1970.
9. Barnett, W. T., "Microwave Line of Sight Propagation With and Without Frequency Diversity," *Bell Syst. Tech. J.*, Vol. 49, Oct. 1970, pp. 1827–1871.
10. Vigants, A., "Number and Duration of Fades at 6 and 4 GHz," *Bell Syst. Tech. J.*, Vol. 50, March 1971, pp. 815–841.

11. "Radiation Pattern Envelopes for Horn Reflector Antenna," Bulletin 1285, Andrew Corporation, 10500 W. 153rd Street, Orland Park, IL 60462.
12. Hartmann, P. R., and Crossett, J. A., "135 MBS-6 GHz Transmission System Using 64-QAM Modulation," *ICC '83 Conference Record*, Publication No. 83 CH1874-7, Boston, MA, June 19–22, 1983, pp. F2.6.1–F2.6.7, © IEEE, 1983.
13. "MDR-2000 Series Digital Radio Systems," Product Description, Collins Transmission Systems Division, Rockwell International, P.O. Box 10462, Dallas, TX 75207.
14. See note 4.
15. See note 5.
16. "Private Communication Networks," Macomnet, Inc., Rockville, MD.
17. Williams, D. S., "Local Distribution in a Digital Communications Network," *ICC '81 Conference Record*, IEEE Publication No. 81CH 1648-5, Denver, CO, June 14–18, 1981, pp. 66.3.1 to 66.3.5, © IEEE, 1981.
18. "Gemlink LSD-112A/122A Microwave Radio," General Electric Microwave Link Operation, Owensboro, KY, 1982.
19. "Video Microwave Systems," Bulletin 9236A, MA-23CC, M/A-COM MVS, Inc. 63 Third Avenue, Burlington, MA 01803.
20. *Telecommunications Transmission Engineering*, Vol. 2, *Facilities*, Bell System Center for Technical Education, Western Electric Company, Winston-Salem, NC, pp. 619–638, © 1977.
21. Di Zenobio, D., and E. Russo, "The New SONET-Compatible Radio Systems: Broadband Local Radio Systems," *ICC '91 Conference Record*, IEEE Publication No. 91CH2984-3, Denver, CO, June 23–26, 1991, pp. 2.6.1 to 2.6.6, © IEEE, 1991.

9

Satellite Transmission*

9.1. INTRODUCTION

Communications satellites, as we know them today, had their beginning with *Syncom*, the first synchronous satellite, in 1963. Syncom soon was followed by *Intelsat I (Early Bird)*, the first commercial satellite, in 1965. Numerous military satellite systems parallel the development of the commercial systems. The first domestic commercial satellite was Telesat's *Anik* (Canadian), followed by Western Union's *Westar* (United States). Because of the predominance of analog voice communication during those years, however, satellite earth stations all over the world were built based upon the analog frequency-division multiplex (FDM) hierarchy. Only with the advent of Satellite Business Systems (SBS) was an all-digital satellite system developed, with the initial objective, however, of data rather than voice communication.

Satellite transmission is unique in that it makes large bandwidths (that is, hundreds of megahertz) available for intercontinental communication. Moreover, with the use of polarization and spot-beaming techniques, a considerable amount of frequency reuse is feasible, which is a significant fact in view of the increasingly crowded radio spectrum.[1]

Satellite transmission is capable of providing global communication, including transmission to and from moving terminals such as ships, and considerable system development effort has been devoted to extending such capabilities to land and aeronautical vehicles.[2,3] Only the polar regions beyond 81° latitude remain beyond line-of-sight of the geostationary satellites. Satellites that are not geostationary,

---

* Much of the material of this chapter is taken from B. E. Keiser, "Broadband Communications Systems," Course Notes, © 1992.

however, can provide voice coverage of the polar regions, as is discussed in Section 9.6.

The satellite may be regarded as a microwave repeater at an elevation high enough that most circuits require only that one repeater. Where a second such repeater is needed, the overall time delay (if the satellites are geostationary) may produce undesirable effects in attempts to carry on conversational voice transmission.

Satellite transmission developed initially on an analog (FDMA) basis, and further developments centered on the use of compandored single-sideband modulation.[4,5] While the use of such techniques provides considerable spectrum economy (7200 voice channels in a 36-MHz transponder), the need exists in these systems to operate amplifiers well below (for example, 10–15 dB) their full power-output capabilities. In addition, signal-to-noise ratios on the order of 27 dB and higher are required, making such transmissions "fragile" with respect to interference. Such advances notwithstanding, the intermodulation problem resulting from transponder nonlinearities makes digital (TDMA) operation highly desirable, as will be explained in Section 9.3.5.2 of this chapter, and current developments are centered on digital technology.

9.2. CHARACTERISTICS OF SATELLITE PROPAGATION

The propagation medium for satellite transmission tends to be a highly stable one, because most of the path is through free space. The only atmospheric effects occur near the earth stations and these effects generally are limited by the fact that propagation usually is upward at a relatively significant angle relative to the horizon. Only at the higher latitudes, or for satellites at a considerably different longitude than the earth station, is a low angle path required.

9.2.1. The Satellite Orbit

For a satellite to remain in a fixed position in the sky relative to the earth stations it is serving, two conditions must be met. First, the satellite must be over the equator. Second, its altitude must be 35,794 km. The latter value results from the fact that the period of an orbiting satellite is given by the equation

$$T = 2\pi \sqrt{\frac{A^3}{\gamma}} \tag{9.1}$$

where

A = semi-major axis of ellipse (earth's radius, 6378 km, plus altitude for a circular orbit)
γ = gravitational constant = 3.99×10^5 km^3/s^2.

For a circular orbit to have a period identical with that of the earth's rotation, the period must equal one day, which is 23 h, 56 min, 4.09 s.

From a geostationary altitude of 35,794 km, three satellites can cover most of the earth's surface.

The planned *Iridium* system will orbit the earth at an altitude of 765 km, and provide 37 cells per satellite, each cell covering an area with a 660-km diameter. Eleven such satellites will be placed in each of seven polar orbits.

9.2.2. Time Delay on Satellite Paths

Although electromagnetic waves travel at the speed of light, 3×10^8 m/s, the distance to a geostationary satellite may range from 35,794 km for an earth station at the sub-satellite point (on the equator) to 41,677 km for an earth station at the horizon relative to the satellite. Accordingly, the one-way propagation time, neglecting electronic circuit delay, from one earth station to another may range from a minimum of 0.239 s to a maximum of 0.278 s. The total round-trip delay time then ranges from 0.478 s to 0.556 s. In conversational voice, such delays begin to be noticeable, but not adversely so. In fact, each party to the conversation gets the impression that the other party is pausing a moment to collect his thoughts before replying!

The same is not true, however, if double-hop satellite transmission is required, as might be the case if a domestic satellite is used in tandem with an international satellite. In such a case, the total round-trip delay time is on the order of one second. When a speaker has finished talking and does not hear a reply forthcoming, he may say, "Did you hear me?" just as the reply is on its way. For such reasons, international satellite transmission is being implemented increasingly on a one-way basis only, with the return path being terrestrial, that is, via submarine cable. This allows a domestic satellite to be placed in tandem with an international satellite for an overall delay of less than 0.6 second in one direction, with the delay on the return path being kept to less than 0.1 second. The combined round-trip delay, on the order of 0.6 to 0.7 second, is tolerable in most cases.[6]

An advantage of the low-earth orbit is that the delay time is short enough to be nearly unnoticeable to the users in most cases.

9.2.3. Atmospheric Attenuation

The attenuation caused by the atmosphere depends upon both the frequency and the elevation angle. This one-way attenuation is given by Fig. 9.1 for standard atmospheric conditions. It results from the presence of oxygen and water vapor in the air. Rainfall causes increased attenuation, especially at frequencies above 10 GHz.[7]

9.2.4. Rain Depolarization

As will be seen in Section 9.3, frequency reuse is an important aspect of satellite-system engineering because of the increasing use of the radio spectrum. One frequency reuse technique is dual polarization. For its successful use, however,

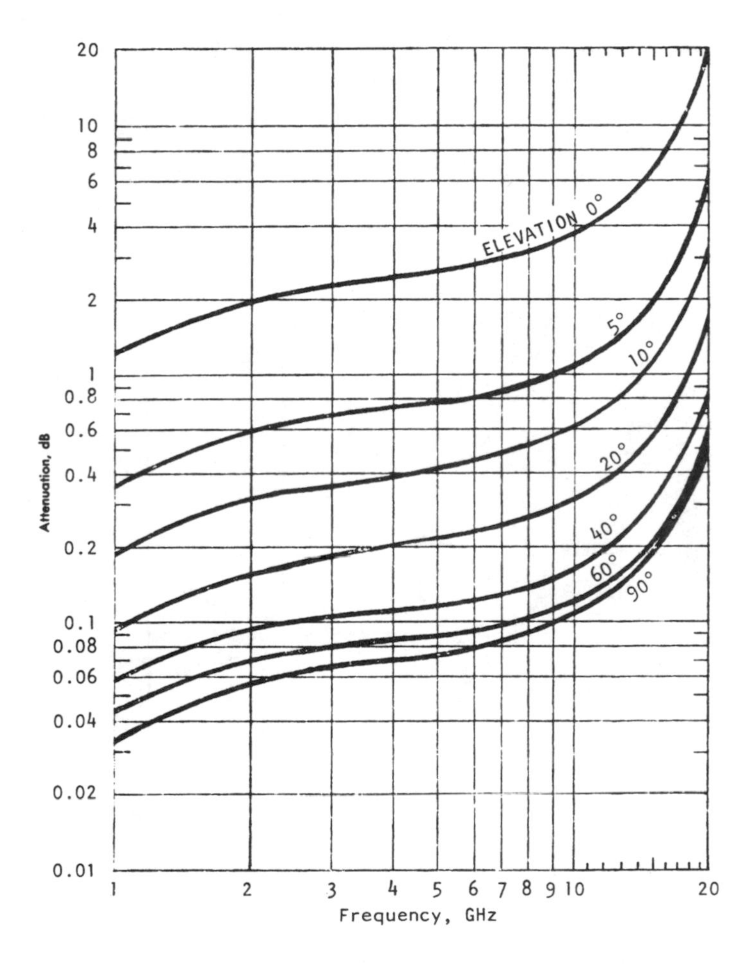

FIGURE 9.1. Atmospheric attenuation over the earth-to-satellite path. (Reprinted by permission from *Microwave Journal*.)

reception of a co-channel or partially co-channel signal must be such that the level of the unwanted (cross-polarized) signal is sufficiently low. In other words, the ratio of carrier to interference, C/I, must be adequate for the planned bit-error ratio. Typically, this calls for $C/I \geq 25$ dB. Two factors allow a cross-polarized signal to be present. One is the polarization purity of the hardware, mainly the antennas, and the other is the extent to which the signals become depolarized during propagation, mainly because of precipitation.

Figure 9.2 illustrates the effect of rain on cross-polarization isolation for 5-km path lengths. The curves are shown for both differential phase and differential attenuation over 36-MHz transponder bandwidths for 4, 6, and 11 GHz. Differential attenuation is the difference in attenuation between the horizontal and vertical polarizations, whereas differential phase is the difference in phase shift between the horizontal and vertical polarizations.[8] As can be seen from Fig. 9.2,

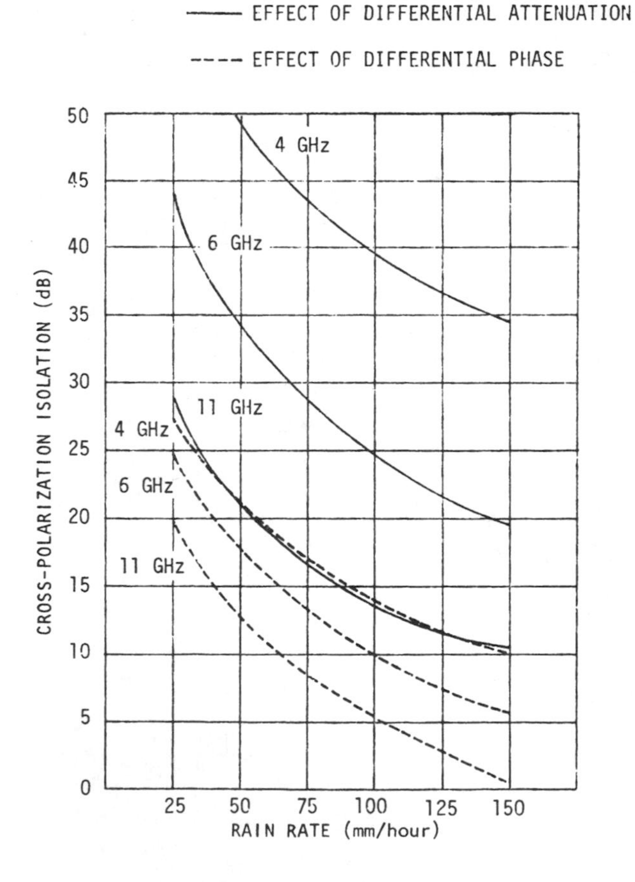

FIGURE 9.2. Cross-polarization isolation versus rain rate for 5-km path lengths.

differential attenuation and phase are key parameters in the production of cross-polarization by precipitation.

9.3. SATELLITE SYSTEM DESIGN

Numerous factors, technical, regulatory and economic, affect the design of a satellite system. This section discusses the major system design factors. Section 9.4 deals with equipment characteristics and limitations. For several topics not covered here, such as echo control, diversity and reliability as well as availability, the reader is referred to the corresponding Sections 5.6, 8.3.4, and 8.3.5 of earlier chapters. The echo control problem (impedance mismatches) and solution (echo cancelers) are quite similar to those of digital microwave radio relay systems. Diversity is required mainly by systems operating in the 11- and 14-GHz bands in regions of heavy precipitation. Here perhaps three earth stations, each separated 10–15 km, may be required to dodge local rain cells. If the three are adequately spaced and interconnected terrestrially, outages due to heavy precipitation can be reduced to an extremely low percentage of the total annual operating time. With respect to reliability/availability, the equipment reliability problem is similar to that of other terrestrial systems, assuming proper spacecraft design and operation. Propagation outages are confined almost entirely to low-elevation-angle paths which are subject to precipitation fading, as well as to the other problems that affect terrestrial microwave propagation, such as multipath.

9.3.1. Frequency Allocations and Usage

Table 9.1 lists the frequency bands allocated to satellite communication and designates the type of usage in each case. In some of these bands, the regulatory authorities have specified maximum power flux densities to minimize interference to existing terrestrial services, since many of these bands are not used exclusively for satellite service, but are shared with terrestrial users.

The term *allocation* means that the frequencies listed in Table 9.1 are available for the usage listed, but many of the bands above 17.7 GHz are currently being used only experimentally in the western hemisphere.

The frequency allocations listed in Table 9.1 might seem to be perfectly adequate for satellite communications purposes in view of the large bandwidths available; however, a serious shortage of spectrum exists for several reasons. First, only one commercial frequency pair, the 3700–4200-MHz downlink and the corresponding 5925–6425-MHz uplink, is suitable for use under all possible weather conditions. The 11- and 14-GHz bands are suitable under most conditions, and rapidly increasing use is being made of these bands. The bands above 20 GHz are more severely limited by precipitation. Their use may depend upon the development of some of the special techniques discussed in Section 9.6.

TABLE 9.1. Satellite Frequency Allocations

Band (MHz)	Usage	
225.0–400.0	military	
1535.0–1542.5	downlink	maritime
1635.0–1644.0	uplink	
1543.5–1558.5	downlink	aeronautical
1645.0–1660.0	uplink	
3700–4200	downlink	commercial
5925–6425	uplink	
7250–7750	downlink	military
7900–8400	uplink	
10950–11200	downlink, international	commercial
11450–11700	downlink, domestic	
11700–12200	uplink	
14000–14500		
17700–20200	downlink	commercial
27500–30000	uplink	
20200–21200	downlink	military
30000–31000	uplink	
43500–45500	uplink	

Considerable use is being made of the existing allocations below 14 500 MHz both domestically and internationally for video and multichannel telephony, as well as for data transmission. Frequency reuse techniques include polarization diversity, used mainly in the 4- and 6-GHz bands, and spot beaming, usable in all bands. Other spectrum conservation techniques include the use of multilevel modulation, as discussed in Section 6.2.

With polarization diversity, the number of transponders occupying an overall band can be doubled, since the isolation between dual-polarized beams can be as great as 30 dB, compared with the *C/I* requirement of 25 dB noted in Section 9.2.4.

With respect to spot beaming, a satellite's antenna can be designed to cover only a portion of the U.S., or a region the size of the U.S. or western Europe. Likewise, beams can be devised to cover only hemispherical regions. In all cases, the beamwidth of the satellite antenna is significantly smaller than the 17.3° width which would cover the entire surface of the earth that is within view of the satellite.

Interest also exists in placing more satellites into the orbital arc. Crowding of the orbital arc is most critical in the 3700–4200-MHz downlink band. To allow a maximum number of satellites to operate in this band, the spacing in the satellite arc over the Western hemisphere has been reduced to 2.5°, and adjacent satellites generally use alternate polarization plans. In addition, earth station antennas are being designed with very low sidelobes to enable them to receive one satellite without interference from its neighbor in the orbital arc.

Sidelobe interference thus constitutes one problem that satellite system planners must face. Both earth station and satellite antennas may exhibit side lobes that are at levels on the order of -20 dB relative to the main lobe. An earth station side lobe may interfere with the signals intended to reach an adjacent satellite. Likewise, side lobes from a satellite may interfere unintentionally with reception at an earth station.

Terrestrial microwave systems on the horizon with respect to a satellite may interfere with, or receive interference from, a satellite system. Such microwave systems might be located in the far north or far south, and have beams in a northerly or southerly direction, or they may be at medium latitudes and have beams in a direction toward a satellite at a significantly different longitude.

International frequency allocation matters are handled by the International Telecommunications Union (ITU), which holds regular international World Radiocommunication Conferences (WRCs), formerly known as World Administrative Radio Conferences (WARCs).

9.3.2. Link Budgets

Link budgets for the satellite service differ from those in the terrestrial microwave service in several ways. First, fading on a satellite link below 10 GHz is almost nonexistent. Accordingly, large fade margins are not needed. Second, the terrestrial earth station receives a signal from a satellite in a relatively quiet sky, unlike a terrestrial receiver, which receives a signal in the presence of numerous terrestrial noise sources. This means that a satellite earth station can benefit from the use of a very-low-noise receiver. Another difference is the fact that the satellite is limited in power to the power available from the sun, and satellite transponders often are quite limited in their peak power-output capabilities. These differences are mentioned to allow the reader to appreciate the differences in approach used in structuring a satellite link budget compared with a terrestrial microwave link budget, as was discussed in Chapter 8.

A link budget provides a complete description of a radio transmission path from the transmitter power output to the detected signal, and often includes an expression of reception quality, expressed in bit-error ratio for digital systems. For a satellite system both an uplink and a downlink budget must be calculated. The uplink budget usually is the less critical of the two, however, since plentiful amounts of power normally are available to the earth-based transmitter. Uplink stations also may have significant amounts of antenna gain. Generally the satellite receiver will have a rather high noise temperature (for example, over 300 K) because the earth itself is noisy, and no value would be gained with a low-noise receiver. In fact, low-noise receivers cannot be made to operate as reliably as high-noise (less sensitive) ones, so are not used on spacecraft.

The criticality of the downlink budget results from several factors. The satellite is power limited by the amount of power available from the solar array and the number of transponders and pieces of control equipment among which this power must be divided. In addition, the physical size of the spacecraft antenna, and thus its gain, also is generally smaller than those on the ground. Another limit may be the available channel bandwidth, as well as the way in which a given transponder's bandwidth and power are allocated. A further limit is that in the 3700–4200-MHz band, international regulations state that the power flux density shall not exceed 12 dBW *effective isotropically radiated power* (EIRP) in any 4-kHz band.

In a purely digital time-division multiplexed system, each user has the entire transponder for the duration of his time slot. Many systems, however, allocate only a portion of the transponder to digital transmission. For example, a single channel per carrier (SCPC) system may transmit a single PCM channel using only a 45-kHz wide portion of an overall 36-MHz wide transponder. While as many as 800 of these SCPC channels may use the same transponder, a smaller limit may be established, with the remainder of the transponder's bandwidth being used for another application.

The basic downlink power budget may be expressed in terms of the carrier-to-noise power normalized to a 1-Hz bandwidth, $(C/kT)_A$, available from a transponder or a portion thereof as follows:

$$\left(\frac{C}{kT}\right)_A = (\mathrm{EIRP})_{\max} - (\mathrm{EIRP})_0 - (\mathrm{TD})_d - (\mathrm{TPL})_d - k + \left(\frac{G}{T}\right)_{\mathrm{eff}} \ \mathrm{dBHz} \qquad (9.2)$$

where

$(\mathrm{EIRP})_{\max}$ = maximum center beam transponder effective isotropic radiated power in watts at its normal operating point, chosen to be backed off from saturation by a specified amount

$(\mathrm{EIRP})_0$ = power in watts devoted to carriers in the transponder other than the carrier of interest

$(\mathrm{TD})_d$ = tilt differential for the downlink in dB, the degradation from the maximum EIRP resulting from the position of the earth terminal relative to the boresight of the satellite antenna

$(\mathrm{TPD})_d$ = total path loss in dB for the downlink
= free space loss (FSL) + miscellaneous loss (m).

The expression for *FSL* is

$$FSL = 32.5 + 20 \log_{10} f_{\text{MHz}} + 20 \log_{10} d_{\text{km}} \tag{9.3}$$

where

f_{MHz} = frequency in MHz
d_{km} = slant range in km
m = loss caused by pointing error, radome loss, atmospheric attenuation and polarization loss
k = -228.6 dBW/HzK (Boltzmann's constant).
$(G/T)_{\text{eff}}$ = effective earth terminal figure of merit in dB/K, including the effect of thermal noise radiated from the satellite to the earth terminal

Equations (9.2) and (9.3) allow a determination to be made of the available C/kT. The next question is "what C/kT is required?" Let this value be designated $(C/kT)_{\text{R}}$. It is expressed as follows:

$$\left(\frac{C}{kT}\right)_{\text{R}} = R + \left(\frac{E_{\text{b}}}{N_0}\right) + L + M \tag{9.4}$$

where

R = $10 \log_{10}$ (bit rate, b/s)
E_{b}/N_0 = ratio of energy per bit to noise power spectral density to achieve required bit error ratio
L = losses resulting from use of channel filters (about 1–2 dB)
M = satellite link margin

Typical values for M are 4 dB for the 3700–4200-MHz band, 6 dB for the 7250–7750-MHz band, and 8 dB for the 10 950–12 200-MHz frequencies.

The required bit-error ratio for the link will determine the E_{b}/N_0 required. This value then is augmented by the required digital rate, as well as an allowance for link margin and losses.

9.3.3. Earth-Station Siting

Satellite earth stations may be found in a variety of places, including building roof tops, but they are best located in valleys or other low terrain, such that their surroundings provide some shielding from interference that may reach them from terrestrial microwave relays. Before attempting to establish a new satellite earth station, a careful examination must be made of the surrounding electromagnetic environment to determine what main beam or sidelobe interference may exist between the new earth station and microwave repeaters.

An important consideration in earth station siting is the beam direction to the satellite. For a fixed installation, the manufacturer provides the pointing angles with the station equipment. For a portable earth station that may be used in various places, computer programs are readily available which compute the required pointing angles, based on the geometry illustrated in the chart of Fig. 9.3.[9] Knowledge of the earth station's latitude and of the relative longitude of the satellite allow the determination of the azimuth and elevation angles of the satellite.

Knowledge of the earth station's main beam direction is important in avoiding obstructions in that direction, as well as in determining the direction of side lobes. Side lobes may cause interference to adjacent satellites or to other satellite earth stations or nearby microwave systems.

Another type of interference that can occur is from sources outside the allocated bands. Such interference can be produced by *environmental intermodula-*

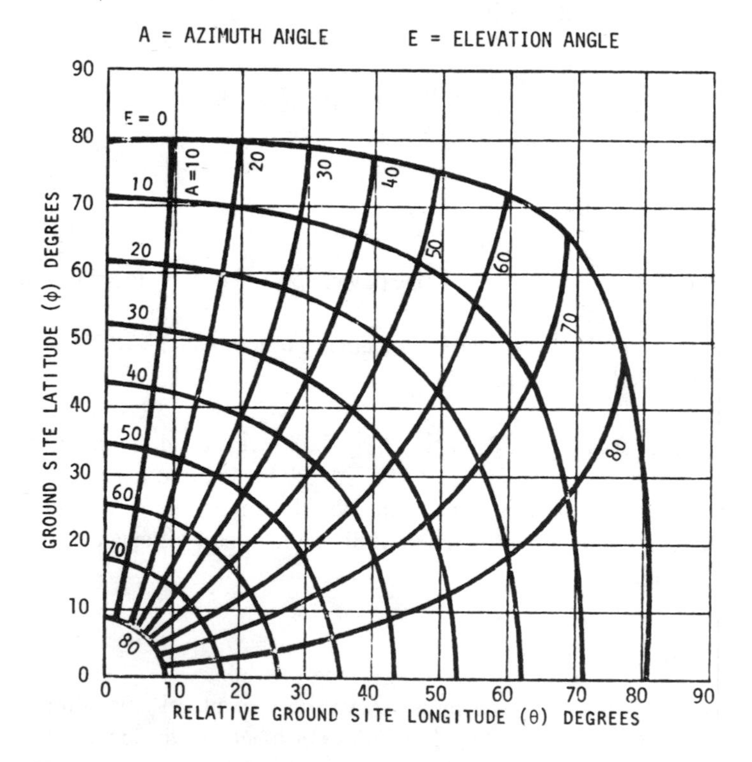

FIGURE 9.3. Direction of a geostationary satellite. (Courtesy F. L. Smith, Ref. 9 © IEEE, 1972.)

tion, resulting from nonlinearities caused by such sources as oxide layers in the waveguide feed system. Alternatively, interference can be caused by *receiver* or *transmitter intermodulation*, resulting from circuit nonlinearities. Any intermodulation products that fall inside the downlink bands may have to be eliminated by suitable treatment of the nonlinearities. In the case of the environmental nonlinearities, oxide layers may have to be removed, or straps may have to be installed around poor joints. In the case of circuit nonlinearities, filtering may have to be added to prevent the unwanted signals from reaching the nonlinear elements.

9.3.4. Multiple Access

Multiple-access techniques allow variously located earth terminals to use portions of a satellite transponder on either a frequency- or time-division basis. In FDMA, each terminal uses some of the bandwidth and transmits some of the power all of the time, whereas in TDMA, each terminal uses all of the bandwidth and produces all of the power, but only part of the time. Finally, in CDMA, each terminal uses all of the bandwidth all of the time, but only transmits part of the power that is on the channel.

9.3.4.1. Frequency-Division Multiple Access

In FDMA, the available spectrum within a satellite transponder is divided among carriers, each of which may carry a digital stream. This digital stream, in turn, may convey messages from one or more sources to one or more destinations. An arrangement often used, especially by small remote earth terminals, is SCPC. Such channels are assigned only when actually needed, and also can be voice activated, so that each channel contributes to transponder power usage only when it is actually required. This helps to minimize intermodulation interference, a problem often encountered by satellite transponders using FDMA. A detailed analysis of intermodulation in FDMA satellite transponders is provided by Spilker.[10]

Many SCPC systems use DAMA.[11] In DAMA, each carrier is modulated by a bit stream that comes from the voice channel of an individual user. COMSAT has devised a form of SCPC known as the *Single-channel per carrier Pulse-code modulation multiple-Access Demand-assigned Equipment* (SPADE). The SPADE system assigns transponder capacity for use on a call-by-call basis. The overall 36-MHz transponder bandwidth is divided into 800 channels, each of which is individually accessible via an earth station. These 800 channels can serve a much larger number of telephone subscribers, the actual number depending on how often the subscribers make calls, and for what durations.

The SPADE system message channels use PCM-encoded voice at 64 kb/s, sent via QPSK at 32 kBd. The bandwidth per voice channel is 38 kHz and the channel spacing is 45 kHz. The bit-error ratio provided is less than 10^{-7}. Frequency stability is ± 2 kHz and is held to this level using *automatic frequency control* (AFC). Correspondingly, oscillator phase noise and incidental FM generated in the oscillators and power amplifiers must be held to a minimum. Also, a part of the SPADE system is a common signaling channel which operates at 128 kb/s using two-phase PSK. This channel allows 50 accesses (49 plus a reference), each of which is allotted a 1-ms burst within a 50-ms frame. The bit-error ratio is less than 10^{-7}. A central network processor controls channel assignments, passing the appropriate commands to a microprocessor at the station at each end of the link. Each microprocessor interprets the commands and sends them on to its channel units via the common channel. The remote terminals can be simplified through the use of a net control terminal, which places the more complex control equipment at a central location. Fig. 9.4 is a block diagram of a SPADE terminal.

In the SPADE system, the channels are spaced uniformly. Such spacing makes efficient use of the transponder spectrum, but causes the intermodulation products to fall directly on the channels being used. Since each carrier is voice activated, however, only about 40% of the carriers are on at any given time. This is equivalent to a 4-dB reduction (back-off) of the transponder's output. This reduction results in operation in a more nearly linear portion of the transponder's power amplifier characteristic, with a consequent decrease in the intermodulation products generated. In addition the time activation of the transponder by each voice signal gates the intermodulation products generated. Moreover, the time activation of the transponder by each voice signal gates the intermodulation products in a random manner, reducing the worst intermodulation noise by 3 dB.[12]

Intermodulation effects can be reduced significantly if the carriers are not

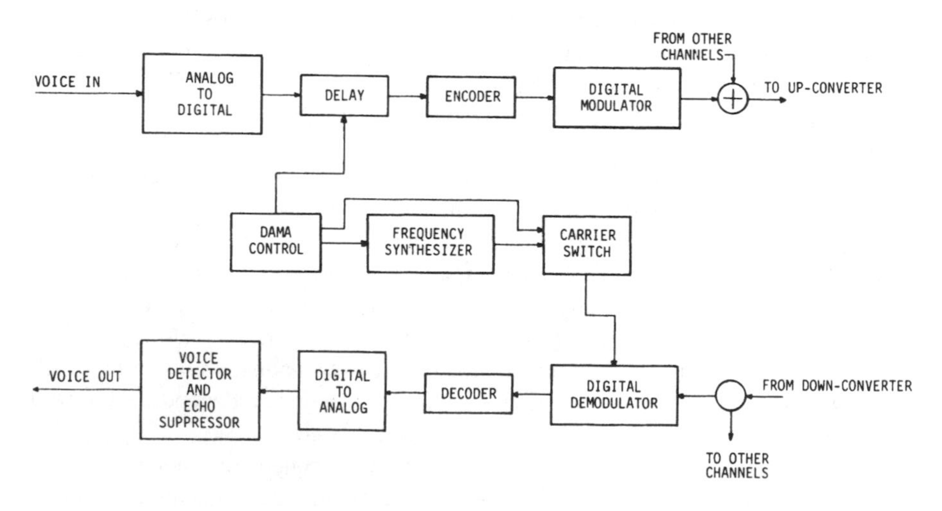

FIGURE 9.4. Block diagram of SPADE terminal.

uniformly spaced. One arrangement involves placing the carriers in clusters of bandwidth W.[13] The spacing between clusters is then made to exceed 3 W.

An alternative approach to the choice of carrier frequencies is one which determines the minimum bandwidth needed for complete avoidance of third- or third- and fifth-order cross products. These results and the frequency spacing required are reported in Ref. 14.

For large numbers of available channels N, randomly spaced channels provide a better overall ratio of carrier-to-intermodulation noise power (C/IM) than the previously mentioned techniques. If W is the bandwidth of the cluster but the carriers are spread out randomly to occupy a bandwidth B, only a fraction (W/B) of the intermodulation power is passed through the receiver. Then, as shown in Ref. 3, the improvement in (C/IM) in the center channel due to the carrier spacing is

$$\left(\frac{C}{IM}\right)_{\text{imp}} = 10 \log_{10}\left(\frac{B}{W}\right) + 7.7 \text{ dB} \tag{9.5}$$

For 100 38-kHz bandwidth carriers in a total bandwidth of 36 MHz, W = 3800 kHz = 3.8 MHz, while B = 36 MHz. Then $(C/IM)_{\text{imp}}$ = 17.5 dB, that is, the use of randomly spaced channels yields a 17.5-dB improvement in C/IM.

9.3.4.2. Time-Division Multiple Access

Section 9.3.4.1. noted the need for transponder back-off in FDMA operation. Typical back-off values range from 3 to 6 dB and constitute a significant decrease in the power which otherwise would be available from the transponder. By comparison, TDMA operates with only one modulated carrier in the transponder at a time. The result is that efficiency in terms of power utilization can reach 90% or more, since the power amplifier can be operated closer to saturation. Moreover, frequency guard bands are not needed in TDMA operation, and time guard slots can be kept to a very small percentage of the total on-time. Thus, while frequency guard bands and allowance for delay distortion near band edges may occupy 10–15% of the frequency band, typical time guard slots are on the order of only 1–2% of the total on-time.

In TDMA, each earth terminal transmits to the satellite in its own time slot, based on a determination of the propagation time to the satellite. Guard time is used to allow for the decay of the pulses so that intersymbol interference is not caused, as well as to allow for timing inaccuracies. Pulse decay is a function of the transient responses of the filters used in both the earth stations and the satellite. These filters must have good phase linearity; that is, low group-delay distortion.

Slight motions of the satellite will cause corresponding changes in propagation time; however, satellite motion is periodic and predictable. Accordingly, elastic buffers, rather than pulse-stuffing buffers, can be used to compensate for satellite path delay variations. This presumes that highly stable clocks are used at all terminals and that the frame rate is held constant at the satellite. The sizes of the elastic buffers must be adequate to prevent overflow or underflow. This can be determined based upon the anticipated orbital characteristics of the satellite, plus a safety factor. In addition, the buffers must be reset regularly because of clock drifts at the earth terminals.

Digital transmission using TDMA is accomplished by the use of modulation during the carrier bursts. The needed phase reference must be derived during each burst since no means exists to synchronize the transmitted phases as received at the satellite. The phase reference can be obtained[15] either by (1) using a phase-locked loop or narrow-band filter which can acquire each carrier rapidly in sequence, or (2) using multiple time-gated carrier-recovery loops or a single time-multiplexed phase-locked loop with phase memory from frame to frame. The latter approach allows the use of narrow-band carrier-recovery loops operating on each time-gated carrier.

With the development of large array antennas for use on communication satellites, narrow spot beams covering only a metropolitan area and its vicinity will become operational, probably at frequencies above 20 GHz. The receiving and transmitting spot beams can be made independent of one another. Such antennas, together with switching on board the satellite, can allow the transmission of speech packets to and from specific ground locations, thus achieving not only full transponder power utilization, but also spectrum economy through frequency re-use in various spot-beam areas. Time-slot coordination can be achieved by having one earth station serve as a central timing source and sending time through the satellite to all the other earth stations.[16]

Satellite-switched transmission can be accomplished using FDMA also, by using frequency bands within the transponder rather than time slots. The transponder, however, then must be operated with its power level backed off from maximum.

9.3.4.3. Code-Division Multiple Access

The discussion of CDMA in Chapter 7 pointed out that in CDMA, the signal is spread over a bandwidth much larger than the information bandwidth. Accordingly, the amount of spectral energy per hertz of bandwidth becomes quite small. In fact, a CDMA signal may raise the effective noise background in a given band by only an imperceptible amount. On this basis, satellite transmission can be done without interference to adjacent satellites, even though the transmitted beams are relatively broad. This means that much smaller dishes can be used for CDMA transmission than are required for FDMA or TDMA, where interference clearly would be caused under the same circumstances.

CDMA is fundamental to the feasibility of the *very small-aperture terminals* (VSATs) used in C-band. Commercial transmission of data at a wide range of rates, from 45 b/s to 19 200 b/s is being done by Equatorial Communications, Inc., based on this principle. Dish sizes as small as 0.6 to 1.5 m are used.

9.3.5. Multiplexing

The reader is referred to Section 5.3 for a general discussion of digital multiplexing. As applied to satellite transmission, PCM voice can be either on an FDM carrier or on a TDM carrier burst. The modulation usually is QPSK. The access technique may be either TDMA or FDMA. The overall designation thus might be PCM/TDM/QPSK/TDMA. The SPADE system, described in Section 9.3.4.1, is designated PCM/QPSK/FDMA. Although FDM generally is used with FDMA and TDM with TDMA, hybrid systems also exist.[17] One is the FDM-master group codec for use in the Telesat TDMA system.[18] Because TDM is used more extensively in digital transmission than is FDM, a more detailed discussion of TDM is provided in Section 5.3.

A low-loss multiplexer has been developed for satellite earth terminals to eliminate the need for a broadband high-power transmitter[19] with its reliability and efficiency limitations. Each 36-MHz channel in the 5925–6425-MHz uplink band is amplified by an individual solid-state power amplifier. Modular units allow the addition of channels as needed. Waveguide equalizers provide for amplitude and time-delay compensation. Such modular units are well-suited to small, unmanned earth terminals.

9.3.6. Demand Assignment

Many earth terminals use large numbers of dedicated channels devoted to continuous or nearly continuous traffic; however, increasing recognition is being given to the fact that many channels are not really needed on a full-period basis. Demand assignment thus can achieve economies by maintaining circuits connected only for the duration of a call, or for even shorter periods of time, such as a talk spurt, as described in Section 5.7.

Demand assignment (DA) is a very old and well-established concept in telephony. A user demands the assignment of a circuit by taking the telephone instrument off-hook. Receipt of dial tone then allows him to access the public switched network by dialing the desired number. A fully variable DA network pools all channels (trunks). A given channel then may be used by any station (end instrument) based upon the instantaneous traffic load. DA is most advantageous for destinations having light traffic loads. DA thus not only uses the space segment more efficiently than fixed assignment, but also allows for more efficient use of the terrestrial interconnect facilities. A result is that direct satellite connections can be made more often, with their improved quality of service as compared with the use of more tandem terrestrial connections.

A semivariable demand-assignment system is one in which blocks of channels are reserved for originating and/or destination stations, but still used only on demand.

A DA system is said to be DAMA when carriers (in FDMA systems) or bursts (in TDMA systems) are assigned on demand. Alternatively, when channels on existing carriers (FDMA) or time slots in existing bursts (TDMA) are assigned on demand, the system is said to be *baseband demand assigned* (BDA). BDA is best suited to networks with many users but only a few large earth stations, whereas fully variable DAMA is best suited to a network with many earth stations, each of which has only low traffic requirements. Mixed approaches also are useful, depending on the requirements to be met by a system.

DAMA systems provide for access on an individual user basis. The SPADE system, described in Section 9.3.4.1, is an example of a DAMA system. Other DAMA approaches are *single-channel per-burst* (SCPB) TDMA and code-division multiple access (CDMA). These systems do not transmit a carrier (FDMA or CDMA) or burst (TDMA) until it is required. The carrier or burst is established only for the duration of a call. For control, a common TDM signaling channel can be used on which all stations can call each other and be aware of the available channels. These available channels can be seized on a first-come-first-served basis.

BDA systems can utilize DSI (see Section 5.7) readily since BDA involves the application of a large number of trunks (usually 40 or more) in the provision of a larger (for example, doubled) number of two-way voice conversations.

9.3.7. Echo Cancellation

Section 5.7.2 discussed the principles of echo cancelers and their importance in long-haul circuits. Echo cancelers are especially important on satellite circuits, since the round-trip delay time is on the order of 600 ms, thus making existing echoes especially noticeable. Usually an echo canceler will be used at each end of a circuit. This is called *split echo control.*

9.4. CHARACTERISTICS OF SATELLITE-SYSTEM EQUIPMENT

A satellite system is comparable to a microwave radio-relay system with a single very-high-altitude repeater. Accordingly, satellite and microwave system equipment characteristics have many similarities. These include the fact that the amplifiers do not have highly linear characteristics. Thus, angle modulation (for example, QPSK or 8-PSK) often is preferred to amplitude modulation. Section 8.4 discusses the characteristics of microwave amplifiers and antennas in more detail.

Figure 9.5 shows the basic elements of a satellite system. They include the satellite (space segment) itself, earth stations (terrestrial segment) and a control station, whose functions are tracking, telemetry, and command.

9.4.1. Space Segment

A spacecraft for telecommunications purposes consists not only of the transponders, but also of a control subsystem, solar array, batteries, monitoring systems, and a small propulsion system, as well as temperature control devices. Power

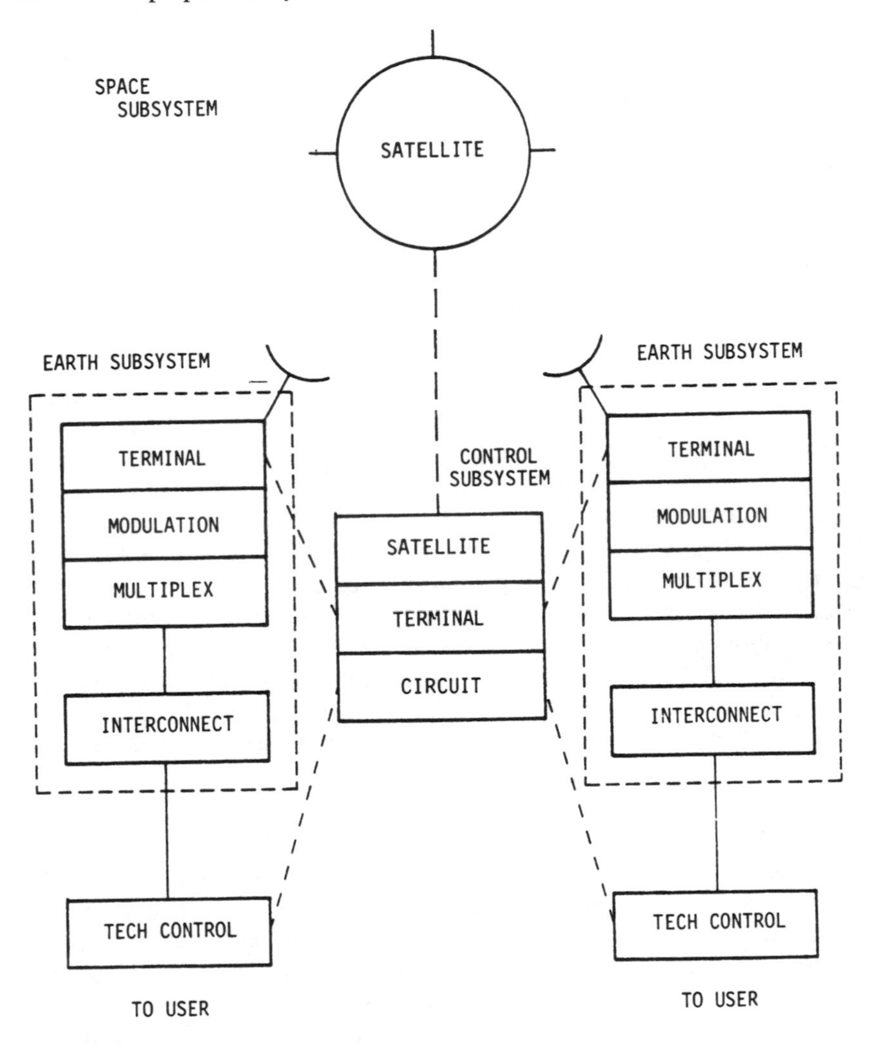

FIGURE 9.5. Satellite system equipment.

for the spacecraft is furnished by a solar array which maintains a charge on nickel-cadmium or nickel-hydrogen batteries. Solar cells provide efficiencies on the order of 15% or more, and thus can produce at least 300 watts dc per square meter of surface area for sun rays striking the cells perpendicularly. The size of the solar array thus limits the power available on board the satellite.

For on-board power storage, nickel-cadmium batteries have been used successfully in space for many years, but their state of charge must be monitored closely. If they are allowed to discharge to less than 50% of full charge, their lifetime may be impaired. They also must be exercised, that is, their state of charge must be varied from time to time. Nickel-hydrogen batteries do not exhibit these limitations, but are relatively new in space. They are, however, being used to an increasing extent.

Twice a year, at the equinox, which occurs in the latter parts of March and September, the spacecraft enters eclipse, that is, the sun sets on the spacecraft. This condition may last up to seventy minutes per day. During such times, the spacecraft must rely entirely upon its batteries for power.

The propulsion system on a spacecraft allows for the correction of its orbital position from time to time to compensate for position drift that may occur.

The remainder of this section describes the transponders and the control subsystem.

9.4.1.1. Transponders

Each satellite contains a number of transponders whose function is to receive, amplify, and retransmit the signals within a given frequency band. The retransmitted signal must be at a different frequency from the received signal to prevent the oscillation that would occur if the transmitter output were fed back to the receiver input. Because of satellite power limitations, the downlink is more critical than the uplink, so the downlink frequency band usually is lower in frequency than the uplink frequency band. A single satellite dish with multiple feeds, however, may be used for both receiving and transmitting.

A large number of signals from various earth stations usually arrives at a given satellite. These signals may be of various frequencies and polarizations, and may occur in different time slots. The antenna feed arrangement is used to separate the signals into two orthogonal polarizations, often linear for domestic satellites and circular for international satellites. Following this, the different frequency bands are separated by the use of channel filters so that each transponder receives the frequency band for which it is designed. Separation of signals in the time domain also may be done. Dual polarization can be achieved such that interference from the oppositely (or cross) polarized signal is about 40 dB below that of the desired signal over relatively small bandwidths (for example, 2%) and beamwidths (e.g., 2°). Over broader bandwidths and beamwidths, however, the oppositely polarized signal reaches higher levels. One *Intelsat* design achieved a −27-dB value for the opposite polarization over the entire 3700–6425-MHz band within a spot beam.

Precipitation degrades polarization isolation by converting linear or circular polarization to elliptical polarization. Thus both the antenna characteristics and the propagation effects must be included in a determination of interference at a receiver from an unwanted transmitted polarization. In the 11–14-GHz bands and at higher frequencies, depolarization caused by precipitation is more severe than in the 4- and 6-GHz bands.

In general, access to a given transponder from the ground is based on the polarization and frequency of the uplink signal. Selection of a given transponder in this manner may be used to control the downlink frequency, and possibly the covered area on the ground, since different transponders may be connected to different spot beams. In addition, transponders may serve different classes of users, with some transponders being dedicated to low-level mobile users while others handle large earth stations with supergroup or master group levels of traffic, or television transmissions.

A transponder frequency plan is shown in Fig. 9.6, using staggered frequencies to aid in isolating the two polarizations. Staggered plans are useful for transponders having small numbers of carriers, that is, for television or master group applications. For SCPC applications, staggering is of only limited value.

A single-conversion transponder is shown in block diagram form in Fig. 9.7. The preamplifier provides gain at the received radio frequency, but usually has an equivalent noise temperature of several hundred Kelvin. Lower noise temper-

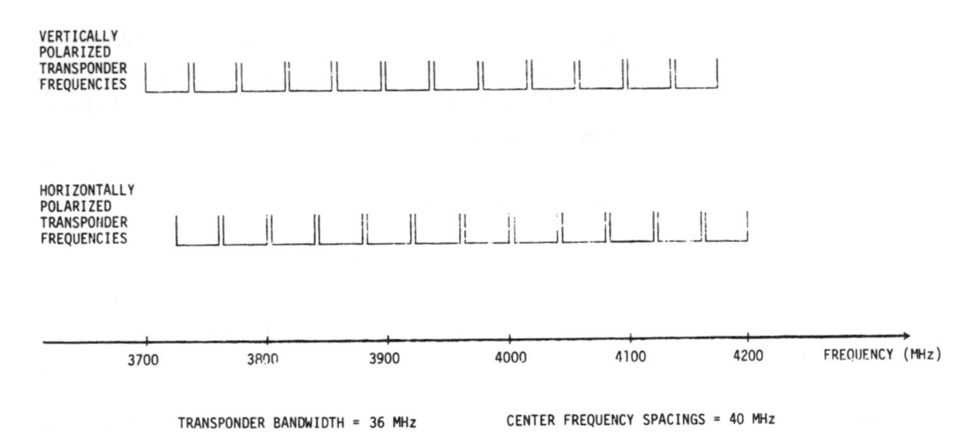

FIGURE 9.6. Downlink satellite transponder frequency plan.

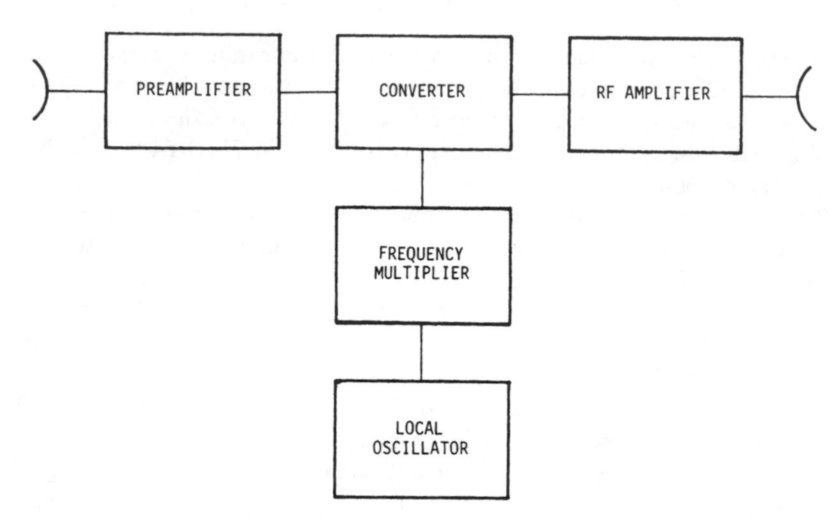

FIGURE 9.7. Single conversion transponder.

atures are not needed because appreciable noise enters from the earth itself. Moreover, amplifiers with low noise figures tend to be less reliable than the noisier type typically used.

The total bandwidth of a given transponder may be as little as 36 MHz or less, as in many domestic and international satellites, or as large as 72, 77, or 241 MHz, as in some of the international satellites. This transponder bandwidth may be subdivided, as is done, for example, in the SPADE system, which provides up to 800 carrier frequencies within a single transponder's bandwidth (see Section 9.3.5.1).

Most transponders provide preamplification of the uplink signal, conversion by a fixed amount to a lower frequency, and amplification at the lower frequency, which then is transmitted to the ground. In the 4- and 6-GHz commercial bands, 2225 MHz usually is subtracted from the uplink frequency to produce the downlink frequency, whereas in the 7- and 8-GHz military bands, 725 MHz or 200 MHz is subtracted. These uplink-downlink frequency differences are the maximum ones available within the given frequency allocations,[20] except for the 200-MHz difference. This relatively small difference results from the fact that the 7900–7950-MHz uplink is received on board the satellite through an earth coverage horn, and is adjacent in the uplink spectrum to other channels received on the earth coverage horn. Signals received in the 7900–7950-MHz transponder, however, are cross-strapped to a 7700–7750-MHz downlink, which is used with a narrow-coverage downlink (transmit) antenna. The frequency assignment was established to place the downlink frequency band adjacent to the other downlink narrow-coverage frequency bands.

Automatic gain-control amplification and limiting are used to maintain uniform signal levels, but transponder gain often is commandable from the ground as well.

The downlink beam produces a coverage area on the ground known as the *footprint.* An example of a footprint is shown in Fig. 9.8.

Most transponders are built on a highly redundant basis to achieve high reliability. The block diagrams in this chapter, for simplicity, do not illustrate these redundancies.

The transponders described thus far involved a single conversion of the uplink frequency directly to the downlink frequency, along with the needed amplification. Various applications exist, however, in which conversion to baseband on board the satellite is advantageous. On-board processing of weather photograph information and earth resources data from multispectral scanners has proven feasible already. Concepts for commercial applications of on-board switching and processing include on-board demand access, also known as the *switchboard in the sky* concept, and TDMA packet transmission via spot beams. On-board demand access allows the transponder input channels to be switched by command from the ground to the correct downlink channel. TDMA packet transmission, or satellite-switched TDMA, uses a preprogrammed switching sequence in which

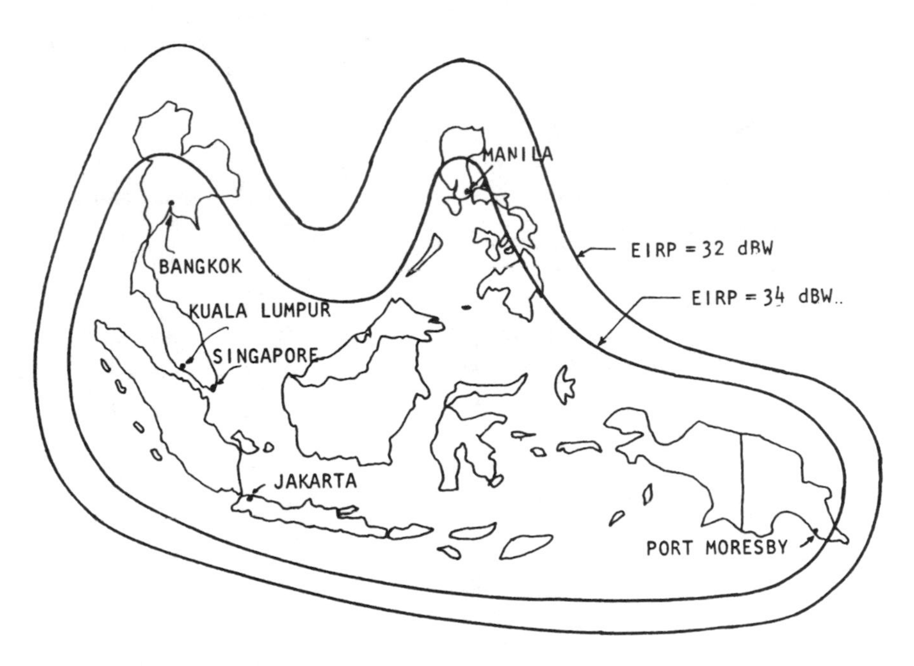

FIGURE 9.8. Footprint coverage of Palapa B satellite (Indonesia). (Reprinted by permission of *Satellite Communications* magazine.)

the packet addresses are used on board the satellite to switch each packet to its proper destination via the corresponding spot beam. A packet typically is 1024 bits long, including address information.

In addition to on-board switching, signal processing can be done to regenerate uplink digital signals for downlink transmission. Such processing is useful both in commercial and military applications. Other military uses include the removal of interference accompanying the uplink signals. In this manner, the repeating of interference can be minimized, and the tendency of strong signals to "capture" the transponder AGC because of nonlinearities can be reduced.[21]

Other concepts for processing transponder applications include the use of uplink single sideband to conserve bandwidth, and its conversion on board the satellite to PCM/FM for the downlink to minimize the power required from the satellite.

Signal-processing transponders are built to perform specific types of conversions. The result is a lack of flexibility with respect to changes that may be desired after the satellite has been launched. The advantages of processing transponders, however, are such that their use in the future probably will increase significantly, especially with the problems of an increasingly crowded spectrum.

Figure 9.9 is a block diagram of a processing transponder. As shown there, the entire transponder input is converted to baseband, at which such functions as switching or signal processing can be performed. Then the processed baseband signals are remodulated and converted to the downlink frequency band.

9.4.1.2. Control Subsystem

The control subsystem includes a special earth station (often called the tracking, telemetry, and command station), whose function is to track the spacecraft's position, with the help of an on-board transponder, to monitor spacecraft telemetry,

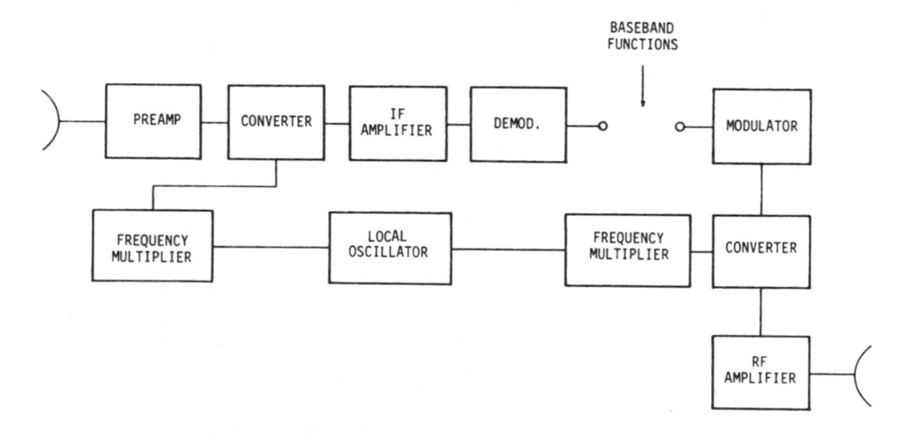

FIGURE 9.9. Switching or processing transponder.

and to send corrective commands to the spacecraft as necessary. Telemetry from the spacecraft includes data on the solar array output, the state of charge of the batteries, the amount of propulsion fuel remaining, on-board temperatures, and the operating level and load of each transponder, including automatic gain-control (AGC) level.

9.4.2. The Earth Station

Figure 9.10 is a simplified block diagram of a satellite earth terminal for digital transmission. Digital streams at the DS1 rate (that is, T1), each conveying 24 voice channels or the equivalent, are multiplexed to form a stream at the DS3 rate (that is, T3) or higher. A single DS3 stream can be handled using QPSK in a 36-MHz transponder. Alternatively, two DS3 streams can be transmitted using 8-PSK, and correspondingly higher rates can be sent through larger transponder bandwidths.

The frequency standard generally is of the cesium type. The tunable frequency synthesizers used in the up and down converters are phase locked to the frequency standard. The required local-oscillator frequencies then are obtained by the use of phase-locked multipliers. Although Fig. 9.10 shows only a single up-conversion step, two actually may be used, with the first IF being centered at 70 MHz and the second at perhaps 700 MHz. Smaller earth terminals may use only a fixed-frequency local oscillator rather than a frequency synthesizer. Earth terminals operating with satellite transponders having appreciably larger bandwidths than 36 MHz use IFs higher than 70 MHz.

The portions of the block diagram described thus far are repeated for each satellite transponder through which the earth station works. A power combiner then adds the outputs of the various up converters, each of which is in a separate band of frequencies. If dual polarization is used, there will be two power combiners, one for the transponders using vertical or right-hand circular polarization and another for the transponders using horizontal or left-hand circular polarization. Following the power combiner is a broadband intermediate-power amplifier (IPA) and a broadband high-power amplifier (HPA). Earth-station power output may be on the order of 500 W to 2 kW or more, depending on the number of channels being handled. Usually the IPA and HPA are dual redundant because of the large number of channels they carry.

A diplexer is used for each of the two polarizations handled by the earth station, as well as for each frequency-band pair (that is, 4–6 GHz, 7–8 GHz, 11–14 GHz, etc.). Dual polarization operation generally is not used in the frequency bands above 14 GHz because of the extensive depolarization caused by precipitation at such frequencies. The diplexers and the transmit and receiver filters must provide the high degree of isolation needed to prevent the transmitter output power from feeding back into the receiver input and thus producing desensitization.

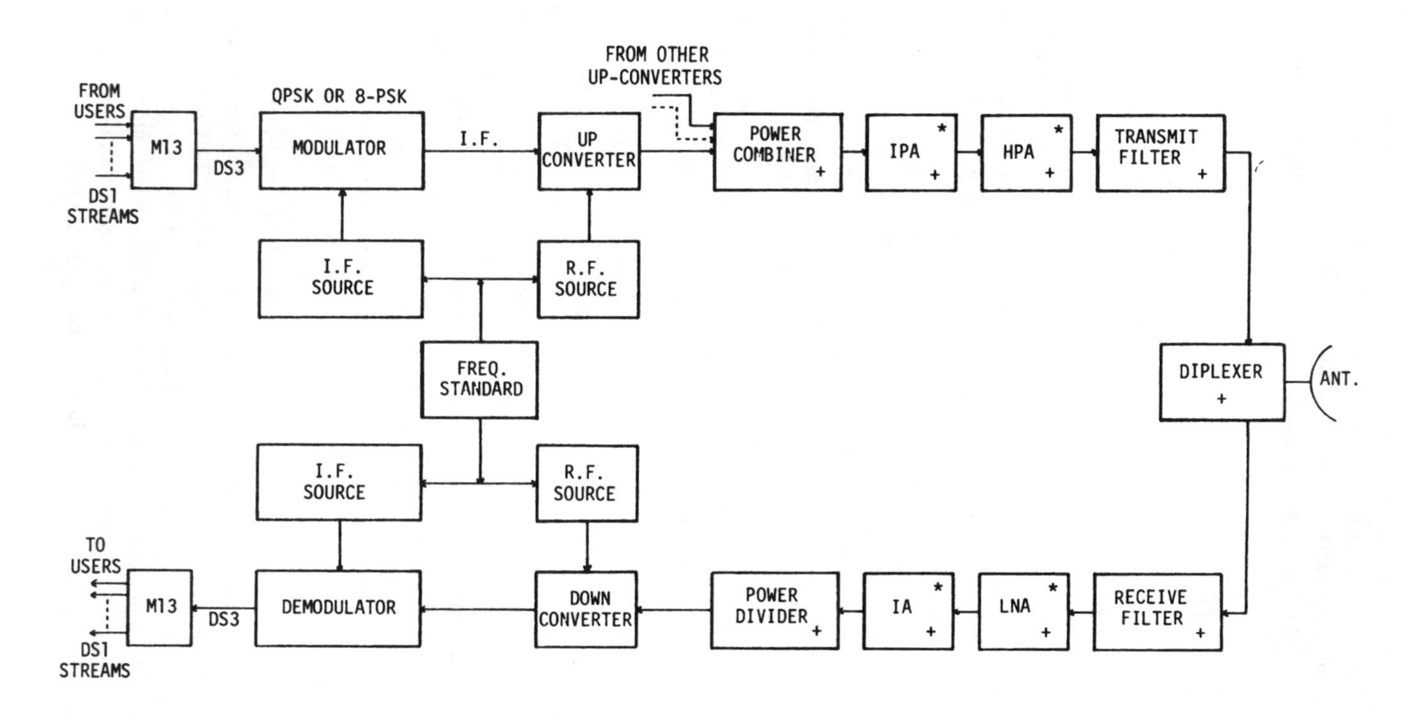

FIGURE 9.10. Satellite earth terminal for digital transmission.

An important characteristic of the earth station is the ratio of the antenna gain G to the effective noise temperature T, where G is expressed in dB and T in dBK (decibels relative to one Kelvin). The ratio G/T thus is expressed in dB/K. It is a receiving system figure of merit. Large values of G/T allow lower-power satellites to be used, but require more expenditure for the earth station because of the cost of larger antennas and low-noise-temperature amplifiers (LNAs).

A separate set of LNAs and intermediate amplifiers (IAs) is used for each received polarization and each frequency band (for example, 4, 7, 11 GHz, etc.). In addition, the IAs and LNAs are redundant for reliability. The power divider separates the channels within the frequency domain based on the satellite's transponder frequency bands. The down conversion, demodulation, and demultiplexing steps then are the reverse of those described for the transmitting direction. All transmitted sidebands must be attenuated sufficiently to prevent them from disturbing the receiving system. Isolation must be on the order of 140 to 180 dB depending on transmitted power level, bandwidth, and receiver noise level. This requires filters with good skirt selectivity, but yet with well controlled delay distortion at the band edges to keep intersymbol interference low.

Earth stations used as part of the Intelsat system are classified as follows:

The Standard A earth station operates in the 4-GHz downlink and 6-GHz uplink bands with $G/T \geq 35$ dB/K. This can be achieved using a 13-m dish and a 45-K LNA. The standard B earth station differs from the Standard A only in that $G/T \geq 31.7$ dB/K. The Standard C earth station operates in the 11-GHz downlink and 14-GHz uplink bands with the following G/T values:

$$\text{Clear sky } \frac{G}{T} \geq 39.0 + L_1 + \log_{10}\left(\frac{f}{11.2}\right) \text{ dB/K} \tag{9.6}$$

where L_1 is the additional attenuation over clear sky all but 10% of the time, and f is the carrier frequency in GHz.

$$\text{Degraded sky } \frac{G}{T} \geq 29.0 + L_2 + \log_{10}\left(\frac{f}{11.2}\right) \text{ dB/K} \tag{9.7}$$

where L_2 is the additional attenuation over clear sky all but 0.017% of the time (1.5 hours per year). These equations are intended to keep the basic telephone-channel noise power to less than 8000 pW0p for 90% of the time.

Equations (9.6) and (9.7) are applied next to illustrate the receiving antenna gain required in a particular case. The example chosen applies to the Intelsat receiving system at Etam, WV.[22] Weather statistics and propagation studies showed that clear sky attenuation, caused by mist, fog, and high stratocumulus clouds, that is, factors not measurable by a rain gauge, is 0.4 dB. Rain conditions for all but the worst 10% of the year contribute an additional 0.55-dB attenuation. Thus $L_1 = 0.40 + 0.55 = 0.95$ dB. Rain conditions for all but the worst 1.5 hours of the year contribute an additional 9.5-dB attenuation. Thus $L_2 = 0.40 + 9.5 = 9.9$ dB. Corresponding contributors to the temperature T are 17 K for the 18° elevation angle at Etam for all but 10% of the time and 419 K for all but 0.017% of the time.

Meeting the Standard C earth station requirements may necessitate the use of a space-diversity pair of receiving sites. For example, the use of a second receiving site at Lenox, WV, 35 km to the northeast of Etam, operated on a diversity basis with Etam, resulted in a reduction in L_2 to $0.4 + 2.5 = 2.9$ dB and a reduction of the noise temperature all but 0.017% of the time from 419 K to 307 K. Because an unacceptably high transmit power (22 kW) would have been required at Etam without diversity, the use of the diversity pair was implemented. A nonstandard earth station requires approval by Intelsat on a case-by-case basis. A "station accessing leased facilities (transponders)" is one that produces less than 400 pWp on an adjacent satellite carrier ($\geq$3° away), in accordance with CCIR recommendation 465–1. Its emission limit is 20 dBW/4-kHz carrier. The abbreviation pWp means "picowatts psophometric" and refers to the use of a voice-band weighting curve used by the ITU/RS. This curve approximates the response of a standard telephone set.

The side lobes produced by the antennas of the above earth stations must all meet the criterion $G_S = 32 - 25 \log_{10} \theta$, where θ is the antenna's half-power beamwidth in degrees. Many newer earth stations must meet more stringent sidelobe criteria, such as $29 - 25 \log_{10} \theta$ or $26 - 25 \log_{10} \theta$.

9.5. MAJOR OPERATIONAL COMMUNICATION SATELLITE SYSTEMS

The largest international satellite system is *Intelsat*, whose earth-station characteristics have been described briefly in Section 9.4.2. Other international satellite systems include the Pan American satellite (*PanamSat*), and *Orionsat*. PanamSat is located at 45° West with C-band transponders for coverage of the Caribbean, as well as Central and South America. K_u-band transponders cover portions of Western Europe, the U.S., Central and South America. Orionsat has locations at 37.5° and 47° West, with K_u-band spot-beam coverage of portions of North America, Europe and Africa.

The International Maritime Satellite Consortium (*Inmarsat*) provides telephony and data services to ships at sea, to off-shore drilling rigs and to some mobile users. The uplink band from the ships or other users to the satellites is 1636.5–1645 MHz, while the downlink band is 1535–1543.5 MHz. The connection between the satellites and the earth stations is via C-band or K_u-band. As is the case with Intelsat, the satellites are located over the Atlantic, Pacific, and Indian Oceans for global maritime coverage.

Many currently operational satellite systems provide both analog and digital transmission at the present time, although the use of digital transmission is increasing. Most systems had their origins in the mid-1970s, and built a sizable number of analog earth-station facilities. The exception was Satellite Business Systems (SBS). SBS started later than the others, and built an all-digital system. Each of its satellites was built with ten transponders of 43-MHz bandwidth, with uplinks in the 14.0–14.5-GHz band and downlinks in the 11.7–12.2-GHz band. For the uplink, SBS has a $G/T = 1.8$ dB/K at the beam edge. For the downlink, it has an EIRP of 37.0–43.8 dBW at the beam edge. Many other satellite systems are implementing digital transmission in one or more transponders, as well as at many of their earth stations. They include Telesat's *Anik*, GE's *Satcom*, American Satellite and *Intelsat*. U.S. domestic satellites which also may be used for digital telephony include Hughes' *Galaxy*, GE's *GSTAR*, SP Communications' *Spacenet*, and AT&T's *Telstar*.

9.6. SATELLITE SYSTEMS FOR DIRECT SERVICE TO THE USER

Low-earth orbit (LEO), medium-earth orbit (MEO), and geostationary-earth orbit (GEO) satellites are expected to play a role in the achievement of worldwide mobile radio telephony, especially in connection with the use of error-correcting codes combined with digitized voice. Error-correcting codes may allow the severe bit-error rates (for example, 10^{-1}) resulting from mobile radio multipath propagation (at low satellite elevation angles) to be reduced to values on the order of 10^{-2} to 10^{-3}, at which most predictive voice-coding techniques give satisfactory voice quality.

While the use of satellites to provide service to portable telephone users over a wide area may seem to be an ideal application of satellite technology, certain limitations are present. First is the matter of frequency reuse. In a terrestrial system, each of two operators has 416 30-kHz-wide channels, totalling 12.5-MHz bandwidth in each direction, or a total of 25 MHz for two operating companies. A cell with a 10-km radius has a coverage area of 314 km^2. On the other hand, the size of a cell produced by a GEO satellite using a 10-m dish diameter at 1.5 GHz (1.5° beamwidth to the -3-dB points), based on a 39 000-km slant range, will be 785,000 km^2 (1000-km radius). This area is 2500 times the area of a typical 10-km radius terrestrial cell. MEO and LEO satellites will produce smaller cell areas, but even these smaller areas are large enough that service must be limited to users who are unable to obtain terrestrial service.

Proposed LEO satellite systems include *Iridium*, a 66-satellite system being developed by a consortium headed by Motorola, Ellipsat's 24-satellite system,

Ellipso, and two 48-satellite systems, Constellation Communication's *Aries*, and Loral/QUALCOMM's *Globalstar*. All of these systems would operate above 1 GHz, in the 1.5–1.6-GHz band. Another system, Inmarsat's *Project 21*, would use about 35 LEO satellites in combination with GEO satellites.

The widely publicized *Iridium* system will feature 66 polar-orbiting satellites, 11 in each of 6 orbital planes. Their 765-km altitude will result in an approximately 90-minute orbit for each satellite. Subscriber sets will search for, and use, a terrestrial channel if one is available. A satellite circuit is opened only if a terrestrial channel is not available. Subscriber sets will feature not only voice service, but also a data port for fax or data. These sets will have a 3-watt transmitter power and operate via an 8-cm whip antenna. Each channel will use QPSK in an 8-kHz bandwidth, with voice being VSELP at 4.8 kb/s, and data being transmitted at 2.4 kb/s. Hand-off normally will occur not because of subscriber motion, but because of movement of the satellites, which will provide 660-km diameter cells, 48 per satellite. Hand-off may be from satellite to satellite in the same orbit via crosslinks above 12 GHz, using fast packet switching, or via one or more of 20 gateways (30-GHz uplinks and 20-GHz downlinks) or 5 interplane links to a satellite in another orbit.

In addition to the LEO systems for voice, several LEO systems have been proposed for such nonvoice services as data, two-way messaging, position detection, etc. These nonvoice LEOs would operate below 1 GHz. They include systems proposed by Orbital Communications, Inc., Starsys, Inc., Volunteers in Technical Assistance (VITA), and Leosat, Inc.

TRW, Inc., has proposed an MEO satellite system, *Odyssey*, consisting of 12 satellites offering both voice and nonvoice services.

Two GEO systems have been planned for both voice and nonvoice services. They are those of (1) the American Mobile Satellite Corporation (AMSC) and its Canadian counterpart, Telesat Mobile, Inc. (TMI), and (2) Celsat, Inc. The AMSC and TMI systems will use separate but identical spacecraft with identical standards, and thus will be capable of backing up one another via their use of opposite circular polarizations. Voice and data services will be provided to closed user groups through private and shared earth stations, with public access through gateways and data hubs. The services provided to commercial and general aviation will have priority of access for safety as well as to promote regularity of flight. The primary access method will be FDMA/SCPC.

The AMSC and TMI systems will provide one beam for each of the four time zones of the contiguous 48 states plus the Canadian provinces.[23] Figure 9.11 displays the mobile satellite (MSAT) coverage. Feeder links in the 11–13-GHz range will connect the satellites to the network control center, as well as gateway and base stations. Speech will be coded at 4.8 kb/s, with a 40% activity factor being assumed. Each channel will have a width of 5 kHz, and the b.e.r. is estimated at 10^{-3} or less. A total of 1400 channels is to be available, or 465 per beam.

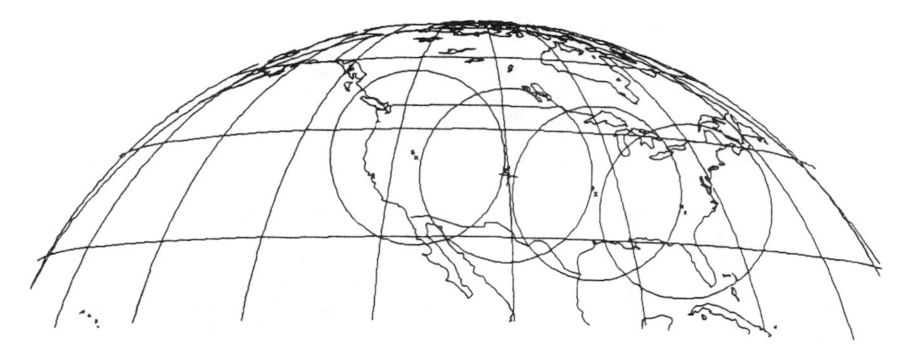

FIGURE 9.11. MSAT coverage (Courtesy Yan, et al. Ref. 23, © IEEE, 1989).

This requires a bandwidth of 7 MHz for both the uplink (near 1.65 GHz) and the downlink (near 1.55 GHz). Since the eastern-most and western-most beams do not intersect, frequency reuse between them will be possible, for a total of 1860 channels for each system. Each mobile will be able to access either satellite through the use of polarization discrimination.

The design of the systems has accounted for such propagation factors as Doppler shift and multipath, which influence the selection of packet lengths, modulation and speech-coding techniques. An integrated-adaptive mobile access protocol (I-AMAP) was developed to optimize performance between two conflicting goals: (1) maximizing the number of subscribers that can access the available bandwidth, and (2) minimizing the delay between a subscriber's request for channel access and completion of service.[24] The MSAT network configuration is shown in Fig. 9.12.

9.7. FUTURE TRENDS IN COMMUNICATION SATELLITE SYSTEMS

In spite of the rapid development of communication satellite systems since the mid-1970s, and the increasing implementation of digital transmission via satellite since the early 1980s, many more developments are on the horizon. The use of on-board processing was described in Section 9.4.1.1. It is expected to allow increased implementation of packet switching, as well as on-board controlled TDMA switching, possibly combined with scanning spot beams.[25] The addresses carried by individual packets might be used for beam and time-slot control. Packet voice has been found to be feasible for telephone conversations, and this concept relates directly to the achievement of on-board controlled TDMA switching.

Intersatellite links involving an up, over and down mode of transmission have already been shown to be feasible through work done by the U.S. Air Force with

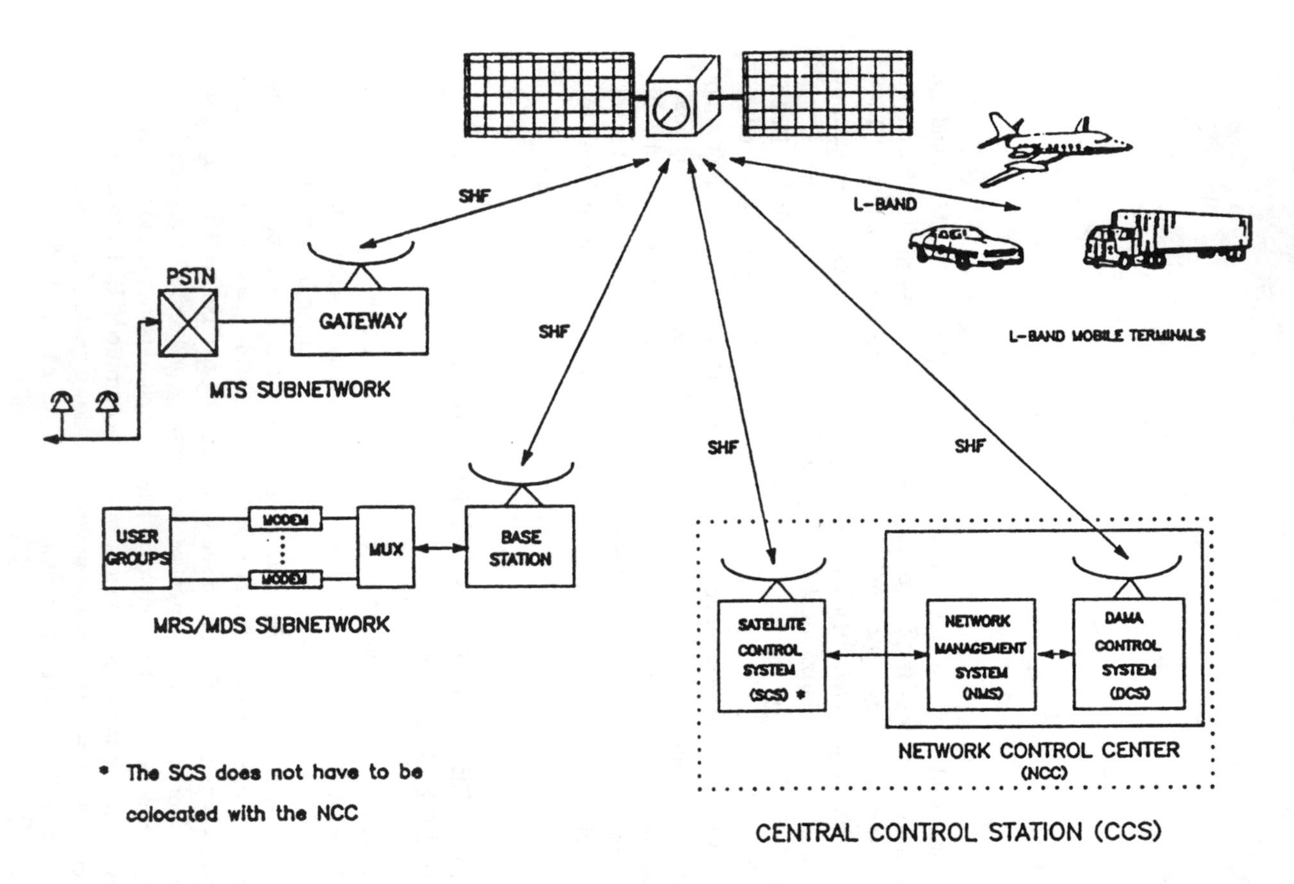

FIGURE 9.12. MSAT network configuration. (Courtesy Wachira, Bossler and Skerry, Ref. 24, © IEEE, 1989).

its *Lincoln Experimental Satellites* 8 and 9 (LES-8 and LES-9), which used a 36–38-GHz crosslink. Such intersatellite links, when implemented commercially, may help to further global telephony as well as to provide more widespread use of land, maritime, and aeronautical radiotelephony. The coming low-earth orbit *Iridium* and *Ellipso* systems will use intersatellite links.

Early satellite-system designs attempted to maximize satellite lifetime by keeping the satellite as simple, and therefore reliable, as possible, while allowing the earth stations to be relatively complex. The trend now, however, is toward more complex satellites but simpler earth stations. This trend is being aided by several major factors. One is the widespread use of large-scale integrated circuits, with their inherently high reliability. Another is the accumulation of many years of experience on the part of spacecraft designers. In addition, experience is being gained in bringing geostationary satellites down to low-earth orbit where they can be serviced by a shuttle and then boosted back to geostationary orbit. The complex satellite and the relatively inexpensive earth station make good economic sense, just as do the present complex central office and the relatively inexpensive end instrument. Many earth stations may use a single satellite. By putting more of the complexity into the satellite, the savings in earth station hardware are multiplied many times.

More complex satellites may provide not only higher EIRP (in the bands above 10 GHz, where such EIRP is allowed) but also more narrow satellite beams for improved orbit-spectrum utilization. More on-board processing is another feature of the more complex satellite. Such on-board processing, in addition to the applications discussed in Section 9.4.1.1, can allow transmission to and from cities experiencing good weather in the 20- to 30-GHz bands, while using the 4–6-GHz bands to and from those cities experiencing precipitation conditions. The smaller earth station sizes will be advantageous primarily from an economic viewpoint.

Satellite technology is applicable to two-way communications for truck fleets, as is being shown by the *OmniTRACS®* System. Two K_u-band transponders on *GStar I* are used to provide encrypted messaging at 9600 b/s, as well as vehicle location using a Loran receiver on each truck. This system has been developed and is being operated by QUALCOMM, Inc., San Diego, CA.

A future satellite system development may be the *large space platform* (LSP), in which a few giant space stations in geostationary orbit replace many present communication satellites. Such an LSP could alleviate interference caused by the crowding of the geostationary orbit, thus permitting smaller earth stations. It could permit the mounting of very large and complex antennas and feed systems with very narrow spot beams. Other features would be the interconnection or "cross-strapping" of various satellite circuits and the achievement of economies of scale by providing common support functions (for example, power, as well as propulsion for orbit corrections) for the multiple satellite systems involved. The LSP would be implemented with the help of the space transportation system (STS) or "shuttle" in its construction and maintenance. The LSP would be serviced and maintained in operation through the use of high-energy upper stages operating from the STS.

Since the LSP concept was originally suggested, thinking has moved toward the idea of a platform consisting of a single STS-borne spacecraft with its mated transfer vehicle.[26] Such a scaled-down platform is now visualized as serving *Intelsat* requirements in the 1990s. Use of the STS can allow spacecraft deployment and checkout in low-earth orbit where minor malfunctions can be corrected manually, something not yet possible at geostationary orbit.

PROBLEMS

9.1. Dual polarization is more commonly used in the 4–6-GHz satellite bands, than in the 11–14-GHz bands. Why?

9.2. Why is a satellite system's downlink budget more critical than its uplink budget?

9.3. A satellite transponder centered at 3950 MHz is used totally for TDMA digital transmission and has a 5-W power output. For earth stations that are at the beam edge (-3 dB) and that have a 45° latitude (satellite and earth stations at same longitude), what is the received level in a 50-dB gain dish if the satellite antenna has a 1.5-m diameter? (Neglect circuit and waveguide losses.) Assume a 55% antenna efficiency.

9.4. In the system of Problem 9.3 the effective earth terminal figure of merit is 20.0 dB/K. What is the available C/kT?

9.5. In the system of Problems 9.3 and 9.4, what symbol rate can be transmitted if the required $E_S/N_0 = 18$ dB, assuming losses and link margin total 4.0 dB?

9.6. You have a choice between placing a satellite earth terminal dish on a building rooftop or near a parking lot on the ground. The path to the satellite is unobstructed in both cases. Which location would you choose and why?

9.7. Discuss the advantages and disadvantages of uniform carrier-frequency spacing in a satellite transponder being used on an FDMA basis.

9.8. Explain how the operating level of a satellite transponder is actually determined by the operation of systems on the ground.

9.9. Are intersatellite links among geostationary satellites useful in *significantly* reducing the end-to-end delay time on intercontinental satellite links? Illustrate your answer numerically.

REFERENCES

1. Keiser, B. E., "Broadband Communications Systems," Course Notes, © 1992.
2. Carr, F. S., "Aerosat—Current Status and the Test and Evaluation Program," *1975 Eascon Record*, Sept. 1975, p. 13.
3. Anderson, R. E., "The Mobilesat System," *Satellite Communications*, Vol. 8, No. 3, March, 1984, pp. 16–18.
4. Brown, R. J., et al., "Companded Single Sideband (CSSB) Implementation on Comstar Satellites and Potential Application to Intelsat V Satellites," *ICC '83 Conference Record*. IEEE Document 83 CH1874-7 June 1983, pp. E2.1.1–E2.1.6.
5. Edwards, Robert D., "An Improved Syllabic Compandor for Satellite Applications," *ICC '83 Conference Record*, IEEE Document 83 CH1874-7, June 1983, pp. E2.3.1–E2.3.5.
6. Helder, G. K., "Customer Evaluation of Telephone Circuits with Delay," *Bell Syst. Tech. J.*, Vol. 45, Dec. 1966, pp. 1749–1773.
7. Benoit, A., "Signal Attenuation Due to Neutral Oxygen and Water Vapour, Rain and Clouds," *Microwave Journal*, Nov. 1968, pp. 73–80.
8. Fang, D., "Attenuation and Phase Shift of Microwaves Due to Canted Raindrops," *Comsat Technical Review*, Vol. 5, No. 1, Spring 1975, pp. 135–156.
9. Smith, F. L., "A Nomogram for Look Angles to Geostationary Satellites," *IEEE Trans. Aerospace and Electronic Systems*, May 1972, p. 394.
10. Spilker, J. J., *Digital Communications by Satellite*, Prentice-Hall, Inc., Englewood Cliffs, NJ, 1977.
11. Puente, J. G., Schmidt, W. G., and Werth, A. M., "Multiple Access Techniques for Commercial Satellites," *Proc. IEEE*, Feb. 1972, pp. 218–229.
12. McClure, R. B., "Analysis of Intermodulation Distortion in a FDMA Satellite Communication System with a Bandwidth Constraint," *Trans. IEEE Intl. Conf. on Comm.*, 1970.
13. See note 10.
14. Babcock, W. C., "Intermodulation Interference in Radio Systems," *Bell Syst. Tech. J.*, Vol. 32, Jan. 1953, pp. 63–73.
15. See note 10.
16. Assal, F., Gupta, R., Apple, J., and Lopatin, A., "Satellite Switching Center for SS-TDMA Communications," *Comsat Technical Review*, Vol. 12, No. 1, Spring 1982, pp. 29–68.
17. Pritchard, W. L., "Satellite Communication—An Overview of the Problems and Programs," *Proc. IEEE*, Vol. 65, No. 3, March 1977, pp. 294–307.
18. Kaneko, H., Katagiri, Y., and Okada, T., "The Design of a PCM Master-Group Codec for the Telesat TDMA System," *Conference Proceedings, ICC '75*, Vol. 3, June 1975, pp. 44-6–44-10.
19. Gruner, R. W., and Williams, E. A., "A Low-Loss Multiplexer for Satellite Earth Terminals," *Comsat Technical Review*, Vol. 5, No. 1, Spring 1975, pp. 157–177.
20. Huang, R. Y., and Hooten, P., "Communication Satellite Processing Repeaters," *Proc. IEEE*, Vol. 57, Feb. 1971, pp. 238–251.
21. See note 17.
22. Gray, L. F., and Brown, M. P., Jr., "Transmission Planning for the First United States

Standard C (14/11 GHz) Intelsat Earth Station," *Comsat Technical Review*, Vol. 9, No. 1, Spring 1979, pp. 61–89.

23. Yan, T.-Y., et al, "A FD/DAMA Network Architecture for the First Generation Land Mobile Satellite Services," *IEEE Globecom '89 Conference Record*, Nov. 1989, pp. 21.2.1–21.2.12.
24. Wachira, M., D. Bossler, and B. Skerry, "FDMA Implementation for Domestic Mobile Satellite Systems", *IEEE Globecom '89 Conference Record*, Nov. 1989, pp. 21.4.1–21.4.6.
25. Reudink, D. O., and Yeh, Y. S., "A Scanning Spot-Beam Satellite System," *Bell Syst. Tech. J.*, Vol. 56, No. 8, Oct. 1977, pp. 1549–1560.
26. Cohen, N. L., and Stone, G. R., "DBS Platforms: A Viable Solution," *Satellite Communications*, Vol. 6, No. 13, Dec. 1982, pp. 22–27.

10

Fiber-Optic Transmission[a]

10.1. INTRODUCTION

Fiber-optic transmission can be implemented wherever coaxial cable or wire-pair transmission is used. Its areas of present and future application extend all the way from the subscriber loop[1] to transoceanic cables.[2]

The basic concept in fiber-optic transmission is illustrated in Fig. 10.1. As shown there, a modulated light source such as a light-emitting diode (LED) or a laser has its output coupled to a fiber which then transmits the light to a remote photodetector, which demodulates the light, thereby providing the desired signal output.

Fiber-optic transmission has numerous advantages over the use of metallic cables.[3] First, the weight and bulk of the cable plant are reduced significantly, thereby allowing more efficient use to be made of duct space. A significant improvement in bandwidth also may be obtained, as is discussed in Section 10.2. This may allow future expansion without plant refurbishment.

Accordingly, fiber optics allows significant cost savings in some applications, while having a cost comparable to that of wire cable in others. The cost of a 50-km transmission system using fiber optics has been shown to be less than the cost of microwave-radio or coaxial-cable transmission for a capacity of 240 duplex channels (15.44 Mb/s) or more. System costs, based upon a 30-km repeater spacing, have been estimated[4] by Future Systems, Inc., Gaithersburg, MD. The costs range from $10,500 for 45-Mb/s transmission to $13 600 for 270-Mb/s transmission. These estimates assume rural installation costs and one-for-one backup pro-

[a] Much of the material of this chapter is taken from B. E. Keiser, "Wideband Communications Systems," Course Notes, © 1990.

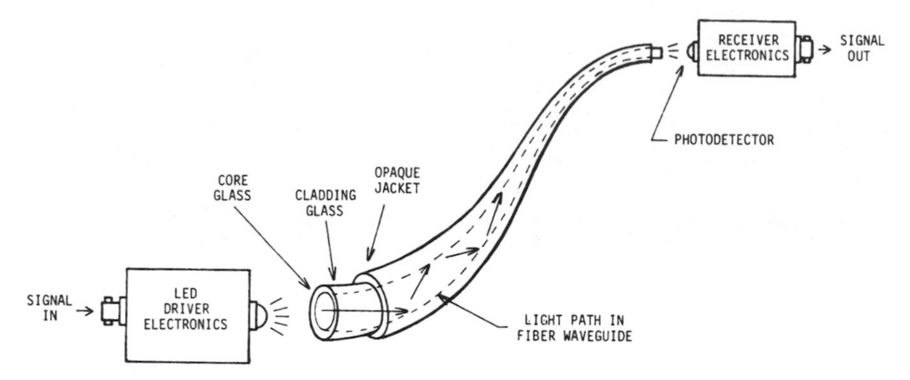

FIGURE 10.1. Basic fiber-optic communication system. (Courtesy Dr. Leonard Bergstein.)

tection. Such factors as legal, property, maintenance, and tax expenses have not been included, because of their wide variability from one place to another.

Fiber-optic systems are not troubled by short circuits or current leakage to ground, although "open circuits" (fiber breakage) can occur. Crosstalk is extremely low, and its effects can be eliminated completely by proper design. Problems of electromagnetic interference and noise do not occur along the optical portion of a fiber system, but such problems can exist on the wire side of the modems that may be used in getting a bit stream on and off a fiber. Correspondingly, there is no electromagnetic pulse problem along the fiber itself, although the fiber does exhibit a response to the ionizing radiation that directly accompanies a nuclear blast, as discussed in Section 10.2.

Optical fibers allow a very high degree of transmission security, which can be extended directly to the user. This results from the fact that optical fibers are very difficult to tap without the user becoming aware of the tap in the form of reflection level changes. In fact, for secure applications, a triaxial fiber system can be built with a noninformation region on the outside to monitor the reflection level. This noninformation region can contain a pulse sequence used only for monitoring purposes, but designed to look like a real bit stream.

10.2. FIBER TRANSMISSION CHARACTERISTICS

The two major factors that limit the transmission distance along a single unrepeatered section of fiber are attenuation and dispersion. Dispersion is important in terms of its limiting effect on bandwidth, and is affected significantly by the process used in manufacturing the fiber. The VAD process[5] provides a high degree of bandwidth-distance capability. The modified chemical vapor deposition

(MCVD) process is popular because it allows the mass production of commercial fibers.

10.2.1. Attenuation

The attenuation characteristics of silicon fibers[6] are shown in Fig. 10.2. Because the lowest attenuations are obtained at near infrared wavelengths, infrared emitting diodes (IREDs) are of major interest. The attenuation peak at 1.4 μm is a hydroxyl (OH) absorption band. Just 1 ppm of OH produces 30- to 40-dB/km attenuation at 1.39 μm.

Note that the attenuation increases appreciably above 1.6 μm. This increase results from phonon absorption, which is caused by interactions between the infrared electrical field and the ions that constitute the glass. The heavier the ions, the longer the wavelength at which phonon absorption occurs. Fibers based on fluoride and chalcogenide may exhibit ultra-low loss in the 2- to 5-μm range, possibly allowing repeaterless transoceanic fiber systems.

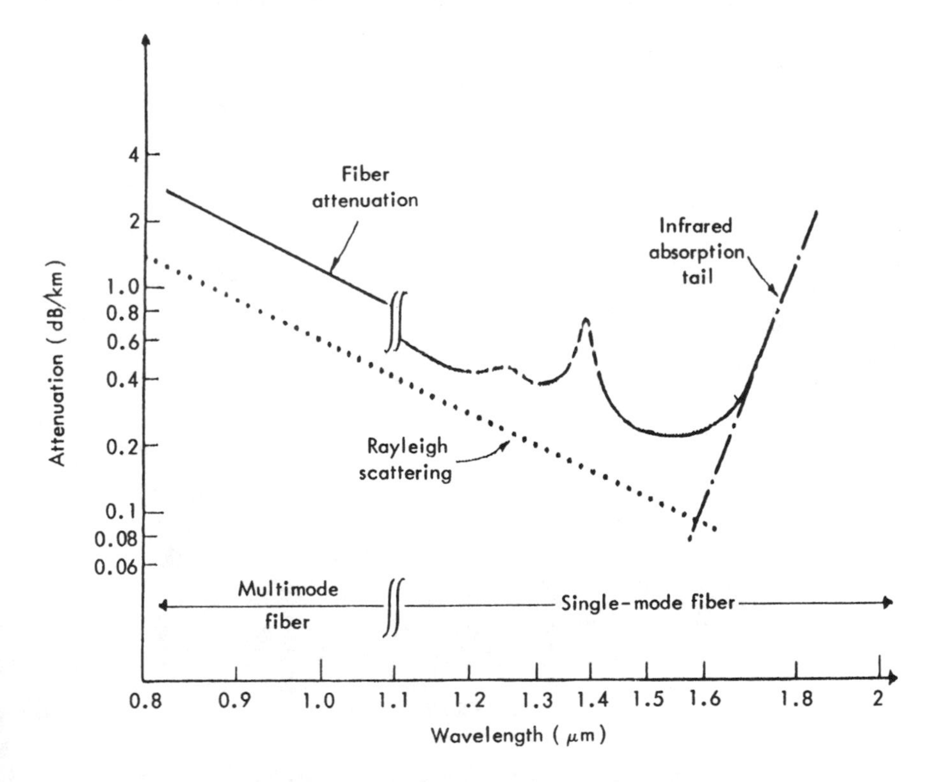

FIGURE 10.2. Fiber attenuation. (From P. S. Henry, Ref. 6, ©IEEE, 1983.)

Of concern in defense programs is the behavior of an optical fiber in the presence of ionizing (nuclear) radiation. When a burst of such radiation occurs, there is a temporary generation of luminescent energy within the fiber-optic waveguide itself. This is followed by a temporary increase in attenuation and then by a smaller permanent amount of attenuation beyond the amount previously exhibited. The use of lead sheath around the fiber cable and its underground installation provide the best protection against this problem.

How does optical-fiber attenuation compare with the attenuation of conventional coaxial cable? Figure 10.3 attempts to answer this question. To establish a comparison, however, note that the entire coaxial cable bandwidth up to a certain limit is used, whereas the optical fiber is operated at a specific wavelength at which it is modulated. Accordingly, the fiber-optic attenuation appears flat as a function of bandwidth, whereas the coaxial-cable attenuation increases steadily with bandwidth. Another difference is noteworthy. A relatively large outer diameter is required for a coaxial cable to exhibit low losses. On the other hand, the optical fiber has only a 0.4-mm nominal outer diameter, compared with a 2.5-mm outer diameter for the smallest (but highest loss) coaxial cable of Fig. 10.3.

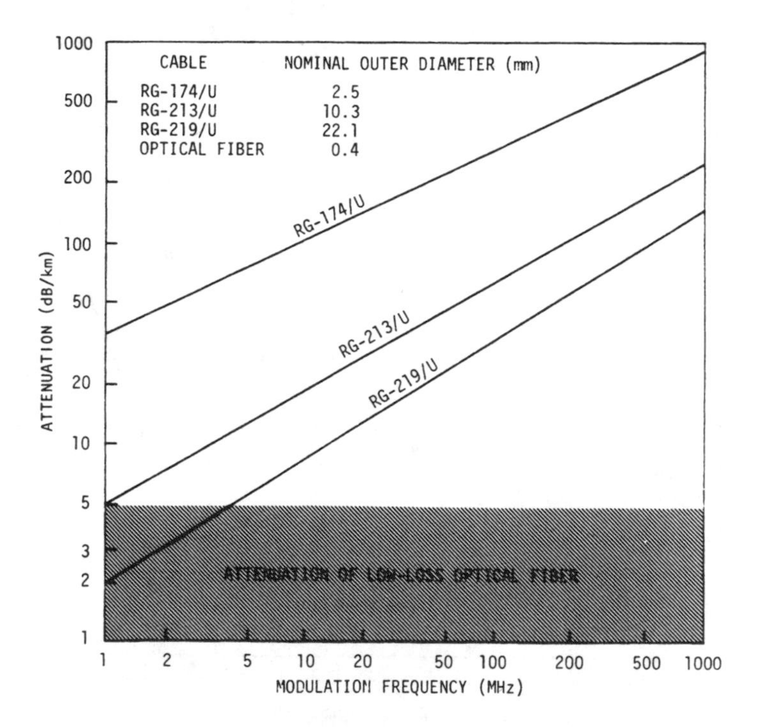

FIGURE 10.3. Fiber-optic cable attenuation versus coaxial cable attenuation. (Courtesy Dr. Leonard Bergstein.)

10.2.2. Dispersion

If the optical energy passing through a fiber were to take only a single path along the length of the fiber, each light pulse would be received exactly as it was transmitted. Pulse-broadening occurs, however, because of the multipath nature of wave propagation in a fiber. This problem is discussed in Section 10.3. Dispersion is measured in nanoseconds/kilometer of fiber length, and may be as low as 0 or as high as 10 ns/km or more, depending on the fiber type. A dispersion of 1 ns/km is equivalent to a 1-Gb/s data rate over a 1-km length of cable for a 50% RZ line code, or to a 500-Mb/s data rate for an NRZ line code. A 10-km length of such a cable thus will have a 100-Mb/s bandwidth if it uses a 50% RZ line code. The intersymbol interference resulting from dispersion thus may be a greater limitation on cable length than is attenuation.

Other types of dispersion also exist. Material, or chromatic, dispersion results from the nonlinear dependence of the fiber's refractive index on wavelength. Thus, chromatic dispersion produces pulse spreading that is proportional to wavelength. All pulses, of course, have a finite bandwidth proportional to their duration. The higher-frequency components of the pulse travel faster than the lower-frequency components. In addition, fibers exhibit a degree of amplitude nonlinearity. The index of refraction is higher at the pulse's peak and lower at its less intense tail. As a result, the peak is retarded compared with the tails.

10.2.3. Soliton Transmission

A *temporal soliton* is a pulse of light that resists a medium's chromatic dispersion,[7] that is, the tendency of the medium to broaden the pulses in time. This ability to resist is achieved by using the fiber's chromatic dispersion to cancel its amplitude nonlinearity. Thus, the faster travel of the higher frequencies, tending to spread the pulse, is countered by the fiber's nonlinearity, which tends to slow the higher frequencies. The result is cancellation of the chromatic dispersion.

The silica glass of optical fibers has properties that suit it well to soliton transmission. Moreover, solitons can be amplified repeatedly, if the distance between amplifiers is short compared to the distance over which dispersion and nonlinearity change the pulse's shape.

10.3. FIBER TYPES

Optical fibers may be made, not only from glass, but also from plastic. The plastic fibers are useful only for short distances. They have diameters on the order of 1 mm or more and are well suited for use with LEDs. They are useful in rugged environments (for example, automotive) but are not found, in general, in telecommunications applications. Plastic-clad silica fibers generally have a diameter less than 600 μm and are used with inexpensive emitters such as LEDs. Because of their large diameter, they allow a large number (>1000) of propagation paths, or modes, and thus are used only for relatively short distances.

Glass fibers have a diameter of less than 200 μm. They are classified as multimode, graded index, and single mode, with the single-mode fibers having a diameter on the order of only 5 μm.

An important factor in fiber optics is the index of refraction n of the fiber. It is defined as the ratio of the velocity of light in vacuum to the velocity of light in the medium. Thus, if a fiber has an index of refraction of 1.5, it will carry light at a speed of $3 \times 10^8/1.5$ or 2×10^8 m/s. Since the velocity of propagation also equals $1/\sqrt{\mu\varepsilon}$, where μ is permeability and ε is permittivity, the value of n also equals $\sqrt{\varepsilon_r}$, where ε_r is the relative permittivity, or dielectric constant. In the foregoing example, the dielectric constant is 2.25.

Light-wave propagation in a step-index fiber occurs as illustrated in Fig. 10.4. The core glass has an index of refraction of n_1 whereas the cladding has an index of refraction of n_2. For total internal reflection at the core-cladding interface, the light flux must have an entrance angle θ_0, relative to the axis of the fiber, that is less than the critical angle θ_c, where

$$\sin \theta_c = \sqrt{n_1^2 - n_2^2} = \text{numerical aperture (NA)} \tag{10.1}$$

A fiber with a small NA thus requires a source with a narrow-output beamwidth, such as a laser, for low coupling loss, whereas a fiber with a high NA can use an LED (broad-output beamwidth) as a source. Single-mode fibers tend to have low NAs while graded-index and multimode fibers tend to have high NAs.

The path length of a ray that passes through the fiber's axis, called a *meridional ray*, then is $L \sec \theta$, where L is the axial length of the fiber and θ is the entrance angle of the ray relative to the fiber axis. The problem of differential delay, and thus pulse dispersion, results partly from the dependence of path length on θ.

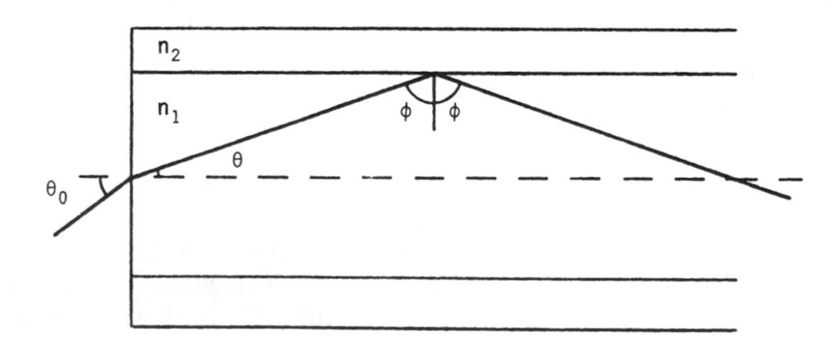

FIGURE 10.4. Propagation of rays in a step-index fiber.

Figure 10.5 shows the cross-sections of different fiber types as well as profiles of their refractive indexes. The single-mode step-index fiber has the lowest dispersion and is very useful for long-haul telecommunication applications. It works well with laser sources. The multimode step-index fiber is most useful for distances up to several kilometers in systems using light-emitting diode sources rather than lasers. The graded index fiber finds common use in systems from several kilometers length to repeatered systems, such as cable television, as well as some long-haul systems. The doubly clad fiber is useful in secure applications.

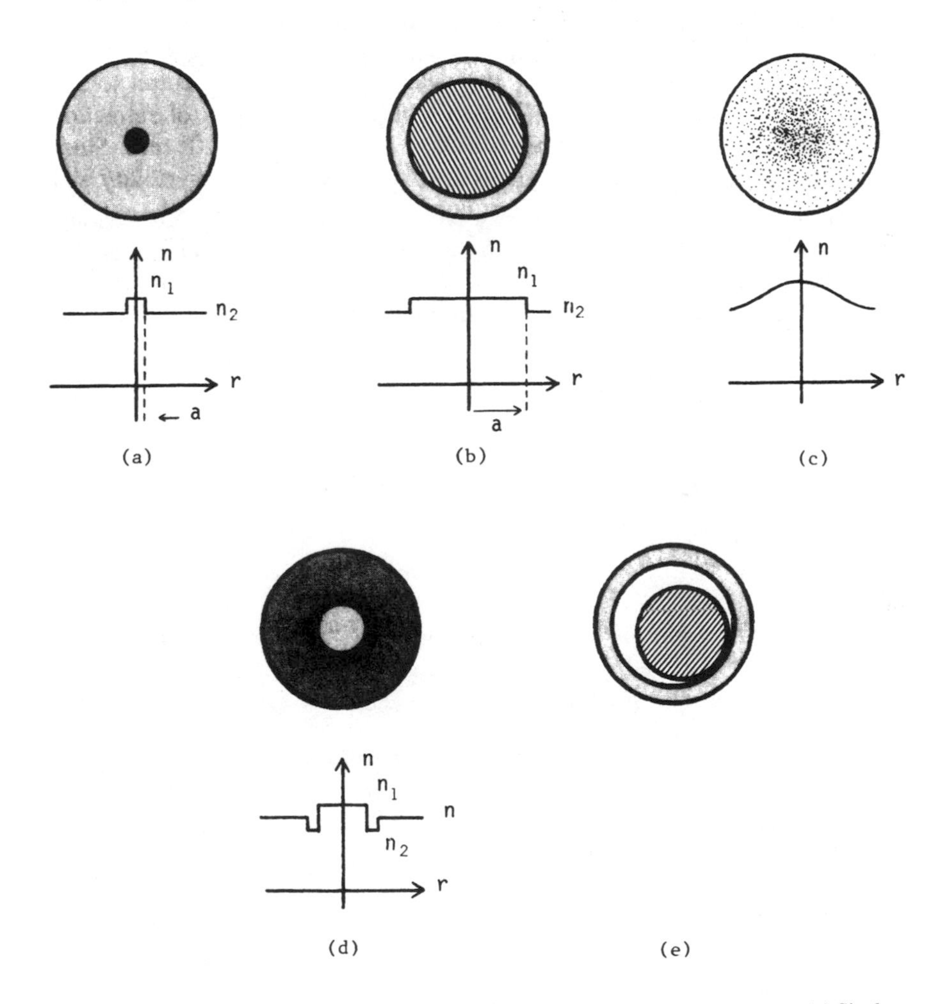

FIGURE 10.5. Cross-sections of fiber types and profiles of their refractive indexes. (a) Single-mode step-index fiber (b) Multimode step-index fiber (c) Graded index (parabolic fiber) (d) Doubly clad fiber (e) Plastic clad fiber. (Courtesy Dr. Leonard Bergstein.)

The plastic-clad fiber is used where rugged physical characteristics are important over short distances, as in automotive applications.

10.3.1. Multimode Fibers

The multimode fiber has a diameter in the 600-μm range. Figure 10.6 illustrates the light paths in a multimode step-index fiber. The group-delay dispersion in a multimode fiber results from four factors: intermodal dispersion, material dispersion, amplitude nonlinearity, and waveguide dispersion. Intermodal dispersion is the group delay spread that results from a variation in group delay between the different propagating modes. This is the dispersion that occurs because the path length of a ray that propagates at an angle θ relative to the fiber axis is $L \sec \theta$ rather than L.

Chromatic dispersion is the group delay resulting from the nonlinear dependence of the fiber's refractive index on wavelength. This chromatic dispersion is proportional to the spectral width of the optical source. Here again the laser is preferable to the LED because the laser has a more narrow spectrum and thus less chromatic dispersion. Typical spectral widths are:

White light:	4000 Å
LED:	350 Å
Injection laser:	15 Å

where 1 Å $= 10^{-10}$ m.

Amplitude nonlinearity causes the fiber's index of refraction to be higher at the pulse's peak than at its tails. The peak thus is retarded compared with the tails. Waveguide dispersion is the group delay resulting from the differing dispersions of each propagating mode. It is a relatively small effect. The foregoing dispersion types also are found in single-mode and graded-index fiber, but inter-

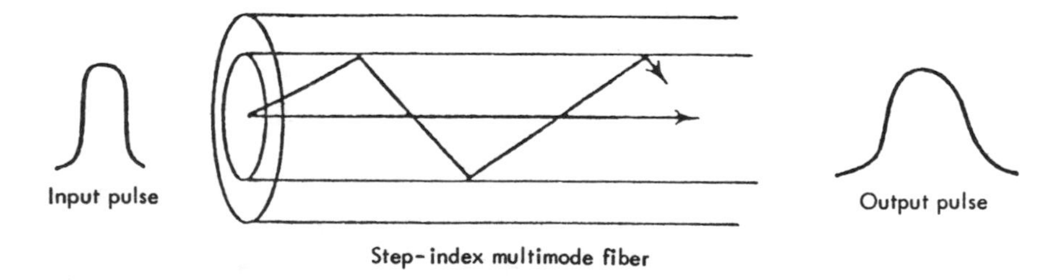

FIGURE 10.6. Paths of light rays in multimode step-index fiber. (Courtesy Dr. Leonard Bergstein.)

modal dispersion and waveguide dispersion are less in magnitude than in the multimode fiber. Chromatic dispersion is less for a laser source than for an LED.

10.3.2. Single-Mode Fibers

The single-mode fiber, also known as monomode, has a diameter of only 5 to 10 μm. This makes it more difficult to splice than the larger types. Its dispersion is very low, however, allowing rates of 10 Gb/s and higher to be sent over distances well beyond 100 km. It has a very low NA, however, which means that a highly coherent high-radiance source, such as a laser or a laser diode, must be used to drive it. Fig. 10.7 illustrates an idealized light path in a single-mode fiber.

Because the single-mode fiber has a greater data rate–distance product than the other fiber types, it is the preferred fiber for long-haul telecommunication applications. Loss as low as 0.2 dB/km has been achieved at 1.55 μm using a silica-based single-mode fiber.[8] The actual choice of wavelength, however, is governed not only by loss but also by dispersion characteristics and the availability of sources and detectors for a given wavelength. For example, a chromatic dispersion null can be achieved at 1.30 μm, making this wavelength attractive even though the losses are somewhat higher than at 1.55 μm.

Single-mode optical fibers are being used in the submarine cable systems installed across the Atlantic and Pacific Oceans. The Atlantic Ocean system is known as the SL Undersea Lightguide System,[9] or TAT-8; it uses single-mode fibers to carry 280 Mb/s at 1.30 μm with repeater spacings of 35 km. The use of DSI with ADPCM allows over 35,000 two-way voice channels to be obtained. The cable contains a total of 12 fibers. The Pacific Ocean system[10] was planned by the Kokusai Denshin Denwa Company (KDD) and extends from Japan to the west coast of the United States.

10.3.3. Graded Index Fibers

The path of the rays in a graded-index fiber is shown in Fig. 10.8. The key feature of the graded-index fiber is essentially the same travel time for the various modes. Thus the effects of dispersion are minimized. This is achieved by causing the rays

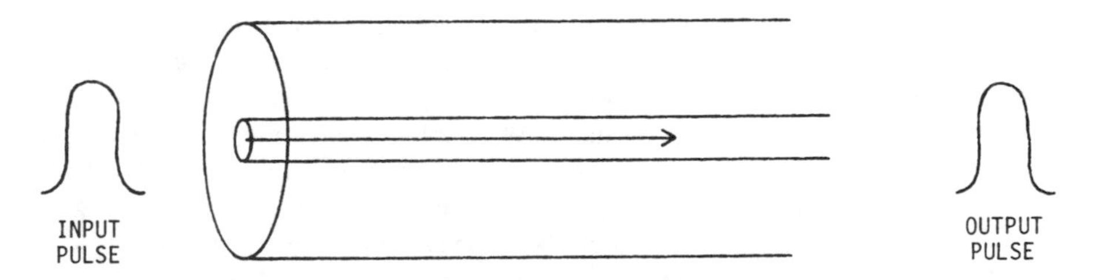

FIGURE 10.7. Path of light rays in single-mode step-index fiber. (Courtesy Dr. Leonard Bergstein.)

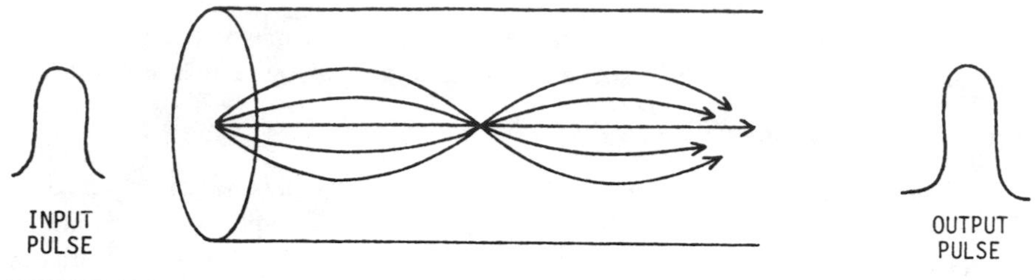

FIGURE 10.8. Path of light rays in graded-index fiber. (Courtesy Dr. Leonard Bergstein.)

to travel faster near the edge of the fiber by making the refractive index lower there. The result is a light path that is almost sinusoidal. With a graded-index fiber, the transmission of 1 Gb/s to a distance of 40 km is possible. Graded-index fibers are made of pure silica and silica doped with boron, germanium, or phosphorus. The core diameter typically is 50–60 μm, with a cladding diameter of 125 μm. As many as 700 modes may propagate in such a fiber.

10.4. OPTICAL SOURCES

10.4.1. Types

The most commonly used sources in optical-fiber transmission are the laser and the light-emitting diode. Fig. 10.9 is a sketch of an injection laser. These devices are capable of output power levels of 5–100 mW on a continuous basis. The modulation of laser diodes is considered feasible at rates as high as 10 GHz for the GaAlAs type, and to 20 GHz[11] and higher for the InGaAsP type. Above 2.5 GHz, however, performance is limited by "chirp," a condition in which the frequency shifts slightly as the laser is modulated, due to a minute shift in its dielectric constant.

The line width of a laser diode is less than 20 Å. Their projected lifetime is on the order of 100,000 to 700,000 hours.[12] Lifetime depends on operating-current density, with the degradation rate increasing with current density. Ambient temperature also is a factor. Degradation is a bulk phenomenon. Over a period of time, strain fields result in the formation of nonradiative recombination centers.

Injection laser diodes are available using the materials shown[13] in Table 10.1.

The narrow beamwidth and narrow bandwidth output of the laser suit it well to long distance (>100 km) and high-data-rate (>2 Gb/s) applications.

Light-emitting diodes (LEDs) are in widespread use for short-distance low-data-rate applications. Their beamwidth is appreciably larger than that of the laser. Table 10.2 compares lasers and LEDs with respect to several basic parameters,

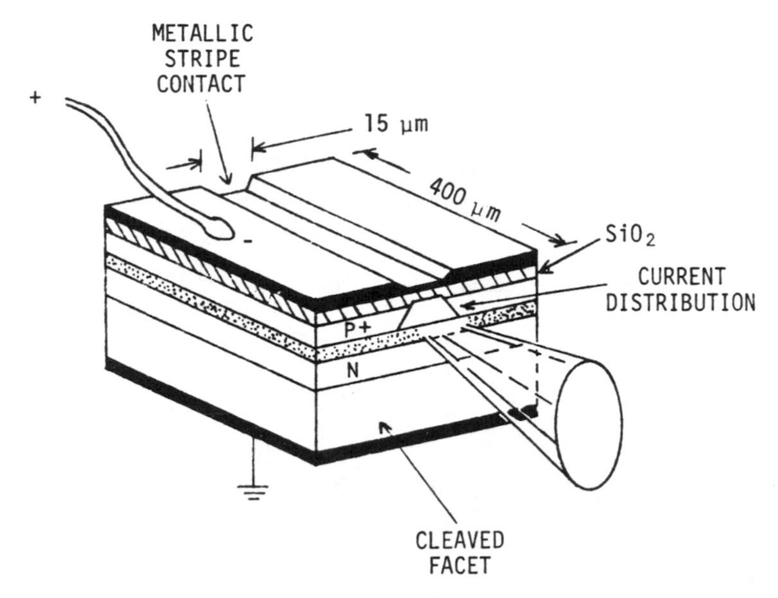

FIGURE 10.9. Injection laser. (Courtesy Dr. Leonard Bergstein.)

for a wavelength of 800–850 nm. The ranges of values indicate variations among units of differing designs and output levels.

During the 1980s, commonly used lasers operated on a noncoherent basis, that is, with a line width greater than the information bandwidth. Coherent lasers now are appearing, as discussed in Section 10.4.2.

10.4.2. Coherent Sources

A coherent optical source, together with a monomode fiber and a coherent receiver, constitute a coherent optical communication system. Light is treated as a carrier that can be modulated in amplitude, frequency and/or phase. Figure 10.10[14] shows the basic configuration of a coherent optical system. Note the use of a laser control function at the transmitter as well as a local oscillator control function at the receiver. An optical amplifier (see Section 10.7) may be added to the transmitter output to compensate for modulator power loss as well as to provide a suitable amount of input power to the fiber being used.

Coherent optical systems exhibit numerous advantages. First, their relatively narrow source bandwidth allows a correspondingly narrow receiver bandwidth with its obvious link budget advantages. In addition, the wavelength-division multiplexing (WDM) of a large number of carriers becomes feasible,[15] as illustrated in Fig. 10.11. Moreover, the power-handling level in single-mode fibers

TABLE 10.1. Injection Laser Diode Materials

Wavelength Region	Material
800–900 nm	Alloys of gallium arsenide and aluminum arsenide
1300 nm	Alloys of indium, gallium, arsenic, and phosphorus
1550 nm	Alloys of indium, gallium, arsenic, and phosphorus

TABLE 10.2. Comparison of Laser and Light-Emitting Diodes[a]

Parameter	Laser	LED
Power	5–100 mW	1–10 mW
Modulation	1–25 GHz	≤200 MHz
Projected lifetime	10^5–10^6 hours	10^5–10^6 hours
Coupling loss to 0.14 NA fiber	–2 to –3 dB	–10 to –17 dB
Line width, $\Delta\lambda$	1–20 Å	200–400 Å

[a]Wavelength: 800–850 nm

using angle (frequency or phase) modulation is improved. Further advantages are discussed in Section 10.5.2.

Fig. 10.12 illustrates the basic differences between a noncoherent ("direct detection") and a coherent ("FSK heterodyne") lightwave system. Coherent optical systems are more complex than noncoherent systems because the transmitter and local oscillator lasers must have very narrow spectral linewidths. In addition, the polarizations of the two must match at the receiver.

Fiber nonlinearities set an upper limit on the power levels that can be transmitted without encountering such impairments as crosstalk between channels, power losses, and unwanted phase fluctuations.[16] Accordingly, nonlinearities set a limit on the transmitted power, the number of allowed channels, and the channel allocations that can be used. As a result, transmitted powers generally are on the order of milliwatts.

The capacity (Gb/s-km) of fiber systems has doubled approximately every year.[17,18] Fig. 10.13 shows the trend in coherent system capacity as a function of time.

10.4.3. Modulation

Intensity modulation of a laser can be used to transmit information digitally. A threshold dc bias is used on the laser to obtain a modulation response in the subnanosecond range. Thus two amplitude levels are used, one to represent 0 and the other to represent 1. This is called direct modulation. Alternatively, a laser diode can be operated at constant output and an electro-optic crystal such as

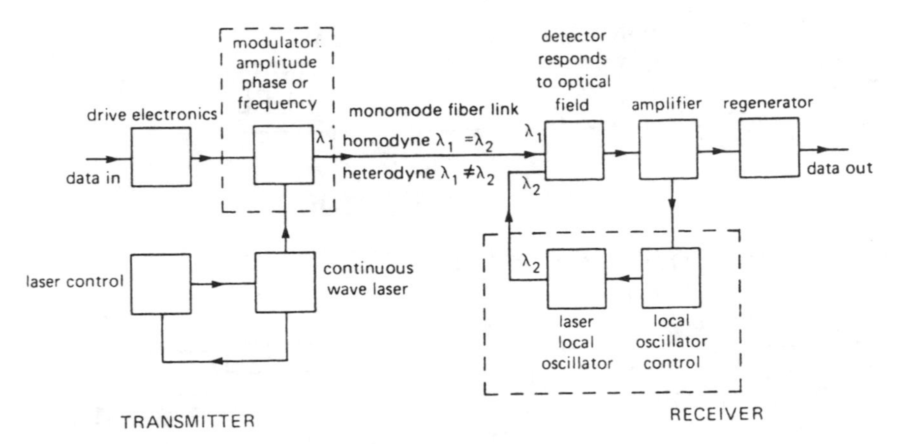

FIGURE 10.10. Coherent optical communication system. (Courtesy I. W. Stanley, Ref. 14, ©IEEE, 1985.)

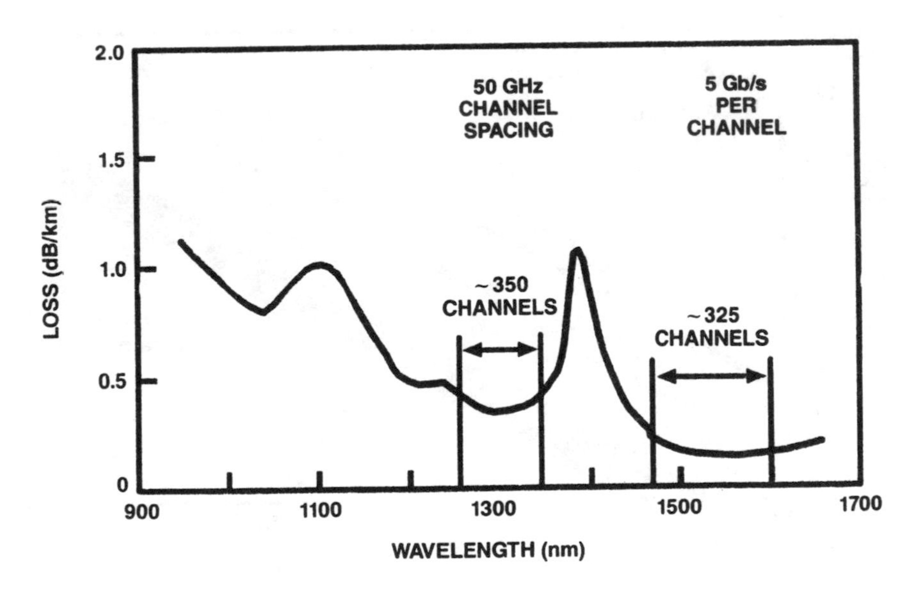

FIGURE 10.11. Information capacity of single-mode fiber. (Courtesy R. S. Vodhanel and R. E. Wagner, Ref. 15, ©IEEE, 1989.)

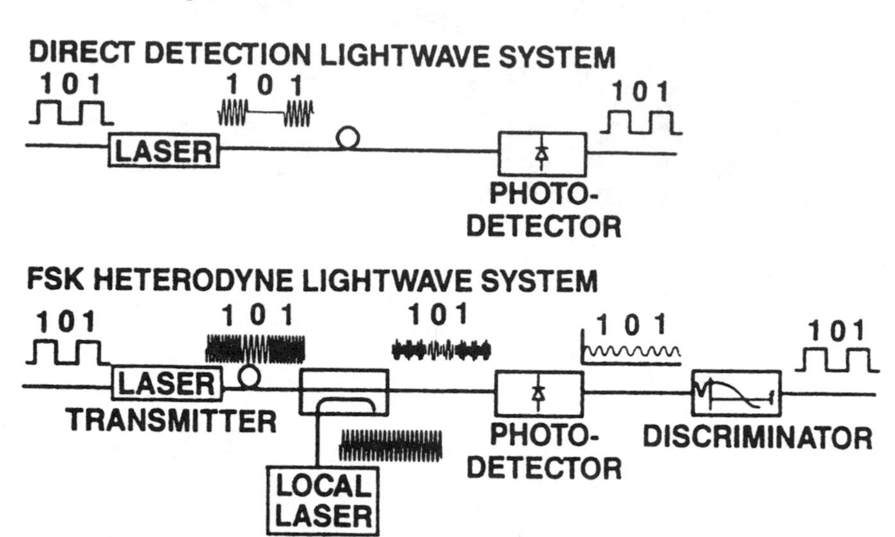

FIGURE 10.12. Comparison between direct detection and coherent lightwave systems. (Courtesy R. S. Vodhanel and R. E. Wagner Ref. 15, ©IEEE, 1989.)

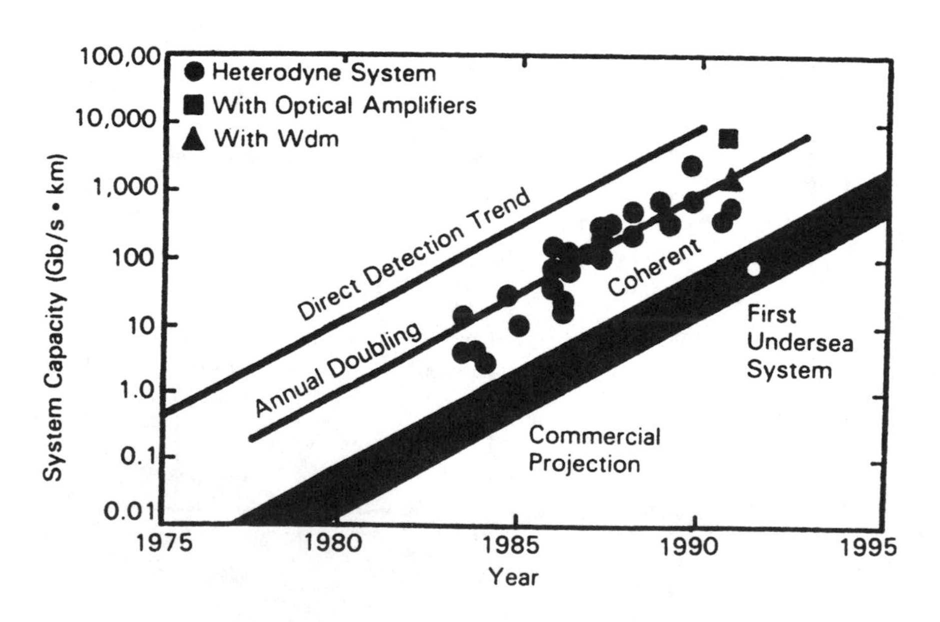

FIGURE 10.13. Trend in coherent system capacity. (Courtesy R. E. Wagner and R. A. Linke, Ref. 17, © IEEE, 1990.)

lithium niobate ($LiNbO_3$) or lithium tantalate ($LiTaO_3$) can be used to modulate the polarization of the light output externally. The bits 0 and 1 then can be represented by the two orthogonal linear polarizations or by the opposite-sense circular polarizations.

Fig. 10.14 illustrates the linewidths of optical sources.[19] Shown at the top of the figure is the spectrum of a multilongitudinal mode (MLM) laser, as has been used for systems operating at rates below 565 Mb/s. Higher-rate systems generally use a single longitudinal mode (SLM) laser with distributed feedback (DFB) resonators. The spectrum is as illustrated in the middle sketch of Fig. 10.14, but the line width may be as large as 0.7 nm (≈85 GHz). Such systems can support a 100-km link at 10 Gb/s.[20]

Coherent modulation of a laser is defined as modulation in which the laser line width is less than the width of the modulation sidebands. Actually, the laser line width can be up to 50% of the modulation bandwidth with essentially no degradation in detection sensitivity. A laser using a very long, that is, 10-cm

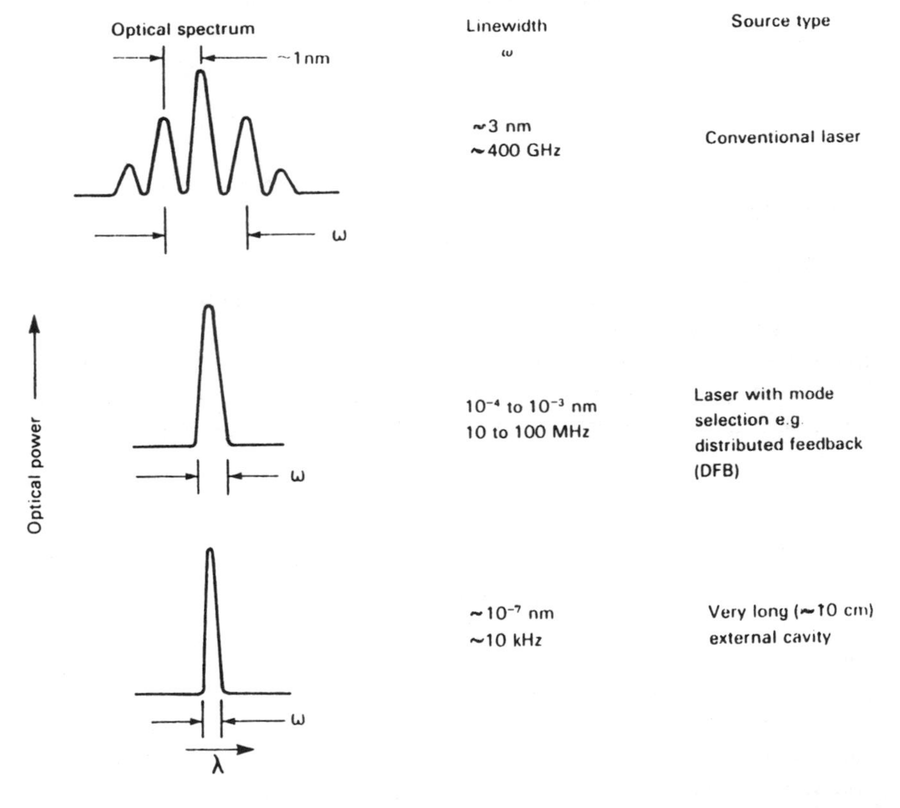

FIGURE 10.14. Linewidths of optical sources. (Courtesy I. W. Stanley, Ref. 14, ©IEEE, 1985.)

external cavity (see Fig. 10.14), typically meets such a requirement. These lasers often are directly FSK modulated, or but may be externally ASK or DPSK modulated instead. Angle modulation is the preferred approach for coherent optical systems because higher-level amplifiers can be used than for amplitude modulation and because of the good receiver sensitivities possible.

A significant limit on modulation rate occurs in the fiber in the form of chromatic dispersion. For example, a carrier that has been coherently amplitude modulated will have two sidebands in phase. As the two sidebands and the carrier propagate down the fiber, chromatic dispersion causes these components to drift apart in phase. Once the sidebands are 180° apart, the amplitude components have cancelled, preventing recovery of the baseband information.

10.4.4. Coupling of Sources to Fibers

The coupling loss from a light source to a fiber is a function of the NA of the step-index fiber and the beamwidth of the laser. Thus the injection laser, with its relatively narrow beam directed along the fiber axis, exhibits the smallest coupling loss, often only several dB. Edge or surface emitters, as well as LEDs, however, produce a relatively broad beam, and thus their coupling losses to the fiber are greater.

10.5. PHOTODETECTORS

A photodetector must have a high response to incident optical energy, an adequate instantaneous bandwidth to respond to the information bandwidth on the optical carrier, and a minimum of internal noise added to the detected signal. In addition, a photodetector should not be susceptible to changes in environmental conditions, especially temperature. Photodetectors are used as both receivers, and in the monitoring of source output to control source level through feedback.

10.5.1. Noncoherent Detectors

A noncoherent optical detector may be a photoemitter, such as the photomultiplier; a photoconductor; or a photodiode, such as an avalanche device. The photodiodes are basically reversely biased *p–n* diodes. The *p–i–n*-type detectors convert optical power directly into electric current. Avalanche photodetectors (APDs) provide an additional internal gain on the order of 10 to 100. For both types, the minimum detectable signal is determined by the level of internally generated noise. In the case of the *p–i–n* diodes, the noise is primarily *shot noise*, resulting from random thermally generated currents. The main source of noise in APDs is the probabilistic avalanche process, in which pairs of photogenerated carriers undergo multiplication. Signal amplification results from this process.[21] Fig. 10.15 compares the minimum detectable *p–i–n* and APD optical power levels (time average) as a function of data rate, with minimum power being defined as the value needed to obtain $p_{be} = 10^{-9}$. The minimum power increases with data rate because binary coding is assumed, and the noise level increases with bandwidth.

10.5.2. Coherent Detectors

A coherent detector uses the *heterodyne* or *homodyne* principle, illustrated in Fig. 10.16. Heterodyne detection involves the use of a local oscillator at a frequency different from that of the signal, whereas homodyne detection uses a local oscillator whose frequency is identical to that of the incoming signal. Heterodyne detection is the type being implemented for operational systems because it imposes less stringent laser linewidth requirements. As modulation rates continue

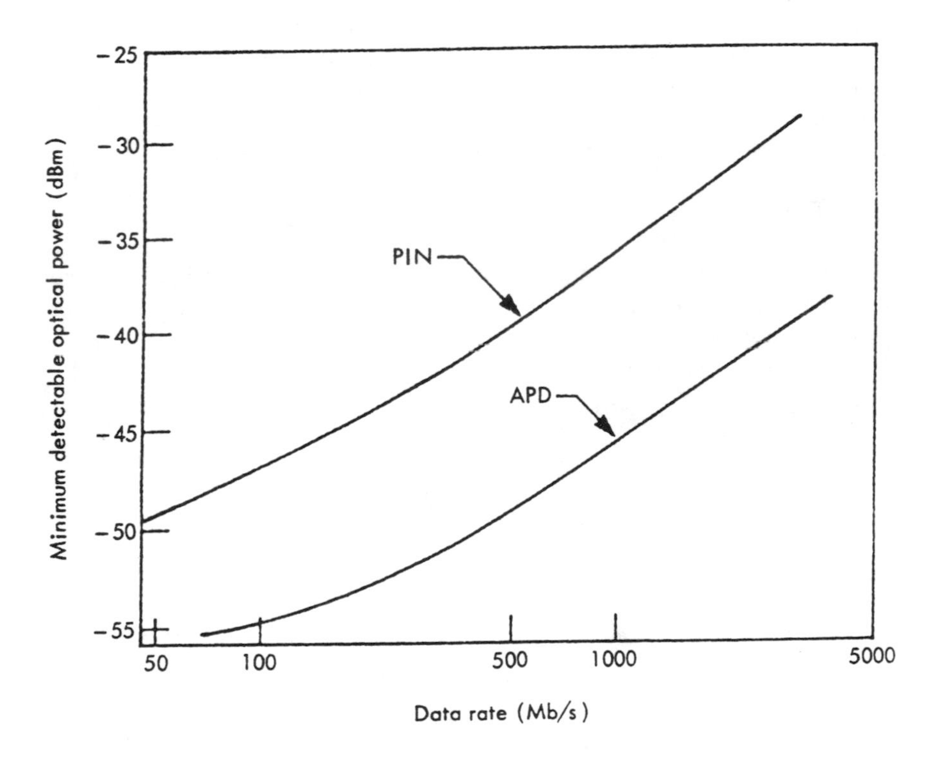

FIGURE 10.15. Minimum detectable optical power for $p_{be} = 10^{-9}$. (Copyright © 1987 AT&T. Reprinted with permission.)

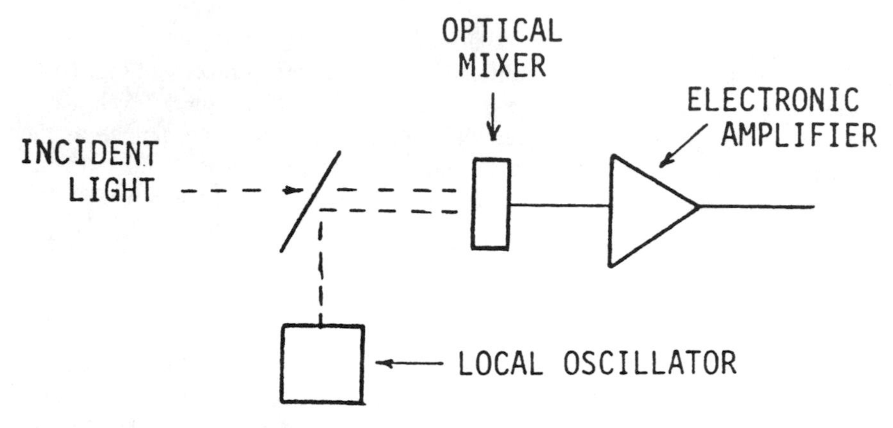

FIGURE 10.16. Heterodyne or homodyne detection. (Courtesy Dr. Leonard Bergstein.)

to increase, however, homodyne detection becomes more attractive, with its more limited optical bandwidth (baseband operation).

To accommodate polarization shifts that may occur within the fiber as a function of time, the receiver is designed for polarization diversity. The local oscillator is tunable, and frequency-locked to the desired channel. This can be achieved through the use of a three-section distributed Bragg reflector (DBR) laser. The intermediate-frequency electronics consists of microwave devices.

10.6 REPEATERS

Repeaters for fiber-optic systems operate by detecting the optical pulses, regenerating them, and then modulating a new light source, as indicated in Fig. 10.17.

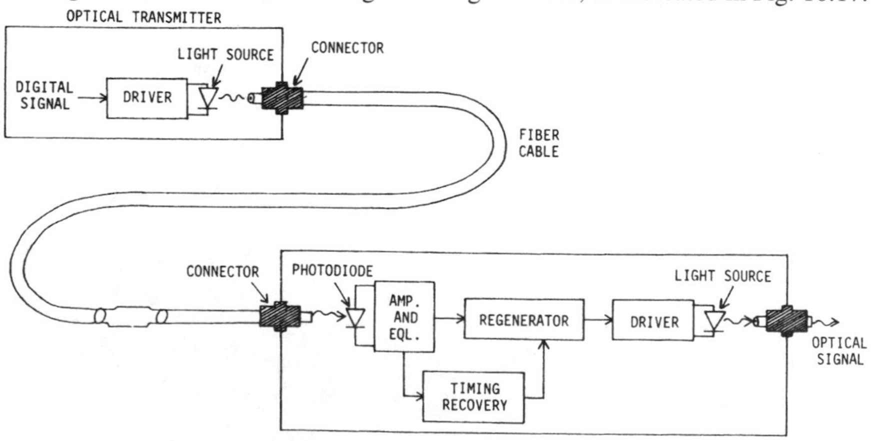

FIGURE 10.17. Use of repeater in fiber-optic link. (Courtesy Dr. Leonard Bergstein.)

Power for such a repeater must be obtained from the central office by means of wire cable paralleling the optical fiber, or may be delivered to the repeater site by the local power company; in this case, an on-site battery backup is required for use in the event of a power outage. Solar powering of the repeater is another alternative, but on-site solar arrays of sufficient size may be subject to vandalism. The distance between repeaters is governed by power budget considerations. Specifically, Figs. 10.2 and 10.15, respectively, indicate the fiber attenuation and receiver sensitivity that may be encountered. From this information, the maximum tolerable attenuation, and thus the maximum distance between repeaters, can be determined. For a high-data-rate system, a fiber's dispersion rating in GHz-km also is a factor in establishing the allowable distance between repeaters.

Regenerative repeaters have been built to operate at rates as high as 10 Gb/s.[22]

10.7. OPTICAL AMPLIFIERS

Regenerative repeaters allow a system to be extended over considerable distances, as has been described in Section 10.6; however, the regenerative repeater is designed for a specific data rate and line code. Thus, its use is not conducive to system upgrades. The optical amplifier, while doing nothing to improve the waveform, provides gain to counter the fiber losses incurred over a long span, especially important in high-data-rate systems whose large bandwidths result in significant system noise levels. In systems whose dispersion is adequately low, use of the optical amplifier is attractive because it does not act as a deterrent to system upgrades in data rate or in the addition of more channels at different wavelengths (see Section 10.10). Flexibility in such parameters is especially important if the repeater is not readily accessible, as is the case in a submarine cable.

An optical amplifier can be built by using a section of erbium-doped fiber (for example, seven to ten meters long in the optical path), to which a pump laser is coupled. The pump wavelength may be in the 980–1500-nm wavelength range. Pump-power levels range from 10–50 mW. The gain and noise figure of the amplifier depend on the pump wavelength and power. Figure 10.18 shows an erbium-doped fiber amplifier and the energy levels associated with its operation. The signal at a 1.55 μm wavelength encounters erbium atoms within an optical fiber that are excited to higher energy levels by a pump at 1.48 μm. The result is that the power of the signal gradually increases along the optical fiber by stimulated light emission. The pump may be a laser diode. Single-mode fibers are used, and a multilayer thin-film filter, coupler, and isolator complete the amplifier.

In general, the pump causes the energy level in the erbium atoms to be increased above the ground state, and then to decay to an intermediate metastable state. Then, the arrival of a photon, for example, in the 1525–1560-nm range, causes coherent radiation of the metastable electrons, with consequent optical

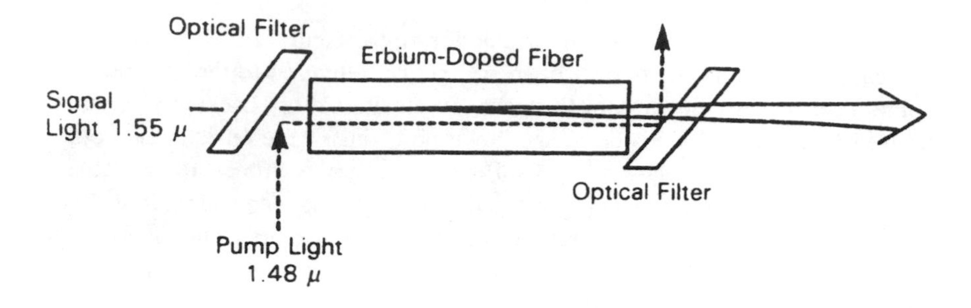

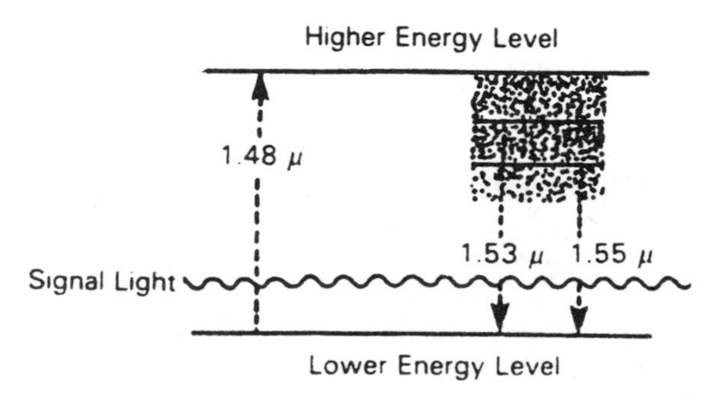

FIGURE 10.18. Erbium-doped fiber amplifier. (Courtesy K. Nakagawa and S. Shimada, Ref. 24, © IEEE, 1990.)

amplification.[23] Gains of 30 dB and noise figures of 3 dB have been achieved. Coupling loss is less than 0.5 dB. The amplifiers are polarization insensitive.

Other optical amplifiers are the semiconductor laser diode, the Raman amplifier, and the Brillouin amplifier. Table 10.3[24] summarizes the key features and application areas of the various optical amplifier types.

Twenty-five erbium-doped amplifiers have provided transmission of 2.5 Gb/s over 2200 km of single-mode fiber. An undersea fiber cable using erbium-doped amplifiers and linking the United States and Japan is planned for 1996. New transoceanic fiber-optic systems planned for the mid-1990s will use optical amplifiers and have capacities of 300,000 simultaneous calls per fiber pair.[24]

10.8. NOISE SOURCES

The primary noise type affecting infrared systems is *photonic*, or *quantum* noise. This noise is proportional to hf, where h is Planck's constant, equal to 6.626×10^{-34} joule-seconds and f is the frequency in Hz. For example,[25] a 1.55-μm signal in fiber has a frequency of $2 \times 10^8/1.55 \times 10^{-6} = 129$ THz, and the quantum noise is $6.626 \times 10^{-34} \times 129 \times 10^{12} = 8.547 \times 10^{-20}$ W/Hz. If the receiver must accept a 100-MHz line width, the noise level is $8.547 \times 10^{-20} \times 100 \times 10^{6} = 8.547 \times 10^{-12}$ watts = 8.547 pW.

By comparison, thermal noise at room temperature (290 K) decreases significantly above 3 THz, at which it equals quantum noise. At 29 K, thermal noise decreases above 300 GHz, at which it equals quantum noise. Fig. 10.19[26] shows how thermal noise at various temperatures compares with quantum noise up to 10 THz, beyond which quantum noise is the predominant type. At and below room temperature, the noise limit clearly is quantum rather than thermal.

Direct-detection receivers usually require received-signal levels 15–20 dB above the quantum limit. This is especially true of the p-i-n diodes, which suffer from dark-current noise, as well as quantum noise. Other noise types include shot noise generated in the preamplifier and speckle or modal noise resulting from source polarization fluctuations. Modal noise is negligible for a single-mode fiber if no polarization-dependent components are in the line.

Coherent detectors can provide sensitivities close to the quantum limit.

10.9. TRANSMISSION SYSTEMS AND SUBSCRIBER LINES

Optical-fiber transmission systems exhibit two fundamental types of limits: loss and dispersion. Fiber loss is discussed in Section 10.2.1, and dispersion in Section 10.2.2. The loss problem must be viewed in connection with the system noise level. Section 10.8 treated the noise types found in photodetectors, together with the basic quantum and thermal noise limits.

In a *loss-limited system*, the losses of the fiber, connectors, and splices equals the difference between the transmitter output and the effective receiver sensitivity,

TABLE 10.3. Optical Amplifiers and Applications

Optical Amplifiers	Features	Application Area
Semiconductor laser-diode	Compact (<mm) Varying wave-band (0.8–1.6 μm) Wide bandwidth (10 THz)	Optical signal processing Optical integration Future multichannel system
Erbium-doped fiber	High gain at 1.5 μm Large relaxation time Matched to fiber	Optical signal transmission In-line amplifier Ring laser
Nonlinear fiber Raman amplifier	Distributed amplifier High pump power required (>100 mW)	Soliton transmission
Brillouin amplifier	Narrow bandwidth ($\approx$100 MHz)	Fiber characterization

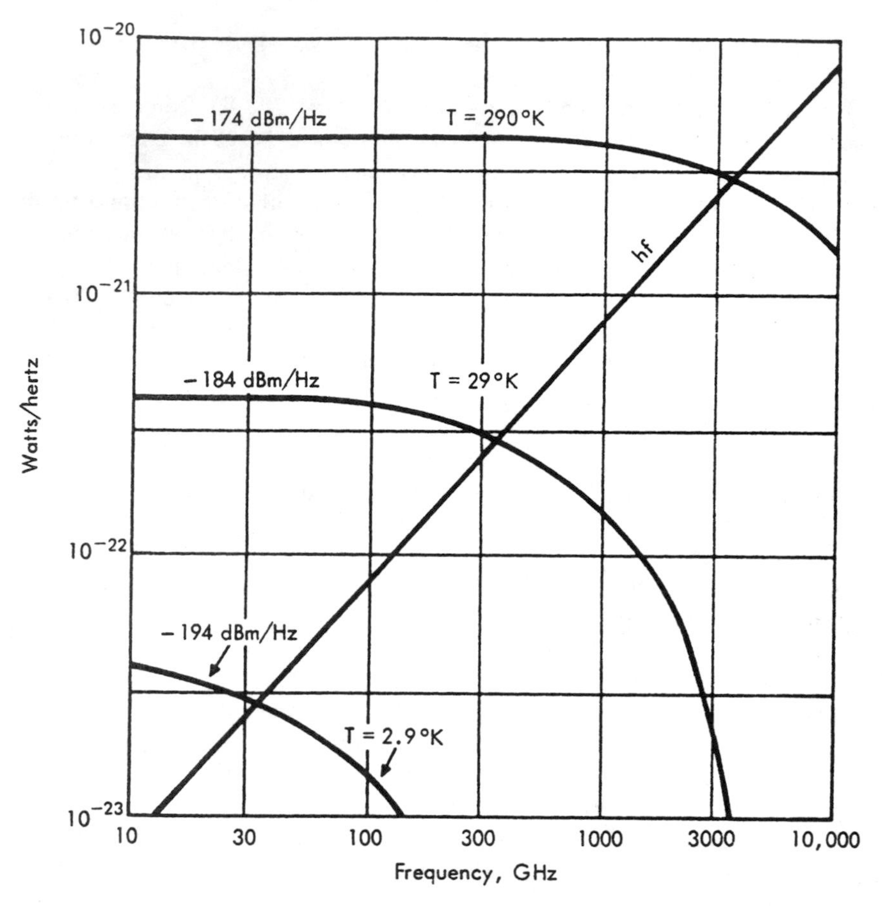

FIGURE 10.19. Comparison of thermal and quantum noise levels. (Copyright 1971, American Telephone and Telegraph Co. Reprinted with permission.)

which depends upon system noise. In a *delay-distortion-limited system*, pulse-spreading causes unrecoverable interference between pulses such that the loss limit is not reached.

At the wavelength of minimum loss, 1.55 μm, a DFB laser with a 0.7-nm line width can provide a 100-km distance between repeaters at a 10-Gb/s rate. Above this rate, the fiber's dispersion limit is reached and narrower line widths are required to maintain the distance limit. At the wavelength of zero dispersion, 1.31 nm, link losses become limiting instead; this is also true for a dispersion-shifted fiber at one of its zero dispersion wavelengths.

Span lengths for coherent detection systems can be much greater because of their better sensitivity and narrow spectrum. Coherent systems may achieve a 50-Gb/s rate over a 150-km span.[27]

10.9.1. Typical Transmission Systems

Fiber telecommunication lines now connect most major metropolitan centers with one another throughout North America, Europe, parts of the Mideast, Japan, and Australia, including an increasing number of transoceanic lines. Extensions to metropolitan centers elsewhere in the world are either under construction or being planned.

10.9.2. Subscriber Systems

The replacement of major portions of copper subscriber-loop plant with fiber is under active consideration throughout the developed world, for the removal of existing bandwidth limitations. Such replacement largely is a matter of economics. While fiber often can be justified economically in interoffice applications, for backbone feeders, and for serving large businesses, the same is not true, at least on a first-cost basis, of extensions to residential and many small business subscribers. The problem is that fewer opportunities for economies of scale exist in the distribution plant.

Bell Communications Research (Bellcore) has developed plans whereby LECs can deploy fiber in the loop (FITL) systems in the distribution plant.[28] Figure 10.20 is a generic view of an FITL system supporting existing telecommunications services with a potential overlay of broadcast video capability. The remote digital terminal (RDT) hosts multiple optical network units (ONUs). The ONUs are positioned at or near customer locations. The RDT and ONUs communicate over an optical path through a passive distribution network (PDN). The PDN consists of optical-fiber cables and possibly some passive optical multiplexers laid out by the telephone company based on local requirements. The ONUs and the corresponding terminations on the host RDT may be viewed as a form of small digital loop carrier (DLC) system supported over an optical path through the PDN. An RDT and all connected OCUs (via the PDN) is defined as an FITL system.

The FITL system must meet the subscriber, the fiber-optic distribution network, and the larger telephone company transmission networks. FITL systems now being installed or planned take the form of fiber to the curb (FTTC), which offers savings similar to those of DLC, which allows dedicated facilities to be replaced with lower-cost shared facilities. The ultimate objective, of course, is fiber to the home (FTTH). Several additional years will be required, however, before FTTH can compete economically with copper for existing services. With

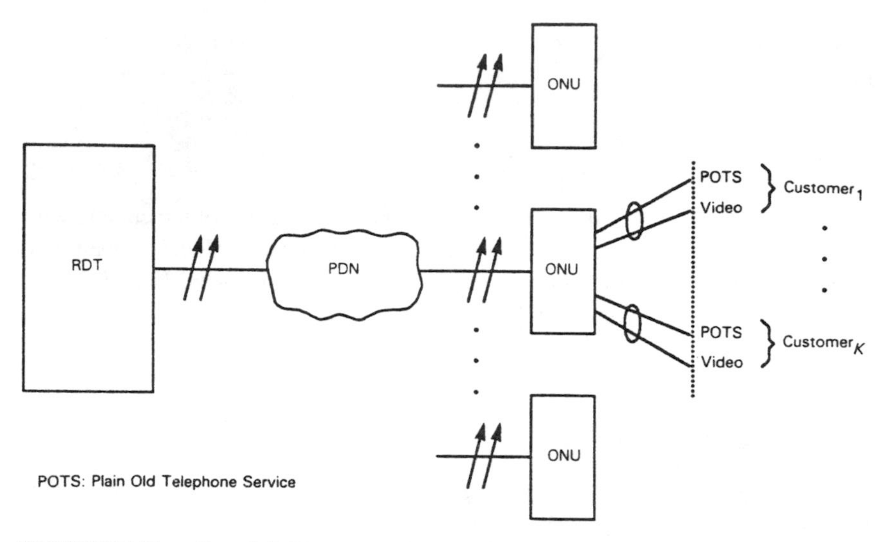

FIGURE 10.20. Generic FITL system. (Courtesy G. R. Boyer, Ref. 28, ©IEEE, 1990.)

FITL, residential broadband access becomes feasible at such internationally standardized rates as 155.52 Mb/s.[29]

Once FTTC has been installed, the cost of upgrading the drop, typically only 30–50 m in length, becomes manageable, and can be done as broadband service needs arise. Moreover, if such fibers are implemented on a single-mode basis, coherent transmission to the home can be installed when available.

Fiber splicing methods have been refined for field use. Splices generally are permanent joints whereas connectors are used where disconnection may be required. The major concern in the use of splices and connectors is the loss that results from such factors as fiber alignment, the end conditions of the fibers, and fiber-core parameters, such as core diameter.

The information carried on the fiber in infrared form is detected by a silicon semiconductor device, such as a P-I-N diode or an APD, as described in Section 10.5. This detection places the information in a standard form for transmission on copper wire.

A further issue relevant to subscriber systems is the desirability of reliability via redundancy, especially since a single construction accident can produce a total service disruption. While ring topologies can be useful in guarding against such outages, subscribers with critical needs may wish to use alternative local access.[30] Alternate-access carriers allow their subscribers to transmit over highly secure fiber, often with diversity routing, to other network locations or to long distance carriers. They provide service via a route that is physically separate from that used by the local telephone company to assure continuity of service by one carrier if the facilities of the other should fail.

10.9.3. Network Compatibility

Not all fiber-optic links are compatible with the digital hierarchy, even though they are used for digital transmission. Many, instead, are designed to provide a connection between a host computer and its remote peripherals. As a result, they will be compatible with transistor-transistor logic (TTL) or emitter-coupled logic (ECL) rather than the bipolar-return-to-zero (BRZ) format used in telephony. Thus a fiber-optic link's compatibility needs to be determined before it is ordered.

Most long-haul optical-fiber systems in the U.S. operate at a multiple of the DS3 rate and interface with other facilities at the DSX-3 cross-connect. European systems generally interface at the European fourth level (139.264 Mb/s), and Japanese systems at the Japanese fifth level (397.20 Mb/s). Generally the North American DSX-4 cross-connect has not been used for optical-fiber systems.[31]

10.10. WAVELENGTH-DIVISION MULTIPLEXING

All fiber-optic transmission involves the modulation of an optical source whose unmodulated output is at a specific wavelength. In the case of a laser, this wavelength may be rather narrowly defined, as was illustrated in Fig. 10.14. Just as microwave and satellite systems use radio channels centered at different frequencies, so optical fibers can carry light of different wavelengths simultaneously. Each wavelength can be modulated separately from the others. This is called WDM. Fig. 10.21 illustrates this concept. The wavelength multiplex coupler consists of wavelength-selective components such as gratings, prisms, or thin-film filters for combining the signals from the input fibers onto the single transmission fiber. The wavelength-demultiplex coupler uses similar components. The couplers work on the prism basis by causing the different wavelengths to be spread out in angle, and thus separated from one another or combined.

WDM offers a simple way to start making effective use of the tremendous capacity of the optical fiber, which is about 30 THz, since the low-loss region of a single-mode fiber extends from about 1.2 to 1.6 μm. Moreover, WDM helps to mitigate the effects of fiber dispersion.[32] By using several relatively low-data-rate channels in parallel, a high overall information rate can be achieved without the dispersion impairments that would be experienced with a single high-rate channel. This approach thus avoids the limits that otherwise may be experienced by a high-rate TDM system.

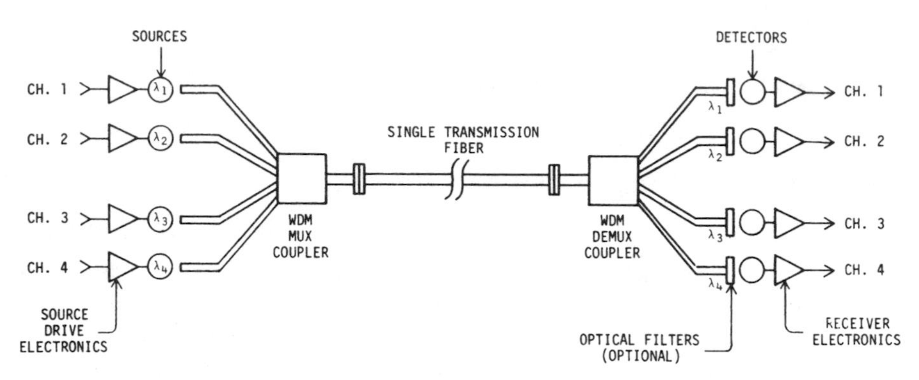

FIGURE 10.21. Wavelength-division multiplexing.

Another advantage of WDM is that it allows users to carry significantly different types of signals on the same fiber. For example, both digital voice and analog video can be delivered to the home on a single fiber. In business applications, numerous users can be interconnected, as is done in local-area networks (LANs). Such an application may involve rapid wavelength switching, and place significant demands on WDM technology.

Also, WDM allows major economies to be achieved in long-haul systems. For example, the capacity of a 1.7-Gb/s system can be doubled to 3.4 Gb/s by adding a 1.5-μm channel to an existing 1.3-μm system.

Advances in optical amplifier technology, as outlined in Section 10.7, make WDM even more attractive in that complicated combinations of regenerators can be replaced with a simple chain of erbium-doped optical amplifiers. Moreover, a system upgrade to a higher data rate can be done with the addition of one or more wavelengths.

Fig. 10.14 illustrates the fact that a variety of optical-source linewidths are available. Use of a linewidth on the order of 1 nm (100 GHz) or less allows the implementation of a *dense WDM* system.[33] The dense WDM system is one whose linewidth, however, still is greater than twice the information bandwidth. For example, if the carrier spacing can be set at 10 GHz, more than 1000 carriers can be transmitted by a single fiber in the 1.55-μm low-loss region. Multichannel frequency stabilization and low-chirp modulation techniques are necessary for the carriers to be densely spaced.[34] (Chirp is a slight shift in line wavelength that occurs upon modulation; it is caused by a slight shift in dielectric constant.) Some authors refer to dense WDM as FDM, but others reserve that term for the use of coherent techniques, as were described in Sections 10.4.2 and 10.5.2.

Dense WDM uses distributed-feedback DBR lasers or other narrow-linewidth lasers, and relatively inexpensive passive components, such as star couplers and fused-fiber multiplexers. It readily allows the transmission of bit rates on the order of 2 Gb/s and higher, and as much as 20 Gb/s in the laboratory. Further increases in such rates are becoming difficult with TDM, but appear relatively easy with dense WDM.

A major feature of dense WDM is that the discrete wavelengths can be regarded as an orthogonal set of carriers which can be separated, routed, and switched without interfering with each other, provided the total light intensity is kept sufficiently low. Such use of wavelength and its processing in passive-network elements distinguish optical networks, in general, from other network technologies.

10.11. THE FIBER DISTRIBUTED-DATA INTERFACE (FDDI)

The fiber distributed-data interface (FDDI) is a local area network (LAN) standard dedicated to the effective use of fiber optics as the transmission medium. The protocol is based on the token-ring access method, in which only one of the attached terminals, the one possessing the "token," is allowed to transmit at any given time, thus avoiding collisions. FDDI has been developed by American National Standards Committee X3-T9, which is chartered to develop computer input/output (I/O) standards.[35]

10.11.1. Basic FDDI

A FDDI network is made up of two counter-rotating token-passing rings, each of which operates at 100 Mb/s.[36] A FDDI LAN can easily extend to 100 km in total length, and include thousands of stations. The use of multimode fiber allows adjacent nodes to be separated by as much as 2 km. The FDDI network's second ring can provide backup in the event of the primary ring's failure. When the second ring is not required for backup, the overall network can support rates up to 200 Mb/s.

FDDI conforms to the open system interconnect (OSI) model of the International Standards Organization (ISO). This model is described in Section 13.3.1.2. The initial version of FDDI uses multimode fiber with LEDs transmitting at 1325 nm. Connections between stations are made using a polarized duplex connector.

Fig. 10.22 shows the functions to be performed in a FDDI station.[37] The components are related in a manner that conforms to both the OSI concept of layering and to IEEE Standard 802, dealing with LANs. Station management (SMT) specifies the local portion of the network management application process, including the control required for the proper internal configuration and operation of a station in a FDDI ring. Media access control (MAC) specifies the lower sublayer of the data-link layer, including access to the medium, addressing, data checking, and data framing. Physical layer protocol (PHY) specifies the upper sublayer of the physical layer, including the encode/decode, clocking, and framing for transmis-

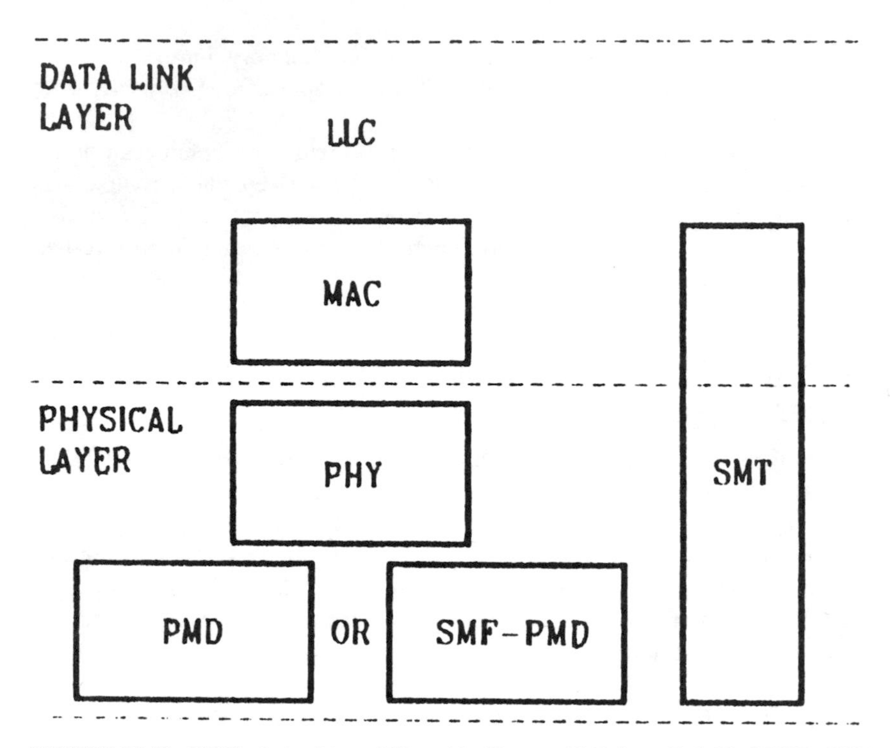

FIGURE 10.22. FDDI relationship to OSI model. (Courtesy F. E. Ross, Ref. 35, ©IEEE, 1989.)

sion. Physical layer medium dependent (PMD) specifies the lower sublayer of the physical layer, including power levels and characteristics of the optical transmitter and receiver, interface optical signal requirements, the connector-receptacle footprint, the requirements of confirming optical-fiber cable plants, and the permissible bit-error ratios. Two alternative versions of PMD are shown: basic PMD and SMF-PMD, which allows the use of single-mode optical fiber.

10.11.2. FDDI-II

FDDI-II is an upward-compatible enhancement of FDDI, adding a circuit-switched service to the existing packet capability. By adding an isochronous data-transmission capability to the network, voice as well as data can be handled. In packet service, the data bytes are placed in frames. Packet lengths may vary. Each packet has beginning and ending marks, as well as the destination address. FDDI packets are called frames.

By contrast, a circuit-switched service provides a continuous connection between two or more stations that has been established by a form of prior agreement.

This prior agreement may take the form of knowing the location of a time slot, or slots, that occur regularly relative to a readily recognizable timing marker. One such common timing marker is the 125-μs clock period used by the public switched network. FDDI-II uses this period.

The data transferred in the circuit-switched mode is treated as a stream with a rate suitable for the service being provided, for example, 64 kb/s for digital voice. By handling both voice and data, FDDI is able to support integrated service applications and provide an efficient interface to a variety of networks, including the public switched analog and digital networks, as well as packet-oriented networks, such as X.25.[38]

Fig. 10.23 shows how FDDI-II is implemented using an additional standard,

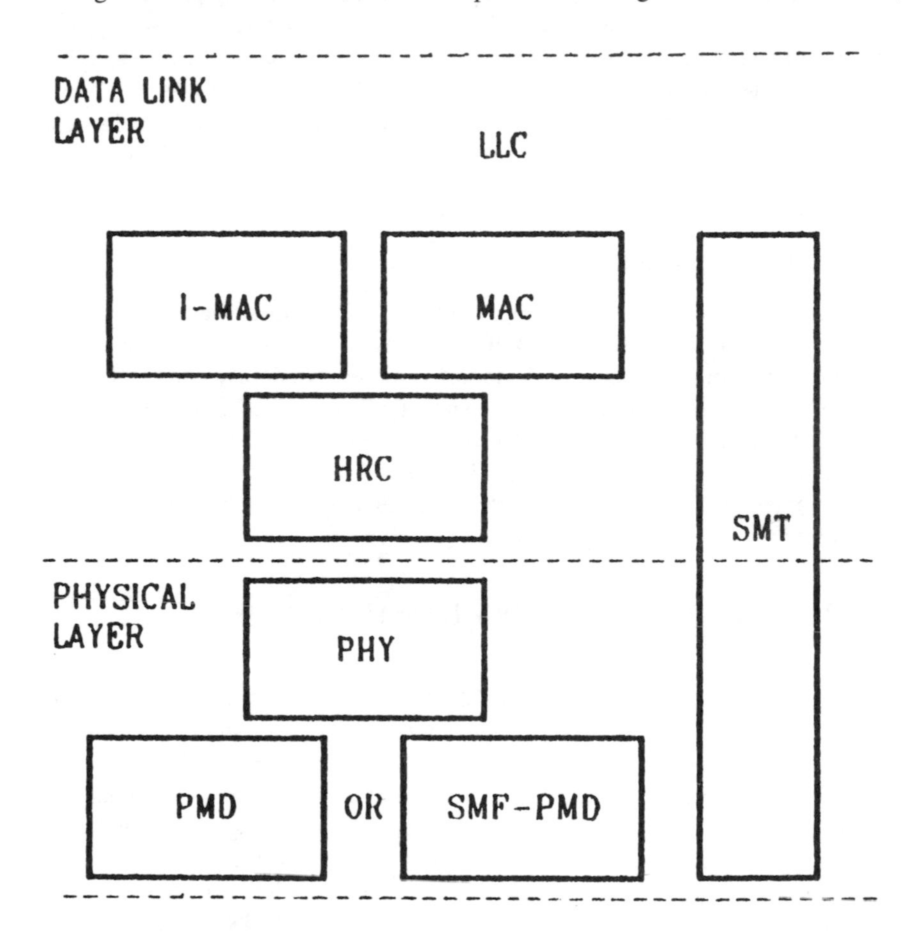

FIGURE 10.23. FDDI-II relationship to OSI model. (Courtesy F. E. Ross, Ref. 35, ©IEEE, 1989.)

hybrid ring control (HRC). HRC becomes the new lowest sublayer of the data-link layer, being situated between MAC and PHY. HRC multiplexes data between the packet MAC (P-MAC) and the isochronous MAC (I-MAC). This means that the P-MAC must be able to transmit and receive data on a noncontinuous basis since packet data are interleaved with isochronous data. Because of the HRCs importance, the FDDI-II standard also is known as the HRC standard.

To accommodate rates above 100 Mb/s on a ring, the ANSI X3-T9 committee has defined 100 Mb/s as a base rate, scalable in increments of 6.144 Mb/s (DS2). Thus, 16 DS2 isochronous channels plus a 768 kb/s packet channel will constitute a 100-Mb/s basic rate. In addition, a single-mode fiber PMD will provide a node-to-node distance of at least 40 km.[39] Work also is in progress to specify how FDDI will run over the SONET services offered by public-network carriers.[40] (SONET is discussed in Section 10.12). Using SONET to extend a FDDI network over long distances will allow a user's LAN to connect machines in different cities while still providing the high performance associated with FDDI LANs.

10.11.3. FDDI Applications

FDDI applications include connecting computer mainframes, memories, peripherals, and controllers. Because FDDI is an industry standard, the connection of equipment from various manufacturers becomes possible without the interface problems associated with proprietary hardware. In addition, FDDI will be used to interconnect other LANs, such as ethernet, [IEEE 802.3], MAC, [IEEE 802.4], token ring, and others. The FDDI data rate of 100+ Mb/s allows large numbers of attached LANs to intercommunicate through gateways on the network.[41]

10.12. SYNCHRONOUS OPTICAL NETWORK (SONET)

The SONET is the ANSI designation for the network standard for synchronous optical transmission. Internationally, the CCITT developed similar, ANSI-consistent synchronous digital hierarchy (SDH) standards: G.707, G.708, and G.709. The new SDH environment provides interfaces that ensure the international connection of digital networks in spite of the independent existing hierarchical rates used by North America, Europe, and Japan.[42] As a result, new services can be offered on a worldwide basis. In addition, synchronous digital networks can be created in existing plesiochronous environments through the use of what are called payload pointers. These payload pointers can be used to lock several 50-Mb/s or 150-Mb/s payloads together to form a broadband payload, on the order of 600 Mb/s. Thus, an SDH-based network can work with existing networks while allowing new services to be developed and introduced.

The multiplexers being recommended by the SDH/SONET standards will allow the multiplexing of a number of plesiochronous signals into an SDH signal.[43] For example, 63 CEPT1 signals (2.048 Mb/s each) might be multiplexed to the STM-1 rate (155.52 Mb/s), or 12 DS3 signals (44.736 Mb/s each) might be multiplexed to the STM-4 rate (622.08 Mb/s). In addition, 16 STM-1 signals might be multiplexed together to form an STM-16 signal. To facilitate the add-drop multiplexing to be done in SDH/SONET, a byte-synchronizing system is used. All bits in a particular block of data are part of a single channel.[44] Such blocks are called "cells." The cells all are of fixed length, and each includes a header with a destination address and other data needed for its transmission. This arrangement is called the asynchronous transfer mode (ATM). Thus, channels may be added or dropped at each node of the network without demultiplexing all signals down to the DS1 rate.

What are called *virtual tributaries* are used to allow streams at rates lower than DS3 to be inserted into the basic STS-1 structure. The lower-speed channels are added in such a way that maintenance overhead bits are added to the payload bits of the resultant channel. This allows the lower-rate payloads to be added to or dropped from the carrier readily.

The high service quality and economy associated with optical transmission are resulting in a rapid expansion of fiber transmission in most segments of telecommunication networks. SONET defines a family of standard optical interfaces for network use. The result will be an environment in which equipment from various vendors can work together, using a common set of standards.

The motivation to deploy SONET on the part of common carriers is threefold.[45] First, many common carriers plan to move existing services, such as DS1s or CEPT1s to fiber because of the improved quality of service and operations benefits that will result. In fact many copper-pair and coaxial services of the interoffice type are being phased out in favor of fiber. The implementation of SONET-based systems is the preferred way to meet these needs. Second, SONET provides such upgraded network capabilities as survivable and self-healing networks, rapid provisioning, networking, grooming and switching capabilities, and network management and control capabilities allowing flexible bandwidth allocation. Finally, new service offerings, including broadband services, will be implemented using SONET.

SONET has been designed mainly as a transport concept, and intentionally structured to make the technology as independent of specific services and applications as possible. Moreover, its attributes and capabilities can further the development and delivery of new services. With new network switching and operations technology, it will serve as a major new delivery platform. With its ability to transmit multiples of a 51.84 Mb/s rate, it will provide a truly broadband transport service for such applications as high-definition television and high-data-rate interconnections between supercomputers. In fact, SONET is viewed by the standards organizations as the foundation for the physical layer of the broadband ISDN (BISDN).

10.12.1. The SONET Standards

The basic SONET building block is the 8-bit byte, repeated every 125 μs. The result, of course, is the 64-kb/s channel. The first SONET level consists of 810 bytes; the corresponding rate is 51.84 Mb/s. This rate transmitted electrically is STS-1; however, no electrical interface is specified. The interfacing is done optically. The optical equivalent is called optical carrier 1 (OC-1). Three OC-1s constitute a synchronous transfer mode 1 (STM-1), with a line rate of 155.52 Mb/s. Multiple STS-1 signals thus are time-division multiplexed together to form the higher rates.

The standardized levels and line rates are the following:

OC Level (STS)	Line Rate	STM Level
OC-1	51.84 Mb/s	
OC-3	155.52 Mb/s	STM-1
OC-9	466.56 Mb/s	
OC-12	622.08 Mb/s	STM-4
OC-18	933.12 Mb/s	
OC-24	1.244 Gb/s	
OC-36	1.866 Gb/s	
OC-48	2.488 Gb/s	STM-16
OC-96	4.976 Gb/s	
OC-192	9.953 Gb/s	

SDH is a synchronous system capable of carrying synchronous or asynchronous tributaries. Correspondingly, the present CCITT hierarchies (European and North American) define asynchronous systems that can carry synchronous or asynchronous tributaries.

Each STS-1 consists of a synchronous payload envelope (SPE), carrying the bits to be transmitted, and DCC, consisting of alarm, maintenance, control, monitoring, and administrative data being sent between the SDH elements and the network management support (NMS) function. The SPE for STS-1 is 50.112 Mb/s, and usually is represented as 87 $\times$ 9 bytes repeated every 125 μs, and includes overhead. The DCC is 192 kb/s per repeater line; 576 kb/s per optical line.

Three types of overhead are used in SDH. *Section overhead* consists of overhead channels that are processed by all SDH equipment, including regenerators.

Line overhead is processed in all SDH equipment except regenerators. *Path overhead* is contained in the SPE along with the payload, and is processed at the payload terminating equipment.

The payload pointer is a number in the line overhead indicating the location of the starting byte for the STS-1 SPE within the STS-1 frame. The pointer level can be adjusted for small frequency differences between STS-1 payloads within an STS-*n* signal.

10.12.2. Virtual Tributaries

As noted previously, a virtual tributary (VT) allows lower rate streams (than DS3) to be combined into the STS-1 rate. A VT is a small envelope within an SPE. The virtual tributary performs mapping into the SPE. The basic VT rates defined by ANSI are as follows:

VT Level	Line Rate	Capacity
VT 1.5	1.728 Mb/s	DS1
VT 2	2.304 Mb/s	CEPT1
VT 3	3.456 Mb/s	DS1C
VT 6	6.912 Mb/s	DS2
VT 6-Nc	N × 6.912 Mb/s	N × DS2
Async DS3[a]	44.736 Mb/s	DS3

[a]Not classified as VT

Virtual tributaries have two modes of operation: *floating* and *locked*. The floating mode allows dynamic alignment of the VT payload within the SPE. Its purpose is to minimize network delay for network elements performing VT switching at ≥1.5 Mb/s. It is suitable for the bulk transport of channelized or unchannelized DSn and for distributed VT grooming.[46] The locked mode requires fixed mapping of the VT payload into the STS. Its purpose is to minimize cost and complexity for network elements performing 64 kb/s switching. It is optimized for bulk (that is, an integral number of STS-1s) transport of DS0s and for distributed DS0 grooming between DS0 path terminating equipment, for example, between DS0 switches.

Table 10.4[47] compares network implications and applications of the two VT modes. The floating VT mode probably will be used extensively in local telephone company networks. Most initial SONET applications involving VT-structured payloads will use the floating VT 1.5 *asynchronous* and *byte synchronous* mappings.

In Table 10.4, *unchannelized* means that the lowest addressable level is a VT, usually equivalent to a DS1, but without the establishment of DS1 framing. An asynchronous mapping requires a minimum timing consistency between tributar-

TABLE 10.4. VT Modes[a]

Mode	Network Implication	Applications
Floating	Minimal delay through the network and at VT crossconnects for DSn transport	Distributed VT crossconnection
	VT path level maintenance and end-to-end performance monitoring possible	Bulk transport and switching of channelized or unchannelized, synchronous or asynchronous DS1s (asynchronous mapping)
		Unchannelized synchronous DS1 transport (bit synchronous mapping)
		DS0 circuit-switched traffic and IDLC (byte synchronous mapping)
Locked	Large delay ($\geq$125 μs) through the network and at VT crossconnects for DSn transport	Unchannelized synchronous DS1 transport (bit synchronous mapping)
	VT path level maintenance and end-to-end performance monitoring not possible	Bulk transport of DS0s and distributed DS0 crossconnection (byte synchronous mapping)
	DS0 level maintenance required at VT crossconnects	
	Cannot transport asynchronous DSn traffic	

[a] Source: Sandesara, et al. Ref. 43. copr. IEEE, 1990

ies and the SONET clock. It is available only in the floating mode of VT operation, and represents unchannelized operation with minimal timing commonality between tributaries.[48] This type of mapping is a technical requirement, due to the unrestricted payload abilities and the support of all DS1 transport needs throughout the SONET network. A bit synchronous mapping establishes a common clock frequency, but assigns an arbitrary phase. Its use is considered undesirable. On the other hand, byte-synchronous mapping represents 64-kb/s channelized operation. A fixed timing reference is established on the DS1 frame. Byte-synchronous mapping is optimized for, and limited to, switch-to-switch service applications.

To summarize the different modes of synchronization:

1. Byte-synchronous is channelized, involves the same frequency and same phase, and the framing is identified.
2. Bit-synchronous is unchannelized, involves the same frequency, but a different phase, and the framing is not identified.
3. Asynchronous is unchannelized, involves different frequencies and different phases, and the framing is not identified.

The term channelized in Table 10.4 means that the DS0 channels are uniquely addressable within the SPE (byte synchronous); both fixed clock-frequency and fixed phase are established based upon the DS1 frame.

With respect to channelization, the locked mode is optimized for channelized tributaries, but may be used to transport unchannelized signals. It is best suited to switch-to-switch service only. The floating mode is optimized for unchannelized transport, but may be configured in a channelized mode.

10.12.3. Architecture

Using SONET, a central office will have the architecture shown in Fig. 10.24,[49] which illustrates the significant simplifications possible in architecture. The main and optical distribution frames are not needed. The all-fiber office terminates connections directly on the SONET digital crossconnect system (DCS). The loop plant consists of SONET digital loop carrier and multiplexers.

As an operational example of the concept of Fig. 10.24, assume that a DS1 signal within OC-M is to be connected to OC-N, where OC-M and OC-N terminate at different locations. The existing central-office architecture would call for that signal to be routed via the DCS. With the SONET capability of including cross-connect and remote operations functions in the SONET multiplexer, the

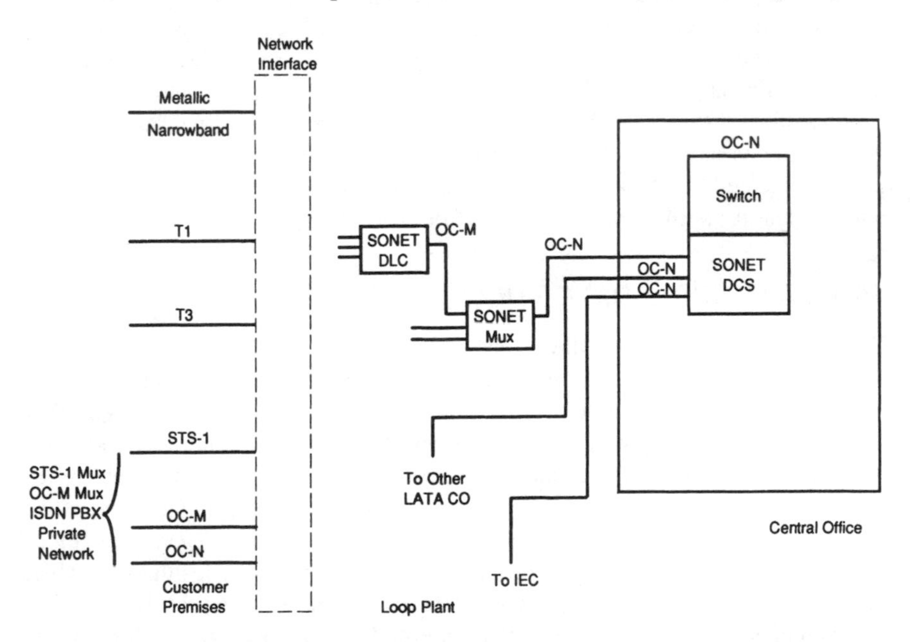

FIGURE 10.24. SONET Central Office Architecture. (Courtesy T. J. Aprille, Ref. 49, ©IEEE, 1990.)

TABLE 10.5. SONET Signal Applications[a]

Interface Signal	Applications
OC-1	Distribution networks
OC-3	Distribution networks, IDLC access, thin interoffice routes, broadband UNI
OC-12	Access rings, loop feeder, interoffice transport
OC-24 OC-48	Interoffice backbone between major hubs, interoffice rings, BISDN feeder
STS-1	Interoffice interconnection
STS-3	Interoffice interconnection, broadband UNI

[a]Source: Sandesara, et al, Ref. 45, copr. IEEE, 1990

need for back-hauling to the central office is eliminated. Thus, the circuit can be routed simply via the SONET mux, resulting in a reduction in loop fiber needs as well as central-office equipment allocations.

10.12.4. Applications

Table 10.5[50] summarizes a variety of applications for the different SONET signals. IDLC is Bellcore's integrated digital loop carrier.[51] UNI is a user-to-network interface. The use of the various signals will depend on such factors as the required traffic capacity, network architecture, and technological factors.

To avoid constraining innovation, the standards do not specify particular technologies, such as lasers or LEDs, or p-i-n or APD receivers. Single-mode dispersion-shifted fiber, however, is called out for certain application categories,[52,53] and attenuation-optimized single-mode fiber for others.[54]

10.13. ATMOSPHERIC OPTICS

Several manufacturers have developed systems using the components and wavelengths of fiber transmission, but without requiring the fiber. The obvious advantage is that an end-to-end connection can be achieved wherever a line-of-sight path exists between the two ends, in the manner of a terrestrial microwave system, but without requiring the regulatory agency authorization or raising interference concerns. Moreover, the cable problems of trenching and right-of-way are avoided. Such systems can be moved readily from one location to another without leaving costly cable behind. Link distance usually is limited to one km for reliable, all-weather operation. Telescopes constitute the transmitting and receiving antennas, and must be kept clean and properly pointed for continued reliable operation.

Applications include ethernet (10 Mb/s), DS1 (1.544 Mb/s), and high resolution video (7.5 MHz).

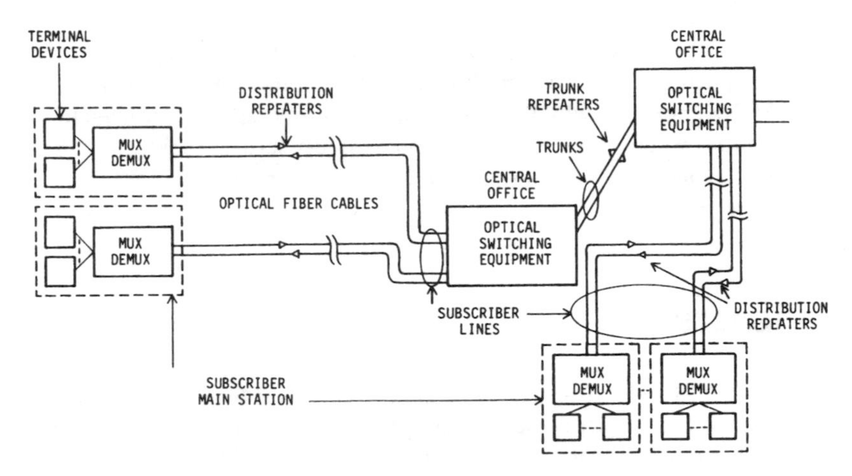

FIGURE 10.25. Future optical telephone network.

10.14. FUTURE OPTICAL TELEPHONE NETWORK

Figure 10.25 illustrates an all-optical telephone network, in which not only the repeaters but also the switches operate entirely on an optical basis. Research now is under way to complete the development of such a system. Advantages will include not only extremely high switching speed, but low-power operation and freedom from electrical interference of various types.

PROBLEMS

10.1. An optical fiber with a 0.2-db/km attenuation is to be used with a 100-km repeater spacing. The fiber is capable of 6 Gb/s-km. If digital video at 18.6 Mb/s per channel is to be transmitted, how many video channels can this system accommodate? If the repeaters were placed 50 km apart instead, what number of channels could be handled?

10.2. Optical fibers often are said to be immune to electromagnetic interference. Does this mean that the use of optical fibers in a telephone network will end all electromagnetic interference problems?

10.3. Lasers often are used with single mode fibers whereas LEDs often are used with graded index or multimode fibers. Will a laser work satisfactorily with a graded index or multimode fiber? Discuss.

10.4. With the common use of simple digital intensity modulation in fiber optic systems, what is the relationship of information rate to symbol rate in such systems?

10.5. Explain why a fiber-optic system designed to interconnect a computer with its peripherals cannot be used directly in a telecommunications application.

10.6. In SONET, explain why the byte-synchronous mode provides channelized operation, whereas the bit synchronous mode does not. Why is the bit-synchronous mode said to result in an "arbitrary phase"?

REFERENCES

1. Kanzow, J., "BIGFON: Preparation for the Use of Optical Fiber Technology in the Local Network of the Deutsche Bundespost," *IEEE Journal on Selected Areas in Communications*, Vol. SAC-1, No. 3, April 1983, pp. 436–439.
2. Fitchew, K. D., "Technology Requirements for Optical Fiber Submarine Systems," *IEEE Journal on Selected Areas in Communications*, Vol. SAC-1, No. 3, April, 1983, pp. 445–453.
3. Keiser, B. E., "Wideband Communications Systems," Course Notes, © 1990.
4. "Cost Comparisons for Long-Distance Communications," *Laser Focus 20*, No. 3, March, 1984, p. 40.
5. Li, T., "Advances in Optical Fiber Communications: An Historical Prospective," *IEEE Journal on Selected Areas in Communications*, Vol. SAC-1, No. 3, April, 1983, pp. 356–372.
6. Henry, P. S., "Introduction to Lightwave Transmission," *IEEE Communications Magazine*, May 1985.
7. Bell, T. E. "Light that Acts Like "Natural Bits", *IEEE Spectrum*, Vol. 27, No. 8, Aug. 1990, pp. 56–57.
8. Nakagawa, K., "Second-Generation Trunk Transmission Technology," *IEEE Journal on Selected Areas in Communications*, Vol. SAC-1, No. 3, April 1983, pp. 387–393.
9. Runge, P. K., and Trischitta, P. R., "The SL Undersea Lightguide System," *IEEE Journal on Selected Areas in Communications*, Vol., No. 3, April 1983, pp. 459–466.
10. Niiro, Y., "Optical Fiber Submarine Cable System Development at KDD," *IEEE Journal on Selected Areas in Communications*, Vol. SAC-1, No. 3, April, 1983 pp. 467–478.
11. Elrefaie, A. F., et al, "10 to 20 Gb/s Modulation Performance of 1.55 μm DFB Lasers for FSK Systems," in *Proc. Conference on Optical Fiber Communication*, Houston, 1989, Paper WN3.
12. Yonezu, H., "Reliability of Light Emitters and Detectors for Optical Fiber Communication System," *IEEE Journal on Selected Areas in Communications*, Vol. SAC-1, No. 3, April 1983, pp. 508–514.
13. Personick, S. D., "Review of Fundamentals of Optical Fiber Systems," *IEEE Journal on Selected Areas in Communications*, Vol. SAC-1, No. 3, April 1983, pp. 373–380.
14. Stanley, I. W, "A Tutorial Review of Techniques for Coherent Optical Fiber Transmission Systems," *IEEE Communications Magazine*, Vol. 23, No. 8, Aug. 1985, pp. 37–53.

15. Vodhanel, R. S. and Wagner, R. E., "Multi-Gigabit/sec Coherent Lightwave Systems," *IEEE ICC'89 Conference Record*, Paper 14.4, June 1989.
16. Waarts, R. G., et al, "Nonlinear Effects in Coherent Multichannel Transmission Through Optical Fibers," *Proc. IEEE*, Vol. 78, No. 8, Aug. 1990.
17. Wagner, R. E., and Linke, R. A. "Heterodyne Lightwave Systems: Moving Towards Commercial Use," *IEEE Lightwave Communication Systems*, Vol. 1, No. 4, Nov. 1990.
18. Nicholson, P. J., "An Introduction to Fiber Optics," *Microwave Journal*, Vol. 34, No. 6, June 1991, pp. 26–43.
19. See note 14.
20. See note 17.
21. Dixon, R. W. and Dutta, N. K., "Lightwave Device Technology," *AT&T Technical Journal*, Vol. 66, Issue 1, Jan.-Feb. 1987, pp. 73–83.
22. Hagimoto, K., et al, "Over 10 Gb/s Regenerators Using Monolithic IC's for Lightwave Communication Systems," *IEEE Journal on Selected Areas in Communications*, Vol. 9, No. 5, June 1991.
23. See note 18.
24. Nakagawa, K. and Shimada, S., "Optical Amplifiers in Future Optical Communication Systems," *IEEE Lightwave Communication Systems*, Vol. 1, No. 4, Nov. 1990, pp. 57–62.
25. Ekas, W., "Photo-nic Finish," *Telephone Engineer & Management*, Vol. 95, No. 23, Dec. 1, 1991, pp. 39–43.
26. *Transmission Systems for Communications*, Bell Telephone Laboratories, Inc., revised fourth ed., 1971.
27. See note 18.
28. Boyer, G. R., "A Perspective on Fiber in the Loop Systems," *IEEE LCS*, Vol. 1, No. 3, Aug. 1990, pp. 6–11.
29. Shumate, P. W., and Snelling, R. K., "Evolution of Fiber in the Residential Loop Plant," *IEEE Communications Magazine*, Vol. 29, No. 3, March 1991, pp. 68–74.
30. Kozak, R., "Alternate Local Access Via Fiber," *Telecommunications*, Vol. 24, No. 7, July 1990, pp. 23–25.
31. Jacobs, I., "Design Considerations for Long-Haul Lightwave Systems," *IEEE Journal on Selected Areas in Communications*, SAC-4, No. 9, Dec. 1986, pp. 1389–1395.
32. Henry, P. S., "Optical Wavelength Division Multiplex," *IEEE Globecom '90 Conference Record*, Paper 803.1, Dec. 1990.
33. Brackett, C. A., "Dense Wavelength Division Multiplexing Networks: Principles and Applications," *IEEE Journal on Selected Areas in Communications*, Vol. 8, No. 6, Aug. 1990, pp. 948–964.
34. Toba, H., et al, "Broadband Information Distribution Networks Employing Optical Frequency Division Multiplexing Technologies," *IEEE Globecom '90 Conference Record*, Paper 803.2, Dec. 1990.
35. Ross, F. E., "An Overview of FDDI: the Fiber Distributed Interface," *IEEE Journal on Selected Areas in Communications*, Vol. 7, No. 7, Sept. 1989, pp. 1043–1051.
36. McClure, B., "FDDI Update: Standards, Testing, and the Future of FDDI," *Telecommunications*, Vol. 25, No. 1, Jan. 1991, pp. 67–69.

37. See note 35.
38. Kadambi, J., "FDDI II Adds Voice Capability to Packet-Switched Token Ring," *Telecommunications*, Vol. 25, No. 4, April 1991, pp. 31–35.
39. See note 38.
40. See note 36.
41. Southard, B. and Kevern, J., "FDDI: A New Era in Computer Communications," *Telecommunications*, Vol. 25, No. 2, Feb. 1991, pp. 58–60.
42. Miki, T., and Siller, C. A., Jr., "An International Perspective on Evolution to a Synchronous Digital Network," *IEEE Communications Magazine*, Vol. 28, No. 8, Aug. 1990, pp. 7–10.
43. Balcer, R., et al, "An Overview of Emerging CCITT Recommendations for the Synchronous Digital Hierarchy: Multiplexers, Line Systems, Management, and Network Aspects," *IEEE Communications Magazine*, Vol. 28, No. 8, Aug. 1990, pp. 21–25.
44. Stoffels, B. and Stewart, A., "SONET: Building Connectivity," Supplement to *Telephone Engineer and Manager*, White Paper, © 1991, Alcatel Network Systems.
45. Sandesara, N. B., et al, "Plans and Considerations for SONET Deployment," *IEEE Communications Magazine*, Vol. 28, No. 8, Aug. 1990, pp. 26–33.
46. See note 45.
47. See note 45.
48. Langmeyer, P., "Considerations for SONET Deployment," NEC America, Inc., Herndon, VA.
49. Aprille, T. J., "Introducing SONET into the Local Exchange Carrier Network," *Communications Magazine*, Vol. 28, No. 8, Aug. 1990, pp. 34–38.
50. See note 45.
51. "Integrated Digital Loop Carrier System Requirements, Objectives and Interface," TR-TSY-000303, issue 1, Sept. 1986, revision 3, Mar. 1990.
52. Recommendation G.652, "Characteristics of a Single-mode Optical Fiber Cable," *CCITT Blue Book*, 1988.
53. Recommendation G.653, "Characteristics of a Dispersion-shifted Single-mode Optical Fiber Cable," *CCITT Blue Book*, 1988.
54. Recommendation G.654, "Characteristics of a 1,500 nm Wavelength Loss-minimized Single-mode Optical Fiber Cable," *CCITT Blue Book*, 1988.

11

Digital Switching Architecture

11.1. INTRODUCTION

In digital switching system design, architectural considerations involve much more than simply replacing an analog switching network with a digital switching network. Several fundamental questions must be answered, and each must then be followed by a large number of decisions, most of which involve trade-offs. Some of the most prominent questions that require major decisions up front are:

1. What is the intended application and maximum size of the switching system?
2. What type of voice encoding should be adopted?
3. How should lines and trunks be interfaced and conditioned for the switching network?
4. What network architecture should be selected?
5. How should signaling and service circuits be handled?
6. What control concept should be used?
7. How should maintenance diagnostics, traffic management, and switching-system administration be accommodated?
8. What recent technological advances should be considered for implementation in the system?
9. What future potential innovations should the design be planned to accommodate?

Not all of these can be answered with finality in the beginning, but all should be addressed. Tentative decisions should be made, and then reexamined as the design process continues. The application and approximate maximum size of the system are determined by a perceived view of the marketplace. Recent advances in solid-state technology should be evaluated, and those most promising for the intended application should be identified for close scrutiny.

If the intended application is for use as a PBX, several encoding schemes are available for consideration. A public network switching system, however, should be able to accommodate the world standard of 64 kb/s PCM. The geographic area of the anticipated market will determine whether μ-law or A-law PCM is selected. For a world market, both should be available as options.

Basic architectural decisions must consider all operational features and capabilities to be provided by the switching system, not only in its initial design capability but also for retrofit of possible future innovations, such as provision for data switching.

11.2. TERMINAL INTERFACE TECHNIQUES

The functions required to be performed in interfacing lines and trunks with the switching network apply to all digital circuit-switching systems, but implementation techniques vary widely. The trend, however, is clearly visible.

11.2.1. Terminal Interface Functions

The terminal interface functions of a digital switching system include those required to interface lines and trunks and to condition the speech paths for presentation to the switching network. A subscriber line is terminated in the tip and ring terminations of a line circuit.

Line-circuit functions are typically defined by the acronym BORSCHT, derived from the first letters of the seven functions described below. The "C" function is not required for digital subscriber lines, and the "H" function is not required for 4-wire lines.

B = Battery voltage supplied to the line. Most central-office battery supplies are nominally negative 48 V dc with a range of negative 42.5–52.5 V dc, although the voltage applied by some central offices at certain stages of switching can exceed 78 V dc and, in rare cases, can exceed 100 V dc.

O = Overvoltage protection to protect the line-circuit equipment from lightning and power-line induction. The protector is placed between the tip and the ring to ground and must be designed to protect the sensitive solid-state components in the line circuit.

R = Ringing circuits to apply controlled ringing voltages to the line to activate the ringer in the called telephone.

S = Supervisory circuitry to detect on-hook and off-hook conditions and dial pulsing.

C = Coder–decoder (codec) equipment to convert analog speech to PCM words. Codecs either contain or are associated with low-pass filters in the transmit and receive paths.
H = Hybrid transformer to convert the two-wire line to a four-wire line, consisting of separate transmit and receive paths. A balance network reduces undesired feedback from the receive path to the transmit path.
T = Test circuitry which connects test equipment to the two-wire line to test the line for continuity, shorts, and impedance characteristics.

The terminal interface area also contains equipment to terminate analog and digital trunks. The functions performed for trunks are similar in some respects to those performed for lines but are modified for the differences in signaling and supervision.

Since most subscriber lines generate low volumes of traffic, lines generally are concentrated for economic reasons. One or two concentration stages of switching usually are included in the terminal interface area. Finally, one or more stages of digital multiplexing are included to increase the number of time slots in the PCM stream entering the switching network. Trunk circuits usually bypass the terminal equipment concentrator and interface the multiplexer directly.

11.2.2. Implementation Considerations

As digital switching systems began to be designed, implementation architecture was influenced by two major factors. Concepts which had been proved effective in electromechanical and electronic space-division systems tended to be carried over to digital design. Also, the relative costs of different solid-state implementations influenced trade-off decisions which were subsequently changed as larger-scale integration became feasible.

11.2.2.1. Analog Line Interface

Some basic line-interface functions are shown in Fig. 11.1. Not shown are battery feed, overvoltage protection, ringing voltage, supervisory signaling, and test circuits. The hybrid network is shown here as a separate component, and the coder/decoder functions are shown separately to illustrate the functions.

Some early designs used shared codecs. As the state of the art of solid-state crosspoint technology, especially the limitation on voltages that can be switched through them, advanced, it became feasible to combine many of the line-interface functions into a smaller number of components. High-voltage silicon technology now permits the passage of higher voltages used for ringing. Recent technology permitted line-circuit chips to include all line-interface functions except concentration and multiplexing, using a minimum of components, as shown in Fig. 11.2.[1] This was followed by PCM bus formatting and concentration. [This line card has

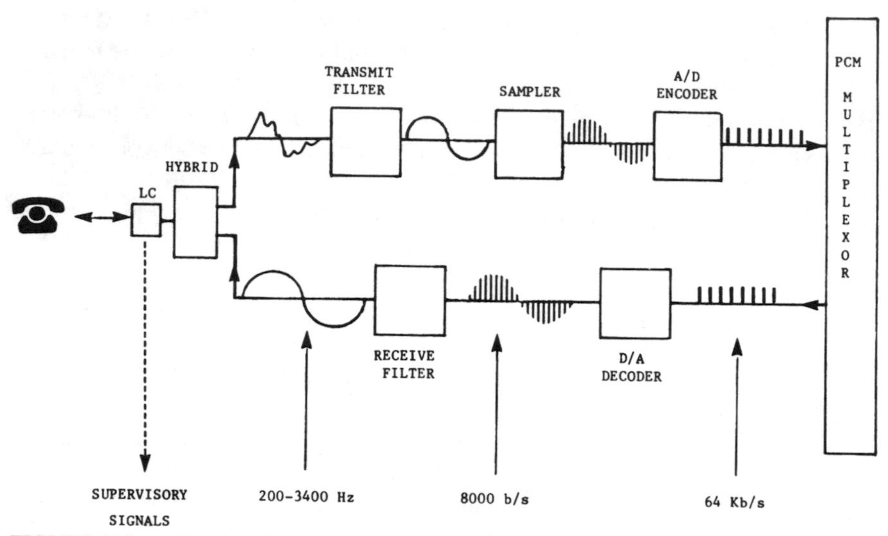

FIGURE 11.1. Terminal interface functions.

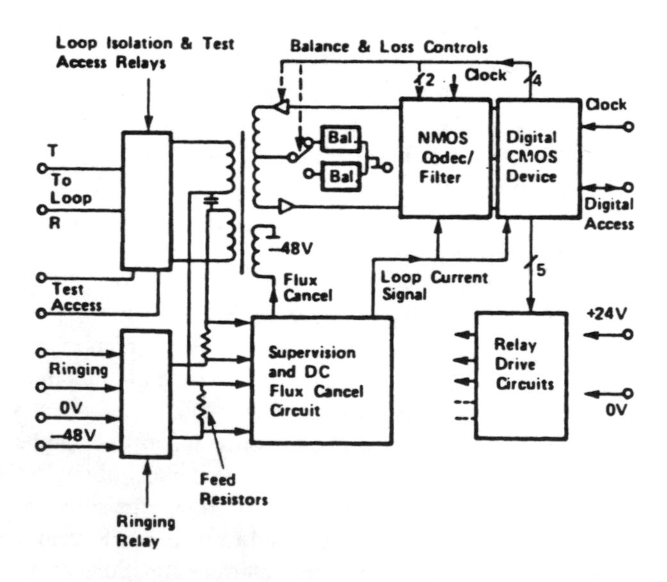

FIGURE 11.2. DMS-100 line-circuit block schematic. (From J. Terry, D. Younge, and R. Matsunaga, "A Subscriber Line Interface for the DMS-100 Digital Switch," NTC '79, © IEEE.)

since been replaced by a software-configurable line card that can be remotely reconfigured to provide a wide variety of user features (see Section 13.5.3.5)].

The concentration function, if performed prior to A/D conversion, is a space-division network. If performed after A/D conversion, switching may be performed in memory (time division), through TDM crosspoints (space division), or both. Functions of a line-interface module are illustrated in Fig. 11.3. The speech paths of 32 lines are converted by the line circuits into PCM streams, which are time multiplexed by the line-access time switch. The outputs of 20 line-access cards, appearing as 20 terminal links, are concentrated through a time-multiplexed space switch into 4 network links at a serial bit rate of 2.56 Mb/s. Each of the 32 time slots contains 10 bits, 8 of which are used for standard PCM words and 2 of which are used for signaling and channel control. The 4 network links are interfaced to the main switching network.[2]

11.2.2.2. Analog Trunk Interface

Analog trunks may be terminated at a switching system in any quantity via metallic paths or in 12-channel groups. The basic treatment is similar to that accorded subscriber lines; however, the implementation is somewhat different. Inband and common-channel signals must be processed separately, and integrity checks are performed as described in Section A.4.2.5. Speech paths are converted to PCM signals and multiplexed to conform to the PCM bit rate required by the switching network. Codecs may be used on a per-channel or multichannel basis. Since trunks carry high traffic volumes compared to lines, they normally are not concentrated in the terminal interface modules.

11.2.2.3. Digital Line and Trunk Interfaces

Digital lines and trunks arrive at the switching system in basic groups of 24 or 30 channels. The North American T-carrier group of 24 channels terminates in a digital terminal unit or in the switch itself. Under clock control, signaling is extracted from the least significant bit in every sixth frame and processed separately by a signal processor. The 24 speech channels can be multiplexed with other T1 channels and placed on a PCM highway to the switching network.

11.2.3. Implementation Trends

The recent advances in large-scale integration (LSI) and very large-scale integration (VLSI) have established a rather clear trend toward consolidation of line interface functions in the line circuit itself. Filters and codecs already had been manufactured on a single LSI chip. Subscriber line-interface circuits (SLIC), containing all BORSCHT functions, have long been produced on a single VLSI chip.[3] Remotely configurable line circuits now can enable digital telephones to be connected to digital transmission facilities.

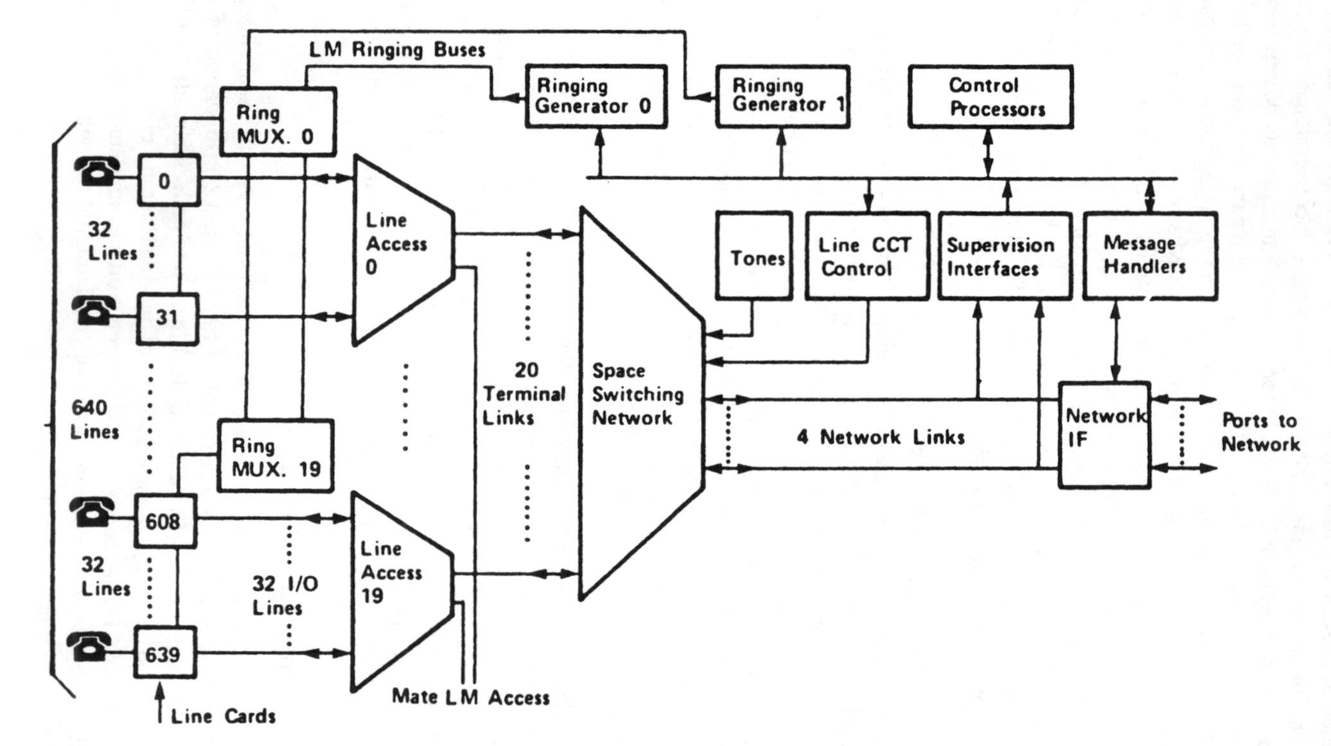

FIGURE 11.3. DMS-100 line-module functional diagram. (From J. Terry, D. Younge, and R. Matsunaga, "A Subscriber Line Interface for the DMS-100 Digital Switch," NTC '79, © IEEE.)

With consolidation of functions in the subscriber line-interface circuit, concentration is performed increasingly on a digital rather than analog basis. The entire terminal interface function can be located either in the local central office or in a remote location. Remote switching units are connected to the host digital switching system by digital transmission facilities over existing cable pairs. The number of connecting digital facilities is determined by the capacity of the concentrator and the traffic load.

11.3. SWITCHING NETWORK CONSIDERATIONS

Selection of switching network architecture involves several considerations. A small or medium-size private branch exchange (PBX) can be designed with only a single switching stage, whereas an economical design for a large system generally requires multiple stages. Time and space switching have somewhat different traffic-handling characteristics. The effect of switching network architecture on control complexity, packaging, and expansion should be considered. The intended application, whether local or tandem switching, and its ultimate size, influence architectural selection. The fact that there are so many different switching network architectures in the marketplace is a clear indication of the many trade-offs which must be considered.

11.3.1. Principles of Time-Division Switching

Analog signals, continuously varying in amplitude, must be altered in order to be switched in time. The speech (or other signals) must be changed into pulses which can be multiplexed into a digital stream, each pulse or group of pulses representing one amplitude sample. Four types of pulse modulation have appeared in telephone switching systems. In all cases, the speech signals are sampled at a controlled rate, each sample pulse representing the amplitude of the signal at the instant of sampling. In *pulse-amplitude modulation* (PAM), the amplitude of the signal sample is represented by the amplitude of the pulse. In *pulse-width modulation* (PWM), the amplitude of the signal sample is represented by the width of the pulse. In *pulse-code modulation* (PCM) and its variations, the amplitude of the PAM pulse is quantized and encoded as explained in Chapter 3. In *delta modulation*, only the polarity of the difference in amplitude between two adjacent speech samples is encoded. PAM, PWM, and delta-modulation schemes have been used in several time-division PBX systems, but PCM technology has been standardized for public network switching systems.

11.3.1.1. Time Switching in Memory

In this concept, a time-division data stream, such as that used in T-carrier systems (see Section 5.3.1), is fed into an *information memory* with each traffic channel occupying a separate time slot position in the buffer. A *control memory* contains the same number of words, or channels, as there are time slots in the information memory, but the content of each word in the control memory is the number of a time slot, or channel, in the information memory. Speech-channel information is read into and out of the information memory under control of clock pulses. A switching processor, under control of the same clock, controls call setup by reading into the control memory a correlation between input and output time slots in the information memory.

Conventionally, the speech time slots are read into the information memory sequentially and read out asequentially in the sequence specified in the control memory. (It can be done in reverse.) In the simplified illustration in Fig. 11.4, the input switches are operated sequentially, but the output switch sequence determines the sequence of the readout. For example, if the first time slot read out is channel 3 during an entire switched connection, then the speech in channel 3 has been switched to channel 1 of the output data stream.

A TSI, such as is used in a PCM switching system, is depicted in Fig. 11.5. In this example, the bus, or PCM highway, has 24 time slots per frame. The 24

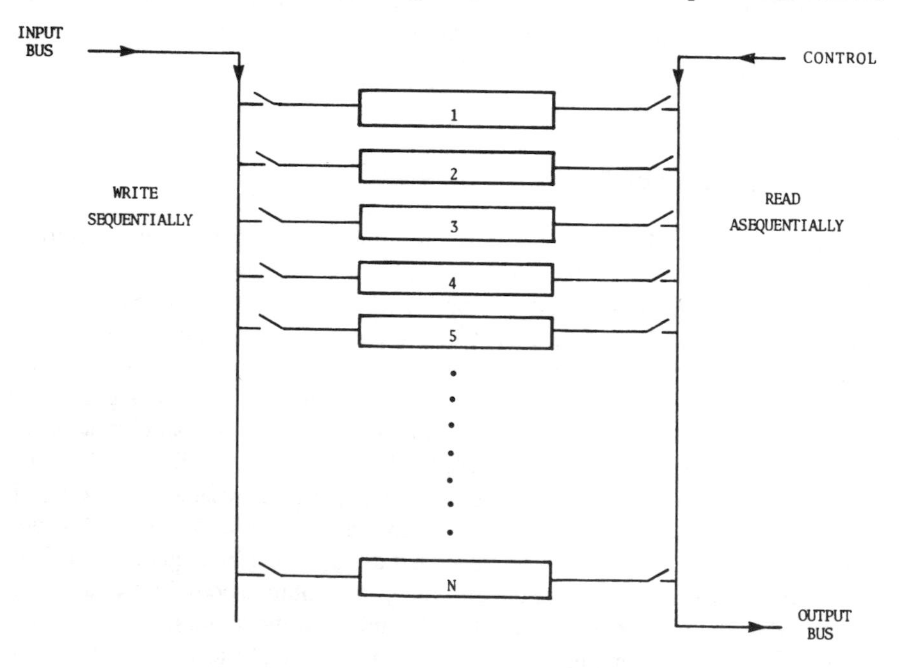

FIGURE 11.4. Switching in memory.

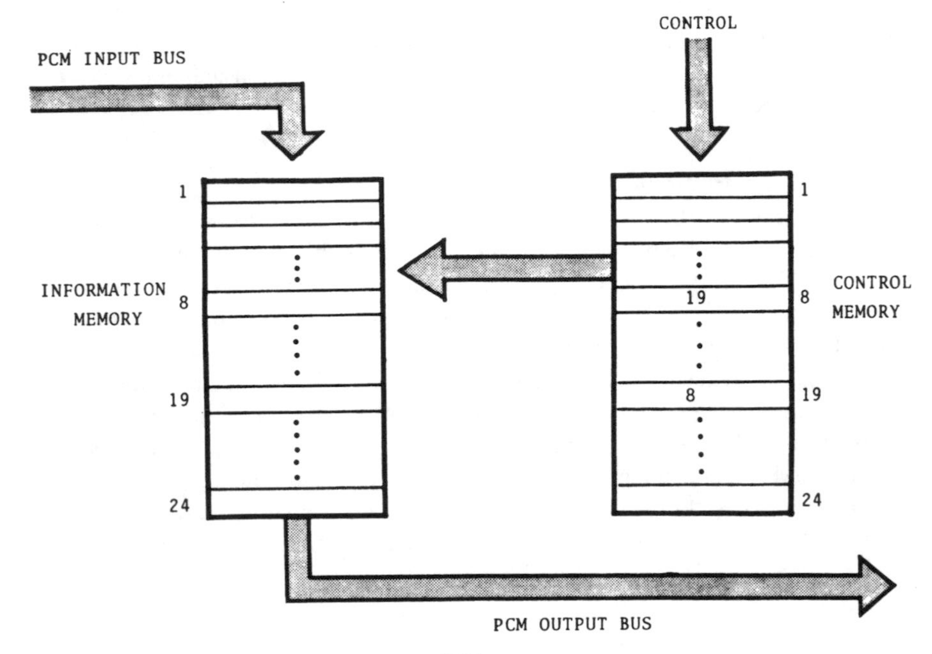

FIGURE 11.5. PCM time-slot interchange (TSI).

voice channels in each frame are read into the information memory in the sequence shown, but the readout sequence is in accordance with the sequence of channels written into the control memory by the control processor. Assume that the illustration represents a 24-port switching system and that ports 8 and 19 are to be connected to each other. When the 8th time slot in the output frame is read out to the output bus, the information memory spills out the word in its time slot 19. Similarly, the 19th time slot in the output frame will contain the PCM word from time slot 8 in the information memory. Since time slot words are being read in and out at clock rate, however, output time slot 19 will contain the PCM word read into time slot 8 in the following frame. Since this word represents only one amplitude pulse, the difference is not detectable.

11.3.1.2. Time Switching in Space

A time-multiplexed signal can be switched in space. A method commonly used with PCM switching systems employs solid-state logic gates as crosspoints, arranged in a square or rectangular matrix. A time-multiplexed PCM bus is connected to each inlet, and an output bus is connected to each outlet. Both input and output buses are controlled by a clock. A call processor causes the logic gates to be opened and closed at clock rate and in the desired sequence to switch the time-multiplexed data streams between the input and output buses. Time switching and space switching generally are combined in large digital switching systems. Either can be used as center-stage switching.

11.3.1.3. Sampling and Coding Rates

Sampling rates in PCM switching systems typically range from 8000 to 16 000 times per second. Higher sampling rates provide higher fidelity and require less complex filtering but cannot interface standard PCM transmission systems directly. Therefore, 8-kHz sampling has been adopted as the worldwide standard for PCM switching and transmission.

As discussed in Chapter 3, 8-bit PCM with $\mu = 255$ companding is the current North American standard. Other techniques are used. CCITT has standardized both the North American system and the A-law companding method used in Europe and most other countries. Some PBX systems have employed more quantization steps and linear coding to facilitate digital signal processing. A disadvantage, of course, is that conversion is required to interface a digital transmission system directly using standard 64-kb/s PCM.

11.3.2. Multistage Digital Switching

Multistage digital-switching networks generally involve both time- and space-division switching. While some switching systems do use multiple time networks only, the time delay involved in time-slot interchanges does place a practical limit on the number of time stages used in a large switching system. Typical PCM switching networks are symmetrical about a central point.

11.3.2.1. Time-Switching Considerations

The principle of time switching in memory was described in Section 11.2.1.1. A major factor in time switching in memory is delay. In a 24-time-slot system, speech can be delayed up to 23 time slots. In a 32-time-slot system, speech can be delayed up to 31 time slots. Some time-switching networks use 128 or more time slots in a 125-μs frame. Therefore, the maximum delay of speech in a single time switch is in the range of 120–124.5 μs or more. When added to other delays inherent in transmission and switching systems, it can add to the effect of echo on the talkers. Consequently, most large switching systems include both time-division and space-division switching stages.

The number of time slots that can be switched in memory during a 125-μs frame time is a function of the cycle time of the information memory. Each time slot requires a write cycle and a read cycle. With 8-kHz sampling, the maximum number of time slots, or channels (C), that can be switched in memory is

$$C = 125/2t_c \tag{12.1}$$

where 125 is the frame time in microseconds and t is the memory cycle time in microseconds.[4] From this, it can be seen that a 256-channel TSI requires a memory with a minimum cycle time of 244 ns, and a 1024-channel TSI requires a memory with a cycle time of 61 ns. Thus, memory cycle time places a practical and economic limit on the size of TSIs.

The impact of TSI size on the control memory should also be considered. The control memory, like the information memory, requires one word for each channel switched in time, but the word length is a function of the number of channels. The number of bits (B) required in each word in the control memory is calculated by

$$B = \log_2 C \tag{12.2}$$

where C is the number of channels.[5] Thus, the maximum number of time slots that can be switched in time with a control memory using 8-bit words is 256, and a 1024-channel TSI requires 10-bit words in the control memory. The number of outgoing time slots in the TSI may be equal to, greater than, or less than the number of incoming time slots, but the most economical arrangement of control memory requires that it control the side of the information memory having the greater number of time slots.[6]

11.3.2.2. Space-Switching Considerations

Time-division space switching involves a matrix which switches m incoming time-multiplexed lines (TML) to n outgoing TMLs having the same number of time slots. The crosspoints in the switching matrix are formed by logic gates controlled by a control memory. The number of outgoing TMLs may be equal to, greater than, or less than the number of incoming TMLs. The control memory controls the logic gates to allow specific time slots in incoming TMLs to be connected to specific time slots in outgoing TMLs.[7] Time slots are not delayed in space-division switching as they are in TSI switching. In space-division switching, incoming and outgoing TMLs are synchronized by the same clock such that channel 1 of an incoming TML is always switched to Channel 1 of one of the outgoing TMLs.

11.3.2.3. Time-Space-Time (TST) Structure

In a time-space-time switching network architecture, time switches are used as the input and output switching stages, and space switches are used as the center stage. The number of internal time slots (that is, the time slots between the input and output stages) is the major factor in the probability of blocking through the switching network. If the number of internal time slots is twice the number of input/output time slots minus one, the network is nonblocking. To achieve a good grade of service (that is, low probability of blocking), the number of input time slots must be significantly less than the number of output time slots in the first time stage. The third stage is a mirror image of the first stage.

In time-division switching with multiple stages, the arrangement of concentration, distribution, and expansion functions shown in Fig. A-14 is reversed. Expansion is performed in the first stage, and concentration is performed in the last stage. The distribution (center) stage or stages always switch on a 1:1 ratio.[8] For example, the AT&T 4ESS input time slot interchange module expands seven TMLs of 120 traffic time slots each into eight TMLs of 105 traffic time slots each for switching. The output time stage then concentrates them back into seven TMLs for transmission.[9] This arrangement does not preclude concentration in the terminal interface module.

Operation of a TST network, shown in Fig. 11.6, involves finding an idle time slot through the space stage which can connect an input time stage to the desired output time stage. The first TSI delays the information in the input time slot 5 until the selected idle space switch time slot 38 appears at the output. Space switching connects the output multiplexed line of the first TSI through the space

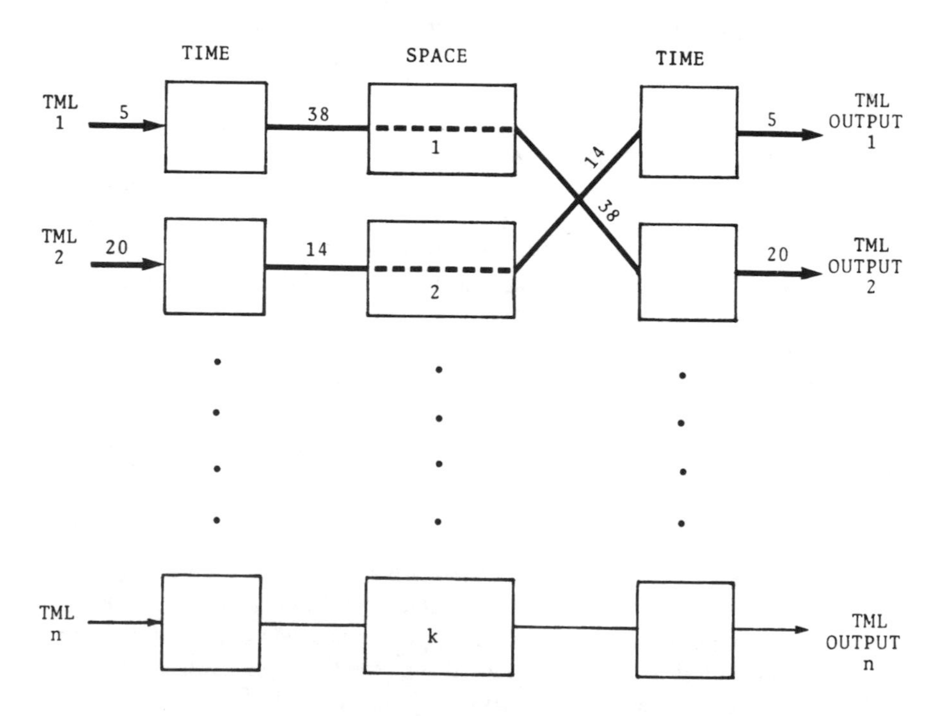

FIGURE 11.6. Time-space-time (TST) switching network.

switch to the input of the third-stage TSI. That time-slot information then is switched in memory to the desired output of the third-stage TSI in time slot 20. A similar connection is established for the other conversation path.[10]

11.3.2.4. Space-Time-Space (STS) Structure

The STS network has a center time stage to effect distribution between the first and third space stages. Time-multiplexed lines function as inputs and outputs to both space stages. The time stage is symmetrical and serves to interconnect time slots in the output TMLs of the first stage to time slots in the input TMLs of the third stage.

Operation of an STS network is shown in Fig. 11.7. The first space stage functions as an expansion stage, having more output TMLs than input TMLs. The last space stage is reversed and is used to concentrate distribution TMLs into a smaller number of output TMLs. To establish a connection, the path controller must find a time switch in the center stage which has an idle input time slot and an idle output time slot corresponding to the time slots used in the space stages. In the example, the first-stage space switch closes a crosspoint path to an internal TML connected to a TSI with an idle time slot 3. The TSI then delays the information until time slot 16 appears in the frame. The third stage then connects that

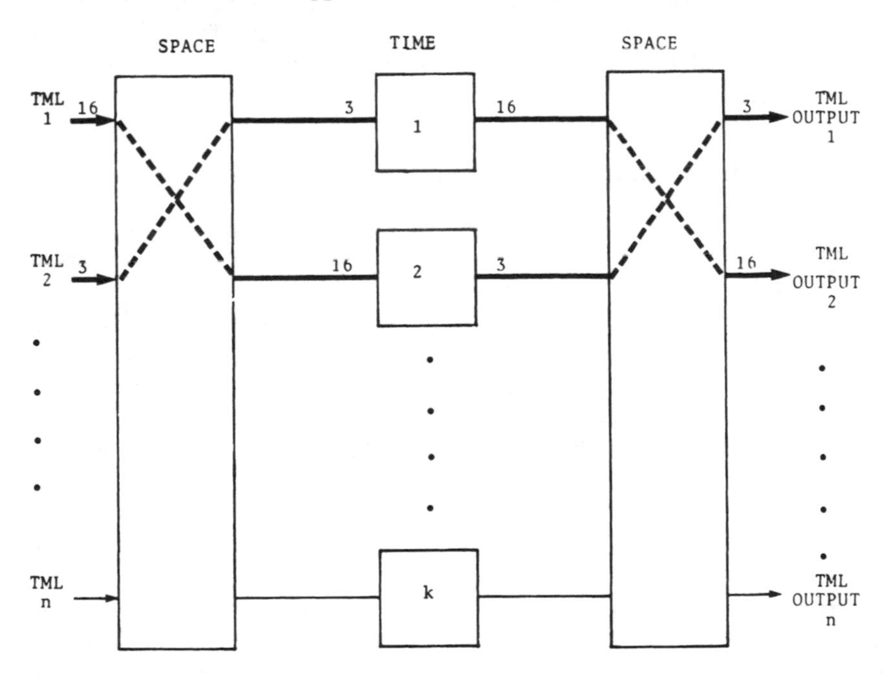

FIGURE 11.7. Space-time-space (STS) switching network.

internal TML to TML OUTPUT 2. In the same manner, time slot 16 in TML 1 is connected to time slot 3 in TML OUTPUT 1 to complete the other half of the conversation.

11.3.2.5. Combined versus Separated Switching

The examples in Figs. 11.6 and 11.7 depict unidirectional paths through the switching networks. Actually the speech highways can be designed to carry two-way traffic. In Fig. 11.6, the two speech paths are shown to use separate time switches, and the two internal time-multiplexed lines are switched through separate arrays in the space stage. This concept is called *separated switching*, and control memories must be provided for each switching array. If both directions of speech are selected identically, however, only one control memory is needed for the bidirectional paths.

In *combined switching*, connections can be established between any two highways connected to a single switching array. Paths may be selected symmetrically or quasi-symmetrically. Symmetrical path selection, in a TST network, uses identically numbered time slots through the center space stage for both speech directions but cannot connect speech paths in the same highway. In an STS network, the two speech directions are switched symmetrically through the same TSI in the center stage. With quasi-symmetrical selection, the two speech directions use two related TSIs controlled by one control memory in an STS network, while they conventionally use even/odd time slots through the space stage of a TST network, permitting connections within the same highway.[11]

11.3.3. Economic and Traffic Considerations

The economics of PCM switching network design must always be weighed against the traffic-carrying capacity of the network. One trade-off is the cost of crosspoints for space switching versus the cost of memory for time switching. For several years, the cost of memory decreased at a faster rate than the cost of logic gates. Currently, however, large-scale and very large-scale integration have significantly reduced the cost of gates, although the effect is limited by the cost of pin arrangements to access the speech paths. Because of the rapid advances being made in solid-state technology, the cost ratio between logic gates and memory depends upon the technological state of the art at the moment and upon the relative costs of control.

The number of time slots (TS) per time-multiplexed line has an effect upon both cost and traffic capacity. Generally, in multistage networks, if the number of terminations and traffic capacity per termination are held constant, switching network costs are reduced if the number of time slots per multiplexed line is increased. This enables the number of multiplexed lines to be decreased, saving logic gates by making the space switches smaller. This also reduces the number of bits required in the space switch control memory but increases the number of bits required in the TSI control memory. This is true for both TST and STS networks. The probability of blocking (see Appendix B) is reduced in TST networks but is increased in STS networks as the number of time slots per multiplexed line is increased. To retain the same probability of blocking, an STS network must be expanded when the number of time slots per multiplexed line is increased. Such action still results in a slight decrease in comparative cost at traffic capacities of about 0.8 erlang per termination or higher.[12] At lower traffic loads, however, a lower loss probability is achieved with fewer time slots per multiplexed line and more multiplexed lines.[13]

Another architectural judgment involves the number of center stages in the switching network, such as T*x*T or S*x*S (x = number of center stages). With a relatively small number (for example, 30) of time slots per multiplexed line, a three-stage PCM switching network is comparable in cost to one having four or more stages up to about 3000 terminations. With 120 time slots per multiplexed line, comparability extends up to about 20 000 terminations. With more terminations, however, multiple center stages are more economical with fewer time slots per multiplexed line.[14] With more time slots per multiplexed line, costs of three-stage and multiple-center-stage networks are roughly equivalent, but the additional center stages may be needed to reduce the probability of blocking.

The most easily recognizable architectural issue is that of the relative merits of TST (or T*x*T) and STS (or S*x*S). This decision is influenced by the factors identified above and others, such as whether data is serial or parallel. There is no clear judgment independent of other factors. Generally, for a given grade of service, expansion in the center stages is somewhat easier and less expensive in T*x*T networks. Since the cost of the switching network, however, rarely exceeds ten percent of the cost of the switching system, other factors generally control the decision on the type of network.[15]

11.3.4. Digital Symmetrical Matrices

The foregoing discussion of switching-network architecture was in terms of separate modules of time switching and space switching. Successful switching-system designers have designed complete switching networks consisting of identical modules of combined time and space switching. Each switch element contains time switching, space switching, and microprocessor control for switching any of 30 channels in any of 16 incoming PCM links (time-multiplexed lines) to any of 30 channels in any of 16 outgoing PCM links. The PCM links are bidirectional, thus serving 16 two-way PCM ports with 30 multiplexed speech channels each. Switch elements can be arranged in multiple stages.[16]

Combined time and space switches have been designed on a single chip with LSI and are called digital symmetrical matrices (DSM). Port capacity is limited by the number of pins on IC cans, chip complexity, and power consumption. Using a 28-pin dual-in-line can, a DSM can support eight bidirectional PCM links and dissipate about 200 mW of power.[17] Analyses have shown that networks composed of DSMs can be designed to be nonblocking and are most cost effective if the least possible number of stages is used.[18]

11.4. SERVICE CIRCUIT TECHNIQUES

Service circuits are common or shared equipment units, implemented in hardware or software, associated with communication paths during the progress of calls. They often are referred to as *pooled common equipment.* Examples of service circuits include tone generators, tone receivers, signaling transmitters, signaling receivers, ringing generators, recorded announcements, conference circuits, and echo suppressors or cancelers.

In analog switching systems, service circuits generally are hardware units that are physically switched into and out of lines and trunks to provide specialized functions. In digital switching systems, service circuits may be implemented in hardware or in memory, or mixed. Incoming tone address signals may be picked off in the terminal interface module and sent to a signal processor, or they may be switched through the digital network to hardware units or to memory locations for interpretation and registration. Outgoing tone signals may be generated by hardware oscillators or in memory. Since most local digital central office switching systems have direct digital interface to remote digital switching units via digital transmission facilities, the trend is to implement service circuits digitally whenever practicable.

11.4.1. Tone Generation

Digital generation of audible and signaling tones in PCM switching systems is constrained by 8-kHz sampling and 8-bit quantization according to the μ-255 companding law. Fundamentally, digital tone generation is accomplished by storing digitally encoded tones in read-only memory (ROM) and then reading out the contents to be decoded for the listener. The PCM constraints result in a unique relationship between the frequency or frequencies of the desired tone signal and the number of samples which must be read out to reproduce the exact tone. One cycle of a presynthesized waveform, representing a single tone or a mixed tone, is stored in ROM. When read out under control of a counter incremented at an 8-kHz rate and decoded either locally or at the distant end of a circuit, the analog tone signal is reproduced. For example, eight words in memory are required to reproduce the digital equivalent of a 0-dBm, 1000-Hz test tone applied at the zero-transmission-level point. This tone is known as a *digital milliwatt.* Digital tone generators have several advantages over analog tone generators. They do not require impedance matching or amplification, and they do not drift in frequency.[19] If tones require interruption, such as busy tone, they are interrupted by turning the ROM on and off at the precise intervals required to produce the desired cadence.

11.4.2. Tone Reception

In a digital switching system, it is desirable, from a control viewpoint, to treat all incoming signals in the voice path uniformly, bringing all incoming tone address signals through A/D conversion and switching them through the PCM system. Given the conversion of the analog tone signals to PCM, it is equally desirable to interpret and register them digitally to save the hardware and D/A conversion necessary to use conventional tone receivers. Several approaches to digital tone detection have been used with varying degrees of success, but most designers have settled on either digital filtering or the discrete Fourier transform (DFT).

Digital signal detection involves single-frequency (SF) tone signals used for supervision (see A.4.2.2), DTMF address signals on subscriber lines (see A.4.2.2), and MF address signals on trunks (see A.4.2.3). Analog signal receivers have rather wide tolerances to compensate for distortion caused by aging transmitters, variations in subscriber keying characteristics, and transmission line impairments. Such distortion compounds the problem of digital recognition.

In the digital filtering method, the PCM samples are linearized and passed through a series of digital bandpass filters centered at each signaling frequency. The filtered signal power at each frequency is measured repeatedly to detect the presence or absence of power for each tone. A signal processor then interprets the signals and translates them for call setup. This method requires a substantial amount of memory and rather elaborate arithmetic calculations.[20]

The DFT method also extracts a measure of signal power at each signaling frequency, but must exercise care to detect signals of minimum specified duration. When used with tone supervision, protection against false operation by speech power must be provided. One system, which can handle tone supervision for 128 analog channels and satisfies CCITT standards, contains 333 ICs and consumes 44 watts of power. The same principle can be used for recognition of DTMF and MF address signals, but must be far more elaborate to cope with the multiplicity of signals.[21]

11.4.3. Digital Conferencing

Analog conference calls typically involve a conference bridge which adds all signals together so that each conferee is heard by all others. With PCM switching systems, the same conference bridge can be used, but it requires converting all signals back to analog and then encoding the composite conference signals for transmission to all conferees. With two-wire transmission on subscriber lines, all signals are passed through a hybrid network for conversion to separate transmit and receive paths required for the digital switching system. When several conferees are thus connected through a digital switching network to an analog conference bridge, the multiplicity of hybrid network impedance mismatches can cause severe instability. This can result in severe echo or even singing, irrespective of the introduction of some attenuation. Therefore, in a digital switching system, digital conferencing is preferred.

There are various methods of performing digital conferencing in a companded PCM system. Some early conferencing systems used an instant speaker algorithm in a switching-type bridge. During conversation, the probability of speech power in any random 8-kHz sample is quite low. If the speech level in each sample of several conferees is compared to identify the sample having the largest binary number, the sample is generally that of the active speaker. The conference processor then transmits that sample to all conferees during the next frame while blocking all other samples. This method involves rather simple circuitry for three-way conference bridges but becomes much more elaborate for larger conferences. The effect of the switching-type conference bridge is similar to that of a long trunk connection with echo suppressors[22] and may be further impaired in the presence of background noise.

Another method involves a summing-type conference bridge using a summing algorithm. Since standard PCM uses a companding algorithm, simple binary addition is not possible. The compressed PCM samples first must be expanded to linear samples. Then all samples in a single conference frame can be summed arithmetically and compressed for transmission to conferees. During the summing process, the sample representing each speaker's own voice signal is removed from the summed data before it is compressed and sent back to that conferee's receiver. Depending upon the number of conferees, some attenuation may be added to the composite signal before compression. The effect of echo is reduced if the sign bits of one-half the incoming samples are inverted, producing a phase inversion to give the conference better stability.[23]

One designer combined elements of the two conferencing methods described above as shown in Fig. 11.8. A loudness-analysis circuit estimates the loudness of each conferee's signal. The loudest signals are expanded and summed in pairs before recompression and transmission to conferees. Attenuation over a range of 0–20 dB is selected for each speaker, based upon that signal's actual loudness as well as its relative loudness compared to others. This is designed to provide a gradual blending of voices when one speaker interrupts another and to provide zero loss for the path from the conference speaker to other conferees. Automatic gain control is used to increase gradually the loudness of the active speaker and decrease gradually the sound volume of the background when no one is speaking.[24]

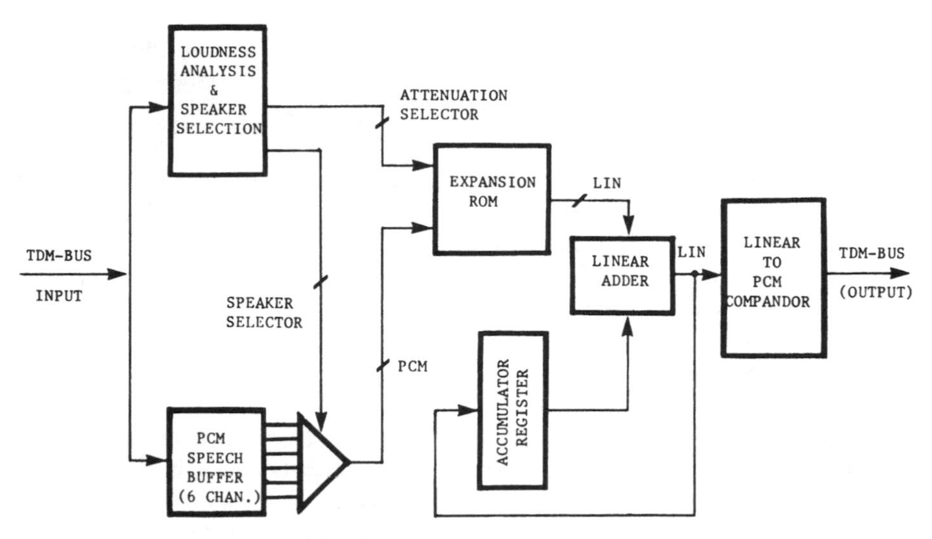

FIGURE 11.8. Digital conference circuit. (From E. A. Munter, "Digital Switch Digitalks," IEEE Communications Magazine, Nov. 1982, © IEEE.)

Audio and data bridge circuits can bridge up to about 60 conference legs using echo cancelers, noise limiters, and automatic gain control on each conference leg.

11.4.4. Digital Recorded Announcements

In the public telephone network, recorded announcements must be available to many subscribers simultaneously. Analog announcement machines generally use a magnetic disk or a continuous magnetic tape. Digital recorder-announcers are an outgrowth of the same techniques used in digital tone generators and digital conference circuits. Speech segments are stored in memory and read out under processor control. Since standard announcements contain substantial redundancy of expression, memory can be conserved by recording the announcements in segments which can be arranged in the sequence needed. Memory requirements can be reduced further if the speech samples are stored in 4-bit ADPCM instead of 8-bit PCM. In such cases, the speech is converted from PCM to ADPCM when it is recorded. An ADPCM-to-PCM decoder is used to provide the PCM-to-ADPCM encoder with prediction values and scaling factors. Speech is read out in segments, including phrases, subphrases, and pauses to multiple output channels.[25]

11.4.5. Digital Echo Suppression

Echo suppressors or echo cancelers are used to control echo on network trunks having a round-trip delay of more than about 45 ms and on certain other trunks.

For many years, echo suppressors had been provided on a per-trunk basis at each end of the trunk. Now, digital technology permits echo suppressor modules to be provided in digital switching systems and inserted into trunk connections as needed.

In normal operation, voice-actuated echo suppressors attenuate signals in one direction (echo) while permitting signals in the other direction (talker) to pass through unattenuated. They can be overridden by both talkers speaking simultaneously. Suppressors can be disabled for data communications by transmission of a 2100-Hz tone in each direction followed by continued presence of a strong data signal.

Digital echo-suppressor modules can be provided on a pooled basis and switched into and out of trunk connections as required under processor control. The processor prevents connection of multiple echo suppressors in tandem. The optional echo-suppressor terminal in the 4ESS toll switching system, for example, can serve up to 1680 trunks simultaneously. Under processor control, echo suppression can be switched into trunk connections in the form needed.[26] Other digital echo suppressors are packaged in various configurations.

When satellite circuits are involved in telephone connections, conventional echo suppressors do not provide sufficient suppression of echo, and the resulting speech clipping is not acceptable to the general public. Digital echo cancelers (see Section 5.6) enable satellite circuits to be used in general-purpose telephone connections. Echo cancelers now are used increasingly on terrestrial circuits in lieu of echo suppressors.

11.4.6. Digital Pads

In telephone networks, the level of signals is closely controlled between the end switching systems by a system of amplifiers to increase signal level and pads to attenuate it. In analog implementation, amplifiers and pads are separate devices. Pads are simple resistors arranged to attenuate signals a specified amount. In digital implementation, digital pads can either amplify or attenuate signals by specified amounts.

A digital pad is a lookup table in ROM in which all desired levels of PCM samples, except the sign bit, are stored. Input PCM samples address locations in the same ROM, and a control input designates the amount of attenuation or gain required. Correlation of the two inputs produces the attenuated or amplified output. Digital pads are shared devices switched into connections which require loss or gain according to a transmission plan.[27] Since digital pads actually obliterate the original bit stream, digital data communication is not possible where they are used. Therefore, digital pads are used only in PBX applications and in local central offices where their use can be restricted to nondata transmissions. Digital pads do result in reduced signal-to-distortion performance comparable to that caused by the use of back-to-back digital channel banks.[28]

11.4.7. Provisioning of Service Circuits

Three different performance criteria are used in the North American network for the provisioning of service circuits. The Ten High Day Busy Hour (THDBH) is the clock hour which has the highest average traffic load for the ten highest-traffic days of the year, excluding Mother's Day, Christmas, and high-traffic days attributable to unusually severe weather or catastrophic events. The High Day Busy Hour (HDBH) is the busy hour of the one day of the ten which has the highest traffic during the busy hour. The Average Busy Season Busy Hour (ABSBH) is the clock hour which has the highest average traffic in the three highest-traffic months of the year. The three months of the "busy season" do not have to be consecutive, and Christmas, Mother's Day, and high-traffic days attributable to unusually severe weather and catastrophic events are excluded.

Typical standards of performance for most service circuits in the North American network are shown in Table 11.1.[29] Tone circuits and ringing circuits have the most stringent requirement and are provisioned more liberally than any other service circuit. General engineering guidelines for pooled common equipment require that the most liberal provisioning apply to equipment necessary to process a call during call setup. Tone circuits, providing call progress tones, and ringing circuits fit that requirement. The least liberally provisioned equipment is that required to originate a call, following the principle that calls which already have entered the network should be given precedence over calls just being attempted. The dial-tone delay requirements and DTMF receiver standards illustrate this principle.

11.5. CONTROL ARCHITECTURES

Three basic control system architectures, each with variations, are used in digital switching systems. All systems use stored-program control, but some use wired logic processors in specialized applications. *Central control, shared control*, and *distributed control* systems have evolved with several variations, including hybrid systems, which make them almost indistinguishable as basic architectures.

11.5.1. Control Workload Distribution

A stored-program control system is an aggregation of several separate programs which work together to process calls. Generally, there is an executive program that coordinates and schedules the work of specific application programs. The tasks performed by application programs are grouped in many ways. One program may be concerned only with the on-hook or off-hook status of lines or trunks, or both. Another may control path setup through all or only a portion of the switching network. A separate program may control service circuits, while another controls

TABLE 11.1. Provisioning of Service Circuits

Service Circuit Element	Time Frame	Probability of All Circuits Busy
DTMF receivers	HDBH[a]	<5%
Interoffice receivers	ABSBH[b]	<1%
Interoffice transmitters	THDBH[c]	<1%
Tone circuits	HDBH	<0.1%
Ringing circuits	HDBH	<0.1%
Dial tone delay	ABSBH	Max. 1.5% >3 seconds
	THDBH	Max. 8% >3 seconds
	HDBH	Max. 20% >3 seconds
Conference circuits	ABSBH	<1% of 3 port attempts
Announcement circuits	ABSBH	<1% with overflow to tone trunks

[a]High Day Busy Hour
[b]Average Busy Season Busy Hour
[c]Ten High Day Busy Hour

call routing. When all control functions are combined into one processor, the executive program controls and schedules the work of the application programs. Not all of the work performed by control systems is related to call processing. The control system may be working at 30–50% or more of its capacity on housekeeping functions even when there is no traffic load at all. Processor time required to run the executive program and to perform the essential functions related to maintenance and administration is known as *fixed overhead*. That time is virtually the same for all switching systems of a specific type, irrespective of the level of traffic and the amount of equipment in the system. *Variable overhead time* is the time spent on tasks performed at a constant rate but at a lower priority than fixed overhead tasks. This time includes such tasks as detecting and responding to line and trunk originations.

The time spent in processing calls is highly variable and is dependent upon the level of traffic. This time is zero if there are no demands for service. Once initiated, however, call processing is performed on a priority basis ahead of variable overhead time. The remainder of the work required by a control system is *deferrable work time*. This includes useful tasks which do not have critical timing requirements, such as administration, audits, and routine maintenance. Audit programs check the status of resources such as hardware subsystems, data blocks, and overall system operation. If the switching system is operating at full traffic capacity, the control system may perform no deferrable work. As a peak traffic load subsides, however, the control system uses the increasingly available time to perform deferrable work. Deferrable work is sometimes known as *fill time*.[30]

11.5.2. Central Control Systems

In a central control system, all work described in Section 12.5.1 is performed by a single processor. In a small or medium-sized PBX, this generally is a microprocessor or minicomputer. In a large central office, however, a more powerful computer is required.

Tasks may be scheduled in different ways. In a timed schedule, tasks are grouped into classes and assigned to fixed time increments, or slots. Scheduled classes in each slot are arranged in priority sequence. The last class of tasks in each slot is comprised of fill work. During each time increment, scheduled classes of tasks are performed sequentially. When all scheduled tasks have been completed, the remainder of the slot time is occupied with fill work. If traffic becomes so heavy that there is insufficient allotted time in the slot to complete all scheduled tasks, the executive program must resolve that situation and decide whether to complete scheduled tasks in the slot or to proceed to the next slot of tasks.

Central control systems can follow a cyclic schedule of tasks. All classes of tasks are arranged sequentially by priority and by required frequency of performance. Thus, one class of tasks may be performed several times as often as another class. There is no fixed cycle time. At low-traffic loads, the time required to cycle through all tasks is short. As traffic builds up, cycle time increases and fewer complete cycles are executed in a given time period. If a particular task takes an exceptionally long time to complete, the time between executions of higher frequency tasks could increase to a point at which it would exceed a threshold established to ensure a particular quality of service. In such cases, the executive program can limit the amount of work done in a class so as not to jeopardize system operation. For example, during traffic overloads, a maximum limit can be placed on the number of new call originations that will be processed during the task of recognizing new calls to be processed. This maximum limit can be set so that new calls that are recognized are processed efficiently, enabling the system to reduce the backlog of work in other critical tasks. As the backlog of work decreases, the task of recognizing new call originations will occur more frequently, and more new calls will be allowed into the system per unit of time. Cyclic schedules function well in heavy traffic and require low overhead.

Hybrid schedules can be designed so that tasks having critical timing requirements are executed according to a timed schedule while other tasks not having such timing requirement are executed according to a cyclic schedule. Various combinations are possible. One arrangement is to perform cyclic tasks when there is no other work to be done and interrupt them at each time interval when time scheduled tasks are to be executed. Another arrangement involves grouping tasks into classes according to their required frequency of performance as in the cyclic schedule. In this case, however, a minimum time between executions of each class is specified. Thus, the timed schedule controls task execution during light-traffic periods. During heavy traffic, the cyclic schedule controls task execution with interrupts to conform to the timed schedule. This arrangement ensures timely execution of tasks having critical timing requirements, while taking advantage of the attractive characteristics of cyclic schedules in heavy traffic.[31]

The complexity of central control systems increases with size, and failures can be catastrophic. Redundancy is essential for high reliability in large switching systems, and is highly desirable in any system. The central control generally is fully duplicated, and continuously monitoring maintenance programs effect automatic switchover in case failures exceed a threshold. Even then, central control systems have been known to fail completely and to require many hours to restore service.

11.5.3. Shared Control Systems

In a shared-control system, multiple processors (generally two) share the load. Each processor works at approximately 40% capacity while the switching system is operating at full-rated capacity. This concept reduces the probability of processor failures during periods of extremely heavy random traffic, and the reserve processor capacity is able more easily to absorb peak traffic loads without jeopardizing critical functions. Each processor can monitor the performance of the other and can handle the full-rated switching system load in case the other processor fails. Processors can be triplicated, in which case two processors share the load while the spare monitors their operation and is automatically switched into service if one fails. The processor complexity, however, is the same as in central control systems because each processor handles all control functions.

11.5.4. Distributed Control Systems

The definite trend in switching-system architecture is toward distributed control systems. Processors may perform specialized tasks only, or they may perform all or virtually all control tasks associated with a group of lines and trunks.

11.5.4.1. Distribution of Control by Function

A popular method of distributed control is to allocate a group of specialized tasks to a controller which does nothing but execute those tasks repeatedly. The application programs for such tasks are relatively simple and are ideally suited to microprocessor technology.

A typical example of a specialized processor is that of a line processor which continually detects on-hook and off-hook conditions of a group of lines, sends that information to another processor, and starts and stops ringing. Most central control systems have allocated signal processing tasks to a separate, stored-program control processor to perform such tasks as line and trunk scanning, reception of incoming address signals, and control of outgoing address signals.

11.5.4.2. Distribution of Control by Block Size

There is an increasing trend toward allocation of all or most processing functions to a controller for a group, or block, of lines and trunks. In a typical distributed control system, a central controller maintains orderly operation of a number of line and trunk controllers and also engages in path selection. In this case, a terminal module processor performs all tasks associated with a group of lines and trunks, including path selection within the module, while a main processor manages centralized data-base information, performs path selection between terminal modules, and executes other centralized programs. A major advantage of terminal-module control is that a switching system can grow modularly without incurring an extremely high initial cost for a powerful central computer designed for ultimate system size. A second advantage is that virtually autonomous terminal modules can be located in outlying areas and operated as remote switching modules with host control of centralized functions.[32]

11.6. MAINTENANCE DIAGNOSTICS AND ADMINISTRATION

In addition to call processing, a digital switching system must have programs to perform diagnostics, collect billing and traffic data, provide for program and database changes, and make traffic-load analyses for network management purposes. These tasks generally are performed by applications processors in communication with call processing controllers.

11.6.1. Maintenance Diagnostics

An ideal design objective is a self-healing switching system which can detect and correct all faults without human intervention. In the absence of such perfection, diagnostic programs must be designed to detect and correct faults whenever possible and to alert maintenance personnel to the existence of uncorrected faults. System effectiveness and maintainability are directly attributable to the consideration given them during architectural design. Functional units that are critical to maintaining continuity of service should be duplicated so that a unit can be removed from service without impairing call-processing capability. Traffic-sensitive common equipment, such as service circuits, typically is provided on the basis of $n + 1$ redundancy.

Equipment should be grouped in such a manner as to facilitate maintenance procedures. This enables a group of hardware units to be removed from service without affecting other portions of the system. This includes partitioning of power supplies such that power-supply faults affect only a limited group of equipment.

11.6.1.1. Maintenance Phases

System maintenance involves seven basic phases: *fault detection, fault analysis, fault isolation, fault reporting, fault localization, fault clearance,* and *service restoration.*[33] While functionally separate, these phases may overlap.

There are two common methods of fault detection, and both are used concurrently. Hardware and software may be designed to activate alarms and initiate diagnostic tests when call failures or other fault conditions occur. Other faults are detected by on-line tests performed routinely as part of the operating program, or they may be performed as audit routines to check blocks of hardware and software.

Ideally, fault-detection processes also should localize the fault to a particular group of equipment units. In case a fault involves more than one equipment group or, for some other reason, is not readily identifiable with an equipment group, further analysis is needed. Failure of path setup, for example, may result from a fault in any of several equipment groups. Analysis may be performed by logical deduction, by historical comparison, or by specific triggering of diagnostic tests.

Once the fault is identified with a specific equipment group, action must be taken to isolate the fault from the rest of the system to avoid further degradation. Traffic may be diverted to a redundant equipment group if one exists, or a faulty group may be locked out of service.

Statistical reporting of faults is essential for effective maintenance. Printed records of fault characteristics and the results of automatic diagnostic tests serve two major purposes. They assist technicians in further actions to clear the fault and restore full service, and they provide a historical record which can be used to develop engineering changes to make the system more reliable. Visual and audible alarms may accompany printed reports when service impairment exceeds a preset threshold.

The next maintenance phase is fault localization. Its purpose is to identify specific faulty equipment within the equipment group. A fault may be localized automatically to a specific printed-circuit board by the diagnostic process, which is the ideal objective, or it may require action by a technician. This is a trade-off between the ideal objective and the cost to achieve it. On-demand diagnostic programs can be initiated by technicians when needed to localize the fault.

Fault clearance involves physical replacement of the defective board or boards. When a board is replaced with a spare, a test program should be run to test the spare board before returning the faulty equipment unit to service. Faulty boards are sent to a repair shop where off-line tests are performed to ascertain the exact trouble and to effect repairs.

The final phase is restoration of service. After the equipment unit with its replacement board is tested, the entire equipment group is functionally tested. When this test is satisfactorily passed, the group can be returned to service in the switching system.

11.6.1.2. Diagnostic Methods

A major advantage of using a 32-channel PCM format, or a multiple thereof, in a switching system is that 30 channels can be used as traffic channels and two channels are available as maintenance and control channels. Some of the 16 bits thus made available in each frame can be used for control messages. Some can be used for parity checks while others can be used for integrity checks.

On-line tests can be used to check control paths between a controller and the equipment units controlled. Detailed checks can be made to verify order execution, data validity, and transmission accuracy. Test transmissions can be looped back and compared. Data can be written into memory, then read out and compared. Network paths can be set up and checked for both continuity and data integrity. Codecs can be checked separately and in connection with network path checks. Signaling transmitters and receivers can be checked by connecting them through the switching network and comparing sent and received signals.

Numerous system-specific parameters are counted as events occur and compared against an alterable threshold. Such events as path check failures, parity failures, signaling timeouts, clock alarms, power fluctuations, loss of framing alignment, and many others are counted and recorded. When a threshold is exceeded, the data printout may be accompanied by audible and visual alarms, depending upon the predetermined urgency of that parameter. Some thresholds, when exceeded, may trigger automatic actions to initiate diagnostic tests, switch to spare equipment units, or curtail certain types of traffic. Periodic polling of equipment units can be performed on either a positive-response or negative-response basis. The positive-response concept looks for a response only from a defective unit. The unit may be so disabled, however, that it cannot respond, or it may be so busy with traffic that a response is seriously delayed. A negative-response concept looks for a positive response from all equipment units which are operating within established parameters. This more positively identifies faulty equipment units but does place an additional workload on the affected control systems. Both concepts are productive when judiciously used.

Since the mean time between failures of any equipment unit in a well engineered switching system is much greater than the mean time to repair that failure, most maintenance strategies are designed around an assumption that a trouble is the result of a single fault. In most cases, the single-fault assumption is correct, but not always. The prioritization of application programs (Section 11.5.1) often results in some maintenance programs being defined as deferred workload. Therefore, most of the maintenance logic of those programs is tested only during low-traffic periods. Units under test are removed from service, tested, and returned to service. When a fault occurs during a high-traffic period, the logic is called upon to isolate the fault while fully duplicated equipment is on line. This usually involves removing a unit from service before testing, removing it from the effects of heavy traffic when actually a unique combination of heavy random traffic may have uncovered a program error which would occur only during such rare conditions. Multiple faults and transient errors do occur and do complicate the maintenance programs.[34]

The most critical maintenance programs are those concerned with severe degradation of control systems which require restart of executive programs to recover system sanity. A common need for system recovery is caused by the mutilation of memory data. Other causes include loss of a vital function, loss of a major facility, problems in control software, multiple failure of redundant units, clock failure, and excessive failures of various types. Most systems have system-dependent recovery phases. When a system is first activated, initialization of all control programs is required. Afterward, required restarts may be applied only to portions of the system which are in trouble unless the system is shut down entirely. The first recovery phase generally saves all established calls and reinitializes temporary memory. The second phase may lose some calls but saves most of them; calls in process of setup are generally lost. The third phase reinitializes major portions of the system; all or most calls are lost. The fourth phase involves reinitialization of the entire system, and all calls are lost. The sequence and effect of the various recovery phases varies with the system.[35]

The criticality of self-diagnostics was illustrated on January 15, 1990 in the AT&T 4ESS network as a result of the insertion of new software concerning sanity restoration procedures. One trunk interface module in a New York switch developed a software problem, and the 4ESS sent a congestion signal to all connected switches that it was not accepting additional traffic. The interface equipment reinitialized itself, correcting the problem within about six seconds, and the 4ESS went back into service and began processing calls. Under the old software, the affected switch would exchange messages via common-channel signaling to verify that sanity was restored before normal traffic was resumed. The new software just provided that the affected switch would begin sending call routing signals associated with calls from subtending switches. Upon receipt of the first signaling initial address message (IAM) over the common-channel signaling system, a connected 4ESS was resetting its internal logic to recognize that the affected switch was back in service when a second IAM was received from the affected switch. The program in the connected switch tried to execute a resulting instruction that was illogical under the circumstances and then shut itself down, sending congestion signals to all of its connected switches. A chain reaction resulted in a major network outage because of receipt of a second IAM before its logic had been reset. Windows are needed at stages to allow time for human examination of the status before progressing beyond the point of effective human intervention.

11.6.2. Administration

Switching-system administration includes database management, generic program changes, and collection of data for billing, traffic engineering, service evaluation, and network management.

11.6.2.1. Database Management

The purpose of database management is to provide the capability of administering the switching system data efficiently on a continuing basis. All insertion, deletion, or modification of data is controlled by the database-management system. The data used in the switching system is organized into formats suitable for processing and for use in system memories, but not for administration. Database information must be available in clearly readable form. Therefore, the database-management system provides translation between the administrative view of the data and the machine-language formats.

The switching system database contains all configuration data pertaining to switching-system equipment and connectivity, including data on subscribers, lines, trunks, service circuits, switching-network equipment, call routing, features, traffic-data collection schedules, and billing information. Access to the database is provided by the database-management system for both administrative and maintenance purposes. The database-management system is provided in one or more of the processors associated with the switching system. An important prerequisite is that it must be designed into the switching system and not added as an afterthought.[36]

11.6.2.2. Generic Program Changes

A generic program is the set of executive program instructions that controls call processing, maintenance, and administrative functions. A means must be provided to modify the program to correct program errors, to provide new features or feature enhancements, to provide system-performance enhancements, or to consolidate accumulated program patches.

Program patches are minor changes to small portions of a generic program to correct program errors, to make minor changes in call processing or features implementation. Program retrofits are major revisions in a generic program, which should be implemented in a working system with minimum impact on subscriber service.[37]

11.6.2.3. Data Collection

Switching systems generate large quantities of statistical data in connection with call processing. Some of the data is used for billing purposes. Billing options for local calls vary according to tariffs that are effective in each locality. Generally, message-accounting systems used for toll billing can also be used for billing local timed message-rate calls. For toll billing, call data includes calling party identification, called telephone number, date, time call is answered, duration of call, and time call is disconnected. Switching system designs must include not only the capability to record all data required for long-distance calls but also options to record data needed for billing of services under many tariff variations and for special studies. Timing of calls must be adjusted for inaccuracies in timer operation, path setup, time delays after called party answer, and clock variations. If conference bridges are provided for three-way calling or larger conferences, timing must include the usage of conference bridges in billing data. Since billing is tariff-dependent, the data-collection system must be easily alterable.

Data collection also includes service measurements which are used to evaluate the quality of service being provided. Service measurements include total switching system traffic counts segregated by type and disposition of calls, customer access statistics, switching system irregularities, network service statistics, and maintenance measurements. While many of the measurements are standardized, some are dependent upon switching-system design. Schedules for data collection vary from intervals of a few seconds to every 24 hours.

11.6.3. Traffic Administration

Much of the collected data is used for multiple purposes, and traffic administration is one of the dominant users. Traffic administration comprises those actions necessary to ensure that equipment and circuit quantities are provided in sufficient number to provide a specified grade of service for assumed traffic loads. It includes both traffic measurements (data collection) and traffic engineering. Traffic considerations for both analog and digital switches are covered in Appendix B.

11.6.4. Network Management

Network management is covered in Section B.4.3.2. Digital switching systems typically have automatically applied controls to maintain a high level of call processing efficiency in the presence of brief periods of peak traffic loads and manually applied controls to protect the system when it is subjected to extended periods of overload. Traffic measurements and overload indicators should automatically alert network managers to service degradation. Indicators and measurements should be designed to identify as accurately and as soon as possible the causes of overload.[38]

An important element of the design of switching processors is to minimize the use of resources expended on processing ineffective attempts. Many telephone users lift the handset and begin dialing immediately without listening for dial tone. During switch overloads when dial tone may be noticeably delayed, those users may become "early dialers," resulting in partial digits being dialed before call abandonment or timeout. Historically, switching-system design has awarded dial tone on a first-in-first-out (FIFO) basis. During overloads with accompanying dial-tone delay, this treatment means that all attempts will be delayed, thereby exacerbating the problem. A last-in-first-out (LIFO) strategy can avoid the interval when many users will dial before receiving dial tone. This strategy attempts to provide dial tone either very quickly or after dialing is completed. In the latter case, an assumption must be made as to the time from the last digit dialed until call abandonment, typically 10–15 seconds. If, during significant overloads, dial tone is given after dialing has been completed and the subscriber is still present, the result is a successful call attempt (that is, an attempt that is processed to the completion of call processing even though it may find all circuits busy). This minimizes the resources used on ineffective attempts.[39]

PROBLEMS

11.1. What are BORSCHT functions and why are they potentially hazardous to digital switching systems using solid state technology?

11.2. What memory cycle time is required for the information memory in a time-slot interchange to switch 512 time slots? How many bits are required in the control memory?

11.3. When time-multiplexed lines are switched through a space-division stage of a digital switching system, why do input PCM code words retain the same time slot number in the output?

11.4. How does the number of time slots per time-multiplexed line affect the probability of blocking in time-space-time and space-time-space digital switching network architectures?

11.5. What factors complicate digital recognition of analog tone signals?

11.6. How are digital pads used to impart gain to a digital signal?

11.7. When the traffic on any processor increases to a level which precludes that processor from having sufficient real time available to process all demands for service, what is the effect and what can be done to provide immediate relief?

11.8. What are the advantages of distributed control over central control architecture?

11.9. What advantage does a 32-channel PCM format have over a 24-channel format in a switching system?

REFERENCES

1. Terry, J. B., Younge, D. R., and Matsunaga, R. T., "A Subscriber Line Interface for the DMS-100 Digital Switch," *National Telecommunications Conference Record, 1979*, IEEE Press, 1979, pp. 28.3.1–28.3.6.
2. Ibid.
3. Caves, Terry and McWalter, Ian, "Filter Codec and Line Card Chips: the New Generation," *Telesis*, No. 4, pp. 2–7 (Ottawa, Bell-Northern Research, 1983).
4. Bellamy, John C., *Digital Telephony*, second edition, John Wiley & Sons, New York, 1991, p. 251.
5. Ibid.
6. Rothmaier, Klaus, and Scheller, Reinhard, "Design of Economic PCM Arrays with a Prescribed Grade of Service," *IEEE Trans. Comm.*, p. 925 (July 1981) (hereafter cited as "Economic PCM Arrays.")
7. Ibid.
8. Ibid.
9. Huttenhoff, J. H., et al., "Peripheral System," *Bell Syst. Tech. J.*, pp. 1037–1041 (Sep. 1977).
10. Pitroda, Sam G., "Telephones Go Digital," *IEEE Spectrum*, p. 51 (Oct. 1979) (hereafter cited as Pitroda).
11. "Economic PCM Arrays," p. 926.
12. Ibid., pp. 930–931.
13. Lotze, Alfred, Rothmaier, Klaus, and Scheller, Reinhard, "TDM Versus SDM Switching Arrays—Comparison," *IEEE Trans. Comm.*, p. 1455 (Oct. 1981).
14. "Economic PCM Arrays," pp. 931–932.
15. McDonald, John C., "Techniques for Digital Switching," *IEEE Communications Society Magazine*, p. 11 (July 1978).
16. Richards, Philip, C., "Technological Evolution—The Making of a Survivable Switching System," in Joel, Amos E., Jr., ed., *Electronic Switching: Digital Central Office Systems of The World*, IEEE Press, New York, 1982, p. 196.
17. Charransol, Pierre, et al., "Development of a Time Division Switching Network Usable in a Very Large Range of Capacities," *IEEE Trans. Comm.*, p. 982 (July 1979).
18. Jajszczyk, Andrzej, "On Nonblocking Switching Networks Composed of Digital Symmetrical Matrices," *IEEE Trans. Comm.*, p. 2 (Jan. 1983).
19. Pitroda, p. 59; Munter, Ernst A., "Digital Switch Digitalks," *IEEE Communications Magazine*, p. 15 (Nov. 1982) (hereafter cited as Munter).
20. Munter, p. 18.
21. Ikeda, Yoshikaza, and Norigoe, Masamitsu, "New Realization of Discrete Fourier Transform Applied to Telephone Signaling System CCITT No. 5." *IEEE Global Telecommunications Conference Record, 1982*, IEEE Press, 1982, pp. D8.1.1–D8.1.6 (hereafter cited as GLOBECOM '82).
22. Munter, p. 19.
23. D'Ortenzio, Remo J., "Conferencing Fundamentals for Digital PABX Equipments," *IEEE International Conference on Communications Record, 1977*, IEEE Press, 1977, pp. 2.5–29 to 2.5–36 (hereafter cited as *ICC '77*).
24. Munter, pp. 19–20.

25. Ibid., pp. 20–23.
26. AT&T Practice 234-100-000, Issue 11, September 1990, p. 83.
27. Munter, pp. 17–18.
28. *Local Switching System General Requirements*, PUB48501, American Telephone and Telegraph Company, Basking Ridge, N.J., 1980, Section 7.4.11.1 (hereafter cited as *LSSGR*).
29. Ibid., Sections 11.2 and 11.4.
30. Brand, Joe E., and Warner, John C., "Processor Call Carrying Capacity Estimation for Stored Program Control Switching Systems," *Proc. IEEE*, p. 1342 (Sep. 1977).
31. Ibid., pp. 1344–1345.
32. Penney, Brian K., and Williams, J. W. J., "The Software Architecture for a Large Telephone Switch," *IEEE Trans. Comm.*, pp. 1369 (June 1982).
33. Treves, Sergio R., "Maintenance Strategies for PCM Circuit Switching," *Proc. IEEE* p. 1363 (Sep. 1977).
34. Willet, R. J., "Design of Recovery Strategies for a Fault-Tolerant No. 4 Electronic Switching System," *Bell Syst. Tech. J.*, pp. 3019–3040 (Dec. 1982).
35. Penney and Williams, p. 1372; Meyers, M. N., Routt, W. A., and Yoder, K. W., "Maintenance Software," *Bell Syst. Tech. J.*, pp. 1139–1167 (Sep. 1977).
36. *LSSGR*, Section 8.5.
37. Ibid., Section 8.6.
38. *LSSGR*, Section 5.3.8.
39. *Switching System Overload Control Generic Requirements*, Bellcore Technical Advisory TA-NWT-001358, Issue 1, January 1993.

12

Operational Switching Systems

12.1. INTRODUCTION

Chapter 11 discussed five aspects of digital switching architecture: *terminal interfaces, switching network considerations, service circuits, control architectures,* and *maintenance diagnostics and administration.* In this chapter, four operational switching systems have been selected to illustrate how the foregoing functions have been integrated into working systems. All are in service in the North American network. Each of these four systems is discussed in more detail than in the first edition of this book, and the reader will detect the system upgrading that has been accomplished in the interim. The data presented was extracted from information supplied by the manufacturers in various depths of detail.

The first system described is a central-office system designed for local and tandem service in locations of moderate size. Next, we discuss a very large central-office system that can be used in local, tandem, and toll switching and as a large private branch exchange. This is followed by a description of a purely toll (transit) switching system and then by a system that can be used in any network function.

12.2. SIEMENS STROMBERG-CARLSON DCO

12.2.1. DCO System Description

The (DCO)[1] was first placed in service in 1977 as a 1920-port system with line concentration. Four combined systems could support 7680 ports. The current version has been enhanced in several ways. Functionally, the DCO can serve as an Equal Access End Office (EAEO), as a local tandem office, and as a Signaling Point (SP) or Service Switching Point (SSP) within a Signaling System Number 7 (SS7) signaling network. Each switching sector can support 2048 ports, and four sectors can be combined to support 8192 ports. The full system can serve up to 32 400 lines as an end office or up to 70 000 lines as a network host with remote switching units.

The DCO has a network capacity of 220 000 *Hundred Call Seconds* (CCS) during the ABSBH. With enhanced processors, a maximum of 71 000 HDBH call attempts can be processed.

The DCO can be configured in a variety of ways from a single small office serving 3240 lines and 384 digital trunks to a large, host–remote network using remote line switches, remote line groups, and digital subscriber carrier systems.

12.2.2. DCO System Architecture

The overall system architecture of the DCO is depicted in simplified form in Fig. 12.1. Access by subscriber lines is through line circuits in the Local Line Switch (LLS) and thence via Port Group Highways (PGH) to a time-space-time switching network. The Trunks and Service Circuits Module provides the interface for analog trunks and converts them to digital streams for transmission via PGH to the switching network. It also stores PCM samples of service circuit tones for injection by the TSI into the signaling time slots of the PGHs. The digital trunk interface module (DS1M) provides an interface for digital trunks and converts the 24-line T1 format to the 30-channel format used on the PGH and in the switching network. The Control Complex contains the central processor and related message communications modules to control the distributed processors in the LLS and the assignment of PGH time slots by other modules. The Operations, Administration, and Maintenance (OAM) System performs maintenance diagnostics, traffic administration, and operations management of the DCO. The SS7 (see Section 13.3.1.2) can be interfaced either through the DS1M or through the Line Interface DSO Module under control of the Local Line Switch LLS.

12.2.2.1. Interfaces

Subscriber lines are interfaced to Line Groups in the LLS. Two other interfaces are described here.[2]

12.2.2.1.1. Digital Interface (DS1) Module. The DS1 Module (DS1M) provides a direct, digital interface between T1 span lines and a DCO or a Remote Line Switch (RLS). A DS1M contains up to eight T1 interface (TIF) boards and redundant pairs of T1 interface control boards, message assembler boards, and

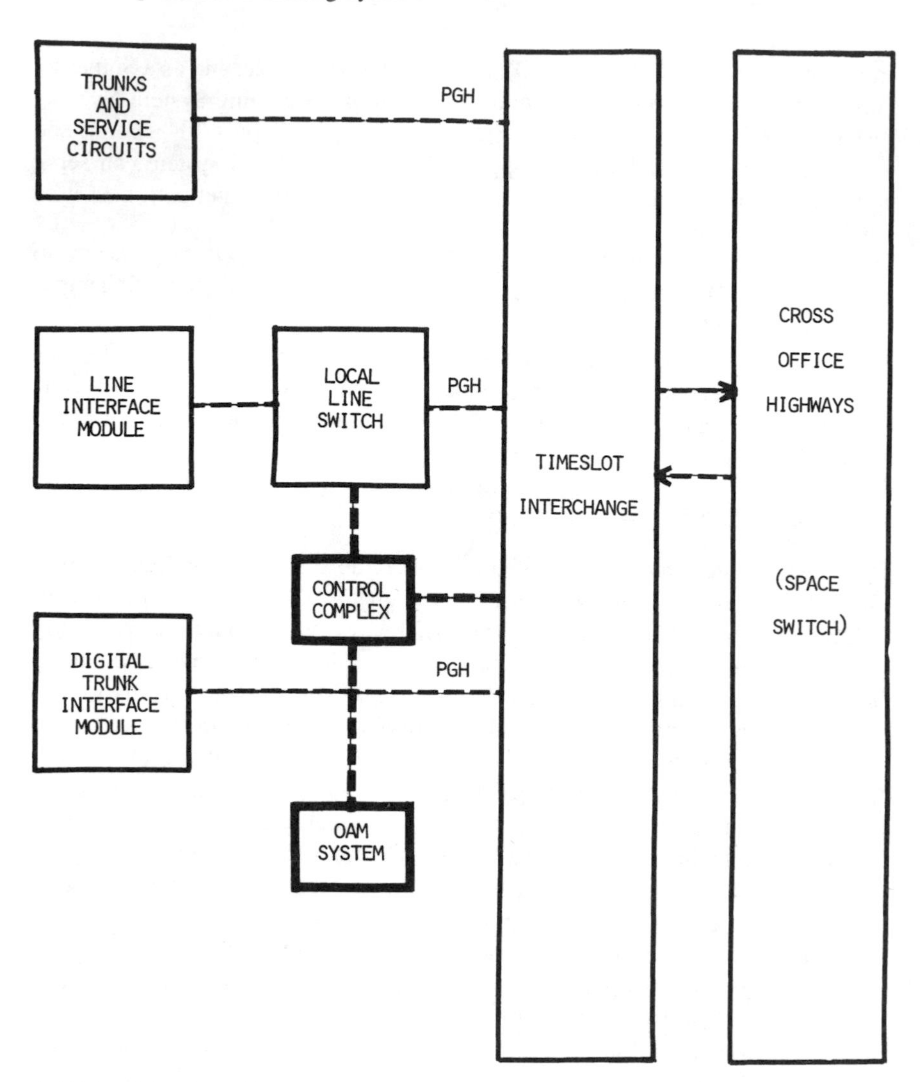

FIGURE 12.1. Simplified DCO block diagram.

power supply boards. The DS1M converts 24-channel T1 signals to the 30-timeslot format of the PGH and converts between the T1 bipolar format and the Non-Return-to-Zero (NRZ) format of the PGH. The DS1M also derives timing from the DCO system clock and provides synchronization for the T1 span lines. Maintenance functions, such as loopback testing, alarm sensing and reporting, are handled by a TIF Control in the DS1M.

The transmit section of each TIF receives data from the PGH and control messages from the message assembler. The data is converted from NRZ to bipolar format, synchronization is added, and the data is transmitted over the T1 span line. The receive section receives data from the T1 span line, regenerates the timing, synchronizes the data, converts the data from bipolar to NRZ format, and applies the data to the PGH and the message assembler.

Two channels on the T1 span line are used for control messages which are buffered by redundant message assemblers at both ends of the T1 span line. The message assemblers at the host office are slaved to the communications buffer controller in the DCO host, while the Line Switch Controller (LSC) at the RLS is slaved to the on-line message assembler at the RLS. With this slaving architecture, a T1 span line or cable failure will not cause a switchover of any redundant equipment at the DCO host or at the RLS.

12.2.2.1.2. Remote Group Interface to SLC-96. When SLC-96 Loop Carrier Systems are connected to a DCO or RLS, The SLC-96 is terminated in an interface contained in a remote group cell that is installed in place of one line group cell in a local or remote line switch, providing an interface between the T1 span lines of the SLC-96 and the Line Group Highways (LGH) of the LLS or RLS. The DCO treats the SLC-96 lines as if they were located in the remote group cell. The remote group cell contains a Line Group Multiplexer (LGM) with three identical circuits, each of which accepts voice and supervisory data from 30 SLC-96 lines and multiplexes the data onto one of three LGHs. Thus, only 90 of the SLC-96 lines can be supported. The LGM can assign any line in a 30-line group to any time slot on the associated LGH.

12.2.2.2. Local Line Switch

The LLS,[3] illustrated in Fig. 12.2, provides up to 7.0 CCS per line to a maximum of 1080 lines. Line interfaces are provided in up to 12 microprocessor-controlled Line Groups (LG) containing 90 line circuits and an LGM. Each line circuit detects the off-hook state and generates a supervisory sense bit that is sent to a Line Group Controller (LGC). Each LGC is the control interface for the 90 lines in the LG. The 90 lines are scanned by the LGC for supervisory data. A codec in each line circuit performs analog-digital conversion. Each LGM groups the 90 lines into three 30-line subgroups and multiplexes them onto three LGHs, each containing 32 time slots.

The 36 LGHs enter the redundant LGHs where: under control of an LSC, they are multiplexed onto one of 2 to 8 PGHs, achieving concentration of 4.5-to-1. The PGH carries voice and supervisory data to and from the TSI so that any line or trunk has access to any PGH time slot. The LGH and PGH data format carries data in sets of 256-bit master frames of 125 μs duration. Each master frame contains eight 32-bit subframes that contain one voice bit for each of the 30 time

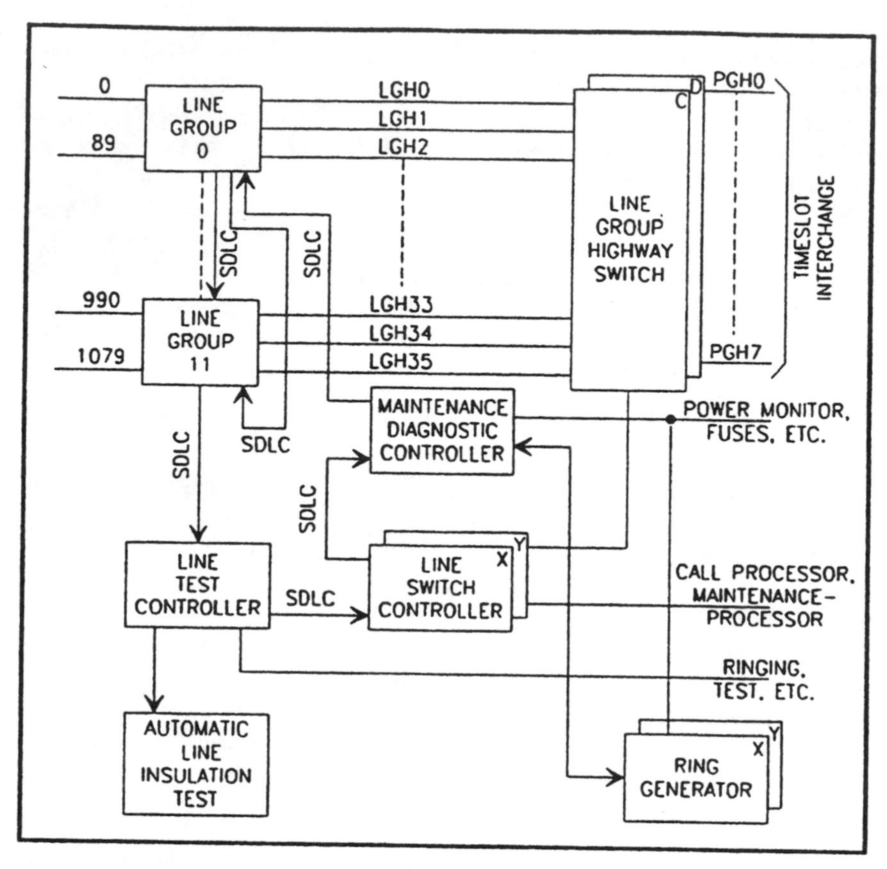

FIGURE 12.2. DCO Local Line Switch. (Reprinted from DCO General Description, Part 1, DCO, Publication 00-020-01, Issue 3, Page 31, Figure 5, August 1992. © 1992 by Siemens Stromberg-Carlson and reprinted by permission of the copyright owner.)

slots plus two control and supervisory bits. Each 15.625-μs subframe contains 32 equal time slots of 488 ns each.

The entire LLS is controlled by the LSC. The LSC monitors the line groups, assigns lines to PGH time slots, controls the maintenance and administration for the LLS, and interfaces the LLS to the call processors and maintenance processor through redundant Communications Buffer Controllers (CBC). The LSC also controls maintenance diagnostics and line testing. Communication between microprocessor controllers is via redundant, unidirectional Synchronous Data Link Control (SDLC) loops that operate synchronous serial data streams in a master-slave arrangement.

12.2.2.3. Common Control

Under direction of the Call Processor (CP), the Common Control[4] performs all telephony-related functions such as detecting requests for service, injecting tones, and switching port-to-port connections. Each common-control sector supports a maximum of 2048 ports. Up to four common-control sectors can be combined to support the maximum DCO configuration of 8192 ports. The common control is composed of three circuit groups: the Telephony Preprocessor TPP, the Time Slot Interchange TSI, and the Differential Bus Interface (DBI), as shown in Fig. 12.3 for one common-control sector. The TSI provides nonblocking connectivity, allowing any port to be connected to any other port for a two-way conversation path. The TSI *send* and *receive* buffers are interconnected by Cross Office Highway (XOH) lines. (The TSI and XOH are described in Section 12.2.2.4.)

The TPP consists of three printed-circuit boards. The TPP0 contains the Port Control Store (PCS), timing and control functions, and supervisory buffers. The TPP1 performs program logic functions, and the TPP2 performs background diagnostics and parity checking. The TPP performs the call-processing functions

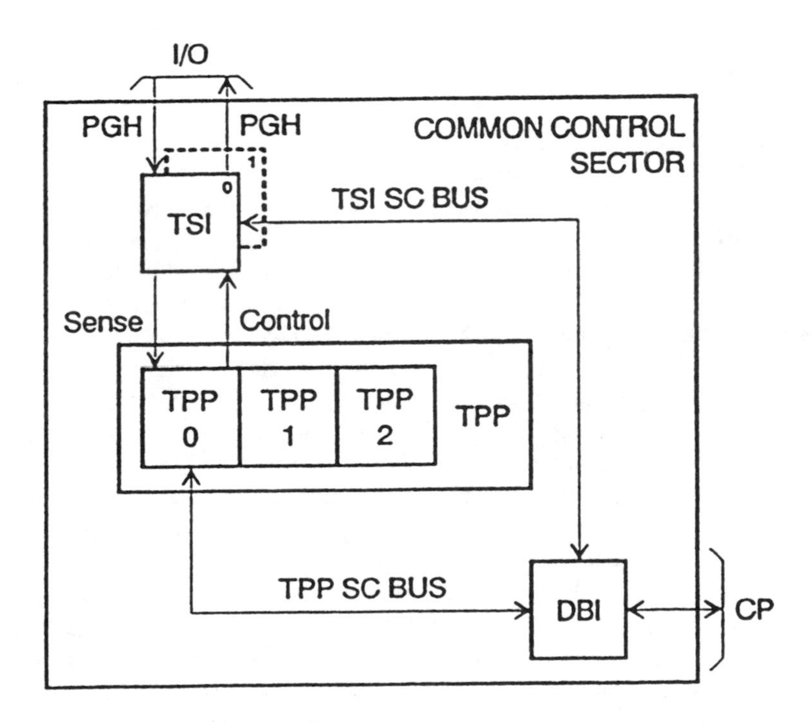

FIGURE 12.3. DCO Common-Control Sector. (Reprinted from DCO General Description, Part 3, Common Control, Publication 00-020-03, Issue 3, Page 3, Figure 1, August 1992. © 1992 by Siemens Stromberg-Carlson and reprinted by permission of the copyright owner.)

for one common-control sector. Random access memory (RAM) in the PCS, within the TPP, contains 2048 Port Store Areas (PSAs). Each PSA maintains the current call state for a specific port. Upon a change of state, the TPP notifies the CP of events that need a higher level of processing. The CP then commands the TPP to direct the port activity toward some general objective, such as "look for seizure" or "ring the line."

The DBI supports bus-to-bus communication among the TPP, TSI, and CP, and generates and distributes digital tones to the TSI, distributes timing signals to the TPP and TSI, and receives and analyzes test patterns routed through the TSI. Bus-to-bus communication paths are shown in Fig. 12.4. One Differential Bus connects the DBI to the A-side Differential Bus Converter (DBC), and one bus connects the DBI to the B-side DBC. Each Differential Bus accommodates 32 unidirectional address bits from the DBC and 32 bidirectional data bits. The transfer protocol is multiplexed address followed by multiplexed data. The TSI SC Bus terminates across the TSI circuits in all common-control sectors and accommodates 32 unidirectional address bits to the TSIs and 32 bidirectional data bits. The TPP SC Bus terminates on the TPP circuit in all common-control sectors and accommodates 32 bidirectional address bits and 32 bidirectional data bits. The CP and the Maintenance Processor (MP) communicate with the TPP and TSI circuits via the DBC and DBI.

12.2.2.4. Switching Network

The switching network[5] is composed of the redundant TSIs and an XOH that functions as a center-stage space switch in a time-space-time network. Each common-control sector can support two TSI boards; eight boards form a fully equipped TSI. A single TSI board is shown in Fig. 12.5. Each board terminates up to 32 PGHs with two-way communication paths. Each board sends data to any of 128 timeslots on any of 8 XOH lines and receives data from any of 128 time slots on any of 64 XOH lines. The XOH connects the send logic of one TSI board to the receive logic of that board and all other TSI boards. Under control of the CP, the receive logic accepts the data in the timeslot directed to it, thus establishing a connection through the TST network.

To establish a connection between two ports, the CP identifies the PGH number and PGH time slot assignments of the calling and called ports and selects an available XOH line and time slot assignment for each port. As the PCM words arrive at the TSI from the PGH, the TSI strips the supervisory sense bits from the signaling time slots and sends them to the TPP, then injects broadcast tone PCM samples into the same PGH signaling time slots. For both transmission directions, the TSI continuously:

- holds the sample until the assigned transmit-end XOH timeslot;
- cross-directs the sample across the assigned XOH, where it is accepted as a receive sample;

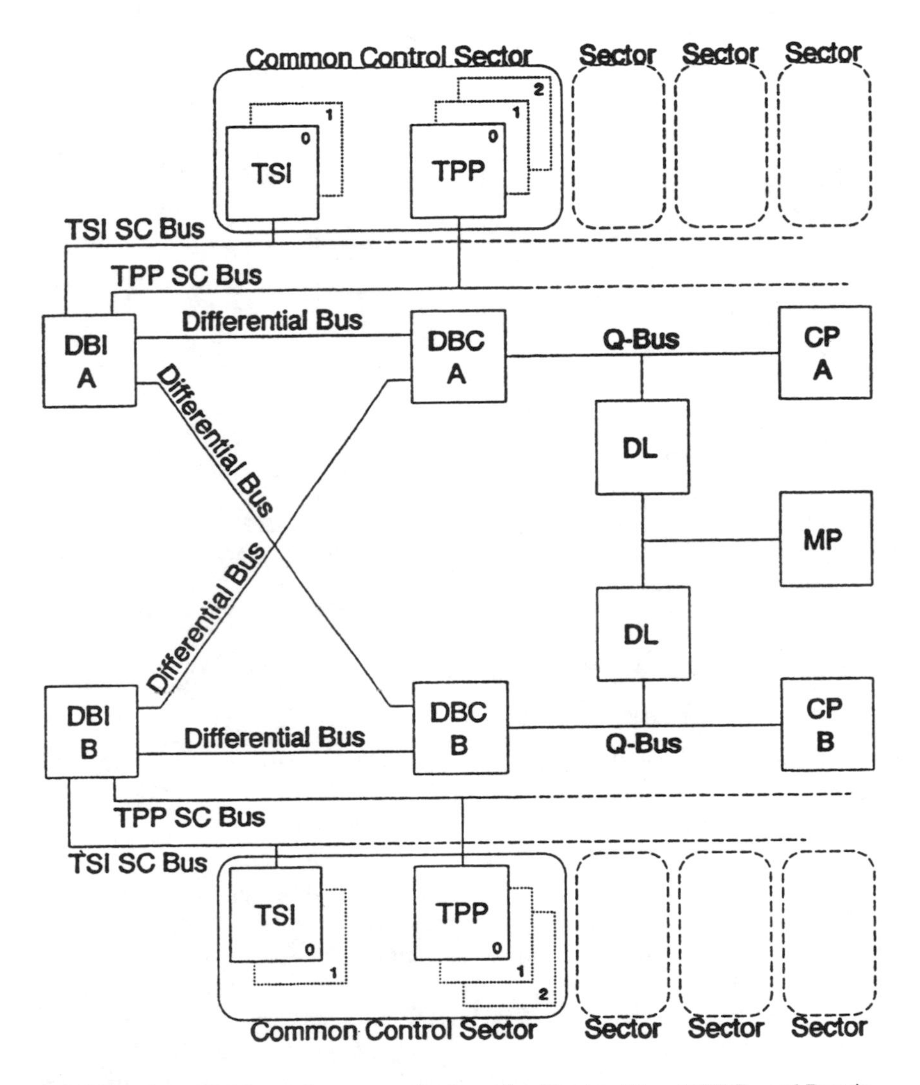

FIGURE 12.4. DCO bus-to-bus communication paths. (Reprinted from DCO General Description, Part 3, Common Control, Publication 00-020-03, Issue 3, Page 11, Figure 4, August 1992. © 1992 by Siemens Stromberg-Carlson and reprinted by permission of the copyright owner.)

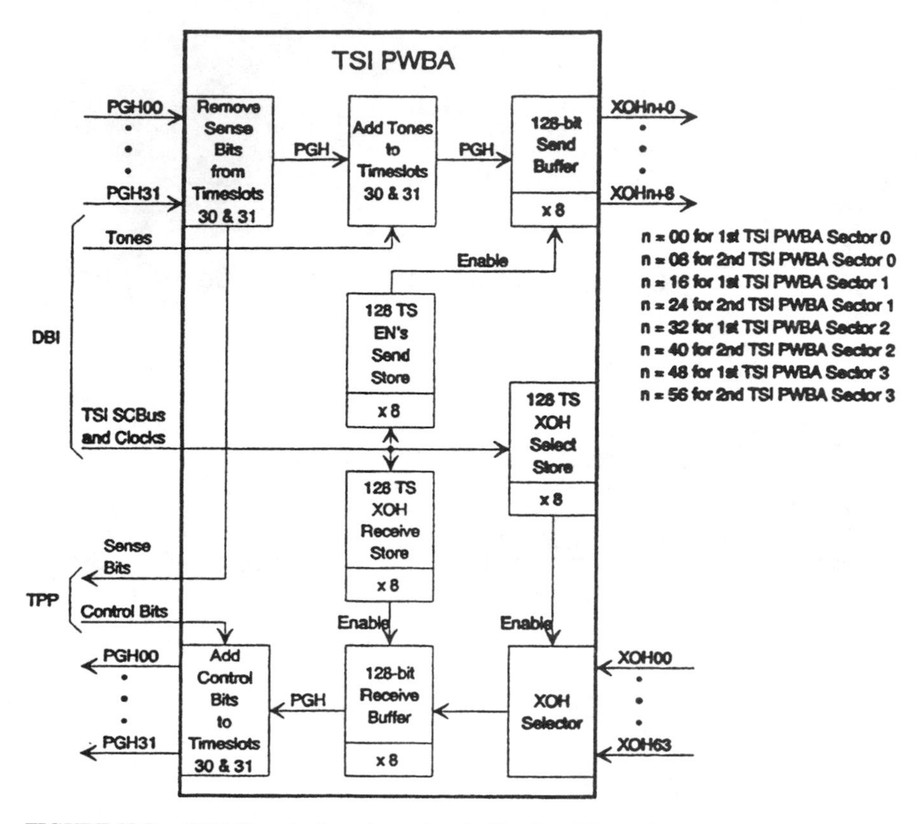

FIGURE 12.5 DCO Timeslot Interchange board. (Reprinted from DCO General Description, Part 3, Common Control, Publication 00-020-03, Issue 3, Page 7, Figure 3, August 1992. © 1992 by Siemens Stromberg-Carlson and reprinted by permission of the copyright owner.)

- holds the receive sample for a time related to the receive-end PGH time slot; and
- delivers the receive sample to the receive-end PGH.

When two to four common-control sectors are combined, XOH connectivity is shown in Fig. 12.6. Each rectangle represents one fully equipped sector containing two TSI boards. Each vertical and horizontal line represents one of 64 XOH lines, physically present in the backplane wiring, that overlays the TSI circuits. The horizontal portion of each XOH line represents the XOH input of the TSI, and the vertical portion of each XOH line represents the XOH output of the TSI. If each of the 128 time slots in each XOH line is used to cross-direct (transmit to receive) a PCM sample, a fully equipped TSI can establish a maximum of 4096 connections.

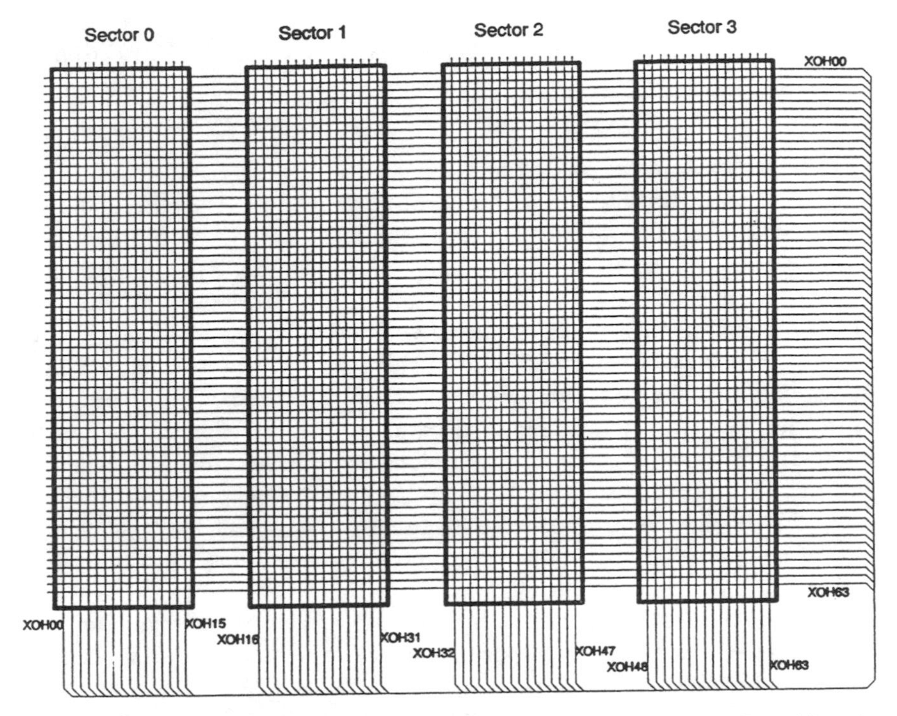

FIGURE 12.6. Cross Office Highway interconnectivity. (Reprinted from DCO General Description, Part 3, Common Control, Publication 00-020-03, Issue 3, Page 6, Figure 2, August 1992. © 1992 by Siemens Stromberg-Carlson and reprinted by permission of the copyright owner.)

Performance of the TSI is monitored by test patterns. The TPP sends repetitive 8-bit test patterns to the TSI. If the test pattern does not change, the TSI simply multiplexes it onto a single return line to send it back to the DBI. The DBI verifies that the test pattern appears in each time slot. If a time slot does not contain the test pattern, the DBI identifies the specific TSI board and reports the failure to the MP.

12.2.3. DCO Remote Operation

The DCO has the capability of operating with various sizes of remote switching equipment.

12.2.3.1. Remote Line Group

The Remote Line Group (RLG)[5] is a 90-line extension of the DCO. The RLG contains 90 subscriber line circuits, a remote TSI, an LGC, and an RLG Control Cell. The RLG Control Cell is the control interface for the RLG and is used at both the remote and the host locations to terminate the connecting T1 span line.

Any one or all of line groups in a DCO host can be replaced by an RLG Control Cell. In the host, the RLG Control Cell connects to the SDLC loops. The remote TSI performs the same function as the LGM in the host LLS.

12.2.3.2. Remote Line Switches

Two basic sizes of RLS can be connected to a DCO host. The RLS-1000 is a host-dependent remote switching system,[7] similar to the LLS, that contains up to 12 line groups serving 1080 lines, redundant LSCs and LGHs, and all ringing and line-testing equipment. Message assemblers and DS1 Modules provide host communications similar to that for the RLG. Each indoor RLS can serve up to three RLS-1000 systems serving up to 3240 lines. Two pad-mounted remote systems are available: the RLS-360 serves up to 360 lines, and the RLS-450 serves up to 450 lines. An optional intranodal switch, consisting of intranodal LGHs, intranodal LSC, and intranodal TSI, all redundant, provides up to 120 switching paths for line-to-line calls within the RLS. After a call is set up through the host office, the CP and LSC determine whether an intranodal switching path can be used. If so, the two lines are transferred to an intranodal line when the call is answered.

The RLS-4000 is a larger remote line switch,[8] capable of serving up to 4500 lines and providing backdoor trunking for up to 288 "backdoor" trunks via 12 T1 span lines to an end office or to a PBX. It contains up to 50 line groups that connect to TSIs via PGHs. The TSIs are interconnected by XOHs, thus producing the time-space-time switching employed by the DCO. Emergency switching capability is available if the host CP is inoperative or if host communications are disrupted.

12.2.3.3. Remote Network Switch

The Remote Network Switch (RNS)[9] is a large remote system that has all the inherent capabilities of a DCO system, including having subordinate remote elements. The basic design and traffic capacity approximate that of the DCO. The RNS can serve up to 7680 common ports serving lines and trunks. Voice is transmitted between host DCO and RNS via trunks, and redundant data links are used for maintenance and administration information and for Automatic Message Accounting (AMA) information.

12.3. AT&T 5ESS SWITCH

12.3.1. 5ESS General Description

The AT&T 5ESS Switch, formerly called the No. 5 ESS, is a large digital switching system capable of processing 300 000 call completions per hour as a tandem office and up to 600 000 POTS call completions per hour as a local central office. It also can function as an international gateway switch.

The 5ESS has been upgraded several times since its introduction in 1982. The 5ESS modular design enables incremental enhancements to individual modules. The distributed control permits call processing to occur in modules affected, thereby relieving the central processor of much repetitive work, and a fault in one module does not necessarily affect the rest of the system. By means of remote switching units, the 5ESS can support hundreds of thousands of subscriber lines. The functional nature of the switch can be changed by adding or changing hardware and software modules.[10]

12.3.2. 5ESS Hardware Architecture

The 5ESS hardware architecture consists of three major parts, as shown in Fig. 12.7.

12.3.2.1. Administrative Module

The Administrative Module (AM) performs administrative and maintenance functions and contains the interface to the SS7 network as shown in Fig. 12.8. The Common Network Interface (CNI) ring provides packet-switching capability for message routing between the switch and the SS7 signaling network.

The AM processor is a duplicated 3B20D Model 3 computer operating in an active/standby mode. The 3B20D performs maintenance functions such as fault detection, diagnostics, and fault recovery. Error-checking circuitry detects and isolates faults. The processor also collects traffic data and billing data, and is the hub of a star network configuration of controllers. It directly interfaces an input/output processor, which connects to external devices, and moving head disk files containing infrequently used programs and data as well as recovery programs.

The input/output processor functions as a front-end processor that controls transfers between the main memory store of the AM processor and the various peripheral units. It interfaces the master control center, remote terminals, and the operations and maintenance center. The master control center provides the local human–machine interface, maintaining system status displays and performing manual control of system operations.[11]

To accommodate the expanded capability of the 5ESS for higher switching capacity and broadband switching, the 3B20D processor is being replaced by the more powerful 3B21 processor, employing the same basic architecture.

12.3.2.2. Communications Module

The Communications Module (CM), shown in Fig. 12.9 is connected to the Switching Module(s) (SM) by fiber-optic Network, Control, and Timing (NCT) links. The Message Switch (MSGS) is a packet switch for transferring call-processing and administrative messages between any two SMs or between the SMs and the AM.

The duplicated Time Multiplexed Switch (TMS) provides three-stage space-

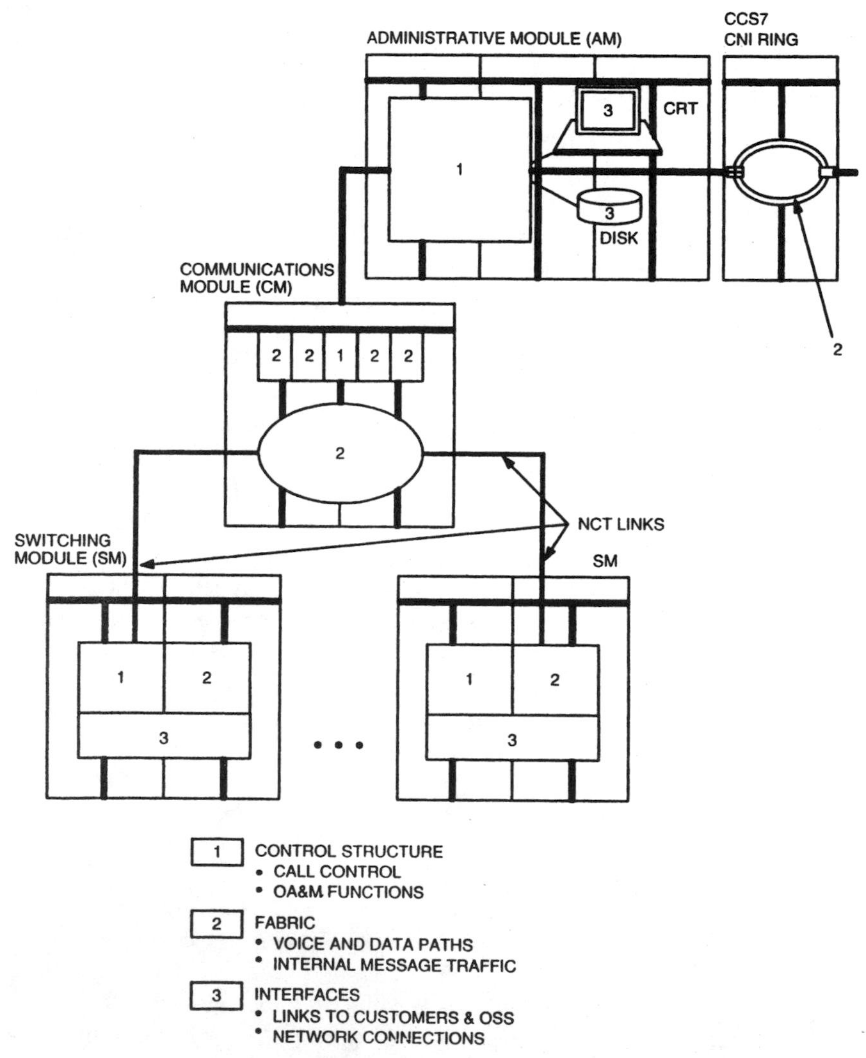

FIGURE 12.7. AT&T 5ESS Switch system architecture. (Copyright © 1993, AT&T Network Systems)

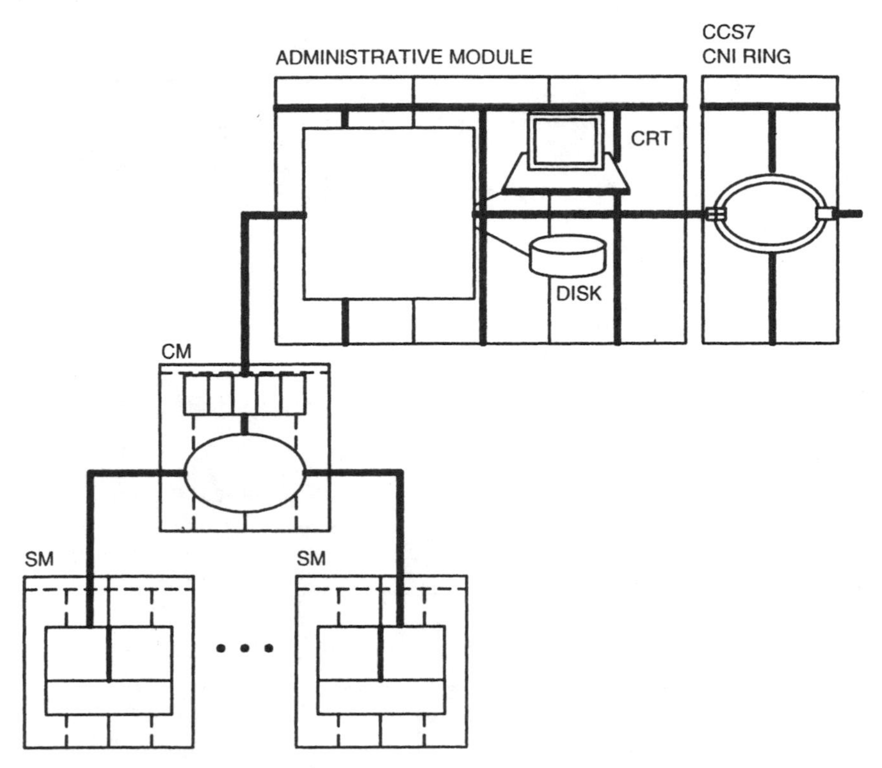

FIGURE 12.8 5ESS Switch Administrative Module. (Copyright © 1993, AT&T Network Systems.)

division switching for digital signals carrying voice, data, and control messages between the SMs and the AM. Each of two planes is connected by two NCT links to the SMs and by one NCT link to the message switch. The AM sends data from the modules to both TMS planes, but the modules accept data from the active TMS only. The standby TMS is activated if the active TMS develops a fault or if one of the NCT links fails.

A Communications Module Processor (CMP) is based on a Motorola MC-68030 microprocessor and performs tasks originally allocated to the AM. It provides real-time and memory relief for the AM and performs centralized control functions such as routing, global resource allocation, and recent changes. It performs call-processing functions, such as TMS path hunt and trunk hunt, and may be augmented in the future as more functions are migrated from the AM to the CMP.[12]

Clock timing pulses are distributed through the TMS. The 5ESS can operate synchronously by deriving timing from an external timing reference over digital links or plesiochronously by using its own frequency reference. The 5ESS net-

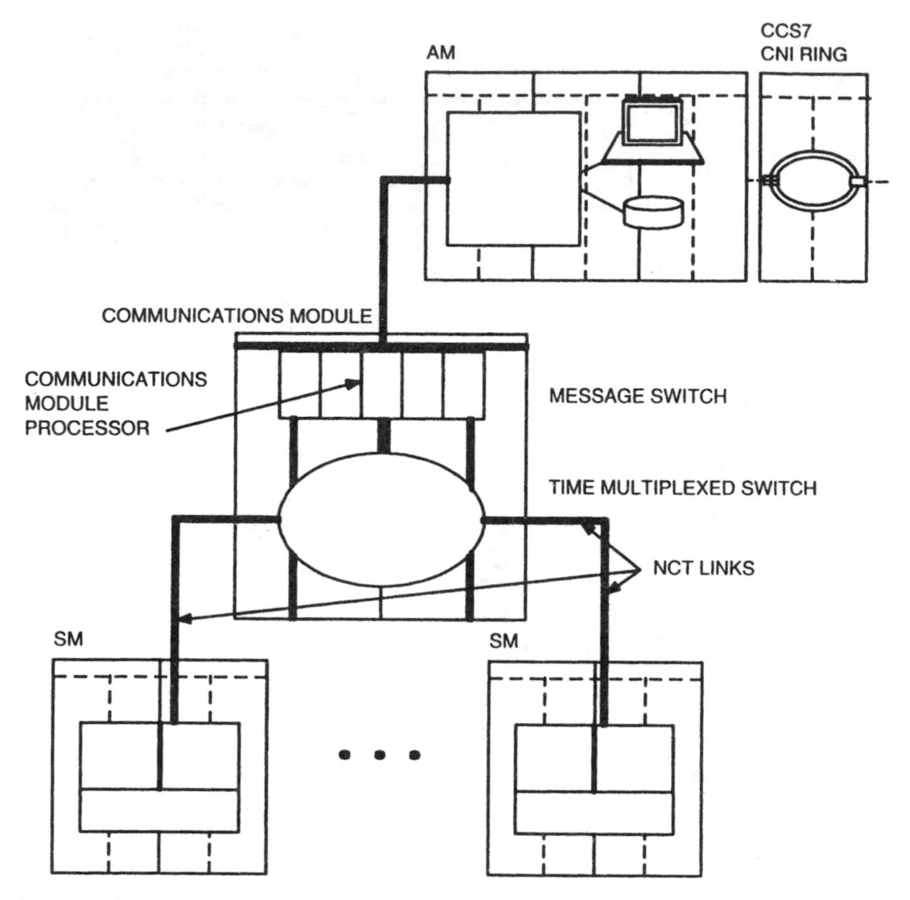

FIGURE 12.9. 5ESS Communications Module. (Copyright © 1993, AT&T Network Systems.)

work clocks are duplicated and operate in an active/standby mode. Medium- or high-stability oscillators are provided, depending upon the network application of the switch. A system with a high-stability clock can be upgraded to a national reference by adding an atomic frequency standard.[13]

12.3.2.3. Switching Module

The SM, shown in Fig. 12.10. provides line and trunk terminations, the first and last stages of switching, and call processing intelligence. Duplicated common equipment in all SMs includes a TSI and a Switching Module Processor (SMP). One or more SMs may be equipped with a Packet Switch Unit (PSU) when the 5ESS is serving digital subscriber lines in an Integrated Service Digital Network (ISDN). The PSU processes D-channel signaling information and switches user

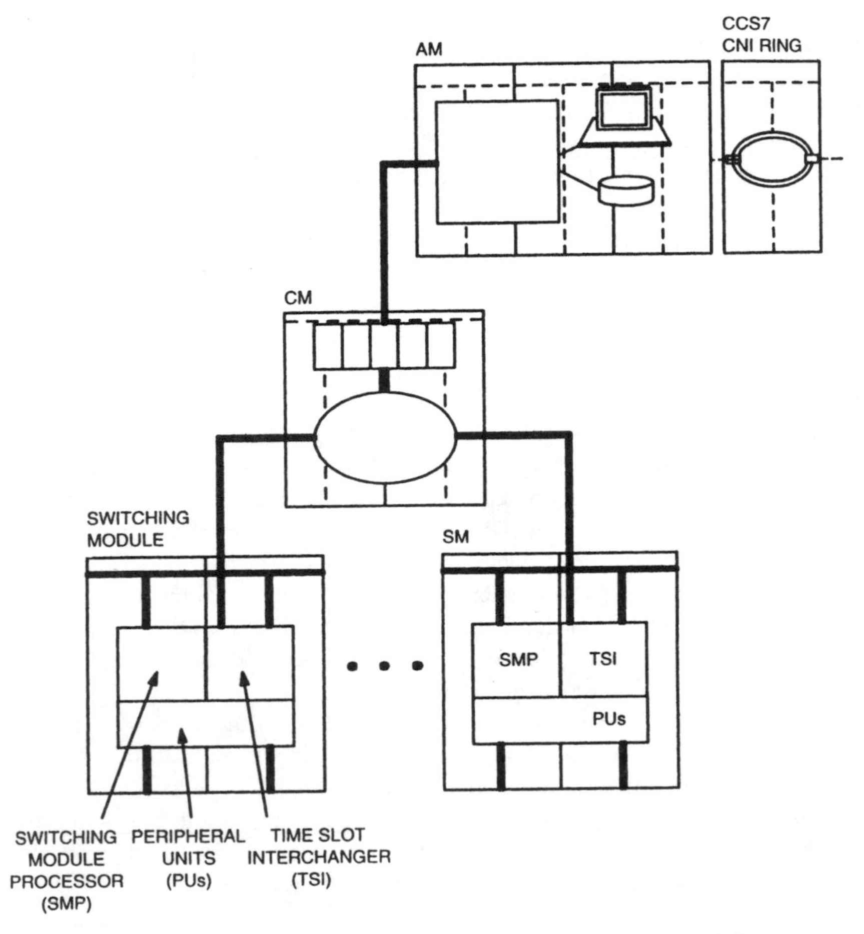

FIGURE 12.10. 5ESS Switching Module. (Copyright © 1993, AT&T Network Systems.)

packet data. The PSU protocol handlers are interconnected by a duplicated 10-Mb/s bus.

The SMP performs call processing for lines and trunks, provides control over peripheral units and the entire SM, and performs maintenance functions for the SM equipment. It is duplicated and operates in an active/standby mode.

The TSI can switch any of its 512 peripheral time slots to other peripheral units in the same SM or to any of 256 time slots in each of a pair of NCT links to the TMS. The TSI provides a special port to permit time-slot access by the local digital service unit for tone generation and tone decoding. A dual-link interface provides a two-way interface between each SM and the TMS. Each link terminates one active and one standby NCT link. The dual-link interface receives data from the TMS through the NCT and delivers the data to the TSI. In the reverse direction, it selects data from time slots in the active NCT link and delivers the data through the NCT link to the TMS. One time slot on each NCT link is dedicated to control messages between the MSGS and the SMP. The TSI and TMS combine to provide time-space-space-space-time (TSSST) switching architecture.

A variety of peripheral equipment is available for use in the SM. Analog line units and analog trunk units provide interfaces with analog lines and trunks, respectively. Each line unit can terminate up to 640 lines or PBX trunks. BORSCHT functions (see Section 11.2.1) are performed by channel circuits which are shared through a concentrator designed into the line unit. The concentrator is a solid-state crosspoint, 2-stage, space switch using high-voltage silicon integrated-circuit technology and provides concentration ratios of 10:1, 8:1, 6:1, or 4:1. The bipolar gated-diode crosspoints can withstand the high voltage required for ringing and line testing. The trunk unit terminates up to 64 analog trunks with no concentration. A Digital Line Trunk Unit (DLTU) contains digital interfaces between the 5ESS and digital transmission facilities. The DLTU provides up to 20 digital facility interfaces to DS1 facilities providing up to 480 digital channels. An Integrated Digital Carrier Unit (IDCU) provides an interface for Digital Loop Carrier (SLC) systems, including the SLC 96, the SLC series 5 systems, and Digital Access and Cross-connect Systems (DACS) facilities.[14]

High-level service circuits are provided uniformly on the basis of 4, 5, or 6 per line unit. They are used for 500 ms at various points in the call setup process and for the 2-second ringing intervals during line ringing. The tone decoder located in the Digital Service Unit (DSU) provides tone generation, cadencing, and decoding of dialed DTMF digits. With later software releases, the DSU also performs recorded announcement and integrated services test functions.[15]

12.3.2.4. Remote Switching Module

The 5ESS can operate with three Remote Switching Module (RSM) configurations, as shown in Fig. 12.11, connected to the host by digital facilities. A single-module RSM contains one SM and digital facility terminating and control units for 2 to 20 DS1 facilities. About 5000 subscriber lines can be served with 10:1 concentration and no trunks or optional equipment that would require trunk units. During normal operation, call processing is controlled by the 5ESS host. If the host is isolated, the RSM continues processing calls to lines and trunks directly connected to it, called stand-alone operation. The RSM can be located up to about 240 km (150 mi) from the host.

A multimodule RSM consists of two to four RSMs connected to each other by dedicated DS1 links. These RSMs must be located within a few hundred feet of each other in the same building. Each RSM also is connected to a host SM by

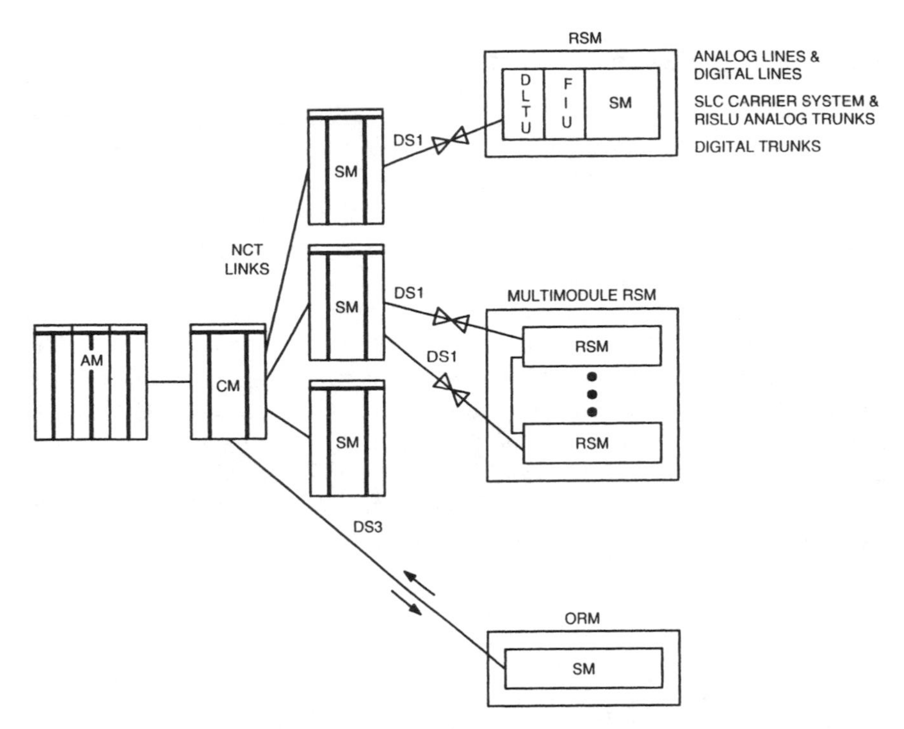

FIGURE 12.11. 5ESS Remote Switching Module architecture. (Copyright © 1993, AT&T Network Systems.)

DS1 links. Typically, the multimodule RSM can serve some 14 000 lines with four RSMs and a concentration ratio of 10:1. This configuration also has a stand-alone capability.

The third RSM configuration is called an Optically Remoted Switching Module (ORM); it provides the same services as the 5ESS host but connects directly to the host TMS rather than to the host TSI in the SM. There are three types of ORMs. A 3.2-km (2-mi) ORM can be remoted over an extended NCT link using multimode optical fiber. A 48-km (30-mi) ORM uses an interface circuit to connect to the CM in the host. This circuit converts the NCT rates to a 45-Mb/s DS3 rate and then multiplexes two of those channels onto a single 90-Mb/s channel. The output drives a repeaterless single-mode fiber link. A 240-km (150-mi) ORM can be remoted over a standard DS3 facility. Transmission converter hardware is required at both the host and the ORM to interface the NCT link to the DS3 facility, which may be either fiber or digital radio.[16]

12.3.3. 5ESS Software Architecture

The 5ESS uses a distributed, layered software system, illustrated in Fig. 12.12. The software architecture is a hierarchy of nested virtual (or logical) machines that span all processors and are structured in sequential layers. Each layer uses the services of any lower layer and provides additional services for the higher layers. The hardware is represented as another layer, consisting of the processors and their peripherals, at the bottom of the hierarchy.

An operating system for each processor runs on the base machine, while virtual machines running on the operating system provide more specialized services. A virtual machine above the operating system is the database manager. Databases also are distributed, with each having its own database manager. The abstract switching machine provides and controls logical entities such as terminal, port, connector, and path to provide switching functions. The application software is the highest layer and contains the call processing software and other subsystems for operations, maintenance, and administrative features.

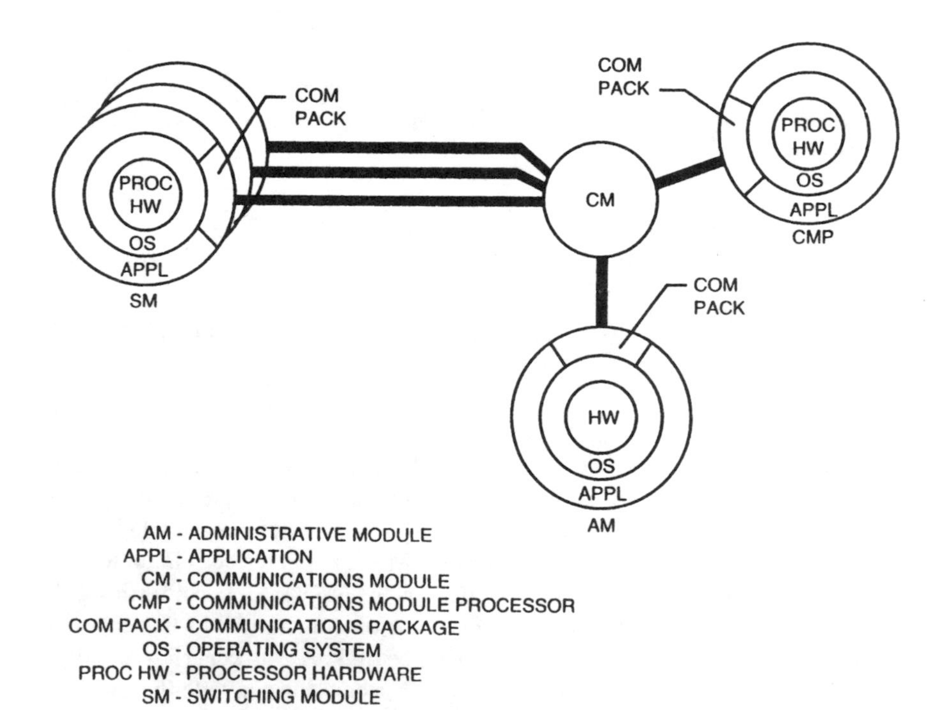

FIGURE 12.12. 5ESS layered software system. (Copyright © 1993, AT&T Network Systems.)

12.3.3.1. Operating System Software

The operating systems provide an interface between the software subsystems and the 5ESS distributed processors. The operating systems manage the software processes, provide communication among the different processes, allocate and control switching hardware, and provide access and control for bulk memory devices, associated work stations, and data links.

The UNIX Real-Time Reliable (RTR) operating system runs on the 3B20D computer in the AM. The Operating System for Distributed Switching (OSDS) runs on the RTR and provides a distributed processing environment in which the various software processes operate to share system resources.[17]

12.3.3.2. Call Processing Software

Call-processing functions are distributed among the processors in the SM, the CM, and the AM. Most functions are performed in the SM, relieving the CM and AM to perform mostly centralized functions. Call processing software is composed of three software subsystems. Peripheral control software performs functions necessary for switching and establishing call connections. It detects an off-hook state and calls the RTA programs to create a feature control terminal process to control the call. Peripheral control connects required service circuits and provides the required transmission characteristics, such as switchable pads. It maintains the status of switching resources and provides maintenance and other data to the administrative subsystems.

The feature-control subsystem controls the sequencing of call-processing functions. It collects and interprets dialed digits and requests the RTA subsystem to provide routing information. Feature control interprets all inputs in the context of the current call state and sends instructions to peripheral control to take the necessary actions. Call-processing functions distributed to the modules are performed by terminal processes. A terminal process controls only a single terminal. The terminal process consists of a series of subroutines, each performing a specific function. The feature-control portion of the process receives all external inputs and sequences call processing actions by distributing call-control actions to other software entities.

The Routing and Terminal Allocation (RTA) software is distributed between the SM and the CMP. Requests for routing are received from feature control, and the terminal allocation process in the SM selects line and trunk subgroups and requests the operating system to create an active terminal process.[18]

12.3.3.3. Maintenance Software

Maintenance software provides initialization of a 5ESS, provides for service and system availability, and enables monitoring, repair, and control of the switching system. Its functions are distributed among subsystems.

Switch maintenance maintains and updates availability status of hardware, detects faults, notifies other software processes, and gives the system the ability to locate faults, identify the faulty component, control replacement of the unit, and verify the repair. Each SM is responsible for its own maintenance functions, reports all maintenance actions to the AM, and follows instructions from the AM for those actions that require manual intervention, scheduled maintenance coordination, reconfiguration, or initialization.

Terminal maintenance software detects problems in lines, trunks, service circuits, and signaling paths. Faults are identified by special fault-detection circuits and related software. Per-call tests, such as power cross, continuity, and leakage resistance, detect most terminal- and circuit-related faults. Routine tests are scheduled to run at low-traffic periods to test line and trunk circuits, time slots, and transmission links. Maintenance personnel can direct diagnostic tests to further isolate and correct faults.

System integrity software is responsible for detecting and correcting faults in system software through audits, integrity monitors, overload detection, defensive checks, and software initialization. Auditing checks system resources for internal consistency. Integrity monitors check the functioning of the entire system and detect overload conditions. Defensive checks compare data relationships and specific data values to detect software faults or bad data as soon as possible and initiate recovery. Initialization is a last-resort recovery process when other corrective measures fail and may be implemented automatically or manually.[19]

12.3.3.4. Administrative Software

Administrative software is distributed in the SMs and the AM. Based upon information from the switching software, the AM administrative software performs data analysis and input/output of the data via data link or office data channels. Administrative software is functionally organized into subroutines that receive and correlate per-call data, generate, collect, and report traffic, plant, and service measurements, handle billing, perform network management surveillance and control, provide data link communications capability, and perform service evaluation.[20]

12.3.4. 5ESS ISDN Architecture

The 5ESS has provision for connecting subscriber services to the ISDN, the evolving network standard for the world (discussed in detail in Chapter 14). The 5ESS ISDN architecture is shown in Fig. 12.13. ISDN is accommodated by adding a Packet Switch Unit (PSU) and an Integrated Services Line Unit (ISLU) to the SM and by additions to the call-processing functions to handle additional feature control and peripheral control. Network signaling for ISDN utilizes CCITT Signaling System No. 7 (SS7). The remote-switching module provides the same ISDN capabilities as the host switch.

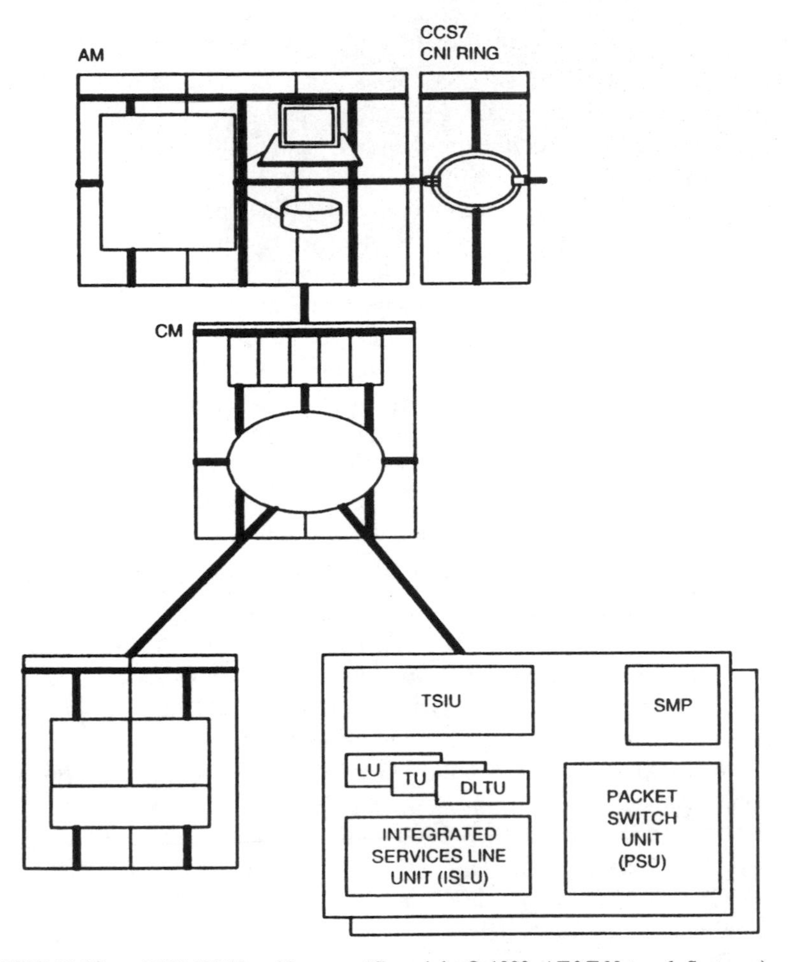

FIGURE 12.13. 5ESS ISDN architecture. (Copyright © 1993, AT&T Network Systems.)

The Basic Rate Interface (BRI), the ISDN line serving low-volume subscribers, is provided by a digital subscriber line card within the ISLU. The line card supports a user data rate of 144 kb/s for full duplex operation on a 2-wire loop using adaptive echo cancellation. The BRI follows the 2B + D format, providing two B-channels at 64 kb/s and a signaling D-channel at 16 kb/s. The D-channels are multiplexed at the line interface and connected directly to the PSU. For circuit-switched connections, B-channels are connected to the TSI in the same manner as other calls are set up. Packet data calls on the B-channel are connected from the ISLU to the PSU via a directly connected peripheral interface data bus.

The Primary Rate Interface (PRI) is provided by the DLTU. In this service, the D-channel is one of the 24 DS1 channels and operates at 64 kb/s. The B-channels for circuit-switched calls are switched in the same manner as for BRI, but packet-mode calls requiring packet-switched functions in the 5ESS are switched to the PSU.

The 5ESS provides fully integrated circuit switching and packet switching. Circuit-switched calls are switched transparently through the TSI and TMS. The PSU handles all signaling messages and data packets. Protocol handlers can handle a variety of protocols for the exchange of packets and signaling messages. The 5ESS can connect packet-switched calls to a Public Packet Switched Network (PPSN) in the same Local Access and Transport Area (LATA) through its X.75 Gateway feature. A 5ESS feature enables packet-switched traffic to be carried on the two BRI B-channels and on the D-channel.[21]

12.3.5. 5ESS Broadband Architecture

Development of the SN-2000 network architecture by AT&T includes an enhanced switching module, called the SM-2000. Initially switching only DS0 channels, the SM-2000 circuit switch is being upgraded to switch *n* DS0 channels. The processor is a Motorola 68040, more powerful than the 68020 used in previous SMs. The SM-2000 has a capacity of 500 000 CCS per hour or 100 000 calls per hour. It can serve some 20 000 lines. In conjunction with the SM-2000, a DACS IV-2000 Digital Access and Cross-connect System can cross-connect fiber-optic channels as well as other digital channels.[22]

The AT&T packet switch BNS-2000, currently being used to provide frame relay service (see Section 14.6.1), runs at 200 Mb/s and likely will be replaced by the GCNS-2000, running initially at 20 Gb/s. This packet switch may be considered as an additional switching module for the 5ESS, according to company sources. There does not appear to be any technological barrier to incorporating a Broadband ISDN cell switching module into the 5ESS architecture, although the GCNS-2000 has been announced as the core network switch for Broadband ISDN (see Section 14.7).

12.4. AT&T 4ESS SWITCH

The AT&T 4ESS was designed strictly as a trunk switch to switch tandem calls. Initially placed in service in 1976, it was the first fully digital switch ever built in North America. There are over 160 switches in service in the United States and foreign countries, some 118 owned by AT&T. A fully equipped 4ESS can connect up to 107 520 trunks and can process some 700 000 calls per hour. All major hardware components are redundant, and others have switchable spares.

The initial and enhanced versions were described in the first edition of this book, and there have been some additional enhancements since.[23]

12.4.1. Overview of the 4ESS Switch

The AT&T 4ESS is a 4-wire, electronic toll or tandem switching system using a solid-state, digital switching network, as shown in Fig. 12.14. The central processor is the same 1A Processor used in the analog 1A ESS switch. The disk file in the original version of the 4ESS switch has been replaced by the 3B Computer (see Section 12.3.2.1) as an Attached Processor System (APS) for storage of file data and other specific functions.

In the United States, the 4ESS is used throughout the AT&T toll network, at international gateway switching centers, and by local exchange carriers as tandem switching centers. In the AT&T network, the 4ESS uses Dynamic Nonhierarchical Routing (DNHR). Calls are routed from the originating toll switch (OTS) to

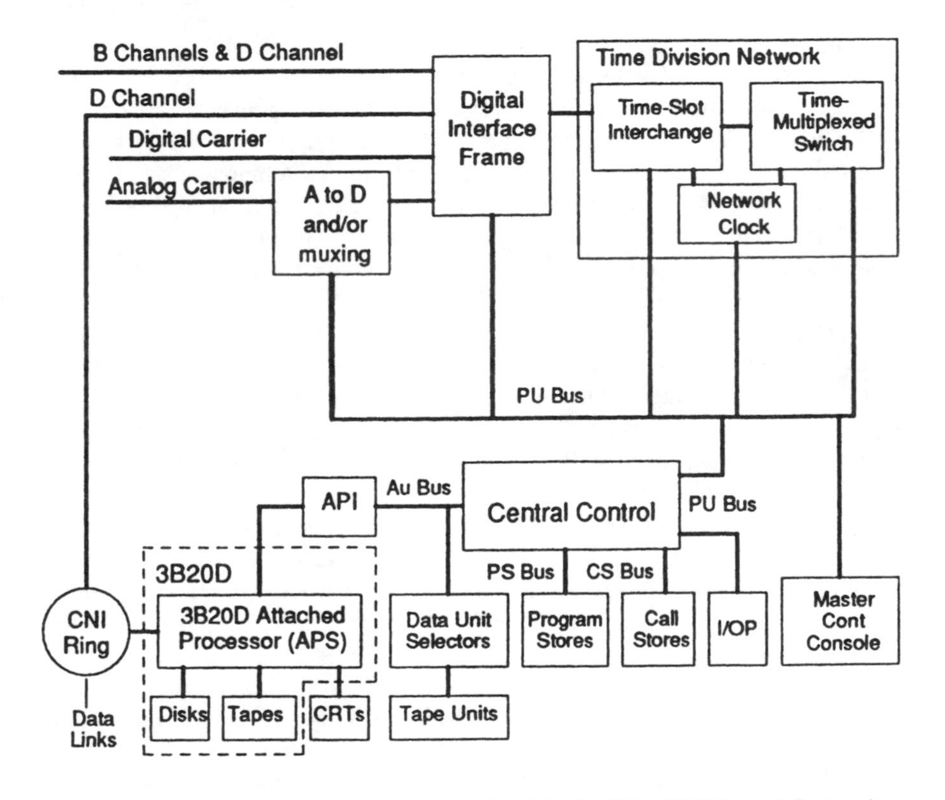

FIGURE 12.14. AT&T 4ESS block diagram. (Copyright © 1993, AT&T Network Systems.)

the terminating toll switch (TTS) over a direct trunk or by way of no more than one DNHR "VIA" switch. The 4ESS can function simultaneously as a DNHR OTS, VIA, and TTS node, depending upon the point of entry of the call into the network. Routing may be changed according to the time-of-day and day-of-week. A 4ESS can have a maximum of 256 routes in effect, but each reroute is limited to a maximum of 7 VIA routes.[24]

12.4.2. 4ESS Switching Network

The 4ESS switching network is composed of three major components: a TSI frame, a TMS frame, and a Network Clock frame (see Fig. 12.14).

12.4.2.1. Architectural Concept

Terminal interface equipment converts all incoming trunk signals to the DS120 format, with 128 time slots comprising a serial stream of PCM signals from 120 trunks, plus 8 time slots for maintenance. Each time slot contains a PCM word from a separate trunk, with certain time slots containing maintenance words. Seven DS120 data streams containing trunk data terminate on the TSI frame, where the data time slots undergo one stage of time switching and one stage of space switching. Eight output DS120 streams of the TSI frame are then subjected to two stages of space switching in the TMS frame. The output DS120 streams from the TMS frame then are switched in the TSI frame through one more space-switching stage and a final time-switching stage before being sent to their respective outgoing trunks. Thus, the architecture of the 4ESS is TSSSST. (*Note*: AT&T documentation describes the time switching and space switching in the TSI frame as "first-stage switching and time-slot interchanging" rather than two stages of switching; thus, it describes the switching network as having four stages.[25] This is explained further in Section 12.4.2.2.) The switching network functions in a folded configuration.

A 4ESS of maximum size has 896 coaxial cables carrying DS120 data streams being switched through 32 TSI frames. Additionally, 128 more coaxial cables carrying null words terminate on the 32 TSI frames. Each TSI frame contains 4 time/space switching entities, each having a capacity to switch 840 trunks and service circuits.

Each TMS frame contains one 256-by-256 two-stage space switch. A fully equipped TMS contains four such frames. With 1024 input cables and 1024 output cables, the TMS provides 131 072 PCM paths through the network, which reduces the probability of blocking within the network. All TMS frames are provided in pairs, with one active and one on standby.

12.4.2.2. Time-Slot Interchange (TSI) Frame

Each TSI frame in the current configuration contains four pairs (active & standby) of time/space switching entities. Each entity provides a *switching and permuting circuit* that can accommodate 840 trunks and service circuits applied through seven DS120 coaxial cables from terminal interface equipment. A fully equipped 4ESS can accommodate up to 128 switching and permuting circuit pairs. Each switching and permuting circuit has a receive portion and a transmit portion. Each portion contains two sets of buffer memories and an 8×8 integrated circuit, serial, space switch for each DS120 formatted data stream.

The receive portion of the switching and permuting circuit, shown in Fig. 12.15, performs four major functions: (1) time buffering, (2) deloading, (3) decorrelation, and (4) time-slot interchanging and first-stage space switching. The resulting output signals are sent to the TMS for two stages of space switching. The output signals of the TMS are applied to the transmit portion of the switching and permuting circuit which performs the fourth stage of space switching and time slot interchanging, recorrelation, and reloading into the transmit DS120 data streams.

The first major function is *time buffering*. Even though the DS120 PCM data generated by the terminal interface equipment is synchronized by timing signals from the network and system clock, minor differences in cable lengths and the fact that some trunk data is routed through an echo suppressor terminal causes some variation among the DS120 frames, up to seven or eight time slots. Buffer memories A0 through A6 in the receive portion of the switching and permuting circuit store the PCM data up to 125 μs (or one frame) to establish frame synchronization.

The second major function is *deloading*, a method for distributing the total traffic load on the switching network. The seven DS120 PCM streams, briefly stored in buffer memories A0 through A6, are mixed with the null words in buffer memory A7 by the deloading and decorrelating circuit. Disregarding the null words, used as spacers in the expanded switching network, the PCM data for 840 trunks on seven cables (and seven buffer memories) is distributed among eight outputs into buffer memories B0 through B7. This lightens the loading on each network path and reduces the probability of blocking by enabling the 840 traffic trunks to be distributed among 960 switching paths in each switching and permuting circuit.

The third major function is *decorrelation.* In some traffic situations, traffic is concentrated among specific groups of trunks. Decorrelation enables concentrated traffic in a DS120 PCM stream to be distributed across the time-division network. Otherwise, blocking could occur if too heavily concentrated traffic were to be switched through the TSI into the TMS. The functions performed by the deloading and decorrelating circuit are similar to those of a multiplexer. The PCM trunk data is moved from one DS120 stream to another, but is kept within the same

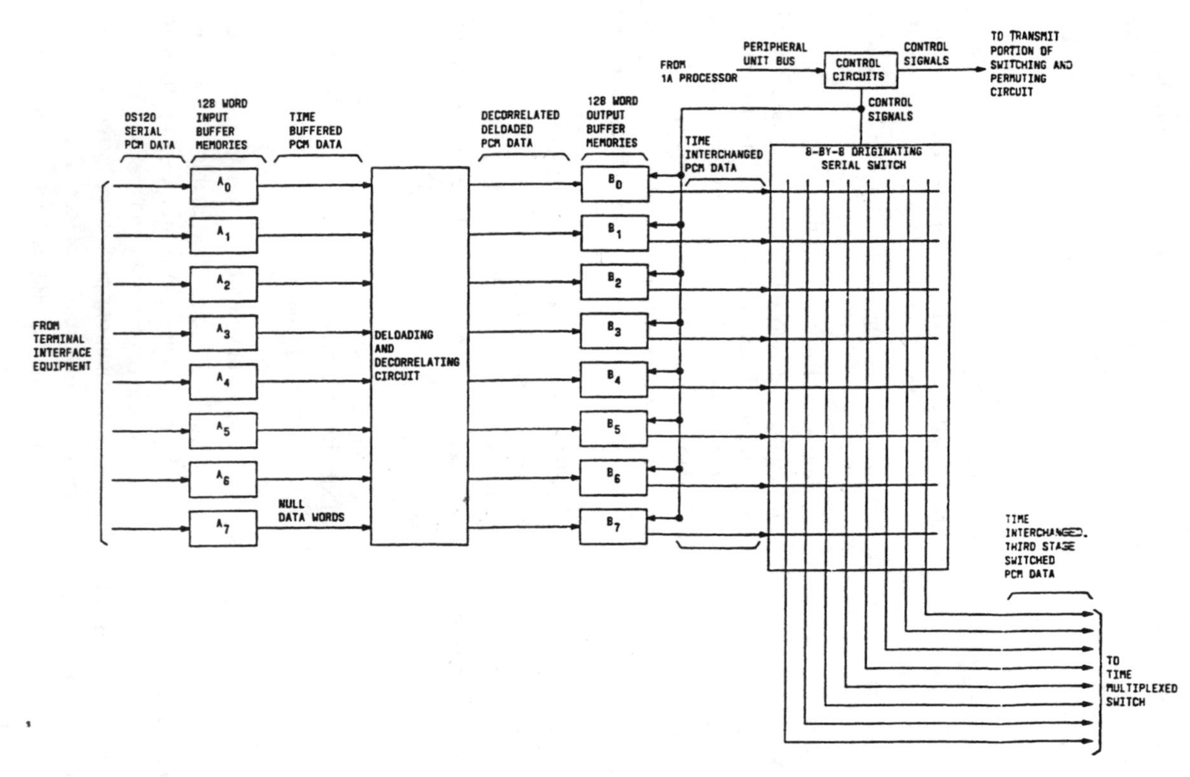

FIGURE 12.15. 4ESS switching and permuting circuit (receive portion). (Copyright © 1993, AT&T Network Systems.)

time slot. The 128 data words in each of the seven input buffer memories and the 128 null words in memory A7 are evenly distributed to all eight output buffer memories according to a precise pattern. Each output buffer memory contains 105 trunk data words and 23 nonmessage words (one-eighth of the 128 null words in memory A7 and one-eighth of the 56 maintenance words in the seven traffic memories).

The fourth major function is *time-slot interchange and first-stage space switching*. The PCM trunk data in each time slot in each output buffer memory B0 through B7 is read out in a different time slot in most cases. The time-slot interchange function rearranges data among time slots in the same DS120 data stream but does not switch the data to different DS120 data streams. Under control of the 1A processor, one time slot from each output buffer memory is read out to the associated level of the 8 × 8 originating serial space switch where it is switched to an output level and sent to an input level of the TMS.

After passing through the two stages of space switching in the TMS, the output levels are applied to the input levels of the 8 × 8 terminating space switch in the transmit portion of the switching and permuting circuit. The outputs are read into buffer memories CO through C7, where they undergo the final time-slot interchange and pass through the reloading and recorrelating circuit to an output circuit, and thence to the terminal interface equipment as DS120 PCM data.[26]

12.4.2.3. Time-Multiplexed Switch (TMS) Frame

The TMS consists of one, two, or four duplicated space-switching grids of two stages each, as shown in Fig. 12.16. Each TMS grid contains 32 integrated circuit 16 × 16 matrices, interconnected to form a 2-stage, 256 × 256 switch array. The grid provides the second and third space-switching stages in the 4ESS. The output levels of the stage 1 are connected to the input levels of stage 2 in a complex pattern, chosen to simplify the establishment of noninterfering paths through the time division network. Since the PCM words are distributed in the DS120 format on the basis of one trunk word per time slot, the 4ESS reconfigures the switching stages 128 times during each 125-μs frame. With 1024 DS120 input and output levels and 128 time slots in each level, the TMS has 131 072 switching paths.[27]

12.4.2.4. Network Clock

The network clock frame is comprised of two identical bays, each containing a duplicated network clock. The network clock provides precise timing signals to control the switching network and all other equipment in the 4ESS. The Network Clock hardware is a Disciplined Rubidium oscillator that provides extremely accurate clock pulses. The oscillators are synchronized to the AT&T Reference Frequency (ARF) and can distribute that master clock signal to other digital network nodes for network timing.

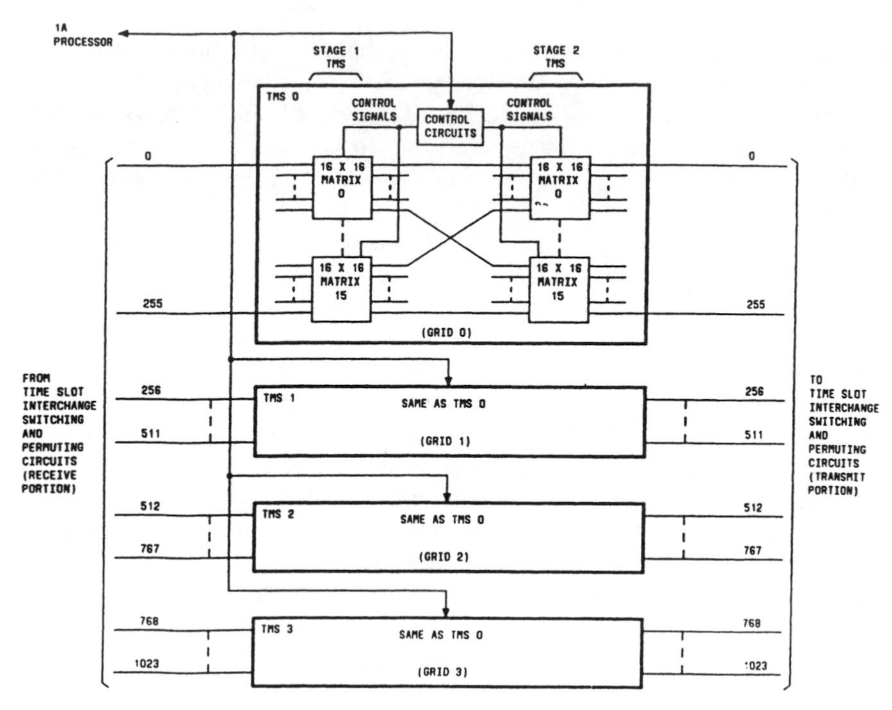

FIGURE 12.16. 4ESS Time-Multiplexed Switch. (Copyright © 1993, AT&T Network Systems.)

12.4.2.5. Call-Switching Process

Tracing a call through the 4ESS will illustrate the call-switching process. Assume that an incoming call on digital Trunk A is addressed to a destination that can be reached by a metallic analog Trunk B. The call on Trunk A arrives in DS1 format in TS 5. The Terminal Interface Equipment combines the PCM data for Trunk A with that of other trunks into the DS120 format. Assume that the PCM data for Trunk A is placed in TS 96 and sent to the TS1 frame for brief storage in Buffer Memory A. Call-processing programs identify and reserve an idle outgoing trunk and time slot to send the call to its destination. Assume that Trunk B is reserved in TS 49 in the DS120 format at the output of the time-division network. The 1A Processor identifies a noninterfering, intermediate time slot in the TMS which is commonly available to TS 96 and TS 49.

Assume that TS "N" is identified. The 1A Processor sends that information to the call processors in the time-division network. The TS1 switching and permuting circuit (receive section) deloads and decorrelates the DS120 input stream and stores the PCM words in Buffer Memory B in TS 96. The PCM words for Trunk A are read out of the time-slot interchanger in TS "N" and are switched through the originating 8 × 8 space switch in the TS1 frame to a predetermined input level in the TMS. The TMS switches TS "N" through two stages of space switching to an output level connected to the terminating TS1 8 × 8 space switch. The terminating 8 × 8 space switch switches TS "N" to the required output level in the switching and permuting circuit (transmit level) and stores the PCM data in Buffer Memory C. The PCM data in TS "N" is time switched (example, read out) into TS 49 in a DS120 stream, which is recorrelated and reloaded into another DS120 stream and sent to the Terminal Interface Equipment associated with Trunk B. That equipment converts the PCM data in TS 49 to an analog signal for transmission on analog Trunk B.

Concurrently with the process described above, the reverse is taking place to establish a path from Trunk B to Trunk A via a mirror image of the path from Trunk A to Trunk B. Trunk B PCM words are being passed simultaneously through the network over the mirror-image path. Those PCM word exchanges occur 8000 times per second until the call is disconnected. Depending upon the input and output time slots involved and the intermediate switching time slot assigned, the data passing through the network may be delayed up to slightly less than three 125-μs frames.[28]

12.4.3. 4ESS Terminal Interface Equipment

The terminal interface equipment comprises a variety of equipment frames that provide an interface between transmission equipment and the 4ESS switch, as shown in Fig. 12.17. The 4ESS can interface high-capacity coaxial multiplex transmission media (L-carrier) through an LT-1 or LT-2 connector. Analog metallic circuits, DS0 circuits, and analog service circuits are connected to D4 channel banks. Analog signals are converted to PCM digital signals in the LT connectors and the D4 channel banks, and then multiplexed into a DS1 signals for processing in the 4ESS equipment. The D4 channel banks also multiplex the DS0 signals but bypass the codec step. Digital signals arriving in the DS1 format, including ISDN primary rate B-channels and D-channels and other signals that have been converted to DS1 format, are passed to a Digital Interface Frame (DIF). In some older configurations, analog trunks terminate on a voiceband interface frame (VIF), and DS1 signals terminate on a Digroup Terminal (DT). Both the VIF and DT perform functions similar to the current production DIF.

The DIF combines five DS1 signals into a DS120 format for switching. A fully equipped DIF can terminate a maximum of 3840 digital circuits and convert them to 32 DS120 streams. (For international analog trunks using CCITT Signaling System No. 5, the capacity of the DIF is 1920.) The 4ESS has a capacity of up to 32 DIFs or equivalent. The output DS120 signals from the DIF can be sent directly to the time-division switching network, except that those requiring echo suppression are sent via an Echo Suppressor Terminal (EST).

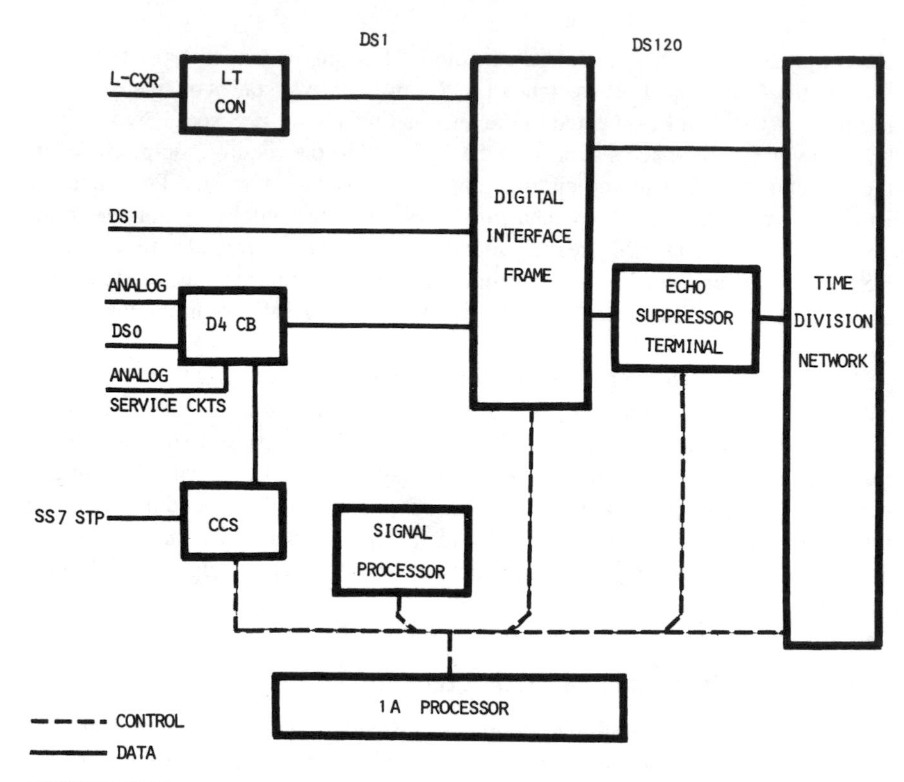

FIGURE 12.17. 4ESS Terminal interface equipment. (Derived from AT&T Network Systems documentation.)

Each EST contains 14 echo-suppressor units. Each unit has one input and one output coaxial cable on the terminal side to connect to a DIF, DT, or VIF, and the same on the switching network side to connect to the TS1. Each echo-suppressor unit can accommodate 120 trunks on a time-sharing basis. Thus, the total capacity of an EST is 1680 trunks. Two EST frames are required to connect the total output of one DIF to the TS1. The four remaining DS120 ports on the DIF are connected to a separate TS1 by using a modulo 8 skewing pattern to keep cable lengths within critical limits and to minimize timing discontinuities. All DS120 cables from a specific DIF are connected either to an EST or directly to a different TS1 frame. If a pair of DIF cables connected to an EST do not require echo suppression, they are connected to an unequipped echo-suppressor unit position through a plug-in unit to the TS1 cables.

Signal Processors (SP) interpret and process inband signals from analog and digital trunks. Inband signaling in the United States can include Dial Pulse (DP), Multifrequency (MF), and Dual Tone Multifrequency (DTMF) signaling methods. Common-channel signals are processed by a CCS7 terminal group. The DIF, EST, CCS7 group, SPs, and time-division network are connected to the 1A Processor via the Peripheral Unit Bus (PUB). On international circuits, the 4ESS can use CCITT No. 5, CCITT No. 6, and CCITT No. 7. Within the CCS7 system, the 4ESS can support the Telephone User Part (TUP), ISDN User Part (ISUP), and the Q.931 protocol (for ISDN PBX signaling).[29]

12.4.4. 4ESS Control System

The 4ESS uses a hybrid control system, as shown in Fig. 12.18. A 1A Processor provides overall central control while major components have their own microprocessor controllers. The 3B20D APS stores file data and provides other functions. The Attached Processor Interface (API) provides the interface between the 1A Processor and the 3B20D and converts data between the 24-bits-per-word format used by the 1A and the 32-bits-per-word format used by the 3B20D. The API also provides direct access between the 1A and 3B memories.

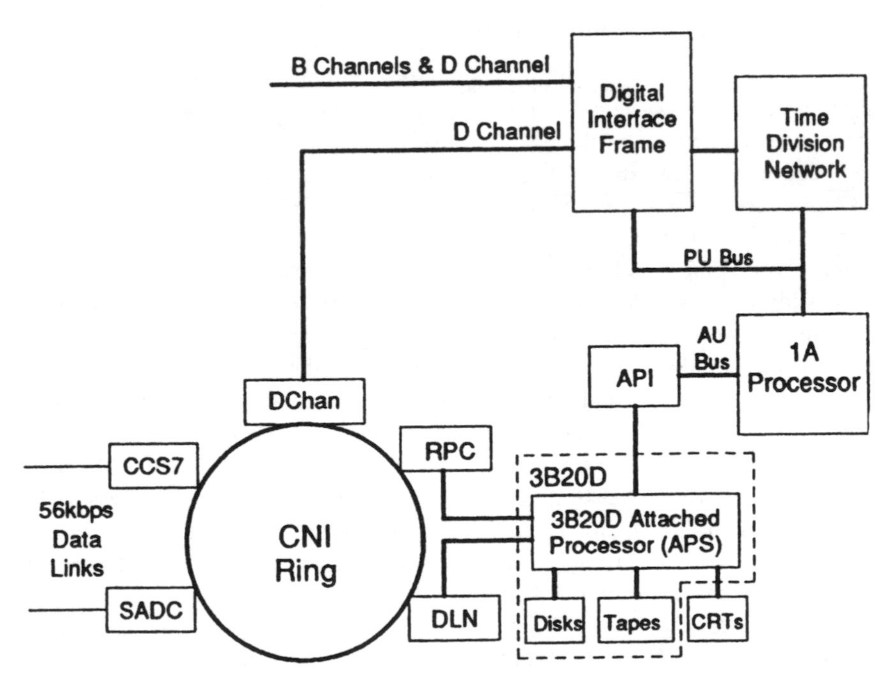

FIGURE 12.18 4ESS control system with CNI ring. (Copyright © 1993, AT&T Network Systems.)

A Common Network Interface (CNI) ring[30] is a dual, token-ring, local area network operating in an active/standby mode. Communication on the CNI ring is by packet-switched messages controlled by an Interprocess Message Switch (IMS). Interfaces to the ring's several nodes are by dual serial channels and DS1 trunks.

- The Ring Peripheral Controller (RPC) node provides booting, ring reconfiguration, and ring maintenance functions.
- The Direct Link Node (DLN) is a high-capacity data interface that provides the direct memory access between the 1A and 3B processors and does checking and reformatting of messages.
- The CCS7 node sends and receives CCS7 messages via 56-kb/s data links to a pair of Signal Transfer Points (STPs) in the signaling network.
- The D-channel Node (DCHN) provides interface to a PBX ISDN D-channel, using the Q.931 signaling protocol between the PBX and the 1A Processor.
- The Special Access Data Channel (SADC) node is used to connect a 4ESS to customer-owned equipment having special data requirements. One feature permits cable TV subscribers to order pay-per-view movies by calling an 800 number. The customer's telephone number is forwarded by the SADC node to the cable company so that the CATV company can enable the customer's cable box and record billing information.

The 1A Processor is a multiple-application computer that can be used with different switching systems. It is dualized, with both processors running synchronously. After each machine cycle, the processors compare a series of test bits and begin error processing if there is a mismatch. The 1A Processor consists of a Central Control (CC) and two major semiconductor memories of two megawords each. The Program Store (PS) contains control programs and infrequently changed office data. Call Store (CS) is used to store frequently changed data related to call processing, such as status of trunks and equipment, records of switching network configurations for calls in progress, address digits received and digits to be outpulsed, programmed diagnostic test data, and translation update data. Call Store also contains an emergency system recovery program to be used to establish a working system if the PS should fail.[31]

The CC connects to the CS memory by a CS bus, to the PS memory by a PS bus, to the API by an AU bus, and to the other 4ESS equipment processors by means of a PUB. In addition, an input/output processor frame within the 1A Processor provides 2-way access between the 1A Processor and processors in other maintenance support and data-collection systems.

All equipment connected to the PUB simultaneously receives all data transmitted on the bus. All such equipment is assigned unique *K-code*. When CC accesses a specific equipment unit, the assigned K-code is transmitted on a peripheral unit enable address bus. The unit whose K-code matches the K-code transmitted responds to the message while units whose K-codes do not match do not respond. The 1A Processor may use a polling technique to obtain status data from up to 24 separate units of the same type. The CC also communicates with certain equipment by means of pulse points to initiate or control specific operations in that equipment.[32]

The APS (3B20D Computer), originally added to replace the 1A file store, is used as an adjunct processor to support specific functions. The directly addressable disk storage available in the 3B20D is 8 388 608 data words of 32 bits each. Additional 3B20D Computers can be attached as adjunct processors to support specific functions.[33]

12.5. NORTHERN TELECOM DMS-100 SWITCH

12.5.1. Overview of the DMS-100 Family

The DMS-100 family of switching systems was developed to provide a basic switching fabric to serve as a local central office switch, as a toll (tandem) switch, as a combination local/toll switch, and as an international gateway switch. The first DMS-200 toll switch was placed in service in January, 1979, and the first DMS-100 local switch followed in September, 1979. Later, a modified version, called the DMS-250, was produced for use by some of the newer interexchange carriers in the United States. When optical-fiber technology was developed sufficiently for SONET use, the switch architecture was modified on a retrofit basis and was redesignated as the S/DMS Product Family. The S/DMS family of products includes *S/DMS AccessNode, S/DMS TransportNode*, and *S/DMS SuperNode* applications. The S/DMS AccessNode functions as a Remote Fiber Terminal connected by a 622-Mb/s SONET (see Section 10.12) link to either a S/DMS SuperNode switch or another S/DMS AccessNode for intraoffice connection to any switch. The S/DMS TransportNode is a SONET-based interoffice connection to any switch. The S/DMS TransportNode is a SONET-based interoffice and long-haul transport vehicle that contains OC-12 or OC-48 SONET network elements that multiplex and provide time switching for STS-1 signals. The SONET links may be over optical fiber or over SONET radio.[34]

12.5.2. Evolution of DMS Architecture

The DMS architecture has been implemented in three major evolutionary design stages while maintaining backward compatibility in each case. The early DMS-100 architecture, depicted in Fig. 12.19, was described in the first edition of this book (pp. 340–345). Line module frames concentrated 640 analog line circuits

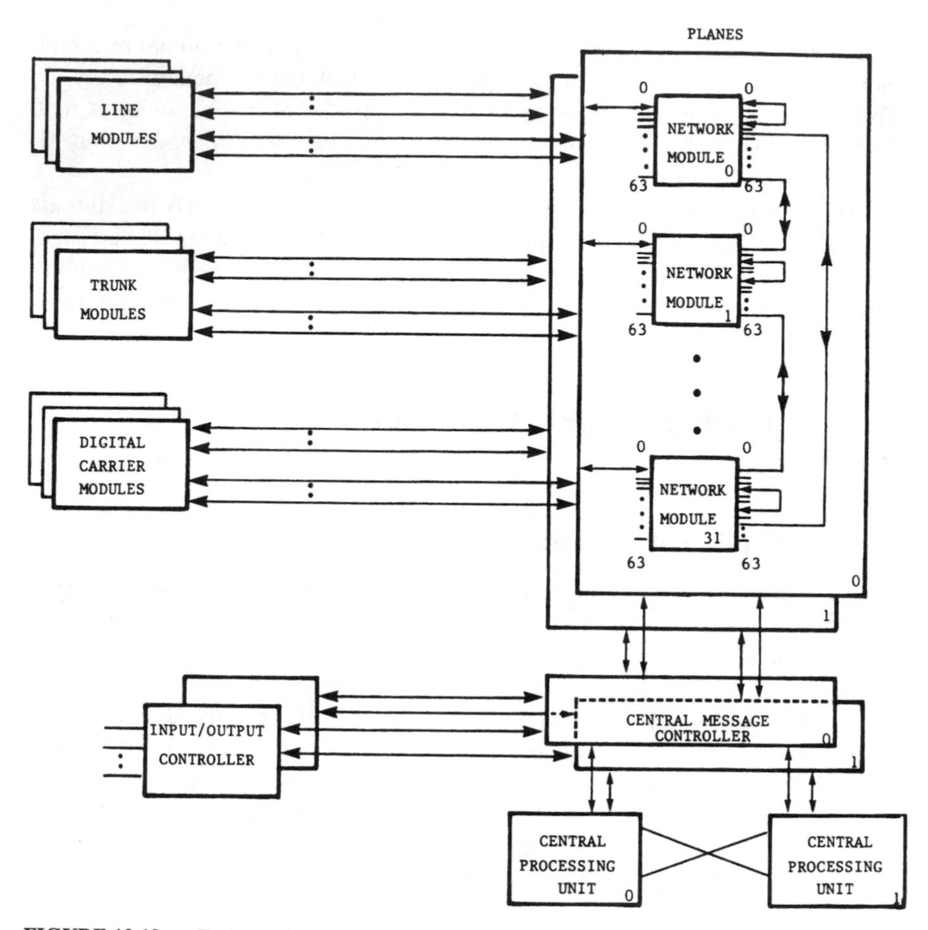

FIGURE 12.19. Early version Northern Telecom DMS-100.

into 60–120 digital channels through a space-switching network. Analog trunk modules and digital carrier modules provided terminations for trunks without concentration. Each of up to 32 switching network modules contained 16 time-space 8 × 8 switch matrices arranged to provide two stages of time-space switching. Each module switched between 64 peripheral ports, each terminating one DS30 digital line containing 30 voice channels and 2 signaling channels, and 64 junctor ports. The serial data rate through the switching network was 2.56 Mb/s. Each module had 64 two-way junctors (wired connections) that connected to other modules in its plane and to itself. The switching network was implemented in two identical *planes*, and the central control complex was duplicated.

The second design stage, called the *DMS SuperNode* (Fig. 12.20), replaced the NT40 central control and central message controller with a *DMS-Core* (see Section 12.5.3.1) and a *DMS-Bus* (see Section 12.5.3.2). A *Link Peripheral Processor* was added to interface various public network protocols such as CCS7, X.25, and Frame Relay, and an Enhanced Network (see Section 12.5.3.4) was added. (Note: Various manufacturers and service providers use different acronyms for CCITT Signaling System No. 7; CCS7 and SS7 are used synonymously.) The first DMS SuperNode system was installed in December, 1987.

The third design stage, called the *S/DMS SuperNode*, modified the architecture so that the basic platform could function as a local central office, as a CCS7 network SP, STP, or SCP, as an SSP in an intelligent network, or as a combination office serving multiple functions, and made provision for SONET interfaces and Asynchronous Transfer Mode (ATM) switching.

12.5.3. S/DMS SuperNode Architecture

A block diagram of the layered architecture of the Northern Telecom S/DMS SuperNode is shown in Fig. 12.21. The highest layer is the Service Processing Layer, containing the DMS-Core, DMS-Bus, and application processors for Line

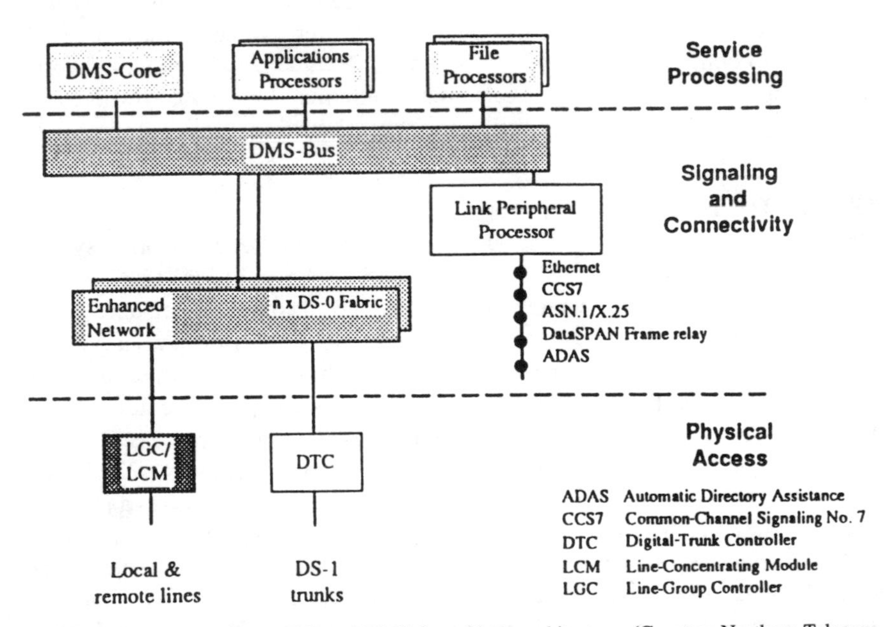

FIGURE 12.20. Northern Telecom DMS SuperNode architecture. (Courtesy Northern Telecom, Inc.)

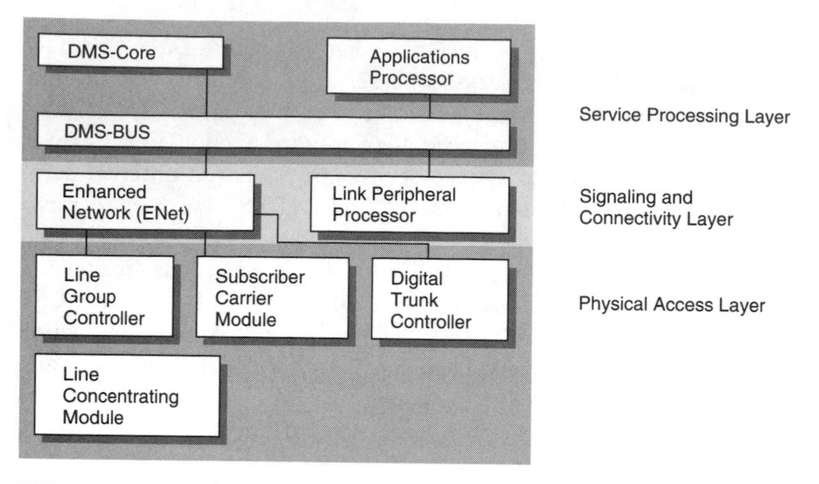

FIGURE 12.21. S/DMS SuperNode layered architecture. (Courtesy Bell-Northern Research.)

and Trunk Services, Network Services, and Operations Management. The second layer, the Signaling and Connectivity Layer, contains the Link Peripheral Processor (LPP), for signaling interfacing and control, and the switching networks. The original switching network has been replaced by an *Enhanced Network (ENet)*. The lowest layer, the Physical Access Layer, contains line and trunk interfaces to both analog and digital facilities.[35] By adding additional networks and call processing functionality, this architecture may be suitable for use as a combined narrowband-broadband, circuit-packet switching platform.

12.5.3.1. DMS-Core

The DMS-Core is the computing and memory resource and performs system management in S/DMS SuperNode applications, as shown in Fig. 12.22. Duplicated computing modules operate in two synchronized computing module planes linked by a Mate Exchange Bus to enable the Master Processor in each plane to compare computations and to check the integrity of the other plane. A semicustom silicon device in each of the single-board computers compares the address and data signals on every memory access attempted by the Master Processor.

The Master Processor in the initial DMS-Core configuration was a 32-bit Motorola MC68020 microprocessor, running at 20 MHz. DMS-Core uses a four-tier memory hierarchy. The first level, a 256-byte, on-chip instruction cache on the MC68020 microprocessor, stores short, repetitive code sequences. The second level, 4 kbytes of zero-wait-state, direct-mapped data cache memory, stores the most frequently and competitively accessed data operands. The third level, up to 512 kbytes of zero-wait-state, static random access memory (SRAM), stores the

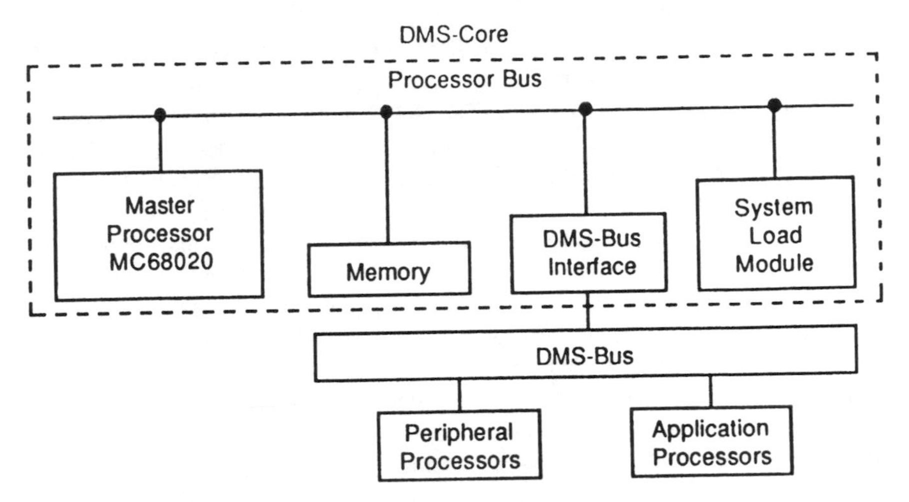

FIGURE 12.22. DMS-Core architecture. (Courtesy Northern Telecom, Inc.)

most frequently accessed software procedures. The main DMS-Core memory comprises the final level in the memory hierarchy. The main memory consists of 10 memory boards, each containing three independent 2-Mbyte memory blocks in a configuration of 256 kbytes by 32 data bits plus 7 error-correction bits and 1 parity bit. The hierarchical memory speeds call processing by allowing the processor to access memory rapidly for the most frequent types of calls. Two System Load Modules, each having a primary and a secondary line to the DMS-Core computer modules, provide redundant paths to load the DMS SuperNode.

Connections between DMS-Core and DMS-Bus are via two optical fibers at 32 Mb/s, operating through rear-mounted paddleboards, containing interface circuitry and other components that replaced the thumbwheel and switches of the DMS-100 NT40 processor core. Other paddleboards provide the remote terminal interfaces used for basic control of DMS-Core.[36]

Later versions of DMS-Core use the Motorola MC68030, running at 33 MHz, or the MC88000, using modified Reduced Instruction-Set Computing (RISC). The MC68020 chip more than doubled the call-handling capacity of the switch compared to the previous NT40 processor used in the earlier DMS-100. The MC68030 is rated at seven million instructions per second (MIPS) and further increased the capacity by a factor of 1.45 times that of the MC68020. The MC88000 microprocessor, called the Series 50 BRISC processor, has demonstrated a capability of handling 1.2 million busy-hour call attempts (BHCA), assuming all attempts were plain old telephone service (POTS) calls. The latest Series 60 processor, using the Motorola MC88100 chip in "burst mode" operation, can process 1.4 million POTS BHCAs. The Series 70 processor, under development, is intended to increase the call-handling capacity to 2 million POTS BHCAs, and future plans are to double that to 4 million POTS BHCAs.[37]

Although upgrading to the Series 50 BRISC processor is accomplished by exchanging some circuit packs in the Computing Module and loading BRISC software, the resulting changes in call processing are rather significant. Using the burst mode, the BRISC processors provide more processing power than traditional computers by using sequences of fast, simple instructions rather than slow, complex instructions. The central processor uses two different types of information: 1. *program code* that contains instructions to the processor and 2. *data* upon which the instructions are to be performed. As shown in Fig. 12.23, internal to the computer, a Code Side Bus handles code traffic, and a Data Side Bus handles data traffic. An Intelligent Prefetcher anticipates the processor's need for certain instructions, retrieves that code from the Dynamic Random Access Memory (DRAM) memory banks and delivers it to a temporary holding area, from which it can be retrieved quickly when needed. This allows a higher proportion of the requested code to arrive at the processor within a single clock period.

Three Motorola MC88200 chips in the Series 50 processor serve as separate cache memory for code and data. One chip is used as cache memory for retrieval of high-speed noncall-processing instructions, and the other two chips are used for high-speed retrieval of data needed for the processing of those instructions. The caches are used for temporary storage of often-used or likely-to-be-used information. The Main Memory Manager coordinates all interactions between the processor board and the DMS-Core main memory, ensuring that the BRISC technology is backward compatible.

Software enhancements also increase the efficiency of the BRISC technology. A new software compiler produces machine code that has more orderly sequences and fewer branches than previous machine code. This enables the large number of memory registers on the MC88000 chip to respond even faster than cache memory and therefore are used to store the most frequently used information. The range of software diagnostics has been increased. All diagnostic testing is done on the inactive computer which, in turn, checks the integrity of the active computer.[38]

12.5.3.2. DMS-Bus

DMS-Bus, shown in Fig. 12.24, is the key architectural element that ties all other elements of the S/DMS SuperNode together. Its most important function is to support multiple application modules with high-speed interprocessor communications. This eliminates the need for interprocessor messages to go through a central control. The internal message network provided by DMS-Bus enables distributed control to be applied to the S/DMS SuperNode. Although DMS-Bus has two redundant planes operating in a loadsharing mode, each plane can support the entire system load.

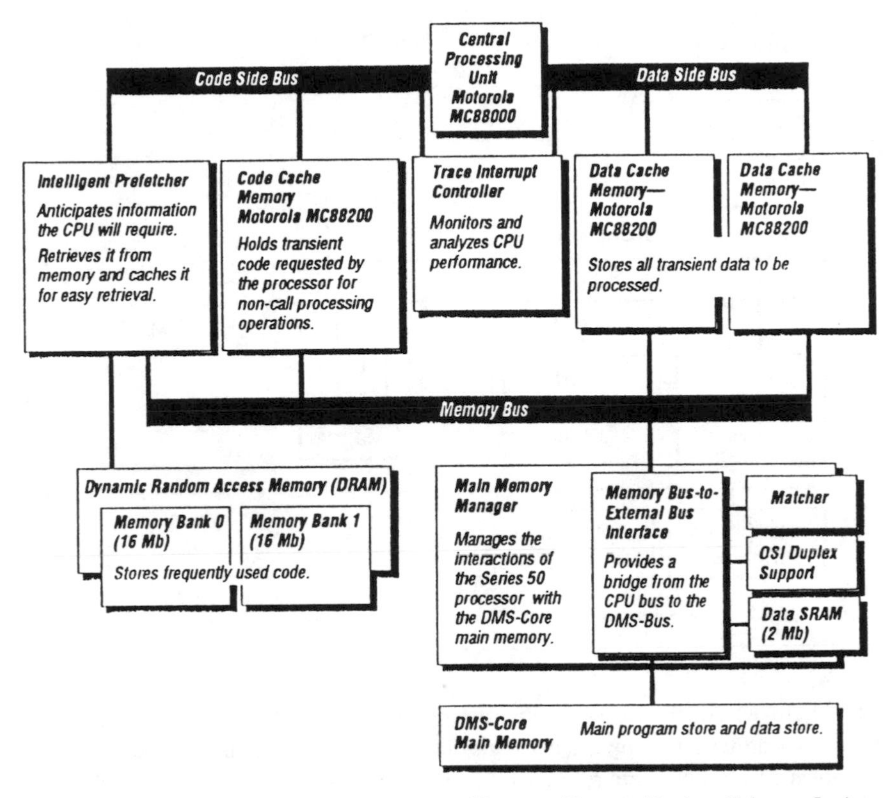

FIGURE 12.23. Series 50 BRISC Processor architecture. (Courtesy Northern Telecom, Inc.)

The Control Processor MC68020, with 6 Mbytes of memory, controls message communications, performs port maintenance and diagnostics, and executes the S/DMS Supernode operating system. The Control Processor uses the Processor Bus to manage and maintain the other components of DMS-Bus. Messages between S/DMS SuperNode components are carried on a 32-bit Transaction Bus. Its capacity of 128 Mb/s enables the bus to support a typical message mix of 100 000 messages per second with a port-to-port delay of less than 100 μs. The Control Processor manages access to the Transaction Bus through the Processor Transaction Bus Interface.

Interfaces from DMS-Bus to Peripheral Processors, Application Processors, and DMS-Core are managed by Port Interface Units (PIU). A PIU contains communication buffers, a microcontroller, and a line handler that functions as a translator between the link data protocol and the messaging protocol used on the Transaction Bus. Messages traveling on the Transaction Bus contain either a physical or logical address header and a message body. A physical address is the

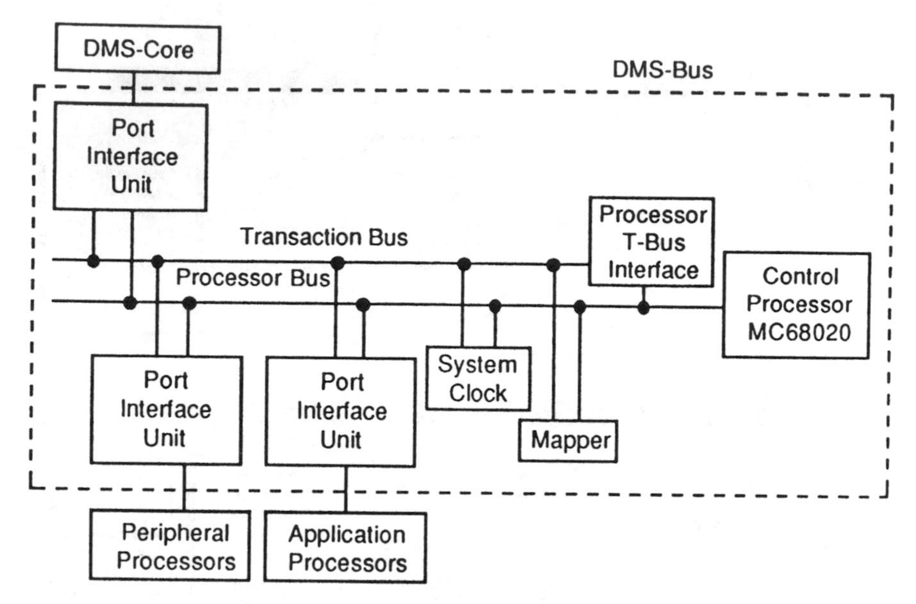

FIGURE 12.24. DMS-Bus architecture. (Courtesy Northern Telecom, Inc.)

actual hard-wired location of a subsystem's interface managed by a specific PIU. A logical address, used by the network modules and the input/output controllers, is a nonspecific subsystem reference that is translated and converted to a physical address by the Mapper. DMS-Bus can be configured to support more than 1400 PIUs, and this capacity is planned to be increased to more than 3000 PIUs.

The built-in System Clock in DMS-Bus is in conformity with Stratum III standards and can be synchronized from an incoming network timing reference. If greater performance accuracy is required in a given situation, an external Stratum I or Stratum II clock can be provided.[39]

12.5.3.3. Link Peripheral Processor

The third major architectural element of the S/DMS SuperNode is the LPP, which provides the protocol structure used on signaling links for internodal communications, as shown in Fig. 12.25. The LPP provides the language syntax that enables advanced services to be delivered to subscribers. A standard set of application and base interfaces include the capabilities for network management and feature programming of an SCP. Protocol sets in the LPP were developed by national and international standards bodies and include the CCS7 set for signaling, Network Operations Protocols X.400 and NOP X.25 for packet transmission, and Q.921 and Q.931 for ISDN access. The CCS7 protocol set in the LPP enables the S/DMS SuperNode to be configured as an SSP by using the Transaction

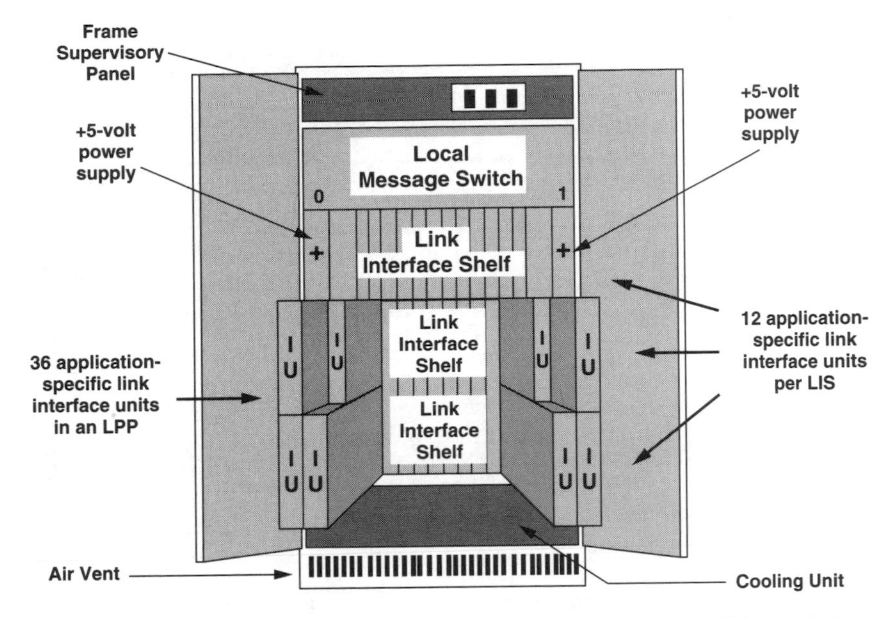

FIGURE 12.25. Link Peripheral Processor architecture. (Courtesy Northern Telecom, Inc.)

Capabilities Application Part (TCAP) to query the SCP database for calls such as those involved in 800 service or automated credit card service. The ISUP is used to set up and break down calls in the public network. Since the calling number and information about the calling party is provided through the ISUP protocol, the LPP facilitates the provision of other advanced subscriber services.[40]

12.5.3.4. S/DMS Switching Network

The ENet is a single-stage, nonblocking, junctorless time switch that provides 64 000 duplicated paths in one cabinet or 128 000 paths in a dual-cabinet configuration. As shown in Fig. 12.26, interconnections between ENet and DMS-Bus are established via DS512 optical links, carrying 512 channels, using 62-micrometer fiber at a wavelength of 1300 nm, operating at 50 Mb/s. ENet interconnects with earlier model peripherals, such as Trunk Module and Line Module, via DS30 copper connections. Extended Peripheral Modules (XPM), Northern Telecom's current series of peripherals, are interconnected via DS512 optical connections. Two types of peripheral-interface paddleboards connect either DS512 fiber links or DS30 copper links with other elements of the S/DMS SuperNode.

The local processor for each ENet shelf uses a Motorola MC68020 microprocessor to control and maintain the shelf and the links to the peripheral modules. It is controlled, in turn, by DMS-Core through DMS-Bus via a DS512 optical link. The processor circuit pack contains 4 Mb/s of memory. A Shelf Reset/RTIF

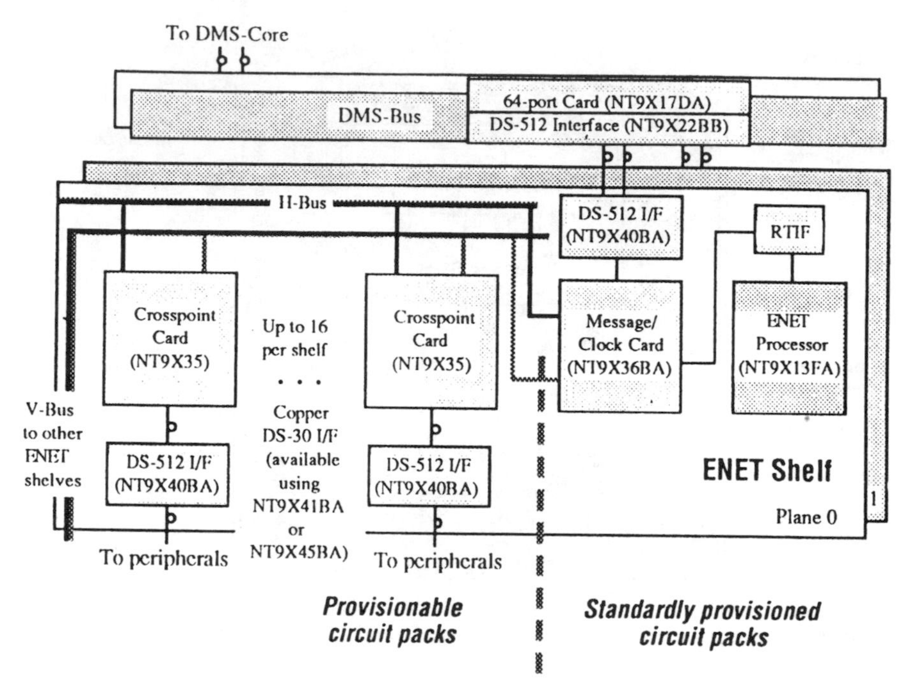

FIGURE 12.26. Block diagram of ENet architecture. (Courtesy Northern Telecom, Inc.)

(Remote Terminal Interface) paddleboard is used to reset the ENet processor in the event of a restart. Out-of-band reset codes from DMS-Core activate and control the RTIF. The Message/Clock card provides clock and framing signals for the ENet shelf and enables the local ENet processor to exchange messages with DMS-Core.

Each ENet shelf contains a 128K-by-32K time switch. The actual switching function of ENet is a crosspoint circuit pack providing switching for 2048 channels. Each shelf can switch 32 000 unidirectional channels. The four shelves in a cabinet can provide one plane of a 128K-channel switching matrix or two planes of a 64K-channel matrix. The switching matrix in the ENet cabinet can be visualized in Fig. 12.27. Channels in either DS512 or DS30 format from originating peripherals enter through the ENet paddleboards, where they are converted from serial to parallel data, buffered in an elastic store, multiplexed onto the Vertical Bus (V-Bus), and are written sequentially into a memory on all crosspoint cards associated with that V-Bus. Alternate frames are written into separate memories. Under direction of messages from DMS-Core, the connection-memory control causes the stored data words to be read out in the desired time slot to the appropriate Horizontal Bus (H-Bus), from which they are sent to a paddleboard, converted back to serial data, and transmitted to terminating peripherals.[41]

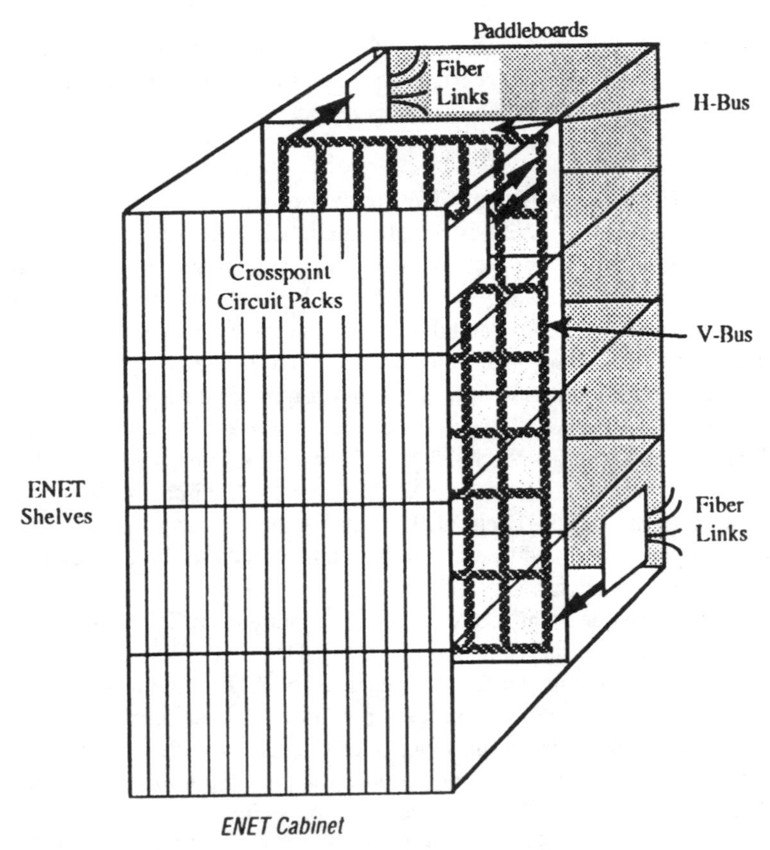

FIGURE 12.27. Illustrative view of ENet single-stage matrix. (Courtesy Northern Telecom, Inc.)

12.5.3.5. S/DMS Peripheral Modules

Subscribers can access the S/DMS SuperNode via conventional copper cables and the traditional peripheral modules shown in Fig. 12.19 or via special modules for ISDN. It is advantageous, however, to shorten subscriber lines by moving the line circuits into remote equipment units such as Remote Switching Centers (RSC) or S/DMS AccessNodes and carrying the access lines to the S/DMS SuperNode switch via optical fiber at the SONET OC-12 (622 Mb/s) rate. Formerly, five major types of line cards were used in the DMS-100 central office, one each for POTS, coin phones, Northern's Electronic Business Set, Datapath data service, and ISDN. A new service-adaptive, programmable line card supports virtually all services that can be extended to residential or business subscribers over a two-wire line. The placing of ringing and generation of other high voltages, formerly generated centrally and distributed to specialized line cards, directly on the service-adaptive line cards permits the use of a single circuit to generate the variety of signals needed in line-card operation.[42]

12.5.3.6. S/DMS SuperNode ISDN Architecture

The LPP, shown in Fig. 12.21, is a multifunctional processing unit that enables the S/DMS SuperNode to function in the SS7 signaling network and also to provide ISDN and other advanced services. The LPP cabinet contains a duplicated Local Message Switch (LMS) and three Link Interface Shelves (LIS) containing modular, add-in cards known as Interface Units (IU). The IUs perform message transfer and control in the SS7 network, support wideband network data services, or provide a DMS Packet Handler (DMS PH) for packet-switched data transfer. The DMS PH is known as the X.25/X.75/X.75′ Link Interface Unit (XLIU). A minimum configuration requires one XLIU to handle X.25 protocols and another to handle X.75 and X.75′. Each XLIU contains three cards to perform Level 2 and Level 3 processing and to provide channelized data to the high-density line controller frame processor.

S/DMS SuperNode architecture for ISDN is shown in Fig. 12.28. Each subscriber line terminates on a line card in the ISDN Line-Concentrating Module (ISDN LCM) that supports up to 480 line cards for ISDN and other services. The module performs low-level call processing functions and provides some fault isolation. The ISDN Line Group Controller (LGCI) provides the interface between the LCME and the switching modules, using DS30A links to support up to 480 ISDN lines. Up to 24 B-channels can be switched together to provide a wideband channel. The Digital Trunk Controller (DTC), ISDN DTC (DTCI), or ISDN Line Trunk Controller (LTCI) provides the interface to ISDN trunks. B-channel and D-channel traffic are switched through the switching network via different paths. D-channel packets are split off, statistically multiplexed, and routed via dedicated channels through ENet using DS30 links to the DMS PH Interface Units in the LPP. B-channel packets are routed through available paths in the switching network to the DMS PH using DS30 links. A Network Interface Unit (NIU) terminates the DS30 links from the network module and distributes data to the DMS PH units in a Link Interface Shelf under control of a channel bus controller.[43]

12.5.4. DMS SuperNode Software Architecture

Restructuring of DMS SuperNode software architecture is being accomplished in two phases.[44] In the first stage, software is layered. A Base Layer and a Telecom Layer form a common software platform for all applications. The Base Layer functions in a manner similar to the operating system of a personal computer,

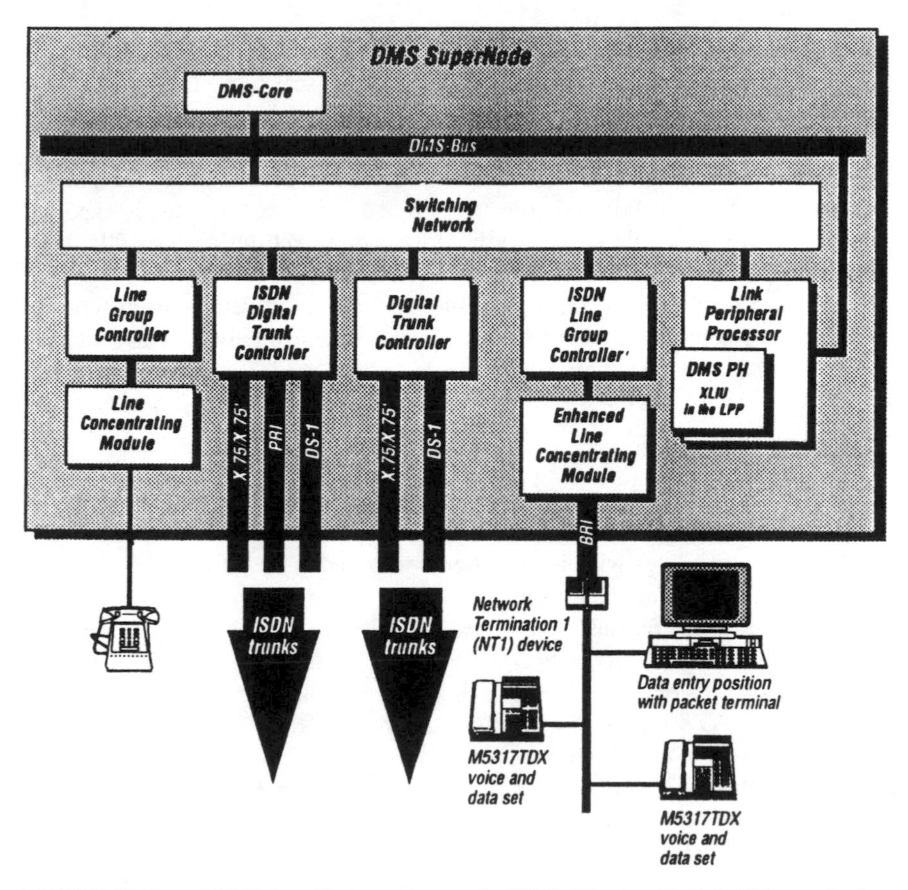

FIGURE 12.28. S/DMS SuperNode architecture for ISDN. (Courtesy Northern Telecom, Inc.)

performing as the hardware controller and scheduler. The Telecom Layer contains the call-processing framework for services offered by today's network. The next higher layer, the Product Layer, enables Telecom Layer functions to be applied to specialized applications in specific markets and contains application-specific software not required by other components. This includes, for example, business applications specific to the North American market. Above the Product Layer, the Customer Layer contains custom-developed features for individual customer users of the switch.

The second stage of software evolution involves adaptation to the SuperNode of the Generic Services Framework (GSF) Call-Processing architecture currently used in Northern Telecom's switch based on Global System for Mobile Communications standards. This software architecture separates the originating and terminating halves of a call. In current Double-Ended Call-Processing architecture, software for each service type contains all the instructions it needs to handle both the originating and terminating ends of a call. In the GSF software processing, the separate halves of a call communicate with each other via a GSF Agent Interworking Protocol (AIP). Communications within each call half are handled by Event-Driven Call-Processing instruction sets. This second stage is expected to be generally available in the North American market during the 1995–1996 time frame.

Advantages of the new architecture include shortened development time for new features, simplified software interaction between relatively self-contained software modules, and simplified interworking between features.

PROBLEMS

12.1. Explain why a T1 span line failure does not cause a switchover of redundant equipment in the Siemens-Stromberg Carlson DCO?

12.2. How is a space-switching function performed in the DCO using the Cross Office Highway interconnectivity?

12.3. What are the residual capabilities of a Remote Line Group, a remote Line Switch, and a Remote Network Switch if communications with the DCO host is severed?

12.4. In the AT&T 5ESS Switch, how many and what type switching stages are used in switching a call arriving via a digital line interface on one Peripheral Unit to a digital interface on another Peripheral Unit in the same Switching Module? In a different Switching Module?

12.5. In the 5ESS, how many and what type switching stages are used in switching an analog line to another analog line in the same Switching Module, assuming 10-to-1 concentration? In a different Switching Module?

12.6. How does the connection of a Remote Switching Module to a 5ESS host differ from the connection of an Optically Remoted Switching Module to the same host and why?

12.7. In the AT&T 4ESS Switch, analyze and explain the two stages of switching in the TSI frame.

12.8. Since incoming digital channels are synchronized to the AT&T Reference Frequency and DS120 streams are synchronized to the network clock, why is it necessary to resynchronize the DS120 streams in the buffer memories of the TSI frame?

12.9. Why are some DS120 cables from a Digital Interface Frame connected to an Echo Suppressor Terminal even though those trunks do not need echo suppression?

12.10. How does the Northern Telecom S/DMS SuperNode architecture facilitate the addition of future switching fabrics?

12.11. In the S/DMS SuperNode, how does the four-tier memory hierarchy in DMS-Core enable faster call processing?

12.12. How does Reduced Instruction-Set Computing (RISC) enable still faster call processing in the S/DMS SuperNode?

REFERENCES

1. *DCO General Description, Part 1, DCO*, Publication 00-020-01, Issue 3, Siemens Stromberg-Carlson, August 1992. (Hereinafter cited as *Part 1, DCO*.)
2. *DCO Remote Line Equipment, Part 2, Interfaces*, Publication 00-300-02, Issue 3, Siemens Stromberg-Carlson, April 1992.
3. *Part 1, DCO.*
4. *DCO General Description, Part 3, Common Control*, Publication 00-020-03, Issue 3, Siemens Stromberg-Carlson, August 1992.
5. Ibid.
6. *DCO Remote Line Equipment, Part 3, Remote Line Group*, Publication 00-300-03, Siemens Stromberg-Carlson, April 1992.
7. *DCO Remote Line Equipment, Part 4, Remote Line Switches 360, 450, and 1000*, Publication 00-300-04, Issue 4, Siemens Stromberg-Carlson, August 1992.
8. *DCO Remote Line Equipment, Part 5, Remote Line Switch 4000*, Publication 00-300-05, Issue 3, Siemens Stromberg-Carlson, April 1992.
9. *Part 1, DCO.*
10. *5ESS Switch Information Guide*, 235-300-010, Issue 4, AT&T Network Systems, 1992. (Hereinafter cited as *AT&T Guide*.)
11. Ibid., pp. 9–10.
12. Ibid., pp. 12–14.
13. Ibid., p. 16.
14. Ibid., pp. 15–16, 43–44.
15. Ibid., p. 44.
16. Ibid., pp. 23–26.
17. Ibid., pp. 29–30.
18. Ibid., pp. 30–31.
19. Ibid., pp. 31–33.
20. Ibid., pp. 33–34.
21. Ibid., pp. 61–67.
22. Ibid., pp. 17–21.
23. *4ESS Switch Training Program, 4ESS Switch Overview*, AT&T, January 1992, pp. 1.1-3 to 1.1-4 (hereinafter cited as *4ESS Overview*).
24. *4ESS Switch General Description*, AT&T Practice 234-100-000, Issue 11, September 1990, p. 26 (hereinafter cited as *AT&T Practice*).
25. *AT&T Practice*, pp. 84–85.
26. Ibid., pp. 85–87.
27. Ibid., pp. 88–89.
28. Ibid., pp. 89–90.
29. Ibid., pp. 74–84.

30. *4ESS Overview*, pp. 1,3–19 to 1.3–21.
31. *AT&T Practice*, pp. 64–65.
32. Ibid., pp. 64–67.
33. Ibid., p. 67.
34. Mellor, F. and Wood, R., "The S/DMS Product Family," *Telesis*, 1990 one/two, pp. 21–32.
35. Ibid.
36. Perry, J., "DMS SuperNode: Technology overview," *Telesis*, 1988 two, pp. 8–10.
37. Northern Telecom Product/Service Information 50058.16/06-92, June 12, 1992.
38. Ibid.
39. *DMS SuperNode System Planner*, May 1988, pp. 8–10 (hereinafter cited as Planner); Ref. 36, pp. 7–8.
40. *Planner*, pp. 10–13.
41. Northern Telecom Product/Service Information 50041.16/08-92, Issue 2, August 7, 1992.
42. Mein, G. and Terry, J., "The evolution of DMS-100 line card technology," *Telesis*, issue no. 92, July 1991, pp. 83–85.
43. Northern Telecom Product/Service Information 50062.16/07-92, Issue 1, July 31, 1992, pp. 4–10.
44. Northern Telecom Feature Planning Guide 4Q93, Product Service Information 50004.11/11-93, Issue 11, November 1993.

13

The Evolving Switched Digital Network

13.1. INTRODUCTION

The North American public switched telephone network (PSTN) has been evolving from a fully analog network toward one that is expected to approach a fully digital network.

The Canadian portion of the network is operated by a group of telephone companies that independently provide local and intraprovincial services, but which associate together as Stentor (formerly Telecom Canada) to provide interprovincial and cross-border services. Interprovincial and cross-border services also are provided competitively by Unitel Communications, Inc.

In the United States, approximately 80% of the telephones are served by the operating companies that were part of the Bell System prior to January 1, 1984. On that date, 22 Bell Operating Companies (BOCs) were divested as a result of a consent decree that settled a government antitrust suit against the American Telephone and Telegraph Company (AT&T). The divested BOCs were formed into seven Regional Bell Operating Companies (RBOCs) and function as holding companies for BOCs and, in some instances, as operating companies themselves. Long distance communications is provided by AT&T, U.S. Sprint, MCI Communications, and a large number of other primarily regional companies, many of which resell services leased from other carriers.

While the switching systems and major transmission links in these networks are largely digital, there still are sufficient analog portions to designate the composite network as *mixed*. For simplicity, unless otherwise specified, these networks or portions thereof are referred to as "the network." The reader is referred to Appendix A for a detailed discussion of the mixed network, including the network number plan, network routing, and signaling principles, but excluding the digital evolution covered in this chapter.

Section 13.2 describes the local telephone networks, operated by the divested BOCs and some 1400 independent telephone companies, in terms of network configuration, transmission, and signaling. Section 13.3 deals with the North American (that is, the United States and Canada) network in terms of signaling and transmission. Section 13.4 discusses synchronization of the digital network.

13.2. INTRA-LATA NETWORKS

Concurrently with the divestiture of the BOCs, a new network configuration was implemented for handling local-area calls and connections between local and competing long distance carriers. The consent decree required separation of exchange and interexchange functions. Former exchange areas were reconfigured into local access and transport areas (LATAs) in BOC territories. Calls within a LATA typically are handled by the BOC serving that LATA. Calls between LATAs are carried by inter-LATA carriers (ICs). Non-BOC local telephone companies generally are not required to conform to these requirements, but may participate on an optional basis. General Telephone was made subject to similar requirements resulting from its own consent decree.[1] In practice, the LATA concept applies throughout the United States.

13.2.1. Local Access and Transport Areas (LATAs)

The continental United States is divided into over 180 LATAs. Other LATAs cover off-shore states and territories. Each LATA is roughly equal to a standard metropolitan statistical area (SMSA) or a standard consolidated statistical area (SCSA). Deviation from SMSA and SCSA boundaries was permitted primarily to preserve existing wire-center boundaries, service arrangements, and communities of interest, to minimize subscriber impact, or to avoid disruption of end-office toll trunking.[2] As established by the consent decree, each LATA must include at least one AT&T toll switching system.

Under the terms of the consent decree, divested BOCs are to provide service for local exchange traffic, intra-LATA toll traffic, and access by subscribers to inter-LATA carriers (ICs). To carry out those functions on a nondiscriminatory basis, changes in network configurations, signaling, and transmission plans were required.

13.2.2. Intra-LATA Network Switching Plan

Intra-LATA networks may serve a small geographic area with a high concentration of subscribers, such as a large metropolitan area, or they may extend for up to about 650 kilometers and include several small cities and towns as well as rural areas. Most LATAs employ a two-level hierarchical network configuration. Central offices, known as *end offices*, are connected to each other by direct *interend-office trunks* (IOTs) when justified by traffic loads, and *tandem offices* are used in the second hierarchical level. A LATA may have one or more tandem switching systems, each connected to subtending end offices in a sector. Very large LATAs may have a third-level tandem office or may designate one or more second-level tandems as a principal *sector tandem*. Fig. 13.1 illustrates a skeletonized two-level hierarchy in a large LATA containing both a metropolitan network and an intra-LATA long distance network.[3]

At divestiture, some switching systems were shared between AT&T and the BOCs during a transition period. Traffic routing, signaling, and transmission plans

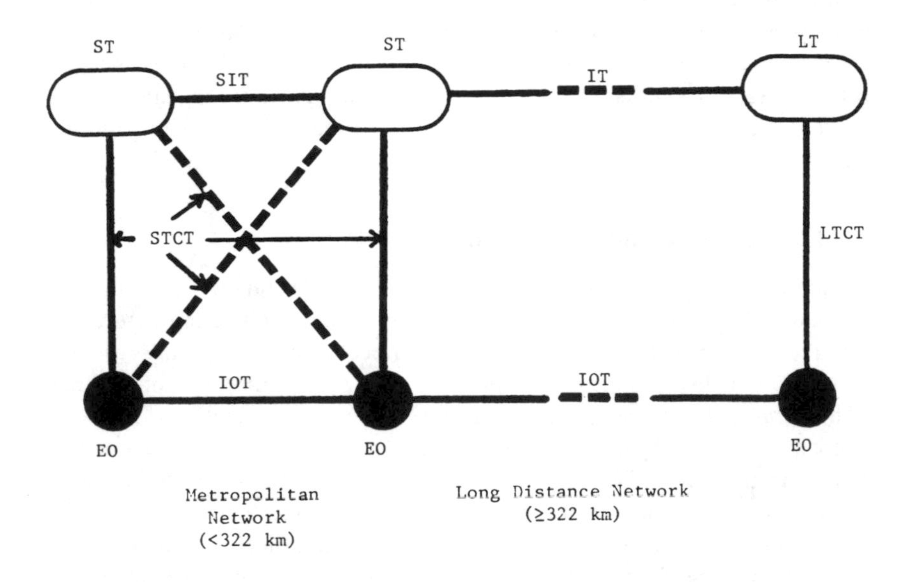

Legend

ST - Sector Tandem
EO - End Office
LT - LATA Tandem
IT - Intertandem Trunk
SIT - Sector Intertandem Trunk
STCT - Sector Tandem Connecting Trunk
LTCT - LATA Tandem Connecting Trunk
IOT - Interend Office Trunk

FIGURE 13.1. Intra-LATA network configuration.

were uniformly established in all BOCs; however, these plans have been modified at different rates according to the growth and modernization within each BOC.

Intra-LATA calls are routed in a modified hierarchical manner. A maximum of four routes can be provided for calls from any end office to any other end office in the same LATA. No more than three trunks are used in any intra-LATA connection. As Signaling System No. 7 with nonassociated signaling is fully deployed, however, dynamic routing is expected to be used throughout the intra-LATA networks. Several dynamic routing alternatives have been investigated. One system called *dynamic routing with 5-minute updates* (DR5), has demonstrated superiority over hierarchical routing. In a trial network with 7% of the high usage (HU) trunks removed, overflow traffic to the tandem switching center was less than half of the overflows under hierarchical routing. Also, over 80% of the DR5-recommended reroutes in the busy hour were successful.[4]

A combination of inband signaling and common-channel signaling is used in the intra-LATA networks. Common-channel signaling, where installed, is used for intra-LATA calls, and is in the process of replacing inband signaling for inter-LATA connections.

13.2.3. Intra-LATA Network Transmission Plan

All transmission systems are imperfect and impair the quality of information transmitted. A major objective of transmission-system engineering is to control impairments in such a way as to achieve an acceptable compromise between transmission quality and the cost of providing the system (see Section A.3.2.3).

The intra-LATA transmission plan is based upon a combination of trunk loss and connection loss objectives. In metropolitan networks, digital connections via IOTs are designed for a 3-dB net loss from end office to end office. Digital tandem connections are designed for a 6-dB net loss. Analog and mixed connections in a metropolitan area may encounter losses between end offices ranging from 3 dB to about 9 dB, depending upon the type of facility and office balance parameters.[5]

Some LATAs cover large areas, such as an entire state. Intra-LATA long-distance transmission standards are patterned after the former Bell System network standards. For all-digital connections, the loss objective is 6 dB between end offices, whether or not a tandem office is used. Mixed connection loss objectives are shown in Table 13.1. The additional loss on connections over 322 kilometers compensates for added echo susceptibility.[6]

13.2.4. Intra-LATA Signaling

Since digital transmission systems can be "proved in" economically more readily than can digital switching systems, transmission-system conversion has progressed more rapidly in some areas than switching-system conversion. This has

TABLE 13.1. Mixed-Connection Trunk-Loss Objectives (dB)

Trunk Type	<322 KM	≥322 KM
Interend office trunks		
Digital	3	6
Combination	3	6
Analog	3	VNL + 6
LATA tandem connecting trunks		
Digital	3	3
Combination	3	3
Analog	3	VNL + 2.5

resulted in network areas in which there are mixed digital and analog switching and transmission systems. Growth of all-digital islands in LATAs has accentuated the need for common-channel signaling to replace inband signaling (see Appendix A). The North American version of CCITT Signaling System No. 7 (SS7), discussed in Section 13.3.1, is rapidly being installed in LATAs for intra-LATA traffic, however, connection of LATA SS7 systems with IC SS7 systems has progressed more slowly. SS7 is an entirely new signaling system that requires a learning curve for both manufacturers and network operators. Both intra-LATA carriers and ICs have been appropriately hesitant to run ahead of that learning curve for fear that catastrophic network failures could occur (this has, indeed, happened on occasion). Additionally, both types of carriers have been reluctant to interconnect their early SS7 systems with those of other carriers until sufficient testing could be completed to generate a high level of confidence in the reliability and compatibility of the separately owned and managed systems.

The FCC mandated the portability of "800" numbers and the reduction of 800-number access delays to five seconds or less by March, 1993, and to two seconds or less within two years; however, this time was extended to May 1, 1993. Therefore, the intra-LATA carriers and ICs have had to interconnect their SS7 signaling systems. Although carriers had to make some interim arrangements to satisfy the FCC mandate, further integration of their respective SS7 systems is continuing. It is expected that the larger independent local exchange carriers will be connected fully to SS7 systems within a reasonable time period.[7]

13.3. THE NORTH AMERICAN NETWORK

13.3.1. Network Signaling

Discussions of the principles of both inband and common-channel signaling are in Appendix A. This subsection discusses the application of those systems to domestic and international networks.

13.3.1.1. Inband Signaling

Inband signaling techniques are covered in Sections A.4.2.3 through A.4.2.7 and A.4.4 through A.4.6.

13.3.1.2. Common-Channel Signaling

Common-channel signaling principles are covered in Section A.4.7. Two common-channel signaling systems have been defined for worldwide use. CCITT System No. 6 uses analog voiceband transmission, and CCITT System No. 7 uses the standard 64-kb/s digital transmission link. Common-channel interoffice signaling (CCIS), formerly used by AT&T as a variation of CCITT No. 6, has been replaced by SS7 in its public network. SS7 contains options for offered services and national network differences.

13.3.1.2.1. OSI Reference Model. SS7 is adapted from the *Reference Model of Open Systems Interconnection for CCITT Applications* (Recommendation X.200), which provides a layered structure for modeling the interconnection and exchange of information between users in a communications system.[8] The OSI reference model is shown in Fig. 13.2.

- The *physical layer* (layer 1) provides the electrical, mechanical, functional, and procedural capability to activate, maintain, and deactivate a physical connection in a communications medium, permitting transparent transmission of a bit stream (at 56 kb/s or 64 kb/s in SS7).
- The *link layer* (layer 2) provides error control and synchronization for the bit stream transmitted over the physical link, and bit stuffing is used to suppress strings of 1s longer than five.
- The *network layer* (layer 3) transfers data transparently by performing routing and relay of data between end users.
- The *transport layer* (layer 4) provides end-to-end control and information interchange with the quality of service needed for the application.
- The *session layer* (layer 5) supports coordination and interaction between communicating application processes; for example, full and half duplex dialog.
- The *presentation layer* (layer 6) provides services to allow the application process to interpret the information exchanged by syntax conversion when necessary.
- The *application layer* (layer 7) provides a window through which a user's application process is managed, and is the sole means for access to the OSI environment.

13.3.1.2.2. Signaling System No. 7 Protocol Model. Signaling system no. 7 uses a 4-layer protocol model[9] that does not correlate exactly with layers of the 7-layer OSI model, as shown in Fig. 13.3. Layers 1 and 2 of the SS7 protocol

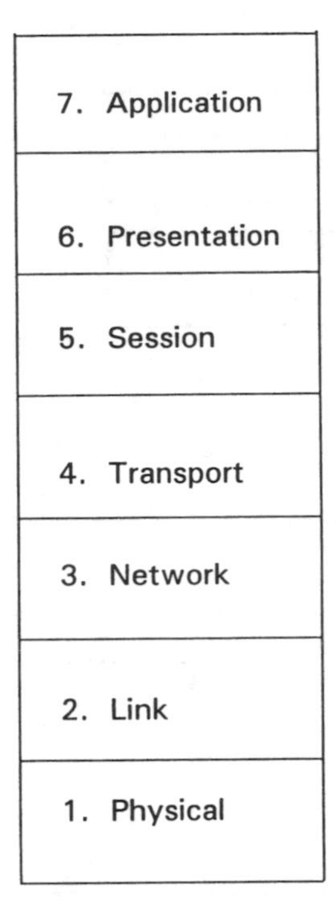

FIGURE 13.2. OSI 7-layer Reference Model.

model are functionally equivalent to OSI layers 1 and 2. The functions of OSI layer 3, however, are subdivided into two sublayers for use in SS7 layer 3 and part of layer 4. OSI layers 4, 5, and 6 and are not defined for SS7. In SS7, layers 1, 2, and 3 together are known as the *Message Transfer Part* (MTP). Its function is to provide reliable transfer of signaling messages between the users or application functions. In this context, the term *user* refers to any functional entity that utilizes the basic transport capability provided by the MTP. Two such users, the *Telephone User Part* (TUP) and the *Data User Part* (DUP), are part of the basic model but are not used in the North American network.

All other SS7 functions are in SS7 layer 4. *The Signaling Connection Control Part* (SCCP), provides specialized routing capabilities for communication between signaling nodes via virtual circuits or via connectionless messages. *A Transaction Capabilities Application Part* (TCAP) provides application-level features for special services provided by the switched network. The *ISDN User Part* (ISUP) provides application-level functions for signaling services, including call setup and disconnect. The ISUP is discussed further in Chapter 14 in connection with the ISDN.

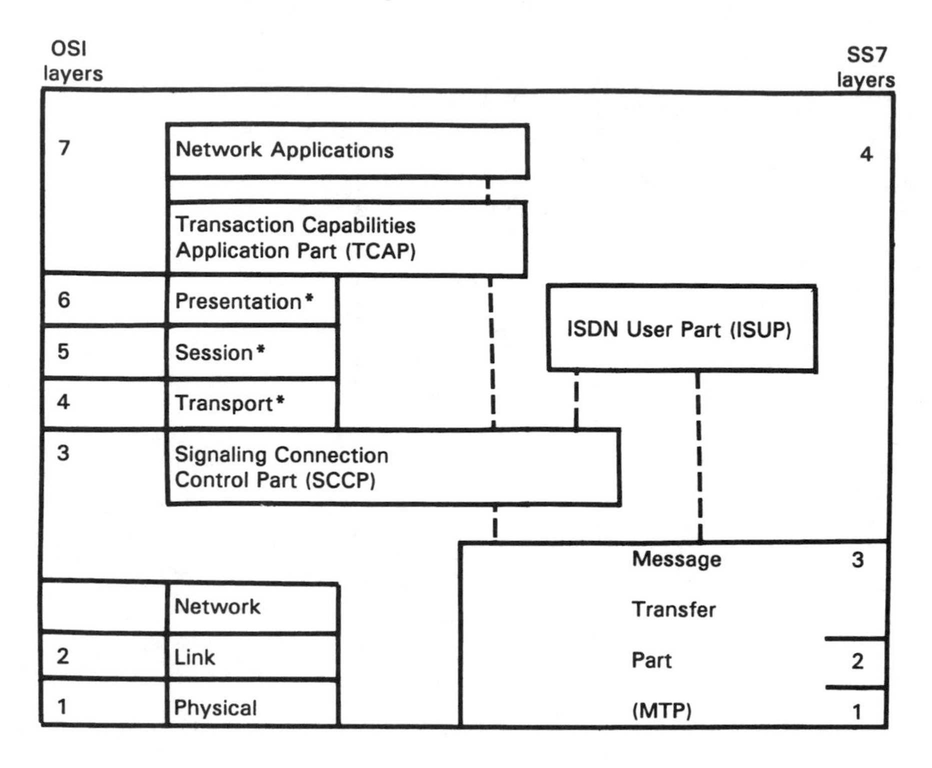

FIGURE 13.3. Signaling System No. 7 protocol model.

13.3.1.2.3. The Signaling Network Configuration. Signaling nodes in SS7, including network switching centers, are referred to in the signaling network as *signaling points* (SP). Signaling messages between two SPs are exchanged over *signaling links* operating at 56 or 64 kb/s. In the *associated mode*, signaling messages between two SPs are exchanged over a directly interconnected signaling link, called an *F* link. When signaling messages between two SPs are exchanged over two or more signaling links in tandem, signaling is said to be in the *non-associated mode*. In such cases, signaling messages are relayed through *signaling transfer points* (STPs), provided in mated pairs. Figure 13.4 depicts a two-level hierarchy signaling network. Each network switching center in the group is connected to the pair of STPs by data links, called *A* links. Each STP in a mated pair is connected to the other STP by a *C* link, and mated pairs of STPs are interconnected by a quad of *B* links. Data bases are established at locations called *network control points* (NCP), alternatively called *service control points* (SCPs), which

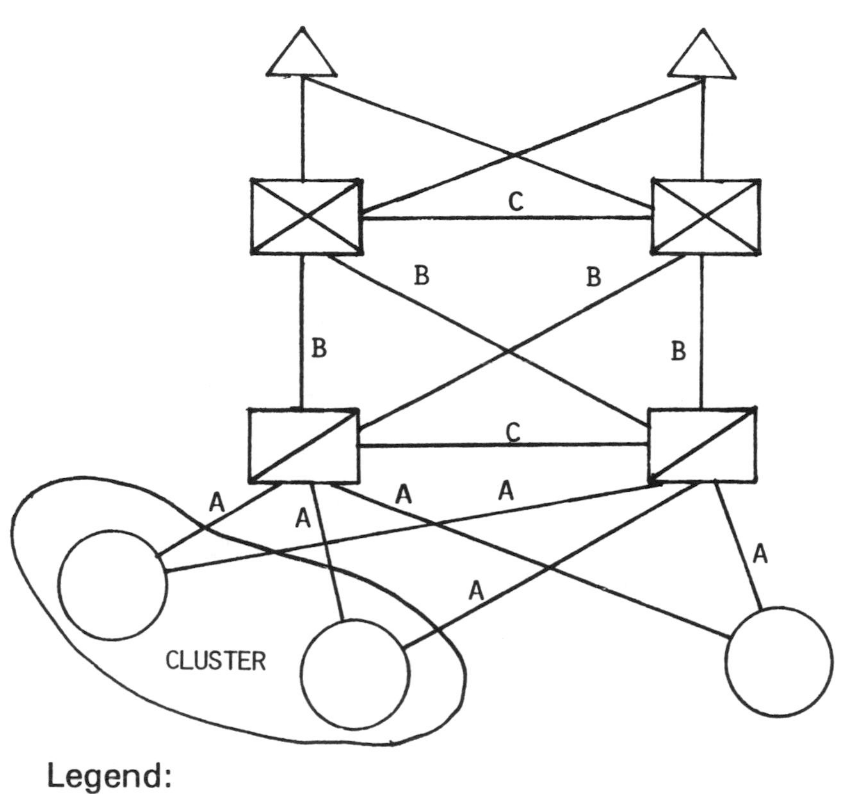

NETWORK CONTROL POINT

PRIMARY SIGNALING TRANSFER POINT

SECONDARY SIGNALING TRANSFER POINT

ORIGINATING SIGNALING POINT

FIGURE 13.4. Signaling System No. 7 network configuration.

may or may not be collocated with STPs. The NCPs provide information for calls requiring special routing.

In a two-level hierarchy, the STP mated pair at the top level of the hierarchy is designated as a *Primary STP pair*. Multiple Primary STP pairs are interconnected via quads of *B* links. The number of Primary STP pairs depends upon traffic volumes, STP capacity, and economic analysis. The lower level of STP pairs in the hierarchy, called *Secondary STP pairs*, are always homed to specific Primary STP pairs via quads of *B* links. If two or more Secondary STP pairs are connected to one Primary STP pair, the connection is made via quads of *D* links. Such Secondary STP pairs in a region may be connected to each other via quads of *B* links; otherwise, all signaling between Secondary STP pairs is routed through the home Primary STP pair.

Signaling points may be grouped into *clusters* to facilitate routing of signaling messages. A small cluster contains an STP pair and all SPs that are directly homed on that STP pair, either Primary or Secondary. Small clusters may be grouped into a large cluster, in which case one of the contained small clusters must contain a Primary STP pair.

A links, providing access for SPs to STPs, should be provisioned on physically diverse routes to each STP in the mated pair. In addition to these *A* links, SPs may be connected to another pair of STPs via *E* links, provided on diverse routes, to provide more efficient handling of signaling traffic. Because signaling traffic over *F* links can be alternate-routed via the overlay signaling network, single *F* link sets are permitted.

13.3.1.2.4. Signal Units. Signaling messages are composed of *signal units* (SU) containing both fixed-length and variable-length fields.[10] The basic format of a message signaling unit is shown in Fig. 13.5a. The least significant bit (bit 1) of each SU is the first bit transmitted. The first octet in each SU contains a flag having the bit pattern *01111110*. To ensure that text bits are not misinterpreted as a flag, the transmitting signaling link terminal inserts a *0* after every sequence of five consecutive *1*s. At the receiving terminal, each *0* that directly follows five consecutive *1*s is deleted. In transmission of consecutive SUs, the opening flag of one SU is considered to be the closing flag of the preceding SU. Within each field or subfield, bits are transmitted with the least significant bit first, except that the 16 *check bits* are transmitted in the order generated. Signal units are sequence numbered using a binary code from a cyclic sequence ranging from 0 to 127. The *forward sequence number* is the sequence number of the SU in which it is carried. The *backward sequence number* is the sequence number of a signal unit being acknowledged. The *length indicator* indicates the number of octets following the length indicator octet and preceding the check bits. The *service information octet* is used at layer 3 to distinguish between different types of messages. Bits 4–1 comprise a service indicator that indicates the MTP-user part involved in the

H	F	E		D	C		B		A	
CK	SIF	SIO	spare	LI	FIB	FSN	BIB	BSN	F	First bit transmitted ------>
16	8n, n>2	8	2	6	1	7	1	7	8	field length (bits)

a) Basic format of a message signal unit

H	G		D	C		B		A	
CK	SF	spare	LI	FIB	FSN	BIB	BSN	F	First bit transmitted ------>
16	8	2	6	1	7	1	7	8	

b) Format of a link status signal unit

H		D	C		B		A	
CK	spare	LI	FIB	FSN	BIB	BSN	F	First bit transmitted ------>
16	2	6	1	7	1	7	8	

c) Format of a fill-in signal unit

Legend:

BIB	Backward indicator bit
BSN	Backward sequence number
CK	Check bits
F	Flag
FIB	Forward indicator bit
FSN	Forward sequence number
LI	Length indicator
SF	Status field
SIF	Signaling information field
SIO	Service information octet

FIGURE 13.5. Signal unit (SU) formats. (© 1993, Bellcore. Reprinted with permission. The information reprinted from Bellcore was deemed current at the time of publication of this work, but since then, the information may have been amended or updated. For more current information, refer to the current Bellcore publication.)

message. Bits 6–5 indicate the message priority, and bits 8–7 identify the network. The *signaling information field* is a variable-length field consisting of up to 272 octets in the North American networks and conveys the actual message content. The first seven octets comprise a routing label containing codes representing the originating and destination points and the signaling link selection.

Three types of signal units are defined for use in SS7.[11] *Message* SUs are used to transmit information generated by MTP users. *Fill-in* SUs are transmitted when no information is contained in a link transmission buffer and message retransmission has not been requested by the remote signaling link node (Fig. 13.5c). *Link status* SUs are used to indicate status of signaling links, such as conditions of busy, out of alignment, normal or emergency alignment, out of service, or processor outage (that is, a problem affecting a functional level above MTP level 2) (Fig. 13.5b). The type of SU being received is identified by the length indicator. Fill-in SUs have a length indicator of zero; link status SUs have a length indicator of one or two to indicate link status; and message SUs have a length indicator greater than two.

13.3.1.2.5. Signaling Link Performance. Obviously, no calls can be completed unless the signaling network is functioning. The signaling network has characteristics similar to those of a store-and-forward message network. The distinguishing feature is the timeliness of delivery required by SS7. Normal loading of signaling links is 40% of transmission capacity (that is, 0.4 erlang of traffic), and links are considered to be in failure condition at twice the normal loading.

A signaling link can be taken out of service by either automatic or manual controls. The tendency is to rely largely upon automatic controls in an attempt to create a self-healing network. Upon detecting signaling-link congestion, signaling-network management messages effect signaling-link changeover to shift signaling traffic to a serviceable link.

Signaling messages are assigned four levels of priority, zero to three, to indicate which message should be discarded in case of signaling-network congestion. Network-management messages, except signaling-route-set-congestion-test messages, are assigned priority level 3, the highest of the priorities. Congestion status is reflected in three types of congestion thresholds (congestion onset, congestion discard, and congestion abatement) at each of three overlapping levels of transmit buffer occupancy. Congestion-onset thresholds detect overload conditions, and congestion-abatement thresholds detect recovery from congestion. Congestion-discard thresholds are used to determine whether a message should be transmitted or discarded during overload conditions.

As transmit buffers begin to overload (that is, exceed 40% occupancy), network-management messages are automatically generated to other nodes to inform them of the congestion status of particular signaling links. As the overload increases, it can trip a threshold that causes nodes to be instructed to discard certain priorities of signaling messages. When call-related signaling messages are discarded, the related calls fail to complete. If congestion continues to build up, higher priority messages are discarded. In the event of severe congestion (that is, 80% or more), no call-related signaling messages are transmitted over the affected links. Failed call attempts habitually generate more call attempts and can contribute to switching equipment congestion by excessive use of call-processing capability.[12]

13.3.1.2.6. Signaling Message Routing. Message Transfer Part users place a routing label in the signaling information field of each message signal unit to direct the signaling message to its destination point. The label contains a Destination Point Code (DPC), an *Originating Point Code* (OPC), and a *Signaling Link Selection* (SLS) field. Point codes are 24-bit addresses that identify signaling points, or nodes, in the signaling network. The point code is composed of three octets designating network identification, cluster identification, and cluster member number, a *cluster* being a homogeneous group of signaling points, any of which can route a signaling message to its intended destination. The routing label format is shown in Fig. 13.6 Signaling protocol is designed to ensure that all messages sent over the same signaling link will be received at the destination point in the order transmitted.[13]

13.3.1.2.7. Signaling Route Management. Orginating SPs should route signaling messages by the most direct route available, whether via *F* links, *E* links, or *A* links. *F* links are the preferred route if they exist, with *E* links the second choice, and *A* links the final choice. Preferred routing choices for STPs to a destination point are *A* links, *E* links, *D* link quads, and *B* link quads. The routing principle, in all cases, it to send the signaling traffic to its destination SP in the most direct route available.

The originating switching system may route signaling messages by point code, by cluster, or by a combination of both. Cluster routing is used only for signaling points that are served by a single STP pair. In a very large network, routing by full point code requires a large amount of memory to maintain status of all signaling points in its routing table. Cluster routing reduces the amount of information required to be stored. To maintain status of signaling points that have a more restrictive status than the cluster of which they are a part, switching systems maintain dynamic exception lists for the clusters in their routing tables. For each signaling point or cluster in a switching system's routing table, routing status is maintained as *allowed, restricted,* or *prohibited.*

Status messages are sent by an STP to indicate its capability to route signaling messages to signaling points and clusters. Although cluster routing saves memory, it can cause additional routing status messages to be transmitted through the signaling network. For example, if an STP sends *Transfer-Cluster-Restricted*

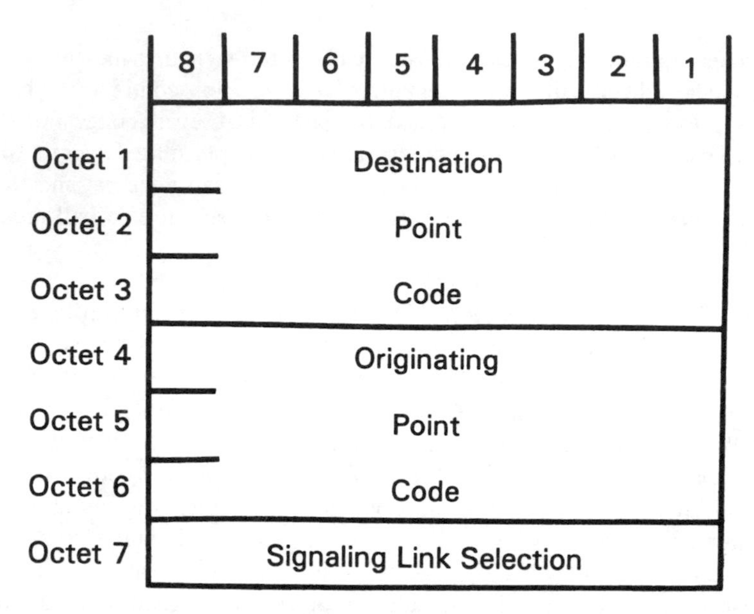

FIGURE 13.6. Routing label format. (© 1993, Bellcore. Reprinted with permission. The information reprinted from Bellcore was deemed current at the time of publication of this work, but since then, the information may have been amended or updated. For more current information, refer to the current Bellcore publication.)

(TCR) message for a given cluster, each message-switching system marks each member of the cluster as restricted on a given route. Then the STP may send a *Transfer-Prohibited* (TFP) message for one member of the cluster, so that signaling point is marked prohibited, and no traffic can be sent to that destination over that route. Then, if congestion in the cluster abates and a *Transfer-Cluster-Allowed* (TCA) message is sent, the message-switching system marks the cluster allowed, but the specific signaling point is still prohibited from receiving traffic on that route. Therefore, the STP must immediately send another status message for each cluster member that does not have the same status as its cluster.[14]

13.3.1.2.8. Noncircuit-Related Signaling. The principles of circuit-related signaling, covered in Section A.4.7.3 in connection with an earlier signaling system, are similar to those of SS7. Noncircuit-related signaling comprises signaling messages that do not pertain to any particular message circuit. Such messages are handled by the SCCP of the SS7 signaling protocol, and the SCCP is indicated by the service information octet as the MTP user. The SCCP data field is provided by the SCCP user, such as the TCAP Four protocol classes are designated for the SCCP.[15]

- *Class 0-Basic Connectionless Class* provides for simple datagram transport with no attempt to guarantee any relationship between messages.
- *Class 1-Sequenced (MTP) Connectionless Class* is similar but provides that message sequencing will be maintained under normal conditions.
- *Class 2-Basic Connection Oriented Class* provides explicitly to associated messages that are part of the same transaction and to ensure message sequencing.
- *Class 3-Flow-control Connection-Oriented Class* provides a sequence-numbering scheme to allow message flow control and to provide for efficient delivery of transaction information contained in a single message SU.

The TCAP is used to format messages containing user data destined for some application at a different node in the signaling network. The TCAP uses the SCCP to route such messages from the originating application to the destination application. If needed, the SCCP provides some specialized routing capabilities before passing the information to the MTP for transport. If the user data does not exceed 254 octets, the message is transferred as a single packet of data, called a *Unitdata* (UDT), as protocol class 0. Larger amounts of user data are segmented into multiple packets, called *Extended Unitdata* (XUDT), and are transferred as protocol class 1 to include the sequencing function. The XUDT messages are reassembled by the destination SCCP and passed to the TCAP user as a complete message.[16]

13.3.1.2.9. Signaling Node Restart Procedures. A signaling node, such as a STP, SCP, or message-switching system, can become isolated because of its own hardware or software malfunction or because all of its connected signaling links are unavailable. When the causing problem has been corrected, the unavailable signaling node is brought back into service through a restart procedure. If a restarting signaling point activates its signaling links serially, the first link that becomes available will enable the other node to start load sharing its signaling traffic on that link. Until sufficient links become available to switch signaling messages to other destinations, however, the restarting node will be receiving traffic that it cannot transfer to the intended destinations. Thus, the initial messages would be lost unnecessarily, calls would be blocked, and retrials would increase call attempts in the network. The solution is to bring the signaling links up to operational status, but not report them as available for traffic until the restarting node has sufficient links to handle the traffic.[17]

13.3.2 Digital Network Transmission Considerations

As digitalization of the network progressed, it became increasingly apparent that the mixed analog/digital network would require rethinking of the transmission plan. A transmission plan for the future fully digital network was necessary before a transmission plan could be developed for the mixed network.

13.3.2.1. Loss and Level Considerations

Digital transmission and switching involve loss and level considerations which were not factors in a purely analog network. The TLP concept, which functions well in an analog environment, requires modification for use in a digital environment. In an analog system, test tones used to line up (adjust levels in) voiceband channels generally are applied at a power level of one milliwatt (0 dBm). Digital systems, however, involve the encoding and decoding of analog signals, and the direct use of the one-milliwat analog test tone is not feasible. Therefore, *a digital milliwatt* has been standardized. For 8-kHz sampling and the standard 64-kb/s channel rate with $\mu = 255$ companding, the digital milliwatt is defined by the CCITT through the bit stream of an eight-word sequence shown in Table 13.2.[18] A consequence of this definition is that the bit stream represents a 1000-Hz sine wave at a power level 3.17 dB below the overload level of the encoder. (The telephone industry in North America has agreed to use 1004 Hz as a test tone to avoid frequency beating of an exact 1000-Hz signal in the presence of PCM multiplex systems.)

In order to utilize the TLP concept in digital systems to the extent possible, encode and decode level points have been defined at the inputs and outputs of encoders and decoders, respectively.

- The *encode level* (EL) in dB at the encode level point (ELP) is the power level in dBm of a sine wave at that point which will result in coded values equivalent to the digital milliwatt.
- The *decode level* (DL) in dB at the decode level point (DLP) is the power in dBm of a sine wave at that point which will result from decoding the digital milliwatt.

TABLE 13.2. Digital Milliwatt[a]

		Bit Number							
		1	2	3	4	5	6	7	8
Word number:	1	0	0	0	1	1	1	1	0
	2	0	0	0	0	1	0	1	1
	3	0	0	0	0	1	0	1	1
	4	0	0	0	1	1	1	1	0
	5	1	0	0	1	1	1	1	0
	6	1	0	0	0	1	0	1	1
	7	1	0	0	0	1	0	1	1
	8	1	0	0	1	1	1	1	0

Note: Bit numbers 4, 6, and 8 individually duplicate the system framing sequence used in T-carrier channel banks. Therefore, to prevent a receiving channel bank from framing on bit numbers 4, 6, or 8 without giving an alarm, the digital milliwatt should not be transmitted over in-service digital facilities.

For example, at 0 EL, a 0-dBm signal produces the digital milliwatt, and encoder overload occurs at 3.17 dBm. (It is common practice to use 3 dBm rather than the more accurate 3.17 dBm.) A D/A converter operating at 0 DL decodes the digital milliwatt into a 1000-Hz sine wave at a level of 0 dBm.[19] A/D and D/A converters can be operated at levels other than zero. By changing the reference voltage of the encoder, a −3 dBm signal can be encoded into a digital milliwatt, resulting in a 3-dB relative gain. This is equivalent to encoding at zero level and inserting 3 dB of analog gain before the encoder. Both result in a 3-dB gain. One also could use 3 dB of digital gain following a 0 EL encoder, but this would destroy bit integrity for transmission of digital data. A decoder also can operate at levels which will result in a specified amount of loss. For example, a −3 DL decoder will decode a digital milliwatt at a level of −3 dBm, equivalent to a 3-dB loss. In summary, an encoder operating at 0 EL results in no gain or loss. An encoder operating at −3 EL, for example, results in a 3-dB equivalent gain. A decoder operating at 0 DL results in no gain or loss. A decoder operating at −5 DL, for example, results in a 5-dB equivalent loss.

13.3.2.2. Digital Network Transmission Plan

In an all-digital network, loss is not calculated on a per-trunk basis. Rather, it is allocated on a connection basis. Current plans specify a fixed loss of 6 dB between end offices for voice connections and zero loss for data connections. Inter-LATA digital trunks operate at 0-dB loss. In an all-digital network, the voiceband channel need be encoded only once and decoded only once. All loss can be inserted at the receiving end of the connection. The DL at the terminating office has no universal value but corresponds to the amount of loss needed for each type of connection. In the connection shown in Fig. 13.7, the terminating office would use a −6 DL. Local exchange end offices will operate at 0 EL and have sufficient DL options so that the proper amount of loss can be inserted for any type of connection.[20] Echo cancellation is used on long digital trunks. The long-term objective is for the network to provide zero loss on all connections and for any

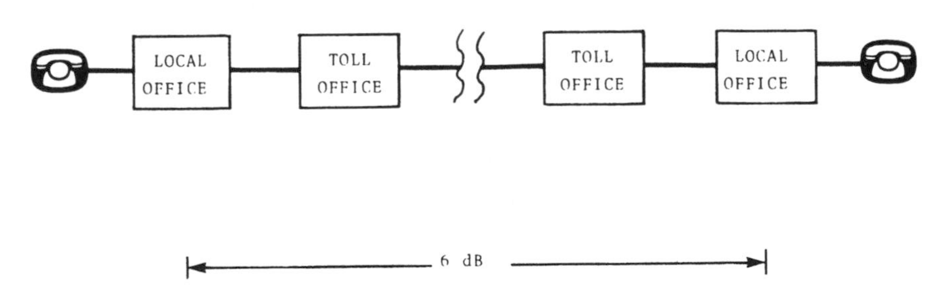

FIGURE 13.7. Digital network transmission plan.

needed loss to be injected at user premises. Until the network is completely digital, the mixed-network transmission covered in Section A.3.2.5.2 applies.

13.4. DIGITAL NETWORK SYNCHRONIZATION

A single digital switching system, operating in isolation from other digital switching systems and digital transmission facilities, functions under control of its own internal clock. As digital switching systems are interconnected by digital facilities, however, a need for synchronization and for a plan to satisfy that need becomes apparent.

13.4.1. The Need for Synchronization

Digital transmission can be either synchronous or asynchronous. When multiple digital streams are being transmitted and received by a single transmission node or a digital switching system controlled by a single clock, synchronization is required.

The need for synchronization can be seen by examining the configurations illustrated in Fig. 13.8.[21] In Fig. 13.8(a), a digital channel bank transmits a bit stream at clock rate F_0. The other channel bank receives at the same bit rate but transmits at clock rate F_1. Each bit stream is received satisfactorily because each receiver aligns itself with the timing of the transmitter at the other end of the channel. Thus, an isolated digital transmission link can use different clock rates in each direction of transmission without impairing transmission quality. This is called *split timing*.

If a digital switching system is substituted for one of the channel banks, as in Fig. 13.8(b), the digital switching system transmits at a bit rate determined by its internal clock, F_0, and needs to receive at the same bit rate. If the distant transmitter sends at a bit rate higher than the switching system clock, the receiver buffer will eventually overflow, causing one frame to be lost. If the received bit stream is at a lower bit rate, the buffer will underflow, causing a frame to be repeated. Either occurrence is called a *slip* (or *controlled slip*). In Fig. 13.8(b), slips are prevented by forcing the distant channel-bank transmitter to operate at the same clock rate as its receiver, which derives its timing from the bit stream transmitted at the switching system clock rate. This is called *loop timing*.

If the remote channel bank also is replaced by a digital switching system, as shown in Fig. 13.8(c), and if each switching system transmits at a bit rate determined by its own clock, slips will occur if the two clocks are not running at the same average rate. When only two digital switching systems are interconnected, synchronization can be achieved by operating two systems in a master-slave mode, whereby each system derives its receive timing from the other. In a network

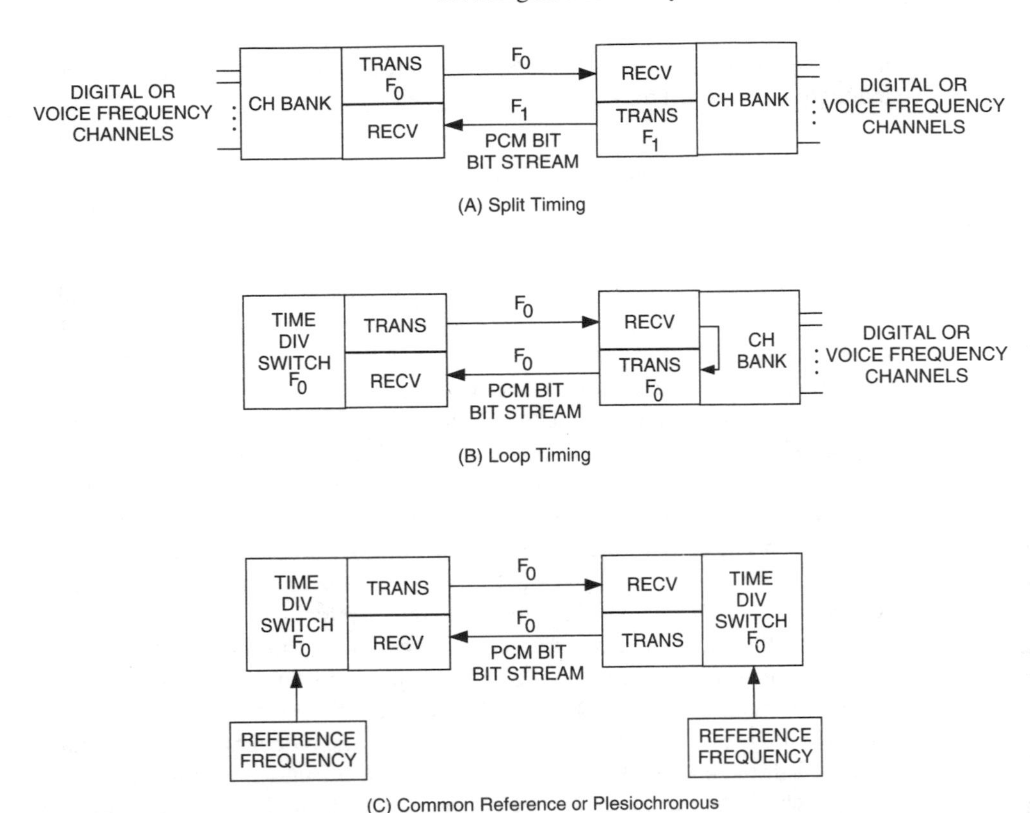

FIGURE 13.8. Need for digital synchronization. (Courtesy TIA. Ref: ANSI/EIA/TIA-594-1991, *Private Digital Network Synchronization.*)

of digital switching systems, however, use of a common reference clock is a better method for minimizing slips.

Line bit rates can vary because daily temperature variations alter the electrical length of digital transmission facility conductors and because of phase jitter and wander. Digital switching systems have receive buffers, or *elastic stores*, to store one complete frame (125 μs in a DS1 frame) plus additional capacity to absorb timing jitter and wander.

The impact of slips depends upon the type of information being transmitted. In voice transmission, slips cause an occasional audible click. In encrypted speech, a slip causes the loss or degradation of data, results in every slip being an audible click, and can cause loss of the encryption key, requiring the encryption key to be retransmitted. A single slip in Group 3 facsimile transmission can cause up to eight horizontal scan lines to be missing.[22] Slips can cause data transmission to be mutilated, or they will require retransmission when error-correction techniques are used. The impairment of video transmission depends upon the encoding algorithm and degree of compression used.

13.4.2. Hierarchical Network Synchronization

The North American networks use a hierarchical method of synchronization involving four stratum levels of clocks.[23] The highest-level clock in the hierarchy is a *Primary Reference Source* (PRS) that maintains a long-term accuracy of 1×10^{-11} or better with optional verification to *universal coordinated time* (UTC). A PRS may be a duplicated *stratum 1* clock, such as a cesium-beam frequency standard, or it may be controlled directly by standard UTC-derived frequency and time services, such as a Loran-C or *Global Positioning Satellite* (GPS) System radio receiver, which are themselves controlled by cesium standards. Thus, since primary reference sources are stratum 1 devices or are traceable to stratum 1 devices, each digital network controlled by a PRS will have stratum 1 traceability.

Stratum 2 clocks form the second level of the synchronization hierarchy; they require a long-term accuracy of $\pm 1.6 \times 10^{-8}$ and have a free-run holdover accuracy of 1×10^{-10} per day when all timing references are lost. Stratum 3 clocks, the third level in the hierarchy, require a long-term accuracy of $\pm 4.6 \times 10^{-6}$ and a holdover accuracy of ≤ 255 DS1 frame slips in the first 24 hours after loss of all timing references. Stratum 4 clocks, the fourth level of the hierarchy, require a long-term accuracy of 3.2×10^{-6} with no holdover requirement. Each stratum 2 through 4 clock must be able to pull itself into synchronization with other clocks at the maximum input-frequency offset from the nominal clock rate.

Hierarchical synchronization is illustrated in Fig. 13.9. The PRS is transmitted to selected stratum 2 nodes. Stratum 2 nodes can provide synchronization to other stratum 2 nodes or to any lower stratum level node. Likewise, stratum 3 nodes can provide synchronization to any stratum 3 or stratum 4 clock. The synchronization plan provides for a secondary synchronization facility to increase reliability of network synchronization in the event the primary synchronization facility is lost. Existing digital transmission facilities between digital switching nodes can be used for synchronization without reducing the traffic-carrying capacity of the transmission system. Digital streams cannot just be selected at random; however, care must be taken to ensure that bit streams transmitted via a Synchronous Optical Betwork (SONET) (see Section 10.12) and used for synchronization timing, are not affected by SONET virtual tributary (VT) pointer adjustments.[24]

Note: SONET quantizes input wander into 8 *unit interval* (UI) steps. When the 8–UI threshold in the buffer is reached, a pointer adjustment is generated.

Global digital networks[25] and multiple national networks derive synchronization timing from multiple primary reference sources. Each network will operate plesiochronously. Two signals are plesiochronous if their corresponding significant instants occur at nominally the same rate, with any variation being con-

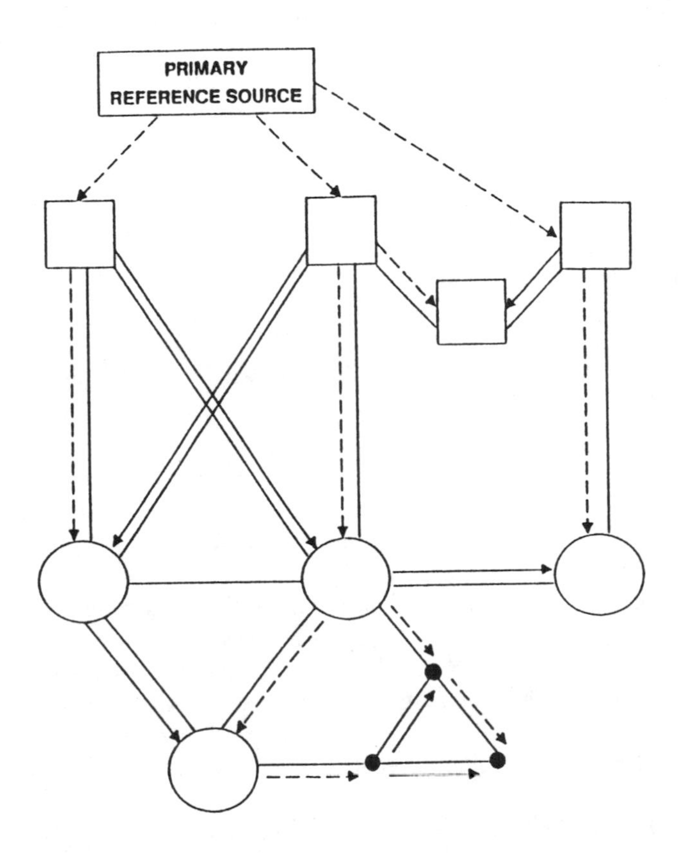

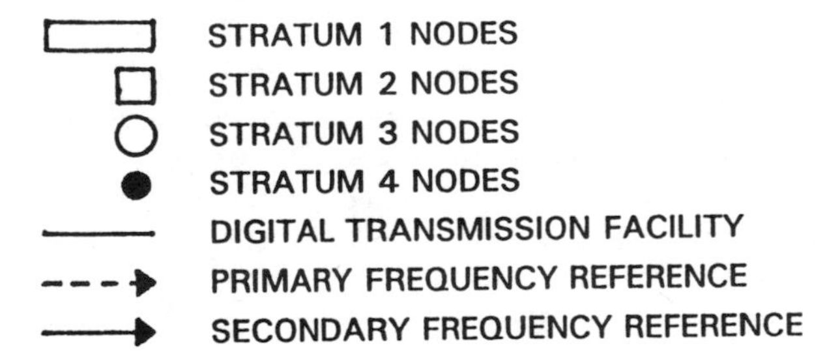

FIGURE 13.9. Hierarchical synchronization network. (Courtesy TIA. Ref: ANSI/EIA/TIA-594-1991, *Private Digital Network Synchronization.*)

strained within specified limits. Digital nodes deriving synchronization from standards of equivalent accuracy can operate synchronously with each other. Global digital networking assumes that national digital networks interconnected by digital links each will be controlled by reference clocks that meet the accuracy requirements specified for stratum 1 clocks.

The slip objective for end-to-end national connections is one slip or less in five hours. For international connections, the objective is not more than one slip in 70 days.[26]

13.4.3. Private Network Synchronization

Large *private digital networks* (PDN) are quite numerous in North America and are becoming increasingly so internationally. Conceptually, a PDN should derive its synchronization timing from the public switched digital network (PSDN). This can cause timing problems if the PDN synchronization layout is not very carefully designed. For example, the PDN could derive timing from one PSDN switching system interface and supply synchronization timing to all PDN switching systems. If the PSDN synchronization link were to fail, however, excessive slips could occur at multiple PSDN interfaces. A better configuration would be to derive synchronization timing from each PSDN interface and carefully distribute that timing to other digital devices.[27] In all cases, secondary timing sources are desirable.

Additional complications can result if the PDN also is connected directly to multiple inter-LATA carriers (IC) operating plesiochronously with the local exchange networks. The PDN may choose to derive its synchronization timing from one of the IC switching systems traceable to a PRS and operate plesiochronously with all other networks, in which case the PDN should obtain a secondary synchronization source traceable to the same PRS.

13.5. DIGITAL IMPLEMENTATION PROGRESS

Switched digital service is available nationwide but not at every location. The digitalization race has been concentrated in long-distance networks and metropolitan areas. The rapidly diminishing cost of optical fiber is enabling digital transmission to become more widespread. As of 1992, about 77% of the local switching systems in the United States were digital, and about 12.5% were still using electromechanical technology (that is, step-by-step or crossbar). Thus, the availability of digital services is concentrated in the larger metropolitan areas, and it could be several years before it will become ubiquitous.

PROBLEMS

13.1. What is the purpose of the additional 3-dB loss on intra-LATA connections over 322 kilometers in length?

13.2. What intra-LATA loss is assigned to an incoming inter-LATA connection to a subscriber served by an access tandem office, and how is that loss provided?

13.3. Why do simultaneous seizures have less adverse impact on international calls using CCITT Signaling System No. 5 than on North American domestic calls using inband signaling?

13.4. How do the lower three layers of CCITT Signaling System No. 7 differ from the lower three layers of the OSI Reference Model?

13.5. How is the start of signal units (SU) in SS7 identified?

13.6. How is signaling traffic overload controlled in SS7?

13.7. What precaution must be taken when restarting an STP in SS7?

13.8. If a standard digital milliwatt is decoded by a -3 DL μ-255 PCM decoder, what is the level of the resulting analog signal?

13.9. What are the likely effects of slips in a digital network, and how are they controlled?

13.10. What precaution must be taken in extracting synchronization from a DS1 bit stream that is carried by SONET?

REFERENCES

1. *Notes on the BOC Intra-LATA Networks*, American Telephone and Telegraph Company, 1983, Section 3 (hereinafter referred to as *BOC Notes*).
2. Ibid., Section 2.
3. Ibid., Section 13.
4. Chaudhary, V. P., Krishnan, K. R., and Vishnubhatla, S., "Technology-trial of Dynamic Traffic Routing with 5 Minute Updates," *GLOBECOM '91 Conference Record*, pp. 48.3.1–48.3.5.
5. *BOC Notes*, Section 7, pp. 33–35.
6. Ibid., pp. 35–37.
7. Powell, D., "Signaling System 7: The brains behind ISDN," *Networking Management*, March, 1992, p. 37.
8. Henderson, B. B., "Helping LANs make ends meet." *Networking, Management*, March, 1992, pp. 30–34.
9. *LATA Switching Systems General Requirements, Common- Channel Signaling, Section 6.5*, Bellcore Technical Advisory TA-NWT-000606, Issue 2, March 1993, pp. 6.5–2,4.
10. Ibid., p. 6.5-10.
11. Ibid., p. 6.5-10-11.
12. Ibid., pp. 6.5-16-17, 6.5-31.
13. Ibid., pp. 6.5-18, 6.5-20.
14. Ibid., pp. 6.5-32-41.
15. *LATA Switching Systems General Requirements, Common- Channel Signaling, Section 6.5*, Bellcore Technical Advisory TA-NWT-000606, Issue 1, Jan. 1992, pp. 6.5–30,31.

16. *Generic Requirements to Support Signaling Connection and Control Part (SCCP), Connectionless Segmentation and Reassembly*, Bellcore Technical Advisory TA-NWT-001269, Issue 1, Feb. 1992, pp. 1–2.
17. *Signaling Transfer Point, Stored Program Controlled Switch, and Service Control Point Note Generic Requirements to Support Node Restart Procedure*, Bellcore Technical Advisory TA-NWT-001229, Issue 1, Oct. 1991, pp. 2–4.
18. CCITT Recommendation G.711 (1984), *Pulse Code Modulation of Voice Frequencies*, Table 6.
19. *Local Switching System General Requirements*, PUB 48501, American Telephone and Telegraph Company, Basking Ridge, NJ, 1980, Section 7.1.7.4.
20. Ibid., Section 7.1.4.2.
21. *Private Digital Network Synchronization*, ANSI/EIA/TIA-594-1991, pp. 7–9.
22. Ibid., p. 9; Bell Laboratories Study on Network Performance, 1983.
23. *Synchronization Interface Standards for Digital Networks*, ANSI T1-101-1987.
24. ANSI/EIA/TIA-594-1991, p. 13.
25. CCITT Recommendation G.811 (1988), *Timing Requirements at the Output of Primary Reverence Clocks Suitable for Plesiochronous Operation of International Digital Links.*
26. CCITT Recommendation G.822 (1988), *Controlled Slip Rate Objectives on an International Digital Connection.*
27. ANSI/EIA/TIA-594-1991, Appendix B.

14

The Integrated Services Digital Network (ISDN)

14.1. THE ISDN CONCEPT

In 1972, the plenary assembly of the International Telegraph and Telephone Consultative Committee (CCITT) approved Recommendation G.711 for Pulse Code Modulation (PCM) of Voice Frequencies. This provided for a sampling rate of 8000 per second and 8-bit encoding/decoding with either A-law or μ-law companding (see Chapter 3). In 1980, the plenary assembly approved Recommendation G.705 which recognized that substantial agreement had been reached in studies of integrated digital networks (IDN) for telephony and that studies leading to an Integrated Services Digital Network (ISDN) were needed. Some conceptual principles to guide the studies provided for:

- the ISDN to be based upon and evolve from the telephony IDN by adding functions and features;
- a layered functional set of protocols;
- gradual transition over one or two decades;
- new services to be compatible with the 64-kb/s switched digital connections of the IDN; and
- arrangements during transition for the interworking of services on the ISDNs and services on other networks.

The CCITT guidelines recognized that countries will develop separate ISDNs, but that compatible international connections will be provided. Several study groups have been involved and continue to be involved in standardizing refinements for supplementary services. The original concept that envisioned the ISDN evolving from a ubiquitous IDN has been overtaken by events and user demand.

Analog technology still is used in portions of telephone networks, while digital technology has been employed in ever-widening areas, particularly in backbone networks and in metropolitan areas.

As ISDN began to be deployed, it was natural that it be used first in metropolitan areas where it could gain acceptance and begin to produce revenue for the providers. Soon, the ISDN "islands" began to be connected by long-haul carriers using Signaling System No. 7. Users began to demand higher bandwidth for higher data speeds, and standards developers began working on Broadband ISDN (BISDN). Meanwhile, carriers developed other techniques to meet the demand for more bandwidth.

This chapter describes the technical characteristics and implementation progress of ISDN in Sections 14.1 through 14.5, existing broadband services in Section 14.6, and the future BISDN in Section 14.7. The emphasis is on the networks in North America.

14.1.1. User Perspective of the ISDN

Not every user will need or desire all the services envisioned for the ISDN. As more sophisticated services develop, however, market stimulation will bring more of those services into the home. A primary user requirement is simplicity of operation at reasonable cost. This mitigates against continuation and proliferation of the multiple access arrangements currently necessary for several existing services which require separate access to separate networks. Users will demand a limited number of standard interfaces, differentiated primarily by bit-rate capability. This will require a single-access arrangement to a local serving office enabling users to access various services. Advances in microprocessor technology and large-scale demand will permit physical realization of these requirements at reasonable cost.

High on the list of ISDN user demands will be a single-billing system for services. This is compatible with the concept of all services passing through a single-access arrangement to a local serving office. This does not mean that the provider of the local serving office must provide all services. It means only that the local serving office must provide network interfaces with the different service providers. It also requires a high level of cooperation among users, local service providers, network providers, enhanced service providers, equipment manufacturers, and government entities to develop the technical standards and agreements necessary for economic implementation.

The network should be capable of fast call setup and disconnect times, a wide range of calling rates and holding times, and variable bit rates for user transmissions. There is increasing demand for cryptographic security for business and industrial transmissions, and the network must make provision for such security.

Efficient directory service for all network services is essential, and users should have the ability to control their costs.

User-interface protocols should be oriented toward the standard family of interfaces being developed rather than toward the location or type of local serving office. Flexibility will enable terminals to be connected directly to a user interface at home or at an office, operated behind a PBX, or plugged in anywhere that ISDN standards are used. The main criterion should be service compatibility.[1]

14.1.2. Narrowband ISDN User-Access Arrangements

CCITT recommendations have specified a limited number of ISDN access arrangements. The principal access and channel structures are shown in Table 14.1.

A block diagram of the standard ISDN access arrangement is illustrated in Fig. 14.1. Two customer premises terminations, NT1 and NT2, are defined. NT1 terminates the subscriber loop and interfaces the line termination (LT) in the local serving office. NT1 and NT2 may be combined or may be provided separately, as determined by the serving common carrier and country regulatory constraints. Four interface reference points are defined. Subscriber ISDN terminal equipment (TE1) connects to the *s* or *t* reference point. Other terminal equipment must connect through the *r* reference point to a terminal adapter (TA). In the United States, the Federal Communications Commission (FCC) requires a distinction between network equipment and customer premises equipment (CPE). Although there is no technical requirement for a reference point on the access line, the *u* reference point and a U-interface have been defined for use in the United States, Canada, and Japan. As competition in telecommunications services increases in Europe, it would not be surprising for the U-interface to find application there.

TABLE 14.1. Access Structure and Channel Bit Rates

Access Structure	Channel Combinations	User Loop Information Bit Rates (kb/s)	Channel Bit Rates (kb/s)
Basic	2B + D	144	64 + 64 + 64
	B + D	80	64 + 16
	D	16	16
Primary rate	23B + D	1544	64
	30B + D	2048	64
	H_0	1544 or 2048	384
	H_{11}	1544	1536
	H_{12}	2048	1920

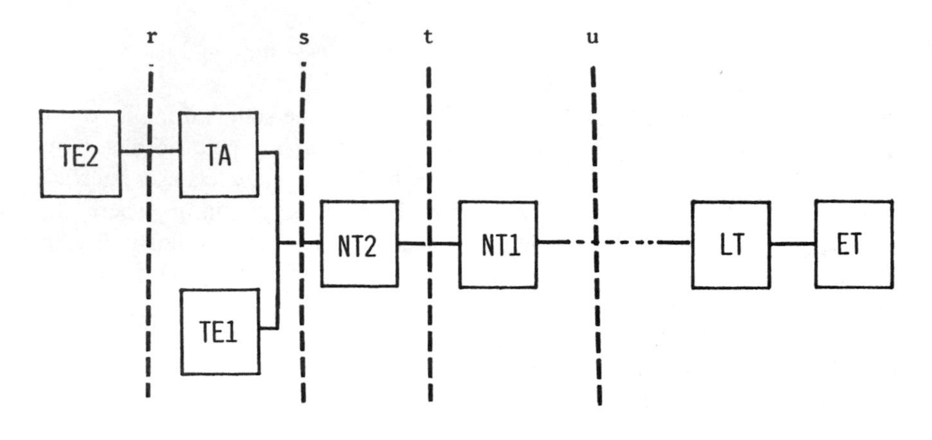

ET - EXCHANGE TERMINATION

LT - LINE TERMINATION

NT1 - NETWORK TERMINATION 1

NT2 - NETWORK TERMINATION 2

TE1 - ISDN TERMINAL EQUIPMENT

TA - TERMINAL ADAPTOR

TE2 - NON-ISDN TERMINAL EQUIPMENT

FIGURE 14.1. ISDN standard access arrangements.

14.1.2.1. The Basic-Rate Access

Basic-rate access (BRA) includes two B-channels at 64 kb/s and one D-channel at 16 kb/s (2B + D). The B-channel will carry circuit-switched PCM voice at 56 or 64 kb/s or packet-switched data at rates adapted to 64 kb/s. Alternatively, a user can switch between voice and data on a B-channel, and a B-channel can be built up by assembling subrate channels. The D-channel will carry signaling information to control the B-channel or packet-switched data up to 16 kb/s. Some users will not require the full basic access, 2B + D; therefore, the user loops can be arranged to support only one B-channel and one D-channel (B + D) or simply one D-channel if the local provider offers those services.

14.1.2.2. The Primary-Rate Access

The Primary-rate access (PRA) will support integrated services digital PBXs or wideband digital terminals at bit rates of 1.544 kb/s (23B + D) or 2.048 Mb/s (30B + D). The bit stream can be arranged as an assembly of 64-kb/s B-channels plus one D channel operating at 64 kb/s. The primary rate also will support wideband digital channels at the net channel bit rates of the primary-access structure or at a specified subrate. In a 24- or 32-channel digital system, the H_0 channel will support channel rates of 384 kb/s. In a 24-channel system, the H_{11}, channel will support a rate of 1536 kb/s, while the H_{12} channel in a 32-channel system will support a rate of 1920 kb/s.

14.1.2.3. The 7-kHz Audio Access

One possible ISDN feature, a family of new voice services providing high-fidelity speech transmission, called *7-kHz Audio*, has been partly standardized.[2] Applications include such services as high-fidelity telephony, audio teleconferencing, audio/graphic services, 7-kHz program audio, and audio/data services. The 7-kHz analog signals are converted to 14-bit precision linear PCM at the 16-kHz sampling rate. A subband adaptive differential pulse code modulation (ADPCM) coding algorithm can convert between the linear PCM and the 64-kb/s format (see Section 3.8). It also can operate at 56 kb/s or 48 kb/s without significant impact on voice quality. Operating at these reduced rates permits simultaneous data transmission on one B-channel. The 7-kHz audio terminals begin transmission in standard PCM; once the transmission path is cut through to the destination, the 7-kHz audio terminals recognize the specified frame alignment, exchange signals for bit-rate allocation, and switch to 7-kHz audio or audio/data operation. The 64-kb/s channel is organized into a frame of 80 octets, a submultiframe of 160 octets, and a multiframe of 1280 octets. A frame-alignment signal of 8 bits is transmitted in the least significant bit of 8 successive octets, every 80 octets. An 8-bit bit-rate allocation signal is transmitted in the least significant bit of the 8 octets following the frame alignment signal.[3]

The nominal 3-dB bandwidth is 50 Hz to 7000 Hz. The coding algorithm restricts the maximum possible frequency to 8000 Hz. The 7-kHz audio/data terminals can send and receive asynchronous data at up to 6.4 kb/s and up to 14.4 kb/s using a rate adaptation protocol. Calls are established via the D-channel, but the parameters of the call can be changed within the B-channel after the 7-kHz audio/data mode is established. Conference calls are possible by the common carrier cross-connecting each DS0 channel into a 7-kHz audio/data bridge at the DS1 rate. Users should force the transmission into PCM prior to disconnect.

Adaptation of this service to the North American ISDN has not progressed rapidly in standards organizations.

14.2. SUBSCRIBER-LOOP TECHNOLOGY

Virtually all subscriber loops are two-wire facilities with multiple wire sizes and frequent bridged taps (see Section A.3.1.1). A *basic-rate interface* (BRI) for ISDN requires full duplex transmission of synchronous digital signals. Two transmission modes have emerged as leading candidates to support the BRI: *echo-canceling hybrid* (ECH) mode and *time-compression multiplexing* (TCM) mode. The ECH mode has better impulse noise immunity and better bandwidth utilization, but requires complex echo cancellation and generates near-end crosstalk (NEXT), while the TCM mode is immune from NEXT because a terminal is never receiving while transmitting. Both can be adversely affected by bridged taps in certain loop configurations. Much research and development has been in progress to develop both the hybrid balancing and TCM concepts.

In TCM, a continuous bit stream at a given information rate is divided into equal segments, compressed in time to a higher transmission rate, and transmitted in bursts which are expanded at the other end to the original information bit rate. A sufficient burst interval is provided to allow the burst energy to decay and to enable a burst to be sent in the reverse direction. To allow for bits needed for overhead and synchronization, the transmission rate for a typical loop length must be about 2.25 times the information rate. Thus, the ISDN full basic access structure of 2B + D will require a burst transmission rate of about 324 kb/s to achieve an aggregate information rate of 144 kb/s. Link bit streams are scrambled to decrease the probability of repetitive zeros, prevent discrete tones from being generated, reduce crosstalk in other systems in the same cable, and simplify extraction of timing information. The TCM method was used in several early, proprietary ISDN systems.

The ECH mode transceiver requires a complex echo canceler at each end of the 2-wire subscriber loop. Advances in very large scale integration (VLSI), used in semiconductor technology, have permitted enhanced designs of echo cancelers that make the ECH mode very attractive. A simplified block diagram of one such transceiver is shown in Fig. 14.2. The ECH method has been standardized for the BRI U-interface. The U-interface permits simultaneous, two-way transmission at a rate of 160 kb/s, with 16 kb/s being used for framing, control, and overhead to produce the 144 kb/s required for the 2B + D channels. Each frame contains 216 bits of payload data, 18 framing bits, and 6 overhead bits organized in an 8-frame superframe.

The line code 2B1Q has been chosen for the U-interface, basic-rate access in North America. This is a 4-level line code that converts blocks of two consecutive signal bits into a single 4-level symbol for transmission. The symbol rate is one-

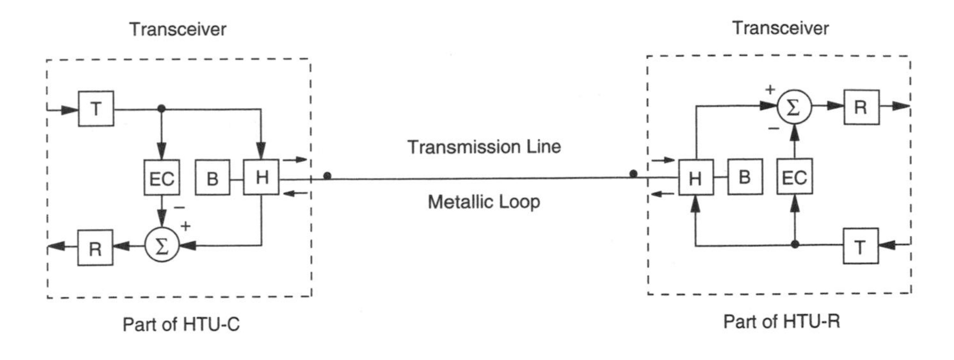

LEGEND:
B = Balance Impedance
EC = Echo Canceler
H = Hybrid
R = Receiver
T = Transmitter
Σ = Summer

FIGURE 14.2. Simplified echo-canceler hybrid transceiver structure. (©1991, Bellcore. Reprinted with permission. The information reprinted from Bellcore was deemed current at the time of publication of this work, but since then the information may have been amended or updated. For more current information, refer to the current Bellcore publication.)

half the information rate. Compared to other codes considered, 2B1Q (see Section 5.9.2) provides better performance in the presence of NEXT and *intersymbol interference* (ISI).[4] Signal bits are scrambled to produce a more random mixture of bits. Each pair of scrambled bits is mapped into a quaternary symbol, called a *quat*. Quats are named $+3$, $+1$, -1, or -3, depending upon their relative position above or below the *zero* line, as illustrated in Fig. 14.3. At the receiver, each quat is converted to a pair of bits, descrambled, and formed into a data stream.

The primary rate interface (PRI) in North America is a conventional DS1 signal comprising 24 channels of 64-kb/s each. A bipolar line code (alternate mark inversion) is used to transmit at the rate of 1.544 Mb/s in each direction over a 4-wire path. To comply with the "ones" density requirement for T-carrier transmission (limitation on number of consecutive "zeros"), a B8ZS (binary 8-zero substitution) line code is used. Any all-zeros octet will be altered by substituting a specified pattern of bipolar violations.

On user premises behind the NT1, transmission across the *s* and *t* reference

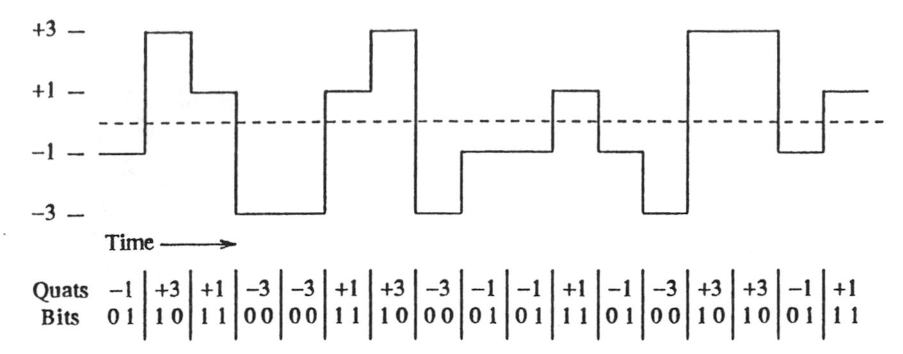

FIGURE 14.3. Example of 2B1Q quaternary symbols. (©1991, Bellcore. Reprinted with permission. The information reprinted from Bellcore was deemed current at the time of publication of this work, but since then the information may have been amended or updated. For more current information, refer to the current Bellcore publication.)

points occurs over a 4-wire facility not to exceed 1 km in length. The data rate is 192 kb/s with 48 kb/s used for framing, synchronization, and control, leaving 144 kb/s to provide the 2B + D. The line code is alternate space inversion. Multiple terminals can be accommodated by a passive bus arrangement.

14.3. NARROWBAND ISDN NETWORK TOPOLOGY

Narrowband ISDN employs circuit-switched 56-kb/s or 64-kb/s (DS0) digital channels, or multiples thereof, via digital transmission systems using CCITT G.711 PCM encoding. Multiple ISDNs can be interconnected to permit worldwide ISDN continuity. In North America and Japan, 24 DS0 channels, along with a framing bit, are multiplexed into a DS1 bit stream at 1.544 Mb/s. In Europe and other countries, 30 message channels, plus 2 channels for frame alignment and signaling, are multiplexed into a bit stream at 2.048 Mb/s. These primary rates are further multiplexed into higher bit-rate streams. Circuit-switched digital networks are arranged into various topologies. The most common topology appears to be hierarchical, although common-channel signaling permits dynamic routing variants to be used.

ISDNs can be interconnected plesiochronously so long as each network derives its timing from a primary reference standard having long-term accuracy of 1×10^{-11} (see Section 13.4).

14.4. NARROWBAND ISDN SIGNALING PROTOCOLS

CCITT Recommendation X.200 has adapted the International Organization for Standardization (ISO) Open System Interconnection (OSI) reference model for

ISDN application. Signaling System No. 7, described in Chapter 13, provides the signaling mechanism for ISDN, using the ISDN User Part (ISUP). For basic-rate access, the lower three layers are related as shown in Fig. 14.4. The physical entity supports the D-channel entity and the two B-channel entities in the data link layer. The B-channel entities in the data link layer service the corresponding entities in the network layer. The D-channel entity in the data link layer serves the signaling entity and, where provided, multiple packet-switching and telemetry entities in the network layer.

The application of the ISDN protocol reference model to a circuit-switched connection is shown in Fig. 14.5. Two functional planes are shown. The signaling plane depicts a connection through two ISDN nodes involving the three lower functional levels to establish and maintain the switched network connection via

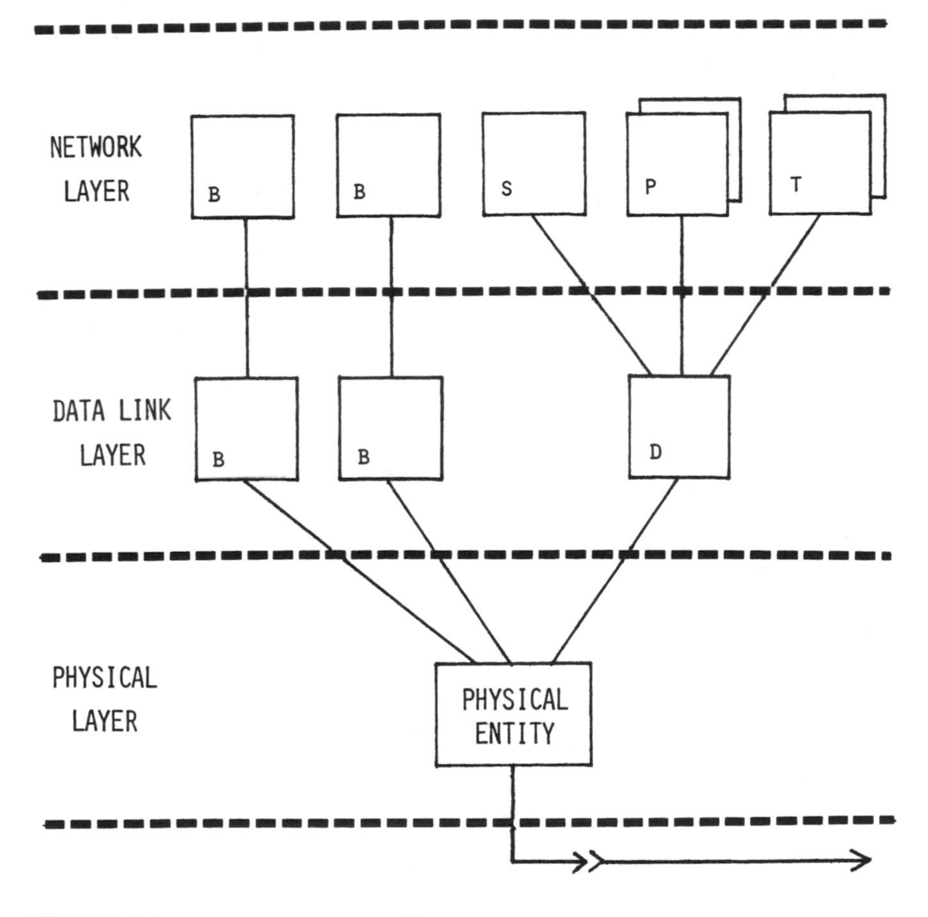

FIGURE 14.4. ISDN basic access model.

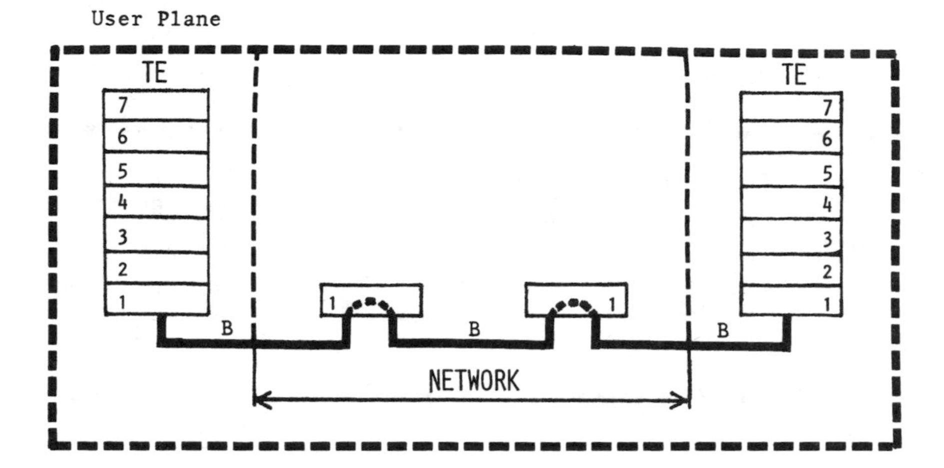

Signaling Plane

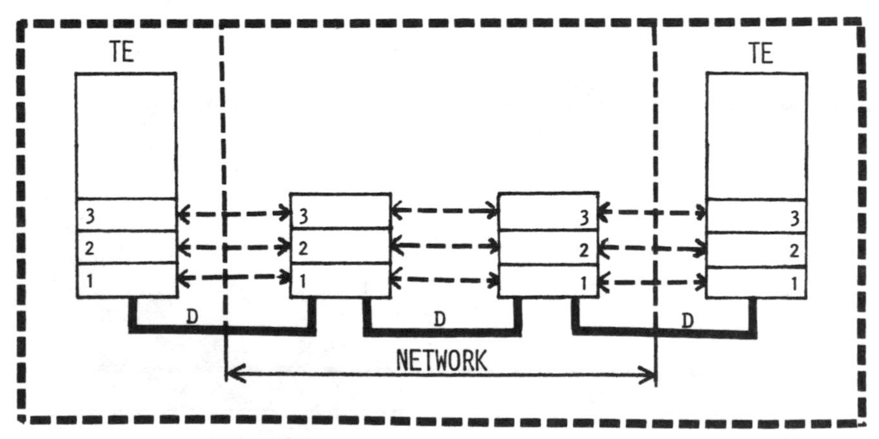

FIGURE 14.5. ISDN circuit switched connection.

the D channel. The user plane shows the functional call set up and release of a B channel. Only the physical layer is involved at the network nodes in this example.

14.5. ISDN IMPLEMENTATION PROGRESS

Network providers generally have been slow to make the large investments necessary to install widespread ISDNs while standards were being developed. As a result, most implementations have been on a trial basis, and some components have used specifications not accepted as standard. Now that worldwide standards have been accepted, there is an increased interest worldwide in network implementation. Several national telephone monopolies have been opened to competition to a varying extent, and international ISDN interconnections have materialized.

The primary deterrent to increased demand for ISDN has been the recognition that, while the basic standards to connect one ISDN terminal with another have existed for several years, the system is not legitimately marketable without all or most of the feature "bells and whistles" that both business and residential customers have become accustomed to using in their *plain old telephone service* (POTS). Provision of those *supplementary services* is quite different and more complex than similar features in POTS. Therefore, as expected, full implementation of ISDN will occur in phases over a period of three to five years because both network providers and manufacturers require some degree of return on their initial investments while the using public is accustomed to a radically different system of communication.

14.5.1. ISDN in Europe

In 1986, the Council of European Communities (CEC) adopted a concept and recommendations for a coordinated introduction of ISDN in Europe. Two major actions resulted. First, the European Telecommunications Standards Institute (ETSI) was created in 1988 to coordinate the technical tasks necessary to harmonize the CCITT standards with national telecommunications systems. ETSI has developed hundreds of standards to assimilate the historically separate and widely diverse individual country systems into a Pan-European system. Second, a Memorandum of Understanding signed in 1989 by telecommunications administrations from 18 countries, resulted in development of a migration plan from the then existing situation into Pan-European ISDN. Agreement was reached on a minimum set of ISDN service offerings, based upon uniform standards, to be provided in each country:

- unrestricted, 64-kb/s, circuit-switched bearer service
- 3.1-kHz, circuit-switched, audio bearer service
- direct inward dialing
- calling-line identification
- calling-line identification restriction
- terminal portability
- multiple subscriber numbers

Other services may be offered optionally as the capabilities are realized. Basic-Rate Acess and Primary-Rate Access are required, along with Signaling System No. 7 (SS7) for international ISDN connections. It was agreed that ISDN service would be introduced by all signatory countries by December, 1993.[5] Projections for 1993 ISDN coverage in Europe included countrywide systems in Belgium, Denmark, and United Kingdom (UK). France, Germany, Italy, The Netherlands, Norway, and Spain had some ISDN service in 1990 and were expected to offer greatly expanded service by the end of 1993.[6]

14.5.1.1. Deregulation and Competition in Europe

Historically, telecommunications has been a total government-owned monopoly in European countries. The barriers to competition have begun to fall in some countries, and other countries are likely to follow in the light of economic realities.

The British Government already has deregulated much of the telecommunications industry. Most of the legal restrictions separating the telephone and cable-television industries have been abolished, opening telecommunications-network access to other than British Telecom and Mercury Communications, Ltd. Under the plan, cable-television companies can offer local telephone service, and the two major telecommunications companies will be able to provide cable television and other entertainment over their telephone lines at least by the end of the decade.[7]

Effective in January 1992, The French Government split the telecommunications and postal functions of its Post & Telecommunications Ministry PTT into two independent organizations. France Télécom will retain monopoly rights to the basic network and voice telephone services, but terminal equipment, data communications, and value-added services will be open to competition.

Faced with the tremendous task of upgrading and expanding telecommunications services in the former East Germany, the Deutsche Bundespost suspended its monopoly on the basic telephone service for an interim period of three to four years. The Bundespost already had completely deregulated data transmission and loosened the restrictions on leased lines by private companies.[8]

14.5.1.2. ISDN in France

France was the first country to place a digital switching system in operation, and was the first to place ISDN in commercial service in 1987, although final standards had not been developed at that time. ISDN *Numeris* basic and primary-rate services were made available nationwide in December 1990. By the end of 1991, some 140 000 B-channels had been placed in service, and the France Télécom goal is one million by 1995. That three-year effort has resulted in all 800 digital exchanges being upgraded to ISDN, reducing the waiting time in urban areas to 15–20 days. In 1992, D-channel access to the X.25 network Transpac was made available, and a virtual private networking capability was expected to be available through an intelligent network concept in 1994. In addition, many ISDN applications have been developed, including computer networking, LAN interconnection, call distribution, document archiving, electronic mail, and image or multimedia transmission. France Télécom concentrated its development of ISDN applications into three major areas: development of the microcomputer as ISDN terminal equipment, computer networking, and upgrading PBXs to ISDN capability. (*Note*: In this book, the term PBX is synonymous with the term PABX, commonly used in Europe.)[9]

Development of microcomputer applications has been the key to ISDN expansion in France. France Télécom has published more than 400 applications using ISDN. By adding specific boards to microcomputers (IBM® PS/2, Macintosh®, or other PCs), ISDN customers have capabilities for audio coding, file transfer protocol handling, image processing, and telephony management. Thus, the microcomputer has become an all-purpose ISDN terminal. Five types of TAs provide access through ISDN to a variety of audio and digital services (X.21, V.35, and V.24) using existing non-ISDN terminals.

Computer networking has been expanded, not only by the enhancements to the microcomputer, but also through partnerships with private industry and by attachments on mainframes and hosts for ISDN access. Some larger companies have private networks, mostly of the X.25 low-bit-rate type, but costs preclude connection to many of their smaller branch offices. Numeris is a cost-effective means of enabling those very small locations to access the private networks. For example, the low $50 monthly charge for Numeris favorably compares with monthly charges of $284 for a 9.6-kb/s Transpac line or $500 for a 50-km leased analog line, excluding the modem. The transmission quality of end-to-end digital connections on ISDN generally provide a bit-error ratio better than 10^{-8}, and call setup time usually is less than 4 seconds. Numeris is able to accommodate the variety of protocols used by its customers.

The average service life of a PBX in France is about seven or eight years. All large PBX manufacturers provide PBXs with digital technology and ISDN interfaces for their large and medium sizes of equipment, and the very small sizes are expected to be upgraded. As replacements occur (about 12% per year), more companies will be able to expand their use of ISDN and interconnect their private networks through PBX access to ISDN.

14.5.1.3. ISDN in Germany

After a trial pilot project, DBP Telekom began commercial ISDN service in eight West German cities in 1989. Thus, ISDN islands have developed with primarily business customers. As those islands become interconnected and expanded, additional applications will become feasible. Following France Télécom's strategy on partnership agreements, DBP Telekom is expanding ISDN through applications involving airline, medical, travel, and retail companies. DBP Telekom expected some 260 exchanges to be equipped for ISDN by the end of 1993.[10]

14.5.1.4. ISDN in Belgium

The Belgium administration (RTT) began commercial ISDN service in June 1989 with three ISDN exchanges, located in Brussels, interconnected with bundles of 2.048-Mb/s systems providing 64-kb/s trunks with associated CCITT SS7 using the ISDN ISUP. The ISDN exchanges were connected to the existing telephone network with CCITT SS7, using the Telephone User Part (TUP), and to the packet-switched network, DCS, using the CCITT X.75 protocol. Basic-rate subscribers were connected over 2-wire loops to concentrators and thence to one of the ISDN exchanges. The most common subscriber installation for basic-rate access is comprised of a 4-wire bus connecting up to 8 different terminals to the NT1 provided by the RTT. Concentrators can be placed all over the country to connect subscribers according to demand. Signaling between the concentrators and the ISDN exchanges is over the D-channels.

The second phase of ISDN implementation was initiated in 1992. It increased the number of ISDN exchanges and greatly increased the bearer services and supplementary services available to subscribers. Applications include slow-scan video transmission, videophone, and video conferences at 64 or 128 kb/s on one BRA and 384 kb/s on three BRAs. Phase 1 subscribers have the option to upgrade to Phase 2, with a change of telephone numbers, or stay on Phase 1 with its reduced basket of services available. Phase 2 ISDN uses Pan-European ISDN specifications for terminal equipment, resulting in terminal portability between countries.[11]

14.5.1.5. International ISDN Connectivity

In 1990, AT&T in the United States established ISDN links to Japan and the United Kingdom. International ISDN dialing to those countries was possible from 37 cities in the United States. The same year, ISDN connectivity was established between Kokusai Denshin Denwa (KDD) in Japan and France Télécom's Numeris network, with primary usage for data and Group 4 facsimile transmission.[12]

France Télécom led other European countries in international connectivity with connections to most of the world's telephone and packet-switched networks. It also has connections to several pre-ISDN specialized networks. Its international transit exchanges are equipped with CCITT SS7, allowing Numeris to be connected to any other ISDN in the world using that system.[13]

The ISDNs in most countries in the CEC are interconnected plesiochronously, using CCITT #7 ISUP. Not every country is connected to every other country, but ISDN communication is possible throughout the CEC via ISDN gateways. Some of the CEC countries are connected to ISDNs in non-CEC European countries.

14.5.2. ISDN in Japan

Nippon Telegraph and Telephone (NTT) introduced commercial ISDN service in Japan in 1988. The ISDN network, called *INS-Net*, was implemented in three phases, the first phase being INS-Net 64, providing circuit-switched services at 64 kb/s. Access for 2B + D uses existing metallic cable pairs with TCM transmission. The following year, INS-1500 provided circuit switching at 64 kb/s, 384 kb/s (H_0), and 1.536 Mb/s (H_{11}) using optical-fiber cable. In 1990, packet switching was added for B-channels and D-channels.

INS-Net transmission is provided over three networks. B-channels are transmitted over the standard digital transmission channels allocated to ISDN. INS-Net 1500 uses a high-speed digital network for H_0 and H_{11} transmission, and packet-switched services are transmitted over the DDX-P packet-switched network. The D70 switching system has been augmented by addition of an ISDN module, providing interfaces for BRA and PRA and providing trunk connections to the D70 switching network for B-channels and to the special INS-Net 1500 network for H_0 and H_{11} channels. The ISDN module connects B-channel and D-channel packet services to a packet module for further processing and connection to the DDX-P network. A block diagram is depicted in Fig. 14.6.

With initial service available only in large cities, NTT faced the same paradoxical problem that other providers have faced in providing a radically new service: demand is slow until wide connectivity is established, and terminal equipment is costly until a greater demand is sufficient to push prices down. In 1990, NTT began a rapid expansion throughout Japan to provide INS-Net service wherever there was subscriber demand. By March 1991, the service area covered about 80% of the population of Japan. NTT's goal is to exceed one million INS-Net 64 subscriber lines by 1995.

Subscriber growth is directly related, not only to availability of service and cost, but also to available applications. The main applications are for business use. Data communications accounts for about 40% of INS-Net 64 use, Group 4 facsimile accounts for another 40%, and all other uses comprise the remaining 20%. Increasingly popular floppy-disk data-transfer terminals can transmit one megabyte of data in three minutes. INS-Net 64 is used by one convenience store chain to transmit point-of-sale (POS) data and order data from about 4400 stores

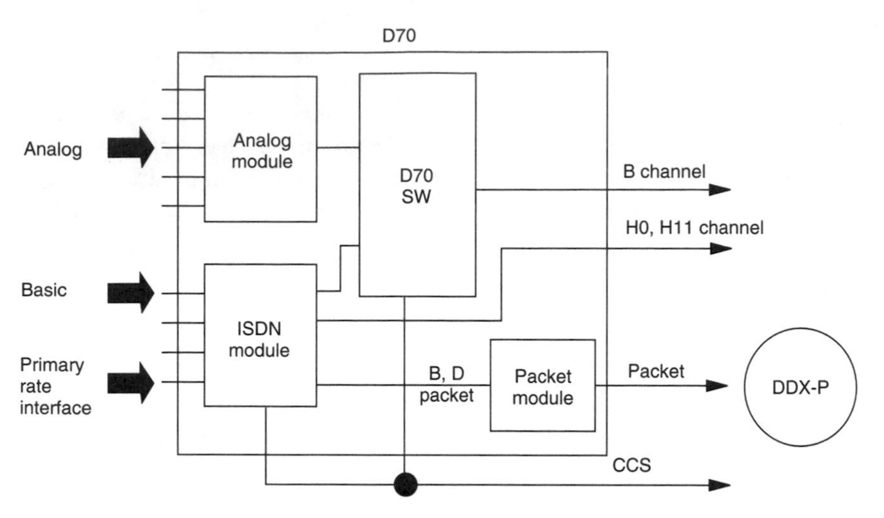

FIGURE 14.6. Block diagram of INS-Net switching system. (From Osamu Inoue, "Implementation in Japan," *IEEE Communications Magazine,* August 1992, © 1992, IEEE)

to its headquarters. Large volumes of data use the circuit-switched service while smaller volumes of data use the packet mode. An accident insurance company transmits visual data of accident scenes taken by a digital still camera to a telephotographic terminal to facilitate cost assessment of automobile repairs.[14]

14.5.3. ISDN in North America

Several ISDN "islands" had been established by both local exchange carriers (LEC) and inter-LATA carriers (IC) in the late 1980s. Interface specifications used were based upon incompletely developed standards and were implemented proprietarily by equipment manufacturers. Therefore, interconnection of those islands was not feasible.

The key to such interconnection is consistency of interfaces. Industry segments collaborated on development of such interfaces and applications to be provided in the first phase of National ISDN-1. Proposed generic requirements were issued by Bellcore in February 1991, and generic guidelines for terminal equipment followed in June 1991. Plans for National ISDN-1 were developed for an initial configuration of 20 local switching systems, manufactured by AT&T, Northern Telecom, and Siemens Stromberg-Carlson, in 17 cities in the United States and Canada (see Fig. 14.7). Inter-LATA connections were provided by AT&T, MCI, and Sprint. Some 13 additional central offices in the initial configuration did not

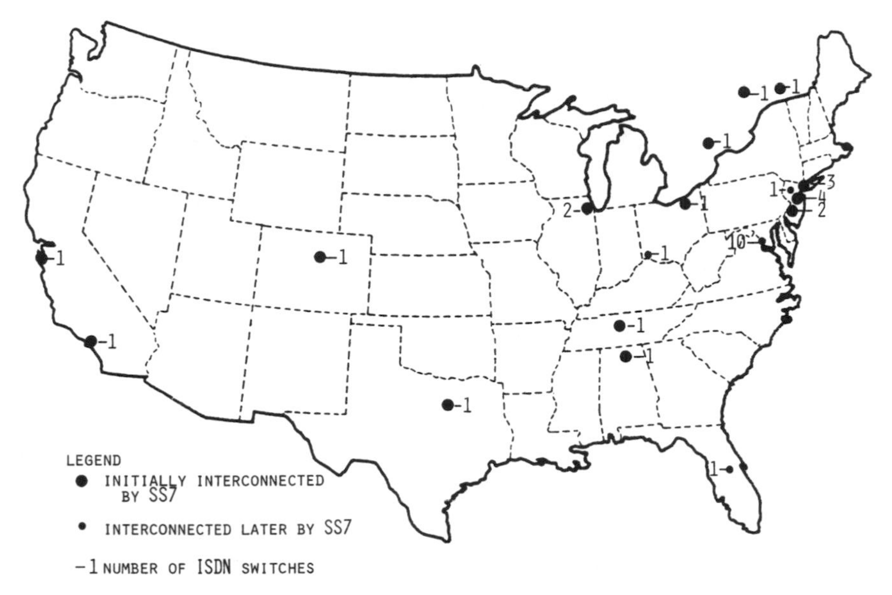

FIGURE 14.7. Initial North American ISDN configuration.

have inter-LATA connections until later. Terminal equipment was provided in approximately 150 locations in 26 states and 7 Canadian provinces.

After extensive testing, the network was cut over on November 16, 1992 with North American and international demonstrations involving ISDN connections to parts of Europe, Asia, and Australia. The initial complement of features included desktop videoconferencing, LAN interconnection, telecommuting, remote access to databases, joint editing, computer-based distance education, financial reporting and review, transmission of still video images, and integrated desktop communications. The main components of National ISDN-1 included:

- a standard BRA interface;
- support for vendor-specific PRA interfaces;
- access to POTS and centrex features;
- selected supplementary services for BRA;
- selected ISDN packet-mode services;
- standard recording for billing support;
- some operations support features.

The second phase of ISDN implementation, National ISDN-2, began in 1993 with enhancements in user customer services and features for both BRA and PRA, improved data capabilities, uniform recording capabilities for billing, and generic operations capabilities. The North American ISDN Users Forum prioritized 123 possible services and applications and indicated the 17 features it most desired to be included in National ISDN-3 by late 1994 or early 1995. About two-thirds of those were included in National ISDN-1 and -2. Deployment plans of the seven Regional Bell Holding Companies project that, by the end of 1994, more than 60.7 million access lines in over 1400 wire centers will have ISDN service available.

14.6. BROADBAND ISDN (BISDN)

14.6.1. Overview of BISDN

A key objective of BISDN is to integrate all services on a common platform using integrated internal network trunk interfaces and a multiservice user-network interface (UNI). There are three major differences between narrowband ISDN and BISDN.

- ISDN uses the infrastructure of the telephony network as it exists today with copper wire pairs; BISDN uses optical-fiber cable and radio.
- ISDN is primarily a circuit-switched network with packet switching on the D-channel, mainly for signaling, and some packet switching on the B-channel; BISDN is completely a packet-switched network.
- ISDN channel bit rates are specified; BISDN uses virtual channels with no specified bit rate. The only bit-rate limitation of the virtual channels is the physical bit rate of the UNI.

In 1988, a decision was made by CCITT to base the development of BISDN on the asynchronous transfer mode (ATM) switching technique. The fixed 53-octet cell format permits processing to be performed by a simple hardware circuit, enabling faster processing speeds than would be possible with software implementation. Cells are multiplexed asynchronously and queued before being interleaved for transmission. The bit rate is adapted to the transmission rate by inserting idle cells that are discarded by the receiver at the destination. Since the ATM is independent of the various services provided by the network, it does not address problems that are specific to applications. ATM cells can be transferred over SONET in a virtual channel with no prespecified bit rate except that of the specified UNI rates of either 155.520 Mb/s or 622.080 Mb/s. Switching is performed by switching ATM cells in fast packet-switching systems.

14.6.2. Existing Broadband Services

Even before ISDN becomes ubiquitously available, standards organizations are developing standards for Broadband ISDN (BISDN). Whereas ISDN has developed slowly while awaiting high user demand, the demand for high-speed data communications, requiring either temporary or fixed-channel bit-rate greater than

1.5 Mb/s, is pushing the development of BISDN through a concentrated standards development effort. That demand, particularly evidenced in North America by user desires to interconnect local area networks (LANs) and metropolitan-area networks (MANs), has resulted in development and implementation of services to accommodate higher bandwidth requirements. Three such services have received significant acceptance among users, and their use is expected to continue to grow.

14.6.2.1. Frame Relay Service

Frame relay (see Section 6.5.3.1) protocols provide for the transport of "bursty" data traffic across digital networks using fast packet technology. Frame relay is a connection-oriented communications service for high-speed bursty data. It is not suitable for constant-bit-rate services such as voice or video. Communication is real-time with no store-and-forward capability and no error-correction capability. The very low error rates inherent in the modern digital network environment permit the error-correction processes of X.25 protocols within the networks to be eliminated. Frame relay relies on end-to-end protocols in user equipment for flow control and for error detection and correction. Frame relay uses variable-length rather than fixed-length frames.

Frame relay is transparent to protocols used in the LANs or MANs. The terminal equipment begins and terminates each frame with a flag and adds address and control information and a *cyclic redundancy check* (CRC). This added overhead encapsulates the payload containing the protocol-oriented informational data, as shown in Fig. 14.8. The public or private frame relay network sets up a

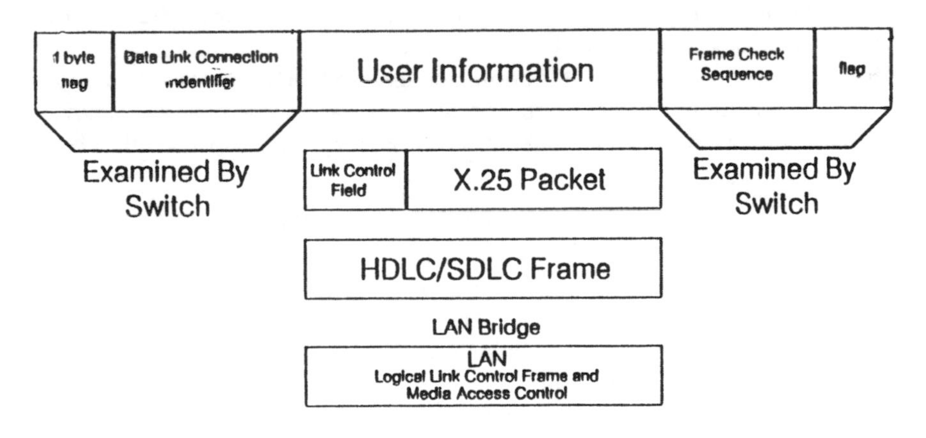

FIGURE 14.8. Frame relay packet. (From Bell Atlantic Seminar, "Partners in Solutions" October 2, 1992)

virtual circuit, comprising a *permanent virtual connection* (PVC), between each origin and destination pair. Based upon the address information in the packet, the originating station transmits the frames at transmission rates of 56 kb/s to 1.544 Mb/s over the PVC. *Switched virtual connections* (SVC) via public networks provide more flexibility in applications. Frame relay traffic is suitable for subdividing payloads into fixed-length payload cells for transmission over cell relay networks (see Section 6.5.3.2).

14.6.2.2. Fiber Distributed-Data Interface (FDDI)

The Fiber Distributed-Data Interface (FDDI) (see Section 10.11) is a set of four standards developed by the Computer and Business Equipment Manufacturers Association (CBEMA) Committee X3T9 and approved by the American National Standards Institute (ANSI). The standards provide interface specifications for a 100-Mb/s token-ring LAN employing fast-packet transmission over optical fiber that can extend to 200 km in perimeter provided that attached devices are no more than 2 km apart on the LAN. The standards cover four different functional areas:

- The *Physical Media Dependent* (PMD) portion specifies the electrical and physical shape of the fiber connectors, including the fiber link, power levels, jitter requirements, bit-error ratios, optical components, and connectors.
- The *Physical Layer Protocol* (PHY) defines physical-layer characteristics that are independent of the actual physical medium, including clocking requirements, data encoding/decoding, framing, smoothing, repeat filter functions, and the elasticity buffer.
- The *Media Access Control* (MAC) layer defines access to the FDDI physical-layer medium and describes packet formation, token handling, addressing, cyclic redundancy checking, and recovery mechanism.
- The *Station Management* (SMT) portion defines the FDDI station configuration, ring configuration, and ring controls, including station insertion and removal, initialization, fault isolation and recovery, scheduling, and statistics collection.[15]

The standards define a dual-fiber, token-ring LAN topology with counter-rotating rings that supports a data rate of 100 Mb/s. An illustrative FDDI network is shown in Fig. 14.9. The network can accommodate up to 500 stations, each terminating both fiber rings. Stations can be repeaters, bridges to other FDDI rings, or gateways to other networks, such as Ethernet or Token Ring. This makes FDDI appropriate for use as a backbone in a high-rise office building or in a campus environment. The specified fiber is multimode, 62.5/125-micron, grade-index fiber with a 0.275 numerical aperture. The end-to-end cable loss, including all fiber, connectors, splices, and optical switches should not exceed 11 dB.

Media access to the FDDI ring is established by a station capturing the circulating token. The station then can transmit frames until a timeout from a previous token capture expires. The token is released immediately following frame

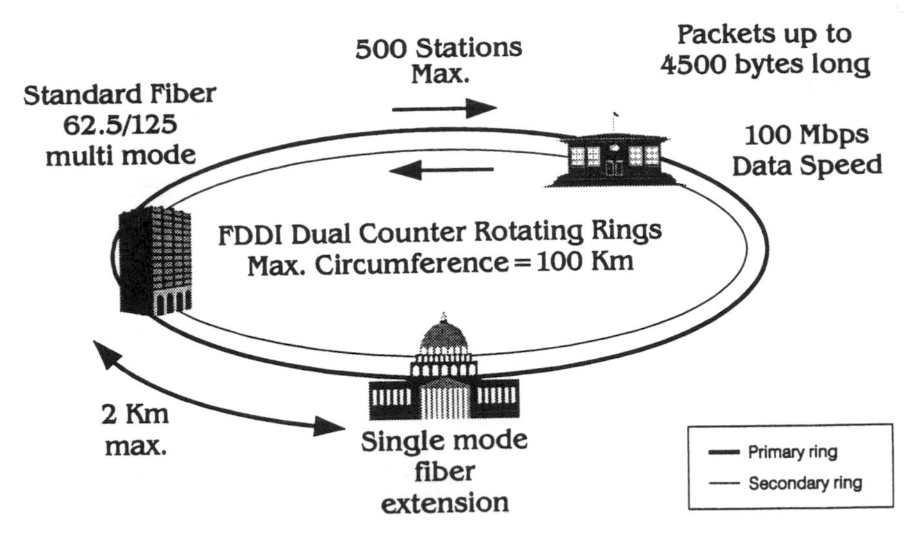

FIGURE 14.9. Illustrative FDDI network. (From Bell Atlantic Seminar, "Partners in Solutions," October 2, 1992.)

transmission to minimize idle time. The frame is transmitted over each ring in opposite directions. Each node converts the optical signal to an electrical signal, amplifies it, copies it if addressed, and converts it back to an optical signal to be transmitted to the next station. The transmitting station strips its own packet from the ring by not repeating the packet. Packets may be up to 4500 bytes in length. Stations may have direct access to the rings (Class A stations) or access via a wiring concentrator (Class B stations). If a node fails, an optional bypass switch can bypass the node optically, removing it from the network. In the event of a cable break, the remaining ring is used. If both rings are broken, Class A stations on each side of the failure automatically loop the data around from one good ring to the other in the reverse direction, as shown in Fig. 14.10. If the cable to a Class B station fails, the wiring concentrator switches the station out of the network.[16]

Industry has been working on standards for FDDI-II, but it is not expected that standards will be completed before 1994 or 1995, unless the tempo of the activity increases. FDDI-II will allow data users to share the bandwidth with voice and video applications. It is forward compatible but not backward compatible with FDDI; for example, if an FDDI network is upgraded to FDDI-II, *all* stations on the FDDI network will have to be upgraded. FDDI-II provides a circuit-switched service embedded in FDDI technology. It allocates isochronous bandwidth for up to 16 channels of 6.144 Mb/s each, with any unallocated bandwidth available for asynchronous packet traffic. Each channel can accommodate up to four 1.544-Mb/s or three 2.048-Mb/s pipes or other subchannels down to 64 kb/s. The 16 channels use 98.3 Mb/s of the total 100-Mb/s bandwidth, leaving a 768-

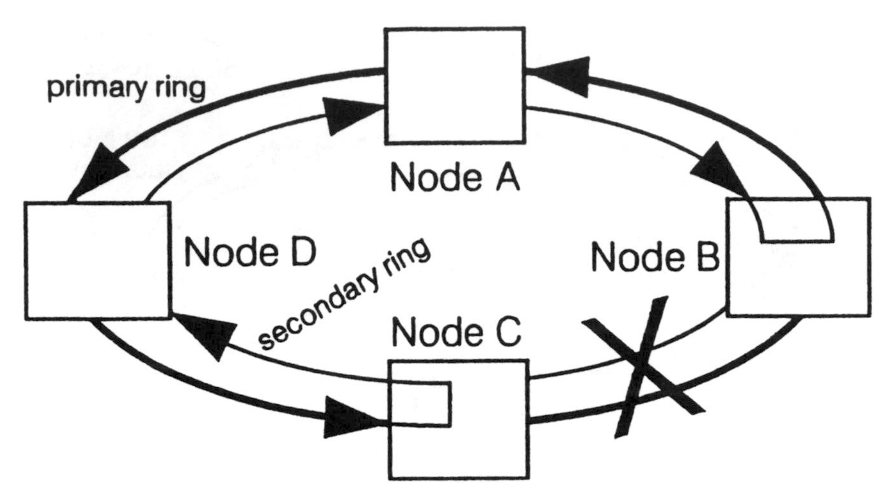

FIGURE 14.10. FDDI fault-tolerant capability. (From Bell Atlantic Seminar, "Partners in Solutions," October 2, 1992.)

kb/s packet-switched channel and 928 kb/s for headers and other overhead. According to a survey of major corporations, the primary attraction of FDDI-II is for videoconferencing, other video, voice, and multimedia applications.[17]

14.6.2.3. Switched Multimegabit Data Services (SMDS™)

SMDS can be defined as a high-speed, connectionless, cell relay, fast packet, public data networking service. Initially, it is being used for bursty traffic to transmit segments of data for short periods of time, using address data rates from 34 Mb/s down to 56 kb/s. Data rates less than 1.2 Mb/s are carried over channelized DS1 facilities, while the higher rates are carried over unchannelized DS1 facilities and over DS3 facilities. SMDS network switches are interconnected by 45-Mb/s (DS3) trunks. Higher access and trunk data rates are planned for the future. When SONET (see Section 14.7.3) is widely used in the networks, the SMDS network access data rate limit could be increased to 155 Mb/s. SMDS is capable of handling both single-address messages and multiple-address (multicast) messages.

There are significant technical differences between SMDS and frame relay in addition to the transmission rates. SMDS is designed to be a multicarrier, nationwide service. SMDS uses fixed-length cells of 53 octets. The specifications for SMDS are being developed to include usage billing, interfaces with other carriers, universal addressing, and network management. Further development will make SMDS suitable, not only for LAN-to-LAN traffic, but also for a wide variety of network applications, such as medical imaging, electronic libraries, computer-aided design (CAD), publishing, multimedia conferences, and scientific visualizations.[18]

Two types of customer premises equipment (CPE) functions are required to execute the 3-layer *SMDS Interface Protocol* (SIP). A router receives a packet, including its headers and trailers from the LAN or other *data terminal equipment* (DTE) and assembles a *Layer 3 Protocol Data Unit* (L3PDU), containing the payload, called the *SMDS Data Unit* (SDU), of up to 9188 octets by wrapping a 36-octet header and a 4-octet trailer around the customer data packet. The source and destination address each contain 8 octets. The router then encapsulates the entire L3PDU inside a 4-octet header, to indicate the payload length and the starting point of a new L3PDU, and 2-octet trailer, and sends it to a CSU/DSU. The CSU/DSU reads the header, calculates the frame check, then strips off the outside header and trailer, and breaks the entire L3PDU into 44-octet payload segments known as *segmentation units* (SU). After further enclosing the SUs in Layer 2 PDU (L2PDU) 7-octet headers and 2-octet trailers to comprise the 53-octet cells, the CSU/DSU adds 4 octets of framing information, called *Physical Layer Convergence Protocol* (PLCP), between cells, and sends them to the network over the physical layer, a DS1 or DS3 pipe. Two octets of the L2PDU header and the 2-octet trailer, along with the 44 octets of the SU, comprise the 48 octets of customer data that remain intact to the destination.[19]

The total overhead of SMDS service is substantial. The PLCP framing and the L2PDU header and trailer use about 23% of the usable bandwidth, and the L3PDU header and trailer comprise 40 bytes of additional overhead for each SDU. The SMDS cell, structured according to the IEEE 802.6 standard for MANs, is compatible with the cell structure of the ATM cells to be used in BISDN.

14.6.3. Broadband Access Technology

Broadband ISDN will be based upon SONET (see Section 10.12). User information will be transmitted across the UNI in ATM cells. Optical interfaces have been defined at 155.520 Mb/s and 622.080 Mb/s, both using two single-mode fiber links suitable for distances up to approximately 15 km. Alternative interfaces are provided. Early implementation of BISDN will employ electrical interfaces at DS1 and DS3 rates to support an initial set of fast packet services, such as SMDS and frame-relay service.[20]

The 53-octet ATM cells include a 5-octet header and a 48-octet payload. The cell header includes routing fields and fields used to (1) indicate the type of information contained in the cell's payload, (2) assist in controlling the flow of traffic at the UNI, (3) establish priority for the cell, and (4) facilitate header error control and cell delineation functions.[21] The cell header format at the UNI is shown in Fig. 14.11. The *generic flow control* field will be used for end-to-end

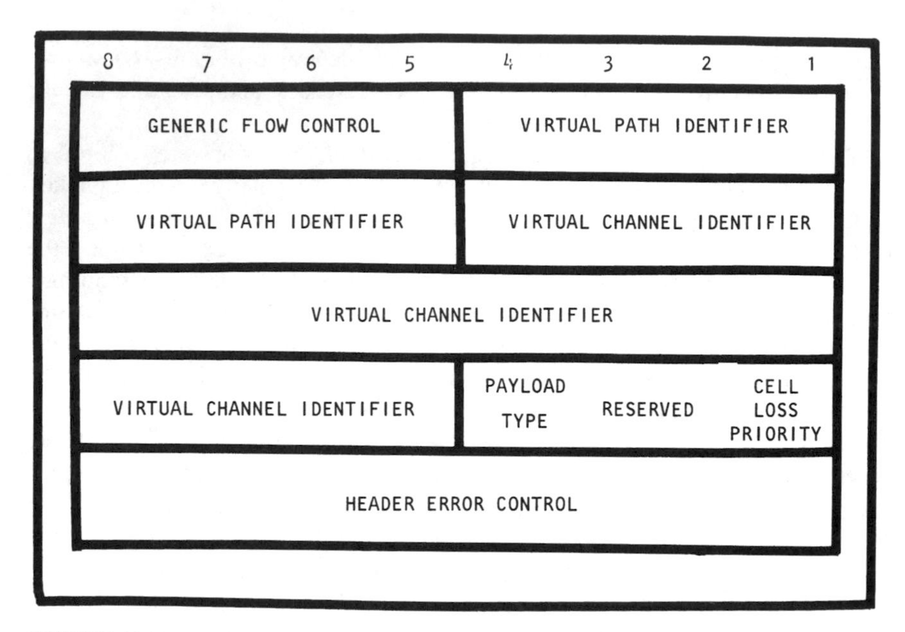

FIGURE 14.11. Cell header format at user-network interface. (©*Telecommunications,* April 1992. Reprinted with permission.)

flow control between ATM users. The *virtual path identifier* identifies the path or route between source and destination. The *virtual channel identifier* defines a logical connection between two ATM users. The *payload type* identifies user or other type of information, such as network management or maintenance messages. The *cell loss priority* indicates whether the cell can be discarded in the event of high network congestion.

The functional reference configuration for BISDN follows the same structure as that for narrowband ISDN depicted in Fig. 14.1 except that the suffix B is used to identify reference points as broadband and the prefix B identifies broadband functionality, as shown in Fig. 14.12. Interfaces for reference points S_B, T_B, U_B and SSB have been standardized.[22] The interface for the SSB reference point is between two TEs and is identical to the interface specification for the S_B reference point. Interfaces for the R reference point may have broadband or narrowband capabilities. A B-NT1 provides functions equivalent to the Physical Layer in the CCITT ISDN Reference Model, such as SONET transmission termination and interface handling for T_B and U_B. A B-NT2 provides functions equivalent to the Physical Layer and higher layers, such as layers 1, 2, and 3 of Broadband Private Branch Exchanges (BPBXs), Layers 1 and 2 of terminal controllers, Layer 1 of simple time division multiplexers, and LAN control functions. Depending upon the implementation, $S_B = T_B$[23].

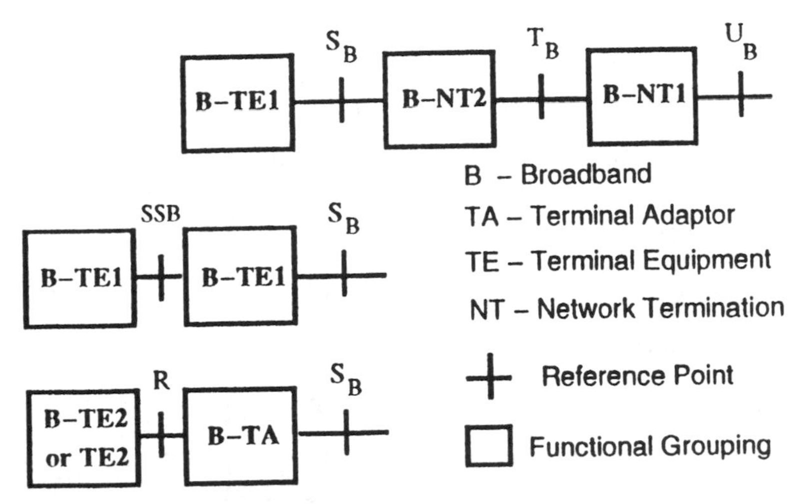

FIGURE 14.12. BISDN functional reference configuration. (©1992, Bellcore. Reprinted with permission. The information reprinted from Bellcore was deemed current at the time of publication of this work, but since then the information may have been amended or updated. For more current information, refer to the current Bellcore publication.)

The concept of separated planes for segregation of user, control, and management functions is introduced in the BISDN protocol architecture model illustrated in Fig. 14.13. The *User Plane* provides for user information flow and associated controls. The *Control Plane* provides call control and connection control functions necessary for set up, supervision, and release of calls and connections. The *Management Plane* provides both plane management and layer management functions necessary for overall network supervision and management.

The *Physical Layer* is based upon SONET. The *ATM Layer* provides a cell-transfer capability for all services by means of statistical multiplexing of virtual channel connections and virtual path connections for both the User Plane and the Control Plane. The cell payload is transferred transparently with no error control or other processing performed at the ATM Layer. The *ATM Adaptation Layer* (AAL) provides service-specific functions to the layer above the AAL for both the user and control planes. The AAL controls the type of service being offered. Class A service is a constant-bit-rate service for voice, video, or clear-channel applications. Class B service is a variable-bit-rate service for packet video applications. Classes C and D services are variable-bit-rate services for data applications. Classes A, B, and C are connection-oriented services, while Class D is a connectionless service.

Service information is mapped by the AAL into ATM cells at the transmitting end and reassembled or read out from ATM cells at the receiving end. The AAL could be terminated at the user end in a B-NT2, TA, or TE, and within the network

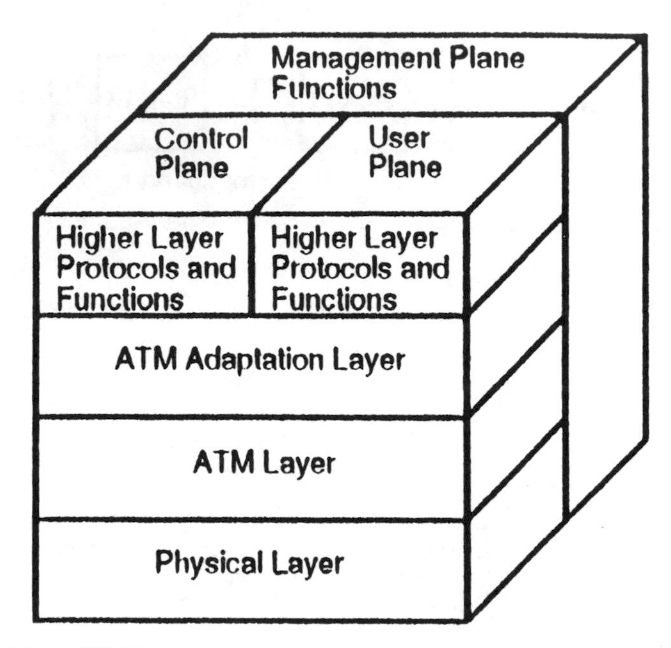

FIGURE 14.13. BISDN protocol architecture model. (© 1992, Bellcore. Reprinted with permission. The information reprinted from Bellcore was deemed current at the time of publication of this work, but since then the information may have been amended or updated. For more current information, refer to the current Bellcore publication.)

at exchange terminations (ET), in service-specific modules, bridging equipment, or Interworking Units (IWU). The layers above the AAL provide call control and connection control in the Control Plane and network supervision functions in the Management Plane.[24]

14.6.4. Broadband Transport Technology

Broadband ISDN provides a method of transporting the ATM cells between two or more points over the Synchronous Digital Hierarchy SDH, or SONET as it is known in North America. The basic signal in SONET is the Optical Carrier (OC), the electrical equivalent of which is the *Synchronous Transport Signal* (STS). The STS-1 signal of 51.840 Mb/s is the basic building-block signal in SONET. Higher-rate signals are formed by byte interleaving of STS-1 signals. Hence, the low-access signaling rate of 155.520 Mb/s is an STS-3 signal (equivalent to STM-1 in the SDH, which is the lowest rate that can accommodate a CCITT level 4 signal of 139.264 Mb/s). In asynchronous systems, multiplexing is performed one step at a time up the hierarchy as in progressing from DS0 to DS3.

In SONET, multiplexing to any level of the SDH can be accomplished in one step because all higher-rate signals are integer multiples of the basic STS-1 building block of 51.840 Mb/s. This enables payloads to be cross-connected easily without completely demultiplexing the higher-rate signal. This permits "Add-Drop" functionality.[25]

Figure 14.14 illustrates the various transmission segments of the fiber-optic network that interconnects *Network Elements* (NE). A *path* is a logical connection between the point at which a standard-frame format for the signal at a given bit rate is assembled and the point at which it is disassembled; for example, a *Terminal Multiplex* (TM). A *line* is a transmission medium and its associated *Line Terminating Equipment* (LTE) required to transport information between two consecutive line terminating NEs, one of which originates the line signal and the other terminates it. A *section* is a portion of a transmission facility, including terminating points, between two regenerators or between a terminating NE and a regenerator. These definitions are illustrated in more detail in Fig. 14.15.

The STS-1 frame of 125 μs, illustrated in Fig. 14.16, consists of 90 columns and 9 rows of octets for a total of 810 octets, producing the STS-1 bit rate of 51.840 Mb/s. The order of transmission is row-by-row from left to right. The most significant bit in each octet is transmitted first. The first three columns are transport overhead, containing 9 octets for section overhead and 18 octets for line overhead. The remaining 87 columns comprise the STS-1 *Synchronous Payload Envelope* (SPE) capacity. Column 1 contains 9 octets comprising the STS *Path Overhead* (POH), leaving 774 octets for net payload. The SPE may begin any-

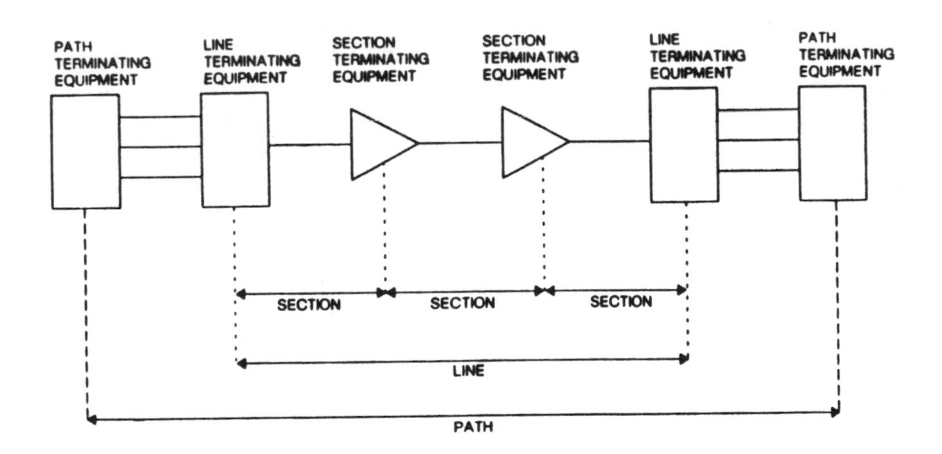

FIGURE 14.14. Simplified diagram depicting SONET section, line, and path definitions. (©1990, Bellcore. Reprinted with permission. The information reprinted from Bellcore was deemed current at the time of publication of this work, but since then the information may have been amended or updated. For more current information, refer to the current Bellcore publication.)

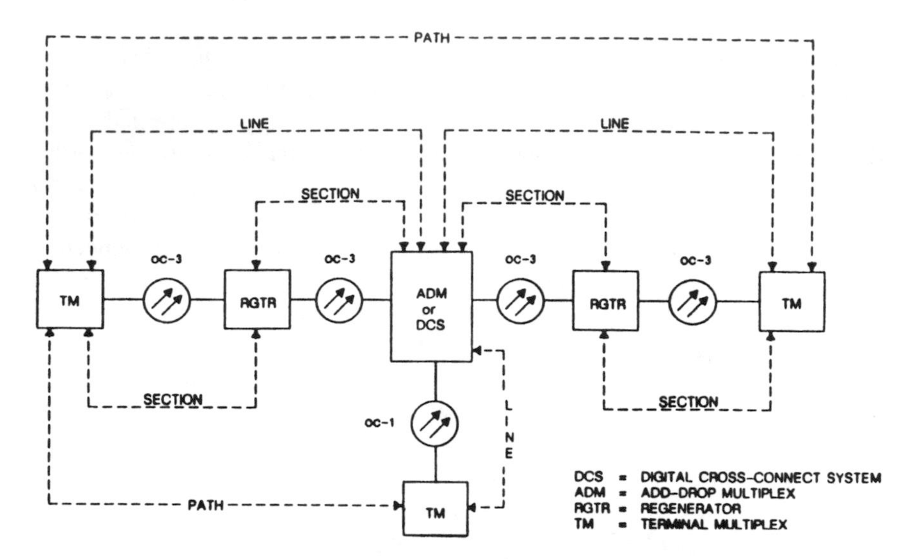

FIGURE 14.15. Illustrative SONET sections, lines, and paths. (©1990, Bellcore. Reprinted with permission. The information reprinted from Bellcore was deemed current at the time of publication of this work, but since then the information may have been amended or updated. For more current information, refer to the current Bellcore publication.)

where in the STS-1 envelope capacity; it may be contained in one frame but typically begins in one frame and ends in the next.[26]

Payload pointers, consisting of three octets in the line overhead, are used to designate the location of the SPE relative to the frame. These pointers indicate the number of octets offset between the pointer octet in the line overhead and the beginning of the SPE. Because the offset can be any value between 0 and 782, the SPE does not need to be frame aligned with the line overhead. The pointer value is adjusted to accommodate slight frequency differences between SONET network nodes, having the effect of moving the beginning of the payload one octet forward or backward relative to the frame.

In response to the demand for existing broadband services and in preparation for BISDN services, both IC and LECs have been installing optical-fiber and SONET transmission systems at an increasingly rapid rate. Canada has a SONET transport system spanning 2200 km, connecting areas that include Edmonton, Calgary, Regina, and Winnipeg. Network providers have been installing fiber more rapidly than previously forecast. Current estimates are for annual sales totaling about $4.5 billion worth of fiber-optic cable and equipment by 1996.[27] The largest United States ICs have implemented all-fiber networks, and are rapidly entering into joint ventures with foreign carriers to provide a wide variety of international network services via underwater fiber-cable and satellite links.

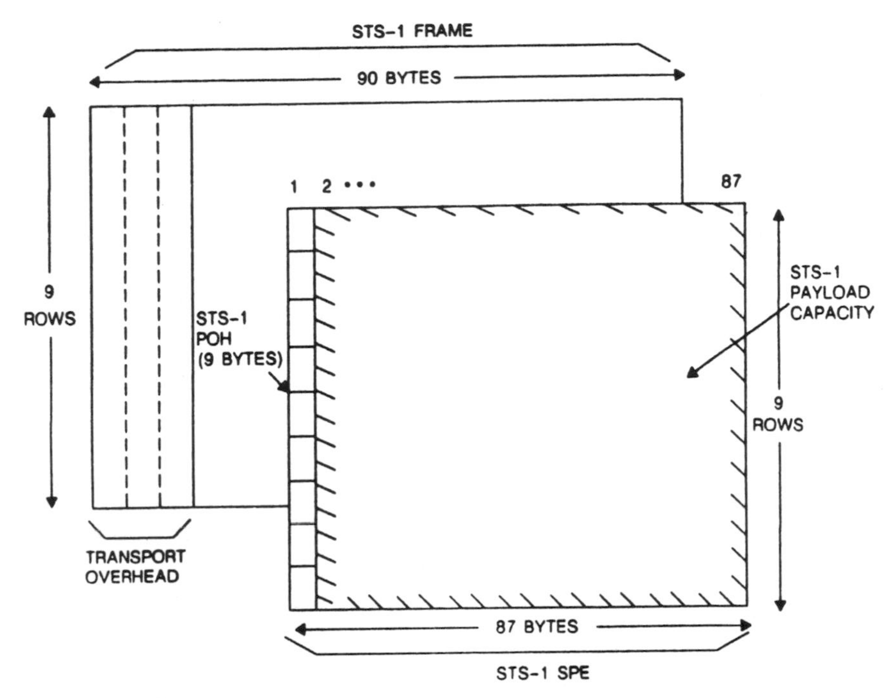

FIGURE 14.16. STS-1 SPE, illustrating overheads and net payload. (© 1990, Bellcore. Reprinted with permission. The information reprinted from Bellcore was deemed current at the time of publication of this work, but since then the information may have been amended or updated. For more current information, refer to the current Bellcore publication.)

14.6.5. Broadband Switching Technology

Since the BISDN concept, selected by CCITT, includes ATM switching and SONET (SDH) transmission, designers have been busy exploring various design techniques for ATM switches. Fast packet ATM switches will receive high-speed data streams from interfacing circuits. Such input packets may be addressed to many destinations, some requiring transmission in a given sequence, such as voice or video, while others can be reassembled at their destination without concern for minor delays. All packets are the same length, use the same number of address bits and other header functions, and need to be transported through the ATM switch using a common routing technique. The only differences are user content, destination, and the requirement for sequencing. Several considerations are important:

- What kind of switching fabric produces the best trade-off to achieve high throughput, low delay, and small probability of cell loss?

- How should the switch be designed for cost-effective modularity to enable systematic growth with optimal efficiency?
- How do multiplexing and queueing arrangements affect the prime considerations listed above?
- What is the most efficient way to route packets through the switching fabric?

14.6.5.1. Switching Fabric Considerations

Since BISDN will carry both constant-bit-rate and variable-bit-rate traffic, switch throughput is one of the most important considerations, and a blocking switching network is highly subject to congestion. That mitigates in favor of a nonblocking switching fabric. The inherent delay in time-division switching is unacceptable in fast-packet switching. Therefore, space-division packet switching has emerged as the key component for integrating voice, data, image, and video in high-performance networks.

The fast-packet switch interfaces will have different data rates of input to the switch. The internal core fabric of the switch needs to be at least as large as the maximum interface data rate. Packets arriving from multiple sources must be buffered briefly to synchronize their frames exactly with the ATM switch clock. The address of each packet must be examined to ascertain which output module within the switch has to receive it in order to send it on its way to its destination. Each packet has to be routed through a switching network to the appropriate output port associated with the SONET transmission system that leads toward the addressed destination. Multiple packets received in the same time slot could be addressed so that they would need to be switched to the same output port at the same time. Therefore, some buffering must be placed somewhere in the ATM switch.[28]

One possible switch fabric is a *shared-memory switch*, illustrated in Fig. 14.17, with N inputs and N outputs, each operating at R b/s, and an $N \times N$ random-access memory (RAM) shared memory. All incoming cells, arriving in a time-

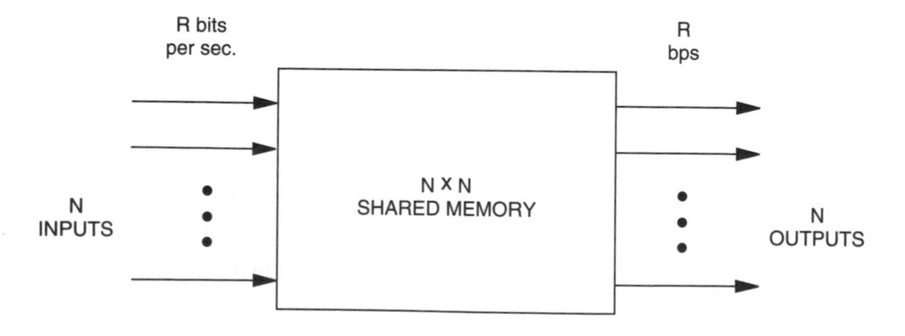

FIGURE 14.17. Simplified shared-memory switch. (From Karol & Eng, "Performance of Hierarchical Multiplexing in ATM Switch Design," *ICC '92 Conference Record,* ©1992, IEEE.)

slotted manner, are stored in the RAM. In later time slots, the stored cells are read out according to their destination addresses. As the incoming cells are written into the RAM, the memory address of each cell is written into a separate first-in-first-out (FIFO) buffer associated with the destination output port of that cell. The FIFO buffer for each output port records the RAM locations of all incoming cells destined for that output port. This produces an excellent throughput-delay performance with minimal buffer size for a small cell switch. As N becomes large, the complexity of such a completely shared memory switch increases. To offset this, one can use hierarchical multiplexing to multiplex a group of input lines to a higher SDH level before entering the core fabric operating at that higher SDH level. Each output of the core fabric is then demultiplexed to the incoming bit rate. The addition of the mux/demux functions simplifies the control function. For example, 16 input lines, operating at 155.520 Mb/s, can be multiplexed to a 2.4-Gb/s core fabric rate and then demultiplexed back to the input rate at the output. As the switch grows in size, however, the required memory size becomes quite large.[29]

An alternative concept is the *knockout switch*, with all inputs and outputs operating at the same bit rate. All input packets arrive in time slots, as shown in Fig. 14.18, and are stored in a short-term buffer. The output port address of each packet is determined through a look-up table, and that address is used by the packet switch to route each packet to its addressed output port. Each input port has a broadcast bus, and each output port has access to the packets arriving on all inputs as illustrated in Fig. 14.19. The nonblocking switch ensures that the only congestion occurs at the interface to each output where, as indicated previously, packets can arrive simultaneously on different inputs destined for the same output. This is unavoidable and is the greatest source of complexity in fast-packet switching. If the number of such simultaneous packets were large, sufficient buf-

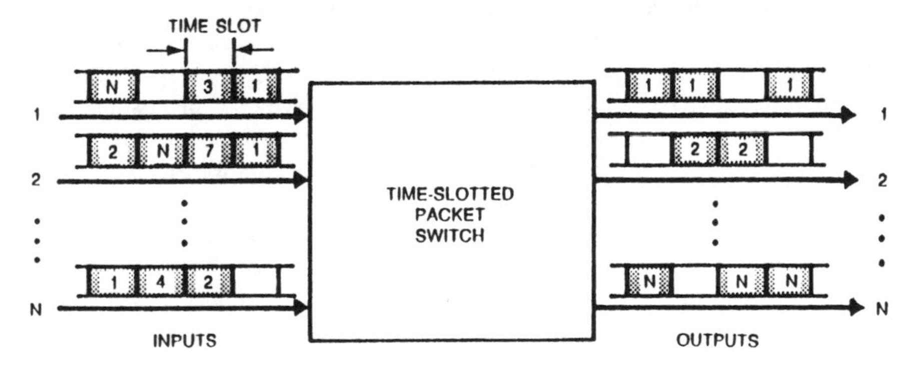

FIGURE 14.18. A simple knockout switch. (From M. J. Karol, M. G. Hluchyj, & S. P. Morgan, "Input Versus Output Queueing on a Space-Division Packet Switch," *IEEE Transactions on Communications*, Vol. COM-35, No. 12, December 1987. ©1987, IEEE)

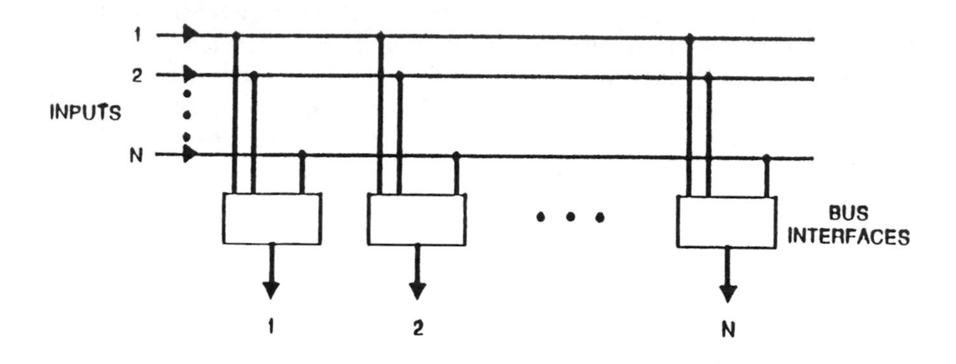

FIGURE 14.19. Knockout switch broadcast bus. (From Y. Yeh, M. G. Hluchuyj, & A. S. Acampora, "The Knockout Switch: A Simple, Modular Architecture for High-Performance Packet Switching," *IEEE Journal of Selected Areas in Communications,"* Vol. SAC-5, No. 8, October 1987, ©1987, IEEE.)

fering to avoid packet loss would cause unacceptable cell delay. If, however, the packet arrivals on different input lines are statistically independent, then the probability that more than a few packets would arrive simultaneously at a given output is very small. Hence, a well known packet switch design principle is based upon the assumption that only a given number of packets are allowed to be accepted for a given output in a single time slot, and the excess packets destined for the same output are dropped, subject to the condition that the dropped packet probability must be sufficiently below all other loss mechanisms such as buffer overflow, link failure, etc. This design principle is known as the *knockout principle.* To avoid excessive packet loss, each output bus interface contains a simple filtering element for each input, and queues the packets addressed to that output into a small set of shared, parallel, FIFO buffers. Contention for the output port in each time slot is conducted like a tournament. Since only one packet can be placed in the first buffer, the losers contend for the second and succeeding buffers. When all buffers are full, any remaining losers are discarded. With a very large number of input lines and a 90% load, 8 parallel buffers produce a packet loss probability of less than 10^{-6}, and 12 buffers guarantees a packet loss probability of less that 10^{-10}.[30]

Queueing of packets may be done basically at the input, throughout the switching fabric, or at the output. Input queueing, or contention, controls the traffic load on the core fabric of the switch but does not grow well. For a large packet switch, input queues with FIFO buffers saturate at moderate traffic loads. Internal queueing increases the packet delay through the switch. Therefore, output queueing achieves the best throughput-delay performance.[31]

The architecture shown in Fig. 14.20 ensures that the permissible number of packets arrive at the desired output group in a time slot. The structure is similar to other three-stage networks except that the output stage consists of packet-switch modules. The intelligent path assignment algorithm prevents path conflicts as packets are routed through the first two stages, called the interconnect fabric. Each switch in the interconnect fabric is a simple, self-routing, memoryless cross-connect device.[32]

Packets must be routed through the switch in an efficient manner to minimize delay and to maximize throughput. The routing function is to direct the input packets through the switch to their addressed output ports without path conflicts. At very-high data rates, the routing algorithm must run extremely fast. Since conventional look-up tables require excessive delay, some other method needs to be devised. A completely self-routing system is very efficient but requires many broadcast buses and knockout concentrators. A simpler, self-routing switch without buffering and a simple, fast, path assignment controller, using a distributed routing (or scheduling) algorithm, provide intelligent load balancing and is simple to implement and easy to expand in size. Packets from all input lines enter the

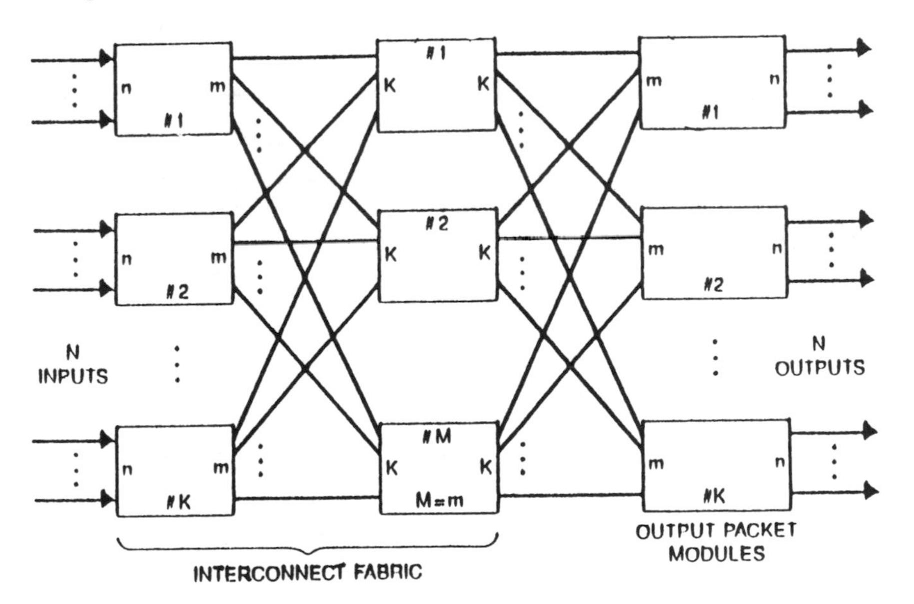

FIGURE 14.20. A growable fast packet-switch architecture. (From Eng, Karol, & Yeh, "A Growable Packet (ATM) Switch Architecture: Design Principles and Applications," *IEEE Trans. Comm.* Vol. 40, No. 2, February 1992, © 1992, IEEE.)

interconnect fabric in synchronized time slots. The task of the routing algorithm is to assign a path through the interconnect fabric connecting a specific time slot in an outlet of an input module to an available time slot in an inlet of the addressed output module through a matching process.[33] Simulation has indicated that cell loss probability from a combination of scheduling loss and knockout loss can be held below 10^{-9} for traffic loads between 70% and 100% by proper dimensioning of the packet switch modules.[34]

14.6.5.2. A Prototype ATM Switch Architecture

A prototype ATM switch operating at 2.5 Gb/s has been developed by AT&T, using the configuration shown in Fig. 14.21. The switch supports multiple standard interfaces for STM-1 (155 Mb/s), STM-4c (622 Mb/s), and STM-16c (2.5 Gb/s). The core switching fabric, using 8 × 8 ATM switch modules and 32:8 concentrators in the output switch modules as shown in Fig. 14.22, can support up to 128 STM-1 interfaces. The switch is expandable up to 1024 STM-1 interfaces. There is no memory in the Cell Distribution Network, which functions similarly to the interconnect network described in Section 14.7.4.1. All incoming cells (packets) are delivered instantaneously to their destination output group addresses in the Output Packet Switch Modules. The switch uses the generalized knockout principle to restrict the maximum number of cells accepted into each output group to the number of input ports to the concentrators. The concentrators are 32-input, 8-output FIFO buffers.

The 8 × 8 ATM Switch Module uses a shared-memory design consisting of

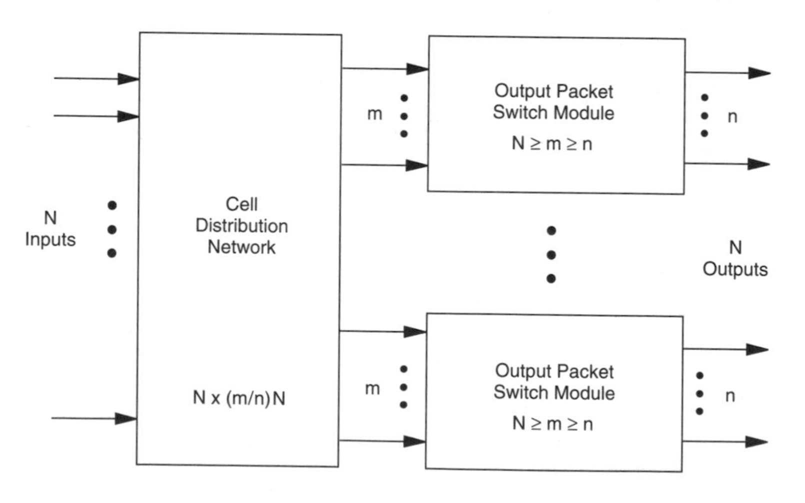

FIGURE 14.21. AT&T prototype ATM switch architecture. (From Eng, et al., "A High-Performance 2.5 Gb/s ATM Switch for Broadband Applications," *GLOBECOM '92 Conference Record,* December 1992, ©1992, IEEE.)

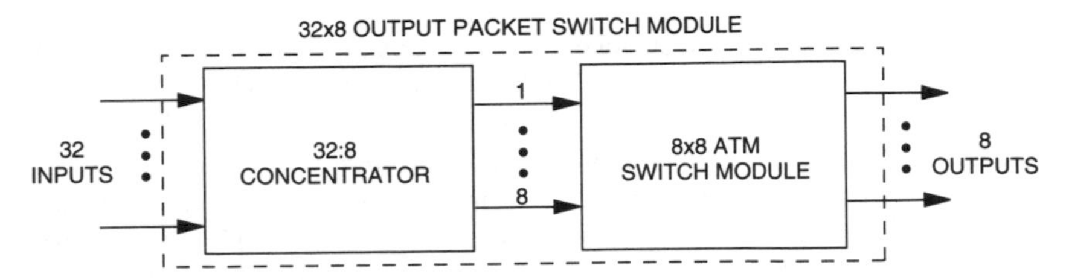

FIGURE 14.22. Concentrator-based Output Packet-Switch Module. (From Eng, et al., "A High-Performance 2.5 Gb/s ATM Switch for Broadband Applications," *GLOBECOM '92 Conference Record,* December 1992, ©1992, IEEE.)

a data path comprised of a converter, a wide data bus, and a large data RAM, and a control section, as shown in Fig. 14.23. Cells are multiplexed by the converter onto a wide data bus and written into and read out of the data RAM under control of the control section. The control section maintains a FIFO queue, containing the address of all cells stored in the data RAM, and a buffer containing all unoccupied locations in the data RAM. Since a cell period is approximately

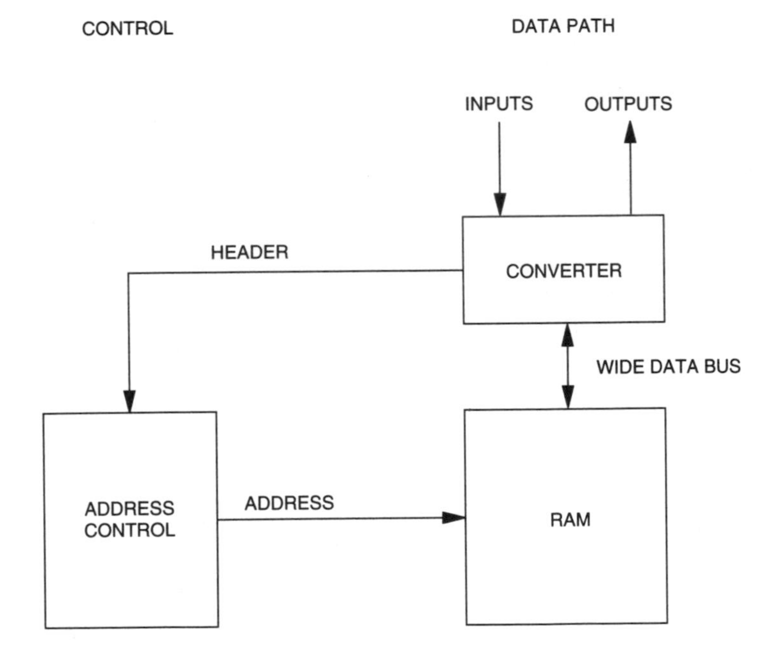

FIGURE 14.23. ATM Switch Module. (From Eng, et al., "A High-Performance Prototype 2.5 Gb/s ATM Switch For Broadband Applications," *GLOBECOM '92 Conference Record,* December 1992, ©1992, IEEE.)

160 ns at 2.5 Gb/s operation, the data bus is wide enough to permit the 16 WRITE or READ operations to be performed within 10 ns per cell. The 8 × 8 ATM Switch Module can function as a stand-alone switch, as shown in Fig. 14.24. Line interfaces are flexible, depending upon line card selection. The line card performs all of the optical interface and SONET functions and some of the ATM cell-based functions. The mux/demux board multiplexes the incoming ATM data streams hierarchically up to the 2.5-Gb/s rate for switching and then demultiplexes them back down to the output data rate for transmission.

For the initial 32 × 32 switch configuration, the Cell Distribution Network broadcasts cells to address filters as shown in Fig. 14.25. Four parallel Output Packet Switch Modules switch packets to four output groups. Each address filter examines the packet header to determine whether the cell is intended for its output group. The cell then is either discarded or passed to a 32 × 8 concentrator and queued for one of the inputs to the associated 8 × 8 ATM Switch Module for

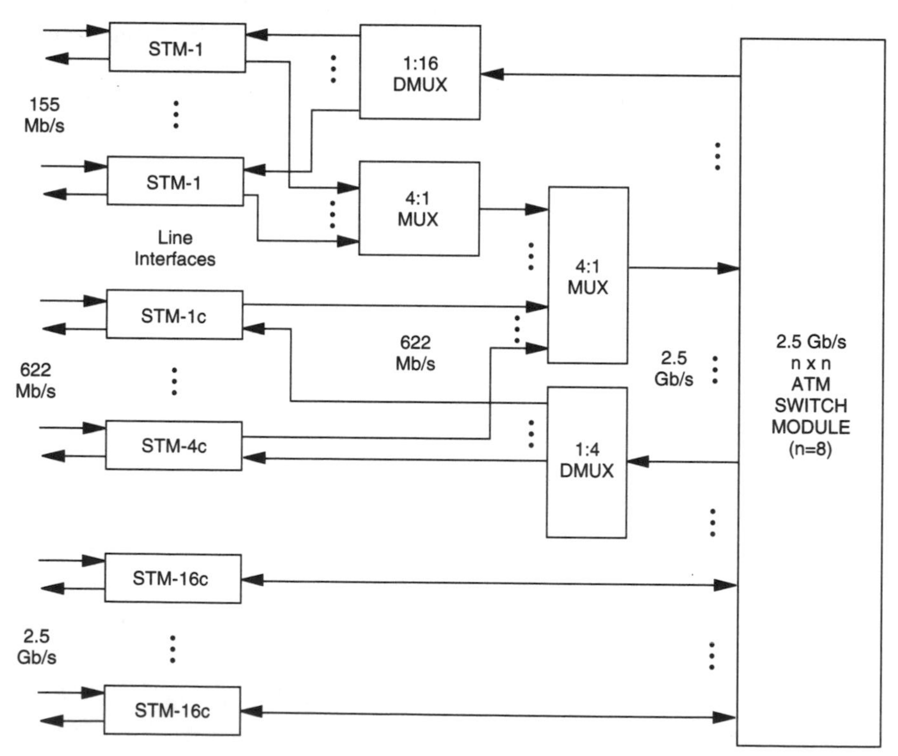

FIGURE 14.24. Stand-alone 8 × 8 ATM switch. (From Eng, et al., "A High-Performance 2.5 Gb/s ATM Switch For Broadband Applications," *GLOBECOM '92 Conference Record,* December 1992, ©1992, IEEE.)

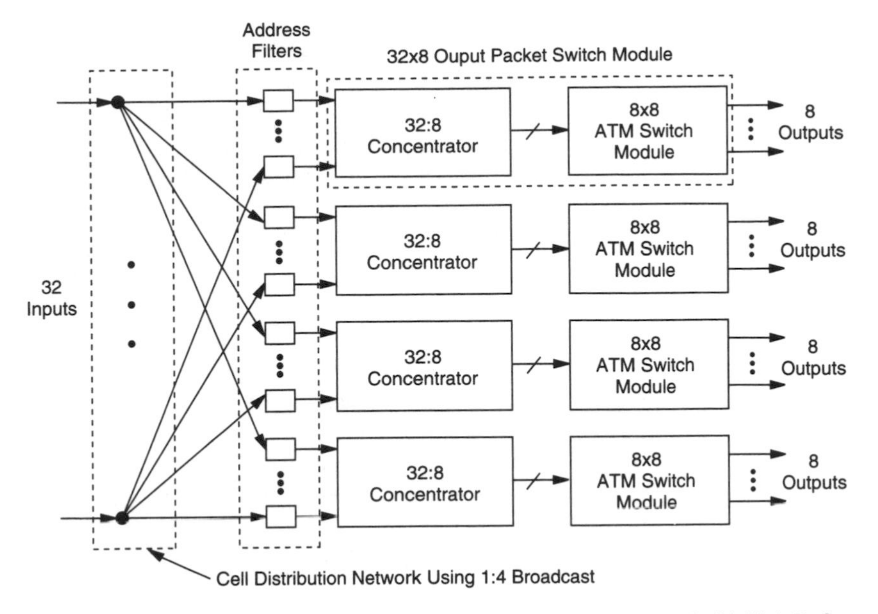

FIGURE 14.25. A 32 × 32 ATM switch output section. (From Eng, et al., "A High-Performance 2.5 Gb/s ATM Switch for Broadband Applications," *GLOBECOM '92 Conference Record,* December 1992, ©1992, IEEE.)

switching to the appropriate output port. The 32 × 32 switch can be expanded to 64 × 64, and further expansions are being researched for development.[35]

AT&T's prototype switch has evolved into the GCNS-2000 ATM switch, said to be capable of speeds up to 20 Gb/s. The GCNS-2000 (marketed as GlobeView-2000™) is the core switch in AT&T's initial ATM network. It initially provides only Category A (constant-bit-rate) and Category C (connection-oriented, variable-bit-rate) point-to-point, virtual services, pending completion of switched ATM standards. Other manufacturers have manufactured initial models of ATM switches and are engaging in ATM trials and in some commercial service. All such switches, however, may require retrofit or redesign to be in conformance with ATM standards when finalized.

14.6.6. Broadband Network Synchronization

Synchronization of an optical-fiber transport network is more complex than that of a channelized digital network employing electrical signals. As indicated in

Section 14.7.3, DS1 bit streams transported via SONET can be displaced in time by operation of SONET virtual tributary (VT) adjustments. Many central offices use a *Building-Integrated Timing Supply* (BITS) plan for distribution of synchronization timing to all NEs in that office. The highest stratum clock in the office is designated as the BITS, or master, clock for that office. Two types of signals are used for intraoffice timing distribution. DS1 provides a frequency-only reference used by digital switches and DCS 1/0 digital cross-connect systems. A composite clock (CC) provides both phase and frequency reference for NEs that have DS0 interfaces. BITS may be implemented by either conceptual BITS or by *Timing Signal Generator* (TSG) BITS. In conceptual BITS, the BITS clock is composed of a DS1 source and a TSG as a CC. The DS1 source uses a terminating DS1 from another office as its frequency reference. The TSG is bridged off a DS1 output signal from the DS1 source and supplies CC to all equipment that requires it. In TSG BITS, the TSG, bridged off a terminating DS1 from another office, functions as the sole BITS clock for the office. TSG BITS is preferred because of the simplicity of having only one timing source in the office and because digital switches may degrade their clock functionality to improve their traffic-handling capability.

As SONET transport is integrated into digital networks, a terminating DS1 bit stream, used for synchronization timing in an office, may have been carried by an intermediate SONET transport system. In SONET, DS1 signals are mapped into Virtual Tributary 1.5s (VT1.5s). In multiplexing 28 VT1.5s to create the STS-SPE, STS path overhead is added. Asynchronous mapping of DS1 into the VT1.5 provides for clear-channel transport of DS1 signals that meet the requirements for DSX-1 signals. Stuff bits accommodate timing variations of the DS1 signal.

There are two modes for VT1.5 transport. The *locked* mode locks the VT payload structure directly to the STS-SPE rather than using the VT pointer. This improves transport and cross-connection of DS0 signals but requires slip buffers between the electrical and optical at every VT1.5 cross-connect point. The *floating* mode uses a VT pointer. This minimizes buffering at VT cross-connect points, but still requires slip buffers at points that cross-connect DS0 channels. Bit-synchronous mapping for DS1 signals is defined only for locked VT1.5s to provide clear-channel transport for synchronous DS1 signals. Byte-synchronous mapping for DS1 signals is defined for both locked and floating modes. In the floating mode, byte-synchronous mapping of DS1 signals allows SONET NEs direct identification and access to the 24 DS0 channels carried in the DS1. The floating mode likely will be used in most cases.

There are four timing modes for SONET NEs. In *external timing*, a SONET NE derives its timing from a DS1 reference from a BITS clock. In *line timing*, a traffic-carrying optical signal (OC-N) is used as a reference to time all OC-Ns transmitted from the office. *Loop timing* is used as a special case of line timing when there is only one OC-N interface in the office. *Through timing* is used specifically for regenerators and *Add-Drop Multiplexers* (ADMs) in the add-drop mode. Each transmitted OC-N derives its timing from each corresponding terminating OC-N. Where BITS is available, external timing is preferred. As SONET is integrated into the existing digital networks, timing derived from the existing DS1 synchronization network ensures that SONET NEs are timed from the same reference source. If SONET NEs have DS0 interfaces, however, they must be timed by a TSG-provided CC. The current synchronization network clock requirements do not specify synchronized short-term stability, and short-term instability in SONET causes pointer adjustments that, in turn, cause jitter on the payloads, correctable by filtering. Existing digital clock requirements are being reviewed to determine whether additional specifications are indicated.

In the absence of a method to identify whether DS1 bit streams carried by SONET transport systems have been impacted by pointer adjustments, such DS1s should not be used for distribution of synchronization timing. SONET-based timing distribution will avoid the serious short-term stability problem of DS1s carried on SONET. In addition, OC-N signals are not likely to be rearranged as frequently as DS1 signals. SONET clocks have tighter short-term stability requirements than current stratum clock specifications, and SONET provides overhead bandwidth for synchronization status messages. There is no optical equivalent, however, to bridging repeaters used to bridge DS1 timing from incoming DS1s to BITS clocks, and current BITS clocks cannot accept OC-N signal references. Therefore, to effectively distribute synchronization timing from OC-N signals, the continued provision of DS1 timing references to BITS clocks is essential. This can be done by having a SONET NE derive a framed DS1 whose frequency is locked to the incoming OC-N signal. The DS1 must be derived directly from the OC-N frame rate without any filtering or processing of the incoming signal. One problem initially is that there may have been line-timed or through-timed ADMs in the incoming OC-N route. A synchronization message set is being defined to identify whether an intermediate ADM is running on its internal clock.[36]

14.6.7. Congestion Control

Congestion control in a SONET-based network with ATM switching is similar to the network management functions described in Section 11.6.4, but the nature of the broadband network creates some major differences. Network management in an existing channelized network is based upon the capability to apply controls on groups of analog or digital channels to protect switching systems from overload. In BISDN, once a user payload enters the network, the packets are transmitted to their destination according to the addresses contained in the packet headers. To avoid network congestion, control actions need to be taken at the point of origin to limit the originating traffic to that which can be carried efficiently or at the destination to buffer or discard packets that cannot be delivered.

The two accepted congestion control methods are preventive control and reactive control.

Admission control is designed to accept a new call only if the network performance can be maintained. To make such a decision, the network must know the current volume of traffic in the network and the characteristics of the new call based upon a description, called traffic descriptors, given by the user. One possible solution is for the user to determine the maximum data rate required. This would require that data-rate requirements be grouped into several classes. Other characteristics that can affect the performance of the network are peak and average bit rate, burstiness, duration of the peak bit rate, bit-rate variance, and maximum number of cells arriving during a certain period. Accepting a call that requires a large data rate can increase the blocking probability of calls requiring a lesser data rate. High-data-rate requirements along a particular route might adversely affect other traffic on the same route. One form of data-rate limitation would be to allocate transmission capacity to a virtual path for a particular origin-destination pair.

Once the parameters for admission control are defined, some form of policing is necessary for enforcement to ensure that a change in a user's traffic characteristics will not impair the overall network performance. Since the main consideration is control of traffic capacity, the leaky-bucket principle has been proposed. A cell would be accepted only when it can draw a token from the leaky-bucket token pool. Depending upon traffic conditions, tokens would be discarded rather than placed in the token pool (or "leaked" from the "bucket"). A variation could involve two or more parameters to be examined, increasing the complexity of the control mechanism. This type of control, however, is not adaptive to changing network traffic loads. One proposal is a virtual leaky bucket, where excessive cells are marked rather than discarded unless the network actually is congested.

Effective congestion control may need to involve some form of *reactive control*. Each cell header contains a *cell-loss priority* (CLP) bit which allows a user to indicate which cells are more or less essential than others. In case of congestion, the less essential cells could be dropped. A system of congestion indicators exchanged among network nodes could indicate the presence of congestion in a route and trigger a control action. Congestion at a point along a route could be avoided by routing traffic around that point.

Probably a combination of admission and reactive controls will be developed, and they will evolve over time as did the current network management controls in the public switched networks.[37]

14.6.8. Assimilation of BISDN

Initially, BISDN will be implemented as an overlay network along high-density routes to serve high-volume business traffic. Over a period of years, as older transmission systems are replaced by SONET-based systems, non-BISDN traffic can be expected to be carried over broadband transport systems. Current DS1 and DS3 services can be carried by a SONET-based network without converting them to the ATM format, although an interworking unit can convert DS1/DS3 protocols to ATM cell format for ATM switching. It is evident that the initial broadband traffic volume will rarely support the minimum access rate of 155.520 Mb/s. Standards for lower access bit rates, such as optical interface at STS-1 (51.840 Mb/s) and electrical interfaces at DS3 (44.736 Mb/s) and even DS1 (1.544 Mb/s), may be needed to accommodate users. Further refinement of ATM signaling, network management, and interworking standards is required. During the 1992–1996 CCITT study period, these and other refinements are being developed so that BISDN trials can be conducted and some commercial service offered with some proprietary design elements in the 1994–1995 time frame. Broadband transmission facilities are being implemented rapidly. A transport backbone employing SONET extends more than 6000 km (3700 mi) across Canada, using both fiber and SONET radio technology.[38] Commercial BISDN, employing CCITT standards, is likely to be implemented in the 1996–1998 time frame, but full development of BISDN services is not likely much before the turn of the century.

14.7. INTELLIGENT NETWORK FUNCTIONALITY

Capitalizing on the capabilities of SS7, CCITT has begun to develop recommendations for a worldwide Intelligent Network (IN). According to CCITT objectives, the IN should be applicable to all telecommunications networks, should evolve from existing networks, should enable service providers to define their own services, and should enable network operators to allocate functionality and resources within their networks. Development of CCITT/ITU-T recommendations will be phased in over a period of years. IN capabilities will be grouped into IN capability sets (CSs) to address one or more of the following: service creation, service management, service interaction, network management, service processing, and network interworking. Each IN CS is to be backward-compatible with the previous IN CS and forward on an evolutionary path toward the established target.

In the study period completed in 1992, Study Groups XI and XVIII completed the first phase of IN recommendations for IN CS-1, the initial set of standardized IN capabilities. Recommendation Q.1200 defines the structure of the Q.12xx series of IN recommendations and outlines the future work processes. The IN conceptual model (INCM) is structured into four planes, each providing an abstract view of an IN network, as shown in Fig. 14.26. The *service plane* describes services and service features from a user perspective, independent of how the service is provided. The *global functional plane* describes units of service functionality, referred to as service-independent building blocks (SIBs), independent

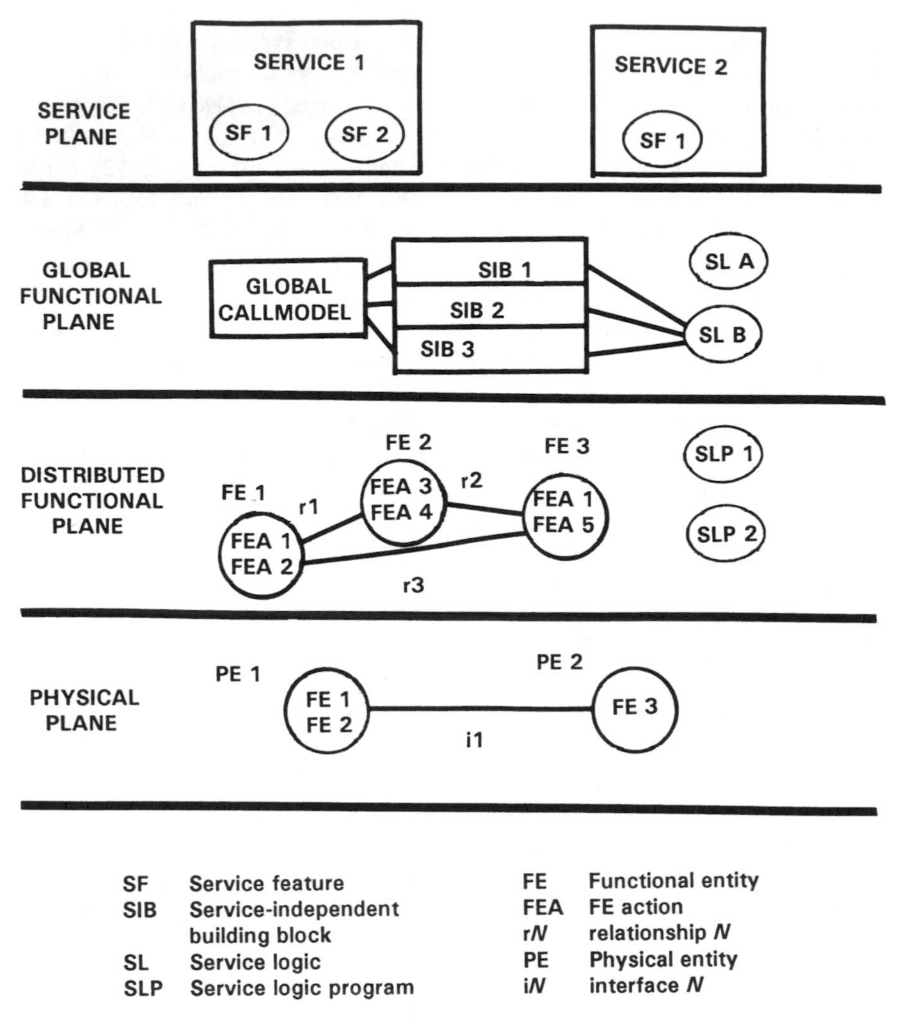

FIGURE 14.26. CCITT IN conceptual model. (From J. J. Garrahan, et al., "Intelligent Network Overview," *IEEE Communications Magazine,* March 1993 © 1993, IEEE.)

of how the functionality is distributed in the network. The *distributed functional plane* describes the functional architecture of an IN-structured network in terms of units of network functionality, referred to as "functional entities," and the information that flows between functional entities, referred to as "relationships." SIBs on the global functional plane are realized on the distributed functional plane by a sequence of functional entity actions and resulting information flows. The *physical plane* describes the physical architecture alternatives in terms of potential physical systems, referred to as physical entities and interfaces between physical entities.

The target set of service drivers for IN CS-1 includes "Freephone," *universal personal telecommunications* (UPT), and *virtual private network* (VPN), with special emphasis on flexible routing, flexible charging, and flexible user interaction. The target set of IN CS-1 services applies only during the call setup phase or during the release phase of the call. Those services are referred to as "single-ended" and generally do not apply to the active phase of the call.

Freephone service enables a subscriber to receive calls at one or more locations from users dialing the subscriber's Freephone number. The calls can originate anywhere, and the subscriber is charged for the call, similar to the well known 800 service. The UPT service provides a subscriber with a unique personal number that is addressable across multiple networks at any network access. The VPN service provides a subscriber with private network functionality without necessarily using dedicated network resources.

CCITT Recommendation Q.1211 describes other target services and service features for IN CS-1. Flexible-routing capabilities route calls to a specific destination or over specific facilities, or they reroute or forward calls under specified conditions. Flexible charging enables a call to be charged to a subscriber's account or credit card, or charged to the calling party, the called party, or split between them, or charged at a premium rate to the calling party for value-added services. Flexible-user interaction provides announcements to call parties to prompt and collect information relative to call processing. These capabilities are realized in a set of 14 SIBs for IN CS-1 that include basic call process (a specialized SIB), algorithm, charge, compare, distribution, limit, log call information, queue, screen, service-data management, status notification, translate, user interaction, and verify.[39]

14.7.1. AIN Overview

The Advanced Intelligent Network (AIN) specifications, being developed under the auspices of Bellcore, harmonize with the IN recommendations under development in CCITT.[40] While the AIN is not part of ISDN or BISDN, its functionality provides the means of implementing features included in those networks. AIN uses the SS7 Message Transfer Part (MTP), SUP, and Transaction Capabilities Application Part (TCAP) to exchange information between network Service Switching Point (SSP) and AIN Service Control Points (SCP). Thus, local and toll switching systems do not each need to store all data required to implement the many features currently planned for the public switched telecommunication networks (PSTN).

Basic AIN architecture is depicted in Fig. 14.27.[41] A originated call needing specialized handling, such as ISDN supplementary services, or any call that needs

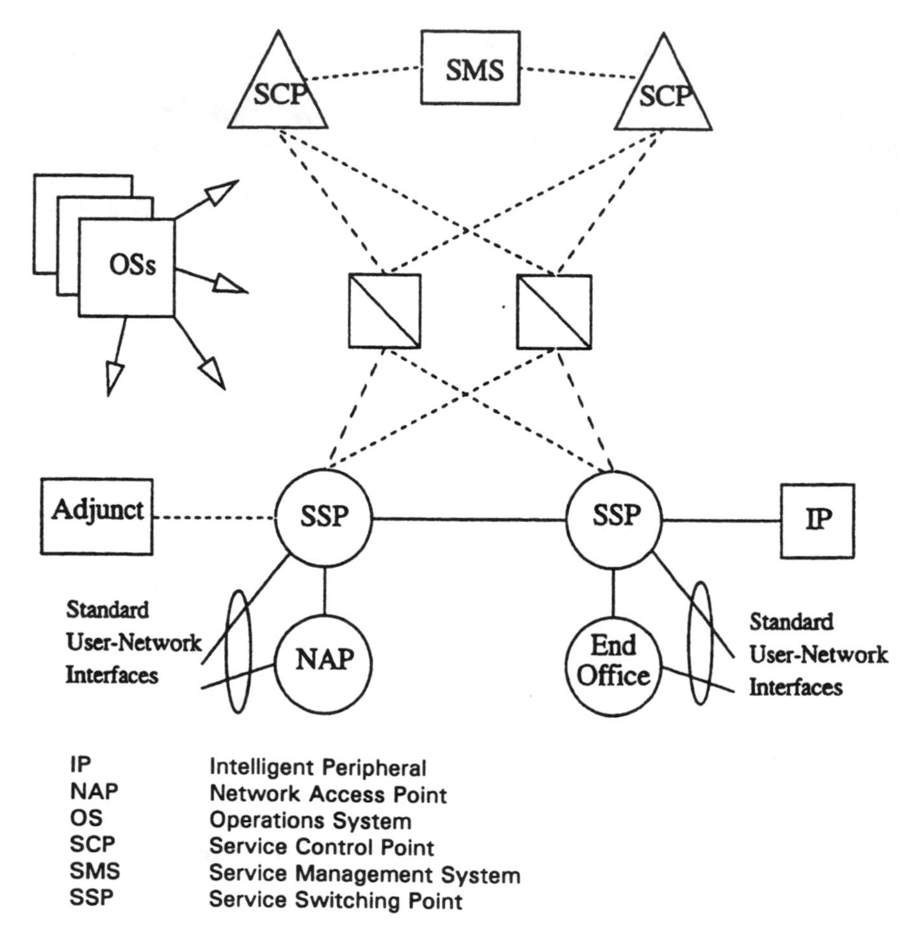

FIGURE 14.27. Advanced Intelligent Network architecture. (© 1993, Bellcore. Reprinted with permission. The information reprinted from Bellcore was deemed current at the time of publication of this work, but since then the information may have been amended or updated. For more current information, refer to the current Bellcore publication.)

interrogation of a specialized data base, is recognized by the local switching system as requiring routing to an AIN SSP. Some SSPs with *Network Access Point* (NAP) functionality recognize calls that require AIN involvement and route those calls to an AIN SSP. Other SSPs without NAP functionality identify such calls by subscriber class marks or normal switching system translations.

The AIN SSP receives the call request, suspends normal call processing, assembles a TCAP Query message, and sends it to an SCP over the SS7 signaling network via an STP after first checking for the existence of network controls that would reveal SCP overload. The SCP processes the query by checking its database to determine call treatment and routing. The SCP sends a TCAP Response message back to the inquiring SSP giving instructions for handling the call. The SCP may request additional information from the SSP to provide more details relative to the call. Service Logic Programs (SLP) containing the specific logic necessary to provide the detailed call-handling instruction to the SSP are resident in Intelligent Peripherals (IP) and Adjuncts so as not to delay normal call processing. For billing purposes, the SCP may request to be notified when the call is disconnected; if so, the SSP sends a TCAP unidirectional message to the SCP giving the time of call disconnect.[42]

The following example illustrates the complexity of programming necessary for an SCP to function in an AIN serving Personal Communications Services (PCS). Assume that a PCS subscriber is assigned a Universal Personal Telecommunications (UPT) number that callers can use to reach that subscriber irrespective of the subscriber's current location. The SCP might contain a "Registration" application program to permit the subscriber to record his current location information. A "UPT Profile Handler" application program could allow the subscriber to customize the subscriber profile information to include special screening information or default routing. A "Call Delivery" application program would be used for the SCP to initiate a TCAP Response message to the TCAP query message received when a caller dials that UPT number. To keep all these programs in working order, there would be a "PCS Maintenance" application program associated with the operation system.[43]

14.7.2. Triggers

A "trigger" occurs when an AIN SSP recognizes that it must query an SCP to continue call processing. Triggers are associated with the services that require AIN involvement, and are grouped into three categories. Users can *subscribe* to triggers related to their subscribed services. Only calls originating on a line or group of lines that subscribes to that trigger or that terminates to a user that subscribes to it will involve AIN processing. Triggers can be *group-based*; that is, they are associated with software-defined groups of users that share a customized dialing plan or routing pattern. If a trigger is *office-based*, it is activated by any call emanating from that office that meets the trigger criteria.

Originating triggers may be assigned to either ISDN users or non-ISDN users, to trunk groups having common authorized features, or to groups of users that share a customized dialing plan or an Automatic Flexible Routing pattern. An SSP detects a subscribed trigger when it receives a setup message from an ISDN user or an off-hook indication from a non-ISDN user. The dialing of NPA codes, Service Access Codes (i.e., 700, 800 or 900), or specific numbers following those codes activate office-based triggers. Some features require extra digits to be dialed subsequent to the initial director number; these features trigger requests from the SCP to collect the additional digits. To enable calls not requiring triggers, such as 911, to be originated by users subscribing to triggered services, escape codes are provided.[44]

14.7.3. AIN Implementation

Initial implementation of AIN was triggered by the FCC order to establish 800-number portability. Subsequent implementation continues in phases to support the many other features requiring AIN involvement. This implementation will continue for several years as network innovations continue to materialize.

PROBLEMS

14.1. Compare the echo-canceling hybrid (ECH) and the time-compression multiplexing (TCM) transmission modes for ISDN basic-rate access (BRA) loops.

14.2. What are the advantages of line code 2B1Q over other line codes for ISDN basic-rate access?

14.3. What factors affect expansion of ISDN services in countries with telecommunications monopolies versus countries with telecommunications competition?

14.4. How does an FDDI ring provide automatic protection in case of failure of a node? of a transmission link?

14.5. How do the three primary pre-BISDN broadband data services differ from one another? What are the advantages of each?

14.6. What are the forces that are driving the rapid development of Broadband ISDN standards?

14.7. What factors enable high-speed data transmission on BISDN to offset the loss of efficiency caused by packet overhead allocation?

14.8. How is SONET multiplexing superior to that of T1 technology?

14.9. Compare payload efficiencies of SMDS and BISDN.

14.10. How is an asynchronous payload identified within a synchronous STS-1 frame in BISDN?

14.11. Explain the knockout principle in cell-switching technology.

14.12. Why is a nonblocking ATM switch subject to loss of cells and how can excessive cell loss be avoided?

14.13. Compare input and output queueing in an ATM switch.

14.14. Why is the synchronization system in the digital circuit-switching network not acceptable for synchronizing BISDN?

14.15. How can congestion caused by traffic overloads be controlled in BISDN?

REFERENCES

1. Hunter, J. and Ellington, W., "ISDN: A Customer Perspective," *IEEE Communications Magazine*, August 1992, pp. 20–22.
2. CCITT Recommendation G.722 (1988), *7-kHz Audio Coding Within 64 kb/s.*
3. CCITT Recommendation H.221 (1988), *Frame Structure for a 64-kb/s Channel in Audiovisual Teleservices.*
4. Lechleider, J., "Line Codes for Digital Subscriber Lines," *IEEE Communications Magazine*, September 1988, pp. 25–32.
5. Liebacher, R., "ISDN Deployment in Europe," *GLOBECOM '90 Conference Record*, pp. 705A.3.1–705A.3.2.
6. Ibid., p. 705A.3.5.
7. Bell, T., "Telecommunications," *IEEE Spectrum*, January 1991, pp. 44–45.
8. Ibid.
9. Temime, Jean-pierre, "Numeris—ISDN in France," *IEEE Communications Magazine*, August 1992, pp. 48–52.
10. Liebacher, p. 705A.3.3.
11. Berenson, B. and David, R., "The ISDN Network in Belgium," *ICC '92 Conference Record*, pp. 201.1.1–201.1.5.
12. Bell, p. 46.
13. Temime, p. 48.
14. Inoue, O., "Implementation in Japan," *IEEE Communications Magazine*, August 1992, pp. 54–57.
15. Howard, M. and McConnell, J., "FDDI-II: A future uncertain," *Networking Management*, Sept. 1992, p. 82.
16. Bender, A., "Getting Fiber to the Desktop," *Networking Management*, April 1989, pp. 24–29.
17. Howard and McConnell, pp. 78–83.
18. Schriftgiesser, D., "SMDS: A Phone Service for Computers," *A Supplement to Business Communications Review*, 1992, pp. 4–9.
19. Strauss, P., "The implications for CPE," BCR Supplement, pp. 18–22.
20. *Broadband ISDN Switching System Generic Requirements*, Bellcore Technical Advisory TA-NWT-001110, Issue 1, August 1992.
21. *Broadband ISDN Operations: Framework Generic Criteria*, Bellcore Framework Technical Advisory FA-NWT-001248, Issue 1, December 1991.
22. CCITT Recommendation I-413 (1992), *BISDN User-Network Interface.*
23. TA-NWT-001110.
24. Ibid.
25. *SONET Synchronization Planning Guidelines*, Bellcore Special Report SR-NWT–2224, Issue 1, February 1992.
26. *Synchronous Optical Network (SONET) Transport Systems: Common Generic Criteria*, Bellcore Technical Advisory TA-NWT-000253, Issue 6, September 1990.
27. *FiberWorld '92, A Progress Report*, Northern Telecom, April 1992.
28. Eng, K., Karol, M., and Yeh, Y., "A Growable Packet (ATM) Switch Architecture: Design Principles and Applications," *IEEE Trans. Comm.* Vol. 40, No. 2, February 1992.

29. Karol, M. and Eng, K., "Performance of Hierarchical Multiplexing in ATM Switch Design," *ICC '92 Conference Record*, pp. 311.4.1–311.4.7.
30. Yeh, Y., Hluchyj, M., and Acampora, A., "The Knockout Switch: A Simple, Modular Architecture for High-Performance Packet Switching," *IEEE Journal on Selected Areas in Communications*, Vol. SAC-5, No. 8, October 1987, pp. 1274–1283.
31. Eng., K., Karol, M., and Yeh, Y.
32. Ibid.
33. Ibid.
34. Karol, M. and I, Chih-Lin, "Performance Analysis of a Growable Architecture for Broad-Band Packet (ATM) Switching," *IEEE Trans. Comm.*, Vol. 40, No. 2, February 1992.
35. Eng, K., Pashan, M., Spanke, R., Karol, M. and Martin, G., "A High-Performance 2.5 Gb/s ATM Switch For Broadband Applications," *GLOBECOM '92 Conference Record*, December 1992, pp. 5.2.1–5.2.7.
36. *SONET Synchronization Planning Guidelines*, Bellcore Special Report SR-NWT-002224, Issue 1, Feb. 1992.
37. Yazid, S. and Mouftah, H., "Congestion Control Methods for BISDN," *IEEE Communications Magazine*, July 1992, pp. 42–47.
38. Gruber, J. and Leeson, J., "Performance in evolving SONET/SDH networks," *Telesis*, issue no. 95, December 1992, pp. 17–18.
39. *Advanced Intelligent Network (AIN) 0.1 Switching Systems Generic Requirements*, Bellcore Technical Advisory TA-NWT-001284, Issue 1, January 1992, Section 1.
40. Garrahan, J. J., Russo, P. A., Kitami, K., and Kung, R., "Intelligent Network Overview," *IEEE Communications Magazine*, March 1993, pp. 30–36.
41. *Advanced Intelligent Network (AIN) Intelligent Peripheral Interface (IPI) Generic Requirements*, Bellcore Technical Advisory TA-NWT-001129, Issue 3, April 1993, p. 1–2.
42. TA-NWT-001284, Issue 1, Section 2.
43. *Advanced Intelligent Network (AIN) Service Control Point (SCP) Generic Requirements*, Bellcore Technical Advisory TA-NWT-001280, Issue 1, August 1992, pp. 2–6 to 2–7.
44. TA-NWT-001284, Section 2.

15

Closing the Loop

15.1. THE LOOP ENVIRONMENT IN NORTH AMERICA

The copper-loop environment is discussed in detail in Section A.3.1.1. Guidelines, developed in the 1980s for Carrier Serving Areas (CSA), produced excellent results for analog services. Additional study was needed, however, to ascertain the extent to which Basic Rate Access (BRA) ISDN could be deployed over copper loops. It was found that about 25% of all copper loops could not carry the required 160 kb/s because of loading coils installed in the loops. It also was found that about 48% of the sampled loops were compatible with CSA guidelines. Other loops in a CSA extend far beyond the maximum guideline length and are provided with Digital Loop Carrier (DLC) in the feeder portion. A Feeder Distribution Interface (FDI) is connected on the subscriber side of the DLC Remote Terminal (RT) and is designed to serve 400 to 600 subscribers.

A more recent survey of CSA loops in five Bell Operating Company (BOC) areas was performed in 1988–1990. The survey focused on two categories of loops: 559 loops served from 101 Potential Broadband (PBB) wire centers that currently were serving user locations having a high potential for future broadband services, and 686 DLC loops served from 126 DLC wire centers covering loop plant in all areas of the wire centers. About half the PBB CSA loops and over 60% of the nonloaded DLC loops sampled were compatible with CSA guidelines.[1] A typical DLC loop configuration is shown in Fig. 15.1.[2]

Of those compatible loops, the average total design loop length was 1552 m

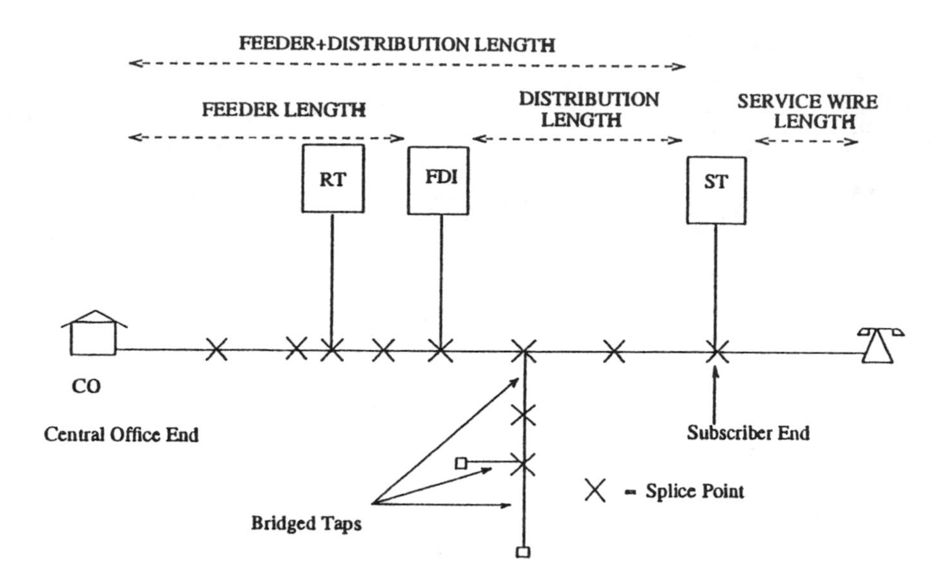

FIGURE 15.1. Typical DLC loop configuration. (©1991, Bellcore. Reprinted with permission. The information reprinted from Bellcore was deemed current at the time of publication of this work, but since then the information may have been amended or updated. For more current information, refer to the current Bellcore publication.)

(5092 ft) with an average working loop length of 1472 m (4828 ft). Bridged taps were evident on 36.3% of the loops, and the average length of bridged taps was 213 m (699 ft). The dc resistance varied around a mean of 364 ohms, and the capacitively reactive input impedance varied widely at frequencies up to 300 kHz. The average length of a CSA-compatible DLC loop is 8924 m (29 279 ft). The average length of the DLC loops from the RT to user location was 1160 m (3807 ft). Nearly 44% of the DLC distribution loops had bridged taps averaging 272 m (892 ft) in length. As with CSA loops, the dc resistance and capacitively reactive input impedance varied widely, primarily because of the concentration of bridged taps in the outer portion of the loops.[3]

15.2. LOOP REQUIREMENTS FOR BROADBAND SERVICES

There is an increasing demand for broadband services to be brought to user premises before BISDN is implemented. Advances in digital compression techniques have enabled video conferences to be conducted on loops with digital capacities of 128 kb/s to full DS1. Other broadband applications include computer-assisted learning over distances, desktop video conferences, videotex, high-speed facsimile, and entertainment television.

Cable-television companies, using coaxial cables to homes, typically provide a choice of up to 60 or more full-color television channels to the home, with selection being made at the television-set control box. If control were shifted to a central location, with adequate digital storage and switching, users could have "video on demand" by selecting channels from storage or from real-time programs available through a centrally located switch. Loop requirements for such a service include:

- the video signal is to be transmitted in a single downstream direction;
- control signals must flow upstream from the user to the information provider;
- the quality of the video signal must be at least that of a video cassette recorder (VCR);
- digital transmission in the loop must be virtually error-free;
- the electrical interface at the user premises must conform to standards.[4]

For primary distribution services, High-Definition Television (HDTV) requires a transmission medium capable of about 150 Mb/s,[5] although digital compression can reduce this sufficiently to fit safely into a SONET STS-3c payload via ATM/BISDN over optical fiber (see Section 10.12). For video in conformance with the National Television System Committee (NTSC) standards, video compression, using either the CCITT H.261 recommendation or the more recent ISO Moving Picture Experts Group (MPEG) recommendation, can provide full-motion video at about 1.3 Mb/s. Thus, the video, an audio channel, and the required overhead can be transmitted within a 1.5 Mb/s signal.

15.3. BROADBAND SERVICES ON COPPER LOOPS

It is apparent that the most efficient means of providing transmission of information, education applications, and entertainment is through network-based services. It is equally apparent that optical fiber to each business and residence would furnish the bandwidth to provide virtually unlimited transmission capacity to satisfy all such requirements and many others that are not foreseen today. That, however, is precluded in the near term by cost considerations, although certain architectures for fiber in the loop are rapidly approaching cost parity with copper for new installations and for major plant rehabilitation.

The divestiture of the BOCs in 1984 resulted in the rapid provision of digital tandem switches in local exchange areas with connecting digital facilities to both local central offices and inter-LATA carriers' points of presence (POP). The initial use of DLC was to provide analog loops by digital means. When digital switching systems were installed, the central-office terminal of the DLC was eliminated, and the line interface of the central-office switch was effectively extended well out into the loop plant. As all-digital loops begin to be realized, transmission performance greatly improves. Removal of analog links virtually eliminates the effect on loop transmission of impairments such as metallic loss, envelope-delay distortion, high-frequency rolloff, and data mutilation through use of digital pads for loss control. This still leaves digital-loop transmission considerations of absolute delay, bit-error rate, slips, jitter, wander, crosstalk, impulse noise, intersymbol interference, and radio frequency interference.[6,7]

A combination of physical media, including the ubiquitous copper wires, and transmission technologies, have the potential to provide information exchange up to DS1 rates. Two such transmission technologies that have appeal for the transition period are High-Bit-Rate-Digital Subscriber Line (HDSL) and Asymmetrical Digital Subscriber Line (ADSL). For higher bit rates, tests already have shown the technical feasibility of transmitting the STS-3 SONET level for short distances on copper wires. In each of these leading-edge technological developments, fiber is likely to be used for some distance in subscriber loops.[8]

15.3.1. High-Bit-Rate Digital Subscriber Line (HDSL)

The HDSL provides an alternative to repeatered T1 lines to support repeaterless DS1 rate over copper loops that conform to CSA loop design rules within a CSA environment. The HDSL provides full duplex 1.544 Mb/s transmission over two nonloaded two-wire metallic cable pairs. Bidirectional data at a rate of 784 kb/s is transmitted and received on each pair. The data from both pairs is combined to form a DS1-compatible format in an architecture called "dual duplex." Each 784-kb/s rate includes performance monitoring, framing, and timing overhead functions. The HDSL has significant advantages over labor-intensive T1 lines:

- no conditioning is required;
- bridged taps do not need to be removed;
- repeaters are not required for use within a CSA;
- binder group separation is not necessary.

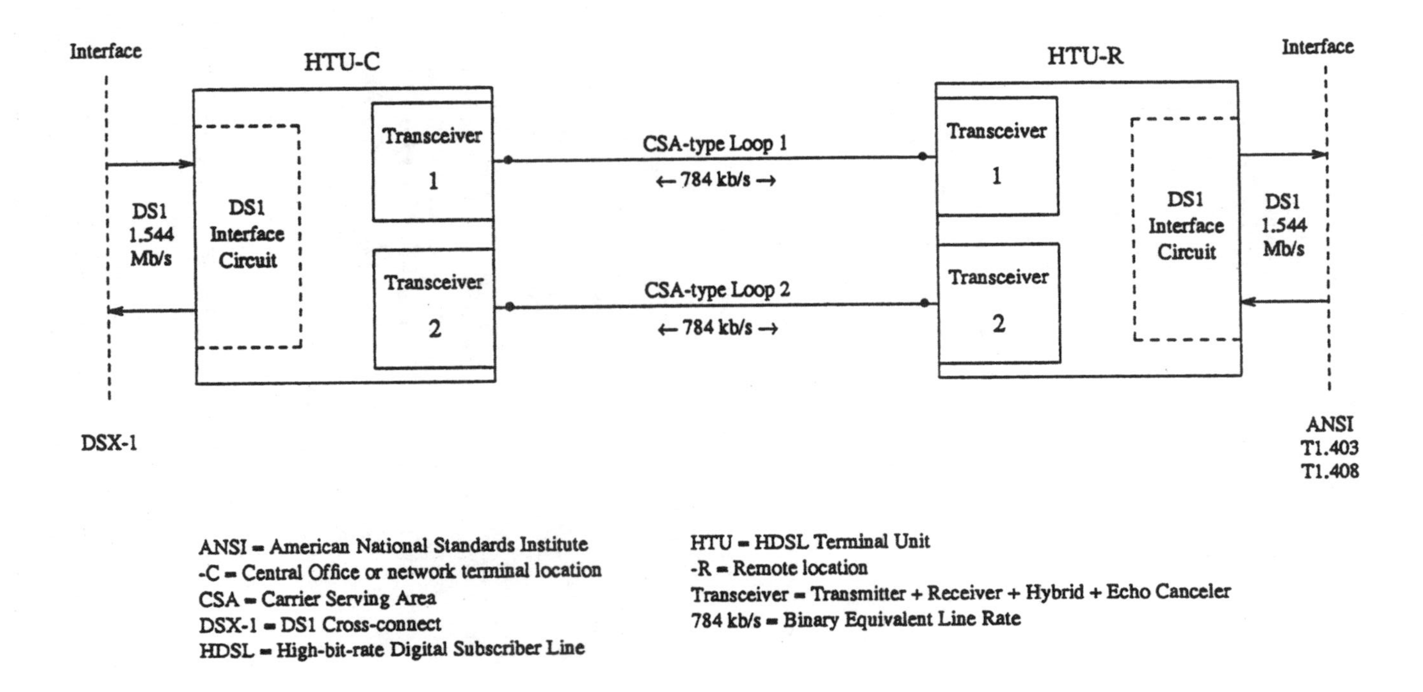

FIGURE 15.2. High-Bit-Rate Digital Subscriber Line (HDSL). (©1991, Bellcore. Reprinted with permission. The information reprinted from Bellcore was deemed current at the time of publication of this work, but since then the information may have been amended or updated. For more current information, refer to the current Bellcore publication.)

A block diagram of an HDSL system is shown in Fig. 15.2. Each High-Bit-Rate Terminal Unit (HTU), at either the central office (HTU-C) or the remote distribution (HTU-R) point, contains a DS1 Interface Circuit and two digital transceivers, each connected to a two-wire CSA loop. Each HTU accepts a DS1 signal, adds HDSL overhead, performs digital signal processing to divide the DS1 payload equally, and generates two 784-kb/s line signals to be placed on the two loops. The HTU at the other end performs digital signal processing to minimize the effects of loop impairments, removes the HDSL overhead bits, and recombines the two payloads into the original DS1 signal. The two transceivers function in a master/slave relationship; the master transceiver is located in the HTU-C, and the slave transceiver is located in the HTU-R. After power is applied to both units, the pair of transceivers establish communication automatically and without field adjustment. The HDSL line uses the principle of echo canceler with hybrid (see Section 14.2) to remove echoes caused by reflections of the transmitted signal from discontinuities, such as bridged taps and gauge changes, from line impedance mismatches, and from transformer hybrid leakage.[9]

The primary source of degradation of HDSL transmission performance is near-end crosstalk (NEXT) between pairs in the same cable, and worst case is self-NEXT, caused by identical transmitters using the same line code. The two main factors that affect NEXT performance are the insertion loss of the loop and the frequency spectrum of the self-NEXT. NEXT coupling in telephone cables increases with frequency at about 15 dB per decade at frequencies above 20 kHz to at least 1 MHz.[10]

Prototype HDSL transceivers were developed using adaptive filtering and equalization to compensate for the variable characteristics of CSA loops and for their tendency to change over time. The 128-tap echo canceler removes echoes of the transmitted signal that have mixed with the received signal. A prototype transmitter sends 4-level PAM symbols through a channel consisting of transmit and receive filters and 3.66 km (12 kft) of 24-ga copper unshielded twisted pair. Samples of the receive filter output, which are provided by a 12-bit A/D encoder, are applied to the input of the adaptive fractionally-spaced equalizer (FSE) portion of the adaptive filter. The resultant equalizer output is applied to a memoryless 4-level slicer, the output of which is proportional to the estimated value of the current symbol. The receiver contains a 30-tap FSE and a 65-tap decision feedback equalizer (DFE) and processes 4-level symbol sequences at rates up to 400 kBd. The DFE feeds back a weighted sum of past decisions to cancel the intersymbol interference (ISI) they cause in the current signaling interval. The performance of the HDSL transceivers is virtually error free so that loops are prequalified in bulk as conforming to CSA design rules and assigned as needed for HDSL use. Since the actual characteristics of the specific loops used with HDSL are not known in advance, the equalizer and transmitter must first go through a training procedure to establish a correct setting for the filter taps.[11]

15.3.2. Asymmetrical Digital Subscriber Line (ADSL)

Another evolving technology that parallels the development of HDSL is that of the ADSL. While HDSL provides DS1 transmission on a two-way basis over copper loops, there is a need for one-way DS1 transmission with a response channel. While HDSL can transmit DS1 over *two pairs* of copper conductors, there are congested loop-plant areas in which two pairs are not available for each residence, and many residences are beyond the CSA range for HDSL. Therefore, the need is for a solution that uses only a single pair of copper wires to each residence. In addition, the user must not be denied access to plain old telephone service (POTS).[12]

15.3.2.1. General Description of ADSL

To accommodate the POTS channel, frequency-division multiplexing can be used to position a digital video channel above the POTS channel similar to the data-over-voice (DOV) services developed several years ago. Early test implementation of ADSL provides a single 1.5-Mb/s video channel, a customer-control channel at 9.6 kb/s, and a POTS or ISDN channel, all over 4572 m (15 000 ft) of 24-ga copper loop.[13] (A proposed upstream signaling channel is capable of supporting a Q.931-based, 16-kb/s data-transmission control channel.)[14]

The encoded video signal is combined with telephony signals within the ADSL device, and the combined signal is transmitted over the two-wire loop facility to the loop ADSL device at the user premises. A low-pass filter directs the telephony signal to a telephony network interface, while the 1.5-Mb/s signal is directed to the DS1 network interface and thence to a real-time decoder which, in turn, delivers the NTSC signal to a standard NTSC television set. In the upstream direction, the loop ADSL device combines the telephony signal with the user signaling information and sends the combined signal upstream to the remote terminal or central office and on to a switch capable of separating the telephony and signaling information.[15]

15.3.2.2. ADSL Technical Challenges

Self-NEXT, the dominant source of interference for HDSL, is eliminated in ADSL because all the transmitters are located at the central office or remote terminal-end of the loop, and all receivers are located at the user ends of the wire pairs. This is the key to permitting ADSL to operate over an extended range. Far-end crosstalk (FEXT), principally self-FEXT resulting from multiple ADSL circuits in the same cable binder group, and thermal noise from terminal equipment are problems to be considered. Also, there are other forms of potential interference to overcome.

- Ringing voltage on the POTS line, applied at the central office, is 86 V rms superimposed on -48 V dc with a pattern of 2 s on and 4 s off. When ringing is interrupted by the user answering the telephone, a condition called *ring trip*, a significant transient can be induced in the line. Other less-significant transients can be induced in the line by circuit interruptions resulting from dial pulsing or from on-hook–off-hook transitions. This could be overcome by encoding the ringing and regenerating it in the user ADSL termination unit, but this would not allow ADSL to be transparent to POTS.

- The precise noise environment that will exist at the users' premises is unknown but is almost certain to be variable. In addition to telephony sources, impulse noise is generated by heavy machinery or smaller electric motors inherent in air-conditioning compressors, vacuum cleaners, grinders, passing motor buses, etc. Although this type of noise in residential environments is anticipated to be slight compared to industrial environments, the actual effect must be determined.
- Radio-frequency interference (RFI) from amateur transmitters, cordless devices, computers, and other electronic equipment can couple into residential wiring and drop wires, and the effect needs to be studied.[16]
- Intersymbol interference from both NEXT and FEXT, induced by HDSL and DSL system using 2B1Q line codes, by T1 systems using alternate mark inversion (AMI), and by other ADSL transmitters are potential problems to be overcome.[17]

15.3.2.3. Prototype ADSL Transceiver Technology

A prototype ADSL transceiver was analyzed through simulation by Bellcore. The prototype used M-ary quadrature amplitude modulation QAM passband signaling. The receiver included a decision feedback equalizer (DFE) comprising either a baud-spaced or fractionally spaced feedforward filter (FFF) and an ideal feedback filter (FBF). Impairments from self-FEXT and additive white Gaussian noise, representing thermal noise and receiver noise, were considered. Simulation was conducted over four loops, including both 24-ga and 26-ga wire, at or close to the extreme range of the loop plant. It used a FEXT model in which 49 pairs in a 50-pair binder group were generating FEXT, using the same type of transmission as the disturbed ADSL signal. Based upon an earlier study,[18] the general assumption was that FEXT varies by 20 dB/decade in frequency and 10 dB/decade in length. It was further assumed that satisfactory performance would be obtained if the bit-error-ratio (BER) did not exceed 10^{-7}.

Simulation results indicated that 16-QAM performed better than 64-QAM or 256-QAM. A study of background noise produced by noise from typical operational amplifiers in the front end of the receiver and thermal noise produced by the cable indicated that background noise is dominated by the receiver noise. It was concluded that satisfactory performance can be obtained if suitable low-noise operational amplifiers are utilized in the design. With background noise power spectral density at -150 dBm/Hz, a signal-to-noise ratio (SNR) of 12 dB, assumed to be sufficient to meet the BER of 10^{-7}, was obtained over 5486 m (18 000 ft) of 24-ga cable and over 4572 m (15 000 ft) of 26-ga cable.[19]

15.3.2.4. Discrete Multitone Technology

Another ADSL concept; Discrete Multitone (DMT) modulation,[20] involves division of a frequency band into small subchannels and optimizes performance of each subchannel by allocating incoming data based upon subchannel signal-to-noise ratios (SNR). Optimum performance across the entire band is achieved without using adaptive equalization. One prototype reportedly has demonstrated the capability to carry the following services at nearly 7 Mb/s over CSA loops:

- a unidirectional channel operating at 6+ Mb/s;
- a bidirectional control channel operating at 16 kb/s;
- a POTS channel, powered from a central office; *plus either*
- one bidirectional H_0 channel operating at 384 kb/s *or*
- one ISDN BRA service with 2 B-channels and 1 D-channel.

The DMT modulator prototype system, illustrated in Fig. 15.3, uses 256 subchannels of 4 kHz each, with each subchannel capable of carrying from 0 to 11 bits out of the data contained within a 250-μs symbol. Subchannels containing no data are turned off. The transmit spectrum will reflect the attenuation and crosstalk noise of a given loop through a loop training process during loop setup. DMT uses Fast Fourier Transforms (FFT) for modulation and demodulation, and the signal processing portion of a DMT modem is dominated by FFTs.

On resistance-design loops, the prototype system can transmit data at 1.544 Mb/s with a full ISDN BRA service underneath. The developer believes that the

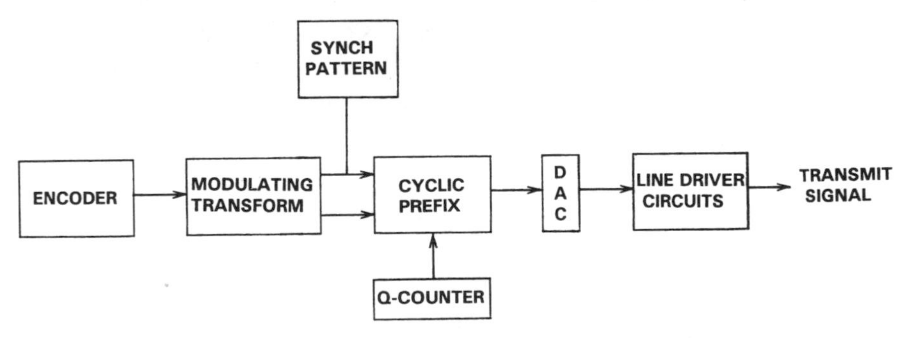

FIGURE 15.3. DMT modulator block diagram. (From Document T1E1.4/93-083, "DMT Specification Overview for ADSL," April 15, 1993, submitted by John M. Cioffi and John A.C. Bingham, Amati Communications Corp., Palo Alto, CA.)

final product will require trellis coding or echo cancellation to provide 6-dB margin on all loops at 1.544 Mb/s, but that both will be needed for robust performance at 6+ Mb/s. Both features are being added to the prototype. On loops with sufficient margin, DMT may be able actually to vacate frequency bands with crosstalk from other services. During one test, DMT had no energy from 120 to 220 kHz, thereby avoiding HDSL interference using 2B1Q line code. Also, energy was eliminated from the band from 750 to 790 kHz, avoiding interference from a local AM radio station picked up by a line looped around a spindle in a parking lot.

The developer expects that gate counts, memory, and speed estimates will lead to a chip that will meet cost targets by about 1996. The final VLSI solution will be programmable and, therefore, able to alter configurations on an individual loop basis. The ADSL product line using DMT is expected to be replaced by a silicon solution in 1996, depending upon standards-development progress. In March, 1993, the T1E1 Committee unanimously adopted DMT as the modulation standard for ADSL standards development. The initial application of ADSL focuses on delivery of prerecorded movies on demand at 1.5 Mb/s over distances up to 5486 meters (18 000 feet). Other services are being developed.

15.3.3. SONET/ATM Signals on Copper Conductors

Experiments conducted by Bellcore have demonstrated the feasibility of transmitting SONET/ATM signals at the STS-3c rate of 155.520 Mb/s over unshielded twisted-pair (UTP) and shielded twisted-pair (STP) cables with BET less than 10^{-13}.[21]

15.3.3.1. UTP and STP Cables

As local area networks (LANs) have proliferated, it has been shown that high-speed data can be transmitted various distances over twisted-pair copper cable. As LAN data rates increased, cable manufacturers have improved the design of copper cables to accommodate those higher rates. To relate those improvements to the higher data rates, UTP-cable characteristics have been grouped into five categories for commercial building uses:[22]

- Categories 1 and 2: used typically for voice and low-speed data;
- Category 3: cables specified up to 16 MHz; used typically for voice and data transmission rates up to 10 Mb/s;
- Category 4: cables specified up to 20 MHz; intended for voice and data transmission rates up to 16 MHz;
- Category 5: cables specified up to 100 MHz; intended for voice and data transmission rates up to 100 Mb/s.

The SONET/ATM experiments were conducted over four types of copper cable:

- single-pair, 24-ga UTP cable;
- Category 3 four-pair, 24-ga UTP cable (LAN grade IEEE 802.3, 10 Base T);
- Category 5 four-pair 24-ga UTP cable (AT&T Type 1061);
- two-pair 22-ga IBM Type 1 STP cable.

The nominal characteristic impedance of the UTP cables is 100 Ω, and that of the STP cable is 150 Ω. The transmission and reflection characteristics versus frequency were measured for various cable lengths. Dispersion was similar in all cables and produced a phase lag asymptotically approaching 90° at high frequencies. Near-end and far-end crosstalk and attenuation were measured and documented for different cable lengths. The study concluded that echo cancellation will be required to reduce interference from NEXT. Category 5 UTP cable performed somewhat better than other UTP cables because of the shorter twist spacing. The measured results suggested that the distorted signal can be reconstructed at the receiver by inserting a real zero in the transfer function to overcome the attenuation and 90° phase shift at high frequencies.

15.3.3.2. Transmission Techniques

A 3-level coding technique was used to evaluate transmission performance. The experimental transmitter in Fig. 15.4 encodes the incoming bit stream into 2-bit code words, using a simple finite-state machine. Correlation between code words results in bandwidth compression. The differential push-pull driver, with a

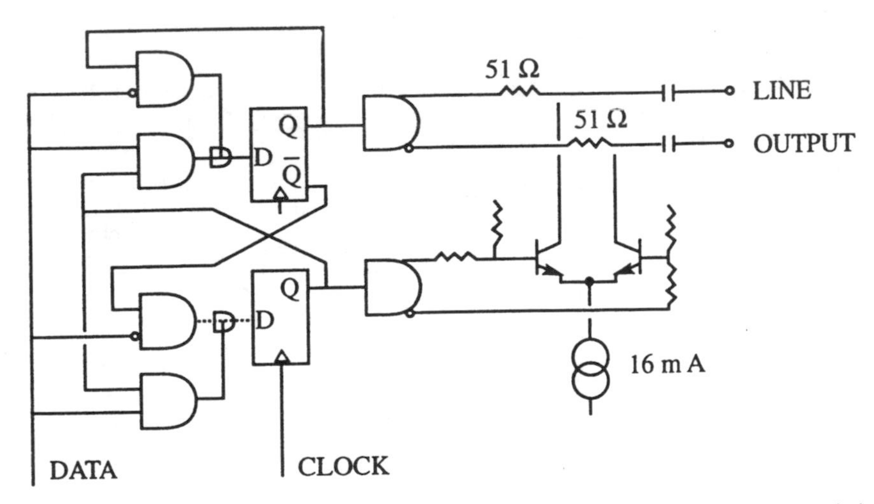

FIGURE 15.4. Experimental 3-level 155-Mb/s transmitter. (From Stephens et al, "Transmission of STS-3c (155 Mbit/sec) SONET/ATM Signals over Unshielded and Shielded Twisted pair Copper Wire," *GLOBECOM '92 Conference Record*, December 1992, ©1992, IEEE.)

100-Ω output impedance in the output stage, converts the correlated 2-bit words at the output of the D-type latches into a 3-level signal. Sequences of logic "ones" are alternately converted into either +800 mV or −800 mV (differential) into 100 Ω. At a logic zero, the output will go to 0 mV (differential). The measured common-mode output current noise can be reduced to acceptable FCC Class B emission limits by using a suitable balun for common-mode rejection.

The receiver in Fig. 15.5 equalizes and decodes the 3-level bipolar signal. The transconductance of the differential amplifier in the input stage is set by the R–C network between the emitters to produce the desired equalization. The collectors drive a wideband current-differencing amplifier whose transimpedance is controlled by negative feedback. For each cable tested, the experimental receiver was manually adjusted for optimum equalization and threshold level. The 3-level bipolar signal is detected and rectified by two high-speed comparators to recover the original binary signal.

15.3.3.3. Transmission Performance

Using the user network interface shown in Fig. 15.6, transmission performance was evaluated by measuring the BER and the noise level for different lengths of the four types of cable. No errors were measured (BER $<10^{-13}$) for 24 to 131 m of single-pair UTP cable, for 100 m of Category 5 UTP cable, or for 213 m of Type 1 STP cable. The effective internal noise, primarily from pattern-dependent baseline wander, was about 5 mV rms for each of those test cables at the lengths

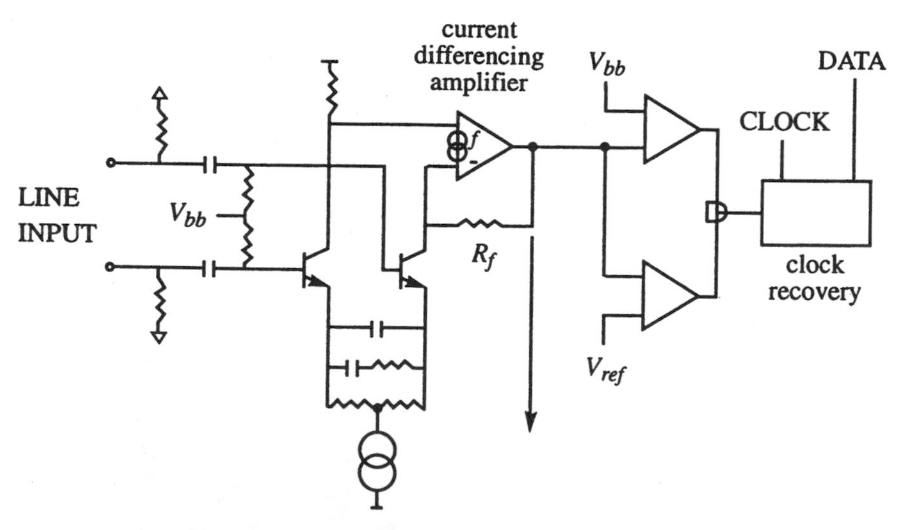

FIGURE 15.5. Experimental SONET/ATM receiver. (From Stephens et al, "Transmission of STS-3c (155 Mbit/sec) SONET/ATM Signals over Unshielded and Shielded Twisted pair Copper Wire," *GLOBECOM '92 Conference Record*, December 1992, ©1992, IEEE.)

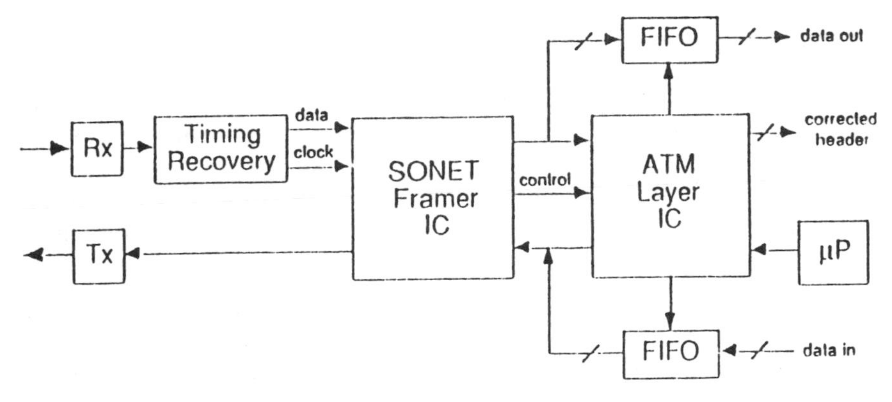

FIGURE 15.6. SONET/ATM prototype user network interface. (From Stephens et al, "Transmission of STS-3c (155 Mbit/sec) SONET/ATM Signals over Unshielded and Shielded Twisted pair Copper Wire," *GLOBECOM '92 Conference Record*, December 1992, ©1992, IEEE.)

indicated. The 213-m STP cable link can tolerate 5-7 mV rms of additional noise, and the 131-m single-pair UTP cable can tolerate 1.4 mV rms of additional noise, at the receiver input, for a BER of 10^{-12}.

Because of the favorable results obtained with Category 5 cable, 2-level coding was investigated. It was found that a 2-level encoder with 1-bit preemphasis (100% overshoot) also can transmit 155 Mb/s over Category 5 cable, but the transmission level must be reduced to satisfy emission restraints. The performance was confirmed by examining both the payload data and the SONET maintenance signals provided by the SONET framer and the ATM-layer ICs for errors. The eye pattern for a typical equalized signal received over 131 m of UTP cable is shown in Fig. 15.7.

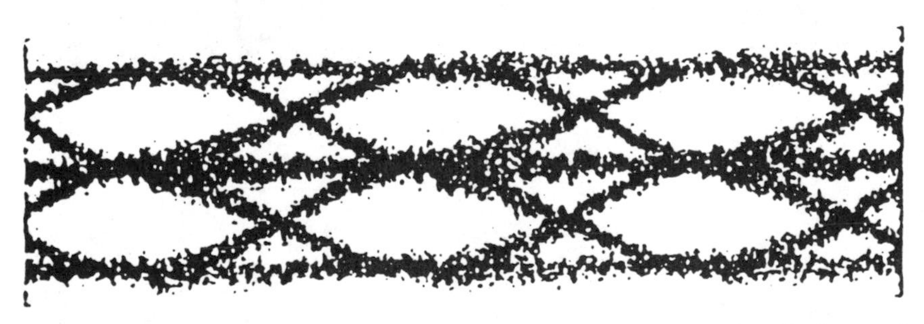

FIGURE 15.7. Eye pattern for typical 155-Mb/s signal over UTP cable. (From Stephens et al, "Transmission of STS-3c (155 Mbit/sec) SONET/ATM Signals over Unshielded and Shielded Twisted pair Copper Wire," *GLOBECOM '92 Conference Record*, December 1992, ©1992, IEEE.)

15.4. FIBER IN THE LOOP (FITL)

Optical fiber has been deployed rather extensively as feeder cables for subscriber services. The only deterrent to deployment of fiber in the subscriber distribution network has been cost, and that deterrent is rapidly disappearing as fiber begins to achieve cost parity with copper. In some situations, cost parity already has been achieved. The initial focus for Fiber in the Loop (FITL) systems is on "Fiber-to-the-Curb" (FTTC), where an Optical Network Unit (ONU) serves several users via copper drops. At such time as the cost of an ONU can be justified to serve a single user, "Fiber-to-the-Home" (FTTH) will have been achieved. A functional model of a FITL system is illustrated in Fig. 15.8.

A Host Digital Terminal (HDT) is connected to subtending Optical Network Units (ONUs) by an optical Passive Distribution Network (PDN). The ONUs are managed by the HDT in a master-slave arrangement. Each ONU terminates fibers from the HDT, performs optical/electrical (O/E) conversions, and sends the electrical signals over twisted-pair wire drops to the network interfaces at user premises. For some services, the drop may consist of coaxial cable. The PDN consists of single-mode fiber and various optical components that may include Wavelength Division Multiplexing (WDM) capability. The HDT performs O/E conversions and interfaces the FITL system to the remainder of the local network.

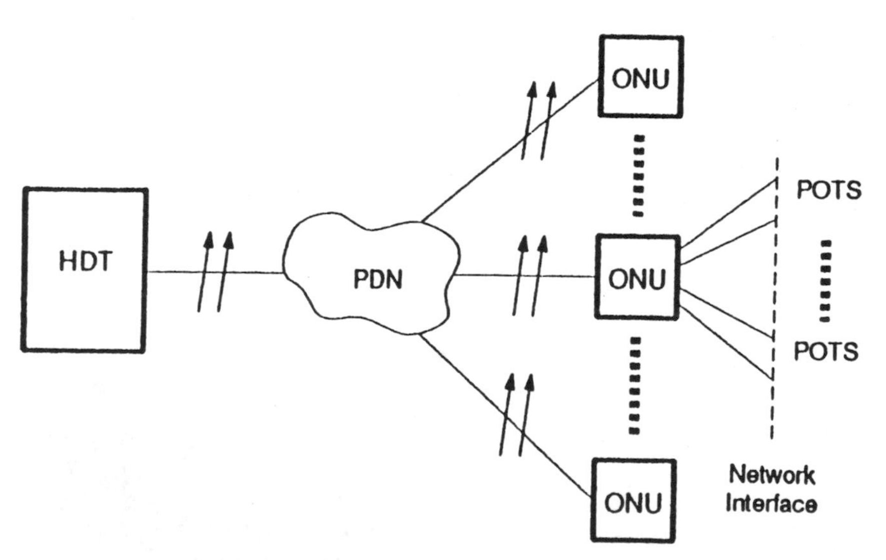

FIGURE 15.8. FITL system functional model. ©1991, Bellcore. Reprinted with permission. The information reprinted from Bellcore was deemed current at the time of publication of this work, but since then the information may have been amended or updated. For more current information, refer to the current Bellcore publication.

The HDT may be located in the outside plant and connected to the digital switch by a Digital Cross-connect and Transmission Facility (DCTF), or it may be located in the central office either as a standalone unit (Universal FITL) or integrated in the digital switch or other network components (Integrated FITL). If the local switch is analog, the HDT is located in the central office and performs D/A conversion.

The HDT performs four generic functions. It *concentrates* traffic from its subtending ONUs for efficient use of transmission capacity to the local switching system. It also separates or *grooms* locally switched traffic from nonswitched and nonlocally switched traffic, so that the nonlocally switched traffic can be efficiently routed away from the local switch. Further, the HDT interprets and translates *signaling* information into an efficient format for processing by the switching system. Finally, it manages provisioning and maintenance *operations* for itself and its ONUs and communicates with both the local switching systems and the appropriate operations system (OS).[23]

The PDN can be arranged in a point-to-point configuration or in a point-to-multipoint configuration. In a point-to-point configuration, a unique fiber facility extends from an optical-line unit in the HDT to each ONU. In a point-to-multipoint configuration, one optical-line unit in the HDT serves multiple ONUs, and a coupler/splitter distributes the fiber facility from the HDT optical line unit to several ONUs. In both cases, the fiber link between the HDT or coupler/splitter and the ONUs could be either a single fiber or two fibers per ONU.[24]

15.4.1. FITL Configuration Alternatives

A functional diagram of an Integrated FITL system is illustrated in Fig. 15.9. The HDT is shown to be connected to the digital switching system via a DCTF through a high-speed digital interface developed for IDLC systems. The interface physical level, on either SONET or DS1 metallic facilities, supports from 2 to 28 virtual tributaries (VT1.5) or DS1 signals for a maximum total of 672 DS0 channels. An Embedded Operations Channel (EOC) is carried in one of the 672 DS0s, using the LAPD protocol, employed on ISDN D-channels at the data link layer, for communication with remote OSs. Call processing involves connection management and services supervision. With common-channel signaling, both functions are combined in messages on a single data channel, but Time-slot Management Channel (TMC) requirements separate the two functions. Connection-management messages are carried over the TMC, but supervision is accomplished within the traffic channels via robbed-bit signaling.[25]

All operations communication between the FITL system and remote OSs passes through the Operations Interface Module (OIM) of the host digital switch. The video block in Fig. 15.10 illustrates a future potential capability for a video dial tone or other broadband feature. In a suburban residential area with predom-

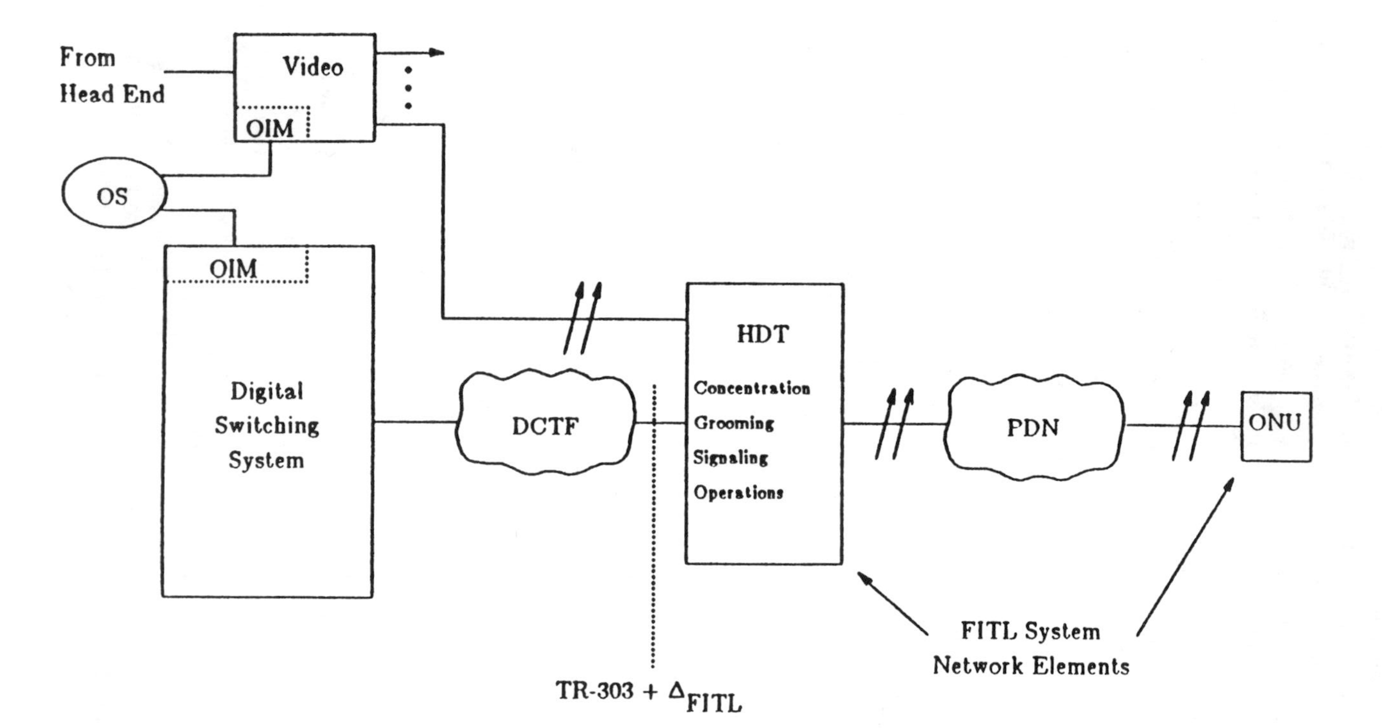

FIGURE 15.9. Integrated FITL system configuration. ©1990, Bellcore. Reprinted with permission. The information reprinted from Bellcore was deemed current at the time of publication of this work, but since then the information may have been amended or updated. For more current information, refer to the current Bellcore publication.

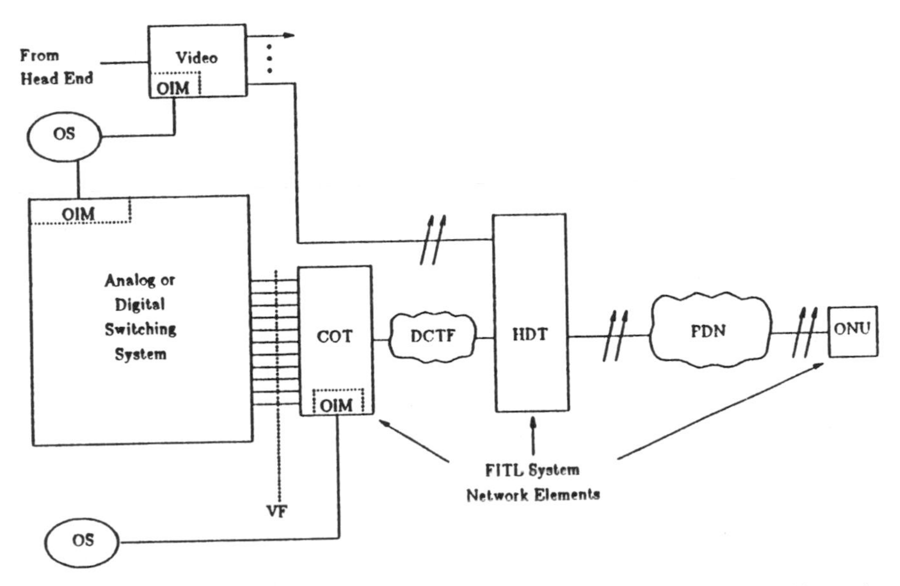

FIGURE 15.10. Universal FITL system configuration. ©1990, Bellcore. Reprinted with permission. The information reprinted from Bellcore was deemed current at the time of publication of this work, but since then the information may have been amended or updated. For more current information, refer to the current Bellcore publication.

inantly single-family homes, the ONU is designed to serve about four homes. In small apartment buildings, it may serve more than four users. In a high-rise apartment building or condominium, the HDT and ONU may be combined into a single network element to provide direct metallic drops to users, thus eliminating the PDN.

Universal FITL systems, consisting of a Central Office Terminal (COT), HDT, and ONUs, can be deployed in areas served by analog switching systems or by digital switching systems that have not been upgraded to support the high-speed digital interface, as illustrated in Fig. 15.10. In these cases, the DCTF terminates on the COT rather than the switch. The COT performs D/A conversions and interfaces the switching system on a voice frequency and baseband basis via a Main Distributing Frame (MDF). The FITL operations communications protocols are converted by the COT to standard interfaces with remote OSs through its OIM function. Since the interfaces on the user side of the ONU and on the switch side of the COT are standard user service interfaces, a Universal FITL system functions as a replacement for existing metallic cable plant.

For large numbers of users located close to the central office, the COT and HDT functions may be combined in a single NE, eliminating the DCTF. Another variation would be to combine the HDT and ONU functions, eliminating the PDN.

The COT could be arranged with a high-speed, metallic, digital interface to serving digital switching equipment equipped for direct interface at the DS1 level to Subscriber Loop Carrier. That interface has limitations that make it undesirable for use in FITL systems. It does not support an integrated SONET interface, full DS0 cross-connect capability, ISDN Basic Access, DS1-based services, or the full range of remote operations capabilities. However, it is more efficient than the alternative of voice frequency and baseband interfaces.[26]

One effect of FITL systems is that digital transmission is extended farther out from the central office into the user access area. When used with traditional digital loop carrier systems, a portion of the metallic cable facilities connecting the DLC to the user premises is replaced by the PDN and ONU. Since ONUs in a suburban residential area are intended to serve four dwellings, signaling, supervision, and transmission requirements have been designed for a maximum nominal drop length of 150 m (500 ft).

15.4.2. Powering Alternatives for ONU

There are three alternatives for supplying power to the ONUs. In a *centralized* power arrangement, dc power from the central office, from the HDT power source, or from a power node is used to power two or more ONUs. Since the power sources may be several kilometers away from the ONUs, Bellcore has proposed the use of − 130 V dc to ensure that enough power is delivered to the ONU and to use voltages already available at many locations. Centralized power has the advantage of using the same backup power that is available at the primary power source. There are several disadvantages to centralized powering. The centralized power source may be insufficient to power all ONUs and may have to be resized. Remote supply of dc power is inefficient and introduces new support requirements for provisioning, maintenance, and trouble sectionalization. There also is no indication at the ONU that it is operating on backup power.

Local powering can derive power from a utility company distribution transformer and convert it to dc at that point. Local powering has the advantage of limiting the distance of transmission to a few hundred meters; therefore, the voltage of the ONUs can be lowered to 48 V dc. Energy is transferred more efficiently to the ONUs, and maintenance of a power network is eliminated; however, disadvantages include the need for a battery backup source at each ONU, the need for ac–dc converters, high-quality batteries for backup, the limited number of hours of backup power available at each ONU, and the additional maintenance required to replace batteries in the field periodically.

In addition, a local-battery backup source may not be sufficient to carry the higher power consumption required for broadband or video services. This likely would require that those services be interrupted during power failures to conserve power for telecommunications services. Thus, a *hybrid* arrangement, using centralized powering for telecommunications services and local powering for broadband/video services, may be the solution.[27]

15.4.3. Spectrum Considerations

Transmission in FITL systems will be confined to single-mode optical fiber. Three wavelengths were considered. Operation in the 800-nm region was rejected because of no industry standard measurement procedures, and its use would impair operations management and maintenance. From the viewpoints of cost, performance, and industry experience, operation in the 1310-nm region was selected for primary use. Synergy with SONET transmitter and receiver use in this region increases the comfort level of operating companies. As digital video evolves, the 1550-nm region may come into use in FITL with WDM technology.[28]

15.5. FIBER TO THE HOME (FTTH)

Extension of optical fiber into the home is the obvious next step after achieving widespread coverage of FITL. Using the FITL architecture, this means that an ONU would be installed in each residence. There have been a number of testbeds installed in the last few years, both in North America and abroad. A description of two of those testbeds is detailed here.

15.5.1. Biarritz, France Optical Network

In the first edition of this book, we briefly described an ongoing project to install a major optical network in the city of Biarritz, France.[29] The first subscribers were cutover in 1984, and some 70 000 subscribers now are connected by optical fiber. The network is constructed in a star topology, with 70 fibers connecting a distribution center to a secondary center, then 10 fibers to a branching box, and 2 fibers to each subscriber premises as shown in Fig. 15.11. About 10 000 km (6214 mi) of multimode 50/125-μm graded-index fibers were installed. Wavelength-division multiplexing is used to serve more subscribers by sharing fibers. Optical emitters in the system include laser diodes, avalanche photodiodes (APDs), and light emitting diodes (LEDs). Low-cost connectors made with thermoplastic molding were developed at a cost of about United States $5.00 each.

Different experiments were carried out, corresponding to two similar networks with some variations in the loops. The distribution center has one LED and one receiving p-i-n diode for each subscriber. A common loop, using cables containing five fibers, channels data from the distribution center to hundreds of branching boxes. The branching boxes are configured in two ways. Network A uses 5 fibers to serve 5000 subscribers in the downstream direction, using temporary mechanical splices (called "bornier") for direct access to each subscriber. An 8-to-1 cou-

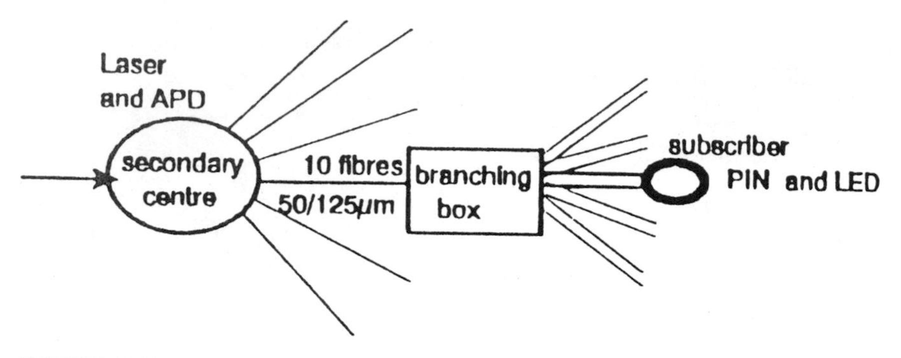

FIGURE 15.11. Biarritz network architecture. (From R. Goarin, "Component Reliability Results from the Biarritz Field Trial and from 'Plan Cable' Volume Deployment," *GLOBECOM '91 Conference Record*, Vol. 1, December 1991, ©1991, IEEE.)

pler collects all upstream selection orders to the distribution center. When the number of subscribers is greater than four, a WDM and two additional borniers are added to the branching box. In Network B, corresponding to about 65 000 subscribers, each subscriber uses a duplexer for bidirectional transmission over the same fiber. Initially, 85/125-μm fiber was used because of its larger core and ease of handling. After improvements in LED output power, 62.5/125-μm fiber has been used since being favored by international standards bodies. The 5 fibers support 10 subscribers using the 0.85-μm first window, and 5 more subscribers can be served by adding a WDM to use the 1.3-μm second window. Experience with the 0.85-μm laser was not satisfactory, because it did not meet reliability requirements.[30]

15.5.2. Heathrow, Florida Optical Network

A later FTTH testbed was installed in Heathrow, Florida in 1988–1989 in a joint project by Southern Bell Telephone Company, Northern Telecom, and Bell-Northern Research, to serve three residential communities with POTS, ISDN, and 54 cable-television channels in cooperation with a CATV supplier.[31] The system uses a star architecture with a dedicated fiber to each subscriber. A feeder cable containing 96 fibers was installed through a splicing manhole to a controlled-environment vault (CEV), a distance of 1.7 km (5600 ft) from the central office. Distribution cables were directly buried in trenches to serve the communities of Brampton Cove and Devon Green and were spliced into the feeder cable 1 km (3300 ft) from the central office to serve the community of Stratford Gardens, as shown in Fig. 15.12. The system uses standard 9/125-μm, single-mode fiber operating at 1300 nm with two fibers to each residence. Although the smaller cable

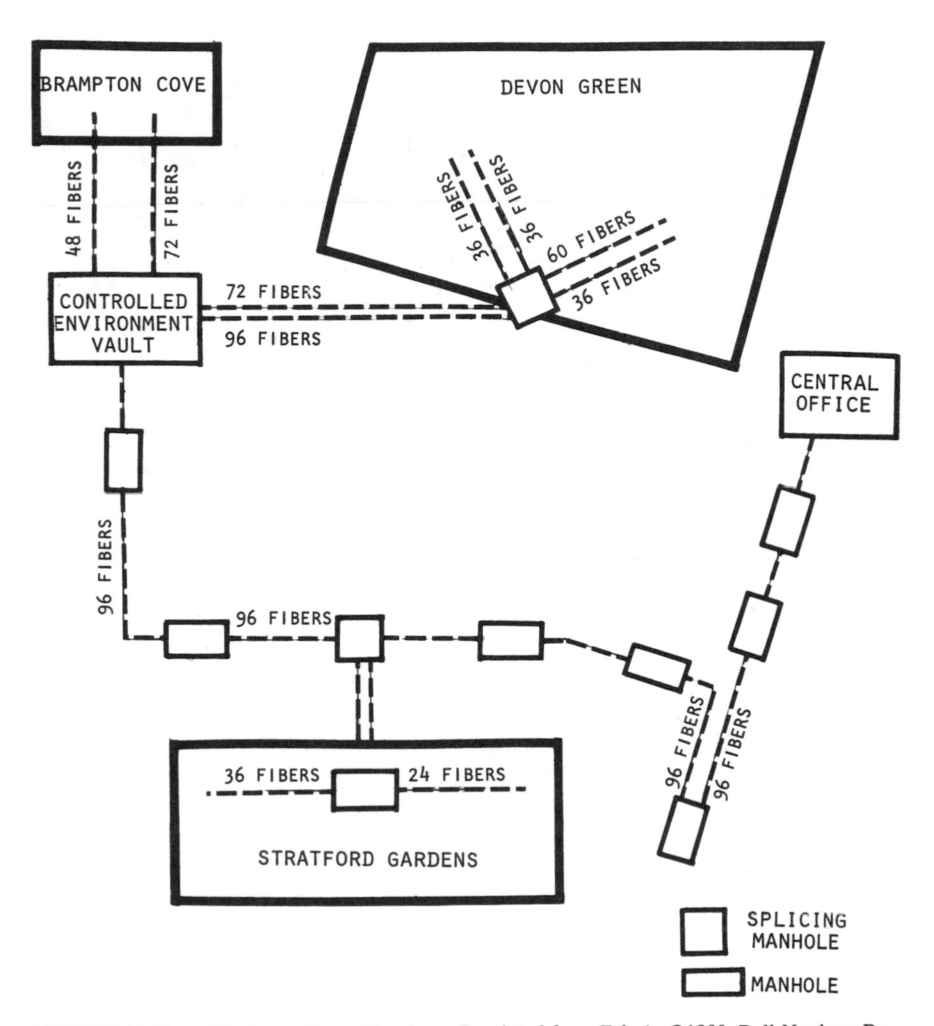

FIGURE 15.12. Heathrow fiber cable plant. (Reprinted from *Telesis*, ©1989, Bell-Northern Research.)

is more difficult to handle than the 62.5/1.25-μm multimode fiber, it is capable of supporting bandwidths in the Gb/s range to accommodate any future services.

In the central office, POTS and ISDN services are provided by a Northern Telecom DMS-100, while a specially developed digital video switch provides selectable CATV channels from the cable operator's headend, as shown in Fig. 15.13. A DMS-1 Urban Digital Loop Carrier was enhanced with optical line cards and placed in the CEV to serve the two more remote communities with custom-calling features and is connected to the central office via a Northern Telecom

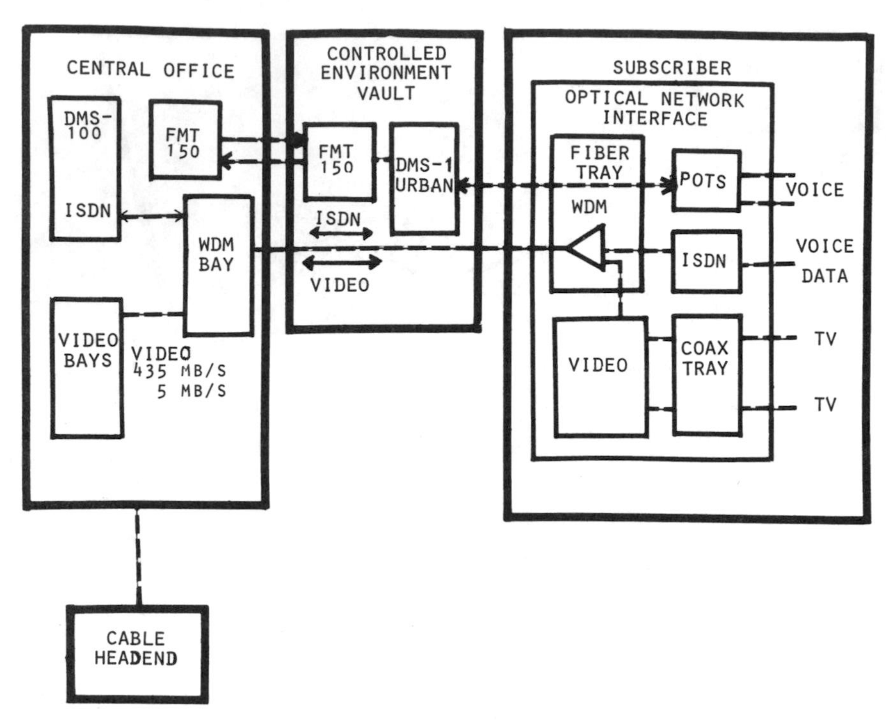

FIGURE 15.13. Heathrow fiber system diagram. (Reprinted from *Telesis*, ©1989, Bell-Northern Research.)

FMT-150 fiber multiplex link. One fiber provides POTS, while ISDN and video services are multiplexed optically on the second fiber through WDM technology. ISDN is carried at 780 and 890 nm, while video and its upstream control channel are carried at 1300 nm. The video downstream channel operates at 435 Mb/s, and the upstream channel operates at 5.12 Mb/s. Both fibers terminate at each residence in an Optical Network Interface (ONI), containing another WDM device, which performs optoelectronic conversion and connects ISDN signals to terminals via a T-bus and video channels to television sets via coaxial cable in a star configuration to as many as six set-top controllers. A POTS module connects two POTS lines to twisted-pair loops.

Transmission performance was excellent. Attenuation because of splices and connectors dominated the link attenuation budget. The impact was such that design allocated 3 dB of splice/connector loss for every 1 dB of fiber loss. Return loss for any splice or connector was specified at -40 dB, to guarantee bidirectional performance. The plan tested fusion and mechanical splices, and both produced results consistently better than 0.2 dB. Fusion splices consistently achieved

−40-dB return loss and typically exceeded −50 dB. Mechanical splices required careful polishing and application of index-matching gel to achieve the specified −40 dB. Mechanical splices were chosen for the local-loop environment, but some splices had to be remade. Subsequent development has resulted in greatly improved splicing procedures.

The video system consists of the video switch at the central office, input from the CATV supplier, and a dedicated optical path to the video module in each subscriber's ONI. The digital video switch can supply up to 256 subscribers with 64 program channels, although 10 channels are held in reserve. Each subscriber has the capability to view four channels simultaneously. The D/A conversion is performed in the set-top controller. The system provides a signal-to-noise ratio of 54 dB. All 64 program channels are available at each subscriber's line card in the video switch, enabling the switch to respond to the viewer's keyed-in 7-digit number for channel selection in less than 350 ms.

The video switch also contains A/D coders that distribute digital signals to the line cards via a digital video bus with 64 tracks, each with 107.52 Mb/s capacity. Each line card supports 2 subscribers and contains 8 video selectors in a single 6300-gate semicustom chip. Current technology would allow the line cards to be placed at a remote network element in the outside plant area, permitting all services to be provided on a single fiber. Since there are about 10 times more line equipment than trunk equipment in the system, cost tradeoffs between video compression and coding were analyzed. One option was to compress the video signal to 45 Mb/s before encoding, while the other was to encode linearly the uncompressed signal for distribution over the virtually unlimited fiber capacity. The linear-encoding option was selected to simplify the design of the D/A converter in the set-top controller. An algorithm encodes linearly a baseband video signal with a superimposed 4.5-Mb/s audio subcarrier into a 107.52-Mb/s digital stream. That bit rate was chosen to be an exact multiple of 2.56 Mb/s to match a digital switching bus. Since SONET bit rates now have been standardized, either 51.84 Mb/s or 155.52 Mb/s likely would be selected to fit into a SONET distribution system.

15.6. WIRELESS-ACCESS COMMUNICATIONS SYSTEM (WACS)

Another subscriber-loop innovation to reduce some of the distribution-cable requirement is the Wireless-Access Communications System (WACS), discussed in Section 7.10. The primary applications provided initially would be 32-kb/s connection-oriented circuit-switched service for voice and voiceband data. The WACS, as proposed by Bellcore, consists of low-power, demand-assigned digital radio links to replace the last portions of distribution loops in residential areas. A radio subscriber unit (SU) in each residence can communicate with a radio port

(RP) via an air interface. As many as 50 to 200 SUs may be served from one RP, depending upon the residential density and other radio systems in the area.[32] Studies have indicated that radio drops may prove to be cost competitive with wireline technologies by the mid 1990s.[33]

15.6.1. WACS System Overview

A proposed WACS configuration is illustrated in Fig. 15.14. The Radio Port Control Unit (RPCU) can be located at the end of the feeder cable or farther out into the distribution plant and provides multiplexing, speech coding, time-slot interchange, privacy coding, and supervision. A Time Division Multiplex/Time Division Multiple Access (TDM/TDMA) radio architecture has been proposed for the radio link. Several fixed-rate bit streams would be multiplexed onto a radio carrier for transmission from a port to several active SUs. In the upstream direction, subscriber transmissions would be time sequenced and synchronized on a common frequency for TDMA. In addition to radio links to residences, an arrangement of radio ports can connect a Personal Communications System (PCS) service provider (PSP) to the telephone network over an open interface.

15.6.2. WACS System Architecture

Radio port antennas need to be mounted 3 to 9 m (10-30 ft) above ground to provide adequate transmission and reception. Density of RPs is dependent upon transmit power and spectrum use. A *Fixed Wireless Access Subscriber Unit*, providing a standard 2-wire interface (tip and ring) to the local exchange network, accesses the network through the TDMA radio link. A *Portable Unit* is a low-power, hand-held unit that allows a PCS user to access the local exchange network via a PCS Access Service for Ports (PASP).

The air interface employs a three-layered protocol. Layer 1 is the physical layer, or radio link, between the subscriber layer and the RP. Its function is to ensure that data and signaling can be established and maintained as free from interference as possible. Multiplexing is a mixture of frequency division and time division, called multifrequency TDM/TDMA/FDD. Duplex channels are spaced 400 kHz apart in a frequency-reusing arrangement. Average RF output power to the antenna in a time slot is 100 to 200 mW, equivalent to 10 to 20 mW per active time slot averaged over time.[34]

Layer 2, the data-link layer, monitors and maintains the performance of the radio link, even when no call is in progress. The SU must maintain phase lock with the RP in order to be alerted and receive broadcast messages. The SU has four Layer 2 states. The SU is in the *Idle Unlocked* state when the SU power is off. When a time slot or bearer channel is seized for registration, call initiation, or response to an alert, the SU is in the *Active Locked* state. It is in the *Active*

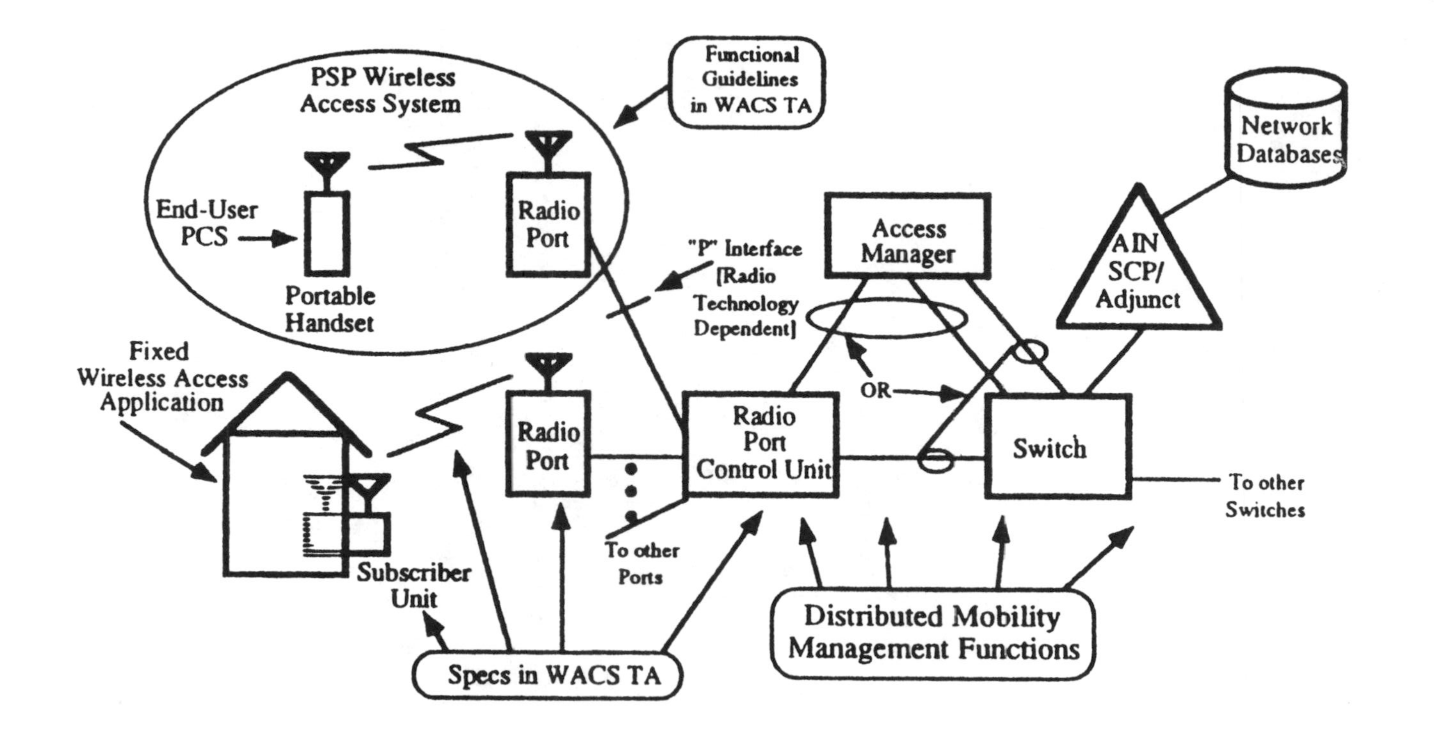

FIGURE 15.14. Proposed WACS configuration. ©1992, Bellcore. Reprinted with permission. The information reprinted from Bellcore was deemed current at the time of publication of this work, but since then the information may have been amended or updated. For more current information, refer to the current Bellcore publication.

Unlocked state when it is synchronized with an RP and is registered to the registration area served by that RP. When the SU is not synchronized to an RP but is actively scanning frequencies to acquire phase synchronization, it is in the *Active Unlocked* state. The initial version of WACS includes five physical channels for signaling:

- a dedicated 16-kb/s or 8-kb/s time slot, primarily for the alerting function;
- an unprotected downlink bit per frame per time slot for power control of the SU;
- an out-of-band channel in each time slot in every frame, consisting of 9 dedicated signaling bits that are not encrypted;
- inband, robbed-bit signaling in the 64 bearer bits in each time slot in every frame, which are encrypted.
- frame-synchronization signaling in the leading 14 bits in each time slot in every frame used for time slot identification.[35]

Layer 3, the network layer, provides the protocols that invoke network services by means of call-processing messages.[36]

15.7. TRENDS AND ISSUES

15.7.1. Multimedia Services via BISDN

There are many factors that must be considered in provision of broadband capabilities in the local loop in North America. Technologies thus far developed have been motivated by the desire of local-exchange carriers (LECs) to be able to offer specific services to residential subscribers to result in near-term revenue. The critical consideration is whether the LECs will be permitted to provide entertainment cable television over their subscriber loops. The FCC already has permitted LECs to provide transport for video services provided by others, but legislation precludes them from having an ownership interest in the video content within their own territories. If the present legislation, imposed to protect cable-TV companies, is lifted, then BISDN and other broadband services will become available to residences via optical fiber. As the cost of videotelephone equipment and service diminishes with large-scale production, that service could be transformed from a novelty into a high-demand service.

Cable-TV plant, primarily coaxial cable, already passes at least 90% of TV households in the United States. By retrofitting their cable plant with switches and electronics for two-way transmission, cable-TV companies can offer a variety of "video-on-demand" services. By employing wireless technology, cable-TV companies also will compete with local telephone companies for telephone network access. As coaxial cable is replaced gradually by optical fiber, provision of BISDN via cable-TV companies becomes even more viable.

To counter this likely trend, LECs are engaging in joint ventures with cable-TV companies to provide video services outside their territories. As competition between LECs and cable-TV companies for local-loop services increases, the case for repeal of the restrictive legislation becomes paramount to provide a level playing field for local-loop competition. If that legislation is not repealed, BISDN services will be provided, but they could be provided primarily outside the plant of the LECs, except for some physical transport.

15.7.2. Multimedia Services via Narrowband ISDN

Perhaps an even more compelling justification for repeal of that restrictive legislation can be seen in the services that can be provided over narrowband ISDN, 2B + D. Experimental research in Bellcore in a project called DEMON, an acronym for Delivery of Electronic Multimedia Over the Network, has proved that many such services can be provided by using both B-channels at 128 kb/s data rate.

The user's end includes a personal computer and a large-screen TV. A conventional television set may be possible in the future. The terminal equipment has the necessary intelligence for the various functions related to media quality and interactive communication. The DEMON project uses media compression, structural analysis, and time shifting to deliver multiple images.

Media compression is used to eliminate unnecessary data before images are transmitted. Some information can be downloaded to the receiving system while other information is displayed on the TV screen. Bellcore has used the international JPEG standard for compressing still images. That standard can enable still color images to be compressed at various ratios from 5:1 to 30:1 for various transmission rates and levels of picture quality. At a digital compression ratio of 5:1, an image is virtually indistinguishable from the original, and very little quality is lost at a 25:1 compression ratio, depending upon the contrast.

Structural analysis enables an image to be transmitted and stay on the video screen for the duration of the multimedia session, if desired. Other images can appear for a short time by scrolling through the information available. The digital data that make up the durable image has to be transmitted only once. The receiving system retains the data and uses it to refresh the screen as required. Multiple viewing windows can be provided by instructing programmed resources in the receiving system.

Time shifting involves transmitting data before it is actually to be displayed. For example, time-consuming data can be interleaved with introductory information as part of a multimedia session such as a travel presentation, real estate housing search, or catalog sales presentation. The receiving system then would reassemble the advance elements in the right order for use at the appropriate time.

Structural analysis breaks a presentation down into components that can be manipulated individually. This enables the peaks and valleys of image transmission to be spread out over time to utilize the communication channel more efficiently. Compression reduces some of the peaks in the data stream so they can be transmitted within the available channel. While this project is in the early stages of research, it has demonstrated the feasibility of transmitting multimedia information over narrowband ISDN.[37]

PROBLEMS

15.1. How do digital loops improve transmission performance?

15.2. How does digital signal processing serve to increase the bit rates that can be transmitted on unshielded twisted pair cable?

15.3. What is the primary source of impairment on High-Bit-Rate Digital Subscriber Lines?

15.4. In order to achieve satisfactory transmission performance in the 800-kHz range on HDSL, why are adaptive rather than fixed filtering and equalization used?

15.5. What are the main differences between High-Bit-Rate Digital Subscriber Lines and Asymmetrical Digital Subscriber Lines?

15.6. Why is near-end crosstalk (NEXT) not a concern in designing ADSLs? What impairments are of concern?

15.7. Why is it more costly to connect a Fiber-In-The-Loop (FITL) System to an analog switching system than to a digital switching system?

15.8. In FITL systems, what are the advantages and disadvantages of centralized powering of the Optical Network Units (ONUs)? of local powering?

15.9. Why is it more advantageous to use the 1310-nm rather than the 800-nm region of single-mode optical fiber in FITL systems?

REFERENCES

1. *Generic Requirements for High-Bit-Rate Digital Subscriber Lines*, Bellcore TA-NWT-001210, Issue 1, October 1991, p. B-21.
2. Ibid., p. 3-2.
3. Ibid., Appendix B.
4. Lawrence, R. W., "Switched Simplex High Bit Rate Services in Today's Residential Environment," *ICC '92 Conference Record*, Vol. 4, p. 211.1.3.
5. Fleischer, P. E., Lau, R. C., Lukacs, M. E., "Digital Transport of HDTV on Optical Fiber," *IEEE Communications Magazine*, August 1991, pp. 36–41.
6. Sibley, L. A., "The State of the Network—1992," *ICC '92 Conference Record*, Vol. 4, pp. 207.1.1–207.1.4.

7. "Copper's high-speed hurdles," *Bellcore Exchange*, March/April 1992, p. 5.
8. "On copper and glass," *Bellcore Exchange*, March/April 1992, p. 2.
9. TA-NWT-001210, Section 2.
10. TA-NWT-001210, p. 4-1.
11. Jones, D. C., "A New Parallel Adaptive Digital Filter Architecture for High Speed Digital Subscriber Line Application," *GLOBECOM '91 Conference Record*, Vol. 3, pp. 56.6.1–56.6.5.
12. Lawrence, R. W., pp. 211.1.1–211.1.6.
13. Bell Atlantic Seminar, Baltimore, MD, October 2, 1992.
14. Lawrence, R. W., p. 211.1.4.
15. Ibid., pp. 211.1.5–211.1.6.
16. Waring, D. L., "The Asymmetrical Digital Subscriber Line (ADSL): A New Transport Technology for Delivering Wideband Capabilities to the Residence," *GLOBECOM '91 Conference Record*, Vol. 3, pp. 56.3.1–56.3.8.
17. Sistanizadeh, K., "Spectral Compatibility of Asymmetrical Digital Subscriber Lines (ADSL)," *GLOBECOM '91 Conference Record*, Vol. 3, pp. 56.1.1–56.1.5.
18. Ahamed, S. V., et al., "A Tutorial on Two-Wire Digital Transmission in the Loop Plant," *IEEE Transactions on Communications*, November 1981, pp. 1561–1562.
19. Barton, M., "On the Performance of an Asymmetrical Digital Subscriber Lines QAM Transceiver," *GLOBECOM '91 Conference Record*, Vol. 3, pp. 56.7.1–56.7.5.
20. "Why DMT Should be Chosen for ADSL Now," T1E1.4 submission by Amati Communications Corp., T1E1.4/93-018, March 8, 1993.
21. Stephens, W. E., et al., "Transmission of STS-3c (155 Mbit/sec) SONET/ATM Signals over Unshielded and Shielded Twisted Pair Copper Wire," *GLOBECOM '92 Conference Record*, pp. 6.6.1–6.6.5.
22. *Additional Cable Specifications for Unshielded Twisted Pair Cables*, EIA/TIA/TSB-36, November 1991.
23. *Generic Requirements and Objectives for Fiber In The Loop Systems*, Bellcore TA-NWT-000909, Issue 1, December 1990, pp. 2.1–2.5.
24. *Optical Source Module for Fiber In The Loop (FITL) Systems*, Bellcore TA-NWT-000786, Issue 2, December 1991, pp. 7–8.
25. TA-NWT-000909, Issue 1, pp. 2.13–2.14.
26. Ibid., pp. 2.5–2.8.
27. Ibid., pp. 2.41–2.45.
28. Ibid., pp. 2.24–2.26.
29. Keiser, B. and Strange, E., *Digital Telephony and Network Integration*, First Edition, Van Nostrand Reinhold, p. 416.
30. Goarin, R., "Component Reliability Results from the Biarritz Field Trial and from "Plan Cable" Volume Deployment," *GLOBECOM '91 Conference Record*, Vol. 1, Dec. 1991, pp. 17.1.1–17.1.6.
31. Balmes, M., Bourne, J., and Mar, J., "The Technology Behind Heathrow," *Telesis*, 1989 two, pp. 31–41.
32. *Generic Criteria for Version 0.1 Wireless Access Communications Systems (WACS)*, Bellcore TA-NWT-001313, Issue 1, July 1992, pp. 1–4.

33. Ibid., p. 19.
34. Ibid., Section 5.
35. Ibid., Section 6.
36. Ibid., Section 7.
37. Rosenberg, J., "Multimedia delivery over public switched networks," *Bellcore Exchange*, January 1993, pp. 20–24.

Appendix A

North American Mixed-Network Technology

A.1. INTRODUCTION

Technological advances in both transmission systems and circuit-switching systems in the 1970s and 1980s were nothing short of revolutionary, and the revolution continues through the 1990s. The thousands of switching systems and millions of miles of transmission systems in the installed network base, however, cannot be replaced in a short time. By the time new technology reaches the implementation stage, even newer innovations are being developed. Replacement of obsolete plant is a function of economics and consumer demand for the new services that can be provided by the new technology.

As a result, when the latest designs of digital switching systems are integrated into the network, they must be able to "talk" to other switching systems of much older vintage when interconnected by transmission systems employing analog or digital technology, or a combination thereof. This means that they must be able to support not only the latest signaling systems but also signaling systems that have been used in the network for over a half century. Because new technology habitually is installed where it will more rapidly increase the return on investment, long-distance networks can be retrofitted much faster than local switching systems, considering the sheer difference in the quantities of installed plant. Thus, some telephone calls can traverse the network coast-to-coast and employ digital technology over all or most of the route, while others use various combinations of digital and analog technology in both switching and transmission systems.

Timing functions involved in connecting and disconnecting circuit-switched calls are critical. Each switching system must be able to use circuit interfaces that are fully compatible with each switching system to which it is connected by direct trunks. Such trunks may be provisioned on various types of transmission media. Therefore, an understanding of the functional characteristics of analog switching, transmission, and signaling systems is a distinct advantage in applying digital technology to a circuit-switching environment that still contains a substantial amount of analog technology.

All Bell operating companies (BOCs) in the United States, except minority-owned Cincinnati Bell and Southern New England Telephone Company, were divested on January 1, 1984 by the American Telephone and Telegraph Company (AT&T) as a result of the settlement of a Government antitrust suit. The BOCs were formed into seven regional holding companies and continue to provide local telephone service as regulated monopolies within defined *Local Access and Transport Areas* (LATAs). There are, however, many examples of direct links between subscribers and long-distance carriers, thus bypassing the local networks.

AT&T and other carriers provide competitive long distance, inter-LATA services (see Chapter 13). The circuit-switching environment described in this appendix is primarily that of the network comprised of AT&T, the BOCs, and many other service providers before extensive market penetration of digital switching equipment. The term *network* in the singular includes multiple, interconnected networks involving different service providers. The term *analog network* is used as a general description even though the gradual introduction of digital technology occurred, resulting in a *mixed network* (see Section A.3.2.5). For simplicity of textual structure, the present tense is used except for cases in which other tenses are specifically intended.

A.2. THE NORTH AMERICAN NETWORK

The public telephone network of Canada and the intra-LATA and inter-LATA networks in the United States utilize a unified numbering plan, coordinated traffic-routing plans, compatible signaling protocols, and similar transmission standards. Several inter-LATA carriers have cross-border connections to the Canadian network. Therefore, each combination of United States and Canadian networks appears to users as one network; the use of country codes for calls between the United States and Canada is unnecessary.

A.2.1. Network Numbering Plan

The numbering plan consists of two parts: a numbering-plan area (NPA) code, known as an area code, and a 7-digit telephone number. The NPA code is a 3-digit number, originally in the form *N* 0/1 *X*, where *N* is any digit 2 through 9, *X* is any digit, and the middle digit was a 0 or a 1. All *N*11 codes were excluded so as not to conflict with local service numbers. The 7-digit telephone numbers consist of a 3-digit central-office (CO) code plus a 4-digit station number. Central-office codes, originally in the form *NNX*, allowing a total of 640 codes per NPA, have been exhausted in some areas, and it became necessary to use codes in the NPA format as office codes. Future growth also may exhaust NPA codes. Therefore, a system has been devised to use the form *NXX* interchangeably for both NPA and office codes. All 63 *NN*0 codes were designated to be the last office codes assigned in any NPA and are to be assigned as both office and NPA codes in reverse sequence. The use of interchangeable NPA codes could occur as early as January 1, 1995. To identify the *NXX* codes as either NPA codes or office codes, long distance calls are prefixed by the digit *1* for direct-dialed, station-to-station calls and by the digit 0 for calls that require "operator" assistance.[1]

A.2.2. Network Routing Plan

Although all inter-LATA carriers use the same numbering plan, each uses its own network routing plan. The former Bell System network used a *hierarchical* routing plan with five hierarchical levels, as depicted in Fig. A.1. (Since divestiture, the AT&T portion of the network has consisted of inter-LATA switching centers and facilities.) Hierarchical levels were connected by *final trunk groups* and *high-usage (HU) trunk groups* as shown. (A HU trunk group was used to route sufficiently large volumes of traffic between two switching centers to justify a direct trunk group. A final trunk group was the last trunk group available to route traffic to any given destination.) Designations of switching centers indicated their positions within the hierarchy. Each of the *regional centers* had a final trunk group to every other regional center. Calls were routed over the most direct route to the destination office subject to strict routing doctrine, illustrated in Fig. A.1.

In Fig. A.1, assume that a call originating at end office A was destined for a subscriber served by end office J. Since there are no interregional HU groups below the class 3 level in this example, the call must progress up the hierarchy via final trunk groups through *toll center* B to *primary center* C. Primary center C has several possible routes to the destination region. The call will be offered first to HU trunk group C–L. If all trunks in that group were busy, alternate routes C–M, C–N, and C–D would be searched in that sequence for an idle trunk. The call would not be offered to HU trunk group C–E because of the doctrine requiring calls to be kept as low as possible in the hierarchy. The HU group C–N could be justified only because of first-routed traffic volume between C and N as a result of regional center N's performance of a primary-center switching function for the toll center homed on it. Once established, however, other traffic is permitted to use it. *Sectional center* D can use any of its routes to the destination region and the final route to regional center E if all HU groups are busy. Proceeding down the destination region hierarchy, the call would use the most direct route available to the destination switching center. *Skip-level* HU trunk groups, such as between

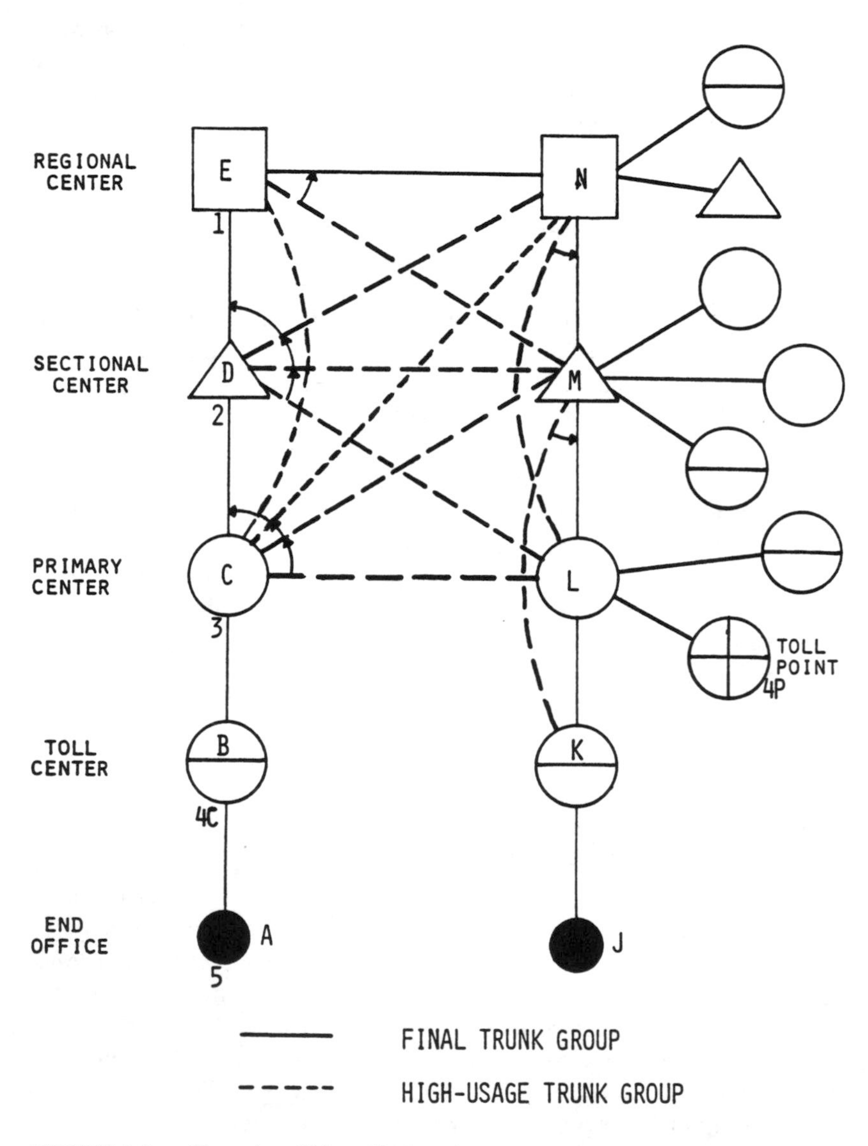

FIGURE A.1. Illustration of hierarchical routing.

regional center N and primary center L, would be used first with blocked traffic overflowing to final trunk groups.[2]

Strict hierarchical routing performed very well for switching systems with limited functionality. With different time zones and unpredictable traffic patterns, some parts of the network could be seriously overloaded with traffic while other parts contained relatively idle capacity. Some nonhierarchical routes could be planned to use capacity to compensate for time zone differences but not for unpredictable traffic patterns. As more sophisticated signaling was deployed in the network, the network could be programmed to identify idle paths to a caller's destination switching center. This resulted in adoption of a *dynamic routing* doctrine, enabling the network to find idle paths, within limits, to destinations in order to complete calls that would have been blocked under strict hierarchical routing doctrine.

AT&T completed installation of DNHR into its network in 1987. DNHR uses several layers of network-based intelligence to identify multiple alternate routes to a destination. In hierarchical routing, the number of alternate routes was limited, and a call could traverse two or three links only to find all trunks busy, blocking the call. With DNHR, calls are routed directly to the terminating AT&T office by way of not more than one DNHR "VIA" switch. Up to 256 reroutes can be in effect for a 4ESS switch, and each reroute can specify up to 7 VIA routes. One of the potential problems with alternate routing is a process called "stitching," the transmission of a call back and forth between two switching centers. In hierarchical routing, when a call was connected to an intermediate switching center, control of that call was transferred, thus allowing the intermediate center to apply alternate routing doctrine to the call. Hierarchical routing doctrine applied strict controls to alternate-routing choices at each switching center to prevent stitching. DNHR uses a feature called *originating office control.* If a call is alternate-routed to an intermediate office and that office is unable to complete the call, control of the call reverts, or is "cranked back," to the originating office so that it may examine other alternate routes.[3]

The superiority of DNHR over hierarchical routing was demonstrated by an analysis of traffic on Thanksgiving Day, one of the three peak traffic days of the year. Average network blocking, which had been 34% on that day in 1986, was only 3% on the same day in 1987. DNHR has permitted AT&T to carry the same traffic load as it could with hierarchical routing with 10 to 15% fewer facilities.[4] Other North American networks use similar nonhierarchical routing plans.

A.3. TRANSMISSION TECHNOLOGY

Transmission technology has developed over the years from a pair of bare copper wires strung between insulators on poles to large-capacity transmission systems.

A.3.1. Subscriber-Loop Transmission

A subscriber loop is that transmission medium connecting a network subscriber to a switching center. Nearly all subscriber loops involve copper cable pairs over all or various portions of the distance. In many cases, analog and digital multiplex systems, called subscriber carrier systems, have been deployed to better utilize existing cable plant.

A.3.1.1. Copper-Loop Environment

A typical subscriber cable contains twisted pairs of copper alloy conductors called *tip* and *ring*. Transmission characteristics are determined by four electrical properties. These are the series resistance of the conductors, the inductance of the conductors, the capacitance between the conductors, and the leakage resistance, or conductance, between the two conductors in a pair. Applying the principles of Thévenin's Theorem[5], a complex network can be represented by a simple equivalent circuit. Thus, a transmission line can be represented by a series of T-sections as shown in Fig. A.2.

Each T-section has the four properties of resistance, inductance, capacitance, and conductance, as shown in Fig. A.2a. Each combination of series and shunt

a.

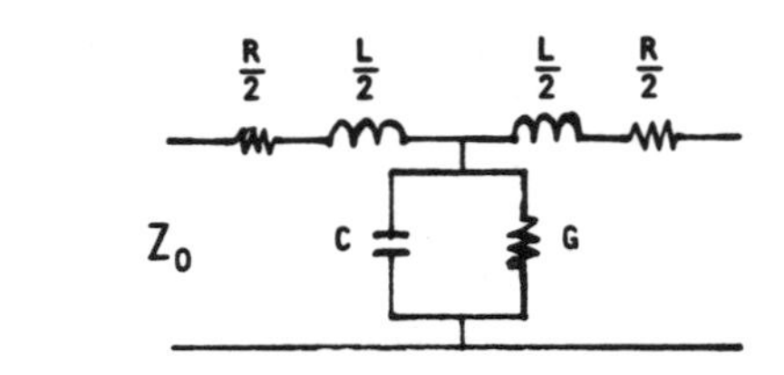

b.

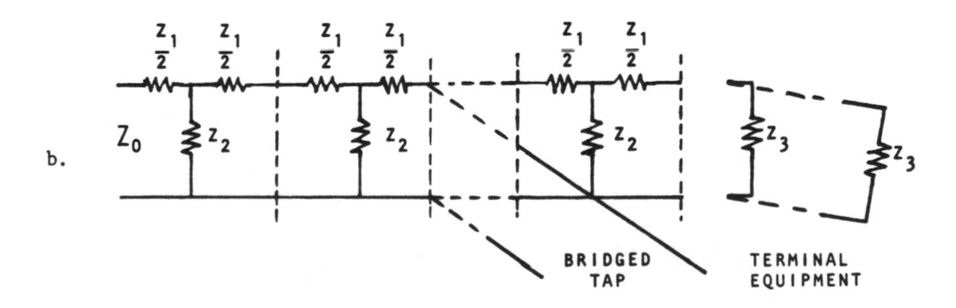

FIGURE A.2. Subscriber-loop characteristics.

properties comprise an impedance, which is repeated as additional T-sections are added to the line, as shown in Fig. A.2b.

Since the characteristic impedance of a transmission line of infinite length terminated in its characteristic impedance is expressed by the equation $Z_0 = \sqrt{Z_1 Z_2}$ the series impedance of a single section may be written as

$$Z_1 = R + j\omega L \tag{A.1}$$

The impedance of the parallel shunt branch is

$$Z_2 = \frac{1}{G + j\omega C} \tag{A.2}$$

Here, per unit length of line, R is the resistance in ohms, L is the inductance in henries, C is the capacitance in farads, G is the conductance in siemens, and $\omega = 2\pi f$, where f is the frequency in hertz.

If all transmission lines were sufficiently long that their input impedance approximated their characteristic impedance, and if all terminal equipment were to have the characteristic impedance of the lines, the transfer of power would be optimum. Metallic conductors in the outside telephone-loop plant serving subscribers, however, have many variables. Loop resistance may vary from near zero to about 2000 ohms or over 3000 ohms with range extenders. On some long loops, loading coils insert lumped inductance at regular intervals to compensate for the additional capacitance. The use of different wire gauges causes impedance mismatches which cause signal reflection. There may be one or more branch circuits, known as *bridged taps*, as shown in Fig. A.2(b), which affect the impedance of the line.

Aerial cable is highly susceptible to power-line interference having magnitudes of up to 50 V rms longitudinal and 5 V rms metallic. Such interference contains primarily the odd harmonics of 60 Hz, and such harmonics in high-tension power lines may have energy values as high as 10 000 watts, compared to telephone-circuit energy which may be only a fraction of a watt. Also, there may be a difference in the ground potential at the telephone company central office and the subscriber's premises of as much as 3 V. Finally, lightning surges may induce voltages up to 5000 V peak and currents up to 1000 A peak with a maximum rise time on the order of 10 μs and a decay time of 10 μs or more.

All such characteristics impair the transmission, not only of voice and data signals, but also of signals used to control the telephone network. The effect of these impairments is increased when multiple sets of terminal equipment are bridged to the line at the premises of the subscriber.

In preparation for the deployment of Basic Rate Access (BRA), Integrated Services Digital Network (ISDN) (see chapter 14), work was initiated in the 1970s to analyze the copper-loop plant and to formulate a plan to utilize as much of it as feasible. Electronic components can overcome the ubiquitous transmission impairments inherent in twisted-copper-wire-pair cable. Advances in VLSI made it possible to amass the large number of needed components on a small number of silicon chips. Plans were made to upgrade the loop plant whenever maintenance or renovation became necessary and by evolving a set of guidelines for new plant. Those guidelines were formalized in the 1980s, and the carrier serving area (CSA) was defined as a plant administration area surrounding a wire center or a remote terminal.[6] Design guidelines require that all loops be nonloaded, and that the length of loops composed of 26-ga conductors and the length of bridged taps be limited. The total length of bridged taps must be less than 762 m (2.5 kft), with no single bridged tap longer than 610 m (2 kft). The total length of multigauge cable containing 26-ga conductors must not exceed:

$$12 - [(3 \times L26)/(9 - LBTAP)] \text{ kft} \tag{A.3}$$

where

L26	is total length of 26-ga cable excluding bridged taps; and
LBTAP	is total length of all bridged taps.

This can be summarized as follows. Loops within a CSA that contain *any* 26-ga cable pairs are limited to a total length of 2.74 km (9 kft). Loops that do not contain any 26-ga cable pairs may be extended to 3.66 km (12 kft).[7] With loop extenders, certain of these loops can be extended to 5.5 km (18 kft). These criteria apply in most local exchange areas in North America.

A loop survey, conducted in 1983 to characterize the loop plant to ascertain the extent to which more sophisticated services could be deployed, found that about 25% of all copper loops were equipped with loading coils and, therefore, precluded the use of adaptive electronics to carry digital data at the required rate. The survey did disclose that about 48% of the sampled loops were compatible with CSA guidelines.

A.3.1.2. Subscriber Carrier Systems

Subscriber-loop plant is categorized as feeder plant and distribution plant. Feeder plant comprises the main cable routes extending outward from central offices to major junctions or to cross-connect points. Distribution plant comprises the subscriber cables, or branch cables, that branch from that point to serve individual groups of subscribers.

As central office serving areas experienced significant growth, both feeder cables and distribution cables reached a very high level of occupancy. Electronic systems were developed to multiplex several subscribers on feeder cable pairs, thereby extending the life of the feeder cable plant and avoiding the cost of some feeder cable expansion. Those subscriber carrier systems were first developed using analog technology but later employed digital technology. Electronic terminals carry multiple subscriber channels from a central office to a remote terminal, at which point distribution cables carry the channels to individual subscribers over copper cable pairs.

A.3.2. Network Transmission

Network transmission involves the transmission of subscriber message content from an originating subscriber through the network to a receiving, or terminating, subscriber. All transmission systems are imperfect and impair the quality of transmitted information to some extent. A major objective of transmission-system engineering is to control impairments in such a way as to achieve an acceptable compromise between transmission quality and the cost of providing the system.

A.3.2.1. Frequency-Division Multiplexing

Frequency-division multiplexing (FDM) (see Section 1.2.2) was standardized to use 4-kHz channels. Filters provided guard bands around effective channel bandwidths of approximately 3.2 kHz. The 4-kHz channels were organized into 12-channel groups. Channel banks provided the frequency modulation and demodulation for the 12-channel groups and provided pilot frequencies for temperature regulation and testing. Five groups were combined into a 60-channel basic supergroup in the frequency band 312-552 kHz with additional pilot frequencies. Ten supergroups were combined into a basic master group of 600 channels in the frequency band 564-3084 kHz, and six mastergroups comprised one basic jumbogroup of 3600 channels in the band 564 to 17548 kHz.

Although FDM systems were designed originally for metallic cable pairs, the cable electrical characteristics limited them to a capacity of 12 channels, later increased to 24 channels. The higher-capacity systems employed two coaxial tubes, one for each direction of transmission. Several pairs of tubes were placed in a single sheath, with one pair being kept in standby status in case of failure of the electronic equipment (terminals and repeaters) associated with any traffic-carrying pair. FDM systems soon were adapted to microwave radio transmission and have been used on satellite links.

Note: Digital transmission systems began to be deployed in 1962 (see Section 1.2.3), primarily on short-haul, high-capacity trunk routes. Known as T-carrier systems (see Chapter 5), they were directly substituted for FDM systems, and the network was still considered to be analog (from a transmission viewpoint) until digital transmission became more widespread, at which time it became known as a *mixed* network.

A.3.2.2. Transmission Impairments

Several characteristics of transmission systems degrade the quality of signals. Among these are noise, loss, amplitude/frequency distortion, crosstalk, echo, and delay distortion. The effects of some impairments can be minimized by quality design of equipment components and circuitry. Others, such as loss, noise, and echo, are inherent in telephone transmission systems (see Section 6.1.2) and can be dealt with only by a carefully controlled transmission plan. Their effects on analog transmission are described in the following paragraphs.

Noise is any signal or interference on a circuit other than the signal being transmitted. There are several types of noise: thermal noise, impulse noise, noise resulting from other multiplexed message channels, quantization noise, and internal noise resulting from a condition of imbalance. Noise always has amplitude and is amplified or attenuated along with the signal in analog systems. The level of inherent circuit noise increases with circuit length; i.e., when the length of a circuit is doubled and the noise per unit length is constant, the total noise level is increased by approximately 3 dB. Design control keeps inherent noise levels well below signal levels. In the event of a weak signal, however, a lower signal-to-noise ratio reduces the quality of the received signal. Noise is measured with a meter that accounts for the interfering effect of noise at different frequencies, the overall interfering effect of combined noise elements, and the effect of noise bursts of different duration. Noise measurements are referred to a reference noise level, defined as 10^{-12} watt, or -90 dBm. A 1000-Hz (or 1004-Hz) tone at a level of -90 dBm would give a meter reading of 0 dBrn. Noise measurements using C-message frequency weighting are expressed in terms of dBrnC and, when referred to 0 TLP, are expressed in terms of dBrnC0.[8]

Loss is another impairment that is inherent in all transmission systems. Loss is the attenuation of power that occurs in traversing a circuit and generally is proportional to circuit length. Amplification can overcome loss, but amplification must be carefully controlled to provide satisfactory end-to-end communications. As signals proceed along a transmission path, they are attenuated, and amplifiers are inserted at intervals to restore the proper levels. If signals arrive at a transmission junction or a switching system at levels that are too high, they are attenuated to a predetermined level by the insertion of attenuation pads. Metallic subscriber loops have varying amounts of loss, depending upon the length and gauge of the conductors used. On extremely long loops, lumped inductance, called *loading coils*, are inserted in the loop at specified intervals to counteract loss due to the capacitance between the conductors.

The third impairment discussed in this subsection is *echo*. Echo is a reflected signal returned with sufficient magnitude and delay to be perceived as distinct from the signal directly transmitted. In telephony, echo occurs generally because of an imperfect impedance match between a 2-wire circuit and a 4-wire circuit. This is illustrated in Fig. A.3.

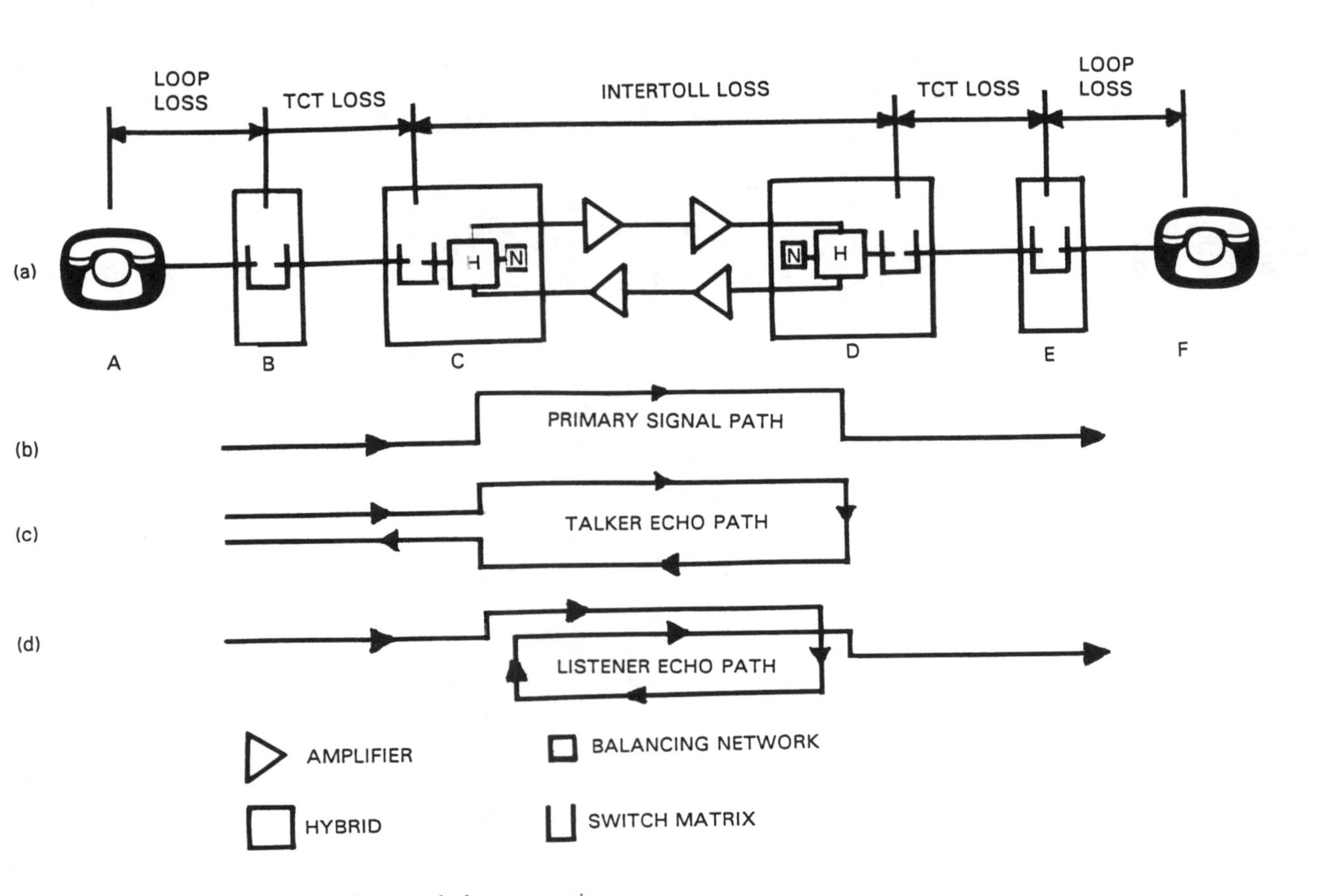

FIGURE A.3. Echo paths in long distance telephone connection.

- Figure A.3a depicts a typical analog toll connection proceeding from telephone A through local central office B, tandem offices C and D, and local central office E to telephone F. The subscriber loops and tandem connecting trunks (TCT) are 2-wire circuits, although TCTs may be either 2-wire or 4-wire circuits. In this example, a single 4-wire intertandem trunk is shown between the tandem switching systems. This requires 2-wire/4-wire conversion at each of the tandem offices. The hybrid transformers habitually use a compromise balancing network that does not perfectly balance the impedance of the 2-wire path. Since the impedances of tandem connecting trunks are closely controlled, the impedance mismatch at the hybrid transformer in a tandem office is less than if it were located at the serving central office, because of the wide variance of subscriber-loop impedances.
- The primary signal traverses the path shown in Fig. A.3b. When the signal arrives at tandem office D, theoretically half the signal power proceeds down the 2-wire path while the other half is dissipated in the balancing network N. In actual practice, however, the imperfect impedance match permits some of the signal power to flow across the hybrid transformer and return to the talker, as shown in Fig. A.3c. If the impedance mismatch at tandem office C is significant, part of the echo signal power can cross the hybrid transformer and be heard by the listener, as shown in Fig. A.3d. The hybrid transformer core and copper losses, added to the approximately 3-dB loss absorbed in the network N, make a total hybrid loss of about 3.5 dB. The principal sources are hybrid imbalance and impedance mismatch.[9]

The effect of echo primarily is a function of transhybrid loss and delay, and delay is a function of the distance between the hybrid transformers and of the propagation velocity of the transmission facility. Tolerance of echo varies with individual people and is influenced by echo level and delay. If circuit amplification is too high, or if the hybrid unbalance is significant, or both, echo signals can circulate around the listener echo path and cause circuit instability, or singing. A near-singing condition causes speech to have a hollow sound as if talking into a barrel.[10]

A.3.2.3. Control of Impairments

It is apparent from the foregoing discussion that transmission impairments must be controlled. A conflicting relationship exists. Amplification overcomes loss but increases the effect of noise, and loss controls echo. Since echo is the most objectionable impairment encountered in telephone networks, control of echo must receive primary consideration. Echo can be kept to a tolerable level through control of loss and transmission delay. The amount of loss needed to control echo increases with increasing delay, but the amount of loss must not reduce the signal below an acceptable level. Since an amplifier affects both signals and noise, noise must be kept sufficiently low that amplification does not cause interference with the signal. Talker echo is controlled by impedance matching and loss insertion in switched connections. Listener echo is controlled by impedance matching and hybrid balancing techniques.

A.3.2.4. Via Net Loss Plan

A plan developed in the early 1950s to allocate loss for echo control in analog connections was called the via net loss plan. *Via net loss* (VNL) was defined as the loss in dB assigned to a trunk to compensate for its added propagation delay, terminal delay, and loss variability. VNL can be calculated for each analog trunk by

$$\mathrm{VNL(dB)} = 0.102D + 0.4 \tag{A.4}$$

where

0.102 is the one-way incremental loss in dB per ms (reduced to 0.1 for practical calculations);
D is the echo path delay in ms;
0.4 is a factor added to compensate for loss variability.

Loss can compensate for echo only to the point at which loss itself begins to degrade circuit quality below acceptable limits.[11]

A.3.2.5. Network Transmission Plans

A.3.2.5.1. Analog Network Transmission Plan. Losses in a long-distance connection were associated with three connection categories: the originating and terminating subscriber loops, all switching system networks, and the interconnecting trunks. Because loop and originating switching losses are highly variable, network transmission plans focused on controllable trunk losses. A loss value, calculated by the VNL technique, was assigned to each trunk to compensate for echo and other impairments. The sum of VNL values for each trunk in a switched connection comprised the transmission loss value for that connection. The loss objective for connections from originating end office to terminating end office was VNL + 5 dB. Because of loss variability, 5 dB of additional loss was assigned to each end-to-end, switched connection. The maximum loss between the originating and terminating switching centers was limited to 9 dB by use of echo suppressors.

In addition to the loss between switching centers, the subscriber loop on each end of the connection also contributed to the overall connection loss. Typical subscriber metallic loops have an average loop loss in the range of 3.5–4 dB. Some rural areas have loop losses in the range of 8.5–10 dB.

A.3.2.5.2. Mixed Network Transmission Plan. With the widespread use of digital switches and digital transmission in the network, the resulting mixed network, comprising a combination of analog and digital trunks in switched connections, required modification of the loss plan. It was no longer practical to assign loss on a per-trunk basis. A fixed-loss plan was developed to simplify echo control as the network evolved. This was accomplished by assigning a fixed loss of 6, 3, or 0 dB to be inserted in the receive path at the terminating switching center. The inserted loss is 6 dB for inter-LATA connections, 3 or 6 dB for intra-LATA connections, and 0 dB for intraoffice connections. VNL loss was assigned to analog trunks remaining in the network.

A.4. SIGNALING TECHNOLOGY

Signaling is the transmission of address and other switching information between subscribers and switching systems or between switching systems. The three principal types of signals used in telephony are supervisory signals, address signals, and information signals.

A.4.1. Supervisory Signaling

Local supervisory signaling is used to determine the busy or idle condition of the subscriber's line. When the line is idle, it is said to be *on-hook*, and a busy line is said to be *off-hook*. Most supervisory signaling systems supply negative battery across the tip and ring conductors of the metallic subscriber line at the central office; i.e., the ring conductor is more negative than the tip conductor. A typical battery voltage is –48 V dc, but it may be as high as 105 Vdc with range extenders. Subscriber carrier systems provide ac signaling between carrier terminals and dc signaling on the subscriber end.

Interoffice signaling systems may employ either dc or ac to provide supervision on metallic trunks between switching systems. When dc signaling is used, it functions in substantially the same way as on subscriber lines, except that the on-hook and off-hook conditions are established under the control of the switching equipment at each end of the circuit. All, or virtually all, interoffice trunks in the toll networks are 4-wire circuits with a separate transmission path in each direction. These trunks generally use ac supervisory signaling. In such systems, a single-frequency, 2600-Hz tone is placed on each side of the trunk when the trunk is idle (or on-hook). An off-hook condition is identified by removing the tone.

Both local and interoffice supervisory signaling are described in more detail in the following subsections.

A.4.1.1. Subscriber Line Supervisory Signaling

There are two categories of supervisory signaling used on subscriber lines: *loop start* signaling and *ground start* signaling.

A.4.1.1.1. Loop-Start Signaling. The principle of operation of loop-start signaling is shown in Fig. A.4. For supervision on loop-start lines, the central office connects the ring conductor through a current sensor to negative battery and the tip conductor to positive battery. The positive end of the battery supply may or may not be grounded. The subscriber interface appears to the central office to be a very high resistance in the idle (*on-hook*) state, presenting a virtual open-circuit condition. When the switchhook contacts in the telephone set are closed, current flow is detected by the sensor, the line is interpreted as being off hook, and dial tone is applied when the central office is ready to receive address digits. In some cases, the central office is not ready to receive digits until 70 ms after dial tone is applied. No current flows in the on-hook state. Loop current varies inversely with the external circuit resistance. Some central-office switching systems open the tip and ring conductors, called open-switch intervals (OSIs), during call set-up and during other changes of call state. OSIs are required to be less than 350 ms in duration, with at least 100 ms between OSIs.[12]

To alert the subscriber to an incoming call, the central office applies an ac ringing signal, superimposed on a dc voltage, usually for 2-second intervals separated by 4-second silent intervals. In some cases, the ringing signal will be applied during the silent interval. In such cases, if the called party lifts the handset, thereby "going off-hook," the central office treats the off-hook as an answer, not as a request for service.

The advantages of loop-start signaling are low cost and simplicity. Its disadvantages are that no disconnect signal is provided to either party when the other party goes on hook, and an occasional OSI may be confused with a disconnect by some sophisticated terminal equipment.

A.4.1.1.2. Ground-Start Signaling. Lines that serve switching equipment, or other automatic equipment at the subscriber's premises, generally operate on a ground-start basis. The principle of operation of ground-start signaling is shown in Fig. A.5. In the idle state, the central office provides battery, negative with respect to ground, through a current detector to the ring conductor of the subscriber line. The tip conductor at the central office is open, and the subscriber terminal equipment presents an open line to the central office. To initiate an outgoing call, the subscriber terminal equipment seizes the circuit by grounding the ring conductor through a resistance. The central office detects current flow in the ring conductor and responds by connecting the tip conductor to ground through a current limiter, applying a dc voltage between tip and ring, and returning dial tone. Some switching systems apply tip ground and dial tone simulta-

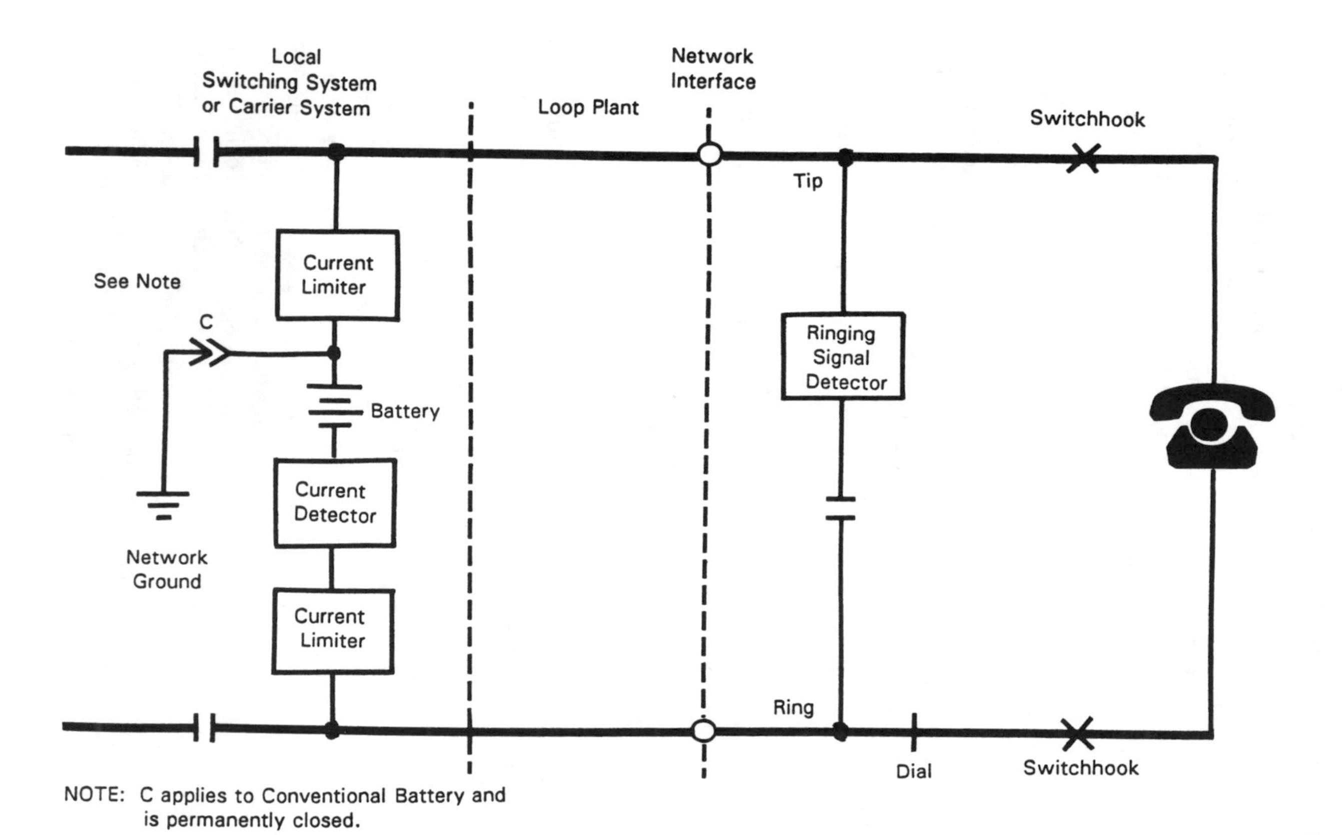

FIGURE A.4. Principle of operation of loop-start signaling.

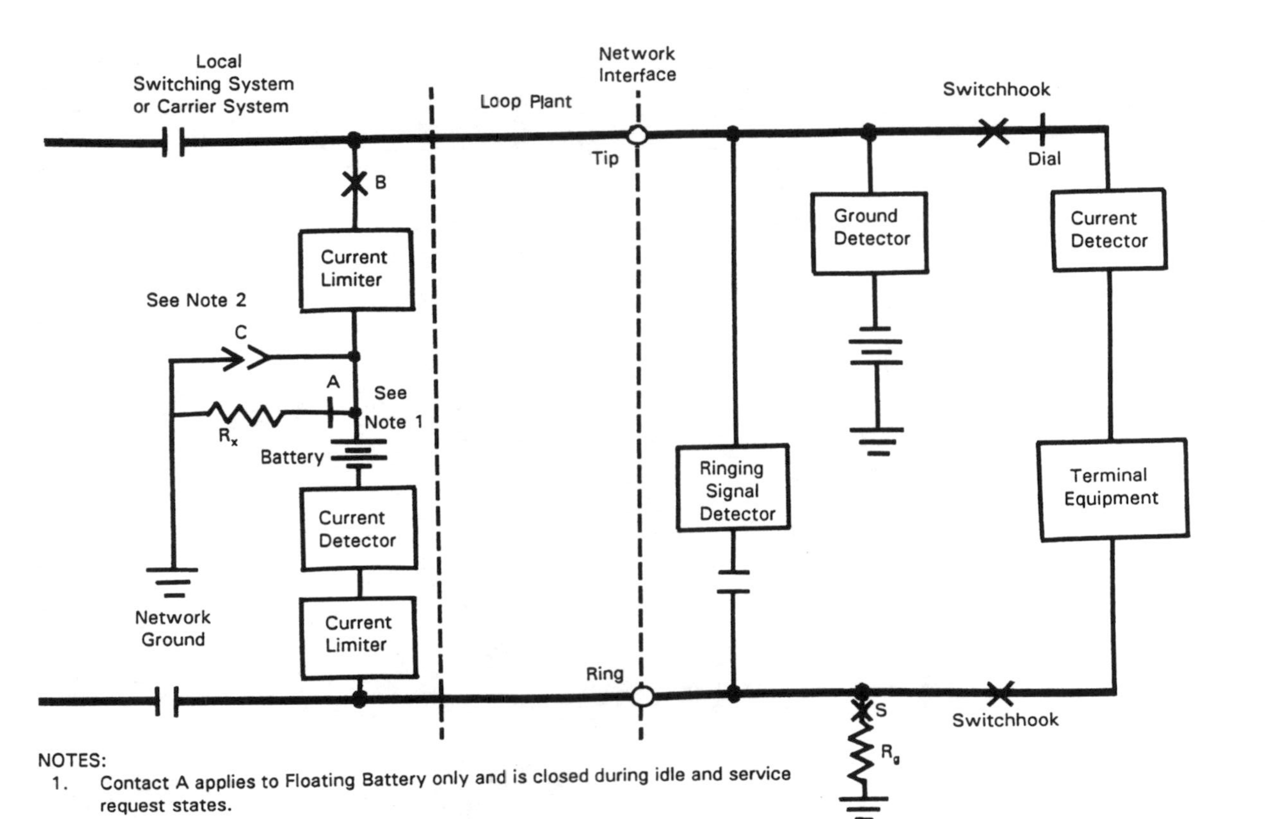

NOTES:

1. Contact A applies to Floating Battery only and is closed during idle and service request states.

2. Contact C applies to Conventional Battery only and is permanently closed.

FIGURE A.5. Principle of operation of ground-start signaling.

neously, or dc voltage and dial tone simultaneously, or both. Upon detecting the central office tip ground, tip-and-ring voltage, or dial tone, the subscriber terminal equipment removes the ring ground and closes the loop, establishing supervision as in loop-start signaling. Incoming calls are detected by the subscriber terminal equipment either by receipt of ringing signals from the central office or by detection of central office tip ground.[13]

Ground-start supervision has three major advantages over loop-start supervision for terminal equipment that dials automatically: (1) it tends to reduce the effect of simultaneous seizures of the line at both ends; (2) the central office tip ground can be used as a start-dial signal, eliminating the need for dial tone detection circuitry; and (3) removal of central office tip ground upon returning to idle state provides a positive disconnect signal to the terminal equipment.

A.4.1.2. Inband Interoffice Supervisory Signaling

Inband supervisory signaling on interoffice trunks uses dc supervision on metallic trunks and ac supervision on multiplexed trunks.

A.4.1.2.1. Loop Supervision on Trunks. Several types of loop supervision are used on metallic trunks between switching systems. In all cases, dc signaling states are superimposed upon the same conductors used for voice transmission. To ensure compatibility, the electrical characteristics, signaling protocols, and sensitivities must be maintained within certain ranges. The maximum loop resistance, or working limit, is dependent upon the sensitivity of a particular trunk circuit to dc pulsing and to the steady-state dc signal. The working limit for any trunk group is the limit that applies to the least sensitive trunk circuit at either end. Loop-signaling trunk circuits should function with conductor leakage resistance as low as 30 000 ohms.

A reverse-battery trunk, in its simplest form, is a one-way trunk that sends opens and closures from the originating end and sends battery and ground reversals from the terminating end. The originating end may substitute series-aiding battery and ground during dial pulsing to increase the working limit of a pulsing-limited trunk. At the originating end, the outgoing trunk circuit indicates an idle (on-hook) condition by maintaining an open condition with at least 30 000 ohms resistance across the tip and ring conductors. An off-hook condition is signaled by bridging the tip and ring conductors with not more than 500 ohms resistance. At the terminating end, the incoming trunk circuit signals an on-hook condition by connecting the tip conductor to ground and the ring conductor to -48 Vdc. To change to an off-hook condition, the incoming trunk circuit reverses the polarity of the tip and ring conductors. During the reversal, the electrical state is not defined. Therefore, the transition period should be kept as short as possible but not longer than 5 ms. To ensure versatility, both incoming and outgoing trunk circuits should have options for reversing the tip and ring polarities. There are several variations of reverse-battery trunks.

A.4.1.2.2. Tone Supervision on Trunks. Inband signaling systems on multiplexed channels use single frequency (SF) signaling units that generate a tone in the voice-frequency band. The SF tone is injected into each transmit side of the 4-wire equivalent path. The presence or absence of tones is transformed into dc signals to and from the switching equipment trunk circuits. In modern systems, the single frequency is 2600 Hz. Formerly other frequencies in the voice band were used. Some SF units could be arranged to send one frequency in one direction while another frequency was used in the other direction, thus enabling the units to be used on 2-wire metallic facilities as well as 4-wire facilities. Conventionally, the 2600-Hz tone is on when the trunk is on-hook and is off when the trunk is off-hook.

A 2600-Hz band-elimination filter in the receive path of some units blocks the tone so that the calling party does not hear the tone when his receive path is on hook. This permits call-progress signals, such as busy tone and audible ring, to be heard by the calling party. The filter is inserted when tone is detected.[14]

A significant problem with SF supervision is its susceptibility to mutual interference between voice transmission and signaling. SF units are subject to false operation, known as "talk-off," from voice sounds that are near the signaling frequency. The selection of 2600 Hz for the signaling frequency tends to minimize the probability of talk-off, but it does occur. Protective measures include specification of a minimum sustained duration of signaling tone to operate the unit and detection of voice-frequency energy other than the signaling frequency to block operation.

A.4.1.3. E & M Lead Control

Some means of communication is required between switching-system trunk circuits and the SF signaling units, used for inband supervisory signaling over interoffice trunks. That means is provided over two metallic leads, designated *E lead* and *M lead.* The E & M designations are derived from their identification on circuit drawings. E & M leads normally are used only within a building or for short distances between buildings and do not appear in the outside plant. They are separate from the transmission paths. E & M leads control the on-hook and off-hook supervisory indications which appear in the transmission paths. The M lead controls the signals from the switching-system trunk circuit to the signaling unit, and the E lead controls the signals from the signaling unit to the switching-system trunk circuit. Some have supposed that the lead designations were derived from the key letters in "transMit" and "recEive;" this is a good memory crutch, irrespective of its validity. The flow of supervisory signals between two switching offices using E & M lead control is shown in Fig. A.6 along with the signaling protocols.

For one switching office to signal off hook to the connected office, the originating-office switching-system trunk circuit operates the M lead toward its SF signaling unit. This causes the SF unit to remove its 2600-Hz tone from the trunk. Removal of the tone causes the SF unit in the receiving office to operate its E lead toward the switching-system trunk circuit, and the off-hook state is recognized.

Five types of E & M lead interfaces have been identified, and all may be found to be in service currently. All types operate the E & M leads by application or removal of battery or ground. Type I interface is a 2-wire interface used with electromechanical switching systems. Battery is supplied by the trunk circuit, and ground is supplied by the SF unit. Type II interface is a 4-wire interface, using separate battery and ground leads, and was designed for use with the AT&T 4ESS switching system. Both battery and ground are supplied by the trunk circuit for

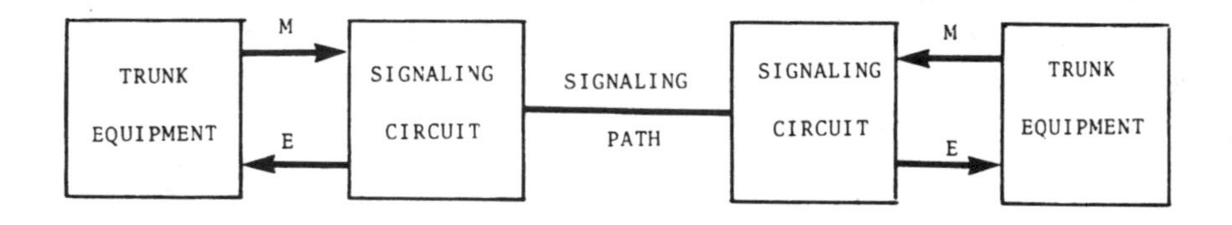

	SENT SIGNALING STATES					
	TRUNK TO SIGNALING CIRCUIT			SIGNALING TO TRUNK CIRCUIT		
TYPE	LEAD	ON-HOOK	OFF-HOOK	LEAD	ON-HOOK	OFF-HOOK
I	M	GROUND	BATTERY	E	OPEN	GROUND
II	M	OPEN	BATTERY	E	OPEN	GROUND
III	M	GROUND	BATTERY	E	OPEN	GROUND
IV	M	OPEN	GROUND	E	OPEN	GROUND
V	M	OPEN	GROUND	E	OPEN	GROUND

FIGURE A.6. E & M lead control signals.

the E lead and by the SF unit for the M lead. Type III interface is an improved Type I interface, using separate battery and ground leads for the M lead, and was designed for use with AT&T analog electronic switching systems. Battery for the E lead is supplied by the trunk circuit, and ground for the E lead and both battery and ground for the M lead are supplied by the SF unit. Type IV interface is an improved Type II interface with a fully symmetrical arrangement but with a slightly different protocol. Battery and ground are supplied by the trunk circuit for the E lead and by the SF unit for the M lead. The fifth type of E & M interface was standardized by CCITT as a symmetrical 2-wire interface. The trunk circuit supplies battery for the E lead and ground for the M lead, while the SF unit supplies ground for the E lead and battery for the M lead.

A.4.1.4. Control of Disconnect

Switched connections normally are held and disconnected under control of the calling party. This concept is commonly known as *calling party control.* Disconnect signals, however, may be initiated by either the calling party or the called party by returning to an on-hook condition at the network interface. Switching equipment must be able to differentiate on-hook signals of different duration. During established calls, temporary on-hook indications can be caused by "hits" on a transmission facility, a legitimate flash to exercise a feature or recall an operator, or a disconnect followed by a reseizure. Hits need to be ignored, but a flash and a disconnect need to be detected. Timers begin timing after the on-hook state is detected by the line circuit or trunk circuit. The timing requirements to differentiate hits, flashes, and disconnects vary slightly with the type of switching system. Any on-hook condition that endures longer than hits or flash timing is interpreted as a disconnect.

Treatment of a disconnect depends upon the type of switching equipment, whether the calling or called party disconnects first, whether the disconnect affects a line circuit or a trunk circuit, the type of trunk circuit, and whether the other party remains off-hook. Time must be allowed to restore the line or trunk to idle state before making it available for reuse. This time varies from immediately to about 37 seconds, depending upon the variables listed above. If both parties have disconnected, however, the guard timing, to ensure trunk release before reseizure, is measured in milliseconds, according to the type of switching equipment and the round-trip signaling time between switching offices.

A.4.2. Address Signaling

When the dialed digits of a called telephone number are transmitted to a switching center, the switching equipment must be prepared to receive and interpret them. It is not cost-effective for a switching system to have physical digit receivers permanently wired to all lines and trunks. Therefore, in analog systems, that equipment is grouped into a pool of common equipment that is connected to each line or trunk only as long as necessary to receive address digits. It then is disconnected and made available for use on another line or trunk. The switching center must signal the calling subscriber or distant switching center when it is ready to receive digits. The calling subscriber is informed by placing dial tone on the line. For trunk signaling, the distant office is notified by protocols described in Section A.4.2.5. Subscribers generally transmit address digits to the central-office switching equipment in either of two ways.

A.4.2.1. Dial-Pulse Signaling

Dial pulses are generated by means of a rotary dial that opens and closes the tip and ring conductors at specified intervals. Some telephones use pushbuttons to generate dial pulses automatically, and some private branch exchange (PBX) equipment automatically generates dial pulses toward the central office. As the line conductors are opened and closed, the line current is interrupted, typically deenergizing and then reenergizing a relay in the central-office digit receiver.

The number of interruptions corresponds to the numerical value of the digit being dialed, except that ten interruptions are used to designate the digit zero. To ensure correct interpretation of dial pulses (and all other address signals) in the presence of transmission line distortion, timing specifications are stringent. Pulsing speed is nominally ten pulses per second, with a tolerance range of 8–11 pulses per second; the pulse period is nominally 0.1 second, which comprises the open-circuit, or *break*, portion as well as the following closed-circuit, or *make*, portion. The structure of dial pulses in a digit train is illustrated in Fig. A.7. The ratio of the break duration to the pulse period, called the *percent break*, typically is a nominal 61%, with a tolerance range of 58–64%.

The interdigital on-hook interval must be long enough to distinguish between two series of pulses representing digits but short enough so that it is not interpreted as a disconnect signal. When pulsing into a step-by-step office, the interdigital

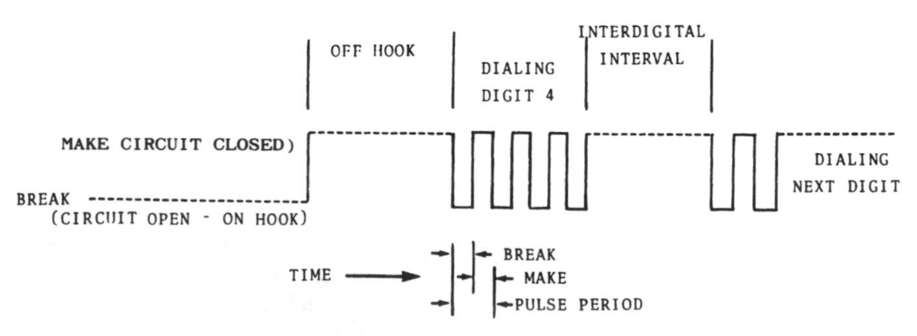

FIGURE A.7. Dial-pulse address signals. (© AT&T, 1980.)

interval must be long enough to allow the selector switch to step to the maximum level and also to allow time to condition a relay in the next switch. That interval should be at least 700 ms. For manual pulsing by subscribers or operators, the interdigital interval is under manual control. (The normal human tendency in using a rotary dial is to space the digits at least one second apart.) To overcome the distortion which is always present on cable pairs, central-office dial-pulse receiving equipment must have greater tolerances than those specified for transmitters.

When dial pulses are formed by dial-pulse generators and transmitted automatically to the central office, more stringent requirements can be designed into the equipment to minimize the holding time of dial-pulse receivers. The first digit should be outpulsed between 70 ms and 5 seconds after receipt of dial tone, and the interdigital interval should be between 700 ms and 3 seconds.

Dial-pulse signaling also is used on some interoffice trunks, especially those involving older switching equipment, such as step-by-step. Dial-pulse transmitters should delay sending dial pulses for at least 70 ms after receiving a start-dial signal and, to avoid registration of transients, the distant office should not register any pulses for 30–60 ms after sending the start-dial signal. Transmitters should send pulses at the rate of 9.8–10.2 pulses per second with a 58–62% break and an interdigital interval of 300–700 ms. To compensate for line distortion, digit receivers should be able to register dial pulses at a rate of 8–12 pulses per second with a 42–84% break.

A.4.2.2. Dual-Tone Multifrequency Signaling

Dual-tone multifrequency (DTMF) signaling is a form of address signaling using pairs of frequencies transmitted over the subscriber line. DTMF shortens dialing time, and thence connection time, and enables longer subscriber loops to be used. It also permits end-to-end signaling after a connection is established. One of the frequencies is selected from a low group of four frequencies, and the other is selected from a high group of three frequencies. A fourth high-group frequency is used in certain private-network applications and is reserved for future use in the public switched-telephone network. Character representation of the tone pairs is shown in Table A.1. The "star" (*) and "number" (#) symbols are used to activate special features.

DTMF transmitters used by subscribers may be powered by the subscriber loop or by a local power source. Table A.2 illustrates some of the more important electrical requirements for signals from DTMF transmitters powered from a local power source. The requirements for loop-powered transmitters are the same, except that greater variations in levels is allowed because of the wide variation of loop current encountered on subscriber lines. Pulse duration can be defined as the time interval that begins when the last tone of the pair exceeds 90% of that tone's steady-state value and ends when either tone of the pair falls below 90% of its

TABLE A.1. DTMF Signaling Frequencies Dual-Tone Multifrequency Signaling

		Nominal high-group frequencies (Hz)			
		1209	1336	1477	1633[a]
Nominal low-group frequencies (Hz)	697	1	2	3	
	770	4	5	6	
	852	7	8	9	
	941	*	0	#	

[a]Not used in the public switched network.

TABLE A.2. Subscriber DTMF Transmitter Requirements

Nominal level per frequency	−6 to −4 dBm
Maximum level per frequency pair	0 dBm
Maximum level difference between two frequencies in pair (high freq. ≥ low freq.)	4 dB
Frequency deviation	±1.5%
Automatic dialer pulsing rate:	
minimum pulse duration	50 ms
minimum interdigital time	45 ms
minimum cycle time	100 ms

steady-state value. Central offices also can be equipped to send DTMF signals to subscriber PBXs arranged for direct inward dialing (DID) (sometimes called Centrex).

DTMF receivers must operate with greater tolerances than transmitters to allow for line distortion. The line impedance should satisfy voice transmission requirements. The receiver should execute a code validity check to ensure that two frequencies are present, one and only one low-group frequency plus one and only one high-group frequency. The receiver should register frequency pairs in which each frequency is within 1.5% of nominal value. The receiver should reject tone pairs having either frequency more than 3.5% from nominal value. It may accept or reject signals between 1.5% and 3.5% from nominal values (this is a *no man's land* that accommodates design and manufacturing tolerances). Speech and other nonsignal energy should be rejected to avoid distortion of signals.

Pulsing speed and duration of pulses must be confined within certain limits. the receiver should register signals with digits and interdigital intervals as short as 40 ms but should ignore those less than 23 ms duration. When pulses between 23 ms and 40 ms are received, digit registration is subject to equipment tolerances. Because the first digit always will be received while dial tone is being sent, the receiver must be able to register that first pulse in the presence of dial tone.

Variable impedances and lengths of subscriber loops (see Section A.3.1.1) can cause the dial-tone level at the input of the DTMF receiver to be approximately as high as 0 dBm.

Problems can be experienced when DTMF signaling is used for end-to-end signaling on either public or private networks. During call processing, some switching systems reverse polarity on the tip and ring conductors. If a polarity guard is not provided on the DTMF generator, the output is disabled. Some single-slot coin telephones use positive, rather than negative, battery on the ring conductor to prevent fraud and to permit proper coin totalizer operation. Again, a polarity guard is needed, or special operational features are required. When a call is made over either a public or private network circuit using echo suppressors and dial tone is sent from the distant switching system, as in dialing off network or dialing into a long off-premises PBX trunk, the first DTMF tone pair can be attenuated. Depending upon the relative levels of the DTMF and dial-tone signals and the type of echo suppressor used, the effect on the first digit may be blockage of the entire digit or attenuation of only the first portion.

A.4.2.3. Multifrequency Signaling

While DTMF signaling generally is confined to subscriber lines, multifrequency (MF) signaling is the primary inband address signaling system used on interoffice, multiplexed trunks and on some DID subscriber lines. MF, like DTMF, uses two simultaneous frequencies, but MF tones are selected from a group of only six frequencies spaced 200 Hz apart. The 15 possible combinations represent the ten digits 0 through 9, signals indicating the beginning (KP) and end (ST) of pulsing, and three signals for special network use. The general-purpose representations used in the North American network are shown in Table A.3.

The major advantages of MF signaling over dial pulsing are accuracy, speed, and signaling distance. A significant disadvantage is susceptibility to fraudulent use because the signaling is in the message channel. Critical transmitter and receiver specifications ensure a very high degree of accuracy of address digits. Frequency tolerance is held to $\pm 1.5\%$, and transmit power is held to within 1 dB of -7 dBm0. The total power of any extraneous frequency components should be at least 16 dB below the power of each signal frequency and should be less than –40 dBm during signal-off time. Each frequency should start and end within 1 ms of each other. The nominal digit rate of seven digits per second in the American network is increased to ten digits per second for international inband signaling, using CCITT Signaling System No. 5.[15] The duration of the start-of-signaling (KP) pulse should be 90–120 ms. All other pulses should be 58–75 ms duration. The interpulse interval, defined as the time when both tone levels are below -35 dBm, must be at least 53 ms to permit accurate registration at the receiver. The total cycle time, including rise time and fall time, should be

TABLE A.3. Multifrequency Signaling

Digit	Frequencies (Hz)
1	700 + 900
2	700 + 1100
3	900 + 1100
4	700 + 1300
5	900 + 1300
6	1100 + 1300
7	700 + 1500
8	900 + 1500
9	1100 + 1500
0	1300 + 1500
Control Signals	**Frequencies (Hz)**
KP-Preparatory for digits	1100 + 1700
ST-End of pulsing sequence	1500 + 1700

at least 110 ms, but a design time of 120 ms is desirable to allow for manufacturing tolerances.[16]

MF signaling is not only faster and more accurate than dial pulsing, it can be used over much greater distances. Distortion of the ac signals, however, is largely a function of distance and the quality of the transmission path, and receivers must be designed to register signals in the presence of some distortion without sacrificing accuracy.

To ensure that each pulse contains two, and only two, valid frequency components, MF receivers are required to perform a code-validity check. Receivers must "unlock" upon receipt of a KP signal of at least 55 ms duration, but may respond to a KP signal as short as 30 ms. Transmission distortion may cause the two frequencies to be shifted in time by as much as 4 ms. Receivers should accept pulses at up to ten digits per second, provided that each frequency component is at least 30 ms in duration. Envelop delay in the transmission path may cause the two frequency components to be shifted in time relative to each other by as much as 4 ms. Interpulse intervals should be at least 25 ms; however, digits comprised of minimum pulse time combined with minimum interpulse time may not be registered accurately.

The presence of circuit noise may affect receipt of MF signals. Receivers should register address signals in the presence of circuit noise at a signal-to-noise ratio of 20 dB and in the presence of impulse noise at a signal-to-noise ratio of 12 dB. Power-line-induced noise at levels of 81 dBrnc0 at 60 Hz and 68 dBrnc0 at 180 Hz should not impair digit registration.[17]

A.4.2.4. Control of User Address Signaling

Control of user address signaling is integral to loop-start signaling and ground-start signaling, as discussed in Sections A.4.1.1 and A.4.1.2, respectively.

A.4.2.5. Control of Interoffice Address Signaling

The three principal protocols used to control the inband transmission of address digits over interoffice trunks are *immediate dial, delay dial*, and *wink start*. On-hook and off-hook supervisory states are described in Sections A.4.1.2.1 and A.4.1.2.2. Seizure of a trunk by a calling office is signaled by changing its on-hook state to off-hook. The called office should recognize the seizure and prepare to receive address signals as rapidly as practicable with reasonable protection against registration of transients (see Section A.4.2.7). An important consideration in interoffice signaling involves a call-by-call integrity check to ensure trunk continuity and, on trunks with loop supervision, the presence of battery and ground with on-hook polarity. On trunks with tone supervision, the receipt of a start-dial signal constitutes an integrity check.

Immediate-dial operation is an uncontrolled method of sending address signals. It must be used when the calling or called office is a nonsenderized step-by-step switching system. In that case, the calling office sends the off-hook seizure signal for at least 150 ms, immediately followed by the entire dial-pulsed address. That ensures sufficient time for the incoming selector in the step-by-step office (see Section A.5.3.1.1) to be conditioned to react to receipt of the first address digit and to overcome distortion in the signaling and trunk equipment. Immediate dial also may be used with common-control offices having fast links and liberally provided digit receivers. A signaling-integrity check, which tests continuity and polarity, is used by common-control offices on immediate-dial calls to progressive-control offices over metallic facilities using loop reverse-battery supervision to identify facility troubles before a connection is completed. When a common-control system uses immediate-dial trunks derived from multiplexed facilities, there is no signaling-integrity check, and the common-control switching equipment outpulses blindly. Any existing trunk trouble usually goes undetected, and a call failure results.[18]

The original method of controlled-address signaling was *delay-dial operation.* The calling office seizure is an on-hook-to-off-hook transition. Upon receipt, the called office immediately returns an off-hook, delay-dial signal. The calling office then does not transmit address digits until a start-dial signal is received from the called office. When the called office is ready to receive address digits, it sends a start-dial signal by an off-hook-to-on-hook transition.

Originally, delay-dial operation did not use an integrity check, causing excessive call failures. The calling office, after sending a seizure (off-hook) signal, timed for 75 ms or 300 ms depending upon the type of trunk, then examined the supervisory state of the transmission-receive path from the called office. If on-hook, address signals were outpulsed; if off-hook, outpulsing was delayed until the receive path from the called office changed to an on-hook state. The called office, upon detection of an off-hook (seizure) signal, sent a delay-dial (off-hook) signal and maintained it until a digit register was ready to receive address digits, at which time a start-dial (on-hook) signal was sent to the calling office. A major disadvantage of this type of controlled signaling is that, if the called office does not send a delay-dial signal, for whatever reason, the calling office, upon finding the supervisory state on-hook, will outpulse address digits that may not be received by the called office. This type of operation is unsuitable for synchronous satellite trunks; it results in excessive call failures on terrestrial trunks, and should be avoided.

To correct this problem, an integrity check was added to delay-dial operation. This requires an incoming response from the called office in the form of a delay-dial signal. Thus, the calling office *expects* a delay-dial signal and does not outpulse address digits without receiving it. The delay-dial signal does not have to be returned within a given time limit (for example, 300 ms); it can be delayed longer (for example, in heavy traffic conditions) because the calling office will not outpulse until it is received. To further increase the effectiveness of this type of controlled outpulsing, the delay-dial signal must be at least 140 ms in duration, and the start-dial (on-hook) signal must not be sent until at least 210 ms after receipt of the seizure (off-hook) signal from the calling office. This added time allows the circuit to stabilize after the supervisory-state transitions. Because of circuit distortion, however, the calling office should be capable of recognizing an off-hook condition as short as 100 ms as a valid delay-dial signal. In most electronic and digital switching systems, the delay-dial signal is generated by the common-control software associated with the trunk equipment.[19]

Wink start operation is similar to delay dial with integrity check except for timing requirements. It can be seen that, if the calling office expects a *timed* delay-dial signal, the called office does not have to send it immediately upon recognizing the seizure signal, but it can wait until it is ready to receive address digits. In this case, the timed off-hook-on-hook signal is called a *wink*. The duration of the wink should be 140–290 ms, and the end of the wink should not occur earlier than 210 ms after receipt of the seizure signal (actual delay varies with switching-system design). To compensate for transmission distortion, the calling office should recognize an off-hook signal of 100–350 ms as a wink. Some switching equipment is capable of starting the wink as soon as the trunk seizure is recognized, and the timing requirement ensures that the wink received by the calling office will be at least 100 ms in duration. On two-way trunks, off-hook signals lasting longer than the intervals specified can result if there is a nearly simultaneous seizure of the trunk from both ends.[20] The latter case is called *glare*.

A.4.2.6. Glare Detection and Resolution

Transmission of the seizure of a two-way trunk by a calling office is received at the called office at a time determined by the propagation delay of the trunk. Time also is required for the called office to detect the seizure, mark the trunk busy in its memory, and condition a receiver to receive address digits. Then, and only then, is the called office ready to send a start-dial signal to the calling office. If the called office should seize the same trunk for an outgoing call before it has marked the trunk busy as a result of detecting the seizure from the other end, each switching system will be sending a seizure (off-hook) signal toward the other. The off-hook condition would appear as a wink of extended duration, and each office would wait for the end of the wink to send address digits.

Protocols for glare detection and resolution depend upon the type of switching equipment terminating each end of the trunk group. The following protocol is typical in North American networks: with delay-dial operation, it is not possible to distinguish between unavailability of a digit receiver and simultaneous seizures; therefore, it requires a minimum of 4 s to assume a glare condition. Wink-start operation is the preferred method of operation for detecting and resolving glare. Any wink signal over 350 ms in duration should be interpreted as glare (with some switching equipment, this can vary up to about 600 ms).

When two offices are connected by a two-way trunk group, one of the offices usually is designated as *control office* for that trunk group. In a hierarchical network, the higher-ranking office typically is the control office to give preference to calls trying to complete down the hierarchy. Whether using delay-dial or wink-start operation, when glare is detected, the control office, if one is assigned, maintains the off-hook condition toward the other office until a start-dial signal is received. The other office, having detected glare, backs out of its outgoing-call attempt, prepares to receive address digits, sends a start-dial signal to the control office, and tries it on another trunk (or returns reorder tone). If a control office is not designated, the first office to detect glare backs out of the connection, sends a start-dial signal, and retries its outgoing call on another trunk. If only one of the two offices can detect glare, it backs out, enabling the other call to be completed.[21]

With reverse-battery supervision (see Section A.4.1.2.1), glare cannot be detected because the loop-closure detector is replaced by a dry bridge (no battery) when the trunk is seized, but glare can be resolved. When an office seizes a reverse-battery trunk, a timer times the waiting period for a wink or delay-dial signal from the called office. It typically times out after 16–20 s and retries the call on another trunk. If the office is not the control office, that timing is reduced to 4–8 s. This enables the office to release the trunk before the control office times out, resolving the glare condition.

A.4.2.7. Signaling Transients

Signaling transients may comprise short changes of on-hook or off-hook state, or they may be short changes of amplitude of tone signals. The rather slow operation of electromechanical switching equipment permits it to be relatively immune to short changes of on-hook or off-hook state during signaling. The much faster switching operations performed by electronic switching systems, both analog and digital, however, increase their susceptibility to such transients, especially when the switching system determines the dc supervisory state by scanning or sampling. As long as older switching systems remain in any part of the network, design of supervisory circuitry should consider the characteristics of older equipment and provide transient-suppression circuitry. Transients resulting from Types I and III M-leads typically can be suppressed by placing a zener diode between the M-lead and ground. If Types I and III M-lead sensors are inductive, a capacitor network generally will provide sufficient transient suppression. If transient suppression is needed for Types II and IV interfaces, however, it must be provided by other than a capacitor network.[22]

Step-by-step trunk circuits and outgoing repeaters generate a spurious pulse just before the first pulse of each digit. The false pulse may begin 6–10 ms before the first valid pulse and may last 2–4 ms. A pulse-repeating relay may cause a momentary *make* contact of about 10 ms during the normal break period, occurring generally about 5–10 ms after the valid make closure is released. Mechanical relays can generate contact chatter for 10–15 ms. A trunk circuit that uses battery-and-ground dial pulsing can cause a false on-hook pulse up to about 10-ms duration when it replaces battery and ground with an inductor bridge during the interdigital period. Step-by-step outgoing-trunk circuits typically use an inductive off-hook bridge with loop reverse-battery supervision. When another office returns answer supervision by reversing battery and ground, the drop in current caused by the inductance may be interpreted by the incoming-trunk circuit as a short on-hook pulse.

One significant transient can be encountered when connected to any type of switching system if loop reverse-battery trunk supervision is used. Some loop reverse-battery outgoing trunks use an idle-circuit termination consisting of resistance and capacitance in series across the tip and ring conductors. Although the capacitance should not exceed 0.5 μF, some outgoing-trunk circuits use as much as 2–3 μF capacitance. If the calling party disconnects first and the idle termination has been bridged across the tip and ring conductors when the called party disconnect is sent, the reversal of battery and ground by the incoming-trunk circuit causes the idle-circuit termination capacitive charge to reverse, giving the appearance of a short off-hook signal. This should not be interpreted as a new seizure.

Short changes in amplitude of tone signals may occur in DTMF transmitters. Such transient that exceed 1-dB deviation from the steady-state signal level should be restricted to the first 5 ms of the tone and should not exceed 12 dB above the absolute peak of the steady-state signal.[23]

A.4.3. Network Information Signals

A wide variety of audible-tone signals and recorded announcements is used in the network. A precise tone plan, comprised of four pure frequencies of 350 Hz, 440 Hz, 480 Hz, and 620 Hz, is used (usually in pairs) to derive some 45 separate indications by varying the tones used and their patterns. These tones are used to provide call-setup progress, indications in connection with optional service features, and for other purposes. The call-progress signals of primary interest to callers are described in Table A.4. The difference between the two frequencies of the tone pair, the *beat frequency*, gives the tone its sound to callers or operators.[24]

Some other tones used in the network have special meaning. A receiver-off-hook (ROH) tone is used by local switching systems, typically with announcements and other tones, to notify a subscriber that the handset has been left off-hook. Though the tones used in the network vary with the local provider and the type of switching equipment used, the most common tone (often called a "howler") consists of four frequencies, 1400 Hz, 2060 Hz, 2450 Hz, and 2600 Hz. Power is applied for 0.1 s and removed for 0.1 s repetitively to provide a 5-pulse-per-second tone. The tones are mixed, amplified, and injected into the subscriber line at a maximum level of 0 dBm total power.[25]

While tones and announcements are used to inform callers of conditions encountered on dialed calls, they also are used by service providers to identify call-failure conditions in the network. Standardized special-information tones (SIT) enable automated call-detection equipment to identify recorded announcements that follow. Such recorded announcements are preceded by a sequence of three precise tone segments with intervening silent intervals. Five standardized codes

TABLE A.4. Call Progress Signals

	Frequency (Hz)					
Tone name	350	440	480	620	Level per Frequency	Cadence
Dial tone	×	×			–13 dBmO	Steady
Line busy tone			×	×	–24 dBmO	60 IPM (tone on 0.5 s and tone off 0.5 s)
Reorder/"no-circuit" tone			×	×	–24 dBmO	120 IPM (tone on 0.25 s and tone off 0.25 s)
Audible ring tone		×	×		–19 dBmO	10 IPM (tone on 2 s and tone off 4 s)

are used in the intra-LATA networks to identify call-failure announcements reflecting vacant code (VC), intercept (IC), equipment reorder (RO′), no circuit (NC′), and ineffective other (IO). Two SIT codes, reorder (RO″) and no circuit (NC″), are used to indicate circuit conditions in the exchange-access or inter-LATA networks.[26]

A.4.4. Inband Signaling Techniques

Inband signaling on North American trunks derived from multiplexed transmission systems uses SF supervisory signals with a tone-on-idle protocol (see Section A.4.1.2.2). Address signaling is either DP or multifrequency (MF) (see Sections A.4.2.1 and A.4.2.3). MF signaling on domestic calls is at the nominal rate of seven pulses per second for most switching systems.

Inband signaling on most international trunks uses CCITT Signaling System No. 5 between foreign and North American gateway switching systems. That system uses *compelled* two-frequency supervisory signals and pulsed MF address signals at the nominal rate of ten pulses per second, using the same six frequencies as are used in domestic signaling. A seizure is indicated by transmitting a 2400-Hz tone to the called gateway switching system. When the called system is ready to receive address digits, it responds by sending a 2600-Hz tone to the originating switching system which then proceeds to outpulse the address digits. The two frequencies also are used for other purposes during call setup and disconnect sequences. Compelled supervisory signaling automatically enables detection of simultaneous seizures by distinguishing between the start-dial signal and the seizure signal. Domestic switching systems, at which international calls are originated or switched through, must outpulse more digits than are required for domestic calls; therefore, the address digits are outpulsed in two stages. The first-stage outpulsing routes the call to an international-gateway switching system. At each switching node, digits may be added or deleted for routing purposes. After a transmission path is established between the originating and gateway switching systems, the second-stage outpulsing occurs on an end-to-end basis. After the gateway office receives the ST pulse at the end of the first-stage outpulsing, it delays at least 700 ms to allow time for switching transients to dissipate, and it then sends a second start-dial signal—an off-hook signal of at least 400–ms duration. The second start-dial signal is followed immediately by a 480-Hz tone that persists until the ST signal is received or until the MF receiver times out. Since some international calls still are dialed by operators, the purpose of the tone is to alert operators to start sending the second-stage address pulses.

A.4.5. Switched Access for Inter-LATA Carriers

Local exchange carriers (LEC) provide means for subscribers to access *inter-LATA carriers* (IC) either directly from an end office to an IC's *point of presence* (POP) in each LATA or via an *access-tandem* (AT) switching system. Access tandems also can perform sector-tandem and LATA-tandem functions. ICs may have more than one POP in a LATA.

Prior to divestiture of the BOCs, competing long distance carriers (other than AT&T) could access their subscribers through *exchange network facilities for interstate access* (ENFIA). Access was provided via line or trunk connections. In the years following divestiture, access to inter-LATA carriers has evolved into two basic methods. The consent decree required that all inter-LATA carriers be accorded *equal access* to the local networks of the BOCs. Thus, equal access has become the dominant access method. Subscribers can presubscribe to the services of a specific inter-LATA carrier, in which case they dial 1 or 0 plus the called telephone number. The BOC then routes that call to the presubscribed carrier of choice. Inter-LATA carriers can provide their own billing services or contract with the BOCs for billing to their subscribers. Subscribers can use dial-pulse or DTMF signaling. A subscriber who has not presubscribed, or one who is calling from a coin telephone or from someone else's telephone, can access his IC by dialing 101XXXX. The expansion from 10XXX has been made mandatory by mid-1995. The XXXX digits comprise a 4-digit code identifying the IC.

Figure A.8 shows the network configuration options and transmission objectives for inter-LATA access. Inter-LATA carriers may optionally access LATA end offices in two ways. Direct inter-LATA connecting trunks (DICT) may be provided between end offices and the IC POP. DICTs are designed for 3-dB fixed loss and have the same quality as predivestiture toll connecting trunks. A BOC AT switching system can be interposed between an end office and an IC POP. The AT is connected to end offices by *tandem connecting trunks* (TCT) designed to the same standards as DICTs. *Tandem inter-LATA connecting trunks* (TICT) between the AT and an IC POP are 4-wire, intertoll-grade facilities designed for 0-dB loss. If subscriber lines also are served by the AT, it must have 3-dB switchable pads to provide the 3-dB loss between the IC POP and the end office.[27]

A.4.6. Equal-Access Dialing and Signaling Plan

An inband-signaling plan was devised for use with equal-access configurations. While the presubscribed subscriber is dialing the called telephone number, the *equal access and office* (EAEO) transmits the calling party telephone number to the IC POP. When the subscriber dials the last address digit, the called number is transmitted.[28]

An illustration of a typical originating signaling sequence on a direct connection is shown in Fig A.9. Subscriber-dialed numbers in parentheses are contingent upon method of calling-party access and type of call. Following receipt of the dialed area code, the EAEO seizes a DICT to the IC POP and, after receiving a wink-start signal, outpulses the identification of the calling party if the IC has arranged to receive *automatic number identification* (ANI). Otherwise, the EAEO

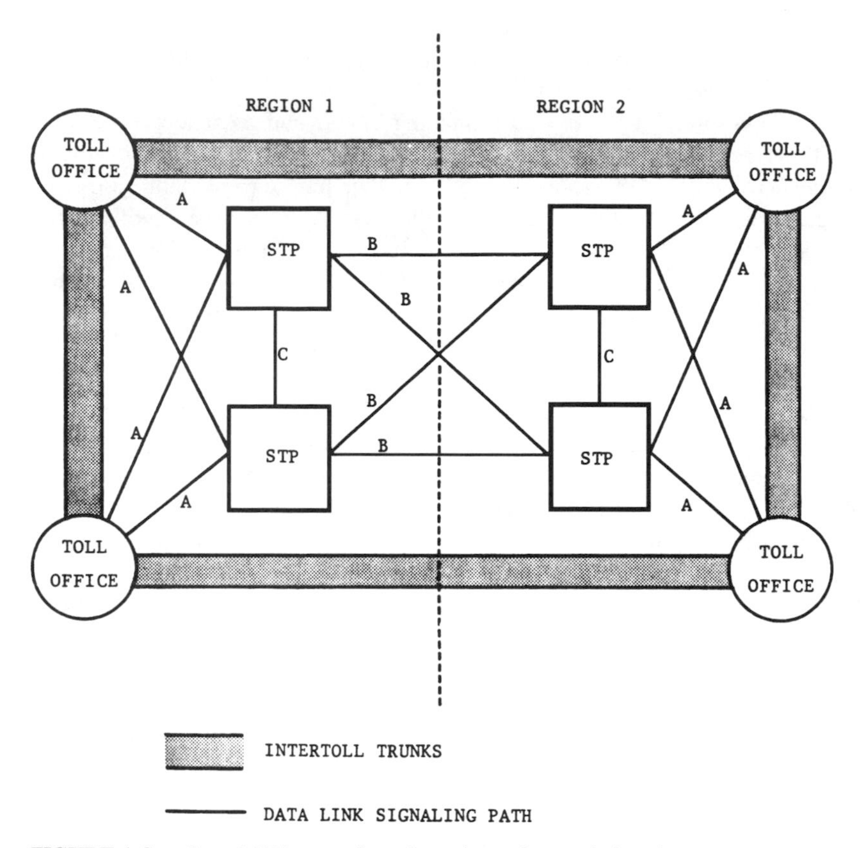

FIGURE A.8. Inter-LATA network configuration and transmission plan.

outpulses KP + ST (see Section A.4.2.3). ANI information digits (II) indicate the category of the call. Upon receipt of the last four digits of the called-party telephone number (address), the EAEO outpulses the address of the called party. The IC POP acknowledges the address by transmitting a wink to the EAEO. When the called party answers, answer supervision is forwarded to the EAEO.[29]

The originating signaling sequence for a tandem connection is illustrated in Fig. A.10. Additional signaling is required for the EAEO to identify to the AT the IC designated to handle the call. That information is transmitted in the form shown. The *OZZ* code designates one of four possible trunk groups between the AT and the IC POP. This is necessary because the IC may desire separate trunk groups to be used for certain categories of calls, such as operator-handled calls, information calls, etc. A 4-digit code in the form XXXX identifies the IC designated to handle the call. The remaining portion of the signaling sequence is the same as for direct connections, except that the AT cuts through a path to the IC

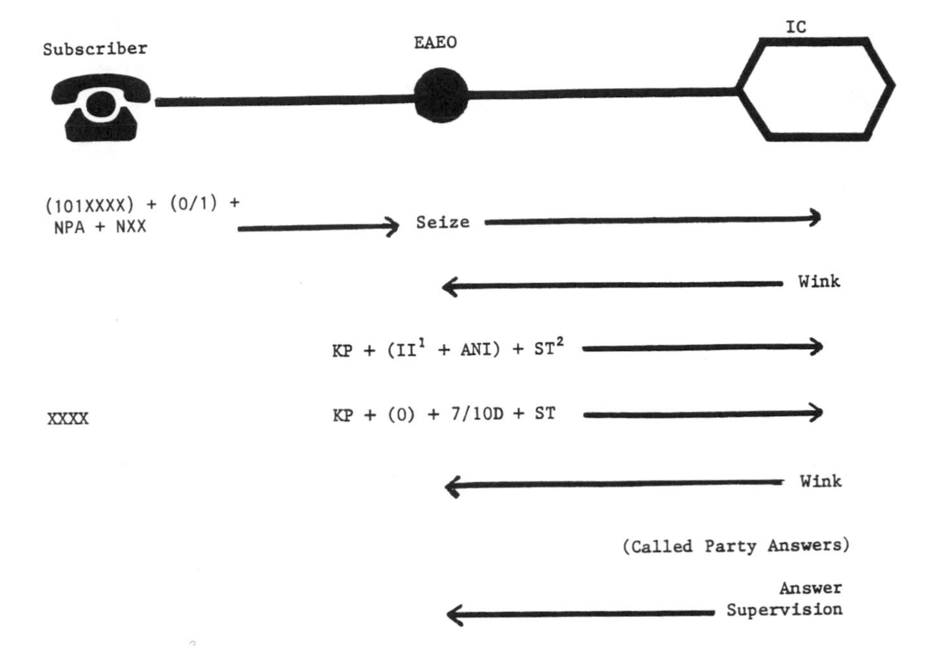

Notes:

1. Information digits to identify type of calling party, e.g., identified direct-dialed call, hotel room call, coin phone, etc.
2. If inter-LATA carrier has not requested ANI, EAEO outpulses KP + ST.

FIGURE A.9. Signaling sequence for a direct connection.

POP after receipt of the carrier identification and a wink from the IC POP. Acknowledgment winks and answer supervision from the IC POP are repeated by the AT to the EAEO.[30]

The terminating signaling sequence is a straightforward application of wink-start control and MF address signaling, in which the IC POP outpulses the called address to either the EAEO or the AT. Called-party answer supervision is returned in the reverse direction.[31]

International calls may be handled directly by *international carriers* (INC) through an INC POP in the LATA or through an INC gateway office and the domestic network of an IC. International calls involve multistage outpulsing to the international carrier. The previous first stage of outpulsing in the form KP + 011 + CCC + ST cannot carry all information required for the international call, carrier, country identification, and call routing. Therefore, a new 3-stage outpulsing format was designed in the form KP + 1NX + XXXX + CCC +

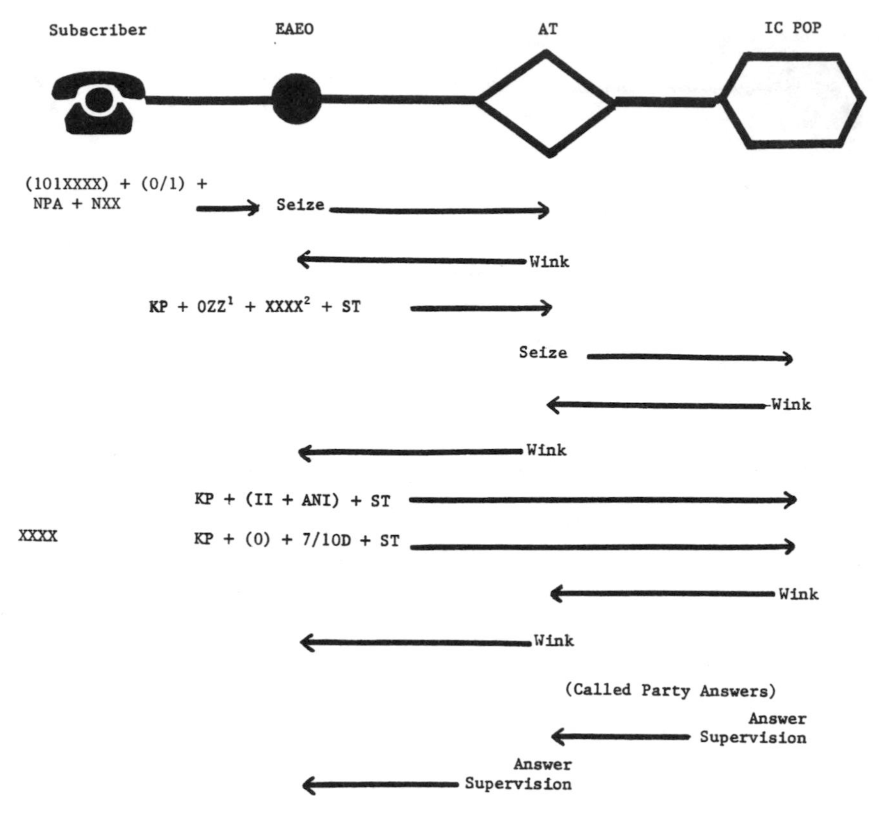

FIGURE A.10. Signaling sequence for a tandem connection.

ST in the first stage. The internal symbols identify call routing, carrier, and country code, respectively. The second stage of outpulsing forwards ANI information, if required, and the third stage forwards the country code and national number of the calling party to the INC directly or through a domestic IC, and possibly through an AT. Winks and answer supervision are the same as for domestic calls.[32]

Some ICs perform combined IC/INC operations. ICs and INCs that do not perform combined operations may negotiate agreements with each other. Subscribers may select an IC or an INC for international calls, and calls are routed accordingly.

A.4.7. Common-Channel Signaling

Common-channel signaling, wherein all signaling is performed over transmission paths completely separate from the voice paths, has become the preferred signaling concept, although inband signaling still is in widespread use. The system installed by AT&T was called Common Channel Interoffice Signaling (CCIS).[33] Other common-channel signaling systems perform similar functions. CCIS differed in message format from the former international common-channel signaling standard, CCITT System No. 6. Although CCIS has been replaced by CCITT Signaling System No. 7 (SS7), described in Chapter 13, CCIS is described here to illustrate the principles of common-channel signaling without encountering the multilevel complexities of SS7.

A.4.7.1. Principles of Common-Channel Signaling

Supervisory signals, address signals, and other signals are exchanged by messages between switching-system processors over a network of signaling links instead of over the speech-transmission paths. (Signaling links in CCIS utilized 4-wire, analog voice-frequency channels with modems and terminals. Initially, the VF channels operated at 2400 b/s but were upgraded to 4800 b/s, and some signaling links subsequently were replaced by 56-kb/s digital channels.) Signaling messages are transmitted over one or more signaling links to a destination switching center (which may not be the final destination of the call) by means of packet-switching technology. Speech paths are checked for continuity and connected through to complete the transmission path. An example of an associated signaling link is shown in Fig. A.11. Since it would be uneconomical to establish such signaling links for all interoffice trunk groups, signaling links are combined into a separate signaling network.

A.4.7.2. Signaling Link Operation

Signaling messages are generated by switching-system processors as signal units (SU). CCIS SUs contain 20 bits of signaling data each and are transmitted over signaling links in blocks of 12. SUs are sent by the switching-system processor to a *terminal-access controller* (TAC) which is custom-designed to be compatible with its associated processor. The TAC arranges the SUs according to priority and routes them to the proper signaling terminal. The signaling terminal stores them in a transmit buffer until 11 have been accumulated. The signaling terminal adds eight check bits to each SU and adds the 12th SU as an acknowledgement signal unit to complete the block of 12. After transmission, the block of SUs is stored until an acknowledgement is received indicating that all SUs in the block have been received correctly. Errors are corrected by retransmission. When no SUs containing signaling data are being transmitted, a synchronization signal unit is transmitted for link synchronization. Signaling reliability is enhanced by link redundancy with automatic transfer and diverse routing.

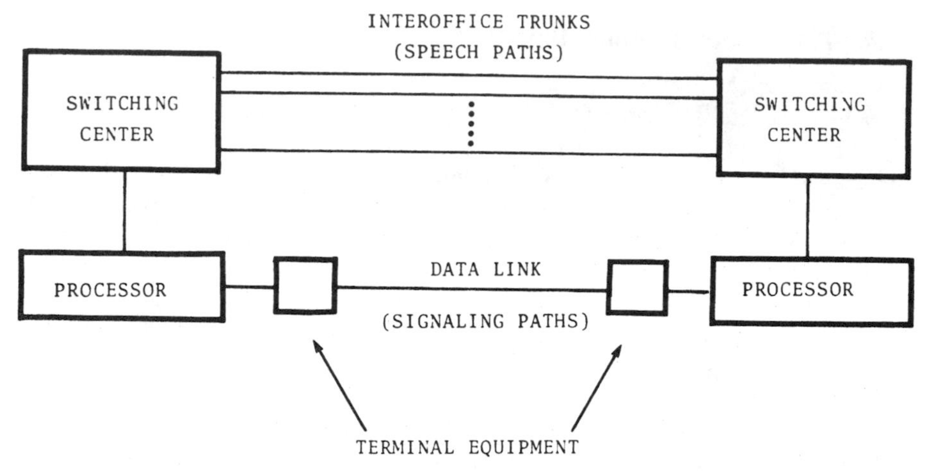

FIGURE A.11. Common-channel interoffice signaling (CCIS).

A.4.7.3. Call Setup with CCIS

To illustrate procedures for call setup by CCIS, an associated signaling link between the originating and terminating switching centers is assumed. The originating signaling office sends an initial address message (IAM) consisting of an initial signal unit (ISU) and up to four subsequent signal units (SSU). Header information relates the ISU and SSUs to each other in the correct sequence. The ISU contains a trunk label consisting of a band number (9 bits) associated with a subgroup of 16 trunks and a trunk number (4 bits) identifying a specific trunk in that subgroup. Routing information and the digits of the called telephone address are contained in the SSUs.

Upon receipt by the destination switching office, the SUs are translated, and a continuity-check circuit is connected to the designated trunk. That circuit simply loops the send and receive paths together through a lossless loop. The originating office connects a transceiver to the trunk, transmits a 2010-Hz tone to the terminating office, and measures the loss of the signal received over the return path. If the loss is out of limits, the failed trunk is locked out of service and subjected to a special test, and the call is reinitiated. If the loss is within acceptable limits, the originating office removes the transceiver and sends a continuity (COT) message over the signaling link. The terminating office acknowledges the COT message by sending an address complete (ADC) message to the originating office.

The call-processing program then translates the called number and tests the called line for busy. If busy, it returns a subscriber busy (SSB) message to the originating office over the signaling link. If the called line is idle, the terminating office applies ringing and returns audible ring tone over the incoming trunk which has been connected to the calling party by the originating office after receipt of the ADC message. When the called party answers, the terminating office connects a cross-office path between the called line and incoming trunk and sends an answer message over the signaling link to the originating office which begins charge timing for billing. When the parties hang up, disconnect messages are send over the signaling link, charge timing is stopped, the cross-office paths are removed, and the trunks and lines are stored to idle state after a specified guard interval.

A.4.7.4. Signaling-Message Formats

A comparison of signaling-message formats used in CCITT System No. 6 and CCIS is shown in Fig. A.12. The principal differences are in the trunk labels and routing information. Three types of SUs are shown.

The formats for a lone signal unit or an initial signal unit of a signaling message containing multiple signal units are shown in Fig. A.12(a). CCITT System No. 6 uses a 5-bit heading, while CCIS employed a 3-bit heading. In the international signaling system, blocks of heading codes are allocated for international, regional, and national use. The 11-bit trunk label in the CCITT System No. 6 permits identification of a total of 2048 trunks. The 7-bit band number identifies a specific group or subgroup of 16 trunks, and the 4-bit trunk number identifies a specific trunk. The 13-bit trunk label used in CCIS enabled the identification of 8192 trunks to accommodate the larger trunk quantities in the AT&T domestic network.

Formats for the first SSUs in a signaling message are shown in Fig. A.12(b). A length indicator in CCITT System No. 6 indicates the number of subsequent signal units in the message. Three bits of route information indicate whether a country code, a satellite, or an echo suppressor is included, and a fourth bit is reserved for future use. The calling party indicator contains other call characteristics, such as whether it is a data call or a test call, and special routing instructions, such as operator language requirement. The CCIS format used 4-bit abbreviated or 16-bit expanded routing, and the choice was indicated by a route bit. If abbreviated routing were used, the first three address digits were included in the first SSU in binary-coded decimal (BCD) format. Additional SSUs were used for address digits, as shown in Fig. A.12(c).

Check bits 21–28 were used for checking the accuracy of signaling messages. Each check bit indicated whether the binary sum of a specific combination of 8–11 bits was odd or even.

A.4.7.5. Datagram Direct Signaling

The addition of a direct-signaling capability permitted messages to be sent from one signaling point to any other signaling point in the network. A portion of the network routing instructions were placed in a database accessible by multiple network switching systems, called action points (ACPs). ACPs sent messages through STPs to the common databases, known as network control points (NCPs),

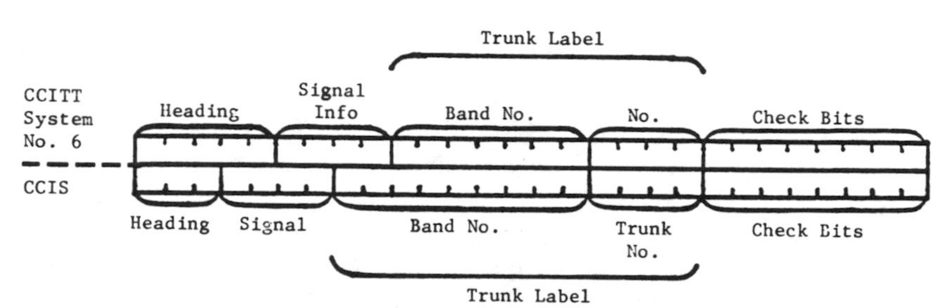

(a) Lone or initial signal unit

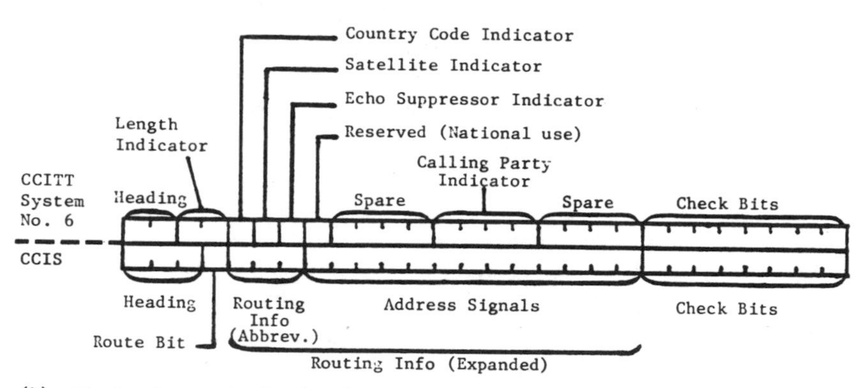

(b) First subsequent signal unit

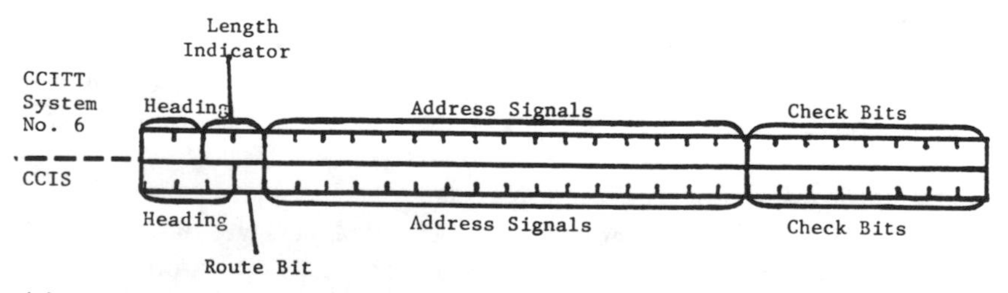

(c) Additional subsequent signal units

FIGURE A.12. Signaling-message formats.

and received routing instructions for calls in return. This allowed the introduction of new subscriber services and routing changes on a time-of-day basis and in response to network traffic conditions.

A.4.7.6. Advantages of Common-Channel Signaling

The primary technical advantages of common-channel signaling are elimination of glare and increased signaling speed, reliability, and flexibility. Common-channel signaling passes signals at a much higher rate than inband-signaling systems. This reduces post-dialing delays and the holding time of trunks. Increased accuracy is achieved by error correction of signaling messages. The increased capacity of signaling messages permits more flexibility in call routing, more efficient network operation, and enhanced subscriber services.

A.5. SWITCHING TECHNOLOGY

A.5.1. Basic Switching Functions

The basic switching functions required to switch telephone calls are supervision, control, signaling, and provision of network paths. A block diagram of a typical local switching system is shown in Fig. A.13. The Terminal Interface Group connects all lines and trunks to the switching system. The Switching Network provides talking paths, and the Control Complex controls all other functions. Service Circuits are peripheral equipment such as tone generators and digit receivers.

Switching systems have other equipment groups, such as power supplies, billing equipment, input/output devices, maintenance and administrative equipment, which support the primary switching functions but which are not shown on the block diagram.

A.5.1.1. Supervision

Supervision involves recognition of the busy or idle condition of circuits (for example, lines and trunks) connected to the switching system. A transition from idle state to busy state is recognized as a demand for service requiring response by the switching system. A transition from busy to idle state is recognized as a termination of connection, or a "disconnect," requiring action by the switching system to restore all associated connections to idle state.

A.5.1.2. Control

The first control function is to recognize and respond to a caller's demand for service. The system control then prepares the system to receive the digits of the called number, or address, and returns dial tone. Upon receipt of the address digits,

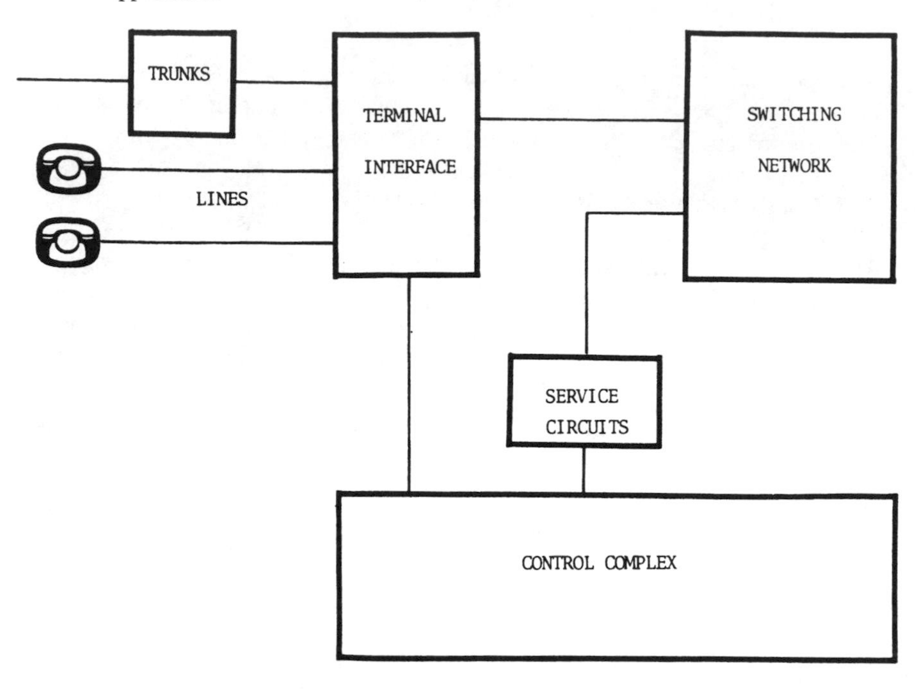

FIGURE A.13. Basic switching-system block diagram.

system control interprets the digits to determine the desired destination in terms of equipment terminations. It then examines the availability of a path through the switching network to the equipment termination representing the destination. If a path is not found, the caller is so informed by a tone or other signal. If a path is found, system control causes a path to be established to the called line or to a trunk termination to another switching system. Ringing is applied to the called line. When the called line is answered by a transition from idle to busy state, ringing is tripped. When conversation is completed, transitions from busy state to idle state are detected, and system control causes the network path to be released.

A.5.1.3. Signaling

Three types of signals are used in basic telephony. Supervisory signals transmit the busy or idle state of lines or trunks (see Section A.4.1). Address signals contain the digits of the called telephone number (see Section A.4.2). Information signals provide information relative to the establishment of a connection through the telephone network.

A.5.1.4. Switching Network

Each switching system has a network of talking paths used to connect lines to lines, lines to trunks, trunks to lines, and trunks to trunks. It also is used to provide access to peripheral equipment such as tone generators and digit receivers.

A.5.2. Control Concepts

The three basic control concepts employed in telephone-circuit switching are operator control, user control, and common control.

A.5.2.1. Operator Control

Telephone operators control the establishment and disconnection of calls under certain conditions. Such control may be exercised by a manual switchboard, with cord circuits to establish connections, or a cordless console, in a standalone configuration or in connection with a switching system.

A.5.2.2. User Control

Older switching systems, such as step-by-step systems, establish calls under user control. Address digits, dialed with a rotary dial, are comprised of one to ten dc pulses that operate stepping switches. Strowger stepping-switch contacts are arranged in horizontal rows, called levels, in a semicylindrical bank. A wiper arm, attached to a vertical shaft, is actuated by dc pulses transmitted by the rotary dial or other dc pulse transmitter on the connected telephone. Each pulse causes the shaft to elevate the wiper arm one level. After reaching the designated level, the wiper arm rotates horizontally across the selected level. Depending upon the application, the wiper arm rotates until it finds an idle circuit represented by one of ten pairs of contacts on the level, or it rotates to a specific pair of contacts directed by the number of pulses dialed. Thus, a connection is established progressively, one switch at a time, through a series, or train, of switches under direct control of the caller. Later improvements have incorporated a *sender* to receive and translate digits. This is called *senderized operation.*

Operation of two-motion stepping switches is controlled by multiple electromechanical relays which require meticulous adjustment. Maladjustment, caused either by wear of metal parts or by deficient maintenance techniques, can result in stepping errors. Some relays are quite large, and the resulting arcing can produce impulse noise in circuits, which causes errors in data transmission. Thus, step-by-step switching systems are not suitable for data transmission. There also is a traffic penalty associated with step-by-step switching. Each train of pulses causes a wiper arm to rotate over a row of ten pairs of contacts representing ten circuits. Large trunk groups must be divided into groups of ten for step-by-step switching. Thus, a call can encounter an all-trunks-busy condition in the group searched while trunks to the same destination are idle in other ten-trunk groups.

Therefore, more trunks must be provided than would be required to carry the same volume of traffic in a single large trunk group to which each call has total access. Stepping of switches normally is at a rate of ten pulses per second, much slower than more modern switching systems. Some step-by-step switching systems still are used in the North American public telephone network.

A.5.2.3. Common Control

Although common-control systems represent current state-of-the-art concepts, existing systems use different design technologies with different characteristics and capabilities. All common-control systems, however, perform substantially the same basic functions.

A.5.2.3.1. Common-Control Functions. Call processing in a local central-office switching system typifies the common-control functions performed in any switching system. The common control receives information from the terminal interface equipment that a caller's line has shifted from idle (on-hook) to busy (off-hook), indicating a demand for service. The calling line is identified for billing purposes, and its equipment designation is transmitted to the common control. The common control establishes a path through the switching network from the calling line to an idle digit receiver and bridges a tone generator to the line to send dial tone to the caller. Upon receiving the first digit of the called party address, the dial tone is disconnected, and the digit is registered. The remaining digits of the address are received, the address is read by the common control, and the digit receiver is disconnected.

The common control translates the address into the designation of the equipment termination of the called line (or trunk if it is an interoffice call) and tests the called line to determine whether it is idle or busy. If busy, the common control connects the calling line to busy tone through the switching network and disconnects that connection when the calling party goes "on-hook."

If the called line is found to be idle, the common control searches for an idle path through the switching network to its equipment termination. If one is not found, the calling line is connected through the network to reorder tone, and the common control proceeds in the same manner as if the line were busy. If an idle path is found, that path is reserved, the called line is connected to a ringing generator, and the calling line is connected to audible ringing tone. When the called party answers, the terminal interface equipment detects the off-hook condition and transmits that information to the common control. The common control then disconnects the ringing generator and audible ringing tone and connects the calling line to the called line through the network path previously reserved.

The network path remains established for the duration of the call. When either party goes on-hook, that transition is detected by the terminal interface equipment, and the common control is informed. The common control disconnects the network path and restores the lines to idle state. If one party remains off-hook, thereby indicating a possible demand for further service, the common control correctly interprets that condition and proceeds to serve the additional call.

If the original call were to a destination that required a connection through another switching system, the common control would search for an idle trunk to the distant (or an intermediate) office. If found, it would cause a transmitter, or sender, to send the address digits to the distant office and connect the calling line to the trunk. If not found, it would connect the calling line to busy tone and proceed as if it were a busy line. On-hook detection and disconnect would occur in substantially the same manner as for a local connection.

A.5.2.3.2. Types of Common Control. There are three basic types of technology used in common-control systems, each reflecting the state-of-the-art at the time of original design.

Electromechanical common control uses electromechanical relays which are hard-wired together in a fixed configuration to perform the control functions. Decisions are made as the result of the operation of particular relays. Variations in the operating program are implemented by wiring options. Many of these systems have not been modernized because of the difficulty and cost of changing the call-handling program. Because of the characteristics of electromechanical relays, timing functions are directly related to the operating time of the relay contacts. This technology is obsolete, but some systems may remain in service for several years. Thus, new digital switching systems must be able to interact with them.

Wired-logic common-control systems use solid-state technology. The solid-state components are hard-wired together in such a way as to perform specified sets of functions, predicated upon recurring routines or contingent upon a finite set of inputs. Such systems require less space, less power, and less maintenance than electromechanical control systems. Memory is used for local database information and for temporary records of calls being processed. A significant disadvantage is the necessity to modify the printed-circuit boards in order to change the generic operating programs. While electromechanical systems can operate satisfactorily in uncontrolled temperature environments, wired-logic systems generate and concentrate heat into a smaller space and require temperature controls. This technology has been overtaken by the dictates of economics, but a few such systems may be in service.

Current state-of-the-art switching systems are designed around stored-program common-control systems. In these systems, both the generic operating program and the local database information are stored in memory. Some of the generic program normally is stored in read-only memory (ROM), while the database information and other program data is stored in programmable random-access memory (RAM). Call set-up information is stored in a scratch-pad memory. Some systems use a combination of wired-logic and stored-program control systems. Stored-program control systems have lower capital and maintenance costs for large switching systems. Maintenance diagnostics are simplified, and power and space requirements are less than for other types of control systems. As in the case of wired-logic systems, however, heat is concentrated into a small area, and environmental controls are needed to maximize the service life of components.

A.5.3. Switching-Network Technology

The switching network is a systematic collection of interconnecting transmission paths which permits connections to be established through a switching system between lines, between lines and trunks, and between trunks. The network also can be used to connect signaling equipment to external circuits. Historically, economic considerations have precluded the provision of sufficient paths such that a path between two terminals would always exist, irrespective of the traffic load. Some lines, such as those in residences, generate very small volumes of traffic, while others, such as those in many business establishments and office complexes, originate and receive very large numbers of calls during the business day. Therefore, switching systems almost universally have used concentration in their switching networks; that is, there are fewer paths than would be necessary to allow all subscribers to participate in talking connections concurrently. Such networks are known as *blocking* networks.

When concentration is used, a large number of subscriber lines is concentrated into a smaller number of switching paths that are distributed in such a way as to be connectible to any desired terminal through expansion switches. Thus, blocking networks can be described as having *concentration, distribution*, and *expansion* stages, as illustrated in Fig. A.14. As can be seen, a connection can be established between the originating and terminating appearance of all subscribers, but only three paths can be established concurrently. Also, when subscriber No. 1 is talking, subscribers No. 2, 3, and 4 are blocked. Large switching networks can be assembled in this manner.

Circuits may be switched in space or in time, irrespective of whether analog or digital technology is used.

A.5.3.1. Space-Division Switching

Conversation paths in a practical switching network can be separated from each other in space. The development of space-division switching networks closely paralleled the development of control concepts.

A.5.3.1.1. Progressive Switching Networks. Establishment of a connection through a progressive switching network is under the direct control of the calling

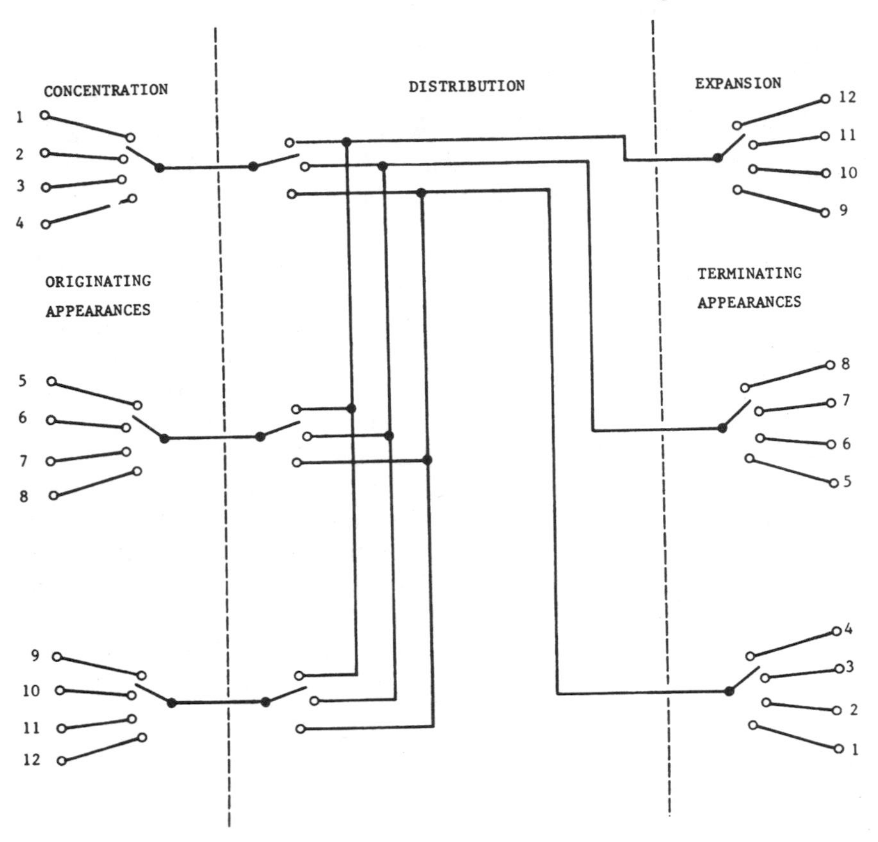

FIGURE A.14. Illustrative switching network.

party and occurs progressively as the address digits are dialed or otherwise transmitted by the calling party.

Using the Strowger two-motion stepping switch as an example, switches are connected in series to form a *switch train*. The first switch in the switch train is called a *linefinder*, and the last switch is called a *connector*. Subscriber lines have appearances on both linefinders and connectors. Intermediate switches are called selectors. Each subscriber line is connected to a group of connectors and to a group of linefinders through a line relay. When a subscriber initiates a call, the off-hook closure causes the line relay to extend battery and ground to specific portions of a preselected linefinder in a linefinder group. The linefinder steps vertically and then rotates its shaft horizontally until the wiper arms find the calling line. Each linefinder is mated to a first selector that returns dial tone. The caller then dials the address digits. First selectors are arranged to drop one, two, or three digits of the 7-digit telephone number depending upon the number of digits required to route the call. A connection established by the last four dialed digits, 2738, of a local call is illustrated in the upper portion of Fig. A.15.

When the digit "2" is dialed, the first selector steps vertically to the second level and rotates its wiper arms horizontally. When its control arm finds an idle path to a second selector, usually indicated by battery and ground on the control contact, its other wiper arm clamps onto contacts, extending the voice path to the second selector. The digit "7" causes the second selector to step in a similar manner to the seventh level and connect to a path to a connector having a terminating appearance of the called line. The last two digits are dialed into the connector. The digit "3" causes the connector to step vertically to the third level, and the digit "8" causes the shaft to rotate its wiper arms to the eighth set of contacts. Finding the line idle, the wiper arm clamps onto the contacts, and ringing voltage is applied to the line. Upon answer, the call is established.

Each linefinder in a group connects to either 100 or 200 lines. Each selector has 100 tip-and-ring paths, and each connector has either 100 or 200 lines connected to it. If the connector serves 200 lines, three digits are dialed into it, the first digit being a "1" or "2" to select the upper or lower bank of 100 line terminations.

The linefinder group comprises the concentration stage of the switch train,

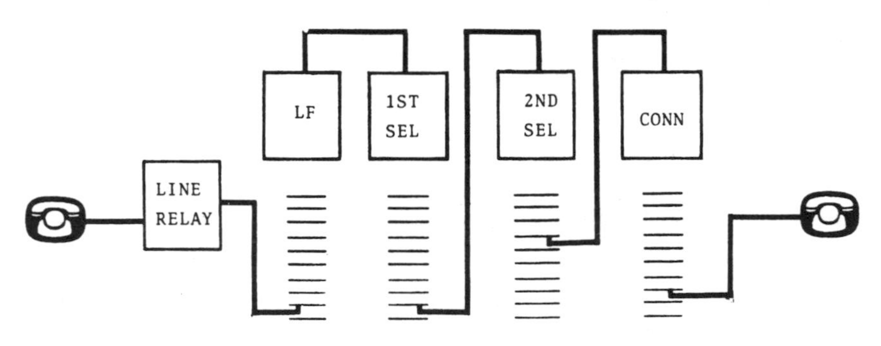

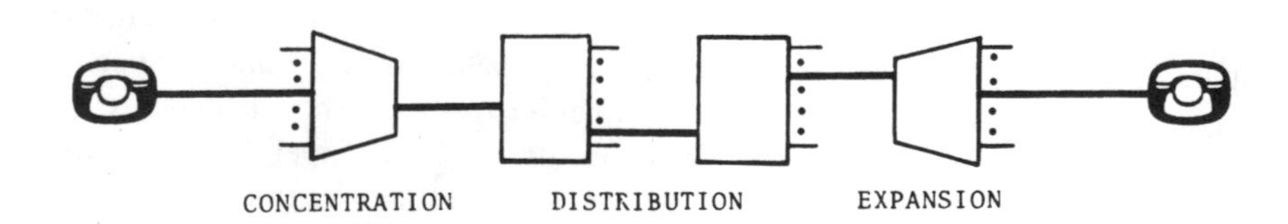

FIGURE A.15. Step-by-step switch-train connection.

concentrating 100 or 200 lines into a smaller number of paths. The connector group forms the expansion stage, expanding the selected path to connect to 100 or 200 lines. The distribution stage is formed by the selectors. The symbolic representation of these stages is shown in the lower portion of Fig. A.15.

The stepping switch provides 100 terminations in a bank of contacts. A separate bank of control contacts has either 100 or 200 contacts, depending upon whether one or two line banks are used. Control-bank contacts are used to indicate busy or idle state of their associated line (or trunk) contacts.

Step-by-step switching systems are characterized by their durability, simplicity, minimum postdialing delay, and lack of an environmental-control requirement. They are, however, feature-limited, require large floor space, and consume large amounts of power. Changes and expansion require major rewiring operations, and directory-number assignments are inflexible. Operation of the large stepping magnets and relays cause electrical impulses that interfere with data transmission, and relay maladjustments are a major source of troubles. A traffic penalty results from the limited availability of paths out of each selector switch.

A.5.3.1.2. Crosspoint Switching Networks. Whereas in progressive networks, the selection of a path proceeds stage-by-stage, with no knowledge of conditions ahead, development of common-control systems made it possible to identify the input and output terminals and then examine the entire switching network for possible paths between them. Crosspoint networks, sometimes known as grid networks or coordinate networks, were developed to take advantage of the characteristics of common-control systems.

A crosspoint-switching network is an assembly of individual coordinate switches arranged into a switching array. Switches may be square with the same number of inlets and outlets, or they may be rectangular with different quantities of inlets and outlets. Individual switches are arranged in the form of a matrix with inlets and outlets on the vertical and horizontal axes. Transmission paths run vertically and horizontally, and connections are made by closing crosspoints at the intersection of the selected inlet and outlet, as shown in Fig. A.16.

In a nonblocking crosspoint switch, there must be a sufficient number of paths and associated crosspoints to ensure that every inlet can be connected concurrently to a separate outlet. As illustrated in Fig. A.16(a), any three inlets can be connected to any of three outlets in the square matrix. The six crosspoints could satisfy the criteria for a nonblocking three-line switching system. If the number of lines is doubled, as in Fig. A.16(b), the required number of crosspoints is increased to 30. For a nonblocking square matrix, the number of crosspoints N_x required can be calculated by $N_x = N(N-1)$, where N is the number of inlet/outlet pairs.

It is readily apparent that nonblocking crosspoint switching networks are very

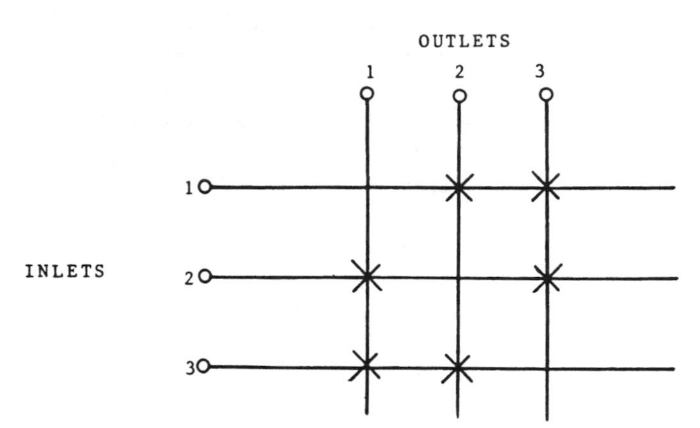

a. Three lines, six crosspoints

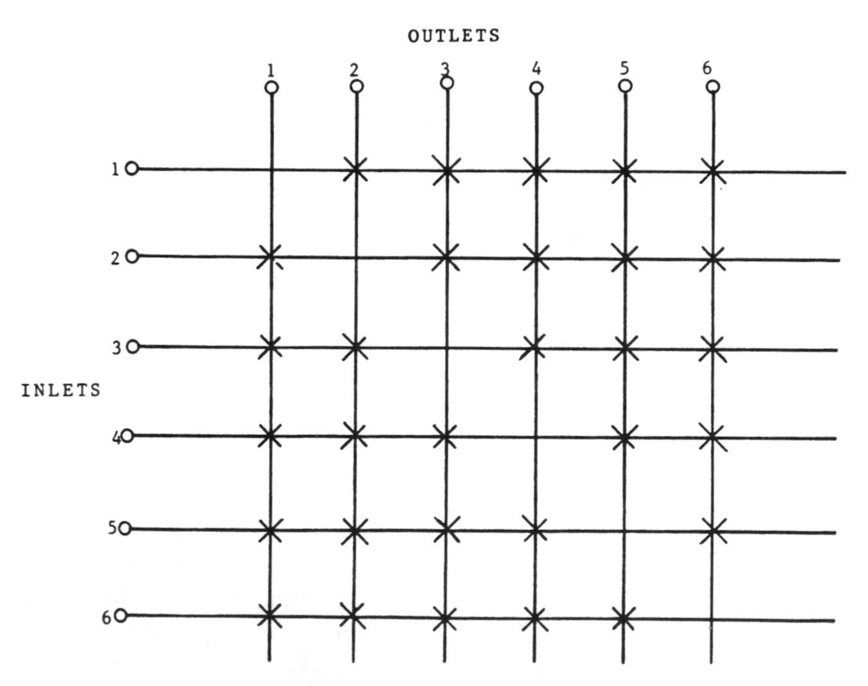

b. Six lines, 30 crosspoints

FIGURE A.16. Illustrative crosspoint switching matrices.

expensive in crosspoints, except in extremely small sizes. For example, a 100-line, nonblocking, three-stage crosspoint network would require a minimum of 5257 crosspoints, according to $N_x(\text{min.}) = 4\,N(\sqrt{2N} - 1)$, where N is the number of lines (ports).[34]

Most crosspoint networks consist of multiple stages and are blocking networks. Two-stage blocking networks can be used for switching systems of small size. A simplified two-stage network is illustrated in Fig. A.17. The six-line network, shown by solid lines, can accommodate three simultaneous conversations for certain combinations of lines; it requires 24 crosspoints. If inlet 1 is connected to

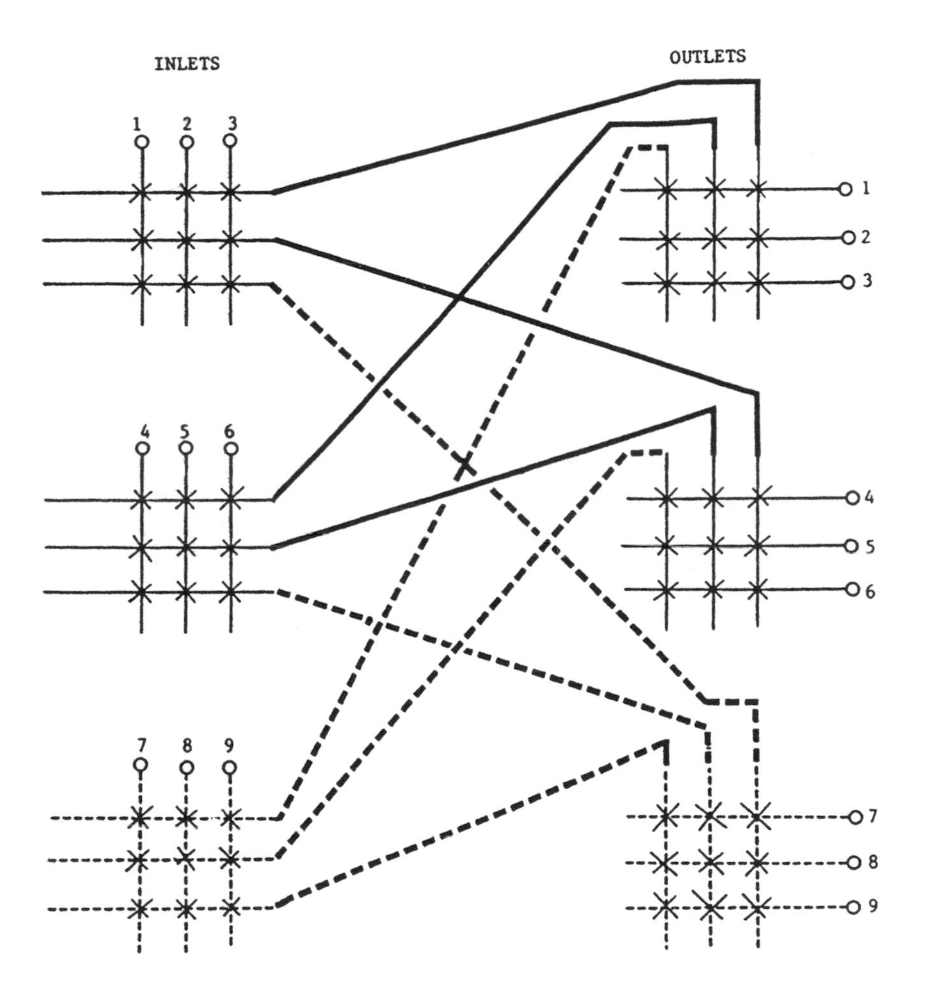

FIGURE A.17. Two-stage crosspoint switching network.

outlet 4, however, inlets 2 and 3 are blocked from outlets 5 and 6. When it is increased to nine lines by the addition of concentration and expansion stages (represented by dashed lines), more conversation paths exist, and 54 crosspoints are required, but the network still is blocking. Although the two-stage array requires fewer crosspoints than nonblocking networks of the same size, they are not efficient switching vehicles.

By adding a center distribution stage, the network shown in Fig. A.18 can accommodate the same number of simultaneous connections as that shown in Fig. A.17 under certain conditions, for both six-line and nine-line configurations, with

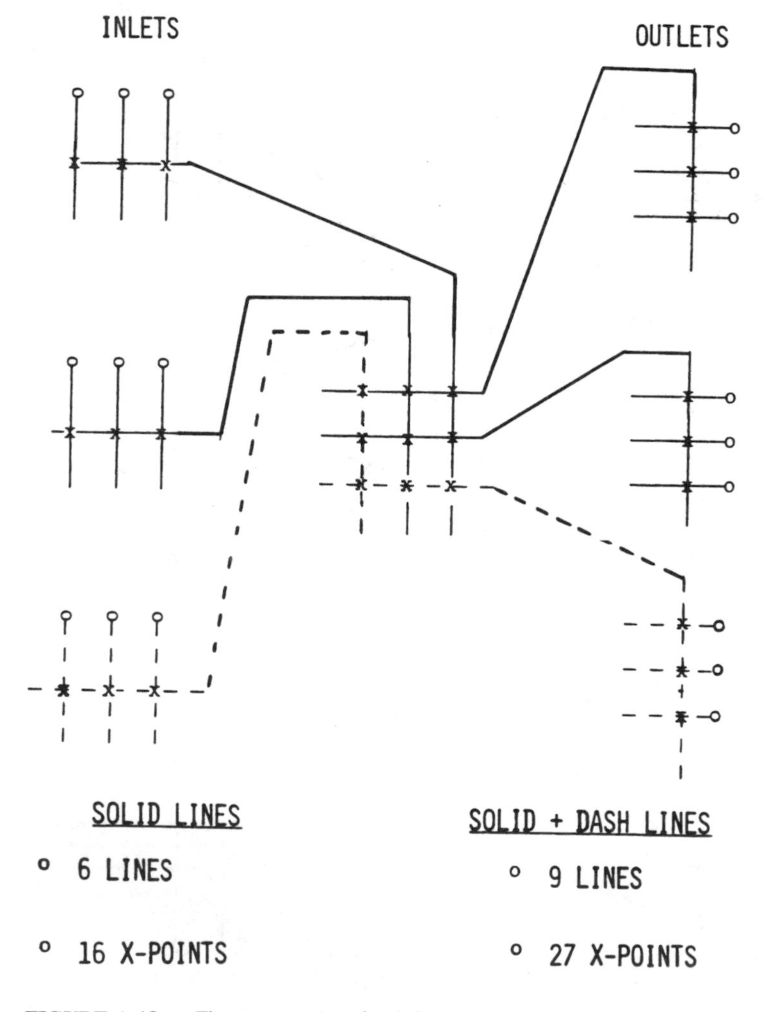

FIGURE A.18. Three-state crosspoint switching network.

fewer crosspoints; 16 and 27 compared to 24 and 54, respectively. A practical, three-stage crosspoint-switching network capable of serving 100 lines, is shown in Fig. A.19. In this example, the A stage is composed of ten rectangular matrices to concentrate ten inputs each into N outputs, where N is the number of center stages. Center (B-stage) arrays are 10×10 square matrices to distribute the concentrated paths to the C-stage switches. C-stage (expansion) switches are the mirror image of those in the A stage and expand the concentrated paths to reach any of the 100 terminations. The traffic capacity and the number of crosspoints required are dependent upon the number of crosspoint switches in the center stage. The number of crosspoints[35] required for a blocking, three-stage network is found by

$$N_x = 2Nk + k(N/n)^2 \tag{A.5}$$

where

N = number of lines (ports)
n = number of lines in each concentration and expansion switch
k = number of center-stage arrays.

A multistage crosspoint network may have several center stages. In planning a system, however, the number of center (distribution) stages should be fixed in the initial installation, based upon the maximum planned size of the system. The maximum attainable size of the system cannot be expanded without taking the switching system out of service. The number of individual switches in each center stage may be increased, along with those in the concentration and expansion stages, as the switching system grows, but the basic architecture fixes the maximum size of the system.

A.5.3.1.3. Crosspoint Technology. In the evolution of switching technology, three distinct generations of crosspoints have become widely used in North America. Each type, coupled with its own common-control system, affects timing considerations during call setup and disconnect operations over trunks to other switching systems.

Crossbar switches generally are configured in 10×10 square or 10×20 rectangular arrays. Each switch has 10 horizontal paths and 10 or 20 vertical paths. A combination of magnets closes and holds connections between vertical and horizontal paths, providing a total of ten possible connections simultaneously on any switch. Five selecting bars are mounted horizontally across the front of each switch, and each bar has one flexible selecting finger attached to it for each vertical path. Each vertical unit has ten groups of contacts—one group of three to six pairs of contact springs for each horizontal path. Each group of contact

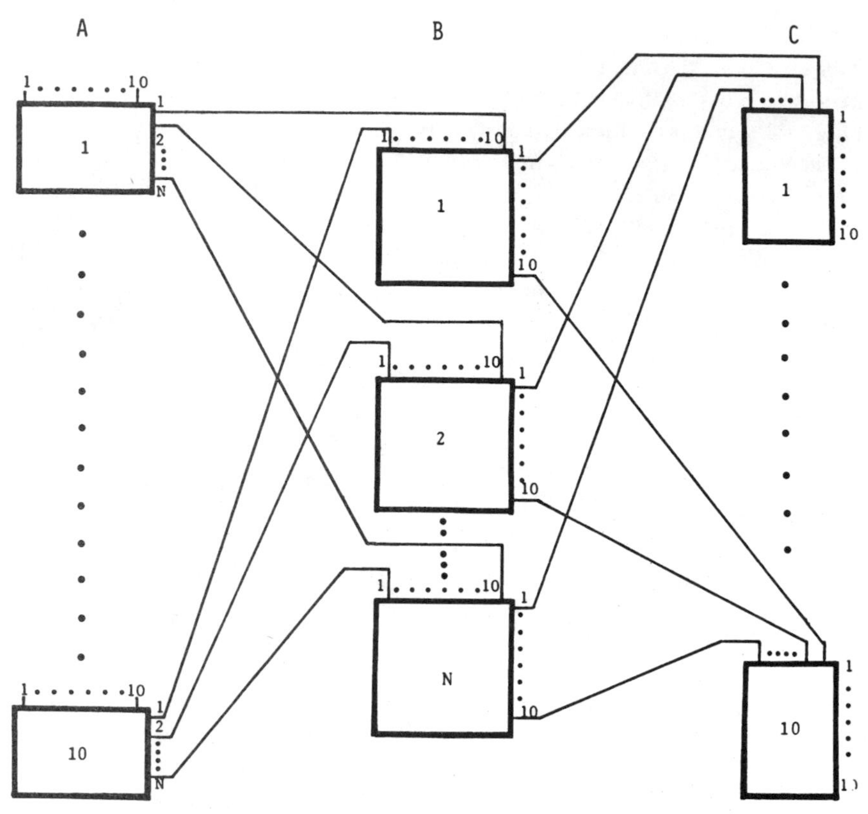

FIGURE A.19. 100-line, 3-stage switching matrix.

springs, when closed by a selecting finger and hold magnet, is considered to be a single crosspoint. Thus, each crosspoint can switch from three to six conductors. To operate a crosspoint, a select magnet causes one of the horizontal selecting bars to be rotated slightly up or down, moving all selecting fingers up or down against a stop; a hold magnet slightly rotates a vertical holding bar, associated with a selected vertical unit, against five selecting fingers (one per horizontal selecting bar), moving them to one side. The selecting finger associated with the operated horizontal selecting bar is pressed against one group of contact springs on the vertical unit, closing the crosspoint. After the hold magnet operates, the select magnet releases, returning all but the operated selecting finger to normal position. When the hold magnet releases, the operated selecting finger returns to normal position, releasing the connection through the crosspoint. A crossbar

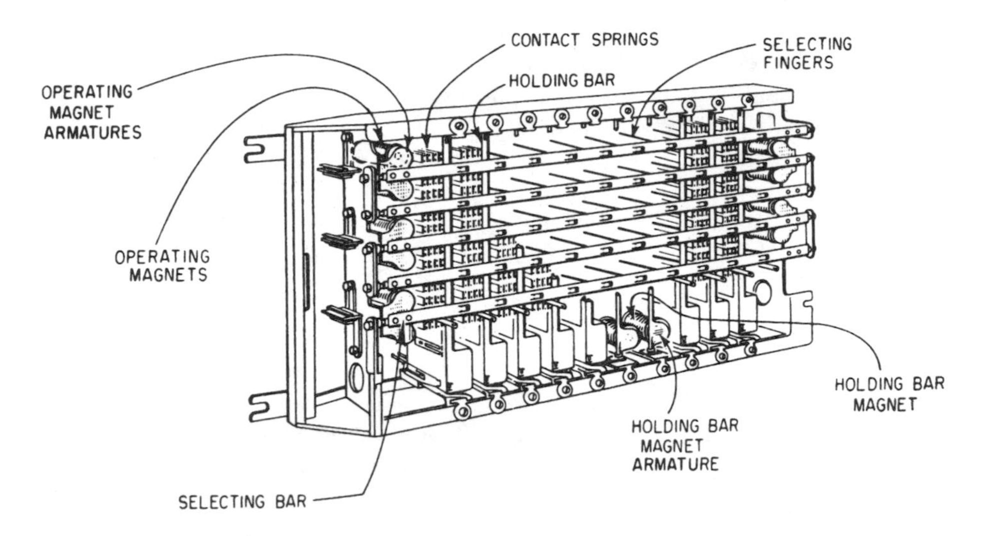

FIGURE A.20. A crossbar switch. (Reproduced with permission of AT&T.)

switch is illustrated in Fig. A-20.[36] Crossbar switches are grouped together to perform the concentration, distribution, and expansion functions of a blocking network.

The second generation of crosspoint technology involved encapsulated *dry reed relays*, consisting of two magnetic reeds sealed in a glass capsule and mounted between plates of a two-state magnetic alloy. When the alloy is subjected to a short pulse of current, the reed contacts close and remain closed magnetically until another pulse of current removes the residual magnetism, opening the contacts. Other techniques use crosspoint coils with operate and hold windings to operate the contacts. Reed crosspoints are arranged into crosspoint arrays. Their use, along with electronic wired-logic or stored-program common-control systems, permitted the design of switching systems with larger capacity, greater reliability, and faster switching times, while requiring less space and less power than crossbar systems.

The third generation of crosspoint technology employs *solid-state* logic devices to switch telephone calls. Diodes, semiconductor-controlled rectifiers, transistors, and integrated circuits have been used as crosspoints. Logic gates are arranged in switching arrays in the same manner as reed relays. Gates are closed to establish a connection and opened to disconnect. Solid-state crosspoints are characterized by long life, low power consumption, high reliability, and very fast switching times. Very-large-scale integrated (VLSI) circuits greatly reduce the size of switching systems.

A.5.3.2. Time-Division Switching

Analog signals, continuously varying in amplitude, must be altered in order to be switched in time. The speech (or other signals) must be changed into pulses which can be multiplexed into a digital stream, each pulse or group of pulses representing one amplitude sample. Time-division switching of amplitude pulses has been used in some analog switching systems, and time-division switching technology quickly became the standard in digital switching systems (see Chapters 11 and 12).

REFERENCES

1. *Notes on the BOC Intra-LATA Networks*, American Telephone and Telegraph Company, 1983, Section 3, pp. 3–7 (hereinafter cited as *BOC Notes*).
2. *Notes on the Network*, American Telephone and Telegraph Company, 1980, Section 3, Appendices 1–3 (hereinafter cited as *AT&T Notes*).
3. Ash, G.R. and Oberer, E., "Dynamic Routing in the AT&T Network—Improved Service Quality at Lower Cost," *GLOBECOM '89 Conference Record*, pp. 9.1.1–9.1.6.
4. Ibid.
5. *Telecommunications Transmission Engineering*, American Telephone and Telegraph Company, 1977, Vol. 1, pp. 69–70.
6. Sistanizadeh, K., "Analysis and Performance Evaluation Studies of High Bit Rate Digital Subscriber Lines (HDSL) Using QAM and PAM Schemes with Ideal Decision Feedback Equalization (DFE) within a Carrier Serving Area (CSA)," *GLOBECOM '90 Conference Record*, Vol. 2, App. A, p. 703.4.4.
7. *Generic Requirements for High-Bit-Rate Digital Subscriber Lines*, Bellcore TA-NWT-001210, Issue 1, October 1991, p. 3-1.
8. *Telecommunications Transmission Engineering*, Vol. 1, pp. 56–58, 413–436.
9. Ibid., Vol. 1, pp. 104–109, 482–486.
10. Ibid., Vol. 1, pp. 486–490.
11. Ibid., Vol. 1, pp. 596–599.
12. *Telecommunications—Interface Between Carriers and Customer Installations—Analog, Voicegrade Switched Access Lines Using Loop-Start and Grand-Start Signaling*, ANSI T1, 401-1993, p. 27.
13. Ibid., pp. 14–15.
14. *BOC Notes*, Section 6, pp. 92–93.
15. Ibid., Section 6, pp. 100–104.
16. Ibid., Section 6, p. 105.
17. Ibid., Section 6, pp. 104–106.
18. Ibid., Section 6, pp. 29–31.
19. Ibid., Section 6, pp. 33–41.
20. Ibid., Section 6, pp. 41–43.
21. Ibid., Section 6, pp. 43–45.
22. Ibid., Section 6, pp. 69–70.
23. Ibid., Section 6, p. 108.

24. ANSI T1.401-1993, Annex D.
25. Ibid.
26. Ibid.
27. *BOC Notes*, Section 13, pp. 5, 9–10.
28. Ibid., Section 6, pp. 198–200.
29. Ibid., Section 6, pp. 201–203.
30. Ibid., Section 6, pp. 203–205.
31. Ibid., Section 6, p. 205.
32. Ibid., Section 6, pp. 205–207.
33. *Local Switching System General Requirements*, American Telephone and Telegraph Company, 1980, Section 6.
34. Bellamy, J.C., *Digital Telephony*, second edition, John Wiley & Sons, New York, 1991, pp. 232–234.
35. Ibid., pp. 230–232.
36. *Local Dial Switching Equipment Reference Guide*, American Telephone and Telegraph Company, 1957, Vol. 4, pp. 2–3.

Appendix B

Traffic Considerations in Telephony

B.1. INTRODUCTION

The purpose of this appendix is to provide readers a brief exposure to traffic considerations applicable to any circuit-switched network, whether analog or digital. Traffic can be defined in different ways, depending upon the particular portion of the network under consideration and whether the context is quantitative or general. Traffic is simply the flow of messages through a communications system. It may apply to the messages sent or received, or both, over one or a group of circuits during a specified period of time, generally one hour, or it may apply generically to the total messages sent and received over an entire network or a significant portion thereof. Useful traffic data pertaining to both switches and circuits is collected and analyzed periodically.

B.2. TRAFFIC ASSUMPTIONS

Teletraffic theory is based upon statistical probabilities and certain assumptions concerning telephone traffic. Four assumptions that are dominant in circuit-switched teletraffic theory are:

- Traffic originated by a large number of sources is random and independent of all other subscribers.
- The holding times (i.e., duration) of individual calls and trunks are exponentially distributed.
- The holding times of certain types of common equipment (for example, signaling units) of a given type are constant.
- There are random variations of traffic loads hourly, daily, and seasonally.[1]

B.3. TRAFFIC MEASUREMENTS

Two types of traffic measurements are used in circuit-switched traffic administration: peg counts and usage. *Peg counts* measure the cumulative number of occurrences of events during a specific time period. *Usage* measures the duration of a specific condition, such as the busy state of a line, trunk, or unit of equipment, commonly called a *server*. Servers locked out of service for maintenance may be recorded as busy for traffic if the two conditions are not treated separately for traffic data collection. Usage may be measured in seconds as it actually occurred, or it may be calculated from a peg count obtained by scanning or sampling a group of circuits or equipment at specified intervals. The scan rate is related to the average holding time of the servers in the group being measured.[2] If the scan rate is at least one-half the average holding time, the result generally is sufficiently accurate. A scan rate of one per second is commonly used for service circuits, while a scan rate of one per 100 seconds for interoffice trunks produces acceptable results. In digital systems, programs can be designed to record traffic data accurately in memory and to read out or transfer that data to other storage media at specific intervals.

B.3.1. Traffic Data Collection

Generally, both peg counts and usage are required for the same system components or events when both measurements are meaningful. Total switching-system measurements are recorded for significant parameters, such as those pertaining to originated calls, terminating calls, outgoing calls, incoming calls, calls initiated but not completed, feature utilization, and processor real-time usage allocation. Typical measurements for trunks and traffic-sensitive components, such as service circuits (see Section 11.4) include peg count of total attempts, peg count of attempts which find all servers busy, usage of servers available for traffic, and usage of servers out of service for maintenance. The traffic-data collection system also should compute dial-tone delay and delay for service circuits. For network management purposes, selected parameters are accumulated for predetermined periods on a continuous basis and are available for readout on demand or when thresholds are exceeded.[3]

B.3.2. Traffic Analysis Considerations

Selection of sensing points for traffic measurements requires a knowledge of how the data will be used. In teletraffic theory, offered traffic load consists of random demands for service from a large number of sources to be connected to a finite number of servers via a switching network. Servers may consist of a group of trunks or service circuits. If all servers are simultaneously busy, the next demand for service, or call attempt, is either blocked or delayed, depending upon the function of the servers. If the servers are a group of outgoing trunks, the attempt is blocked, or lost, unless there is an alternate route, in which case the attempt overflows to the alternate group of trunks. If the servers are subscriber digit receivers, the attempt will queue and wait until a server becomes available or until a specified time limit expires, called *timeout*. All offered traffic which is not carried immediately by the server group is lost traffic, overflow traffic, or delayed traffic. Traffic relationships are illustrated in Fig. B.1.

In determining the quantity of servers to be provided in a group, the following factors need to be considered:

- whether the servers will handle traffic on a blocking basis or on a delay basis;
- the type of access to servers—full or limited availability;
- the statistical characteristics of the offered traffic in terms of randomness and holding times;
- the required grade of service; i.e., the amount of loss or delay that can be tolerated.[4]

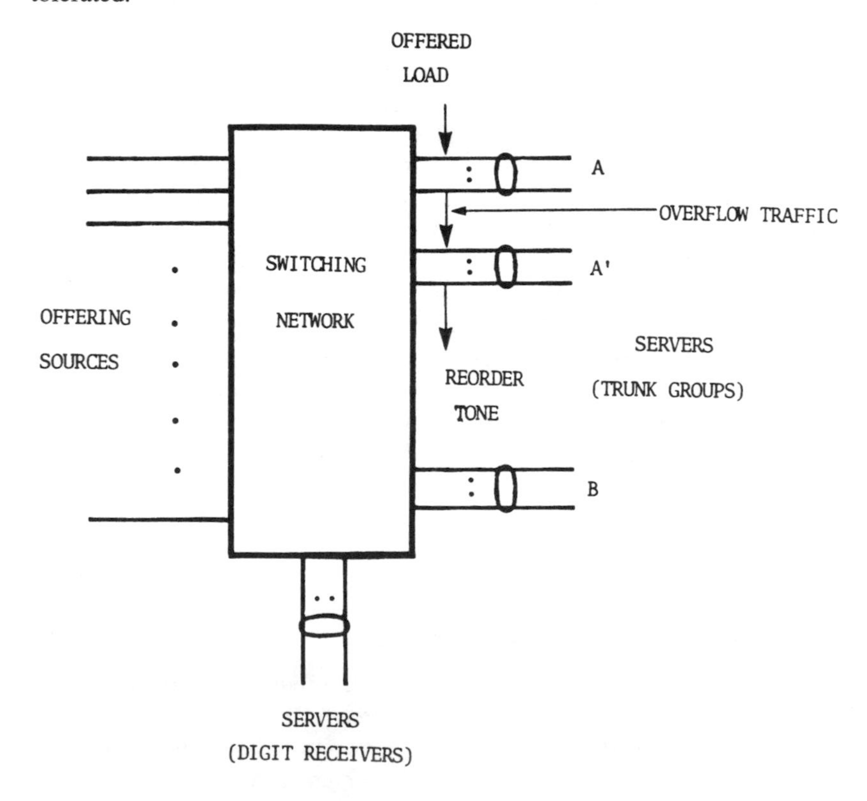

FIGURE B.1. Traffic relationships.

B.3.3. Traffic Loss Probabilities

Grade of service for blocking systems is expressed as probability of loss. For delay systems, it generally is expressed in terms of probability of delay exceeding t seconds and average delay of offered or delayed calls. The grade of service is determined by the offered load, the number of servers, and the underlying assumptions of traffic characteristics. Three basic formulas in prominent use in North America are based upon three different assumptions as to the behavior of call attempts finding all servers busy.

B.3.3.1. Erlang B Formula

The Erlang formula of the first kind, known as Erlang B formula or Erlang loss formula, is based upon the assumption that calls finding all servers busy will be cleared from the system and will have no further effect on it. This is known as the *blocked calls cleared* assumption. The probability of blocking for a call in traffic load a offered to s servers is expressed by

$$B_{s/a} = \frac{a^s/s!}{\sum_{k=0}^{s}\left(\frac{a^k}{k!}\right)} \tag{B.1}$$

where a is the traffic load in erlangs (one erlang = 3600 s). In Fig. B.1, traffic offered to trunk group A behaves according to the assumptions underlying the formula, because calls finding all trunks in group A busy will overflow and be offered to trunk group A′. Traffic offered to trunk group B, however, does not overflow and is blocked. The caller will receive a busy signal and likely will retry the call within a short time or even immediately. For a two-way trunk group, traffic loads offered from each end are added to obtain the total offered load a.[5]

B.3.3.2. Poisson Formula

The underlying behavioral assumption for the Poisson formula is that a call finding all servers busy will wait for its intended holding time, then depart the system, and will not reappear for the remainder of the study period, usually an hour. If a server becomes idle while the call is waiting, the call will seize the server and hold it for the balance of its intended holding time. This is known as the *blocked calls held* assumption. The probability of blocking for a call in traffic load a offered to s servers is defined by[6]

$$P_{s,a} = e^{-a}\sum_{j=s}^{\infty}\frac{a^j}{j!} \tag{B.2}$$

where a is the traffic load in erlangs and e is the natural or Naperian logarithmic base, a constant 2.71828. Although the queue discipline of the Poisson formula is unlike any actual physical behavior of waiting traffic, it does offer some allowance for retrials and just happens to give a slightly better solution than Erlang B or C for blocking on a trunk group which has no alternate route for blocked calls. Though retrial habits of different groups of people vary widely according to circumstances, this formula does happen to interpolate between the Erlang B formula and the delay formula, Erlang C.[7]

B.3.3.3. Erlang C Formula

The Erlang formula of the second kind, known as the Erlang C formula or Erlang delay formula, is based upon the assumption that offered calls which find all servers busy will wait indefinitely for a server to become available and then will occupy that server for its full intended holding time. This is known as the *blocked calls delayed* assumption and implies that the offered load is equal to the carried load. Perhaps the Erlang C formula can best be understood by its relationship to the Erlang B formula. Using the terms of Eq. (B.1), the probability of delay $P(>0)$ for a call finding all s servers busy can be expressed by

$$P(>0) = \frac{B_{s,a}}{1 - \frac{a}{s}(1 - B_{s,a})} \tag{B.3}$$

where a is the offered load in erlangs. This formula is valid only when the offered load is less than the number of servers, when calls are handled in the order of arrival, and for exponentially distributed holding times. For constant holding times, formulas were developed by A. K. Erlang for only single-server applications, but graphs have been prepared for multiple-server groups for both exponential and constant holding times. Values of $P(>0)$ for constant holding times are only slightly lower than the corresponding values for exponential holding times. For holding times which are neither constant nor exponential, the data for exponential holding times should be used because they yield a slightly better grade of service.[8]

B.3.3.4. Comparison of Traffic Formulas

A comparison of the results of the three traffic formulas is shown by the curves derived from them in Fig. B.2. The abscissa denotes the load in erlangs offered to a group of ten trunks, and the ordinate reflects the probability that a call attempt finds all servers busy. For the Erlang C curve, the ordinate reflects the probability of delay $P(>0)$. For low blocking probabilities up to about 1%, the Erlang C and Poisson curves yield the same results, and there is only a slight difference in results obtained from Erlang B. For example, at an offered load of 4 erlangs to a

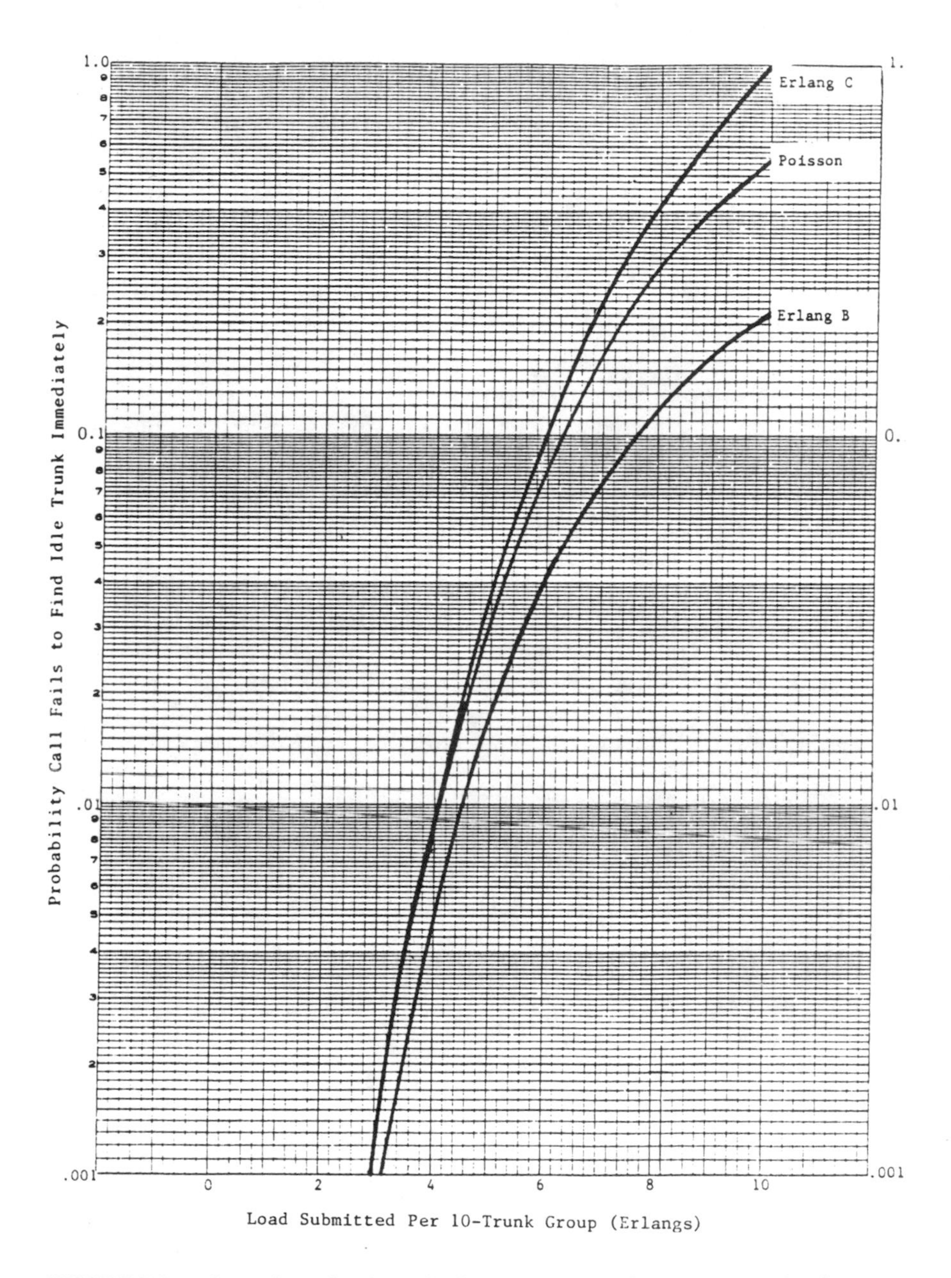

FIGURE B.2. Comparison of major traffic formulas. (From *Defense Communication System Traffic Engineering Practices*, Vol. XII.)

10-trunk group, the Erlang B curve reflects a blocking probability of 0.5%, while the Poisson and Erlang C curves show 0.9% probability of blocking. At higher traffic loads, however, there is some divergence of the Erlang C and Poisson curves because of the Erlang C assumption that calls wait indefinitely, while the Poisson assumption is that they wait only a limited time. The greater divergence of the Poisson and Erlang B curves results from the Erlang B assumption that blocked calls disappear from the system immediately. An offered load of 6 erlangs to the 10-trunk group results in approximately 4% blocking under the Erlang B assumption, 8% blocking under the Poisson assumption, and 10% blocking under the Erlang C assumption.[9] The true blocking probability is a function of retrial habits.

B.3.3.5. Nonrandom Traffic Theories

The assumptions that offered traffic is purely random and that it is from a large number of sources are not always true. For example, in Fig. B.1, the traffic overflowing from trunk group A to trunk group A′ is from a single source and has a peakedness characteristic. This can be seen by examining the pen recording of the number of simultaneously busy trunks in a trunk group during a 45-minute period, shown in Fig. B.3. If the traffic shown represented the characteristics of the traffic offered to trunk group A in Fig. B.1, and if trunk group A had 15 trunks, the overflow traffic offered to trunk group A′ would consist of spurts, or peaks, of traffic overflowing trunk group A at about 9:46, 9:58–10:01, and 10:23. If one were to assume that first-routed traffic, in addition to overflow traffic from other trunk groups, is offered to trunk group A′, it can be seen that *average* traffic

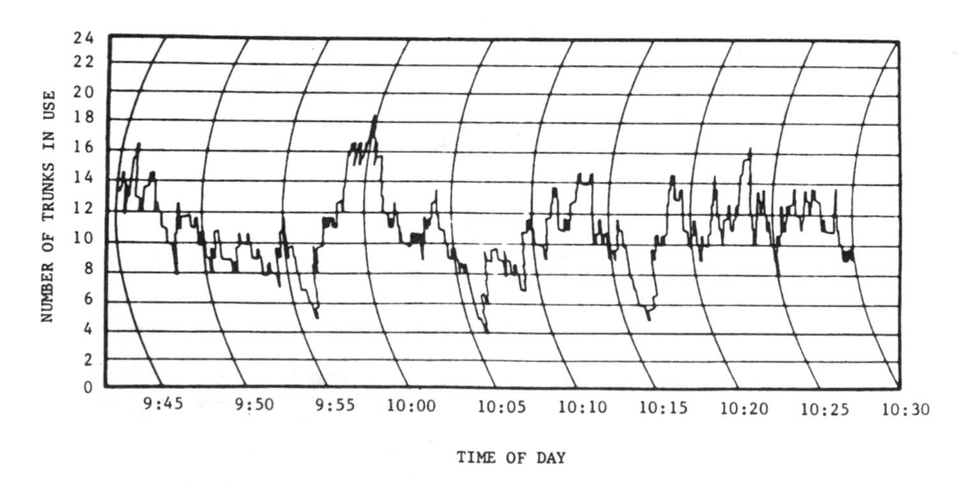

FIGURE B.3. Continuous load distribution during the busy hour. (From *Defense Communications System Traffic Engineering Practices*, Vol. I.)

intensity would not be a true representation of the total offered traffic load. The peakedness factor can be quantified by relating the peakedness of overflow traffic to that of random traffic. If random traffic has a peakedness of one, overflow traffic can have peakedness up to about four.[10]

Two theories, producing about the same results, were developed at about the same time to solve the problem of nonrandom traffic. R. I. Wilkinson, of Bell Telephone Laboratories, published an *equivalent random theory* in 1956, and G. Bretschneider published a *traffic variance method* the same year. The principles are the same; only the methods vary slightly in handling the variance. The principle involves finding a pseudotrunk group with random traffic input which is equivalent to the actual trunk group with peaked traffic. Wilkinson's method uses a series of curves based upon the mean and variance of the traffic, while Bretschneider's method uses tables and curves based upon the mean and random factor characteristics of the traffic to produce the same results.[11]

B.4. NETWORK MANAGEMENT

A modern telephone network using extensive alternate routing and common-control switching makes efficient use of transmission facilities but is subject to degradation during periods of heavy traffic overloads or major equipment failures. As traffic increases beyond engineered capacity, the use of alternate routing involves more trunks and switching systems per call. Under those conditions, more and more calls receive all-trunks-busy signals, and retrials add to the traffic demands on service circuits and control systems. This effect is much more severe with inband signaling than when common-channel signaling is used.

The most degrading effect is switching congestion. If an incoming call, using inband signaling, finds all digit receivers busy, it will queue until a receiver becomes available or until the transmitter at the other switching system times out and routes the call to reorder tone or an announcement. Transmitter holding time has increased, and the control system has had to use real time to route the call twice—once to the trunk group and once to reorder. Under exceedingly heavy traffic overloads, the mean holding time of service circuits and the average call processing time increase. As the overload continues to feed on itself, ineffective attempts build up, switching congestion spreads, and the network-carried load decreases. Finally, so much processor time is spent on ineffective attempts that trunks begin to be idle and trunk holding times become shorter.

The most difficult type of ineffective attempt to cope with is subscriber reattempts after unsuccessful first attempts. The frequency of reattempts and the time before the first reattempt vary with network conditions. Some subscribers will reattempt almost immediately after an unsuccessful attempt. Others will wait for a short time, while still others will wait a considerable length of time. Many factors influence reattempt habits. On normal heavy traffic days, such as Christmas Day or Mother's Day, calls are likely to be reattempted very frequently. The type of indication received during an unsuccessful attempt also is a factor. Announcements tend to result in greater spacing between reattempts than tones. Studies have shown that the probability of reattempting after an ineffective call attempt typically is in the range of 0.55 to 0.85 with a planning assumption of 0.7.

The effect of traffic overloads can be seen in Fig. B.4, which shows the results of an event simulation of a network of about 50 switching systems. As the offered load built up toward the engineered load of 1600 erlangs, the carried traffic in-

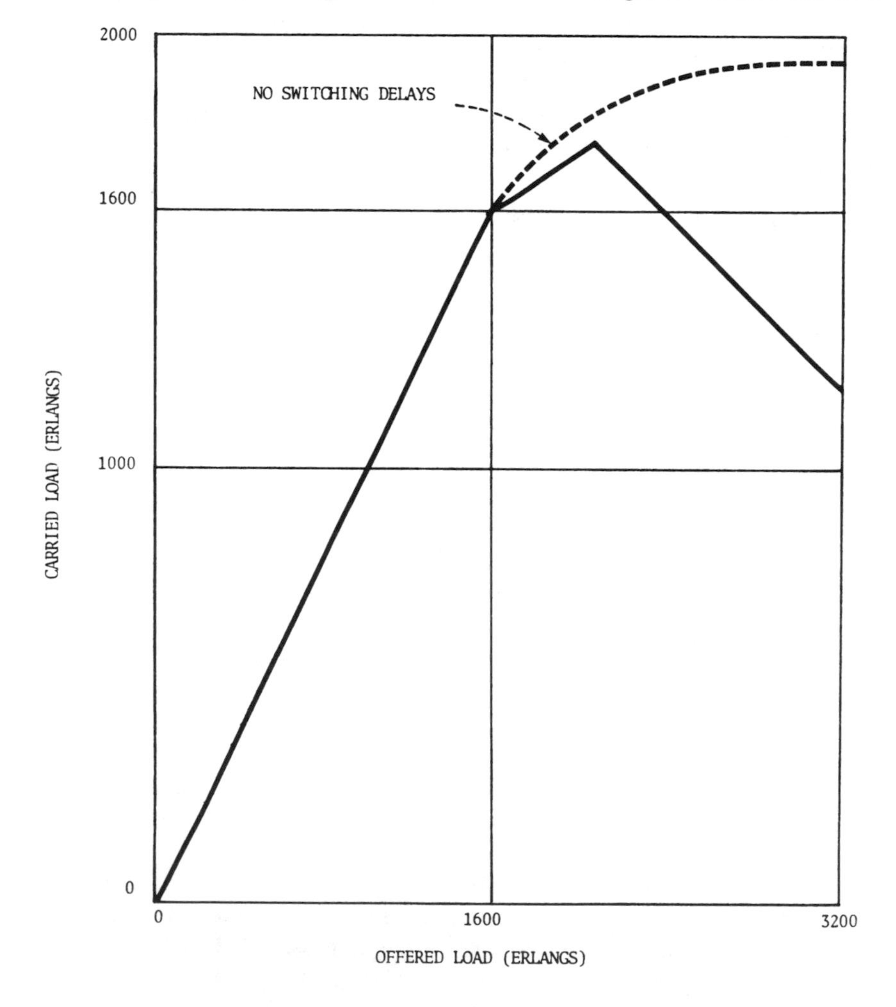

FIGURE B.4. Effect of switching congestion. (© 1977 Bell Telephone Laboratories. Reprinted by permission.)

creased linearly. The dashed line shows how the carried load would be affected with no switching congestion as the offered load increased to twice the engineered load. The solid line, however, shows that the carried load did continue to increase for a while but at a slower rate. A point was reached—generally occurs at about 20–30% above engineered load—where the carried load began to decrease sharply and continued to decrease even though the offered load continued to rise.[12]

Network management is real-time surveillance and control to enable a degraded network to operate as efficiently as possible. Network management is an art—not an exact science. Traffic analysis is used by network managers to take actions to implement controls, but there are an infinite number of possible situations that can be encountered in a large network, and each one must be treated individually.

B.4.1. Principles of Control

The overall objective of network management is to enable as many calls as possible to be completed under any condition of degradation. Certain principles have proved valid for large networks. The overriding principle is to *use the minimum controls necessary to enable the network to operate as normally as possible.* When switch overloads do occur, overload controls are necessary to *inhibit switching congestion.*

B.4.2. Principal Controls Available

There are two major categories of controls available for network management. *Expansive* controls are employed to divert some traffic from a portion of the network that is overloaded to call destinations via other portions that are underloaded. *Restrictive* controls remove some traffic from overloaded portions of the network when that traffic has a low probability of completion. Most controls are of the restrictive type. Controls may be applied manually or automatically, and switching systems should have the capability to implement controls using both methods, as appropriate. Dynamic controls generally are more effective than manual controls because they can be applied and removed instantaneously based upon traffic thresholds.[13]

The following control is applicable primarily to networks using inband signaling and hierarchical routing.

Trunk directionalization changes two-way trunks into one-way operation. It has two major applications. In case of severe switching congestion or a natural disaster which attracts focused heavy calling, trunks can be partially directionalized in favor of the affected switching system. This will allow calls to depart the congested switch, but the number of calls entering the switch will be reduced. In addition, during periods of exceptionally heavy traffic throughout the network, trunk groups can be partially directionalized in favor of traffic proceeding down a network hierarchy. That traffic may already have used several switching systems and trunks and may need only one or two more links to complete. In electromechanical switching systems, trunk directionalization is applied dynamically using manually set thresholds. In stored program systems, it can be applied through software.

The following controls are applicable to any network configuration using any signaling system. In networks using common-channel signaling, they are implemented through the signaling system:

Alternate route cancellation is an effective control to protect overloaded switching systems. Its primary application is in a network employing fixed alternate routing with inband signaling. It may be implemented manually or automatically, but is always applied selectively. All traffic overflowing a specific route may be prevented from advancing to its next alternate route, or all overflow traffic may be prevented from advancing to a specific route. Both variations may be implemented manually, and the latter may be applied automatically upon receipt of a signal from a congested switching system.

Code blocking is used when one or more locations are virtually isolated from the network, generally because of equipment failure, and traffic to the affected destinations has little or no chance to complete. Calls to affected codes should be diverted to a recorded announcement at a switching location as near to the point of origin as practicable.

Recorded announcements are an effective way to increase the spacing between retrials. In one subjective study of calling habits, it was found that the average spacing between retrials was approximately doubled when attempts were diverted to announcements rather than reorder tone.

Call gapping limits the number of calls routed to a specific code or to a specific number. It is very effective in controlling mass calling by blocking some of the calls at or near the point of origin rather than at or near the destination switching system. An adjustable timer in stored-program-control switching systems or in common-channel signaling systems, set at one of several time intervals (for example, from 0 to 600 seconds), allows one call to have access to the network code or number and then blocks all others for the duration of the time interval, after which it allows another call to go through and recycles itself. A call gapping time interval set to infinity can function as a code blocking control.

Reroutes expands routing doctrine to allow calls to be routed in other than the normal manner. This permits heavy traffic in one portion of the network to use idle capacity in another portion. Reroutes are particularly effective in cross-routing through different time zones and in routing calls around a major facility failure.

Common-channel signaling systems reduce the need for controls which have their greatest application in a network with predominantly inband signaling. Common-channel signaling systems, however, can be affected by traffic overloads, failures, and switching congestion. Dynamic overload controls are effective in controlling signaling link overloads and in reducing the traffic load on congested signaling processors.

REFERENCES

1. *Switching Systems*, American Telephone and Telegraph Company, 1961, pp. 91–98, 105.
2. *Local Switching System General Requirements*, PUB48501, American Telephone and Telegraph Company, 1980, Section 7.4.11.1.
3. Ibid., Section 8.2.
4. *Telephone Traffic Theory—Tables and Charts, Part 1*, Siemens Aktiengesellschaft, Munich, 1970, pp. 15–17 (hereinafter cited to as *Telephone Traffic Theory*).
5. Cooper, Robert B., *Introduction to Queueing Theory*, The Macmillan Company, New York, 1972, pp. 65–71.
6. Ibid., pp. 77–80.
7. *Engineering and Operations in the Bell System*, Bell Telephone Laboratories, Inc., 1977, p. 484 (hereinafter cited as Engineering and Operations).
8. *Telephone Traffic Theory*, pp. 351–359.
9. *Defense Communications System Traffic Engineering Practices*, Defense Communications Agency, 1969, Vol. XII, pp. 2–3 (hereinafter cited as *Traffic Engineering Practices*).
10. *Engineering and Operations*, p. 485.
11. *Traffic Engineering Practices*, vol. XII, p. 5.
12. *Engineering and Operations*, pp. 493–495.
13. *Notes on the Network*, American Telephone and Telegraph Company, 1980, Section 11, pp. 6–9.

Appendix C

Analog Cellular Systems

C.1. INTRODUCTION

The purpose of this appendix is to provide the reader an understanding of the analog cellular system used exclusively in the United States and Canada prior to the advent of digital service. This system is known as *Advanced Mobile Phone Service (AMPS)*. It still is in use throughout both countries as the primary system in the less heavily populated areas. In addition, it serves as the alternative system in heavily populated areas for users without digital end instruments, and for users whose end instruments cannot operate on the standard of their selected system.

The reader is referred to Chapter 7 for coverage of those topics that are the same for both digital and analog systems. These topics are the introductory material of Section 7.1, the mobile antenna (Section 7.2.1), signal-fading characteristics (Section 7.2.2), and frequency reuse (Section 7.2.5). In addition, Section 7.3.3, roaming, and Section 7.6, spectrum efficiency, provide additional information that is common to both digital and analog systems.

Much of the material of this appendix is taken from Ref. 1.

C.2. ANALOG SYSTEM OPERATION

In analog system operation, each cell site has a transceiver for each voice channel assigned to it, as well as transmitting and receiving antennas capable of operating on these channels. The cell site also contains scanning receivers capable of measuring the signal strengths of mobiles that may be entering the service area of the cell. Certain designated radio channels serve as setup channels rather than voice channels. On the setup channels, information needed to establish calls is exchanged. When a mobile is turned on it looks for the strongest setup channel and monitors that channel.

C.2.1. Call Setup

The mobile continues to monitor the setup channel unless directed otherwise by the setup channel, or unless poor reception causes it to look for a better channel. When the mobile unit detects that it is being called ("paged"), it quickly samples the signal strength of all the system's setup channels so it can respond through the cell that provides the strongest signal at its current location. The mobile unit then seizes the new setup channel and sends its response to the page. The system next sends a voice-channel assignment to the mobile. Following this, the mobile tunes to the assigned channel, where it receives a ringing signal to notify the user of the incoming call. A similar set of actions occurs when the mobile user initiates a call.

C.2.2. Signal-to-Noise Ratio

Listener tests have shown that at a signal-to-noise (S/N) ratio of 18 dB, most listeners considered the channel to be good to excellent. System designers thus attempt to achieve or exceed this value, with a high probability in 90% or more of a cell's service area.

C.2.3. Signal-to-Interference Ratio

Listener tests showed that at S/I = 17 dB, most listeners considered the channel to be good to excellent. Accordingly, this value must be achieved over at least 90% of a cell's service area. In the high interference environments of Newark, NJ and Philadelphia, PA, a system could meet this requirement at a cochannel separation of 4.6 cell radii with the use of 120° directional antennas. This separation corresponds to a 7-cell repeating pattern (see Section 7.2.5).

C.2.4. FM Deviation

Analog cellular voice channels operate with a peak FM deviation of 12 kHz and a spacing of 30 kHz. Section 7.2.6 describes the channel allocation plan. To minimize adjacent channel problems, the largest possible frequency separation is maintained between the channels of a given set. If N channel sets are required, the nth set ($1 \leq n \leq N$) contains channels n, $n + N$, $n + 2N$, etc. Thus if $N =$ 7, set 2 contains channels 2, 9, 16, etc.

C.3. ANALOG SYSTEM CONTROL

Control techniques relate to the functions of supervision, paging, access, and seizure collision avoidance, the last item referring to the loss of calls because of the simultaneous arrival of two or more control messages. Mobile telephone supervision includes 1. detecting changes in the subscriber's on-hook/off-hook condition, and 2. ensuring that adequate signal strength is maintained during a call.

C.3.1. Supervision

False supervisory indications caused by cochannel interferers must be avoided. To achieve this, the system uses a combination of a tone burst and a continuous out-of-band modulation for supervisory purposes. These are known, respectively, as the signaling tone (ST) and the supervisory audio tone (SAT).

C.3.1.1. Signaling Tone

The signaling tone (ST), 10 kHz, is present when the user is (1) being alerted, (2) being handed off, (3) disconnecting, or (4) flashing for custom services during a call. The ST is used only in the mobile-to-land direction.

C.3.1.2. Supervisory Audio Tone

The supervisory audio tone (SAT) used at a given cell site is 5970, 6000, or 6030 Hz. A mobile receives a SAT from a cell site and transponds it back. The received signal strength from the mobile can be used in controlling its power level. If the cell site should receive a SAT different from the one it sent out, it concludes that interference corruption is present in the path to or from the mobile. The use of three SATs has been shown[2] to multiply effectively the D/R ratio for supervision by $N^{1/2}$. For $N = 3$, this is 1.732. Thus, for $N = 7$, a cell site with both the same RF channel set and the same SAT is effectively as far away as if N were 21. Thus, in addition to its frequency management features, the three-SAT supervisory scheme provides supervision reliability by reducing the probability of misinterpreted interference.

The three SATs are sufficiently close together that a single phase-locked tracking filter can lock onto any one of them. They are high enough in audio frequency that they can be filtered out easily from the voice signal. This also aids in the control of intermodulation products. The FM deviation of the SAT is ± 2 kHz.

C.3.1.3. Locating Function

Locating is done by (1) measuring the RF signal strength of a mobile on the channel it is using, and (2) measuring its range based on the round-trip delay time of the SAT. Analysis of this information determines whether a change of channels and/or handoff is required.

C.3.2. Paging and Access

The *paging* function is performed by the mobile telephone switching office (MTSO) seeking a mobile to complete a call to it. The *access* function is performed by the mobile in placing a call. These functions are performed on *setup* channels.

Paging messages contain the binary equivalent of the mobile unit's telephone number. In addition, paging messages include *overhead word* messages to give the mobile descriptive information about the local system. The contents of the overhead word are described in Section C.3.3. To the 28 message bits required for a page, 12 additional bits are added for parity protection. The encoded message is transmitted at a 10 kb/s rate.

The access function involves (1) informing the system of the mobile's presence, (2) supplying the system with the mobile's identification and the dialed digits, and (3) waiting for a channel designation.

C.3.3. Setup Channels

Setup channels provide the information needed to establish calls. Calls *to* the mobile involve paging, while calls from the mobile involve access. The paging function includes the sending of overhead words. The overhead word includes (1) an "Area Call Sign" to permit the automatic roaming feature, (2) the cell site's SAT identification, (3) a parameter *N* that specifies the number of setup channels in the repeating set, (4) a parameter *CMAX*, which specifies the number of setup channels to scan when a call is made, and (5) a parameter *CPA*, which tells the mobile units whether the paging and access functions share the same setup channels.

Each setup channel has a data-rate capability on the order of 10 000b/s. The actual information rate to and from each mobile is only 1200 b/s, thus allowing for the redundancy needed for forward error correction.

The use of the setup channel by the mobile is as follows:

1. Upon turn-on, and about once per minute thereafter, the mobile scans the top 21 channels and selects the strongest one on which to read an overhead message. This permits the mobile to determine if it is being served by its home system, and to retrieve the frequency reuse factor *N*. To receive pages, the mobile then scans the appropriate set of *N* channels to find the strongest one.
2. When a call is initiated either to or from the mobile, the mobile must repeat the scanning process to select the best cell for access. To do this, it scans the *CMAX* channels.
3. The mobile determines if its chosen setup channel is idle. If so, it attempts an access by transmitting the necessary information to the cell site: (a) if answering a page, its identification, or (b) if originating a call, its identification and the dialed digits. The mobile then turns its transmitter off but remains tuned and synchronized to its chosen setup channel.
4. After the land portion of the system has processed the access information, it sends a channel-designation message to the mobile on its chosen setup channel.

Upon receipt of this message, the mobile tunes to the designated channel, whereupon the voice portion of the call can proceed.

C.3.4. Seizure Collision Avoidance

Since all mobiles compete for the same setup channels, methods are needed to minimize collisions and to prevent temporary system disruption when collisions occur. Several techniques are used for this purpose:

1. The forward setup channels set aside every 11th bit as a "busy/idle" bit. As long as a cell perceives legitimate seizure messages directed toward it, it sees that the "busy/idle" bit is set to "busy."
2. The mobile sends a "precursor" in its seizure message. This precursor tells the land portion of the system which cell site it is attempting to use.
3. Before the mobile attempts to seize a setup channel, it waits a random time. This eliminates the periodicity introduced into the mobile seizures by the format of the setup channel messages.
4. After a mobile sends its precursor, it opens a "window" in time in which it expects to see the channel become busy. If the idle-to-busy transition does not occur within the time window, the seizure attempt is aborted.
5. If the initial seizure is unsuccessful, the mobile automatically tries repeatedly at random intervals. To prevent continual collisions and consequent system overload, however, a limit is placed on the number of automatic reattempts that are permitted.

C.3.5. Error Limits

The error limits for information transfer are the following:

1. The allowed miss rate for messages is $\approx 10^{-3}$ at $S/I = 15$ dB. Since S/I is better than 15 dB in much of the service area, this implies a miss rate of about 10^{-4}. This miss rate is very small compared to the probability that a call will be missed because a mobile is unattended. Moreover, it is consistent with the requirement placed on mishandled calls in the wire-line telephone network.
2. The falsing rate (incorrect data interpretation) should be less than 10^{-7} for a given message. This requirement is needed to assure that less than one transmitted page in 200 will result in a false response.

The mobile propagation environment produces a high probability that bits in long blocks will be lost. Extensive interleaving of bits might be used to reduce the number of message bits lost in a given fade. The effectiveness of bit interleaving in the mobile channel, however, is limited by the high probability of long fades.

Radio channels do not lend themselves to the request for repeat and retransmission techniques that otherwise might be used when an error is detected. Instead, each message is automatically repeated and the repeats are summed to decrease the effective bit-error-ratio.

Each repetition is encoded using a cyclic redundancy check and the sum word is error corrected. Message repetition with majority voting and single-error correction has proved to be a simple way to supply reliable signaling over the mobile telephone channel. The redundancy inherent in this approach, and the self-clocking nature of the Manchester code used, are the basis of the data transmission technique that provides the signaling and control information essential to the operation of the system.

C.3.6. Blank and Burst

When the need arises to hand-off a mobile from the serving cell to another one, the new channel information must be sent to the mobile on the voice channel in use at the time. To do this, the voice signal is blanked and the data is sent in a burst using the channel's bandwidth. The falsing rate requirement for this function is 10^{-8}, because the effect of falsing in this case is similar to that of a mishandled call on the wireline network.

Messages are repeated 11 times in the forward direction but only 5 times in the reverse direction. The reason for the difference is that the handoff message is considered to be a critical function. A false interpretation results in a mishandled call. The transmission usually occurs under poor S/I conditions.

C.3.7. Making Calls

The procedures followed in calling to and from a mobile are described next.

C.3.7.1. Call to Mobile

The actions required in processing a call to a mobile are:

1. Paging: From the calling party's central office, the call is routed by the wireline network to the home MTSO of the mobile. The MTSO receives the digits, converts them to the mobile's identification number, and instructs the cell sites to page the mobile over the forward setup channels. The paging signal thus is broadcast over the entire service area.
2. Cell site selection: The mobile unit, upon recognizing its page, scans the setup channels used for access in the mobile serving area, using parameters derived

from the overhead word, and selects the strongest one. The selected channel usually will be one from a nearby cell site.
3. Page reply: The mobile responds to the cell site it selected over the reverse setup channel. The selected cell site then reports the page reply to the MTSO over its dedicated land-line data link.
4. Channel designation: The MTSO selects an idle voice channel and an associated land-line trunk in the cell site that handled the page reply. It then informs the cell site of its channel choice over the appropriate data link. The serving cell site then informs the mobile of its channel designation over the forward setup channel. The mobile tunes to its assigned channel and transponds the SAT over the voice channel. Upon recognizing the SAT, the cell site places the associated land-line trunk in an off-hook (in use) state, which the MTSO interprets as successful voice-channel communication.
5. Alerting: Upon command from the MTSO, the serving cell site sends a data message over the voice channel to ring the mobile telephone. Signaling tone from the mobile causes the cell site to confirm successful alerting to the MTSO. The MTSO, in turn, provides audible ringing to the calling party.
6. Talking: When the mobile user answers, the cell site recognizes removal of signaling tone by the mobile and places the land-line trunk in an off-hook state. This is detected at the MTSO, which removes the audible ringing circuit and establishes the talking connection so that conversation can begin.

C.3.7.2. Call from Mobile

The actions required in processing a call from a mobile are:

1. Preorigination: The user enters the digits to be dialed into the mobile unit's memory.
2. Cell site selection: After the mobile unit is placed in an off-hook state, cell site selection occurs as in 2. of Section C.3.7.1.
3. Origination: The stored digits, along with the mobile's identification, are sent over the reverse setup channel selected by the mobile. The cell site associated with this setup channel receives this information and relays it to the MTSO over its land-line data link.
4. Channel designation: As in 4. of Section C.3.7.1, the MTSO designates a voice channel and establishes voice communication with the mobile. The MTSO also establishes routing and charging information at this time by analyzing the dialed digits.
5. Call completion: The MTSO completes the call through the wireline network using standard techniques.
6. Talking: When the MTSO establishes a talking connection, communication between the users takes place when the called party answers.

C.3.8. Handoff

The actions common to mobile-originated and mobile-completed calls are:

1. New channel preparation: Location information gathered by the cell site serving the mobile, as well as by surrounding cell sites, is sent to the MTSO over the various cell site land-line data links. The data are analyzed by the MTSO, which may decide that a handoff to a new cell site is to be attempted. The MTSO selects an idle voice channel and an associated land-line trunk at the receiving cell site. It also informs the new cell site to enable its radio. The cell site then sends SAT.
2. Mobile retune command: A message is sent to the current serving cell site informing it of the new channel and new SAT for the mobile. The serving cell site sends this information to the mobile over the voice channel.
3. Channel/path reconfiguration: The mobile sends a brief burst of signaling tone and turns off its transmitter. It then retunes to its new channel and transponds the SAT it finds there. The old cell site, having recognized the burst of signaling tone, places an on-hook signal on the trunk to the MTSO. The MTSO reconfigures its switching network, connecting the other party with the appropriate land-line trunk to the new serving cell site. The new serving cell site, upon recognizing the transponded SAT on the new channel, places an off-hook signal on the associated land-line trunk. The MTSO interprets these two signals (off-hook on new trunk and on-hook on old trunk) as a successful handoff.

C.3.9. Disconnect

Disconnection can be initiated by either the mobile party or the land party. The resulting actions differ to some extent, as described next.

C.3.9.1. Mobile-Initiated Disconnect

The actions that occur when the mobile party goes on-hook are:

1. Release: The mobile unit sends ST and turns off its transmitter. The ST is received by the cell site, which places an on-hook signal on the appropriate land-line trunk.
2. Idle: In response to the on-hook signal, the MTSO idles all switching office resources associated with the call and sends any necessary disconnect signals through the wireline network.
3. Transmitter shutdown: As the final action in the call, the MTSO commands the serving cell site over its land-line data link to shut down the cell-site radio transmitter associated with the call. All equipment used on this call now may be used on other calls.

C.3.9.2. System-Initiated Disconnect

The actions that occur when the land party goes on-hook are:

1. Idle: In response to the disconnect signal received from the wire-line network, the MTSO idles all switching office resources associated with the call.
2. Ordered release: The MTSO sends a release-order data-link message to the serving cell site. This cell site sends this command to the mobile over the voice channel. The mobile confirms receipt of this message by invoking the same release sequence as with a mobile-initiated disconnect.
3. Transmitter shutdown: When the MTSO recognizes successful release by the mobile, it commands the serving cell site to shut down the radio transmitter, as described in 3. of Section C3.9.1.

C.3.10. Summary

The MTSO, cell site, and mobile work together to perform the system control functions. These include the regular telephony control functions plus a set of functions that result either directly from the cellular concept or from the nature of the radio environment. The control functions are partitioned among the control elements as shown in Fig. C.1.

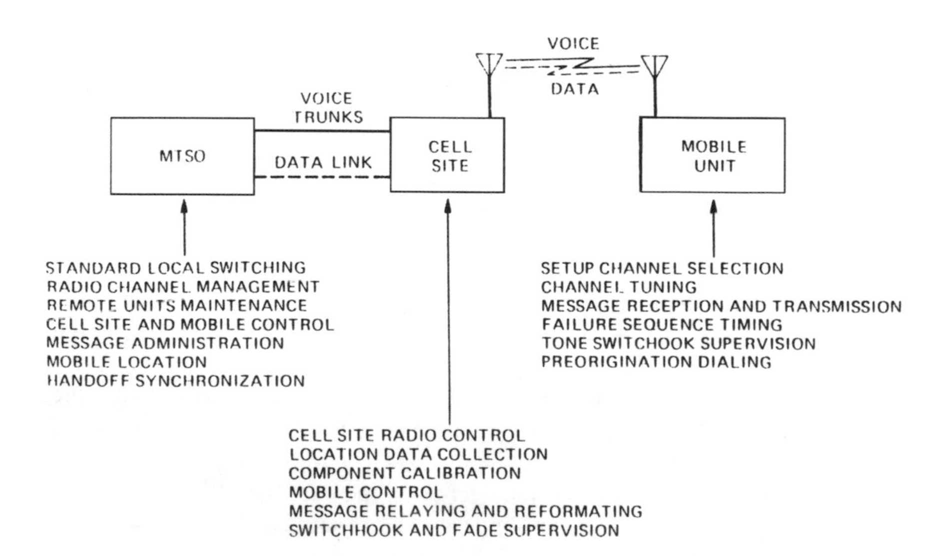

FIGURE C.1. Partitioning of control functions among AMPS control elements.

C.4. NARROWBAND ADVANCED MOBILE PHONE SERVICE (NAMPS)

As a result of the rapid growth of cellular service in many areas, channels have become scarce. Not all communities, however, require the considerable increases in spectrum efficiency afforded by digital systems. In addition, some system operators have been unsure about which of the new digital standards they should adopt. These factors, along with a feeling on the part of some operators that the new digital technologies were untried in a full operational environment, led to their interest in a development by Motorola known as *Narrowband Advanced Mobile Phone Service (NAMPS).*

In NAMPS, the digital-control methods of AMPS are used, but the number of voice channels is tripled by reducing the channel bandwidth from 30 kHz to 10 kHz. This reduction is achieved by using a much lower FM deviation than is used for AMPS. The lower deviation provides less FM improvement, but the smaller bandwidth means less channel noise, and the added number of channels means less co-channel interference, other things being equal. Accordingly, performance from the user's viewpoint is essentially the same as is achieved with AMPS, and the system operator has three NAMPS channels for every AMPS channel that is taken out of service.

REFERENCES

1. "Advanced Mobile Phone Service," Special issue of *Bell Syst. Tech. J.*, Vol. 58, No. 1 (January, 1979).
2. MacDonald, V. H., "AMPS: The Cellular Concept," *Bell Syst. Tech. J.*, Vol. 58, No. 1 (January, 1979).

Glossary

In this glossary, some of the definitions are peculiar to digital telephony, even though in other applications the same terms may have different definitions.

AAL ATM adaptation layer.
A-Law A companding characteristic used in European PCM systems.
ABSBH Average busy season busy hour.
access Actions performed by a cellular mobile telephone in placing a call.
A/D Analog-to-digital conversion.
adaptive delta coding Delta coding using a step size that depends on the magnitude of the input.
adaptive equalization The provision of a uniform amplitude or group delay channel response over a given frequency range by sensing the channel's characteristics, and automatically adjusting an equalizer to achieve the required uniform characteristic.
adaptive predictive coding Coding based on predicting the present input sample using the previous cycle or pitch period.
adaptive pulse code modulation Pulse-code modulation using adaptive quantization.
adaptive quantization Quantization in which the step size varies so that it matches the variance of the input signal.
adaptive transform coding A coding technique in which input subbands are transformed and the resulting coefficients encoded, usually using adaptive PCM.
address 1. The destination of a message in a communications system. 2. The group of digits comprising a complete telephone number.
address signals Signals which are used to convey destination information.
ADC Adaptive delta coding.
ADPCM Adaptive DPCM.
ADSL Asymmetrical digital subscriber line.
AIN Advanced intelligent network.
aliasing A condition in which a frequency Δf above half the sampling rate is reproduced as a frequency Δf below half the sampling rate.
alternate routing The use of a route other than a direct or first-choice route.
amplitude-phase keying (APK) Digital modulation in which both the amplitude and the phase of the carrier are altered to produce the various symbol states.
AMPS Advanced mobile phone service.
AMSC American Mobile Satellite Corporation.
analog signal A signal dependent upon continuous changes in magnitude to express information.
ANSI American National Standards Institute.
APC Adaptive predictive coding.

APCM Adaptive pulse-code modulation.

APD Avalanche photodiode. A detector used in optical-fiber systems

ARDIS Advanced radio data information service.

aries A LEO satellite system proposed by Constellation Communications Corp.

array, switching A matrix of crosspoints forming part of a switching network.

asynchronous communication A mode of transmission characterized by start-stop operation, with undefined time intervals between transmissions.

ATC Adaptive transform coding.

ATM Asynchronous transfer mode. A method of employing fast packet technology to switch multimedia signals.

autonomous registration The process by which a mobile telephone initiates registration on a cellular network serving its current location.

availability The percentage of time during which a system provides its intended service.

average busy season busy hour The three months with the highest average traffic in the busy hour, excluding Mother's Day, Christmas, and extremely high traffic days attributed to unusual events.

AWGN Additive white gaussian noise.

baseband The band of frequencies consisting of an original signal prior to modulation, or after demodulation.

battery The direct-current voltage used to power telecommunications equipment or circuits.

BCH Bose-Chaudhuri-Hocquenghem.

Bell System A consortium of telephone companies, mostly owned by the American Telephone and Telegraph Company, that provided telephone services in most of the United States prior to divestiture on January 1, 1984 as a result of an antitrust consent decree.

b.e.r. Bit error ratio. The ratio of erroneous bits in a bit sequence to the total number of bits in that sequence.

BETRS Basic Exchange Telecommunications Radio Service.

BHCA Busy hour call attempts.

BISDN Broadband ISDN.

block code A code in which the redundant bits relate only to the information bits of the same block.

blocking The inability of a telecommunications system to establish a connection because of the unavailability of a path.

BOC Bell operating company.

BORSCHT An acronym referring to functions performed in, or in connection with, subscriber-line circuits in a switching system: Battery, Overload protection, Ringing, Supervision, Coding, Hybrid, and Test access.

BPSK Bi-phase-shift keying.

BRA ISDN basic rate access.

break The open state of relay or switch contacts.

BRI ISDN basic rate interface, at 144 kb/s, serving low volume ISDN subscribers.

bridged tap A pair of wires branched from a main pair in a telephone cable.

BRZ Bipolar return to zero. A channel code used for digital transmission.
BSC Base-station controller.
busy hour The clock hour during which the most traffic is experienced in a switching system or over a group of circuits.
B3ZS Bipolar with three-zero substitution. A channel code.
B6ZS Bipolar with six-zero substitution. A channel code.
B8ZS Bipolar with eight-zero substitution. A channel code.

C-message weighting A weighting used in a noise measuring set to measure noise on a line that would be terminated by a device having acoustic properties similar to those of a Western Electric Type 500 telephone set.
call progress signals Signals sent to a caller by a switching system to provide information relative to the establishment of a connection.
carrier 1. A company carrying electrically transmitted information for the public. 2. A wave having at least one characteristic that may be varied from a known reference value by modulation.
carrier serving area A geographic area served by a local telecommunications central office, delineated by the defined extent of the transmission system emanating from that office to serve customers in that area.
CCIS Common-channel interoffice signaling.
CCITT International Consultative Committee for Telegraphy and Telephony.
CCS7 *See* SS7.
CDM Continuous delta modulation.
CDMA Code-division multiple access. A technique used in cellular and satellite telephony transmission to provide a degree of protection against interference.
CDVCC Coded DVCC.
CEC Council of European Communities.
cellular Of or pertaining to the use of limited-area repeaters in the provision of mobile telephony service.
CELP Code-excited linear prediction.
Celsat A proposed GEO satellite system for mobile telephony.
central office One or more public network switching systems installed at a single location; the term is often used synonymously with switching system.
cepstrum The Fourier transform of the logarithm of the power spectrum.
channel bank A device which combines a number (example, 24) of voice channels together into a digital stream (for example, at 1.544 Mb/s) based upon sampling each voice channel at a specific rate (for example, 8000 times per second).
channel coder A device that processes a binary input into a multilevel or modified binary signal.
characteristic impedance The impedance a transmission line would present at its input terminals if it were infinitely long.
chip 1. Channel information piece; a signal element in spread-spectrum transmission. 2. An integrated circuit
circuit 1. A transmission path connecting two or more points. 2. A component of equipment.
CLAN Cordless local area network.

clock Equipment providing a time base used in a transmission system to control the timing of certain functions, such as the control of the duration of signal elements, and the sampling (CCITT).

CO Central office.

coding 1. Conversion of an analog function to digital form using a specific set of rules. 2. Using redundant bits for error control. 3. Formatting for transmission. 4. Encryption.

combined switching In time-division switching, the switching of each direction of conversation by a single control.

common-channel interoffice signaling (CCIS) A signaling system used in North America, in which a separate signaling network is used to exchange supervisory and address signals between switching systems.

common equipment Items of like equipment used on a shared basis by a switching system to establish connections.

communications quality Speech quality that is acceptable to users in a mobile environment.

compander A compresser-expander.

compresser A device which reduces the dynamic range of a signal.

concentration The process of connecting any of a number of inlets to one of a smaller number of outlets.

continuous delta modulation Delta modulation (coding) using syllabic adaptation of the step size.

continuously variable-slope delta modulation Delta modulation (coding) using a set of discrete values of slope variations in which the slope changes are made at a syllabic rate.

convolutional code A code in which the redundant bits check the information bits in previous blocks.

cordless Of or pertaining to a telephone end instrument using a radio path to connect it with the wireline PSTN.

correlation Multiplication of an incoming signal by a locally generated function and averaging of the result. If the locally generated function is a delayed form of the received signal, the process may be called autocorrelation.

correlation bandwidth The span between two frequencies that fade within 3 dB of one another along a given path.

correlative coding Coding which uses finite memory to change the baseband digital stream to a form which improves coding efficiency from a spectrum occupancy viewpoint.

COT Central-office terminal.

CPE Customer-provided equipment.

CRC Cyclic redundancy check.

crosspoint A controlled device used by a switching system to connect one path to another.

crosstalk Undesired power coupled to a communications circuit from other communications circuits; may be intelligible or unintelligible.

CSA Carrier serving area.

CSU Channel service unit.
CVSD Continuously variable-slope delta modulation (coding).
CWR Continuous word recognition.

D/A Digital-to-analog conversion.
DAC Digital-to-analog converter.
DADE Diversity antenna delay equalization.
data under voice (DUV) Transmission in which a digital stream is sent using baseband frequencies lower than those used for analog transmission.
dBm Power level in decibels referred to a power of one milliwatt, used in telephony as a measure of absolute power.
dBmO The power in dBm measured at, or referred to, a point of zero transmission level.
dBrnc Decibels above reference noise, with the reference at −90 dBm using C-message weighting.
DCA Dynamic capacity allocation.
DCDM Digitally controlled delta modulation (coding).
DECT Digital European cordless telephone.
delay The amount of time by which a signal or event is retarded, expressed in time or in number of symbols or characters.
delay spread The lengthening of the duration of a pulse caused by the presence of multiple transmission paths from its source to its destination.
delta coding A one-bit version of DPCM in which the output bits convey only the polarity of the difference signal.
demand assignment The assignment of a channel on demand when needed, for the duration of the communication.
DEMS Digital electronic message service.
dial pulsing A means of address signaling consisting of regular, momentary interruptions of the direct current path at the sending end, in which the number of interruptions corresponds to the value of the address digit being transmitted.
differential pulse-code modulation Pulse-code modulation in which the quantization is done on a differential waveform produced by subtracting from the input the previous value of the output, or a weighted combination of previous output values.
digit synchronization The condition in which each digit (usually bit) is correctly sampled by the receiver, thus assuring its proper reception.
digital Information in the form of one of a discrete number of levels.
digital radio A radio that transmits a signal whose informational content is at least partly digital.
DSI. Digital speech interpolation.
DSM Digital symmetrical matrix.
digital termination systems (DTS) Radio local loops provided by competing common carriers in a given metropolitan area.
digitally controlled delta modulation Delta modulation in which step size information is derived directly from the bit sequence produced by the sampling and quantization process, with companding at a syllabic rate.

diphthong A gliding monosyllabic speech item that starts at or near the articulatory position for one vowel and moves to or toward the position for another.

discrete multitone A transmission technique in which a multiplicity of carriers (tones) are used in the transmission of a digital stream over a subscriber line.

dispersion Pulse broadening in an optical fiber caused by multipath wave propagation. The result is intersymbol interference, corresponding to a limitation on the bandwidth that can be transmitted.

distortion The amount by which an output waveform or pulse differs from the corresponding input waveform or pulse. It may be expressed in terms of any signal attribute, such as amplitude, phase, etc.

distribution The switching of traffic between concentration and expansion portions of a switching system network.

distribution cable Telephone cable that distributes circuits to individual subscriber locations.

dithering A variation of the quantization levels to break up signal-dependent patterns in the quantized result.

diversity The use of dual frequencies, paths, or polarizations to minimize fading problems in a microwave radio-relay system.

DLQ Dynamic locking quantizer.

DM Delta modulation (coding).

DNHR Dynamic nonhierarchical routing.

doppler Of or pertaining to a shift in a received frequency caused by motion of the frequency source toward or away from the observer.

DPCM Differential pulse-code modulation.

DRS Digital radio system.

dry contacts Contacts through which no direct current flows.

DS Direct sequence.

DSI Digital speech interpolation.

DSL Digital subscriber line. A loop transmission system utilizing one metallic pair and providing for ISDN basic-rate access (2B + D).

DSU Data service unit.

DS1 Digital stream rate number 1: 1.544 Mb/s.

DS2 Digital stream rate number 2: 6.312 Mb/s.

DS3 Digital stream rate number 3: 44.736 Mb/s.

DTMF Dual-tone multifrequency signaling.

DTS Digital Termination System.

DTX Discontinuous transmission.

dual mode Of or pertaining to a subscriber set capable of operation using more than one transmission standard.

dual-tone multifrequency signaling A method of transmitting address signals by transmitting a pair of discrete audio-frequency tones from a group of eight tones, usually used on subscriber lines.

duobinary A correlative coding technique in which a two-level binary sequence is converted into one that uses three levels. The conversion involves intersymbol interference extending over one bit interval.

DVCC Digital verification color code.

dynamic nonhierarchical routing A routing algorithm for circuit-switch connections that is not in conformity with a network hierarchical routing philosophy.

dynamic locking quantizer A quantizer capable of sensing and handling either speech or data inputs.

dynamic range The difference between the overload level and the minimum acceptable signal level.

E & M signaling A technique for transferring information between a trunk circuit and a separate signaling circuit over leads designated E and M.

ECH Echo-cancelling hybrid.

echo A reflection of a transmitted signal having sufficient magnitude to be perceived as separate from the transmitted signal.

EIRP Effective isotropically radiated power.

elastic store A digital store unit which accepts data under one timing source but outputs it under another. In this manner, jitter related to either the input or the output can be removed.

Ellipso LEO satellite system proposed by the Ellipsat Corporation.

EMX Electronic mobile exchange.

erlang A unit of telephone traffic intensity obtained by multiplying the number of calls by the average length of calls in hours. One erlang equals 60 minutes of traffic.

error coder A device which adds redundant bits to a bit stream to provide for error detection and, possibly, correction, at the receiver.

E-TDMA Extended TDMA.

ETSI European Telecommunications Standards Institute.

expander A device which increases the dynamic range of a signal.

expansion The process of connecting any of a number of inlets to one of a larger number of outlets.

extended framing format An extension of the multiframe structure for use of the framing bit sequence for signaling, cyclic redundancy check, and a data link.

FACCH Fast associated control channel.

F-Bit Framing bit.

fade margin The number of decibels by which a signal can fade before the fade takes the signal's level below the receiver's threshold.

far-end crosstalk (FEXT) Crosstalk that is propagated in a disturbed channel in the same direction as the propagation of signals in the disturbing channel.

FDD Frequency-division duplex.

FDDI Fiber-distributed data interface.

FDMA Frequency-division multiple access.

feeder cable Telephone cable that carries circuits from a central office or remote switching point to a subscriber distribution point.

FEXT Far-end crosstalk.

FH Frequency hopping.

FITL Fiber in the loop.

FHMA Frequency-hopping multiple access.

flat fading Fading that is uniform with respect to frequency.

FM Frequency modulation.

FM deviation The frequency shift imparted to a sinusoidal carrier in the process of frequency modulation.

FMR Follow-me roaming.

footprint The coverage area on the surface of the earth from a satellite beam.

formant A band of speech energy in the frequency spectrum. A resonance frequency of the vocal tract tube.

four-wire circuit A two-way transmission circuit using separate paths for each direction of transmission.

frame synchronization The condition in which each frame or block of received bits is correctly timed with respect to the received signal for the proper identification of the received bits as well as the individual channels.

framing The determination of which groups of bits constitute quantized levels and which quantized levels belong to which channels.

frequency spectrum The complex amplitude (magnitude and phase) versus frequency characteristic of an electrical signal. It is defined as the Fourier transform of the time domain expression of the signal.

Fresnel zone A means of expressing the clearance of a microwave beam over an obstacle. The boundary of the *n*th Fresnel zone consists of all points from which the reflected wave is delayed $n/2$ wavelengths.

fricative A sustained unvoiced sound produced from the random sound pressure that results from turbulent air flow at a constricted point in the vocal system.

FTTH Fiber to the home.

FWA Fixed-wire applications.

GEO Geostationary earth orbit.

G/T The ratio of a receiving system's gain to its noise temperature.

glare The condition resulting from a near-simultaneous seizure of a two-way trunk from both ends, in which the seizure signal appears to each switching system as a signal indicating a readiness to receive address digits.

Globalstar A LEO satellite system proposed by Loral/QUALCOMM for the provision of global service to mobile telephones.

GMSK Gaussian MSK.

graded index fiber An optical fiber that minimizes dispersion effects by providing nearly the same travel time for the various propagating modes.

grooming The separation and rearrangement of digital input signals (for example, VT1.5s or DS1s) into different channels; grooming requires a digital crossconnect capability.

ground start A method of supervision on subscriber lines by which a seizure is indicated by placing ground potential on one of the conductors.

GSM Global system mobile. *Also* Groupe Spéciale Mobile.

guard time The time interval provided between the pulses of two different time-division digital channels to allow the receiver to distinguish between the channels.

HDBH High day busy hour.

HDLC High-level data-line control.

HDSL High-bit-rate digital subscriber line.

HDTV High definition television.

hierarchical routing A standard plan or set of rules by which calls are routed over a network of switches arranged in a hierarchy.

high day busy hour The hour in the one day, among the ten in the ten high days, which has the highest traffic during the busy hour determined from the ten high day busy hour analysis.

HLR Home location register.

holding time The total time that a circuit or unit of equipment is held busy, usually expressed in seconds or other units of time.

HPA High-powered amplifier. The output stage in a transmitting earth station for satellite communication.

hybrid coder A combination waveform and parametric coder. Often a hybrid coder will perform waveform coding of voice pitch but parametric coding of the voice formants.

IC 1. Integrated circuit. An electronic circuit that consists of many individual circuit elements such as transistors, diodes, resistors, capacitors, inductors, and other active and passive semiconductor devices, formed on a single chip of semiconducting material and mounted on a single piece of substrate material. 2. Inter-LATA carrier. 3. Intercept announcement.

IDN Integrated digital network.

impulse function A function that begins and ends within a time so short that it may be regarded mathematically as infinitesimal, although the area described by the function is defined as being finite.

impulse noise Intermittent or spasmodic noise consisting of high-level pulses of short duration.

impulse response The response of a network to an impulse.

index of refraction The ratio of the velocity of light in vacuum to the velocity of light in a given medium (e.g., an optical fiber).

INS-Net An integrated services digital network in Japan.

integrated services digital network (ISDN) A switched network, with end-to-end digital connectivity, which supports a wide range of services.

integrity A condition of a component, system, or subsystem that is operating normally (also known as sanity).

interdigital time The time interval between address digits being transmitted over a circuit.

interface The point at which two systems or two parts of one system interconnect.

intermediate frequency A frequency band to which all frequencies incoming to a receiver are converted prior to demodulation.

intersymbol interference A form of self-interference in which the energy of preceding symbols occupies a time slot reserved for the present symbol.

intermodulation The production of unwanted frequency components as a result of a nonlinearity within a system.

IRED Infrared emitting diode. A source used in optical-fiber systems.

Iridium A system of 66 LEO satellites proposed by Motorola to provide global service to mobile telephones.

ISDN Integrated services digital network.

ISI Intersymbol interference.

ISUP ISDN User Part, used in CCITT Signaling System No. 7.

IWR Isolated word recognition.

jitter Short-term variations of the significant instants of a digital signal from their ideal positions in time (CCITT).

LAN Local-area network.

LAPC Link access protocol C.

LATA Local access and transport area.

laser diode A junction diode consisting of positive and negative carrier regions with a P-N transition region (junction) that emits electromagnetic radiation at optical frequencies. The emitted beam is very narrow, allowing the output to be coupled efficiently into single-mode fibers usable in long-range optical transmission systems.

LD-CELP Low delay CELP.

LDM Linear delta modulation (coding).

LEC Local exchange carrier.

LED Light-emitting diode. A source used in short-range optical-fiber systems. A diode that emits electromagnetic radiation at optical frequencies. The emitted beam is broad, allowing the output to be coupled efficiently only into relatively large cross-section multimode fibers. Such fibers normally are used only in short-range optical transmission systems.

LEO Low earth orbit.

Leosat A LEO satellite system proposed by the Leosat Corporation to provide non-voice global service to mobile users.

linear delta modulation Delta modulation (coding) in which the input time function is approximated by a series of linear segments of constant slope.

linear predictive coding A parametric coding technique in which the perceptually significant features of speech are extracted from its waveform.

LNA Low-noise amplifier. The input stage of a receiving earth station for satellite communication. Its effective noise temperature usually is lower than the ambient temperature.

loading coil An inductance placed in a metallic cable pair at specific intervals to compensate for capacitance between the conductors.

logarithmic compression Reduction of the dynamic range of a signal based upon the logarithm of its instantaneous amplitude.

long haul Transmission over a microwave radio-relay system to distances in excess of 400 km.

longitudinal circuit A circuit formed by one conductor (or by two or more conductors in parallel) of a metallic circuit, with a return path through ground or through other conductors.

loop The transmission path between a central office and a customer's premises.

loop extender A device that enables a metallic loop to operate over a greater distance than would be feasible otherwise.

loop start A method of supervision on subscriber lines by which a seizure is indicated by a closure of the two conductors in the subscriber loop.

loss 1. Power that is dissipated in a circuit without doing useful work. 2. The drop in power of a signal traversing a circuit or a switched connection.

loss variability The statistical variability of signal attenuation, calculated for a group of transmission paths.

LPC Linear predictive coding.

LSI Large-scale integration of circuitry in semiconductor elements.

MF Multifrequency signaling.

μ-Law A companding characteristic used in North American PCM systems.

MAHO Mobile assisted handoff.

make The closed state of relay or switch contacts.

MAN Metropolitan area network.

MEO Medium earth orbit.

metallic circuit A circuit formed by metallic conductors which have no contact with signal ground or earth.

mixed network A telecommunications network comprised of part analog and part digital facilities.

metropolitan area trunk (MAT) A cable designed to minimize crosstalk where large numbers of circuits are required between central offices.

MF Multifrequency signaling.

microcell A cell of limited coverage area for use by a variety of subscriber sets such as personal portables, mobiles, and suitably designed cordless telephones.

modem A modulator-demodulator. This device is used to convert a digital stream to a quasi-analog form (tones) suitable for transmission on analog facilities, and to reconvert to digital form at the receiving end.

modified duobinary A correlative coding technique in which the intersymbol interference extends over two bit intervals.

modulation Variation of the amplitude, frequency, or phase of a carrier wave to convey information.

MPEG Motion pictures experts group.

MS Mobile station.

MSAT Mobile satellite.

MSK Minimum-shift keying. A form of frequency-shift keying in which the peak frequency deviation equals ± 0.25 times the bit rate and coherent detection is used.

MTP Message transfer part, used in CCITT Signaling System No. 7.

MTSO Mobile telecommunications switching office.

muldem A multiplexer-demultiplexer.

multiframe A set of twelve consecutive frames in which the position of each frame can be identified by reference to a multiframe alignment signal for the group of consecutive frames.

multifrequency signaling (MF) A signaling system, normally used on telephone trunks, by which address digits are indicated by a pair of discrete tones.

multimode fiber An optical fiber whose diameter is large enough to allow the transmission of multiple propagation modes.

multiple access Techniques allowing variously located earth terminals to use portions of a satellite's transponder on either a frequency- or a code- or a time-division basis.

multipulse LPC A form of LPC that provides an excitation that is a sequence of pulses at times $t_1, t_2, \ldots t_n, \ldots$ with amplitudes $\alpha_1, \alpha_2, \ldots \alpha_n$.

near-end crosstalk (NEXT) Crosstalk that is propagated in a disturbed channel in the direction opposite to the propagation of signals in the disturbing channel.

network An arrangement of telecommunications switches and interconnecting circuitry.

noise Any unwanted signal or interference on a circuit other than the signal being transmitted.

noise figure The ratio, expressed in decibels, of a system's input signal-to-noise ratio to its output signal-to-noise ratio.

nonblocking The ability of a telecommunications system to establish a connection from any inlet to any outlet irrespective of the amount of traffic.

noise power spectral density The noise power per unit bandwidth, usually expressed as watts/Hz.

NPA Numbering plan area.

NRZ Nonreturn to zero. A channel code in which there are only two states of a signal parameter used to represent data. These are the 0 state and the 1 state.

NT Network termination.

numbering plan A set of rules for assignment of subscriber numbers in a network.

Numbering plan area A geographic area within a network numbering plan within which all subscriber numbers begin with the same area code.

Odyssey A MEO satellite system proposed by TRW to provide global service to mobile telephones.

off-hook (1) In line signaling, the condition indicating that a line is in use (line loop: closed). (2) In trunk signaling, the signaling state which exists, in the forward direction, to indicate a seizure of the trunk by the switching equipment and, in the backward direction, to indicate an answered call or an element of signaling protocol.

office code A three-digit number designating a group of subscriber numbers, usually served by the same central-office switch.

on-hook 1. In line signaling, the condition indicating that a line is idle (line loop: open). 2. In trunk signaling, the signaling state which exists in the forward direction, to indicate that the trunk is not in use and, in the backward direction, to indicate that a call is awaiting an answer, a disconnect signal from the called end, or an element of signaling protocol.

ONU Optical network unit.

open switching interval Momentary opening of the tip and ring conductors of a transmission path for 100 to 350 milliseconds during signaling.

Orbital Communications, Inc. A corporation proposing to provide nonvoice global communications to mobile users.

OSI 1. Open switching interval. 2. Open systems interconnection.

outpulsing The transmission of address digits necessary to establish a switched connection.

packet transmission The transmission of a stream of bits which has been divided into packets of a specific length (example, 1024 bits). Each packet carries its address. The overall stream is reassembled at the receiving location.

pad 1. A resistance or other network inserted into an analog transmission path to provide a controlled amount of loss in the path. 2. In a digital transmission path, a change in the encode/decode level of a digital signal to effect an attenuation of the digital content.

paging Actions performed by a cellular system in seeking a mobile telephone to complete a call to it.

parabolic As used in antenna descriptions, the shape of a reflector whose curvature is that of a parabola.

parametric coder A device which is designed to digitize an input in terms of its parameters, such as frequency bands, amplitudes, periodicities, etc.

parity bit The name given to a redundant bit added to a sequence of information bits so the total sequence adds to either one or zero. If the received sequence does not add to the same number (one or zero), a parity error is said to have occurred.

partial response signaling (PRS) The use of controlled intersymbol interference to increase the transmission rate in a given bandwidth.

PBX Private branch exchange. An automatic switching system providing switched telephone communications at a subscriber's premises and connections between the premises and the public switched network.

PCM Pulse-code modulation.

PCN 1. Personal communications network. 2. Personal calling number.

PCS 1. Personal communications system. 2. Port control store.

PDN Passive distribution network.

PDU Protocol data unit.

peg count The count of the number of traffic attempts made on a group of circuits or equipment elements during a given time period.

phase-shift keying A form of digital modulation in which the bits shift the instantaneous phase between predetermined discrete values. It uses 2^m phases to represent m bits of information each.

phoneme A distinctive sound within a language.

π/4-DQPSK Differential phase-shift keying in which each symbol differs from the previous one by at least 45°.

pitch The fundamental or lowest predominant frequency produced by the human voice.

plant Switching equipment, transmission equipment, circuitry, and related ancillary equipment comprising a telecommunications system or a portion thereof.

plesiochronous operation Independent operation of two interconnecting digital networks, each synchronized to a separate primary frequency standard.

plosive A sound resulting from making a complete closure (usually toward the mouth end), building up pressure behind the closure, and abruptly releasing it. The letters B, K, P and T are plosives.

polarization The direction of the electric vector of a propagating electromagnetic wave. For circular polarization, this vector rotates at a rate equal to the carrier frequency.

polybinary A correlative coding technique in which the intersymbol interference extends over more than two bit intervals.

POP Point of presence of an inter-LATA carrier in a LATA.

POTS Plain old telephone service; for example, voice connections without significantly advanced features.

PRA Primary rate access in ISDN.

prediction The process of estimating a future value by using weighted sums of past values.

PRI Primary rate interface, at 1.544 Mb/s, serving high-volume ISDN subscribers.

private automatic branch exchange (PABX) *See* PBX.

Project 21 An MEO and GEO satellite system proposed by Inmarsat to provide global service to mobile telephones.

protection channel A spare channel for use when channel equipment outages occur, or during deep fades on a microwave radio relay system.

protocol A formal set of conventions governing the format and timing of message exchange between two elements of a telecommunications system or network.

pseudonoise generator A generator of a very long periodic digital sequence. The length of the stream is great enough that it appears to be random.

PSK Phase-shift keying.

psophometric weighting Selective attenuation of voiceband characteristics based upon the use of a filter recommended by the CCITT and calibrated with an 800-Hz tone at 0 dBm.

PSTN Public switched telecommunications network.

PTT Postal telephony and telegraphy authority.

pulse-code modulation The use of a code to represent quantized values of instantaneous samples of a waveform.

pulse period The period of time from the beginning instant of one pulse to the beginning instant of the next consecutive pulse.

pulsing The generation and transmission of pulses to provide signaling information to a switching system.

PVC Permanent virtual connection.

QAM Quadrature amplitude modulation.

QPRS Quadrature partial-response signaling.

QPSK Quaternary phase-shift keying.

QSELP QUALCOMM version of VSELP.

quadrature amplitude modulation The independent amplitude modulation of two orthogonal channels using the same carrier frequency.

quantization noise Noise produced by the error of approximation in the quantization process.
rake receiver A receiver in which multipath echoes are summed so they contribute to the desired signal.
ramp time The time required for a transmitter to reach a designated power level.
range extender *See* loop extender.
Rayleigh A statistical distribution that characterizes the signal fluctuations found on a mobile radio link.
RBS Radio base station.
real time Pertaining to the actual time during which a physical process transpires.
regeneration The process of recognizing and reconstructing a digital signal so that the amplitude, waveform, and timing are constrained within stated limits (CCITT).
reliability The percentage of time during which equipment performs its intended function.
RELP Residually excited linear prediction.
retrial An additional attempt to seize a unit of equipment or a circuit, or to find a path through a switching network, after a previous attempt has failed.
Rician A form of fading which is produced by a combination of a Rayleigh faded (random or scatter) component and a direct (nonfaded or specular) component.
ring One conductor in a pair of wires, as distinguished from the tip conductor.
ringing signal A signal sent over a called line or trunk to alert the called party by audible or visual means to the incoming call.
roaming The operation of a mobile subscriber set through other than its home carrier.
routing plan A set of rules for routing of telecommunications calls over a network to their respective destinations.
RPCU Radio port control unit.
RPE-LTP Residual pulse excitation with long-term prediction.
RPMUX Radio port multiplexer.
RSSI Received signal strength indication.
RT Remote terminal.

SACCH Slow associated-control channel.
SAI Standard air interface.
sampling The process of sensing a waveform's amplitude at specific instants of time. Sampling usually is done periodically.
sanity *See* integrity.
SAT Supervisory audio tone. A tone transmitted by a cellular telephone.
SBC Subband coding.
SCCP Signaling connection control part, used in CCITT Signaling System No. 7.
SCPC Single channel per carrier. A technique used in thin-route satellite communications.
scrambler A device which alters a bit stream using a specific set of rules so that the transmitted stream does not contain long sequences of zeros, but so that the receiver can reconstruct the original stream.
seizure collision avoidance Methods of preventing collisions in the attempts of two or more mobile telephones to seize a given setup channel at the same time.

sender A device in an electromechanical common-control switching system which receives address or routing information and outpulses the correct digits to a trunk or to the local equipment.

separated switching In time-division switching, the switching of each direction of conversation by separate controls.

server A circuit or item of equipment which provides service to a call attempt.

service circuit Any of several groups of common equipment used on a shared basis within a switching system to establish connections.

setup channel A channel used in a cellular telephone system to set up and to disconnect calls.

short haul Transmission over a microwave radio-relay system to distances less than 400 km.

sidetone The signal produced in the telephone receiver by one's own voice or by room noise through the telephone transmitter.

signal-to-noise ratio The ratio of the signal power to the noise power at a given point in a system.

signaling The transmission of address and other switching information between subscribers and switching systems or between switching systems.

single frequency (SF) signaling A method for conveying dial pulse and supervisory signals from one end of a trunk to the other end by the presence or absence of a single specified frequency.

single-mode fiber An optical fiber whose diameter is so small that only a single mode can propagate through it.

SMDS Switched multimegabit data service.

SNI Standard network interface.

SNR Signal-to-quantizing-noise ratio.

soliton A light pulse that does not spread as it travels in a fiber.

SONET Synchronous optical network. A high-speed, synchronous transmission technology using optical fiber or radio as a transmission medium.

source coding The process of digitizing an analog input using a specific algorithm.

SPADE Single-channel per carrier Pulse-code modulation multiple-Access Demand assigned Equipment. A SCPC technique used in thin-route satellite communication.

SPEC Speech predictive encoded communications.

SS7 CCITT Signaling System No. 7.

SSMA Spread-spectrum multiple access.

ST Signaling tone. A tone transmitted by a cellular telephone.

Starsys A corporation proposing to provide nonvoice global communications to mobile users.

stitching Switching of a call back and forth between two nodes.

stop A sound produced by an abrupt release of a pressure built up behind a complete occlusion.

STP 1. Shielded twisted pair. 2. Signaling transfer point: a switching point for SS7 signaling messages.

STS Space-Time-Space switching architecture.

subband coding Frequency-domain coding in which each of several subbands is coded separately.

subrate A digital rate less than 64 kb/s.

supervision The process of detecting a change of state between busy and idle conditions on a circuit.

SVC Switch virtual connection.

switch A telecommunications switching system.

switchhook The combination telephone handset cradle, or other handset rest, and related switch that opens and closes the subscriber loop.

switching array (or network) A matrix of crosspoints forming part of a switching system.

symbol A set of one or more bits transmitted using a unique digital level.

sync character A repetitive bit pattern used by a receiver to establish that synchronization has been achieved.

synchronization A means of ensuring that both transmitting and receiving stations are operating together (in phase).

synchronous communication A mode of digital transmission in which discrete signal elements (symbols) are sent at a fixed and continuous rate.

synthetic quality The quality of computer-generated speech, which often lacks human naturalness. Synthetic-quality speech is intelligible, but the speaker may not be recognizable.

TA Terminal adapter.

tandem 1. A network arrangement in which a trunk from the calling office is connected to a trunk to the called office through an intermediate point called a tandem switching office. 2. To establish a trunk-to-trunk connection through a switching office.

TASI Time-assignment speech interpolation.

TCAP Transaction capabilities application part, used in CCITT Signaling System No. 7.

TCM Time-compression multiplexing. Time-domain separation of two directions of transmission.

TDD Time-division duplex.

TDM Time-division multiplexing.

TDMA Time-division multiple access.

temporal soliton A pulse of light that resists a medium's chromatic dispersion.

ten high day busy hour The ten-day average traffic level for the time-consistent busy hour, excluding Mother's Day, Christmas, and extremely high traffic days attributed to unusual events.

TH Time hopping.

THDBH Ten high day busy hour.

thermal noise Noise produced within a circuit by the thermal agitation of electrons, atoms, and molecules. The noise power is directly proportional to the absolute temperature.

time-division multiplexing The sharing of a transmission circuit among multiple users by assigning time slots to individual users during which any one of them has the entire circuit's bandwidth.

time slot interchange A switching system element that switches between circuits by separating signals in time.

timeout The expiration of a designed time period during which an expected event did not occur, resulting in a designed response to the lack of occurrence.

tip One conductor in a pair of wires used in telephony, as distinguished from the ring conductor.

TMI Telesat Mobile, Inc.

toll quality Speech quality based upon a laboratory test in which the signal-to-noise ratio exceeds 30 dB and the harmonic distortion is less than 2%. The bandwidth is 300–3200 Hz.

traffic 1. The messages sent and received over one or a group of communication channels. 2. A quantitative measure of the total messages and their length, expressed in specified units.

transient A rapid fluctuation of voltage or current in a circuit, usually of short duration, caused by switching, changes in load, momentary crosses, ground, or nearby lightning surges.

transmission The propagation of a signal along a path between system elements.

transmission level The power measured at a given point in a circuit, usually expressed in dBm, over a given range of frequencies or at a specific frequency.

transmission level point (TLP) The reference level point in a transmission system at which signal strength comparisons are made.

transmultiplexer A device used to convert TDM signals to FDM, and vice versa, thus serving as an interface device between digital and analog networks.

trellis coding A combination of digital modulation and error-correction coding in which the error-correction bits are used in selecting the sequence of signal state points.

trunk A communication channel provided as a common traffic path between two switching systems.

trunk circuit A network of circuit elements used to connect a switching system to one of its associated trunks.

trunk group A group of trunks that can be used interchangeably to carry traffic.

TS Time slot.

TSI Time-slot interchange.

TST Time-space-time switching architecture.

UDPC Universal digital portable communications.

UNI User-to-network interface.

usage The intensity of traffic carried by a group of circuits.

UTP Unshielded twisted pair.

VAD Voice-activity detection.

VITA Volunteers in technical assistance.

VLR Visited location register.

VLSI Very large-scale integration of semiconductor elements in circuitry.

VMR Violation monitoring and removal.

vocoder A device designed to digitize voice in terms of its pitch, amplitude, voicing, and formants.

VSELP Vector-sum-excited linear prediction.

WACS Wireless-access communications systems.

wander Long-term variations of the significant instants of a digital signal from their ideal positions in time. (Long term implies lower in frequency than 10 Hz.).

WARC World Administrative Radio Conference.

waveform coder A device that converts samples of an analog waveform to bits.

wavelength-division multiplexing (WDM) The simultaneous transmission of optical carriers of different wavelengths on a given optical fiber.

WIN Wireless information network.

wink A single supervisory pulse used between switching systems to signal a readiness to receive address digits.

WRC World Radiocommunications Conference.

X.25 A CCITT standard for data network transmission and related protocols.

Index

Abate's rule 57
Absolute timing 202
Absorption 271–273
Access 386, 402, 407, 421, 425, 428, 437, 438, 442, 443, 445, 449, 456, 459, 460, 464, 480–484, 487, 490–493, 495, 496, 498–501, 504, 509, 516, 519, 521, 522, 523, 527, 533, 535, 544, 545, 549, 550, 552, 588, 589, 599, 624, 628–629
Adaptation, instantaneous 41
Adaptive coding 11
Adaptive delay 202
Adaptive delta coding (ADC) 58–62, 67, 73–74, 94, 98, 111
Adaptive DPCM (ADPCM) 49–52, 68, 73–75, 86, 94, 111, 391, 483
  subband 68
Adaptive equalizers 260
Adaptive delta modulation – *see* Adaptive delta coding
Adaptive PCM (APCM) 40–41, 74
Adaptive prediction, backward 52
Adaptive predictive coder (APC) 94–97
Adaptive predictor 50
Adaptive quantization, backward 41
Adaptive quantizer 50
Adaptive transform coding (ATC) 68–70, 86
Added digit framing 113
Added variable delay 202
Additive white Gaussian noise (AWGN) 160, 166, 262
Adjacent channel interference (ACI) 160, 237, 275–276, 284
Advanced Intelligent Network (AIN) 266, 521–524
Advanced Mobile Phone Service (AMPS) 256, 626
Advanced Radio Data Information Service (ARDIS) 263
Alarm 149
  blue 149
  red 149
  resupply 149
  yellow 149
Alarm indication signal (AIS) 149
Aliasing 66–67
Alignment 112
Alignment signal 113
  frame 113, 133
  multiframe 114, 133
All-optical telephone network 369
Alternate channel 239
Alternate digits inverted (ADI) 116–117
Alternate mark inversion (AMI) 29, 294
American Mobile Satellite Corporation (AMSC) 326
American National Standards Committee X3T9 359
American National Standards Institute (ANSI) 362, 498
American Satellite 325
American Telephone & Telegraph Company

(AT&T) 5, 325, 384, 400, 416, 428–430, 433, 455–457, 460, 492, 494, 512, 515, 537, 558, 559, 561, 577, 589, 593, 595
Amplitude alignment 29
Amplitude distribution, Gaussian 29
Amplitude probability distribution 27
Amplitude shift keying (ASK) 177–179
Amplitude-phase keying (APK) 180–181
Analog FM 226, 266
Analog system control 627–634
Analog trunks 377, 407, 422, 440, 569, 570
Analog differencing and integration 45
Analysis, speech 88–89, 94
Andrew Corporation 286–288
Anik 298, 325
Antenna feed 316
Antenna gain 214
Antennas 285–288
  horn reflector 285–288
  parabolic 285–286
Aries 326
Assignment (AS) 242
Asymmetrical Digital Subscriber Line (ADSL) 145, 530, 533–536
Asynchronous communication 24
Asynchronous mapping 365–366
Asynchronous time division multiplexing (ATDM) 202
Asynchronous transfer mode (ATM) 6, 204, 363, 441, 496, 501, 503, 504, 507, 508, 512–515, 517, 519, 529, 536, 538, 539
Atal-Schroeder adaptive predictor 51
Atmospheric optics 368
Attentuation 217, 280–281, 300–301, 307, 390, 392, 535, 548, 566, 581
  atmospheric 280–281, 300–301, 307
  foliage 217
Autocorrelation function 46
Autocorrelation vocoder 87, 90
Automatic frequency control (AFC) 310
Automatic gain control (AGC), digital 42
Automatic line buildout circuit 139
Automatic number identification (ANI) 589, 590, 592
Automatic roaming 234–235
Autonomous registration 230–231
Availability 282–283
Avalanche photodetectors (APDs) 348–349
Avalanche process 349
Balance network 375
Balanced mixers 171
Bandpass filter bank 126
Bandwidth 159, 480, 483, 484, 497, 499, 501, 517, 529, 537, 547, 565
Bandwidth efficiency 189, 254
Barnett, W. T. 282
Base station 229
Base Station Controller (BSC) 240–242
Baseband demand assigned (BDA) 314
Baseband repeaters 289
Basic Exchange Telecommunications Radio Service (BETRS) 98, 263–264
Batteries 316
  nickel-cadmium 316
  nickel-hydrogen 316
Battery voltage 374
Baud 165, 532, 534
Bell Communications Research (Bellcore) 355, 494, 521, 534, 536, 544, 549, 553
Bell System 455
Bell Telephone Laboratories 282
Binary phase-shift keying (BPSK) 170, 173–174
Bipolar return to zero (BRZ) 29, 30, 132, 134
Bipolar violation (BPV) 130, 149, 485
Bipolar with six-zero substitution (B6ZS) 123–124, 132
Bipolar with three-zero substitution (B3ZS) 123–124, 132, 293
Bit error ratio (BER) 233, 491, 535
Bit synchronous mapping 366
Bits 131, 377, 383, 387, 390, 399, 412, 438, 439, 443, 464, 466, 483–485, 507, 532, 535, 552, 593–595
  null 131
  stuff 131, 516
Blank and burst 631
Blind delay 201
Block coding 191, 196–197
Block companding 42
Blocking 383, 384, 387, 390, 430, 431, 508, 518, 561, 602, 607–609, 616–618, 620, 624
BORSCHT 374, 377, 422
Bose-Chaudhuri-Hocquenghem (BCH) code 197, 230
Brick wall filters 189
Brick-wall channel 161
Bridged tap 484, 528
Brillouin amplifier 352, 353
British Telecom 490

Broadband CDMA 251
Broadband ISDN (BISDN) 6, 304, 480, 496, 497, 501–504, 506, 508, 517, 518, 521, 529, 552, 553
Busy hour 393
Byte synchronous mapping 366
Bytes 23

Call from mobile 632
Call setup 230–232, 627
Call to mobile 631–632
Carrier phase recovery 171
Carrier Serving Areas (CSA) 527, 528, 530, 532, 533, 535, 564
Carrier-to-interference ratio, C/I 224
Carriers 293–294
  extended service 294
  limited service 294
Cell 202–203, 226, 228, 229–231, 233, 428, 496, 498, 501–503, 508–510, 512, 513, 514, 518
  adjacent 230
  boomer 229
  neighbor 230, 233
  sector 228
Cell relay 204, 498, 500
Cellular channel assignments 228
Cellular radio systems 146, 211–269, 626–635
  analog 224, 226, 626–635
  TDMA 224, 226, 229, 237–245, 256–258
  CDMA 226, 229, 244–253, 258
Cellular Telecommunications Industry Association (CTIA) 212
Celsat, Inc. 326
Central control 393–397, 438
Central Office Terminal (COT) 263–264, 543, 544
Cepstrum vocoder 87, 90
Channel, memoryless 166
Channel access protocol, Aloha 261
Channel assignments 228
Channel augmentation by bit reduction 137–138
Channel bank 117–118, 124, 289
  digital 289
Channel encoders 289
Channel filters 316
Channel noise 191
Channel roll-off factor 164
Channel service unit (CSU) 73, 134
Channel vocoder 87, 90
Channels
  A, B 228
  setup 230
Character available flag 117
Characteristic impedance 537, 563
Chip rate 199
Chirp 358
Chromatic dispersion 348
Circuit spectrum efficiency 254
Circuit-switched environment 200
Circular orbit 300
Class 1 bits 219
Class 2 bits 219
Classen-Mecklenbräuker method 127–128
Clock 377, 381, 393, 399, 400, 402, 408, 419, 420, 430, 433, 444, 446, 472, 473, 476, 508, 516, 517
Co-channel interference (CCI) 160, 262
Code 197
  cyclic 197
  polynomial 197
Code division multiple access (CDMA) 5, 198–199, 212–213, 224, 244–253, 260–262, 312
Code excited linear prediction (CELP) 86, 98–103
  low delay CELP (LD-CELP) 100–101, 111
Codec 237, 239, 242, 375, 377, 399
  full rate 237
  half-rate 239, 242
Coded digital verification color code (CDVCC) 232
Coded mark inversion (CMI) 134
Coder 111–112, 130–135
  channel 112, 130–135
  error 112
  source 111
Codevectors 70
Coding
  block 196
  convolutional 197
  stochastic 98
  tree 100
  trellis 100
  waveform 23–88
Coherent detectors 349–350
Coherent optical communication system 343–344
Coherent optical sources 343–344
Coherent PSK 184
Common control systems 3

Communication efficiency 255
Communications quality 19
Companding 390
  A-law 32, 34
  instantaneous 31
  nearly instantaneous 39, 42–43
  syllabic 31
  μ-law 32–34, 382, 388
Companding advantage 34
Compandored single-sideband modulation 299
Competitive clip 138
Compression 29
Compressor 31
Concentration 375, 377, 379, 384, 409, 422, 440, 541, 602, 605, 608, 609, 611
Congestion 466, 467, 502, 509, 517, 518, 621, 623, 624, 625
Consent decree 455, 456
Consonant recognition test (CRT) 80
Constellation 168, 178
Constellation Communications 326
Continuous phase FSK (CP-FSK) 184
Continuous word recognition (CWR) 151–152
Continuously Variable Slope Delta (CVSD) 60–62, 76–80, 111
Control memory 380, 381, 383, 386, 387
Control subsystem, satellite 322
Controller (C) Interface 266
Convolutional code 191, 196–197
Cordless booster 260
Cordless local area network (CLAN) 262
Cordless PBX 261
Cordless systems 146
Correlation 199
Correlation bandwidth 218, 242
Correlation function 46
Correlative techniques 184–188
Coupling loss 348
Crest factor 35
Critical angle 338
Crosspoints 375, 381, 383, 385, 386, 448, 605, 607–611
Cyclic redundancy check (CRC) 114

Data above video (DAVID) 208
Data above voice (DAV) 208
Data compression 181, 203
Data compressor 203
Data link connection identifier (DLCI) 204
Data link layer 204
Data service unit (DSU) 73, 134
Data transmission 254, 359–360
  isochronous 360
  mobile 253–254
Data under voice (DUV) 206–207, 287
DBP Telekom 492
Decimate 64
Decimation 127
Decision feedback equalizer (DFE) 237
Decode level (DL) 470, 471
Delay 382, 385, 393, 402, 435, 511, 530, 566, 568, 569, 605, 615–618
Delay dial 583–585
Delay distortion limited system 354
Delay spread 217–224
Delta coding 52–62, 145, 379
  adaptive 58–62
  continuous 59–62
  linear 52–58
Delta modulation—*see* Delta coding
Demand access, on-board 319
Demand Assigned Multiple Access (DAMA) 264, 309, 314
Demand assignment 237, 313–314
Department of Trade and Industry (DTI) 262
Despreading 199
Detection
  coherent 171
  differential 171
Detectors
  Coherent 349–350, 353
  noncoherent 348–349
  optical 348–351
Deutsche Bundespost 490
Deviation ratio 182, 184
Diagnostic rhyme test (DRT) 89
Differential 12
Differential attenuation 302
Differential detection 171
Differential PCM (DPCM) 44–58
Differential phase 302
Differential quaternary phase-shift keying (DQPSK) 171–174
Digital color codes (DCCs) 231
Digital Conferencing 389, 390
Digital Echo Suppression 391–392
Digital electronic message service (DEMS) 294
Digital European Cordless Telephone (DECT) 262

Digital filtering 389
Digital hierarchy 120–122, 124
  European 133
Digital Loop Carrier (DLC) 355, 527
Digital loop plant 142
Digital microwave radio 285–295
Digital milliwatt 388, 470, 471
Digital noninterpolated interface (DNI) 138
Digital pads 392
Digital recorded announcements 391
Digital repeaters 139–142
Digital speech interpolation (DSI) 39, 86, 117, 137–139, 212, 239–244
  gain 138–139
Digital stream rates 120
DS0 131, 483, 486, 504, 516–517, 541, 544
DS1 121, 123, 125, 128, 131, 132–134, 407, 416, 422, 423, 428, 434, 435, 438, 473, 474, 483, 485, 486, 501, 516–517, 519, 529, 530, 532, 533, 541, 544
DS1 substitution signal 206
DS1C 132
DS2 121, 123, 132, 187
DS3 122, 123, 132, 149, 500, 501, 504, 519
DS4 122, 123, 134
Digital symmetrical matrices 387
Digital telephones 377
Digital termination system (DTS) 293–294
Digital verification color code (DVCC) 238
Digital-to-analog converter 35–37, 471, 549
Digitally controlled delta coding 62
Diphthong 150
Diplexer 321
Direct detection 344
Direct sequence (DS) 199, 235, 244–245
Directed retry 231
Disconnect 232–233, 402, 461, 523, 571, 574, 577–579, 586, 588–601, 609, 611, 633–634
  mobile-initiated 633
  system-initiated 634
Discontinuous transmission (DTX) 242
Discrete Fourier transform (DFT) 389
Discrete multitone (DMT) modulation 145, 535–536
Discrete Multitone (DMT) Systems 145–146, 535–536
Dispersion
  chromatic 337, 340
  intermodal 340
  material 337
  waveguide 340
Distributed antenna system 259, 260
Distributed Bragg reflector (DBR) laser 350
Distributed control 393, 396, 417
Distributed feedback (DFB) 347
Distributed feedback DBR laser 354, 358
Distribution 384, 385, 602, 605, 608, 611
Diversity 224, 282
Diversity Antenna Delay Equalization (DADE) 292
Doppler shift 218
Double hop-satellite transmission 300
Double-talk 136
Downlink 306–307
Dual mode 239, 244
Dual polarization 301, 316
Duobinary 184–186
Dynamic capacity allocation (DCA) 251, 262
Dynamic locking quantizer (DLQ) 51
Dynamic routing 200, 458, 561

Earth station 321–324
  earth station accessing leased facilities 324
  nonstandard 324
  standard A 323
  standard B 323
  standard C 323–324
Echo 382, 390–392, 427, 431, 435, 436, 458, 471, 484, 532, 536, 537, 566, 568, 569, 570, 581
Echo cancelers 135–136, 201, 388, 391, 392, 484, 532
Echo canceling hybrid (ECH) 484, 485
Echo cancellation 146, 314, 471, 484, 536, 537
Echo control 303
  split 314
Echo suppressor 135–136, 201, 388, 390, 391–392, 431, 435, 436, 568–569, 581, 595
Echos 161
Eclipse 316
Economic spectrum efficiency 255
Effective isotropically radiated power (EIRP) 306
Efficiency 130, 202, 311, 313
  antenna 271, 272, 285
  bandwidth 189–190, 202
  coding 184, 195
  data 171
  reuse 247
  sampling 65

spectrum 18, 167, 205, 222, 239, 254–256, 262
transmission 4
Eight-plane shift keying (8-PSK) 175–177
Elastic buffers 311
Electro-optic crystal 344, 347
Electronic mobile exchange (EMX) 228–229
Electronic serial number (ESN) 230
Elevation angle discrimination 248
Ellipsat 325
Ellipso 326, 329
Encode level (EL) 470, 471
Encryption 111, 262
Energy scale factor 94–95
Exchange network access for interstate carriers (ENFIA) 589
Enhanced TDMA™ (E-TDMA) 239–244
Environment 258–263
macrocellular 258–259
microcellular 258–263
Environmental intermodulation 308–309
Equal access 407, 589
Equalization 140
Equatorial Communications, Inc. 313
Equinox 316
Erbium-doped fiber 351, 353
Erlang B formula 617, 618, 620
Erlang C formula 618, 620
Error burst 120
Error control 191–197
Error correction 195–196
Error detection 195
Error detector 187
Error variance 28
Error-free interval (EFI) 120
Error-free seconds (EFS) 120
Euclidean distance 178
European Telecommunications Standards Institute (ETSI) 212, 489
Excitation sequence 100
Expander 31
Expansion 379, 384, 385, 387, 602, 605, 608, 609, 611
Extended multiframe (superframe) structure 114–115
Extended-TDMA (E-TDMA) system 139, 237, 239–244
Eye pattern 165, 539

Fade margin 224, 280
required 282–283
Face rate 218
Fading 214–224, 259, 271, 280–282
Rayleigh 218–219, 259
Rician 259
Failure time 120
Fast associated control channel (FACCH) 233, 238, 244
Fast Fourier Transforms (FFT) 535
Fast packet switching 204
Fast stat-mux 203
Fault location 150
Feeder links 326
Fiber
erbium-doped 351, 353
chalcogenide 335
fluoride 335
graded index 338–339, 341–342
multimode 338–341
plastic-clad 335, 337–338, 340
single mode 338–339, 341
Fiber distributed data interface (FDDI) 359–362, 498–499
basic 359–360
FDDI LAN 359
FDDI-II 360–363, 499, 500
Fiber in the loop (FITL) 355, 530, 540–545
Fiber splicing methods 356
Fiber system 333–342
attenuation 335–336
dispersion 337
triaxial 334
Fiber to the curb (FTTC) 355, 340
Fiber to the home (FTTH) 355, 340, 545, 546
Fiber transmission characteristics 334–337
Fiber types 337–342
glass 338
plastic 337–338
plastic-clad silica 335, 337–338
Figure of merit 307
Filter bandwidth limitations 191
Filter roll-off factor 174
Fixed path routing 200
Fixed station originated call 231–232
Fixed wire applications (FWA) 264
Floating mode 365
Foliage attenuation 217
Follow-me roaming 230, 234
Footprint 319
Formant 51, 88, 94, 150

Formant vocoder 87, 90
Forward Digital Traffic Channel (FDTC) 238
Forward error correction (FEC) 85, 107, 146, 191, 246
Forward-acting timing 116
Fractional interval equalizer 237
Frame alignment 112–116
Frame alignment sequence 114
Frame alignment signal 113–115, 133
Frame definition messages 203
Frame relay 204, 497, 500, 501
Frame synchronization 24, 112–115
Framing, added digit 113
Framing sequence, composite 114
Frame Télécom 490–492
Free space loss (FSL) 273, 306
Freephone 521
Frequency allocations 276–278, 302–305
Frequency division multiple access (FDMA) 146, 197–198, 260, 306, 309–311
FDMA (SCPC) 146, 306
Frequency division multiplexing (FDM) 2, 143, 565
Frequency hopping (FH) 199, 235, 244
Frequency hopping multiple access (FHMA) 253
Frequency reuse 224–228
Frequency-shift keying (FSK) 181–184, 294, 344
Fresnel zone 280–281
Fricative 48–49, 87, 151
FSK heterodyne 344
Fujitsu 293
Future Systems, Inc. 333

Gain
  ADPCM system 51
  DPCM system 48
  DSI 138–139
  on-axis 271
Galaxy 325
Gaussian minimum shift keying (GMSK) 184, 219
General Electric (GE) 325
General Telephone & Electronics (GTE) 171, 456
Geographical spectrum efficiency 255
Geostationary altitude 300
Geostationary earth orbit (GEO) 325–327
Geotek Industries, Inc. 253
Glare 584, 585
Global Systems Mobile (GSM) 106, 108, 184, 219, 257–258
Globalstar 326
Glottis 151
Grade of service 149, 617, 618
Graded index fibers 338–339, 341–342
Granular noise 54
Groupe Spéciale Mobile (GSM) 212
GStar 325, 329
Guard time 311

Hamming code 195
Hand-off 233–234, 248, 326, 633
  soft 248
  softer 248
Harmonic scaling 70
Helsinki Telephone 262
Heterodyne detection 344, 349–350
Hierarchical 457, 458
High Definition Television (HDTV) 529
High Density Bipolar-3 (HDB-3) 133
High-Bit-Rate Digital Subscriber Line (HDSL) 530–534, 536
Home location register (HLR) 229, 257
Homodyne detection 349–350
Huffman principle 86, 203
Hughes Network Systems, Inc. 212, 239
Hutchinson Rabbit 262
Hybrid 93–94, 143
Hybrid CDMA/TDMA system 253
Hybrid coder 93–94
Hybrid Ring Control (HRC) 362
Hybrid speech coding 5, 10
Hybrid transformer 375

IEEE Standard 802, 359
Immediate dial 583
Impulse 161, 163
Impulse response 161
In-building environment 260–263
Index of refraction 338
Indoor ratio systems 260–262
Information bandwidth 199
Information density 254–255
Information memory 380–383
Information rate 165, 167
Information retrieval systems 155

Infrared emitting diodes (IREDs) 335
Infrared transmission 265
Initial address message (IAM) 400
Initialization 400
Injection laser 343
Injection laser diodes 342
Inmarsat 324, 326
Innovation sequence 100
Integer band samping 65–67
Integrated Digital Loop Carrier 368
Integrated Service Digital Network (ISDN) 6, 420, 426–428, 435, 437, 438, 446, 449, 450, 461, 479–482, 486–496, 502, 521, 523, 527, 533, 535, 546–548, 553, 554, 563
Integrated-adaptive mobile access protocol (I-AMAP) 327
Integrity checks 377
Intelligent Network (IN) 519–521
Intelsat 317, 324–325, 330
Intelsat I (Early Bird) 298
Interdigital Communications Co., Inc. 213, 245
Interexchange carriers 204, 439
Interference, sidelobe 305
Interference zone 224
Intermediate-frequency (heterodyne) repeaters 289
Intermodal dispersion 340
Intermodulation (IM) 159, 198, 284, 308–309
  environmental 308–309
  receiver 309
  transmitter 309
International Standards Organization (ISO) 359, 486, 529
International Telecommunication Union (ITU) 276
  Radiocommunications Sector (RS) 276
Interpolate 64
Interpolation 127
Intersatellite links 329
Intersymbol interference (ISI) 160–161, 164, 166, 171, 184, 186, 217, 260, 485, 532, 534
Intersystem roaming 234
Intersystem signaling 234
Ionizing (nuclear) radiation 336
Iridium 300, 326, 329
Isochronous MAC (I-MAC) 362
Isolated word recognition (IWR) 151–152
Isopreference contours 76

Japanese Digital Cellular System 257
Jitter 123–124, 473, 498, 530
Justification 131

Knockout Switch 509, 510
Kokusai Denshin Denwa Company (KDD) 341, 492

Lag 100
Large scale integration (LSI) 377, 386, 387
Large space plateform (LSP) 329–330
Laser 333, 338, 342–343
Laser diode 342–343, 352
  semiconductor 352–353
Leaky bucket 518
Leaky feeders 259
Leosat, Inc. 326
Level 159, 161
Light-emitting diode (LED) 333, 338, 342–343
Lincoln Experimental Satellites 8 and 9 (LES-8 and LES-9) 329
Line card 375, 377, 427, 449–450
Line circuit 374, 377, 409, 449
Line code 532
  AMI 29, 294, 485, 486
  B3ZS 123, 132
  B6ZS 123, 132
  B8ZS 485
  zero substitution 294
  2B1Q 147, 484, 534, 536
Line degradation 194
Line of sight 280
Line overhead 365
Linear prediction model 188
Linear predictive coder (LPC) 87–90, 111
  multipulse 95
  vector quantization LPC (VQ-LPC) 91–93
Link 229
fiber optic 229
microwave 229
Link budget 278–280, 305–307
Lithium niobate 347
Lithium tantalate 347
Local Access and Transport Area (LATA) 428, 456–459, 495, 530, 558, 559, 588, 589
Local area network (LAN) 6, 262
Local exchange carrier (LEC) 204, 264, 494, 552, 553, 588
Locating 628

Locked mode 365
Log PCM 31–43
Logical multiplexing 200
Long term prediction 86
Long haul systems 270
Loop timing 472
Loral/QUALCOMM 326
Loss limited system 353
Low delay CELP (LD-CELP) 100–101, 111
Low-earth orbit (LEO) 325–326
Low-pass filter bank 126

M/A-Com 294
Macrocell 258–259
Maintenance diagnostics 373, 397, 406, 407, 410
Manual registration 234
MCI Communications 494
Mean-square error 27, 34
Media Access Control (MAC) 359
Medium-earth orbit (MEO) 325–326
Mercury Communications, Ltd. 490
Meridional ray 338
Message-switched environment 200
Metropolitan area systems 293–295
Metropolitan area trunk (MAT) 120
Microcell 230
Microcellular environment 258–263
Microwave equipment 283–289
Microwave propagation 271–275
Minimum shift keying (MSK) 184, 212
Mixed network 455
Mobile assisted handoff (MAHO) 233, 238
Mobile identification number (MIN) 230
Mobile station (MS) 242
Mobile switching center (MSC) 228, 232, 234, 235, 241
Mobile telecommunications switching office (MTSO) 228
Modems 2, 491, 535
Modified chemical vapor deposition (MCVD) 334–335
Modified duobinary 185
Modulation 167–191, 344–348
  digital 167–191
  direct 344–346
  intensity 344–345
  16-QAM 181
  64-QAM 181
  256-QAM 181
Modulation technique performance 185–191
Monitoring and maintenance 149–150
Monomode fiber 341
Motion-compensated video 294
Motorola 212, 325
Muldems 122–124
Multiframe alignment signal 114–115
Multilevel PSK (M-PSK) 174–177
Multilongitudinal mode (MLM) laser 347
Multimode fibers 338–341
Multimode step-index fiber 339
Multipath 273–275, 280, 282
Multipath fading 237, 273–274
Multipath interference 199
Multiple access 197–199, 235–253, 309–313
Multiplexer
  fused-fiber 358
  M12 122–123, 132
  M13 122–123, 132
  M1C 122, 132
  M34 122–123
Mutiplexing 313, 375, 505, 508, 509, 516, 533, 550, 564, 565
Multipliers 171
Multipulse LPC 95, 97
Multistate modulation method 126

Narrowband Advanced Mobile Phone Service (NAMPS) 256, 635
National ISDN-I 494, 495
National ISDN-2 495
National ISDN-3 496
Near-end crosstalk (NEXT) 121, 484, 485, 532–534
Near-far problem 261
Nearly instantaneous companding (NIC) 39, 42–43
Network interface 266, 417, 438, 450, 480, 496, 533, 538, 540, 548, 577
Network control point (NCP) 462
Network layer 202, 204, 460, 487, 552
Network management 397, 401, 402, 446, 466, 500, 502, 517–519, 615, 621, 622
Network management support (NMS) 364
Nickel-cadmium batteries 316
Nickel-hydrogen batteries 316
Nippon Telegraph and Telephone (NTT) 493
Noise 391, 534–535, 538–539, 566

impulse 160, 530, 534
level 20, 67, 139, 244, 246, 254, 279–280, 284, 323, 348, 349, 351, 355, 354, 538
photonic 352
quantum 352, 353
thermal 160, 353
Noise figure 279, 284
Noise power 28
Noise sources, optical 352–353
Noise temperature 284
Nonblocking 384, 411, 508, 605, 607, 608
Nonlinear time warping 152
Nonreturn-to-zero (NRZ) 131
Nonwireline carriers 228
Northern Telecom 292, 439, 441, 447, 451, 494, 546–547
Norwegian Telecom 262
Null bits 131
Numbering plan area (NPA) 523, 558, 559
Numerical aperture (NA) 338, 341
Nyquist theorem 13, 117, 161–166

Octonary phase-shift keying (OPSK) 175–177
Odyssey 326
Omnipoint Communications, Inc. 213, 253
OmniTRACS® System 329
Open switch interval (OSI) 571
Open System Interconnect (OSI) model 202, 359, 571
Optical amplifiers 351–352
erbium-doped 358
Optical Carrier 1 (OC-1) 364
Optical detector, noncoherent 348–349
Optical Network Units (ONUs) 355
Optical sources 342–348
Orbital Communications, Inc. 326
Orionsat 324
Orthogonal function vocoder 87, 90
Outage 120
Overhead word 629
Overreach 275
Oversampling 52
Oversampling index 52
Overvoltage protection 374, 375

Packet MAC (P-MAC) 362
Packet switch 203, 417, 420, 426, 428, 508, 509–514
Packet transmission 199–204, 320
Packet voice 200–202
Paging 230, 628–629
Pair-gain systems 145
PanamSat 324
Parametric coding 9, 10
Parametric speech coding 5
Parity check code 195
Partial response classes 188
Partial response coding 149
Partial response techniques 184–185
Passive Distribution Network (PDN) 355
Path overhead 365
Peak clipping 28
Perplexity 151
Personal calling number (PCN) 260
Personal communication networks (PCNs) 259–260
Personal Communications Services 523, 550
Personal communication system (PCS) 265–266
PCS interface to the PSTN 265–266
Phase-locked loop (PLL) 171
Phase shift keying (PSK) 170–177, 212
coherent 171
differential 171, 246
Phoneme 150
Phonon absorption 335
Photodetectors 348–350
Photodiodes 348–349
Physical Layer Medium Dependent (PMD) 360
SMF-PMF 360
Physical Layer Protocol (PHY) 359
Picowatts psophometric 324
P-i-n diodes 348–349, 353
Pitch 88–90
Pitch gain 94
Planck's constant 352
Plesiochronous 474, 476, 493
Plosive sound 150
Pointel 262
Poisson formula 618, 620
Polar binary 133, 134
Polar format 132
Polar ternary 141
Polarization 271, 275, 298, 316–317
circular 317
elliptical 317
linear 317
vertical 213
Polarization diversity 304

Polarization loss 307
Polyphase method 126
Polyvinylidene fluoride (PVDF) 147
Port (P) Interface 266
Power amplifiers, microwave 284
Power budget, downlink 306–307
Power control 248
  closed loop 248
  open loop 248, 251
Pre-origination dialing 232
Precipitation 273, 276, 303, 316–317
Prediction 11
  backward-acting 50, 52
  forward-acting 50
Predictive coding 44
Predictor coefficients 50
Probability of error 191
Processing gain 199
Program patches 401
Project 21, 326
Propagation 221–224, 282
  drop-out spots 224
  effect on system design 221–224
  multipath 282–283
Propulsion system 316
Protection channels 278
Protection line switching 150
Protection switching 114
Pseudo-noise (PN) sequences 21, 246
Pseudo-random sequence 198–199
Public switched telecommunications network (PSTN) 259, 261, 265, 266, 455, 521
Pulse amplitude modulation (PAM) 379
Pulse broadening 337
Pulse code modulation (PCM) 5, 12, 23 25–41, 73–75, 77–78, 80, 84, 111, 379–382, 387–392, 479, 483, 486
  linear 25–31
  PCM 30, 126
  performance equation 28
Pulse dispersion 338
Pulse stuffing 121, 131, 132
Pulse width modualtion (PWM) 379
Pump laser 351

Quadrature amplitude modulation (QAM) 177–181
  16-QAM 181
  64-QAM 181
  256-QAM 181
Quadrature mirror filter bank (QMFB) 67
Quadrature partial response signaling (QPRS) 135, 186
QUALCOMM, Inc. 106, 213, 245, 329
  QSELP 106, 245–246
Quantization 10, 15, 25, 382, 388
  error 28
  nonuniform 29
Quantizer 15, 28
  mid-riser 34
Quantizing distortion 15
Quantizing noise 15, 27, 48
Quaternary phase-shift keying (QPSK) 170–174, 295
Quantizer, optimum 34

Radio Base Station (RBS) 264
Radio common-carrier systems 146
Radio Port (RP) 264
Radio Port Control Unit (RPCU) 265
Radio systems, SONET-compatible 295
Radome loss 307
Rain attenuation 280–281
Rain depolarization 301–303
Raised cosine frequency response 174
Raised cosine function 164, 166
Raised cosine spectrum 164–165
Rake receiver 224, 247
Raman amplifier 352, 353
Ramp time 238
Ray interference 275
Rayleigh distribution 218–219
Received signal strength indication (RSSI) 233
Rectangular pulse 163
Reflective fading 282
Reflectors 217
Refraction 273–275, 280–281
Regenerative repeater 119, 139, 295, 350–351
Regenerator 85, 140, 505, 517
Registration 230–231, 234–235
  autonomous 230–231, 234–235
  forced 234
Registration identification (REG ID) 235
Regular pulse excitation with long term prediction (RPE-LTP) 106–108, 111, 191
Reliability 282–283
Remote channel 239
Remote digital terminal (RDT) 355
Remote electronics (RE) subsystem 295

Remote terminal 263, 417, 443, 448, 527, 533, 564
Repeater siting 280
Repeaters 119, 139, 264, 350–351
  optical 350–351
  regenerative 119, 139, 295, 350–351
Request (RQ) 242
Required fade margin 282
Rescan 231
Residual 95
Residual-excited linear prediction (RELP) 95–99, 111
Residue pitch 88
Restricted bandwidth 187
Retrofit 205–208
Reverse digital traffic channel (RDTC) 238
Ring trip 534
Ringing 374, 375, 393, 598, 600, 604
Roamer 211
Roamer access port 234
Roaming 230, 234–235
Rockwell 289–293
Round trip measurement 201
Row check 195

Sampling 13, 379, 382, 388, 392
Sampling theorem 13
Satcom 325
Satellite Business Systems (SBS) 298
Satellite orbit 299–300
Satellite system design 303–314
Scrambler 112, 128–130
Scrambling 205
Section overhead 364
Sector cells 228
SEGQ parameters 94
Seizure collision avoidance 630
Selective fading 278
Semivowel 150
Service circuits 373, 388, 393, 397, 401, 406, 407, 435, 597, 615, 621
Service Control Point (SCP) 441, 446, 447, 464, 521–524
Service Switching Point (SSP) 407, 441, 446, 521–524
Setup channels 230, 629–630
Seven level partial response coder 205
Seven-level modified duobinary 187
Shannon limit 166–167, 190
Shannon-Hartley law 166
Shared control 393, 396
Shared memory switch 508, 509
Short haul system 270
Shot noise 348
Side tone 147
Sideband, vestigial 169
Sidebands 168
Siemens Stromberg-Carlson 406, 494
Signal impairments in transmission 159–161
Signal level 159, 161, 388, 390, 392, 470, 471, 474, 539, 566, 568, 581, 587
Signal unit (SU) 464–466, 469, 549–550, 593–595
Signal-state space diagram 168, 178
Signal-to-quantization-noise ratio 28
Signaling
  and supervision 112
  call progress signals 587
  common channel interoffice signaling (CCIS) 60, 593
  common channel signaling 400, 458–460, 486, 41, 593, 597, 621, 625
  dial pulse signaling 374, 436, 578, 579, 581–583, 586, 588, 589
  Dual Tone Multifrequency (DTMF) signaling 389, 393, 437, 579–581, 586, 589
  E & M lead signaling 575–577
  ground start signaling 571, 574, 583
  inband signaling 458–460, 575, 597, 621, 623–625
  loop start signaling 571, 574, 583
  multifrequency (MF) signaling 389, 436, 581, 588, 591
  point (SP) 407, 441, 462, 464, 467–469
  reverse battery 583, 585, 586
  tone (ST) 628
  Transfer Point (STP) 438, 462, 464, 467, 468, 522, 595
  System No. 5 437, 581, 588
  System No. 6 437, 593, 595
  System No. 7 (SS7) 73, 407, 426, 437, 441, 446, 450, 458–462, 466, 468, 480, 487, 490, 492, 493, 519, 521, 593
Silicon fibers 335, 337–338
Single channel per burst (SCPB) 314
Single channel per carrier (SCPC) 146, 306, 309
Single-frequency (SF) 576, 577, 588
Single longitudinal mode (SLM) laser 347
Single-mode fibers 338–339, 341

Sixteen-phase shift keying (16-PSK) 177
SL Undersea Lightguide System 341
Slip 472–474, 476, 516, 517, 530
Slope overload distortion 54
Slot 204
Slow associated control channel (SACCH) 233–234, 238
Small signal amplifiers 284
Soft hand-off 248
Solar array 316
Soliton, temporal 337
Soliton transmission 337
Sound 87
  unvoiced 87
  voiced 87
Source coder 10, 111
SP Communications 325
Space diversity 275, 280
Space segment 315–321
Space transportation system (STS) 329–330
Space-time-space (STS) 385, 386, 387
Spacenet 325
SPADE 309–310, 318
Speaker verification system 155
Spectrum efficiency 254–256
Speech 86–90
  analysis 88
  fricatives 87
  synthesis 88
  voiced segments 87
Speech clipping 242, 244
Speech coder ratings 75–80
  objective 75
  perceptual 75
Speech coding 237, 239, 242
  full rate 237
  half rate 239, 242
Speech recognition 150–153
Speech recognition systems 151
Speech recognition techniques 152–153
Speech spurt 241
Sprint Communications 494
Spot beams 298, 304, 312
Spread-spectrum multiple access (SSMA) 198, 235
Spreading loss 271–273
Standard, cellular
  IS-41 234
  IS-54 235–244
  IS-95 244–252
Standard air interface (SAI) 265
Standard Network Interface (SNI) 264
Star coupler 358
Starsys, Inc. 326
Stat-mux 202–203
Station Management (SMT) 359
Statistical multiplexing 202–203, 503
Statistics, speech activity 240–245
Stentor 455
Step size control 39–41
Step-index fiber 339
  multimode 339
  single-mode 339
Stochastic coding 98
Stops 87, 151
Stored program control 394–395, 396–397
Streams
  DS1 121, 123, 125, 128, 131, 132, 134
  DS1C 132
  DS2 121, 123, 132, 187
  DS3 122, 123, 132, 149
  DS4 122, 123, 134
Stoger, Almon B. 3
Stuff bits 131
Subband coder 62–68
  interband leakage 65–68
  variable-band 64–68
Subrate 123
Subscriber line 142, 374, 375, 377, 389, 390, 407, 417, 420, 422, 449, 450, 493, 570, 571, 579, 581, 587, 589, 602–603
Subscriber Loop Carrier-40 (SLC-40) 145
Subscriber loop multiplex (SLM) 145
Subscriber loop systems 145
Subscriber systems, optical 355–357
Superframe 114–115
Supervision 374, 586, 591, 592, 597, 628
Supervisory audio tone (SAT) 628
Supervisory signaling 375, 570, 571, 574, 575, 588, 593, 598
Swedish Televerket (PTT) 262
Switchboard in the sky 319
Switched Multimegabit Data Services (SMDS[SM]) 500, 501
Switching
  combined switching 386
  networks 373, 377, 379, 382, 383, 386, 387, 399, 406, 407, 438, 440, 442, 450, 461, 508, 597–600, 601–602, 605, 615

packet switching 200, 417, 428, 442, 493, 496, 508, 509, 593
remote switching 379, 416, 449
separated switching 386
space division switching 375, 377, 382, 383, 417, 508, 602
systems 406–454, 515
time division switching 382–384
Syllabic adaptation 59–60
Syllabic rate 50
Syllabically companded delta 59–62
Symbol 161
Symbol rate 161, 164–165, 167
Sync search 24, 117
Synchronization 23, 112–115, 408, 409, 431, 456, 460, 472–476, 484, 486, 508, 515–517, 552, 594
frame 112
Synchronous communication 24
Synchronous Digital Hierarchy (SDH) 362–365, 504, 505, 507, 509
SDH/SONET standards 363
Synchronous Optical Network (SONET) 6, 120, 362–368, 439, 449, 474, 496, 500–508, 514, 516–519, 529, 530, 536, 537, 538, 539, 541, 544, 545, 549
SONET central office architecture 367–368, 441–452
SONET compatible radio systems 295, 439
SONET Digital Crossconnect System (DCS) 367
SONET standards 368
Synchronous payload envelope (SPE) 364–365, 505, 516
Synchronous time division multiplexing (STDM) 202
Synchronous Transfer Mode 1 (STM-1) 364
Synchronous Transport Signal 1 (STS-1) 364, 504–506
Syncom 298
Synthesis, speech 88–89
Synthesizer, speech 94
Synthetic quality 19
System identification (SID) 235
System interface arrangements 287–289
System monitoring 292

T-carrier systems 5, 119–120
T1 system 205, 377, 380, 485, 565
T1 Outstate (T1/OS) System 120, 121
T3 system 204
TAT-8 cable 341
Telecommunication Industries Association (TIA) 212, 234, 239, 473, 475
Telepoint 262
Telesat 298, 313
Telesat Mobile, Inc. (TMI) 326
Telesciences 294
Telescopes 368
Telstar 325
Terrain features 226
Terrestrial segment 315
Test circuitry 375
Tilt differential 306
Time alignment 238
Time Assignment Speech Interpolation (TASI) 4, 137
Time compression multiplexing (TCM) 143–145, 484, 493
Time division multiple access (TDMA) 5, 198, 237–244, 260, 299, 311–312, 550
Time division multiplexing (TDM) 2, 117–128, 550
Time domain harmonic scaling (TDHS) 68
Time hopping (TH) 235
Time slot coordination 312
Time slot interchange (TSI) 265, 380, 382–387, 407, 409, 411, 412, 414–416, 420–423, 427, 428, 430, 431, 433, 550
Time slot transfers (TST) 265
Time slots 375, 380–387, 407, 412, 414, 415, 421, 422, 426, 430, 431, 433–435, 448, 508–509, 510–512, 541, 552
Time-space-time (TST) 383, 386, 387, 407, 416
Timing 115–116, 393–396, 402, 408, 411, 412, 419, 431, 433, 436, 446, 472–474, 476, 484, 486, 516–517, 530, 557, 577, 584, 585, 595, 601, 609
Timing advance 237
Timing extractor 116
Timing jitter 205
Timing recovery 116–117
Toll quality 19, 29
Tone generation 388
Tone reception 389
Total path loss 306
Traffic 373, 375, 379, 384, 386, 387, 393–402, 407, 417, 426, 428, 431, 450, 457, 458, 466–469, 500, 511, 517, 518, 541, 559, 561, 565, 584, 599, 600, 602, 605, 609, 614–618, 620–625

Traffic channel structure 237
Trailer 200
Transistor-transistor logic (TTL) 134–135
Transmultiplexers 122, 125–128
Transponder 317–321
  processing 319–320
  single conversion 317
Transponders 316–320
Transport layer 202
Tree encoding 86, 98
Tree structure 72
Trellis coding 146, 536
Trunked system 211
TRW, Inc. 326

Unchannelized operation 365
Unitel Communications, Inc. 455
Universal Digital Portable Communications (UDPC) 262
Universal Personal Telecommunications (UPT) 521, 523
Universal telephone handset 266–267
Unshielded twisted pair (UTP) 536–539
Unvoiced sound 151
Uplink 305
User information 238
User-to-Network Interface (UNI) 368

Vapor-phase axial deposition (VAD) 334
Variable quantizing level (VQL) 42
Variable slope delta coding (VSD) 60, 62, 76
Vector dimension 70
Vector PCM 70–73
Vector quantization (VQ) 70–73, 98
  multistage 72
Vector quantization LPC (VQ-LPC) 91–93
Vector techniques 70–73
Vector sum excited linear prediction (VSELP) 104–106, 191
Very large scale integration (VLSI) 1, 3, 5, 136, 147, 377, 386, 611
Very small aperture terminals (VSATs) 312
Via net loss (VNL) 569
Vigants, A. 282
Violation monitor and removal (VMR) 290
Virtual circuit routing 200
Virtual private network (VPN) 521
Virtual tributary (VT) 363, 365–367
Visited location register (VLR) 257
Vocal chords 151
Vocoders 86–93
Voice activity detection (VAD) 241–242, 253
Voice frequency order wire 187
Voice input system 153
Voice mail 235
Voice response 153–156
Voiced sounds 48, 87, 151
Volunteers in Technical Assistance (VITA) 326

Wander 473, 474, 530, 538
Water vapor 273
Waveform
  bipolar 134
  polar binary 134
  unipolar 134
Waveform coder 10
Waveform coding 9, 23
Waveform regeneration 2, 141, 294
Waveforms, orthogonal 198
Waveguide dispersion 340
Wavelength Division Multiplexing (WDM) 540, 545, 546, 548
Wavelength division multiplexing (WDM) system 343, 357–359
  dense 358
Weaver structure method 126, 128
Westar 298
Western Electric 293, 295
Western Union 298
Wide area network (WAN) 6
Wink start 583–585, 591
Wireless Access Communications System (WACS) 264, 549, 550–552
Wireless In-Building Network (WIN) 265
Wireless local loop 263–266
Wireless PBX 261
Wireline carriers 228
Word 202–203
Word recognition systems 153

X.25 standard 361

Zero code suppression 26, 117, 128